STRATEGIC MINERALS

VOLUME I

STRATEGIC MINERALS

VOLUME I

MAJOR MINERAL-EXPORTING REGIONS OF THE WORLD

Issues and Strategies

W. C. J. van Rensburg
University of Texas at Austin

PRENTICE-HALL, INC., Englewood Cliffs, New Jersey 07632

Library of Congress Cataloging in Publication Data

VAN RENSBURG, W. C. J.
    Strategic minerals.

    (Prentice-Hall international series in world
resources, energy, and minerals)
    Bibliography: v. I, p. 527
    Includes index.
    Contents: v. I. Major mineral exporting regions of the
world, issues and strategies.
    1. Mineral industries—Government policy—Case
studies. 2. Metal trade—Government policy—Case
studies. 3. Strategic minerals—Government policy—
Case studies. I. Title. II. Series.
HD9506.A2V36 1986   333.8'5   85-6569
ISBN 0-13-851387-2 (v. 1)

Cover design: Edsal Enterprises
Manufacturing buyer: Rhett Conklin

PRENTICE-HALL INTERNATIONAL SERIES IN WORLD RESOURCES,
    ENERGY, AND MINERALS

    W. C. J. van Rensburg, Series Editor

Printed in the United States of America

10 9 8 7 6 5 4 3 2 1

ISBN 0-13-851387-2 01

Prentice-Hall International (UK) Limited, *London*
Prentice-Hall of Australia Pty. Limited, *Sydney*
Prentice-Hall Canada Inc., *Toronto*
Prentice-Hall Hispanoamericana, S.A., *Mexico*
Prentice-Hall of India Private Limited, *New Delhi*
Prentice-Hall of Japan, Inc., *Tokyo*
Prentice-Hall of Southeast Asia Pte. Ltd., *Singapore*
Editora Prentice-Hall do Brasil, Ltda., *Rio de Janeiro*
Whitehall Books Limited, *Wellington, New Zealand*

Contents

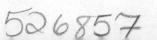

Preface

Several analysts have noticed that a substantial part of the literature dealing with strategic minerals covers only selected aspects, such as technology or politics, whereas we are convinced that the various aspects are very closely interrelated in a complex fashion. The United States, Western Europe, and Japan have become increasingly dependent on imported sources of these materials. There is a difference of opinion about the significance of this development. Some analysts contend that this growing import dependence has little significance for the security of the noncommunist world because the producing countries are so dependent on revenues from the sale of their minerals that they will be forced to continue to sell these minerals to the Western industrial countries, no matter what governments happen to be in power. Others point out that the United States and its allies no longer have political control over the mineral resources of the Third World, as they once did. Also, the United States and its allies no longer have control of the high seas and of the sea-lanes through which these minerals have to pass to reach their markets. By contrast, the USSR is largely self-sufficient in strategic mineral supplies, and has consistently followed a policy of mineral self-sufficiency at any cost. These analysts contend that the mineral import dependence of the West constitutes a real danger, and one that could easily be exploited by the USSR. A major reason for the lack of concern about mineral supply dependence by liberal analysts lies in their belief that the next war between the superpowers will inevitably be a nuclear holocaust, in which case adequate supplies of strategic minerals would be of no consequence. Conservative analysts, on the other hand, emphasize that this has not been the pattern over the past three decades, but that we have had a number of brushfire wars in which the availability of strategic minerals was extremely important. Neglect of conventional weapons almost invites the Soviet bloc to exploit this weakness of the West. The option with the lowest risk and cost would be to deprive the West of its oil supplies from the Middle East and its metal supplies from southern Africa. This would force Western Europe to seek an accommodation with the Soviets. Japan might have to follow suit, leaving the United States isolated. This would dramatically change the world balance of power.

In response to the need for a comprehensive study of the world's strategic minerals, a team consisting of mining engineers, regional planners, economists, geologists, finance majors, and some persons having language skills was assembled. All the team members were Master's degree candidates in the Energy and Mineral Resources Program at The University of Texas at Austin. Their task was to assemble the massive amount of information required for a detailed study of the supply and demand of strategic minerals on a worldwide basis. Our approach was first to undertake a library investigation into all the relevant aspects of the project, and to follow this by a direct approach to individuals, corporations, government departments, and so on, in order to obtain as much information as possible in a firsthand manner.

Our main objectives were to assess the adequacy and security of strategic mineral supplies of the United States, Western Europe, and Japan. The adequacy of resources and reserves, of productive capacity and of actual production, current and potential constraints on production, potential bottlenecks, disruption scenarios, local cost structures, levels of political morality, taxes and tariffs, incentives for further processing, markets for raw and processed products, competition from other producing countries, technological developments affecting both the supply of and the demand for minerals, substitutes, recycling, alternative sources of supply, and past statistical trends were all analyzed to obtain a global picture of the supply and demand picture for strategic minerals. Such a study requires a detailed investigation of a variety of aspects of the mineral industries. We also considered the potential effects of embargoes, cartels, stockpiles, and monopolies. A very important part of our investigation was an analysis of the effects of government policies, laws, and regulations on the availability, reliability, and cost of producing minerals.

ISSUES COVERED IN THE STUDY

Reserves and resources
Location of deposits
Geological types of deposits
Size of deposits
By-products and co-products
Metallurgical complexity
Mining methods
Labor availability and productivity
Physical infrastructure
    Transport
    Water
    Energy
Social infrastructure
    Housing
    Recreational facilities
    Services
    Research facilities
Taxation
Depreciation and depletion
Ownership of mineral rights
Local cost structure
Metallurgical flowsheets
Markets for raw and processed products
Foreign competition
Substitution
Secondary recovery (recycling)
Political climate
Cartels, embargoes, and monopolies
Government policies
Stockpiles (producer and consumer)
Risks (political and financial)
Degree of local processing
Constraints (technical, economic, political, etc.)
Costs (production, processing, etc.)

LIST OF STUDENT PARTICIPANTS

Peter Savinelli	Australia
Phill Ballinger	The USSR, Western Europe, and the one-crop-economy countries
Brian Muehling	Canada and the United States
Ralph Boeker	China
Lee Calaway	Southern Africa
Tom Braschayko	Japan
Steve Taylor	Latin America
Perry Lindstrom	U.S. strategic stockpile

This project has now been completed. The results are being published in two volumes, because the amount of material collected turned out to be far too voluminous for a single volume. It appeared to make much more sense to split it up into two volumes of more or less equal length, dealing with groups of countries with similar concerns. There is no completely satisfactory subdivision for a book of this nature, but a division between countries, on the one hand, with a major concern in finding export markets for their minerals, and on the other, with a major concern for obtaining sufficient volumes of minerals to satisfy their large domestic consumption, appeared to be a logical subdivision. This is to some extent a compromise, because a country such as the United States is, for instance, both a large producer and a large importer of a variety of critical strategic minerals. China, on the other hand, is a very large country, but not at present a very large producer or exporter of most strategic minerals. The two volumes were subdivided as follows:

Volume 1: The Major Exporting Countries

1. Canada
2. Australia
3. Southern Africa
4. Latin America
5. China
6. The one-crop-economy countries

Volume 2: The Major Consuming Countries

1. The United States
2. Japan
3. Western Europe
4. The USSR

The illustrations for this publication were prepared by Barbara Hartman. Editing of this manuscript was undertaken by Dr. W. C. J. van Rensburg, assisted by Dr. Paul Anaejionu, both of The University of Texas at Austin. The editor was also the supervisor of the theses on which these two volumes are based.

MINERALS INVESTIGATED

1. Chromium
2. Manganese
3. Cobalt
4. Platinum-group metals
5. Aluminum
6. Titanium
7. Vanadium
8. Fluorspar
9. Columbium

10. Tantalum
11. Nickel
12. Gold
13. Diamonds
14. Copper
15. Lead
16. Zinc
17. Iron
18. Uranium
19. Coal
20. Oil and gas
21. Phosphates
22. Potash

A number of criteria were used to select these minerals. Among the most important were: scale of use in the industrial countries, level of import dependence, percentage of consumption in defense applications, concentration of production, political reliability of supply sources, ease of substitution, magnitude of reserves relative to consumption, and vulnerability of supply lines to physical disruption. Although several attempts were made to quantify the strategic importance of individual minerals, it appeared that this type of classification ultimately has to contain at least an element of subjective judgment. Furthermore, the rankings in terms of strategic importance will vary from one country to another, based on each country's own resource base, geographical location and political relations with trading partners, and so on. At present, most Western industrial nations are particularly concerned about the possible disruption of supplies from southern Africa. Hence minerals obtained largely or solely from that region would feature prominently on their lists of minerals of greatest concern. However, political factors are by no means the only ones. Australia is regarded as a politically secure source of a wide range of minerals, but one with a high potential for loss of production through strikes, and also as a country with a high cost structure. The relative strategic importance of minerals also changes over time, as a result of technological developments or new discoveries. Therefore, it should be clear that the selection of this list of minerals is somewhat arbitrary. Most analysts agree that the first five minerals on our list are the ones of greatest concern. Four of them have a very high concentration of reserves and production in just two areas of the world: southern Africa and Siberia. Aluminum is second only to steel in terms of tonnage of consumption. While the United States is self-sufficient in terms of aluminum smelting capacity, it is dependent on imports for more than 90% of

its bauxite supplies. Platinum-group minerals are usually regarded as precious metals. However, more than 95% of the consumption of these metals is in essential industrial uses. Chromium is essential in the production of all special alloy steels. However, some 99% of world reserves occur in South Africa and neighboring Zimbabwe. Steel cannot be produced without manganese. The USSR and South Africa supply more than 50% of the world's manganese. More than 50% of the world's cobalt production comes from Zaire, which has been a very unstable country in recent years. Titanium has become essential in defense applications, and Western Europe and the United States are heavily dependent on imports at various stages of processing. The rest of the top 10 on our list are important alloying elements for the steel industry. Fluorspar does not appear on most lists of strategic minerals—a fact that surprises us, since it is essential in the smelting of aluminum and as a flux in the steel industry.

The minerals listed as numbers 11 to 22 are probably not quite as critical at this stage. They include gold and diamonds, which are important as stores of value, and the traditional nonferrous metals, copper, lead, and zinc, the consumption of which has stagnated. However, these metals are still important in many industrial and military applications. The energy minerals, uranium, coal, oil, and gas, were included because of the importance of energy in the processing of many of the most important strategic minerals. Finally, the fertilizer minerals, phosphate and potash, were included because of their importance in food production.

Some attention was devoted to minerals likely to become much more important in the future as a result of their use in the high-technology industries. Included in this are minerals such as germanium, selenium, silicon, gallium, berillium, and lithium. However, these minerals will be analyzed in much more detail in a separate study which is now being undertaken. The emphasis was obviously on those minerals that are of greatest importance at this time, in terms of tonnage of world production or value of world production. Clearly, we are in an era of very rapid technological change and of volatile economic conditions. Undoubtedly, some of the minerals included in our study will become relatively less important in the future, and others not considered may turn out to be more important. The mineral industry is dynamic, and a study of this type should be updated regularly, as we fully intend to do.

The various parts of these two volumes differ somewhat in style and emphasis, reflecting different concerns in various parts of the world, as well as differences in the backgrounds of the students who undertook the individual studies. We trust that this work will contribute to a more informed and objective debate on the issues involved in securing and maintaining adequate supplies of strategic minerals for the United States and its allies.

W. C. J. VAN RENSBURG

Executive Summary

The dependence of the United States on imported strategic minerals has grown substantially since World War II. Imports as a percentage of apparent domestic consumption is close to 100% for a number of these minerals, compared with about 30% for imported oil. Reserves and production capacity are highly concentrated in a small number of foreign countries for several minerals, such as chromium, manganese, cobalt, and platinum-group minerals, resources of which are highly concentrated in southern Africa. The only other major source of these minerals is the USSR.

Sharp increases in transportation and processing costs since the oil crises of 1974 and 1980 have led to a change in the nature of this import dependence. Where previously the United States imported ores and concentrates, strategic minerals are now usually imported in processed forms. Much more stringent environmental regulations since 1969 have also had an adverse effect on domestic mineral-processing industries.

The reasons for America's growing import dependence include increases in the level of consumption and an expanded range of minerals required by modern industries, the depletion of high-grade domestic deposits, neglect of mining and metallurgical research, growing concern with protection of the environment, restrictions on exploration and mining in the federal domain, and growing competition for available supplies from other noncommunist industrial countries. Inadequate domestic mineral policies and tax laws that do not sufficiently encourage domestic exploration, mining, and mineral processing undoubtedly also contributed to the problem. Public ignorance of the importance of minerals as raw materials for industry and for defense did not improve the situation.

Western Europe and Japan are far more dependent on mineral imports than is the United States. By contrast, the Soviet bloc is largely self-sufficient in strategic minerals supplies. The USSR is aware of the fact that the import dependence of the United States and its major industrial allies makes them vulnerable to energy and mineral supply disruptions.

During World War II, a lack of adequate energy and mineral supplies contributed substantially to the defeat of Germany, Japan, and Italy. The United States and the British Empire, together with the USSR, had political control of most of the world's mineral reserves. They also had control over the major sea-lanes. However, the Allies came perilously close to having their mineral supplies severely disrupted by the success of German submarines. The lack of an adequate strategic minerals stockpile proved to be a costly mistake.

Today, the British Empire no longer exists, and with its demise, political control over many mineral reserves has been lost. The United States and its allies no longer have the dominant control over sea-lanes, and are faced with a formidable challenge from the Soviet navy. The West is far more vulnerable to energy and mineral supply disruptions than we were during World War II.

More emphasis on domestic mineral production, recycling, substitution, and stockpiling could significantly reduce the import dependence of the United States. However, for the foreseeable future the United States, Western Europe, and Japan will remain highly dependent on imported energy and strategic minerals. The change toward the import of these minerals in more processed forms usually means that the number of potential suppliers is reduced.

In Volume I of this book the strategic minerals potential of our major suppliers is analyzed. Included in this volume are Canada, southern Africa, Australia, Latin America, China, and the so-called one-crop-economy countries. Canada, South Africa, and Australia are the most important suppliers to the United States, South Africa is Western Europe's most important provider, and Australia is critically important to Japan. Mexico, Brazil, Venezuela, and China are potentially very important producers of a wide range of minerals. Such countries as Zambia, Zaire, Chile, Peru, Gabon, Morocco, Indonesia, and Malaysia are already very important sources of one or two commodities each. They are the one-crop-economy countries. A number of oil-producing countries are also highly dependent on exports of that commodity.

CANADA

Canada is the second largest nation in the world in terms of surface area. Its vast size allowed for the complex interaction of climate, geology, and time needed to form a wide variety of mineral deposits. As a result, Canada has developed into the third largest mineral-producing nation in the world. The industry is quite important because it contributes a great deal to Canada's balance of payments and also encourages a great deal of foreign investment in Canada. Canada's major source of investment funds, and major trading partner, is the United States, which has become more dependent on minerals from Canada than from any other nation. Canada has recently enacted policies to counter what it feels are adverse effects of excessive U.S. influence, and also a dependence on foreign trade, and these policies raise questions about the ability of the United States to continue to depend on Canada as a source of supply. Canada will continue to be a dependable source, however, because of such factors as the evolving patterns of world trade, and its desires to fully diversify its domestic industrial base.

SOUTHERN AFRICA

Countries in southern Africa contribute significantly to the world mineral industry. This is especially true of South Africa, which is the world's fourth largest mineral industry. The mineral industries of Angola, Botswana, Lesotho, Malawi, Mozambique, Namibia/South-West Africa, South Africa, Swaziland, Zaire, Zambia, and Zimbabwe are presented in this book. Important factors that influence individual mineral industries are reviewed, and include geographic and geologic features, supportive infrastructures, and local and regional economic and political policies. The region is particularly important for minerals that are considered critical to industrialized countries. Reserves of these critical minerals ensure that southern Africa will remain an important source of nonfuel minerals in the future.

South Africa is an important producer of gold, platinum-group metals, diamonds, chromium, manganese, vanadium, copper, lead, zinc, iron ore, titanium, zirconium, fluorspar, phosphates, andalusite, and silliminite. It is also a major exporter of steam coal and uranium, but is highly vulnerable to an embargo on oil imports. In addition to being a major supplier of strategic minerals to the United States, Western Europe, and Japan, South Africa is a world leader in mining and metallurgical technology, and the technology for producing oil from coal.

Zaire is an important source of cobalt, copper, and industrial diamonds. Zambia is important for copper and cobalt, and Zimbabwe for chromium. Namibia is a major uranium and diamond producer, and Angola is the only significant oil producer in the whole region.

The political rivalries between South Africa and her black-ruled neighbors are well known. However, the economic realities of the region—the interdependence of all the nations of the subcontinent—are rarely appreciated in the West. All the developing countries of the region are, and are likely to remain, highly dependent on South Africa's physical infrastructure, technology, food supplies, medical and veterinary services, and markets.

Disruption of the South African economy and mineral industries will inevitably lead to serious losses of mineral supplies from all the other countries in southern Africa.

AUSTRALIA

Australia is one of the largest sources of minerals on both a production and an export basis. World trade in iron ore, bauxite, mineral sands, nickel, manganese, lead, zinc, copper, tin, tungsten, diamonds, and gold are all influenced significantly by Australia. In addition, Australia has become an important source of coking and steam coal and uranium, and in the future may become a source of liquefied natural gas.

Since Australia exports roughly 50% of her mineral production, the country depends on trade to keep her mineral industry healthy. Japan is Australia's most important trade partner, consuming around 40% of her mineral exports. While the United States and Western Europe are traditional markets, the industrializing nations of the western Pacific region hold the greatest potential for new trade.

Australia has suffered competitive problems due to high government, capital, labor, and infrastructure costs. Although some of these are unavoidable, many could be controlled through better planning and cooperation between the various sectors of the mineral industry.

To better its position in the future, Australia must become more competitive so that it can gain additional access to markets outside Japan. This could be accomplished through revisions in the tax structure or by altering regulations concerning foreign investment and infrastructure financing. In addition, there are opportunities to increase the level of domestic processing. This would also increase mineral revenue. Perhaps the most important issue that has worked against the Australian mineral industry is the intergovernmental conflict between the state and federal governments. If Australia is to continue to have a prosperous mineral industry, there must be greater cooperation between the two tiers of government.

LATIN AMERICA

Latin America exports large volumes of a wide range of mineral commodities to the United States, Japan, and Western Europe. Among the United States' largest sources of mineral imports, Latin American countries supply over 30 different mineral commodities to the American economy.

The level of mineral development varies widely among the countries of this region. The principal factors responsible for this are the quality of infrastructure, the resource endowment, the historical role of multinational corporations in mineral development, foreign loans, the level of state ownership of mining companies, and political and economic stability. Of these, the first three are the most important. The use of protectionist policies to promote mineral development is another important consideration. Politically stable countries, such as Mexico and Brazil, tend to have more highly developed mining industries.

The minerals industries of Latin America face numerous challenges in the 1980s. In addition to the problems created by the world recession and low commodity prices, ore grades are falling at mines throughout the region. Historically high interest rates have further compounded the serious foreign debt problems of Brazil, Mexico, Chile, and Venezuela, and others. Analysis of past and present mineral production and trade of these countries reveals several trends that will alter the level of their mineral exports to Western industrialized countries. Latin American countries are diversifying the composition and destinations of their mineral exports. Rapidly expanding steel industries in Brazil and Mexico will consume a portion of their iron ore, manganese, and fluorspar production now available for export. Increases in the domestic processing of minerals will have serious implications for the mineral-processing industries in the industrialized countries. Latin America will play a more important role in the world mineral economy as new projects in the region reach full productive capacity in the mid to late 1980s.

Peru and Chile presently have the most favorable climates for foreign investment in the region. Recent liberalizations of their mining laws have increased foreign companies' interest in investing in their mining industries. Large foreign debts, and the lack of infrastructure, will continue to limit foreign investment and mineral development in Bolivia, Brazil, and Mexico. Mineral development in the region's minor mineral-producing countries will be very limited in the 1980s.

Latin American countries have been major suppliers of strategic minerals to the United States in previous wars and international conflicts. Political instability in southern Africa and the Middle East increases the importance of Latin America's mineral and energy resources to the United States. Latin American countries will continue to supply large shares of U.S. requirements of a number of minerals.

CHINA

China is a country with a large potential for energy and mineral development. Although China is a small factor in world trade, there is a stated commitment to increase mineral commodity exports. The extent of China's resources is not readily apparent, due to incomplete exploration, and failure of the Chinese to fully disclose their mineral resources and reserves data to outsiders. The potential for development must be compared with the likelihood of increased commodity demand associated with the growing Chinese population and economy. Resources development, and economic development in general, are constrained by the inadequate social and physical infrastructure. Problems range from transport capacity limitations to insufficient trained personnel to operate sophisticated imported equipment. China has sought outside assistance, and is offering payback-in-kind, including resources, to attract additional assistance. Investment by resource-dependent countries to assure supply is one area offering a great deal of promise. Without outside technical and economic assistance, the chances for China to emerge as a significant mineral exporter are not good.

THE ONE-CROP-ECONOMY COUNTRIES

A large number of countries are dependent on the export of one or only a few mineral commodities. These countries face special problems because of wide fluctuation in the prices of and demand for their particular commodities. For this reason, they have a particular interest in price-stabilization schemes; and in joining forces with other producers of the same commodity in cartels, in order to maximize the benefits derived from the exploitation of their resources. Since minerals are non-renewable resources, these countries are also faced with the ultimate depletion of their minerals, and with the need to use the income derived from these minerals to provide for a new economic base.

Apart from countries such as Chile, Peru, Bolivia, Zaire, and Zambia, covered in other sections of this volume, and the major oil-exporting nations, the mineral position of other one-crop-economy countries are discussed in this section of the book. They include Gabon (manganese), Guinea (bauxite), Morocco (phosphates), Malaysia (tin), Indonesia (tin and oil), Papua–New Guinea (copper and gold), and Thailand (fluorspar).

STRATEGIC MINERALS

VOLUME I

Introduction

The United States is no longer the world leader in mineral production that it once was. That role — one held for years — is being relinquished despite adequate domestic resources for a number of minerals, a skilled work force, and despite a continuing domestic demand for minerals. It has become apparent that an adequate resource base does not ensure adequate domestic mineral production in the face of increasing foreign competition and diminished markets.

As a result, the United States is falling behind in the world marketplace as a source of mineral supplies for other nations as well as for its own industries. In fact, the United States is becoming much more dependent on foreign mineral supplies. That condition has been blamed on many factors. Some have contended that it is the result of environmental regulations, destroying the productivity of U.S. industries. Others have blamed cheap foreign labor, luring away U.S. jobs and contributing to closures of processing plants. Depletion of domestic reserves, or the state of the world economy, or a decline in America's technological lead over other nations, have also been listed as primary reasons for this sad state of affairs.

Western Europe and Japan are far more dependent on imported sources of minerals and energy than is the United States. Western Europe must import about 75% of its minerals, and Japan some 90%. However, these countries have had to deal with a high level of import dependence over many decades, and have developed more sophisticated methods of dealing with the issue than has the United States. Japan in particular has had to develop the means of ensuring adequate and secure sources of supply for virtually all of its raw material supplies. To reach this objective, Japan has had to develop close cooperation between the private and public sectors in all aspects of raw material acquisition. Japan simply could not afford the adversary relationship that exists in the United States. Japanese steel companies routinely cooperate in the acquisition of raw materials such as iron ore and coking coal from abroad. This might be prohibited in the United States in terms of antitrust legislation. Japanese companies, in many instances, acquire a minority share in foreign mines and processing plants. U.S. firms generally prefer either a majority share in foreign plants, or are prepared to take the risk of buying on the spot market. These strategies were acceptable during times of an abundance of mineral supplies, and before the advent of a much more nationalistic posture on the part of resource-rich developing countries.

In the United States, the general public has only barely begun to grasp the reality of this country's dependence on imported crude oil; indeed, most of them are convinced by the present glut that the oil import dependence is the result of a conspiracy between OPEC and the major oil companies. They are blissfully unaware of this country's far greater dependence on a wide range of critically important strategic minerals.

A few examples may serve to illustrate the nature of the problem and the complexity of its causes. The United States has historically been the world's leading producer of copper. Since 1970, however, domestic production of copper has leveled off, and even declined, while production in other countries has increased. In 1982, copper production in the United States declined by 28%. In the same year, production of copper in Chile increased by more than 10% at the same time that U.S. copper mines were closing down. A number of factors played a part in these developments. Declining ore grades in the United States was certainly one. Other forces that adversely affected the U.S. industry were inflation and the strengthening of the U.S. dollar against almost all other currencies. For example, between 1979 and 1983, inflation in the United States amounted to some 28% and the average cost of producing a pound of copper increased by 13 cents. The rate of inflation in Chile during this same period was about three times that in the United States. However, Chile went through a series of rapid devaluations in 1982 and in 1983. As a result, Chilean production costs in U.S. dollar terms were about the same in 1983 as they were in 1979. More important even than the economic factors were a series of political developments that took place during the past decade. In a number of resource-rich developing countries the mineral industries which traditionally had been controlled by private companies were either nationalized or the governments' share of ownership increased sharply.

1

During periods of excess production, and when metal prices are depressed, privately owned mines are likely to shut down operations or to at least decrease production when their marginal costs exceed their marginal revenues. Government-owned mines, on the other hand, are less concerned with profits than with earning foreign exchange to pay for their burgeoning oil import bills. Many of these countries have unemployment rates of 20% or more. This is another reason why the governments are unwilling to cut back on mineral production. It would be political suicide to lay off people under such circumstances. Many of the mineral-producing developing countries also have huge debts to international banks. To avoid defaulting on these loans, the banks in some instances have granted additional loans to the developing countries to enable them to pay the interest on their previous loans. In one-crop economy countries, the only means of paying the interest on their loans — and hopefully one day the principal — is to sell more minerals, even in a depressed market. So U.S. and European banks provide loans to these countries, which are sometimes used to increase productive capacity that is not needed. In other words, excess productive capacity in the developing countries is financed by the international banks, or even by the governments of industrialized countries, to protect their loans. This excess capacity in turn leads to the closure of mines and processing plants in the industrialized countries. It is obvious that normal supply and demand relationships are distorted by these developments.

Depleted resources will soon place the United States at a competitive disadvantage with regard to the production of phosphates. At a price of $30 per ton, in constant 1983 dollars, phosphate production in the United States is likely to start declining quite soon. Large capital investments will be required to develop huge low-grade phosphate resources in the southeastern United States. The U.S. Bureau of Mines estimates that a price of some $50 per ton will be required to replace the anticipated loss of production from existing mines. Under these circumstances, it is very likely that the United States will lose export markets for phosphate to Morocco.

U.S. zinc resources are huge. Yet production of zinc in the United States has declined by 65% since 1965. Primary smelter capacity for zinc has dropped by 60% during the same period, while production has declined by 75%. The United States has become much more dependent on imported sources of refined zinc metal. Previously, we imported less zinc, largely as concentrates. Of the 14 primary U.S. zinc smelters operating in 1968, 10 have closed

down permanently. Increasingly stringent environmental regulations played an important part in this decline in the U.S. zinc industry. Import duties, government stockpile releases, and price controls were also contributing factors.

These three examples, quoted recently by the director of the U.S. Bureau of Mines, illustrate quite clearly that there is no single cause for the decline of the U.S. mineral industry. Rather, an ignorant public, uninformed politicians, and disinterested administrations have all contributed to a situation that now constitutes a threat to the very security of the United States. We must have materials to build weapons for national defense and to provide the goods, services, and technologies that have made life in the United States and in other Western industrialized nations the envy of the world.

There are no easy solutions to this problem of increasing import dependence by the United States, Western Europe, and Japan. The situation is made even more serious by the fact that our major adversary—the USSR—is largely self-sufficient in minerals and energy and has consistently followed a policy aimed at maintaining a high level of production in order to maintain that self-sufficiency. The USSR is completely aware of the fact that the dependence of the West on remote and unreliable sources of strategic minerals constitutes a serious security risk.

It has even been suggested that the USSR is already engaged in a "resource war" with the West in which it is their objective to make it as difficult and expensive as possible for the West to obtain energy and mineral supplies. In Richard Nixon's book *The Real War* he quotes former Soviet leader Brezhnev as stating: "It is our intention to deprive the West of its two main treasure troves: the oil fields of the Persian Gulf, and the strategic mineral resources of central and southern Africa." He went on to state that control over these resources would render Western Europe a hostage to the USSR and would isolate the United States.

The concept of a resource war has been hotly debated by mineral economists. Proponents of the resource war thesis point to developments in southern Africa as clear evidence of a Soviet attempt to gain control of a resource-rich area, producing a variety of minerals that are of vital importance to the United States and Western Europe. The other major source of most of these minerals, curiously enough, is the USSR. Opponents hold that the Soviets merely regard this vast area with its racial conflicts as a convenient political target of opportunity and feel that the presence of rich mineral resources in the area is incidental. There are also differences of opinion even among proponents of the concept about the exact reasons why the Soviets are so active in southern Africa. Some believe that the Soviets, being largely self-sufficient in minerals, merely wish to deny these resources to the West. Others contend that the Soviets have depleted their own high-grade resources of minerals such as manganese, chromium, cobalt, and platinum-group metals, and that their true objective is to seize higher-grade deposits in southern Africa.

These arguments show that the "resource war" debate involves more than just an academic difference of opinion. It is also essentially an argument between liberals and conservatives about the true motives of communism. Additionally, it is a debate between supporters and opponents of the mining industry's position regarding the security and availability of vital mineral commodities to Western industrialized economies. In the process, the debate has become very emotional on both sides, with strong and sometimes unsubstantiated statements being made.

Our objective is not to enter this emotional debate, but rather to point out that an in-depth analysis of the issues involving global supplies of essential minerals will include considering factors ranging from the geography of mineral locations and government economic policies to the ideological orientations of nations. This approach is necessary for us to gain an understanding of global mineral supply and demand problems, especially those that affect the adequacy of mineral supply to the so-called "free world."

The first volume of this study deals with the mineral industries of the major mineral-exporting regions of the world. Included in this volume are Australia, Canada, southern Africa, Latin America, China, and the "one-crop-economy countries." As far as the United States is concerned, Canada, South Africa, and Australia are by far the most important sources of nonfuel mineral imports, with Mexico as a potentially significant supplier. Japan is heavily dependent on Australia, whereas Western Europe obtains the bulk of its strategic minerals from South Africa. The one-crop-economy countries are those whose economies are heavily dependent on one, or at most a few commodities. Examples of such countries are Morocco (phosphates), Gabon (manganese), Guinea (bauxite), Niger (uranium), Malaysia (tin), Bolivia (antimony and tin), Papua–New Guinea (copper and gold), Zaire (cobalt and copper), Zambia (copper), and of course, a number of oil-producing nations. These countries are highly dependent on the markets for

their particular commodity and are very vulnerable to declines in the demand for those minerals. They have a particular interest in commodity stabilization schemes, tend to become involved in attempts to establish cartels, and tend to support the concept of the New International Economic Order.

One of the great misconceptions of our time is that most strategic minerals are produced in developing countries, and that we are in danger of physically running out of minerals. Although there were great hopes immediately after World War II that the developing countries would become a major source of minerals, this did not, for the most part, come about for a variety of reasons. The developing countries, upon obtaining their independence after the war, were initially prepared to grant exploration and mining concessions to multinational corporations under generally favorable conditions in the hope that mining ventures would provide employment as well as generate the capital which they so desperately needed. Unfortunately, many of the developing countries soon decided that they wanted a larger share of the profits derived from the exploitation of their resources. Nationalization of foreign company assets, increased taxation, forced changes in agreements, loading of the payrolls with political cronies, and other forms of harassment of the multinationals became quite commonplace. This is not to suggest that the companies were in any way blameless. Many of them were less than generous in their treatment of the developing countries, and must carry at least part of the blame for what happened.

The result of all this, however, is that over the past two decades most exploration for strategic minerals has not taken place in developing countries but in a relatively small number of developed countries, including Canada, Australia, South Africa, and the United States. If we do not explore in the developing countries, we cannot expect them to produce significant amounts of minerals in the foreseeable future. There are long and lengthening lead times between exploration and development of mineral deposits.

There is little danger that we will physically run out of most minerals on a global basis, but a very real danger that we will not invest enough in exploration and development. In developed countries such as the United States where mining has taken place on a very large scale over many decades, the higher-grade deposits have for the most part been depleted. Most of the deposits that remain to be exploited are likely to be either small, low-grade, complex, remote, or to have a combination of these characteristics.

In the first volume we deal with a number of recently developed countries, such as Australia, Canada, and South Africa, which have sophisticated technology for the discovery, mining, and processing of minerals, and with a large number of developing countries, which, for the most part, will have to obtain the technology from elsewhere.

The second volume deals with the major consumers of minerals. Included in this volume are the United States, Western Europe, Japan, and the USSR. The United States and the USSR are both large mineral producers in their own right. Japan and Western Europe are still very large consumers of minerals.

The problem of increasing mineral import dependence by the noncommunist industrialized countries is complex. Although several attempts have been made to explain this trend toward increasing import dependence, most have tended to offer somewhat simplistic solutions to a problem that is inherently highly complex. The problems did not originate, nor do the answers lie only in geology, mining technology, metallurgy, infrastructure, economics, policies, or politics. All these factors, and others, such as public perceptions, the relations between nations, and changing life-styles, have to be considered. In addition, it should always be appreciated that the mineral and energy industries are international in nature. International trade accounts for a major share of the production of virtually all strategic minerals.

A detailed investigation of world supply and demand of strategic minerals should therefore take into account all the various factors that play a part in global output and consumption of minerals. Although the emphasis of this study is on strategic nonfuel minerals, the importance of energy in the processing and transport of minerals cannot be ignored.

The first aspect that has to be evaluated is the magnitude and distribution of world resources and reserves of all the commodities included in the study. Unfortunately, this is much easier said than done. Published estimates of the magnitudes of resources and reserves of minerals in the United States, Western Europe, and Japan are at best only accurate within an order of magnitude. Those for the recently developed mineral-producing countries, Canada, Australia, and South Africa are somewhat less reliable, because exploration has generally not yet been quite as extensive. The figures for developing countries are little more than guesswork. Immediately, therefore, we are faced with the problem that one of our key indicators about the adequacy

of minerals is by definition incomplete. However, this is a fact of life that we simply have to live with.

We can, of course, get an inventory of existing and planned mines and processing plants. This provides us with adequate information about the supply potential in the short to medium term. Note that knowledge about productive capacity is just as important as information about actual production.

Information about the size, grade, location, and geometry of individual deposits and mines is obviously also very important. Information about the mining methods used at the individual mines provides clues about likely costs of extraction, although this can vary considerably from one location to another.

The occurrence and grades of by-products and co-products may have a material effect on the cost of extraction of the desired strategic mineral. Over time, by-products may, in fact, become the major products. A knowledge of the genetic type of deposit that we are dealing with gives some indication of the types of by-products to expect. For instance, in the case of porphyry copper deposits, we would expect to find molybdenum or gold, whereas in magmatic sulfide deposits we would look for nickel, cobalt, and platinum-group metals.

The mineralogy of ore deposits, grain sizes, types of intergrowths of the ore minerals, degree of alteration, and nature of the gangue minerals all provide useful information about possible beneficiation processes. In the case of existing processing plants, we would, of course, prefer to have detailed information about the actual flowsheets, because this can provide valuable clues about likely production costs—an aspect about which information is rarely readily available.

The availability and cost of physical infrastructure is of critical importance to the mineral industry. Transport facilities; roads, railway lines, waterways, and ports are essential for the provision of plant and equipment to the mine, and for moving the products to the markets—domestic or foreign. In the case of minerals with a high place value and a low unit value, such as the bulk commodities coal, iron ore, manganese ore, chromite, and so on, the cost of transportation may represent more than 50% of the delivered cost of the products.

The capacity, quality, and state of repair of the transport network has a major impact on the viability of any mineral venture. For instance, there are substantial economies of scale in ocean transport. South Africa and Australia, with deep-draft ports capable of handling tankers with a capacity of more than 200,000 tons have a substantial advantage over U.S. coal exporters, who have to operate from shallow-draft ports, capable of handling only small vessels.

The capital costs of physical infrastructure can be substantial. For example, in the development of the iron ore deposits of the Pilbara region in Western Australia, the cost of the physical infrastructure exceeded that of the mines and associated facilities by a factor of 2. In developing countries that ratio can be much higher.

The provision of major transport facilities is not only expensive but takes time. Therefore, the capacity of transport facilities may place a ceiling on a nation's capacity to export minerals. During periods of high inflation, countries with adequate existing transport facilities will generally have a comparative advantage over those that have to provide new ones. Monitoring of new transport facilities can provide useful advance warning about additional mineral export capabilities.

Many strategic minerals require substantial amounts of energy for their processing. This may be in the form of electricity, coke, diesel fuel, or gasoline, depending on the type of processing or the location. The availability and cost of energy can be critical to many mining and mineral processing ventures. Countries with adequate domestic sources of energy at relatively low costs not only have a comparative advantage over those that do not, but are more likely to process their minerals before export.

The cost of water is usually not as important as that of transport or energy in determining the economic viability of mineral ventures, but its availability obviously is. Most mineral deposits tend to occur in remote and inhospitable areas. The industry may therefore be forgiven for tending to believe that water invariably occurs in the wrong places, in insufficient quantities, and of unacceptable quality. Many mines in remote areas have to undertake expensive and elaborate schemes to obtain their water supplies, sometimes from far away.

The cost and availability of physical infrastructure often determines where processing plants are located. Unlike mines, processing plants do not have to be located at the deposit. The cost of transporting large amounts of waste in ores or concentrates may be offset by cheaper energy costs away from the mine and in an urbanized area. Government policies with regard to the financing of infrastructure are very important in this regard. Infrastructure is usually more developed in urban areas than in the remote areas where most mines

are located. This may persuade companies to locate their smelters and refineries away from the mine in an urban area with existing infrastructure. This will, in turn, tend to exacerbate existing environmental problems.

The question of who pays for infrastructure to serve mines and processing plants in remote areas is always a matter of debate. If the government is willing to finance the infrastructure, it greatly reduces the capital costs of the mining companies. However, this also places the government in a much stronger position to determine where processing plants will be located. Governments may agree to finance the infrastructure, but may then charge excessive rates for power, water, and transportation. This can even take the form of additional indirect taxation. The policies of any particular government on these issues will have a material effect on the viability of mining and processing ventures.

Social infrastructure is also very important. Under this heading we include items such as housing, schools, recreational facilities, medical services, geological surveys, metallurgical test facilities, and mining research institutes. Some of these facilities may be provided by governments, others must be financed by the companies. The more the companies have to provide, the less likely they are to invest in any particular project.

The availability and level of sophistication of research and development facilities in any particular country will have an effect on its competitive position on world markets and on the degree of processing of its minerals before export. The availability of local geologists, mining engineers, and metallurgists has become particularly important in this age of increasing nationalism.

The cost structure within a country will obviously have an effect on its competitiveness on world markets. For instance, Australia, with its high wages, industrial unrest, and sparse population, generally has to mine deposits with higher grades than does South Africa, which is regarded as politically vulnerable. Exploration, mining, processing, and transport costs are all important in this regard. Relative cost structures change over time as a result of inflation, currency revaluations, and changes in labor productivity. Of course, production costs of minerals are also materially influenced by the grades of deposits and by by-product, co-product relationships.

The level of economic development and sophistication of a country will give an indication of the availability of goods and services required by the mineral industry. The availability of domestic capital is an important indication of a country's ability to participate actively in the development of its mineral resources.

Political relations and the political system of a producing country are aspects of growing concern to the industry. The level of political morality is important. In this regard the key question is: Will the government honor existing laws, regulations, and agreements? Of equal importance is the stability of the political system. In a two-party system it is very important to know whether there is at least a measure of bipartisanship in the mineral policies of the two parties. There are long lead times in the development of mineral projects. If a major policy change takes place during this period, it can have disastrous consequences for the project. Political stability and predictability are very desirable factors that mining companies and the major consuming nations consider in planning their investment policies.

The propensity of a country to join producer associations (cartels), to support the concept of the New International Economic Order, or to adopt excessively nationalistic or socialistic mineral policies will have a significant influence on the behavior of multinational companies in terms of their decisions to invest in mineral projects in the country. For all its other favorable attributes, these are areas of concern for groups considering investments in Australia. These concerns may explain why several multinational corporations undertake political risk analyses in connection with proposed or existing mining projects. Historically, mining companies have regarded nationalization of assets as the major risk. Because many resource-rich developing countries have learned that such actions invite reprisals, other forms of political risks, such as sudden shifts in mineral policies, also have to be evaluated.

Although most cartels have been unsuccessful in the longer term, the recent success of OPEC has again raised fears about possible disruption of mineral supplies by powerful producer cartels. Organizations such as CIPEC, the bauxite producers association, the iron ore producers association, the International Tin Agreement, the International Lead-Zinc Study Group, and producer associations in the mercury and manganese industries have at various times made attempts to control or manipulate the markets for their commodities, usually with scant success.

Monopolies in the mineral industry have been few and far between, with the notable exception of the diamond monopoly run by the Central Selling Organization of De Beers. Fears about the establishment of other monopolies appear to be unfounded.

The increasing volatility of the world economy, and fears about supply disruptions developing as a result of the growing nationalism of developing countries, have played a part in persuading countries such as the United States to establish national strategic stockpiles of a range of minerals that are not available in sufficient quantities from domestic sources. Although the motivation for building up such stockpiles is obvious, there is no doubt that they constitute a serious overhang on the markets. In the past stockpiles have been used to manipulate mineral markets. For instance, the U.S. stockpile contains more than a year's world production of tin. This greatly concerns the major producers of tin.

In recent years, there have been a number of attempts to place mandatory embargoes on the export of oil and other products to South Africa. These initiatives have had some success. The South Africans have threatened to impose their own counterembargo on the export of strategic minerals, such as chrome, platinum-group metals, and manganese in the event of an all-out oil embargo against them.

Cartels, monopolies, stockpiles, and embargoes have the potential of distorting mineral markets, and of causing higher prices—at least in the short to medium term. The fact that they have seldom had much success in the past should not lead us into a false sense of security.

Clearly, in a study of this magnitude it is impossible to strive for a complete data base. One has to accept that the data base will be incomplete and that it will soon become outdated. In this study we have attempted to obtain statistics to the end of 1982 wherever possible, and in some instances to 1983.

Our conclusions are that shortages of strategic minerals are not likely to come about as a result of physical depletion of reserves for most minerals. However, serious disruption of supplies may well occur as a result of a number of economic and political developments. For instance, failure to provide adequate funding for exploring and developing the mineral resources of the Third World has probably already sharply reduced the potential supplies from those countries, at least in the short to medium term. Of course, the reasons for this lack of investment are largely political. The current fad in the United States is that this country's economy is undergoing a structural change from a manufacturing to a high-technology and service-oriented society. Although there is certainly an element of truth in this, there is a danger that this concept is being greatly exaggerated and that the importance of the primary extractive industries, even the manufacturing sector, is no longer adequately appreciated. Even if we accept the emphasis on high-technology industries, the question must still be raised concerning the availability of mineral supplies for the high-technology industries. Will the United States be less dependent on foreign sources for high-technology minerals than it is for conventional minerals? Will the United States and other industrial countries adopt existing or new policies to ensure adequate and secure supplies of the raw materials required by these new industries? The environmental movement in the United States has made it much more difficult and expensive to extract minerals domestically. Interventions by environmental movements and a growing body of increasingly complex and onerous environmental and other regulations have led to serious delays of some processing plants and to the demise of some industries, such as the production of ferroalloys and the refining of zinc. There are distinct possibilities for massive and extended disruptions of mineral supplies from southern Africa, as a result of political events in that region. We have no right to be complacent about the future adequacy of mineral supplies for the United States, Western Europe, and Japan.

It appears that Canada, southern Africa, and Australia will remain the major sources of mineral supply to the industrial nations of the West. The rich potential of countries like Mexico, Brazil, Venezuela, Indonesia, Zaire, and Zambia has not yet been realized. Unless more investments take place in their mineral industries, this will not happen soon.

The United States appears to have lost its technological lead in mining and in mineral processing. This, together with the depletion of domestic mineral deposits and a continuing decline in the grades of U.S. deposits, has eroded the competitiveness of the U.S. mineral industry and has led to a change in the nature of America's import dependence. Where previously America imported ores and concentrates, and protected her domestic processing industry with stiff import duties, the United States is now increasingly importing minerals in processed forms. This is not only more costly, but it also substantially reduces the number of potential sources of supply for a number of critical minerals, such as chromite. It also means the loss of jobs in mineral processing.

Western Europe has long since depleted most of her economically feasible mineral deposits and has no option but to live with a high level of mineral import dependence. Japan is even more import dependent and is actively moving toward the import of minerals in highly processed forms. This has created real opportunities for mineral-producing countries having advanced technologies and established physical and social infrastructure to increase the benefits that can be derived from the exploitation of their mineral resources. However, it is also likely to make producers of unprocessed minerals in the Third World less competitive. Such developments could distort markets for at least some minerals and may encourage attempts at the establishment of cartels, as has already taken place in the case of the uranium industry.

Whether or not the USSR is engaging the West in a "resource war," as some have suggested, remains debatable. However, it is obvious that extensive disruptions of mineral supplies to the West would benefit the Eastern Bloc. We should certainly not expect them to bail us out in the event of such supply disruptions.

The United States, Western Europe, and Japan will become even more dependent on imports of strategic minerals from politically unstable or hostile countries. These minerals will be imported in increasingly processed forms. This will reduce the number of potential suppliers. This growing import dependence, together with a loss of strategic control over the high seas, leaves the West vulnerable to supply disruptions in the event of a conventional war. The USSR's mineral and energy supplies are far more secure. Soviet control over the natural resources of the Middle East and southern Africa would radically change the world balance of power.

PART I

CANADA

Introduction

The two wealthiest nations of North America, Canada and the United States, are engaged in a truly unique relationship. They have a bilateral trade flow far in excess of that found between any other two nations in the world. One important component of this trade flow is that of mineral resources, both energy and nonfuel. Although both nations are world leaders in mineral production, both must import certain minerals. Fortunately, their mineral import needs are complementary to a great extent. As a result, Canada and the United States are each other's primary source of mineral imports.

As of late, the United States has become somewhat concerned about its dependence on other nations for minerals. Although Canada is very stable and secure as a supplier, compared with most other nations the United States must depend on, certain factors have recently emerged as sources of concern in the United States. These factors revolve around Canada's economic nationalism, which has resulted in controversial government policies concerning foreign investment, energy development, mineral trade, and so on. Given these concerns, the United States is faced with the question of whether it can continue to rely on Canada for minerals to the extent it once did.

In order to investigate this possibility, several aspects of Canada's mineral industry are examined. First is the nature of the geology of Canada and the mineral resources found in Canada. Second is the structure of Canada's mineral industry: its advantages, disadvantages, and future prospects. Third is a consideration of Canada's policies on trade, foreign investment, and mineral development. Finally, the nature of the United States–Canada relationship is discussed with an eye to improving bilateral relations in the future, and thereby assuring the dependability of Canada as a source of minerals for the United States.

Chapter I-2

Canada's Mineral Endowment

Any attempt to describe Canadian production of minerals first requires a discussion of Canada's mineral endowment. The geology of Canada is extremely complex and offers great potential for widely different types of mineral formations. The result is twofold. Canada presently produces over 60 different minerals and, furthermore, has very large reserves of a number of these minerals. However, the gradual depletion of these reserves dictates the need for additional mineral exploration.

GEOLOGY OVERVIEW

An exhaustive discussion of the geology of Canada is beyond the scope of this project, yet a synopsis is in order. The initial factor to consider about Canada's geology is the vast size of the nation. Figure I-2-1 shows a map of Canada. The second largest nation in the world, Canada has a land area of some 3,852,000 square miles. Of that total, 2,964,000 square miles are mainland, 596,000 square miles are islands (mainly arctic), and fresh-

water lakes make up another 292,000 square miles. Adding Canada's offshore areas (to within the limits of the continental slope) increases the total to 5,526,000 square miles, 2.9% of the globe's surface.[1]

This size is important for two reasons. First, it is to be expected that a great deal of geological diversity will exist in such a large area. This raises the possibility of a wide variety of mineral deposits. The second factor of importance is that a landmass so large of necessity occupies different climatic zones. This affects both the process of mineral formation and the physical recovery of minerals. In Canada's case, its location in the temperate and arctic zones of the northern hemisphere has had a significant influence on the type and rate of mining occurring there. This point is discussed at greater length in Chapter 3.

Physiographically, Canada has two major parts. The first is the Canadian Shield, a region of Precambrian age crystalline rocks which is slightly depressed at its center (Hudson Bay). The prominent

FIGURE I-2-1 Map of Canada.

feature of this 1.8-million square mile region is the impact of glaciation. The Shield is surrounded to the north, west, and south by an elevated crescent of younger, folded, and stratified rocks.

Geologic Provinces

Most material in this section is from Douglas[1] and Derry.[2] Thirteen geologic provinces have been identified in Canada on the basis of tectonic features. These provinces are illustrated in Figure I-2-2. The focal point of the Shield is the depressed Hudson Platform, which underlies most of Hudson Bay and much of northern Ontario. Those geologic provinces comprising the Shield are the Bear, Slave, Churchill, Superior, Southern, Grenville, and Nutak provinces. The Arctic, Interior, and St. Lawrence Platforms make up the northern, western, and southern transitions, respectively, from the Shield to the surrounding crescent. The crescent is made up of Phanerozoic and late Precambrian rocks, which form the Cordilleran, Innuitian, and Appalachian Orogens, or belts of deformed rocks.

Cordilleran Orogen. This highly complex orogen is located in far west Canada and underlies the bulk of the land area of British Columbia and the Yukon Territory. The subduction of the Pacific and American tectonic plates in this area resulted in the development of five major north-south-trending folds. From the west to the east, they are the Insular volcanic belt, the Coast plutonic complex, the Intermontane belt, the Omenica crystalline terrace, and the Foreland thrust and fold belt. These folds roughly coincide with a west-to-east rock-type transition of granites, volcanics, and sedimentary rocks. Major minerals produced in this orogen are porphyry copper and molybdenum, lead, zinc, silver, gold, tungsten, and coal.

Innuitian Orogen. This young fold system is located in the northernmost of Canada's Arctic islands. It is composed mainly of deformed sedimentary rocks and some volcanic rocks. Lead, zinc, and coal occurrences are prevalent, but there is little possibility that these will be exploited soon. This is due largely to their remote location and the resulting high costs of development and transportation.

Appalachian Orogen. This geologic province runs along a southwest-northeast trend line from the eastern United States to Newfoundland. Deformed sedimentary rocks predominate to the west, whereas volcanics, sedimentary rocks, and some granites are found to the east. Copper, lead, and zinc deposits are common, but the dominant minerals being recovered at this time are asbestos and coal. The sedimentary rocks in the western part of this orogen presently yield about 80% of Canada's asbestos production.[2]

Interior Platform. Situated just to the east of the Cordilleran Orogen is this region of partly deformed sedimentary rocks. It underlies most of the prairie provinces and part of the Mackenzie District in the western Northwest Territories. The only producing metal mine is the Pine Point lead-zinc mine

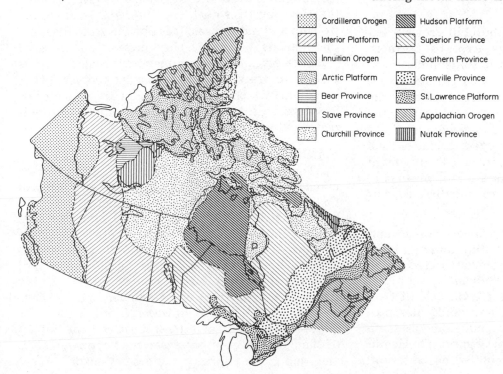

Legend:
- Cordilleran Orogen
- Interior Platform
- Innuitian Orogen
- Arctic Platform
- Bear Province
- Slave Province
- Churchill Province
- Hudson Platform
- Superior Province
- Southern Province
- Grenville Province
- St. Lawrence Platform
- Appalachian Orogen
- Nutak Province

FIGURE I-2-2 Geological provinces of Canada. (From Refs. 1 and 2.)

in the Great Slave Lake region. This geologic province is best known for its dominance in the production of oil, gas, and potash.

Arctic Platform. This platform is situated just south of the Innuitian Orogen and is also made up mainly of sedimentary rocks such as dolomite. Many lead-zinc occurrences have been identified. One such deposit, on Little Cornwallis Island, was recently developed and is now the northernmost operating mine in the world.

St. Lawrence Platform. The platform sporadically appears along a trend line roughly parallel to the St. Lawrence River. It is composed of largely flat-lying sedimentary rocks, and has no producing metal mines. Various industrial minerals, and some oil and gas, are produced here.

Bear Province. Located in the extreme northwest corner of the Shield, this province is nestled by the Interior and Arctic Platforms. Its rock types show a gradual progression from the sedimentary rocks of the platform to the granites of the Shield. Silver, copper, and gold are currently being mined. Uranium mines, once in production, have since been mined out.

Slave Province. Given its location to the southeast of the Bear Province, the Slave Province exhibits a much higher proportion of the volcanic and granitic rocks common to the Shield. This area has historically been dominated by lode gold mining, but has some potential for the development of copper and zinc operations.

Churchill Province. The Churchill Province extends up along the western edge, across the northern edge, and to the northeast of the Hudson Platform. The Province is made up mainly of granitic rocks. Most known mineralization occurs at or near the contact between the Churchill granites and the rocks (usually sedimentary) of adjacent geologic provinces. This generalized pattern includes such important cases as the Thompson nickel district in Manitoba, the Lake Athabasca uranium unconformity in Saskatchewan, and the northern edge of the Quebec-Labrador iron range.

Superior Province. The province wraps around the southern edge and contacts the Churchill Province on both sides of the Hudson Platform. Granitic and volcanic rocks are the dominant types found here. This province produces a good deal of Canada's gold, copper, and zinc, and could also produce some asbestos. Uranium conglomerates are found here, as is the new and very important Hemlo volcanogenic gold deposit near Marathon, Ontario.[3]

Southern Province. South of the Superior Province is the Southern Province, which continues south until it is just under the northern edge of Lake Superior. Sedimentary and volcanic rocks cover older Superior age rocks in this region. This province produces silver, uranium, and many other minerals, but it is best known for the very rich Sudbury (Ontario) nickel-copper-platinum intrusive and the Cobalt (Ontario) silver camp.

Grenville Province. The Grenville geologic province lies between the Superior Province and the St. Lawrence Platform. Although composed largely of granitic rocks, the Grenville Province has a large amount of anorthosites and some deformed sedimentary rocks. Anorthosites north of the St. Lawrence River yield ilmenite, while the bulk of Canada's iron ore production occurs to the north in the sedimentary regions. Molybdenite veins located near granite contacts (as at Val D'Or, Quebec) have been mined at times.

Nutak Province and Hudson Platform. These regions are both largely under water, but land-based areas are made up of granites in the Nutak and sedimentary rocks in the Hudson Platform. No metallic mineral deposits have yet been identified in either area.

MINERAL RESERVES

Table I-2-1 lists the 1980 reserve levels for 17 non-fuel minerals produced in Canada. It also presents the 1981 level for coal, and the 1982 level for petroleum and natural gas. It is obvious from the table that Canada has fairly large proportions of total global reserves of many minerals. It is also obvious that there are some other minerals (cobalt, PGM, vanadium) which are not found in large quantities in Canada. Still, the conclusion is that Canada is richly endowed with minerals.

Variation of Estimates

The figures given in Table I-2-1 are not to be taken as absolute, static levels. Reserves are identified amounts of a nonrenewable mineral which can be extracted at a profit at today's price for that specific mineral. Thus these amounts tend to increase or decrease on the basis of the changing profitability of extraction. This profitability is in turn determined by price variations, technology, the tax structure mining firms operate under, and a variety of other factors.[4] Price, however, is by far the most important factor. Table I-2-2 illustrates the impact of changing prices on the reserves of copper, nickel, zinc, and gold at operating mines in Canada.

TABLE I-2-1   Reserves of Major Canadian Minerals

Mineral	Unit	(1) Canadian reserves	(2) World reserves	(1) as percent of (2)
Asbestos	a	37,000	142,000	26.1
Cadmium	a	75	645	11.6
Cobalt	a	27	2,404	1.1
Columbium	a	122	4,082	3.0
Copper	a	32,000	494,000	6.5
Gold	b	13	1,040	1.3
Ilmenite	a	44,000	199,000	22.1
Lead	a	12,000	127,000	9.5
Molybdenum	a	635	9,480	6.7
Nickel	a	8,700	60,000	14.5
Platinum group	b	9	1,180	0.8
Potash	c	2,700	9,100	29.7
Silver	b	1,600	8,125	19.7
Tantalum	a	1	65	1.5
Tungsten	a	270	2,585	10.4
Zinc	a	30,000	162,000	18.5
Uranium	a	235	2,590[d]	9.1
Petroleum	e	7,020	670,189	1.1
Natural gas	f	97,000	3,023,527	3.2
Thermal coal	g	16,091	1,963,887	0.9

[a]Thousand metric tons.
[b]Million troy ounces.
[c]Million metric tons K_2O.
[d]Western world data only.
[e]Million barrels.
[f]Trillion cubic feet.
[g]Million metric tons minable coal.

Sources: Mineral Facts and Problems 1980; Helmut Schmidt and Manfred Kruszana, *Regional Distribution of Mining Production and Reserves of Mineral Commodities in the World.* Hanover, West Germany: Federal Institute for Geosciences and Natural Resources, January 1982; *BP Statistical Review of World Energy 1981*; *World Coal*, July/August 1981; *Oil and Gas Journal*, 27 December 1982.

TABLE I-2-2   Selected Metal Reserves in Canada, 1975-1981, and Their Relation to Average Prices, 1974-1981

Year	Copper Total (thousand metric tons)	Price (cents/lb)	Gold Total (kg)	Price (dollars/oz)	Nickel Total (thousand metric tons)	Price (dollars/lb)	Zinc Total (thousand metric tons)	Price (cents/lb)
1974		77.3		159.74		1.74		36.0
1975	17,048	64.2	372,371	161.49	7,268	2.03	28,274	39.0
1976	16,803	69.6	354,105	125.32	7,266	2.20	28,083	37.0
1977	16,634	66.8	396,012	148.31	7,326	2.30	27,407	34.4
1978	16,471	66.5	366,421	193.55	7,389	2.08	26,908	31.1
1979	15,840	93.3	409,582	307.50	7,070	2.49	26,452	37.3
1980	16,869	102.4	540,493	612.56	7,179	3.41	28,635	37.4
1981	16,831	85.1	769,890	459.64	8,304	3.43	29,436	44.6

Sources: Canadian Mining Journal, February 1982; U.S. Department of the Interior, Bureau of Mines, *Minerals and Materials/ A Monthly Survey*, various issues; *Mineral Commodities Summaries 1983*.

Ultimately, these current reserves will be mined out. Estimates by Energy, Mines and Resources Canada indicate that very little of the mineral production expected in the year 2000 will come from reserves identified in 1981.[5] It has been estimated that some 200 new mines must be developed in the remainder of this century if Canadian mineral production is to maintain its share of world markets. This would entail the establishment of 10 new mines per year, compared with the annual average of six per year from 1960 to 1980.[6] New reserves must be added to the present amounts; otherwise, Canadian production of minerals will fall.

Addition to reserve levels can occur in two ways. Known and unknown subeconomic resources can become reserves if real metal prices increase or if extraction costs fall. Alternatively, reserves can increase if newly discovered resources are such that they can be profitably recovered at current prices. In either case, some degree of exploration is called for, as presently exploited mine reserves cannot last forever.

THE ROLE OF EXPLORATION

If exploration is to lead to new discoveries that will ultimately increase, or at least maintain, Canada's mineral reserves, several factors must be dealt with. First, exploration will not be as easy as it once was. Surface outcrops of ore that served as the early foundation for the mining industry will rarely, if ever, be found again. The same point holds for most near-surface mineralization. Thus exploration will have to focus on deep mineralization, as well as on those regions of Canada not yet thoroughly examined for surface and near-surface mineral occurrences.

The Northern Shift

Exploration trends in Canada point to an increased emphasis on relatively unexplored regions. Figure I-2-3 shows that the exploration expenditures in the Yukon and Northwest Territories have increased from roughly 5% of total expenditures in 1967 to 20% in 1978. DeYoung[7] has also shown this pattern to hold true on the basis of expenditures corrected for inflation, that is, in constant-dollar amounts. Part of this increase, relative to other regions, can be attributed to higher costs caused by the remote location and extreme climate. Even so, the figures indicate an increased emphasis on exploring regions once largely avoided. The discovery, and recent development of the Polaris lead-zinc mine on Little Cornwallis Island in the Canadian Arctic, seems to confirm this.

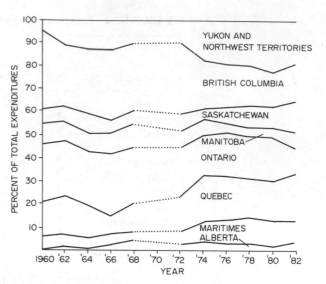

FIGURE I-2-3 Exploration expenditures by province, 1967–1978. (From Ref. 9.)

Expenditure Levels

Even if exploration continues largely in the southern regions of Canada, it will be more difficult because of the need to look for deep ore occurrences. This, plus the forays into the Arctic, will increase the real costs of exploration ventures. Mid- to late 1970s exploration expenditures for all of Canada averaged over C$100 million per year (in constant 1974 dollars). This average annual investment may have to be doubled, or possibly even quadrupled, to replace the reserves now being mined.[6]

This estimate of required investment in exploration should be regarded as an average goal for exploration expenditures. Changing economic conditions tend to create fluctuations in these expenditures. *Canadian Minerals Yearbook* data show that annual exploration expenditures from 1967 to 1978 ranged from C$73.9 to C$194.9 million but with a great deal of variation from year to year. This holds true for both current- and constant-dollar comparisons. Current-dollar expenditures totaled C$400 million in 1980[8] and reached an all-time high of C$596.2 million in 1981,[3] but this was the last recent period of strong growth for the mining sector. Preliminary figures from Energy, Mines, and Resources Canada indicate a 32% drop in exploration outlays from 1981 to 1982, with a 1982 outlay of C$413.7 million.[3] Expenditures in British Columbia, for example, dropped 53% from 1981 to 1982.

The Economics of Exploration

Mineral exploration efforts undertaken by private firms operating in market economies are subject to the same criteria as any other business venture. That is, the business venture must be profitable. It must yield an "adequate" return on investment or the investment funds will be placed elsewhere. However, mineral exploration ventures are far more risky than are most other ventures. Aside from the typical considerations of price, taxation, and so on, faced by all companies, exploration firms must deal with several fairly unique constraints.

The first is the time required to see if an investment will pay off. Few other investments have such long lead times. Any well-run exploration venture must consistently ensure that it acquires all available information on a mineral prospect. Each new level of information is utilized to test further the economic and geologic feasibility of that prospect. This information gathering is never a rapid process. Even with the advent of sophisticated geophysical instruments, progress is slow. The increased emphasis on deep mineral occurrences could further slow down the process of information gathering. In any case, Mackenzie noted that the total time required for a complete, successful exploration program can run from six to 15 years.[9]

The second constraint is the inherent riskiness of the exploration activity. Mackenzie pointed out three important risks.[9] First, there is the question as to whether or not a discovered mineral deposit is economic at today's prices. That is, can the mineral be recovered economically and at a profit? Second, even an economically recoverable deposit can vary in quality. Thus the rate of return on the exploration investment may not be adequate for some companies. Finally, fluctuations in commodity prices can radically alter the expected profitability of a venture not to be exploited for some time.

Taken all together, these factors present an exploration firm with a choice. Does it want to tie down funds for extended periods of time in an operation which, for a variety of reasons, may offer little or no return on investment? Such a decision is obviously influenced by corporate goals for profit and survival. The actions of major Canadian firms during the recent recession indicated that concern about survival was a dominant influence in their investment activities. Major firms such as Noranda, Cominco, Denison, Inco, and Falconbridge all cut back on exploration activities to help ride out the recession.[10] At least one of these firms, Falconbridge, has announced that exploration expenditures for 1983 will be reduced from earlier levels.[3]

Such cutbacks in exploration do have a useful purpose in the short run, but they appear to ultimately hurt the firm over the long run.[10] Cutting back on exploration expenses may impair information gathering to the extent that a potentially feasible prospect is not identified as soon as might otherwise be the case. Thus long-run cash flows decrease, possibly to the extent that the financial strength of the firm may be tested. As Peters[11] pointed out:

> An exploration effort is a sequence of investigations and decisions in which the experience of past failures is just as important as that of past successes. A major requirement, therefore, is a continuity of experience within the exploration department. It is not economical to make drastic reductions in exploration personnel and then expect a totally new department to discover ore bodies quickly in times of brighter outlook.

In addition, the exploration cutbacks undertaken by private firms have a social impact in a nation such as Canada. All Canadians benefit in some measure from the positive trade balances generated by the export of minerals. Should reserves not be kept up through continued exploration, production and exports of minerals will ultimately fall. The trade-off between private and social costs is a touchy issue and is discussed at greater length in another section of this paper.

REFERENCES

1. Douglas, R. J. W., ed *Geology and Economic Minerals of Canada.* Ottawa: Minister of Supply and Services Canada, 1976.
2. Derry, Duncan R. "Canada's Mineral Resources." *Engineering and Mining Journal* 181 (November 1981).
3. See various issues of *The Northern Miner*, for example, 3 March 1983.
4. van Rensburg, W. C. J. "The Classification of Coal Resources and Reserves." Mineral Resource Circular 65-1980. Austin, Tex.: The University of Texas—Austin, Bureau of Economic Geology 1980.
5. Energy, Mines and Resources Canada. *Mineral Policy: A Discussion Paper.* Ottawa: Minister of Supply and Services Canada, 1982.
6. Owens, O. E. "Canada and Mineral Exploration in the 1980's" *CIM Bulletin* 73 (August 1980).
7. DeYoung, John H., Jr., "Effect of Tax Laws on Mineral Exploration in Canada." *Resources Policy* 3 (June 1977).
8. Bonus, John L., Managing Director, The Mining Association of Canada. Personal Communication, 5 November 1982.
9. Mackenzie, Brian W., and Geoffrey G. Snow. "The Environment of Exploration: Economic, Organizational, and Social Constraints." *Economic Geology*, 75th Anniversary Volume (1981).
10. "Exploration-Changing Perceptions." *Mining Journal* 299 (26 November 1982).
11. Peters, W. C. "The Economics of Mineral Exploration." *Geophysics* 34 (August 1969).

The Canadian Mineral Industry

HISTORICAL BACKGROUND

Interest in, and the exploitation of, minerals in Canada can be traced back some 400 years. It is thought that the discovery of silver and copper occurrences near the St. Lawrence River may have encouraged Samuel de Champlain to establish permanent settlements in Canada in the seventeenth century.[1] Coal deposits in Cape Breton were worked in the 1700s, and iron was exploited shortly thereafter in regions now called Quebec. These activities and others helped establish a trend crucial to Canada's history: a mineral discovery, next a settlement, then related economic growth, and finally a move onward.

Even with this pattern, it was not until the 1850s that mining and processing of minerals took on major economic significance.[2] Iron, steel, and coal industries were by then well established in the Maritime Provinces, and things picked up quickly from there. Among the most important examples of mineral discoveries and subsequent exploitation were nickel at Sudbury, Ontario, in 1883, the Sullivan lead-zinc-silver deposits at Kimberly, British Columbia, in the 1880s, the Noranda copper deposit in Quebec in 1921, and the Leduc oil deposit in the late 1940s.

IMPORTANCE TO THE ECONOMY

By the 1920s Canada's mineral industry had established itself as a world-class producer of a variety of minerals. The industry's historical dependence on exports as the basis for its development caused some problems, but it fostered the development of an industry crucial to the entire Canadian economy. This importance can be illustrated in several interesting ways.

Employment

Roughly 135,000 miners were employed in Canada in 1982[3] while some 280,000 overall were employed, in 1981, in mines, mills, smelters, and related operations.[4] Although these figures represent only 1.4 and 2.7%, respectively, of the total labor force,[5] two facts show the significance of the figures. First, the mining industry indirectly accounts for a much larger overall employment level because of its impact on relevant support industries.

Examples would be the Canadian railway system and manufacturers of heavy machinery. All told, over 6 percent of Canada's labor force is based, directly or indirectly, in mining.[1]

Second, this level of employment takes on added significance because of its geographic dispersion. Employment directly related to mining or primary mineral processing serves as the dominant economic base for some 175 communities throughout Canada.[4] Two examples of such communities are Sudbury, Ontario, and Faro, Yukon Territory. These and other communities are the sites for the 260 mines, 230 mills, 16 smelters, and 15 refineries used by the nonfuel minerals sector alone.[6] These facilities and their employees are scattered throughout Canada. Locations range from the arctic north (the Polaris mine on Little Cornwallis Island) to southern Canada (Sudbury), with just as great a dispersion along an east-west axis.

Impact on Gross National Product

Table I-3-1 illustrates the gross value of Canadian mineral production as a percentage of gross national product (GNP) from 1960 to 1982. The annual share of GNP held by nonfuel mineral production (exclusive of coal, oil, and gas) has averaged 4.37% over this time frame. The annual share of GNP held by total mineral output (fuel plus nonfuel) has averaged 7.49% of GNP. The overall impact mineral production has on GNP is somewhat higher because of the GNP share held by industries that support mining. Examples would be heavy-equipment manufacturers and private exploration firms.

An alternative illustration of this point focuses on the importance of mining to provincial economic strength. Table I-3-2 illustrates the share of census value added held by mining out of all goods-producing industries. Several interesting points can be drawn from Table I-3-2. First, as is the case with the Yukon and Northwest Territories, some areas have economies which are almost totally dependent on mining. The converse of this would perhaps be Ontario. Although it ranks second in overall provincial mineral production (by value), Ontario's economy is so large and so diversified that mining holds only a small share of total value.

TABLE I-3-1   Canadian Mineral Production as Percentage of GNP, 1930–1982
(Million Canadian Dollars)

	(1) Mineral production		(2)	(3)	(1) as	(2) as
Year	Nonfuel	Fuel	Total	GNP (current dollars)	percent of (3)	percent of (3)
1982	10,920	22,160	33,080	349,040	3.13	9.48
1981	13,290	19,050	32,340	331,300	4.01	9.76
1980	13,897	16,119	30,016	291,900	4.76	10.28
1979	11,569	14,529	26,098	261,600	4.42	9.98
1978	8,429	11,232	19,661	230,407	3.66	8.53
1977	8,600	9,873	18,473	209,379	4.11	8.82
1976	7,584	8,109	15,693	191,166	3.97	8.21
1975	6,694	6,653	13,347	165,343	4.05	8.07
1974	6,552	5,202	11,754	147,528	4.44	7.97
1973	5,143	3,227	8,370	123,560	4.16	6.77
1972	4,041	2,367	6,408	105,234	3.84	6.09
1971	3,948	2,015	5,963	94,450	4.18	6.31
1970	4,004	1,718	5,722	85,685	4.67	6.68
1969	3,269	1,465	4,734	79,815	4.10	5.93
1968	3,379	1,343	4,722	72,586	4.66	6.51
1967	3,146	1,235	4,381	66,409	4.74	6.60
1966	2,829	1,152	3,981	61,828	4.58	6.44
1965	2,669	1,046	3,715	55,364	4.82	6.71
1964	2,392	973	3,365	50,280	4.76	6.69
1963	2,142	885	3,027	45,978	4.66	6.58
1962	2,070	811	2,881	42,927	4.82	6.71
1961	1,929	674	2,603	39,646	4.87	6.57
1960	1,927	566	2,493	38,359	5.02	6.50
1955	1,381	414	1,795	28,528	4.84	6.29
1950	844	201	1,045	17,955	4.70	5.82
1940	451	79	530	6,713	6.72	7.89
1930	212	68	280	5,720	3.71	4.89

Sources: The Globe and Mail, 26 January 1983; "Annual Industry Review and Forecast," Canadian Mining Journal, various issues; Canadian Minerals Yearbook, various issues; International Economic Indicators, December 1982; The Financial Post, 5 February 1983; Geology and Economic Minerals of Canada, Part A, 1976.

TABLE I-3-2   Census Value Added in Goods-Producing Industries, by Industry and Province, 1977

(1) Province	(2) Total census value added	(3) Mining share	(4) (3) as percent of (2)
Newfoundland	1,402,998	435,980	31.1
Nova Scotia	1,722,661	121,189	7.0
New Brunswick	1,495,598	79,538	5.3
Prince Edward Island	167,212	—	—
Quebec	19,742,501	901,844	4.6
Ontario	34,537,717	1,418,804	4.1
Manitoba	2,801,235	152,665	5.4
Saskatchewan	3,834,553	989,936	25.8
Alberta	14,558,941	7,843,187	53.9
British Columbia	9,894,289	1,154,107	11.7
Yukon and Northwest Territories	261,235	221,943	84.9
Canada	90,418,840	13,319,193	14.7

Source: Ref. 94.

Export Orientation

The mineral sector's export orientation is important in terms of its impact on Canada's trade balance of payments. Simply put, the mining sector is currently one of only a few sectors generating a positive trade payments balance. Daly,[6] among others, noted that the Canadian mining sector still has a competitive advantage in world trade, whereas the opposite is true of the manufacturing sector in Canada. Thus exports by the mining sector are crucial to balancing out the deficits incurred by importing manufactured products of various types.

Table I-3-3 and Figure I-3-1 show several important sides of this issue. First, over the period 1960–1982, an average of 81.3% (by value) of the minerals produced in Canada were exported. Even with the dramatic rise in value of oil and gas after the 1973–1974 period, nonfuel mineral exports made up an average 74.7% of the minerals exported by Canada. This bias in favor of nonfuel minerals, especially base metals, occurs because Canada exports little of its fuel production.[7] Furthermore, mineral exports have averaged 31.7% of the value of all items exported from Canada since 1960. This is impressive for an industrial

sector that accounts directly for only 7.5% of Canada's GNP.

Finally, the Canadian Manufacturers' Association has noted that almost 50% of the jobs in Canadian industries exist because of exports.[8] Since mineral exports account for roughly 32% of all exports, it could be that up to 16% of the jobs in the entire industrial sector are derived from mineral exports.

Transportation Impact

A final example of the importance of the mineral sector to the Canadian economy concerns the dependence of the transportation system on mineral traffic. Half the total railway freight revenues comes from the movement of crude and fabricated minerals, while half the cargoes loaded for export at Canadian seaports is composed of minerals.[9] In fact, it has been argued that mineral freight revenues subsidize the transportation of other commodities on Canadian railways.[10]

This is due in part to the extremely low Crow's Nest Pass rates charged for the transportation of grain. For the past 50 years, that rate has been half a cent per ton-mile.[11] These rates caused 1981 grain transport operating losses of C$335 million for

TABLE I-3-3 Importance of Mineral Exports to Canada's Economy, 1960-1982 (Values in Million Canadian Dollars)

Year	(1) Mineral production	(2) Nonfuel	(3) Mineral exports Fuel	(4) Total	(5) Canada exports	(4) as percent of (1)	(4) as percent of (5)	(2) as percent of (4)
1982								
1981								
1980	33,080	15,551	10,141	25,692		86		61
1979	26,098							
1978	19,661	9,211	5,291	14,503	51,919	74	28	64
1977	18,473	8,081	5,078	13,159	43,684	71	30	61
1976	15,693	7,049	5,026	12,075	37,651	77	32	58
1975	13,347	5,955	5,276	11,231	32,587	84	34	53
1974	11,754	6,374	4,844	11,218	31,739	95	35	57
1973	8,370	4,982	2,310	7,292	24,838	87	29	68
1972	6,408	3,869	1,630	5,500	19,671	86	28	70
1971	5,963	3,797	1,242	5,039	17,397	85	29	75
1970	5,722	4,233	970	5,202	16,401	91	32	81
1969	4,734	3,293	771	4,064	14,498	86	28	81
1968	4,722	3,432	672	4,103	13,270	87	31	84
1967	4,381	2,893	577	3,469	11,112	79	31	83
1966	3,981	2,650	473	3,123	10,071	78	31	85
1965	3,715	2,363	420	2,782	8,633	75	32	85
1964	3,365	2,359	225	2,584	8,094	77	32	91
1963	3,027	1,694	320	2,014	6,799	67	30	84
1962	2,881	1,621	314	1,935	6,179	67	31	84
1961	2,603	2,030	205	2,235	5,756	86	39	91
1960	2,493	2,048	119	2,167	5,266	87	41	95

Sources: Canadian Minerals Yearbook, various issues; Ref. 12.

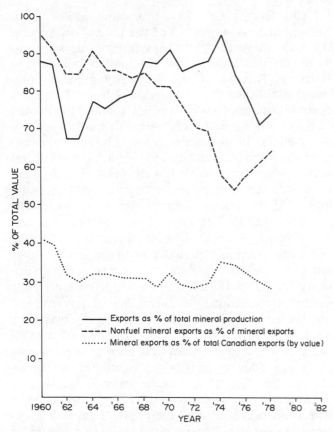

% OF TOTAL VALUE

YEAR

——— Exports as % of total mineral production
- - - - Nonfuel mineral exports as % of mineral exports
········ Mineral exports as % of total Canadian exports (by value)

FIGURE I-3-1 Export orientation of Canadian mineral output. (From "Energy, Mines and Resources," *Canadian Minerals Yearbook*, various issues.)

both the major railways, Canadian National and Canadian Pacific.[12] However, other items such as minerals contributed to a net profit.

The situation changed in 1982 when, for example, Canadian National posted a grain-based loss of C$296 million. Drops in minerals traffic were enough that Canadian National posted a C$120 million operating loss on total rail traffic in 1982.[13] It should be noted that the Crow's Nest Pass rates are now in the process of being changed.

CONCERNS ABOUT THE MINERAL INDUSTRY
Firms operating in the mineral industry are often exposed to risks and costs—in exploitation and development—that most other industries can avoid. These risks and costs manifest themselves in a variety of ways.

Impact of Export Orientation
The industry's export orientation leaves it vulnerable to swings in global economic activity. The fact that raw minerals demand levels react more than proportionately to economic downturns only exacerbates this vulnerability.[14] A good illustration can be drawn from the recent global recession. Canada, which ranks with the USSR as one of the

top nickel producers worldwide, exports the vast majority of its annual production. From 1975 to 1978, an average of only 6% of annual nickel production was consumed domestically, with most of the remainder being exported from Canada.[15]

The decreased demand for nickel during the recent recession caused world mine production of nickel to fall by roughly 25% from 1981 to 1982. Canada's production, however, given its free market export basis, fell by almost 50%. Production in the USSR, on the other hand, fell by only 2.3%.[16] Furthermore, the drop in nickel production (as with other minerals) in Canada resulted in the unemployment of 70,000 Canadian miners by November 1982.[2] Thus mining's share of Canada's unemployment ranks at that time was 5.4% even though mining makes up only 1.4% of the total labor force.[5]

An additional concern about this export orientation is the fact that Canada's exports go primarily to the United States. As discussed more fully in Chapter I-5, the U. S. economy is no longer as vibrant as it once was. Thus Canada is concerned about its dependence on the strength of the U. S. economy. This concern is especially strong given the forecasts for low growth in the U. S. economy. The International Monetary Fund, for example, has projected only a 1.6% increase in the U. S. gross national product for 1983.[17] Prospects beyond 1983 are still uncertain.

Declining Export Competitiveness
Even if Canada accepts the occasional slump brought on by its export orientation, and the resultant dependence on strong global economic activity, Canada may not be able to maintain its share of worldwide mineral trade. There are both domestic and external reasons for this declining export competitiveness. On the domestic level, there are indications that the mining industry is losing its competitive edge over producers in other countries. An earlier study by the Canadian federal government indicated that the industry was competitive in only three out of 10 possible categories.[9] These three were technology, energy costs, and trade revenues. The industry was not competitive in such areas as ore grade, labor costs, and taxation levels.

Energy, Mines and Resources Canada summarized Canada's competitive position in the production of nickel, copper, and zinc. Eleven variables quite similar to those used by the federal study were investigated.[10] Although Canada's position overall was competitive, this position was based generally on only three factors: low energy costs, recovery of by-products and co-products, and favorable currency exchange rates. As noted in the study, the first and third factors are subject to quick variation. The recent drop in the price of oil may significantly reduce Canada's advantage with respect to energy costs. Although the issue of competitive advantage is based on many factors, present domestic trends do offer cause for just concern.

The externally based factor behind Canada's declining competitiveness is the advent of new producers in developing nations. Many of these nations have few sources of foreign exchange other than minerals. Thus these nations place great emphasis on maintaining, if not expanding, the level of mineral production and exports. This is true even when recessionary market forces would normally call for decreased production. For example, nickel production in the Philippines and Botswana, two large exporters of nickel, declined by only 14 and 1% respectively, from 1981 to 1982.[16] This compares with the 50% reduction in Canadian production over the same time period. This tendency for developing nations to continue production during economic downturns will have an adverse impact on Canada's trade competitiveness in the future.[18]

Industry Structure

The third concern about the minerals industry's structure focuses on the role of junior mining companies and Crown corporations. Junior mining companies (JMCs) are small, independent exploration firms with budgets rarely exceeding several million dollars a year. Their significance lies in the fact that they are often willing to take risks typically avoided by larger, more diversified firms. As a result, these companies have historically accounted for a disproportionately large share of total mineral discoveries relative to their capital expenditures. For example, data from Energy, Mines and Resources Canada indicates that one-third of the major mineral discoveries made in the 1970s were made by JMCs. This occurred even though their share of exploration outlays dropped from 50% to 10% of the total from 1969 to 1979.[7]

Freyman[19] noted that 62% of the metallic mineral discoveries in Ontario made from 1951 to 1974 were made by JMCs. This took place even though these firms accounted for only 28% of the exploration outlays. Obviously, these organizations are quite important to the continued success of the Canadian mineral industry. The concern is that present tax policy, combined with declining business and the risky nature of exploration itself, may force these firms out of business. This is discussed at greater length in Chapter I-6.

Crown corporations are provincially or federally owned companies active in the private sector in Canada. It has been estimated that there are 300

to 400 federal Crown corporations alone in all fields, but only a handful of the federal ones are in the mineral sector.[20] Perhaps the two most prominent on the federal level are Petro-Canada and Eldorado Nuclear. There are many such corporations established at the provincial level. Two such corporations are Quebec's SOQUEM and the Potash Corporation of Saskatchewan, both of which are involved in the minerals sector. The incentive behind the establishment of Crown (and other public) corporations varies with the situation. In any case, it is assumed that public ownership and management will be better than its private counterpart.

These public firms have lately become the target of much criticism. The argument is that public firms are given tax breaks and other special operating concessions which allow them to compete unfairly with private firms. One such provision is that Crown corporations on one level of government not be taxed by the other level.[10] This provision is a major incentive for the large numbers of provincially established companies. It allows the provinces to keep a share of profits that would otherwise be lost to the federal corporate income tax.[20] The debate over the usefulness of publicly owned firms will probably continue for some time, but these firms do seem to have a good operating record in the minerals sector.

MAJOR NONFUEL MINERAL COMMODITIES

The Canadian mineral industry is currently the third largest in the world in terms of output. On the basis of exports, however, the Canadian industry is the undisputed world leader.[12] This export orientation, as noted above, has led to the development of an industry that is constantly in a state of flux in response to market forces. This must be so, for the failure to adapt to changing market conditions would lead to a decreased Canadian share of the world mineral trade market. Out of the 60+ minerals produced in Canada, 13 are of particular importance because of their export orientation. Moreover, these minerals are, except for molybdenum, exported to the United States.

Some minerals, such as copper and iron ore, are exported to the United States largely because of the proximity of the two nations. Other minerals are exported because of lower production costs in Canada. This is true, for example, of aluminum. Finally, there are a select number of "strategic minerals" (such as cobalt and the platinum group metals) which Canada exports to the United States. Although these quantities are small, they are important in that they represent a small alternative to already limited U.S. supply-source choices. This point is discussed further in Chapter I-5.

Figure I-3-2 shows the relative share of the total

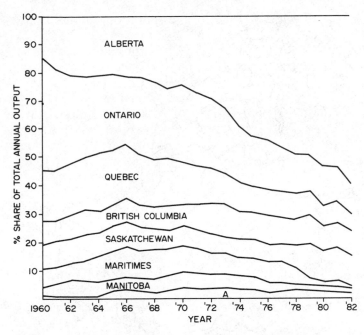

value of Canada's mineral production held by different provinces and territories. Without a doubt, the most prominent feature of the figure is the increase in Alberta's share which began in the early 1970s. This was due, of course, to the world price increases for oil and natural gas. The increase in this value is such that it has overshadowed absolute production increases in other provinces.

The following description of major mineral commodities is grouped alphabetically according to whether the mineral is nonfuel or energy in its use.

Aluminum

As with many other mineral industries, Canada's aluminum industry was developed to a great extent by foreign interests.[21] These developments were initiated in 1899 when Pittsburgh Reduction, the forerunner of Alcoa, began construction of a smelter at Shawinigan, Quebec. The facility was run by Northern Aluminum Company, Limited, an Alcoa branch developed to oversee all Canadian operations. By 1925, Northern Aluminum had changed its name to the Aluminum Company of Canada, or Alcan. Alcan then proceeded to purchase hydroelectric sites and power plants in the Saguenay, Quebec, region and built a smelter there in 1927. In 1928, Alcoa created another Canadian corporation, Aluminum Ltd., to oversee all its foreign holdings. In 1937, the U.S. Justice Department filed an antitrust suit against Alcoa because Alcoa and its subsidiary Alcan controlled the North American aluminum market. In 1950, Alcoa and Alcan were ordered to sever their financial ties.

Aside from Alcan, the only other Canadian aluminum firm is Canadian Reynolds Metals Company Ltd. Formed in 1955 as the Canadian British

Aluminum Co. Ltd., it was initially a joint venture between British Aluminum and the Q.N.S. Paper Company Ltd. By 1970, the company's assets were 100% controlled by Reynolds Metals of the United States—hence the name changed to the present one.

Figures I-3-3A and I-3-3B show the location of Canadian smelting facilities, while Table I-3-4 notes the annual rated capacity of these facilities. Total smelting capacity now stands at 1,115,000 metric tons per year.[22] This will increase to 1,335,000 metric tons per year upon the completion of the Becancour, Quebec, smelter to be constructed by the French firm Pechiney Ugine Kuhlman (PUK).[23] The 1984 completion of the 122,000-metric ton Baie Comeau expansion will bring the total capacity for aluminum production to 1,457,000 metric tons per year.[12] This will comprise roughly 8.2% of world capacity as of year end 1982.[16]

However, Alcan has substantial overseas investments in smelting capacity, as shown in Table I-3-5. Addition of this capacity to Canada's domestic smelting capacity yields a total of 2,426,000 metric tons per year. This means that Canadian operations help account for 12.4% of the world's smelting capacity, effective at year end 1982.[16]

Canada produces no bauxite, so feed for domestic smelters comes from a variety of sources. Canadian firms have diversified their supply sources so as to reduce their vulnerability to supply cutoffs. For example, Alcan has bauxite investments in seven countries and alumina production facilities in seven countries. Actual import sources have varied over time due to a variety of factors. Until 1974, Canada imported most of its bauxite from Guyana, in which Alcan's 100%-controlled Demerara Bauxite Co. Ltd. was located. Imports

FIGURE I-3-3A   Nonferrous smelters and refineries in Canada. (From Refs. 15 and 98.)

FIGURE I-3-3B   Key to Figure I-3-3A Nonferrous Smelters and Refineries in Canada

Number	Location	Product
1	Kitimat, B.C.	Aluminum ingots and alloys
2	Port Coquitlam, B.C.	Tungsten, titanium, and tantalum-columbium powders and carbides
3	Kamloops, B.C.	Copper
4	Trail, B.C.	Lead, zinc, gold, silver, cadmium
5	Fort Saskatchewan, Alta.	Nickel, cobalt, copper sulfide
6	Flin Flon, Man.	Copper anodes, zinc, cadmium
7	Thompson, Man.	Nickel, precious metals, cobalt oxide
8	Timmins, Ont.	Zinc, cadmium
9	Noranda, Que.	Copper anodes
10	Cobalt, Ont.	Silver
11	Falconbridge, Ont.	Copper-nickel matte
12	Brampton, Ont.	Gold, silver, platinum-group metals
13	Port Colborne, Ont.	Nickel products, copper products, platinum-group residues, cobalt products
14	Port Hope, Ont.	Uranium hexafluoride, uranium metals
15	Haley, Ont.	Magnesium and magnesium alloys
16	Ottawa, Ont.	Gold, silver
17	Beauharnois, Que.	Aluminum ingots and alloys
18	Montreal East, Que.	Copper products, gold, silver, nickel sulfates
19	Sorel, Que.	Titanium dioxide slag, iron
20	Shawinigan, Que.	Aluminum ingots and alloys
21	Isle Maligne, Que.	Aluminum ingots and alloys
	Jonquiere, Que.	Aluminum ingots and alloys
22	Baie Comeau, Que.	Aluminum ingots and alloys
23	Murdochville, Que.	Copper anodes
24	Belledune, N.B.	Lead, silver

TABLE I-3-4   Canada's Domestic Aluminum Smelting Facilities, 1982

Location	Owner	Annual capacity (thousand metric tons)
Jonquiere, Que.	Alcan	432
Isle Maligne, Que.	Alcan	73
Shawinigan, Que.	Alcan	84
Beauharnois, Que.	Alcan	47
Grande Baie, Que.	Alcan	177
Kitimat, B.C.	Alcan	268
Baie Comeau, Que.	Canadian Reynolds	154
		1115

Sources: Refs. 12 and 22.

TABLE I-3-5   Alcan Investments in Non-Canadian Smelting Facilities,
January 1, 1981

Location	Percent ownership by Alcan	Annual capacity (thousand metric tons)
Lynemouth, U.K.	100	125
Saramenha, Brazil	100	60
Aratu, Brazil	100	28
Ludwigshafen, West Germany	100	44
Belgaum, India	55.27	73
Hirakud, India	55.27	25
Alupuram, India	55.27	20
Kurri Kurri, Australia	70	90
Valladolid, Spain	42.53	26
Aviles, Spain	42.53	100
San Ciprian, Spain	23.39	180
Kambara, Japan	50	64
Tomakomai, Japan	50	134
		969

Source: Ref. 22.

levels from Guyana, however, have steadily dropped since the 1971 peak, for Guyana at that time nationalized the Alcan subsidiary. Brazil now provides almost 50% of Canada's bauxite imports.[12] Similarly, 80% of Canada's alumina imports comes from Australia, the United States, and Jamaica.[12] Figure I-3-4 shows the level of Canadian imports of bauxite and alumina from 1964 to 1980.

By diversifying their supply sources, Canada's aluminum firms have invested in capacity levels far in excess of their current needs. For example, Alcan's overseas alumina plants have an annual capacity of 5,497,000 metric tons.[22] As shown in Figure I-3-4, Canadian imports of alumina topped the 1,000,000-metric ton level only once from 1964 to 1980. The result of this situation is that additional supplies should be readily available should Canada's domestic smelters decide to increase production levels.

However, Canadian firms have also looked into alternatives to bauxite as a feedstock. This investigation took place even before the Guyana nationalization, so it was at least partly motivated by cost considerations. Alcan, in the 1960s, acted on the cost consideration when it attempted to perfect a process involving monochloride purification of directly reduced aluminous ores. This process seemed promising but could not compete with the Bayer/Hall-Heroult process, so it was abandoned.[24] In the 1970s, Alcan and Pechiney Ugine Kuhlman of France worked on other processes to recover alumina from nonbauxitic sources (such as clays), but with similar results.[21]

It is clear that cheap hydroelectricity has helped Canada establish a very competitive market position for its aluminum. As noted by Skea, the union of an aluminum smelter and a hydroelectric plant is a profitable and sensible one.[25] The aluminum

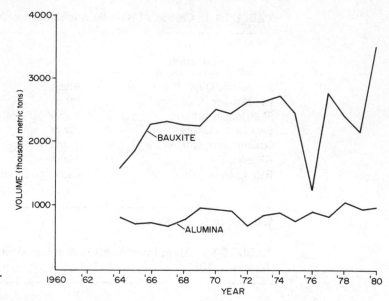

FIGURE I-3-4 Canadian imports of bauxite and alumina, 1964–1980. (From Refs. 15 and 35.)

smelter needs large, steady amounts of electricity generated at the lowest cost possible; this is something a capital-intensive hydro facility is well suited for. The hydro facility, given its capital-intensive nature, needs a guaranteed baseload in turn. The experience in Canada seems to verify Skea's argument.

Hydroelectric plants with a total installed capacity of 3.583 million kilowatts have been constructed by Alcan to power its domestic smelters.[22] Hydro-Quebec provides electricity to Canadian Reynolds's Baie Comeau facility. Hydro-Quebec has also contracted to provide the electricity for the Pechiney Ugine Kuhlman smelter to be constructed at Becancour. In fact, a major incentive for the decision to locate at Becancour was Hydro-Quebec's offer to provide the first five years of electricity at half-price, and a less-than-cost price for the following 10 years.[26]

Figure I-3-5 shows production and export levels for domestically produced aluminum. It is obvious that the bulk of annual production is exported. Although Canada's exports go worldwide, they have lately been focused more and more on the United States. In 1955, Canada exported 460,000 metric tons of aluminum. Of that amount, 176,000 metric tons, 38.2% went to the United States. By 1976, the corresponding figure was 73.5%.[21] Obviously, different smelters export to different locations. Whereas the Quebec-based smelters export largely to the U.S. market, Alcan's Kitimat smelter exports 75% of its metal to Asian markets. In fact, 45% of Kitimat's exports go to Japan.[22]

Future prospects look good for Canada's aluminum industry in terms of export growth potential. For one thing, a large number of world producers have been reducing domestic smelting capacity because of high energy costs. Actual and future

FIGURE I-3-5 Canadian production and exports of aluminum metal, 1960–1982. (From Refs. 15, 16, and 35.)

planned reductions in capacity will reduce output capacity in Japan by almost 50%, for example. Japan's smelters had an average energy cost in 1981 of 70 mills per kilowatt-hour. Canada, on the other hand, averaged 2 to 3 mills.[27] In addition, the United States is expected to continue as a net importer of aluminum.[16]

Asbestos

Asbestos produced in Canada is in the form of chrysotile, a hydrous magnesium silicate found in serpentinized peridotite. This ore is in almost every case mined by open-pit methods. The vast majority of Canada's asbestos production is centered in southern Quebec along the St. Lawrence River (Figures I-3-6A and I-3-6B). The only other currently producing mine in Canada is found in northern British Columbia. However, asbestos was previously

FIGURE I-3-6A Asbestos and potash mines in Canada, 1983. (From Refs. 15 and 98.)

FIGURE 1-3-6B Key to Figure I-3-6A Asbestos and Potash Mines in Canada

	Company	Location
	Asbestos mines	
A	Cassiar Asbestos Corp.	Cassiar, B.C.
B	Asbestos Corporation Ltd.	Putuniq. Que.
C	Advocate Mines Ltd.	Baie Verte, Nfd.
D	Carey Canada, Inc.	East Broughton, Que.
E	Asbestos Corporation Ltd.	Thetford Mines, Que.
	Bell Asbestos Mines Ltd.	Thetford Mines, Que.
	Lake Asbestos of Canada Ltd.	Black Lake, Que.
	Lake Asbestos of Canada Ltd.	Thetford Mines, Que.
F	Johns-Manville of Canada	Asbestos, Que.
	Potash Mines	
1	Denison Mines	Sussex, N.B.
2	Cominco Ltd.	Vanscoy, Sask.
	Potash Corporation of Saskatchewan (Cory Division)	Saskatoon, Sask.
3	Potash Corporation of America	Saskatoon, Sask.
4	PCS (Allen Division)	Allen, Sask.
	Central Canada Potash Division	Allen, Sask.
5	PCS (Lanigan Division)	Lanigan, Sask.
6	International Minerals and Chemical Corporation	Esterhazy, Sask.
	PCS (Esterhazy Division)	Esterhazy, Sask.
7	PPG Industries	Kalium, Sask.
8	PCS (Rocanville Division)	Rocanville, Sask.

mined in Newfoundland, Ontario, and the Yukon Territory, and large resources are scattered throughout Canada.

The dominant producer is Johns-Manville Canada, which operates the Western world's largest known asbestos deposit. The deposit, located at Asbestos, Quebec, produces roughly 545,000 metric tons of finished fibers per year, and thereby accounts for 38% and 11% of current Canadian production and total world production, respectively.[9] The mine has the capacity to produce up to 645,000 metric tons of fiber per year. This comprises over 37% of total Canadian production capacity (in 1981) of 1,725,000 metric tons of finished fibers per year.[12]

Canada was, until recently, the world's leading producer of asbestos. Table I-3-6 illustrates Canada's share of world production from 1960 to 1982. Canada's share has declined, even though Canadian production has generally increased. Canada's relative decline as a producer is largely due to the enormous production increases which occurred in the USSR

from the 1950s to the 1970s. By 1980, the USSR accounted for 44% of world production.[12]

Figure I-3-7 illustrates that Canada exports most of the asbestos it produces. Exports normally represent about 95% or so of total production. Furthermore, Canadian exports, even today, account for some 65% of total world exports.[12] Figure I-3-7 also shows that roughly 40% of Canada's exports go to the United States. These exports have since 1960, accounted for over 90% of total U.S. imports of asbestos fibers. In fact, 97% of U.S. imports from 1977 to 1981 came from Canada.[16] Canada dominates other export markets, too. It has been the largest single exporter to Japan and the European Community nations for some time.[28]

Asbestos producers in Canada (and elsewhere in the free world) are faced with two major difficulties. The first, and more important one for now, is that the major use of asbestos is in construction materials and is thereby drastically affected by market recessions. Canada's exports to the United States and Western Europe, its two largest customers have declined by 700,000 metric tons since a peak in 1973.[29] This is a 41% drop in demand for Canadian exports and has, in turn, helped cause a nearly 50% drop in industry employment levels.[30] However, even if these two markets fail to improve, Canadian producers hope to see continued growth in the developing nations. This market took 20 to 30% of Canada's export during the 1970s and may exhibit 5% real annual growth in demand through the 1980s.[12]

The other difficulty is the increasing level of concern over health problems related to exposure to asbestos. For example, asbestos-exposure lawsuits taken out against the Manville Corporation, the parent of Johns-Manville Canada, forced it to file for Chapter 11 bankruptcy protection in 1982.

TABLE I-3-6 Canada's Share of World Asbestos Production, Selected Years, 1960–1982 (Metric Tons)

Year	Total world output	Canada Output	Canada Percent of world total
1960	2,213,080	1,118,000	50.5
1965	3,101,994	1,259,000	40.5
1970	3,493,085	1,507,000	43.1
1973	4,184,641	1,690,000	40.4
1975	4,138,756	1,056,000	25.5
1980	4,818,369	1,291,371	26.8
1982	4,000,000	880,000	22.0

Sources: Refs. 16 and 35.

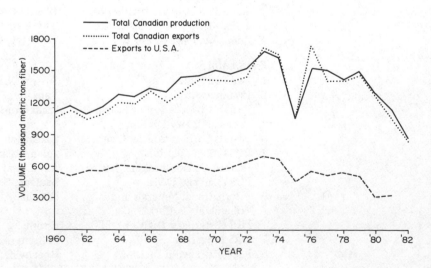

FIGURE I-3-7 Canadian production and exports of asbestos, 1960-1982. (From Refs. 12, 15, 16, and 35.)

Estimates of legal liabilities reached some $5 billion by September 1982 alone, but all other asbestos producers face the same problems as the Manville Corporation.[27]

One other issue of importance remains for asbestos producers in Quebec. The government of Quebec has argued for some time that not enough processing of asbestos fibers takes place in the province. As noted before, Quebec seeks to gain greater economic benefit from all domestic industries. In the case of asbestos, this may be a reasonable goal, given the size of the industry. Only 1200 people were employed at asbestos manufacturing jobs in Quebec in 1975, as opposed to the 60,000 jobs in the United States at the same time.[28]

Some estimates claim 10,000 such jobs could be created in Quebec, yet there is a good deal of uncertainty about the reliability of this estimate.[29] At any rate, the government of Quebec has acted to improve the situation. First, it established a Crown corporation (Société Nationale de l'Amiante, or SNA) to exploit asbestos deposits in Quebec. It also purchased Asbestos Corporation, a private firm with strong overseas export markets. Additional future actions are possible should the government not be satisfied with the results of current actions.

Cadmium

Cadmium is recovered in Canada and elsewhere as a by-product of zinc smelting and refining. This is because cadmium usually occurs as greenockite (CdS), a sulfide ore associated with sphalerite and other zinc sulfides. As such, cadmium production in Canada derives from the demand for zinc, and is virtually inelastic to cadmium price variations. Figure I-3-8 shows cadmium and zinc production levels, and cadmium exports, for the years 1960 to 1982. The vast bulk of cadmium exports goes to the United States and the United Kingdom. Future exports should be slow but steady in their increase, with nickel-cadmium batteries the major demand.[6]

Some 27 mines scattered over most of Canada produce cadmium-bearing zinc ores (see Figures I-3-9A and I-3-9B). Table I-3-7 shows that cadmium metal is then produced at four centers which had a 1978 annual capacity of 1705 metric tons. Final recovery of metal from ore ranges from 32 to 72%.[15]

Cobalt

In Canada, cobalt occurs in arsenides and sulfarsenides such as cobaltite (CoAsS). These are typically low-grade ores. The ore grade at Sudbury, for example, averages 0.07% cobalt.[31] Cobalt ores are associated with nickel and copper ores; thus Canadian production of cobalt is largely determined by the market for copper and nickel and is relatively price inelastic. As an illustration, Figure I-3-10 shows production levels for nickel and cobalt from 1960 to 1982. It is obvious that the by-product relationship is a strong one for nickel and cobalt.

Mines that produce cobalt ore are shown in Figures I-3-11A and I-3-11B. These operations use a variety of mining methods. At Sudbury, ores are recovered using open-pit, block-caving, and blast-hole-stoping methods. Figure I-3-3 and Table I-3-8 illustrate the location and capacity of Canadian cobalt-processing facilities. Inco Metals Co. Ltd. and Falconbridge Nickel Mines Ltd. recover cobalt from their nickel and copper operations. Cobalt-bearing matte is then shipped to Falconbridge's nickel refinery in Kristiansand, Norway. Production capacity for electrolytic cobalt currently stands at 1800 metric tons per year. Inco's cobalt oxide is

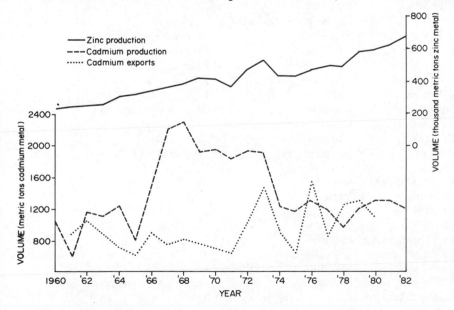

FIGURE I-3-8 Cadmium production and exports in relation to zinc production, 1960–1982. (From Refs. 15, 16, and 35.)

FIGURE I-3-9A Cadmium, lead, and zinc mining areas in Canada. (From Refs. 15, and 98.)

FIGURE I-3-9B Key to Figure I-3-9A: Cadmium, Lead, and Zinc Mining Areas in Canada

Region		Company	Location	Product
Yukon Terr.	1	United Keno Hill Mines	Elsa	Lead, zinc, cadmium
	2	Cyprus Anvil Mining	Faro	Lead, zinc, cadmium
British Columbia	3	Western Mines Ltd.	Myra Falls	Zinc, lead
	4	Northair Mines Ltd.	Brandywine	Lead, zinc, cadmium
	5	Silvana Mines, Inc.	New Denver	Lead, zinc
	6	Cominco Ltd.	Kimberley	Lead, zinc, cadmium
	7	Teck Corporation	Beaverdell	Lead, zinc
Northwest Terr.	8	Nanisivik Mines	Arctic Bay	Lead, zinc
	9	Pine Point Mines	Pine Point	Zinc, lead
	10	Cominco Ltd.	Little Cornwallis Island	Lead, zinc
Saskatchewan	11	Hudson Bay Mining	Flin Flon	Zinc, cadmium
Manitoba	12	Hudson Bay Mining	Snow Lake	Zinc, lead
	13	Hudson Bay Mining	Flin Flon	Zinc, cadmium
	14	Sherritt Gordon	Fox Lake	Zinc
	15	Sherritt Gordon	Ruttan Lake	Zinc
Ontario	16	Kidd Creek Mines	Timmins	Zinc, lead, cadmium
	17	Noranda Mines	Manitouwadge	Zinc, lead
	18	Falconbridge Copper	Sturgeon Lake	Zinc, lead
		Mattabi Mines	Sturgeon Lake	Zinc, lead
	19	Selco Mining	Uchi Lake	Zinc
Quebec	20	Falconbridge Copper	Noranda	Zinc
	21	Noranda Mines	Matagami	Zinc
	22	Lemoine Mines	Lemoine Twp.	Zinc
New Brunswick	23	Heath Steele Mines	Newcastle	Zinc, lead
	24	Brunswick Mining	Bathurst	Zinc, lead
Nova Scotia	25	Esso Resources Can.	Gays River	Zinc, lead
	26	Yava Mines Ltd.	Silvermines	Lead
Newfoundland	27	ASARCO Incorp.	Buchans	Zinc, lead, cadmium
	28	Newfoundland Zinc	Daniel's Harbour	Zinc

TABLE I-3-7 Cadmium Metal Production Centers in Canada

Location	Company	Capacity (metric tons)
Valleyfield, Que.	Canadian Electrolytic Zinc	544
Trail, B.C.	Cominco Ltd.	544
Flin Flon, Man.	Hudson Bay Mining	163
Timmins, Ont.	Kidd Creek Mines Ltd.	454
		1705

Source: Ref. 15.

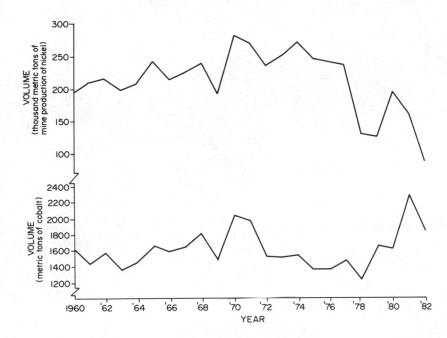

FIGURE I-3-10 Cobalt-nickel production in Canada, 1960–1982. (From Refs. 15, 16, and 98.)

reprocessed into oxide and salt compounds at the company's refinery in Clydach, Wales. Sherritt Gordon Mines Ltd. recovers its cobalt from nickel on a toll basis.[12]

Canada accounted for only 6% of world cobalt production in 1982.[16] This is unlikely to change much, given the dominance of Zaire in production. In fact, given the volatility of the nickel and copper markets, there is a chance that Canadian production of cobalt could fall from its present level. Expected demand levels for cobalt are low and will probably not improve much in the near future. This could change, though, should defense-related production increase in the United States.[16]

However, should the market pick up, a recently discovered copper-cobalt deposit in northwestern British Columbia may be started on the road to exploitation. Located only kilometers from the border with Alaska, the Windy Craggy deposit is surrounded on three sides by glaciers in very rough terrain. Thus production costs would be very high. However, currently identified resources already amount to 334 million metric tons, grading at 1.52%

copper and 0.08% cobalt.[34] This significantly increases Canada's resources of cobalt and could influence actual production levels in the future.

Columbium and Tantalum

The only columbium mine in Canada is the St. Honore, Quebec, pyrochlore mine (Figure I-3-12). The mine is run by Niobec, Inc., which in turn is owned equally by Teck Corporation and Quebec Mining Exploration Company, or SOQUEM. The mine has lately accounted for some 12% of the world's production of columbium, which makes it the second largest producer overall.[9] The mine is particularly important, in that it is the sole source of columbium concentrates for the United States, as Brazil discontinued its exports to the United States at the end of 1980.[12] Figure I-3-13 shows the mine's production levels since its startup in 1961.

The pyrochlore deposit worked by Niobec is located in carbonatite and alkalic rocks of the Canadian Shield. Ore grade is now 1.7% columbium

FIGURE I-3-11A Cobalt, iron ore, and nickel mines in Canada. (From Refs. 15 and 98.)

FIGURE I-3-11B   Key to Figure I-3-11A: Cobalt, Iron Ore, and Nickel Mines in Canada

	Iron ore mines	*Location*
1	Wesfrob Mines	Tasu, B.C.
2	Pickards Mather & Co.	Bruce Lake, Ont.
3	Algoma Steel Corp.	Wawa, Ont.
4	Cliffs of Canada Ltd.	Kirkland Lake, Ont.
5	Cliffs of Canada Ltd.	Temagami, Ont.
6	Inco Metals Co.	Sudbury, Ont.
7	Iron Ore Co. of Canada	Schefferville, Que.
8	Quebec Cartier Mining	Mount Wright, Que.
9	Sidbec-Normines, Inc.	Fire Lake, Que.
10	QIT-Fer et Titane Inc.	Lac Tio, Que.
11	Iron Ore Co. of Canada	Schefferville, Nfd.
12	Iron Ore Co. of Canada	Labrador City, Nfd.
13	Wabush Mines	Wabush, Nfd.
14	Borealis Mines	Roche Bay, Melville Peninsula, N.W. Terr. (not yet in operation)

Nickel-cobalt mining areas

	Company	*Location*	*Product*
6	Inco Metals Co.	Sudbury, Ont.	Nickel, cobalt
16	Inco Metals Co.	Shebandowan, Ont.	Nickel
17	Union Miniere	Pickle Lake, Ont.	Nickel
18	Falconbridge Nickel Mines	Falconbridge, Ont.	Nickel, cobalt

Prospective mine

15	Windy Craggy deposit	British Columbia	Cobalt-copper

TABLE I-3-8   Canadian Cobalt Production Facilities

Location	Company	Product	Annual capacity[a] (metric tons)
Fort Saskatchewan, Alta.	Sherritt Gordon	Cobalt metal powder	907
Port Colborne, Ont.	Inco Metals Co.	Electrolytic cobalt	907
Thompson, Man.	Inco Metals Co.	Cobalt oxide	NA
Sudbury, Ont.	Falconbridge Nickel	Cobalt matte	NA
		Partial total (metric tons)	1814

[a]NA, not available.

Source: Ref. 12.

KEY:
1 Canada Tungsten Mining Corp., Tungsten, Northwest Terr.

2 Tantalum Mining Corporation of Canada, Ltd., Bernic Lake, Man.

3 Niobec Inc., St. Honore, Quebec

4 QIT-Fer et Titane, Inc., Lac Tio, Quebec

0   500 km

FIGURE I-3-12   Columbium, tantalum, titanium, and tungsten mines in Canada, 1983. (From Refs. 15 and 98.)

pentoxide (Cb_2O_5), but this is expected to improve when some higher-grade ore zones are mined. All ore is recovered by underground techniques. Ore is concentrated to an average grade of 61% Cb_2O_5, and is then exported for further processing. Major destinations are Europe, the United States, and Japan, all of which require columbium for high-temperature alloys and superconductors.[33]

Just as with columbium, tantalum produced in Canada comes from only one mine. This is the Bernic Lake, Manitoba, mine run by the Tantalum Mining Corporation of Canada Ltd. (Tanco). Tanco is owned by Hudson Bay Mining and Smelting Ltd., the KBI Division of Cabot Corporation, and the Manitoba government. The mine's location is shown in Figure I-3-12.

The Bernic Lake mine recovers tantalum-bearing minerals from an albitic zone in a large pegmatite. The operation is unique in that it recovers tantalum pentoxide (Ta_2O_5) as a primary product. Two-thirds of world tantalum production comes as a by-product of other operations, usually tin processing. Room-and-pillar mining techniques are used to recover wodginite (at 70% Ta_2O_5), and by-products of beryllium, cesium, lithium, and rubidium.[12] Overall ore grade is 0.113% Ta_2O_5. Current mine and mill capacity is 1000 metric tons per day, which yields roughly 145 metric tons of tantalum per year.[33] The final product is a concentrate grading 35% Ta_2O_5. The mine has lately accounted for some 20% of the free world's tantalum output.[16]

Both the Niobec columbium and Tanco tantalum mines have been shut down in recent years due to low demand levels and high stocks. However,

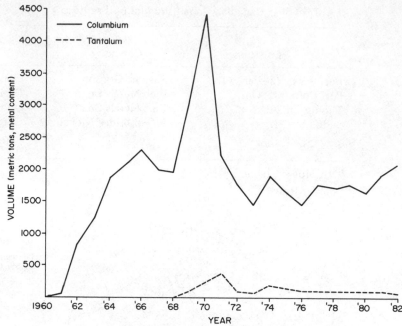

FIGURE I-3-13 Canadian production of columbium and tantalum concentrates, 1960–1982. (From Refs. 15, 16, and 98.)

the Niobec mine was reopened in 1983[34] and the Tanco mine was reopened in 1984.[32]

Copper

Copper is produced in Canada in a variety of geological settings. The two dominant settings are the porphyry copper deposits found in British Columbia, and the volcanogenic copper-lead-zinc deposits found in Quebec and other provinces. The porphyry copper deposits, currently being mined in British Columbia, also contain molybdenum as an important by-product. Nickel-copper and copper-zinc mines often produce cobalt and a variety of precious metals as by-products.

As of year end 1982, there were 40 copper mining areas in operation in Canada. Figures I-3-14A and I-3-14B show the location of these areas. These

FIGURE I-3-14A Copper and molybdenum producing areas in Canada. (From Refs. 15 and 98.)

FIGURE I-3-14B   Key to Figure I-3-14A: Copper and Molybdenum Producing Areas in Canada

Region		Company	Location	Product
British Columbia	1	Canada Wide Mines	Stewart	Copper
	2	Noranda Mines	Babine Lake	Copper
	3	Placer Development	Endako	Molybdenum
	4	Falconbridge Nickel	Tasu	Copper
	5	Gibralter Mines	McLeese Lake	Copper, Molybdenum
	6	Noranda Mines	Hendrix Lake	Molybdenum
	7	Utah Mines Ltd.	Port Hardy	Copper, Molybdenum
	8	Afton Mines	Kamloops	Copper
	9	Bethlehem Copper	Highland Valley	Copper, Molybdenum
		Teck Corporation	Highland Valley	Copper, Molybdenum
		Lornex Mining	Highland Valley	Copper, Molybdenum
		DeKalb Mining Corp.	Highland Valley	Copper, Molybdenum
	10	Craigmont Mines	Merritt	Copper
	11	Western Mines Ltd.	Brandywine	Copper
	12	Brenda Mines Ltd.	Peachland	Copper, Molybdenum
	13	Newmont Mines	Princeton	Copper
	14	Equity Silver Mines	Houston	Copper
Manitoba	15	Sherritt Gordon	Fox Lake	Copper
	16	Sherritt Gordon	Ruttan Lake	Copper
	17	Inco (Thompson mine)	Thompson	Copper
		Inco (Pipe mine)	Thompson	Copper
	18	Hudson Bay Mining	Flin Flon	Copper
		Hudson Bay Mining	Snow Lake	Copper
Ontario	19	Selco, Inc.	Uchi Lake	Copper
	20	Union Miniere	Pickle Lake	Copper
	21	Noranda Mines	Sturgeon Lake	Copper
		Mattabi Mines	Sturgeon Lake	Copper
	22	Inco Metals Co.	Shebandowan	Copper
	23	Noranda Mines	Manitouwadge	Copper
	24	Kidd Creek Mines	Timmins	Copper
	25	Pamour Porcupine	Schumacher	Copper
		Pamour Porcupine	Holtyre	Copper
	26	Inco Metals Co.	Sudbury	Copper
	27	Falconbridge Nickel	Falconbridge	Copper
Quebec	28	Campbell Resources	Chibougamau	Copper
		Patino Mines	Chibougamau	Copper
	29	Falconbridge Copper	Chapais	Copper
		Lemoine Mines	Lemoine Twp.	Copper
	30	Noranda Mines	Matagami	Copper
	31	Falconbridge Copper	Noranda	Copper
	32	Madeleine Mines	Ste-Anne-des-Monts	Copper
	33	Noranda Mines	Murdochville	Copper, Molybdenum
New Brunswick	34	Brunswick Mining	Bathurst	Copper
	35	Heath Steele Mines	Newcastle	Copper
Newfoundland	36	Consolidated Rambler	Baie Verte	Copper
	37	ASARCO Incorporated	Buchans	Copper
Northwest Territories	38	Canada Tungsten Mining	Tungsten	Copper
Yukon Territory	39	Cyprus Anvil Mining	Faro	Copper
	40	Hudson Bay Mining	Whitehorse	Copper

mines have a production capacity of roughly 1,100,000 metric tons per year of contained copper, but lately have been producing well below this level. Mine production in 1982, for example, was 640,000 metric tons of contained copper, 8.2% of world output.[16] Figure I-3-15 shows Canada's mine production of copper from 1960 to 1982. It is clear that production has declined since a peak in the mid-1970s. Capacity, however, is expected to reach 1,300,000 metric tons of contained copper per year by 1985.[16]

Concentrate from the mines is either exported or processed domestically. Figure I-3-3 shows the location of the eight smelters and refineries that process copper in Canada. Table I-3-9 shows their capacity and major product. Canadian smelters had, in 1979, a total capacity, in metal content, of 640,000 metric tons per year. As seen in Table I-3-9, the corresponding figure for the two refineries was 627,000 metric tons per year. These rated capacity levels are estimated to reach 1985 levels of 950,000 and 750,000 metric tons per year for smelters and refineries, respectively.[31]

The vast majority of the copper produced in Canada is exported in a number of different forms. Japan and Norway historically are the two major importers of copper concentrates and matte. Falconbridge processes nickel-copper matte at its

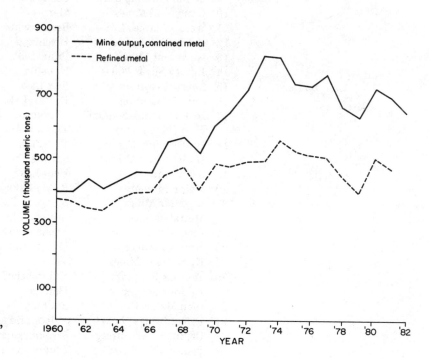

FIGURE I-3-15 Production of copper in Canada, 1960–1982. (From Refs. 15, 16, and 35.)

TABLE I-3-9 Copper Smelters and Refineries in Canada

Company	Location	Major product	Capacity (metric tons/year)
Smelters			
Gaspe Copper Mines Ltd.	Murdochville, Que	Copper anodes	336,000
Noranda Mines	Noranda, Que.	Copper anodes	1,542,000
Falconbridge Nickel Mines, Ltd.	Falconbridge, Ont.	Copper-nickel matte	590,000
Inco Metals Co.	Sudbury, Ont.	Blister copper	3,630,000
Hudson Bay Mining & Smelting	Flin Flon, Man.	Blister copper	522,000
Afton Mines Ltd.	Kamloops, B.C.	Blister copper	22,500
			6,642,500
Refineries			
Canadian Copper Refineries Ltd.	Montreal East, Que.		435,000
Inco Metals Co.	Copper Cliff, Ont.		192,000
			627,000

Source: Ref. 15.

Kristiansand, Norway, refinery. Japan imports concentrates from western Canadian copper mines, many of which were started because of Japanese capital infusions. The proportion of total copper exports made up by concentrates, on a metal content basis, has gradually increased over the past 20 years. Concentrates and matte accounted for 12.2% of exports in 1961, 30.2% in 1970, and 40.6% in 1980. The peak was 1973, when concentrates made up 46.2% of all exports of copper, and when Japan could still process concentrates with cheap oil.[35]

Copper in unwrought shapes, scrap, and semimanufactured forms is exported to a wide range of countries, but the United States is traditionally the primary destination.[36] In 1980, the United States received 37.8% of Canada's exports of refinery shapes, 59.9% of scrap, and 64.9% of Canada's semimanufactured copper items.[35] Canadian exports to the United States have increased somewhat lately, due to falling domestic production in the United States. This is due partly to more stringent environmental controls and partly to higher production costs in the United States.

The future outlook for Canadian copper production and exports is generally favorable because of two factors. First, Canada's pollution control laws are not as stringent as those in the United States, and thereby add less of an additional cost to Canadian copper. Second, Canada has the lowest average break-even production costs of the largest free world producers (while the United States has the first or second highest).[37] The key to Canada's low cost is production credits from by-products recovered along with copper. Because these by-products are typically of medium- to high-unit-value items such as molybdenum, cobalt, and precious metals, Canada's copper producers have a break-even cost of 60 cents a pound.[38] Copper producers in the United States, however, require 90 cents to $1.50 a pound.[39]

Some factors will present a challenge to any increase in Canada's copper output, though. First, many Latin American and other developing nations are not inclined to slow down domestic copper production because of a need for foreign exchange. Thus competition for exports is sure to be strong. Second, Canada may increase the stringency of its air pollution control laws. Third, fiber optics, plastics, and other products have begun to substitute for copper in major markets, such as electronics and plumbing. This is restricted largely to the United States for now, but is expected to occur soon in other developed nations.

Even so, it looks as though traditional Canadian markets will at least be maintained. The United States will probably continue to import large amounts of processed copper because of high costs and declining production in domestic operations. Japan will most likely continue to import large volumes of concentrates. This is expected because Japan needs larger volumes of sulfuric acid for domestic industry, and the sulfur found in copper concentrates has done quite well as a source.[36]

Gold

After an extended slump in production, Canadian gold production has lately risen. Production in 1982 was ranked third in the world, after South Africa and the USSR.[16] Canadian production and exports are expected to increase in the near future, with production topping 3 million troy ounces in 1985[40] (Figure I-3-16).

As of year end 1981, 44 gold mining operations

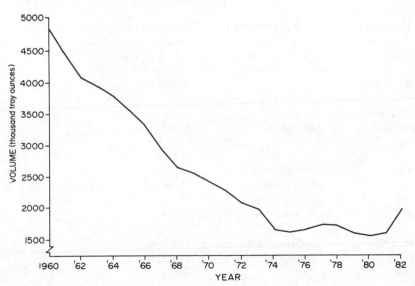

FIGURE I-3-16 Canadian gold production, 1960–1982. (From Refs. 15, 16, and 35.)

existed in Canada, 37 of which were lode gold mines (see Figures I-3-17A and I-3-17B). Table I-3-10 shows the number of mines operating in each province and the distribution between different recovery methods. The amounts of gold recovered have shifted somewhat with time. In 1960, the share of total output held by lode, base metal, and placer operations was 85, 13, and 2%, respectively. These standings had changed to 68, 31, and 1%, respectively, by 1978.[15] A major factor behind the change was the advent of large copper mines in British Columbia, which produced gold as a by-product.

Ontario is currently the leading mine producer of gold, with roughly 34% of total production.[12] In descending order of output, the next most important producers are Quebec, British Columbia, and the Northwest Territories. Gold occurring in base metals is recovered from smelters at Trail, British Columbia, and refineries in Toronto, Ottawa, and Montreal East.

Since 1980, gold has been the only truly bright spot in all of Canada's nonfuel minerals. Its high price has allowed the development and continuation of many mining operations. For example, gold recovered from nickel ores mined by Inco and Fal-

FIGURE I-3-17A Gold, silver, and platinum-group metal mines in Canada. (From Refs. 15 and 98.)

FIGURE I-3-17B   Key to Figure I-3-17A: Gold, Silver, and Platinum-Group Metal Mines in Canada

Region		Company	Location	Product
Yukon Territory	1	United Keno Hill Mines	Elsa	Silver
	2	Cyprus Anvil Mining	Faro	Silver
	3	Hudson Bay Mining and Smelting	Whitehorse	Gold, silver
Northwest Terr.	1	Nanisivik Mines Ltd.	Arctic Bay	Silver
	2	Echo Bay Mines	Port Radium	Silver
	3	Terra Mining	Port Radium	Silver
	4	Cominco Ltd	Yellowknife	Gold
		Giant Yellowknife Mines	Yellowknife	Gold
British Columbia	1	Erickson Gold Mining	McDame Lake	Gold, silver
	2	Noranda Mines	Babine Lake	Gold, silver
	3	Mosquito Creek	Wells	Gold

	4	Wesfrob Mines	Tasu	Silver
	5	Gibralter Mines	McLeese Lake	Silver
	6	Utah Mines Ltd.	Port Hardy	Silver, gold
	7	Afton Mines Ltd.	Kamloops	Gold, silver
	8	Lornex Mining Corp.	Highland Valley	Silver
	9	Western Mines Ltd.	Myra Falls	Silver, gold
	10	Northair Mines Ltd.	Brandywine	Gold, silver
	11	Brenda Mines Ltd.	Peachland	Silver
	12	Silvana Mines, Inc.	New Denver	Silver
	13	Cominco Ltd.	Kimberley	Silver
	14	Teck Corporation	Beaverdell	Silver
	15	Newmont Mines Ltd.	Princeton	Gold, silver
	16	Dankoe Mines	Keremus	Silver, gold
Saskatchewan	1	Hudson Bay Mining	Flin Flon	Gold, silver
Manitoba	1	Sherritt Gordon Mines	Fox Lake	Silver
	2	Sherritt Gordon Mines	Rutton Lake	Silver
	3	Hudson Bay Mining	Flin Flon	Gold, silver
	4	Hudson Bay Mining	Snow Lake	Gold, silver
	5	Inco Metals Co.	Thompson	Platinum group
Ontario	1	Campbell Red Lakes	Balmertown	Gold
		Dickerson Mines	Balmertown	Gold
	2	Selco Mining Corp.	Uchi Lake	Silver
	3	Union Miniere	Pickle Lake	Silver
	4	Falconbridge Copper	Sturgeon Lake	Silver, gold
		Mattabi Mines	Sturgeon Lake	Silver
	5	Noranda Mines	Manitouwadge	Silver
	6	Kidd Creek Mines	Timmins	Silver
	7	Dome Mines	South Porcupine	Gold
		Pamour Porcupine Mines	Schumacher	Gold
		Pamour Porcupine Mines	Pamour	Gold
	8	Pamour Porcupine Mines	Holtyre	Gold, silver
	9	Mining Corp. of Canada	Palomar	Gold
	10	Pamour Porcupine Mines	Matachewan	Gold
	11	Wilroy Mines	Kirkland Lake	Gold, silver
	12	Kerr Addison Mines	Virginiatown	Gold
	13	Agnico-Eagle Mines	Cobalt	Silver
		Canadaka Mines	Cobalt	Silver
		Teck Corporation	Cobalt	Silver
	14	Inco Metals Co.	Sudbury	Gold, silver, platinum group
	15	Falconbridge Nickel	Falconbridge	Silver, platinum group
Quebec	1	Campbell Chibougamau	Chibougamau	Gold, silver
		Patino Mines	Chibougamau	Gold, silver
	2	Falconbridge Copper	Chapais	Gold, silver
	3	Noranda Mines	Matagami	Silver, gold
	4	Agnico-Eagle Mines	Joutel	Gold
	5	Falconbridge Copper	Noranda	Gold, silver
		Noranda Mines	Rouyn	Gold
	6	Camflo Mines	Cadillac	Gold
		Les Mines Est-Malartic	Malartic	Gold
	7	Belmoral Mines	Val D'or	Gold
		Lamaque Mining	Val D'or	Gold
		Sigma Mines	Val D'or	Gold
	8	Madeleine Mines	Ste-Anne-des-Monts	Silver
	9	Noranda Mines	Murdochville	Silver
New Brunswick	1	Brunswick Mining & Smelting	Bathurst	Silver
Newfoundland	1	Consolidated Rambler	Baie Verte	Gold, silver
	2	ASARCO Incorporated	Buchans	Silver, gold

TABLE I-3-10   Canadian Gold Production Distribution

Region	Number of mines	Recovery method
Northwest Territories	2	All lode gold
Yukon Territory	1	Mainly placer, some base metal
British Columbia	9	99% base metal
Saskatchewan	1	All base metal
Manitoba	2	All base metal
Ontario	12	91% lode, 9% base metal
Quebec	14	70% lode, 30% base metal
New Brunswick	1	All base metal
Newfoundland	2	All base metal

Sources: Refs. 12 and 15.

conbridge provided 30% of their gross revenues in 1980, a year when nickel sales started to slump.[9]

Two new gold districts in Ontario promise to sustain the optimism surrounding gold mining in Canada. The Detour Lake project in northeastern Ontario will be the largest gold mine in all of Canada. Gold is to be recovered as a primary product. Reserves are 27.7 million metric tons, grading 3.89 g of gold per metric ton, plus silver and copper. Expected output from the initial open-pit mine is 2000 metric tons of ore per day. An expected switch to underground mining will ultimately double the mine's capacity.

The Hemlo district near Marathon, Ontario, on the northeast shore of Lake Superior, is the first new gold district to be discovered in the last 50 years. Four deposits have been identified already, and reserves are conservatively estimated at 10 million tons of ore, grading at 0.3 oz of gold per ton.[41] The district is crossed by both major highway and railway lines and should, therefore, have low infrastructure costs.

Exports of gold metal alone, from 1976 to 1980, amounted to 105% of production during those years; so gold stocks are being drawn down somewhat. The United States received 84 percent of Canada's gold exports during the years 1976 to 1980.[35] Exports of gold-bearing ores and concentrates had a more diversified list of destinations. Japan consistently is the largest importer, but the United States, the United Kingdom, and the USSR also import large amounts.[31]

Iron Ore

Canada's deposits of iron ore are sedimentary in nature and are of a low to medium grade, averaging 20 to 50% iron.[12] The residual deposits at Schefferville, however, grade 50 to 60% iron. Large tonnages of ore, combined with low ore grade, have meant that the ore is almost always mined by open-pit methods. Those minerals recovered from mining are normally magnetite, hematite, and siderite.

Iron ore mining in Canada started on a large scale only by the 1950s. In 1949, the Iron Ore Company of Canada (IOC) was formed to exploit the deposits in the Labrador Trough, a north-south-trending metasedimentary iron formation. Given its location in an extremely isolated part of eastern Quebec and western Labrador, a great deal was initially spent on just developing the required infrastructure. Whole new towns were built, such as Schefferville, as was a 576-km railway from Schefferville to the St. Lawrence Seaway port of Sept-Iles. Production began in 1954. IOC then built two pelletizing plants to upgrade the ore to a level suitable for export. One plant, in Labrador City, was a 10-million metric ton per year facility, while the other plant, at Sept-Iles, had a capacity of 6 million metric tons per year of pellets.[42]

Not long after IOC began operations, other deposits in the Labrador Trough area were developed. Major firms were the Quebec Cartier Mining Company (QCM), Wabush Mines, and, eventually, Sidbec-Normines, Inc. All these developments also required massive infrastructure investments. For example, QCM built new towns at Gagnon and Fermont, as well as extensive railway links. Wabush Mines built the new town at Wabush, and a 6-million long ton per year pelletizing plant at Pointe-Noire. The capital-intensive nature of this development is exemplified by the fact that some railways had an average cost of $1.24 million per kilometer.[42]

Figure I-3-11 shows the location of the 13 iron ore mines in operation in 1980, as well as a new mine due to start up by 1987. Of the 13 mines open in 1980, three were in Newfoundland, four in Quebec, five were in Ontario, and one was located in British Columbia. The dominant producer was IOC, with roughly 40% of total Canadian output.[9]

Production capacity was rated at 44.5 million metric tons of contained iron in 1979, but capacity is expected to fall to 40.8 million metric tons by 1985 because of economic factors.[31] Even so, Canadian production should continue to fall short of capacity. For example, production in 1982 was only 34.5 million metric tons.[16] Figure I-3-18 shows production and export levels from 1960 to 1982.

Iron ore mining in Canada has several distinct disadvantages it must overcome. First, mining itself is hampered somewhat by the extremely cold climate in the Labrador Trough area. Recovered ore freezes regularly in the Schefferville area, for example, and some of the area's ore bodies are located in permafrost.[31] Second, the extremely

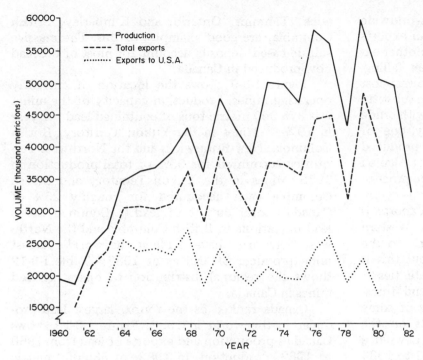

FIGURE I-3-18   Canadian production and exports of iron ore, 1960–1982. (From Refs. 15, 16, and 35.)

isolated work environment prompts strikes which occur regularly whenever contract renewals are due. These two factors often result in production levels far below rated capacity. An additional disadvantage is the relatively low grade of Canadian iron ore. Since it does not measure up to the grades of Australian or Brazilian ore, Canadian iron ore is usually pelletized. This highly energy intensive process puts Canadian output at somewhat of a cost disadvantage relative to other producers. Some ore, though, is sintered in Canada.

Trade patterns for Canadian iron ore show the same pattern as do many other Canadian mineral exports. In other words, most exports go to the United States. This is borne out by Figure I-3-18. The U.S. share of Canadian exports has dropped somewhat, though, as shown in Table I-3-11, and Japanese and European deliveries have risen.

This shift in export destinations can be attrib-

uted to several factors. The initial emphasis on exports to the United States was based both on proximity and also on corporate linkages. Many U.S. firms helped establish Canadian iron ore firms; for example, QCM was established as a subsidiary of U.S. Steel.[42] Thus it was natural for the early export emphasis to be on the United States. The United Kingdom was also a major destination because of historical ties and the size of its steel industry. However, during the 1970s the U.K. and U.S. steel industries eventually started to falter and other, lower-cost producers began to take a larger share of Canada's exports. This trend has continued into the 1980s. The intensity of the slump in the U.S. steel industry was such that Canadian exports of iron ore to the United States in the first half of 1982 were 50% below the level of a year earlier.[43]

Even with the shift in destinations, Canadian iron ore production and exports have decreased re-

TABLE I-3-11   Major Export Destinations for Canadian Iron Ore, Selected Years, 1965–1980

Year	Export levels (thousand metric tons)	Primary destination	Secondary destination
1965	31,293	U.S. (77.1%)	U.K. (9.5%)
1970	39,348	U.S. (61.6%)	U.K. (12.7%)
1975	36,034	U.S. (53.5%)	Japan (11.7%)
1976	44,685	U.S. (54.8%)	Japan (11.4%)
1977	45,060	U.S. (59.1%)	Netherlands (11.4%)
1978	31,929	U.S. (59.1%)	Netherlands (11.2%)
1979	48,849	U.S. (46.8%)	Netherlands (13.4%)
1980	39,021	U.S. (44.3%)	Netherlands (15.9%)

Source: Ref. 35.

cently. This is to be expected, given the worldwide slump in steelmaking. However, Canadian exports are at a cost disadvantage with respect to other exporters. Since steel production has been falling, major producers have turned to their lowest-cost sources of supply. That usually means the substitution of sinter for pellets (which are made with crude oil). However, during 1979–1981, only one of Canada's 15 iron ore processing plants produced sinter. This was the Algoma Ore Division of Algoma Steel, based at Wawa, Ontario. Ten of the processing plants produced pellets.[12]

The future for Canadian iron ore exports is questionable. Low market demand in Western Europe is expected to continue. Exports to the slumping U.S. industry will continue, but this is mainly because of interlocking corporate ties.[42] Furthermore, increasing infrastructure and transport costs will hurt the competitiveness of some Canadian ores. For example, the tolls on the Montreal–Lake Ontario portion of the St. Lawrence Seaway were increased by 17% from 1981 to 1983 to offset declining revenues from the drop in seaway traffic.[12]

There is one bright spot on the horizon, though. Borealis Exploration is expected to receive federal support and approval for the development of a magnetite mine at Roche Bay on the Melville Peninsula, Northwest Territories (Figure I-3-11). Open-pit reserves of over 4 billion tons have already been identified, and the developers strongly believe that the magnetite concentrate produced there will find markets in Europe, Japan, and Southeast Asia. The mine, expected to produce 7.3 million metric tons of concentrate a year, is located 9 km from the deep-water harbor at Roche Bay. It is thought that concentrate production, transportation to, and loading onto ocean vessels at Roche Bay will cost C$9.81 per metric ton. Total cost of the concentrate, landed in Europe, is C$15 per metric ton, which dramatically undercuts the $25.72 charged for Brazilian hematite delivered to Western Europe ports.[44] Should this mine come on-stream by 1987, as expected, Canada's iron ore industry should find itself looking far more competitive than it is now.

Lead

Geologically speaking, lead occurrences in Canada can be grouped into two major settings.[31] The first setting is the stratabound deposits found in such Devonian Age sedimentary rocks as dolomite and limestone. An example of this would be the Pine Point mine in the Northwest Territories. The other major setting for lead occurrences is in massive sulfides. The mines at Bathurst, New Brunswick, Timmins, Ontario, and Kimberley, British Columbia, are good examples of this. The massive sulfide-based deposits account for most of the lead now produced in Canada.

Figure I-3-9 shows the location of currently operating mines. Production capacity of the mines was 375,000 metric tons of contained lead per year in 1979.[31] Mines in the Yukon Territory, British Columbia, New Brunswick, and the Northwest Territories accounted for 95% of total production in 1978. Mines in the Yukon Territory and British Columbia each accounted for roughly 25% of Canada's total output of lead.[15] Cominco Ltd.'s lead operations in British Columbia and the Northwest Territories have made it the world's largest mine producer of lead as of 1982.[9] Table I-3-12 shows the regional distribution of operating lead mines in Canada.

Canada ranked as the fourth largest lead producer in the world in 1982. Figure I-3-19 shows Canada's production and exports of lead from 1960 to 1982. Production, in 1982, of 335,000 metric tons of lead accounted for 9.7% of global output.[23] Most lead production is as a co-product or by-product of copper and/or zinc, with occasional by-products of gold, silver, antimony, bismuth, and cadmium. This co- and by-product relationship means that Canadian lead production is determined to some extent by the economics of copper and zinc production. This is in sharp contrast with the American situation, where many lead deposits are rich enough that lead is mined as the primary commodity.[45]

The two major lead smelting plants in Canada had a 1981 capacity of 217,000 metric tons of lead slag per year. These plants are Cominco's smelter at Trail, British Columbia, and the Bellendune, New Brunswick, smelter run by Brunswick Mining & Smelting Corp. Ltd., a Noranda subsidiary. The

TABLE I-3-12 Distribution of Lead Mining Areas in Canada

Region	Number of mines
Northwest Territories	2
Yukon Territory	2
British Columbia	5
Manitoba	1
Ontario	4
New Brunswick	2
Nova Scotia	2
Newfoundland	1
	19

Sources: Refs. 15 and 98.

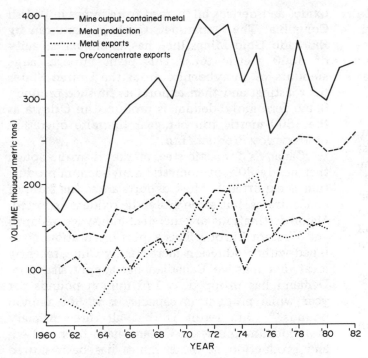

FIGURE I-3-19 Canadian production and exports of lead, 1960–1982. (From Refs. 15, 16, and 35.)

Belledune smelter has a rated capacity of 77,500 metric tons of lead per year. It also has the capacity to produce 110,000 kg of silver, 4500 metric tons of 40% copper matte, and some bismuth and antimony every year. Zinc concentrate is not smelted at this facility but exported to overseas markets. The Trail smelter is due to increase its capacity to 180,000 metric tons per year sometime in the near future.[9]

These same two locations house primary lead refining centers. Both centers use the Betts electrolytic refining process and had, in 1979, a rated capacity of 217,000 metric tons per year.[31] Cominco's expansion will increase capacity at its Trail operation from 145,000 to 180,000 metric tons

per year by 1985.[12] Secondary lead production capacity was 110,000 metric tons per year in 1978.[15]

Figure I-3-19 shows that over 80% of Canada's lead production is exported. Primary uses by importing nations are for batteries and gasoline additives for automobiles. The battery market alone accounts for over 50% of lead demand.[46] Possible increases in export demand will be contingent on the recovery in the United States and other major producers of automobiles, and also on the strength of the recovery of the housing industry.

Table I-3-13 shows that Canada's exports of lead represent a large proportion of total output. Table I-3-13 shows that exports of contained lead

TABLE I-3-13   Canada's Lead Exports as Proportion of Production, 1962–1980 (Metric Tons of Contained Lead)

| | | Total exports | | |
Year	Mine production	Ore/concentrate	Metal	As percent of mine output
1962	191,706	53,973	117,913	81.7
1964	187,205	72,900	93,454	88.9
1966	293,180	102,452	102,576	69.9
1968	326,610	130,501	134,260	81.8
1970	357,195	150,513	149,857	84.1
1972	376,257	162,001	145,144	81.6
1974	320,253	194,088	95,015	90.3
1976	256,324	140,942	133,973	107.3
1978	365,782	142,682	156,076	81.7
1980	296,641	147,007	152,327	101.0

Source: Ref. 35.

in ores and concentrates have gradually come to roughly equal the level of exported metal. This can be attributed to increased Japanese imports of concentrates, for Japan is by far the largest importer of Canadian lead concentrates. The major importer of lead metal from Canada is the United States.

Molybdenum

With 16% of the world's 1982 mine production, Canada maintained its status as the third-largest producer of molybdenum in the world.[16] Only the United States and Chile produced more. Canada produced 12,849 metric tons of molybdenum in 1981 and 16,460 metric tons in 1982.[47] British Columbia's mines accounted for 90.5% of 1982 output and the remainder came from one mine in Quebec.[48] Figure I-3-20 shows Canada's production and exports of molybdenum from 1960 to 1982.

The location of Canada's molybdenum mines is shown in Figure I-3-14. Six of the mines (five in British Columbia, one in Quebec) are porphyry copper-molybdenum operations which recover the latter metal as a by-product. Three other mines in British Columbia are porphyry molybdenum operations which recover the metal as a primary product. These deposits are usually worked by open-pit mining methods, and exhibit the low ore grade characteristic of porphyries. The ore grade for mines worked in 1978 averaged 0.043% molybdenum.[15]

Molybdenite is processed into molybdic oxide at two roasters. The Endako, British Columbia, operation run by Placer Development Ltd., has an annual capacity of 10,800 metric tons molybdic oxide, and derives all its feed from mines in British Columbia. The Duparquet, Quebec, plant run by Eldorado Gold Mines Inc., has an annual capacity of 4500 metric tons. This plant imports large amounts of molybdenite from the United States for roasting, and then exports its finished product. Some ferromolybdenum is produced in Ottawa at the 1800 metric ton per year thermite operation of Masterloy Products Ltd.[15]

Canada's domestic steel market is small enough that nearly 90% of domestic molybdenum production is exported.[15] Most exports are in the form of concentrates, but some molybdic oxide is exported. Major destinations are the steel industries of Japan and Western Europe. This export orientation, combined with production increases by Chile, recently created a massive Canadian oversupply situation. Demand has dropped to 140 million pounds per year while production capacity is at 300 million pounds.[48] One result is that all three primary molybdenum mines in Canada have been closed, and production at other mines has been slowed down or temporarily stopped. The recent price drop of molybdenum has, in turn, had an adverse impact on the economics of base metal operations in British Columbia. These operations have depended on an operating subsidy from molybdenum profits to keep the mines going.[49]

Nickel

Canada has, until recently, been the dominant producer of nickel in the world. Moreover, one firm, Inco Ltd., has dominated both the Canadian and world scene. Inco's dominance can be attributed

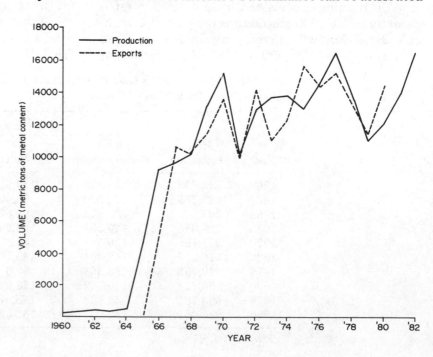

FIGURE I-3-20   Canadian production and exports of molybdenum, 1960–1982. (From Refs. 15, 16, and 35.)

to its possession of the Sudbury intrusive in Ontario, unquestionably the largest and richest nickel-sulfide ore body yet discovered anywhere. The ore body also contains large quantities of many other metals, such as cobalt, precious metals, and copper.

However, Inco's dominance in nickel production has steadily decreased with the advent of new producers in other parts of the world. In the early 1950s, Inco, Falconbridge Nickel Mines (of Canada), and France's Société Métallurgique Le Nickel (SLN) were the only producers of nickel. By the 1970s, production was taking place in the Philippines, Botswana, Australia, Cuba, and many other countries. The result is that Inco's Sudbury operations now account for only 25% of free world production capacity.[9]

By year end 1981, there were 14 operating nickel mines in Canada. Eleven were in Ontario (mainly in the Sudbury region), and three were in Manitoba. Their location is shown in Figure I-3-11.

Rated capacity, in recoverable nickel, is 275,000 metric tons per year, a figure expected to hold until 1985.[31] Figure I-3-21, which shows Canada's share of world nickel production from 1960 to 1982, indicates that output has recently been from one-fourths to two-thirds of rated capacity. Several other companies besides Inco and Falconbridge mine nickel in Canada. The other Ontario-based firm is Union Minière Exploration and Mining Corp. Ltd., while Sherritt Gordon Mines Ltd. is currently the only other operating nickel-mining firm in Manitoba.

The two smelters and four refineries which process nickel in Canada are described in Table I-3-14, and their location is seen in Figure I-3-3. Copper-nickel matte from the Falconbridge smelter is shipped to Falconbridge's refinery at Kristiansand, Norway. Some of Inco's smelter output is refined at Clydack, Wales, while the remainder is refined domestically. Sherritt Gordon's refinery processes

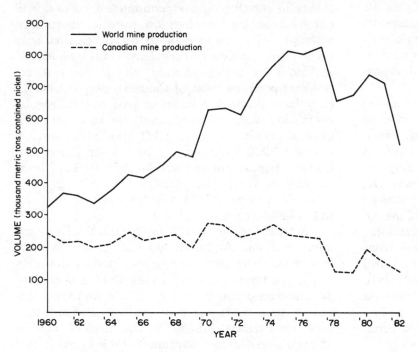

FIGURE 1-3-21 Canadian share of world nickel production, 1960–1962. (From Refs. 15, 16, and 35.)

TABLE I-3-14   Nickel Smelters and Refineries in Canada, 1981

	Company	*Location*	*Product*
Smelters	Falconbridge	Falconbridge, Ont.	Copper-nickel matte
	Inco	Sudbury, Ont.	Nickel sulfate
Refineries	Inco	Port Colborne, Ont.	Nickel oxides, metal, and cathodes
		Sudbury, Ont.	Nickel oxides, pellets
		Thompson, Man.	Nickel metal
	Sherritt Gordon	Fort Saskatchewan, Alta.	Nickel metal

Source: Ref. 12.

feedstock from Lynn Lake and Thompson, Manitoba, as well as imported concentrate and matte from New Caledonia and Australia.[34] The importance of nickel mining and processing is indicated by the fact that nickel comprises roughly 50% of the value of all mineral production in Manitoba, and almost 30% in Ontario. Moreover, nickel comprises roughly 10% of the total value of Canadian mineral output.[50]

Canada's influence on world nickel production is larger than its own domestic production, though. This is because Inco and Falconbridge Nickel have both invested in nickel mining operations overseas. Inco has investments in nickel laterite mines in Guatemala and Indonesia.[50] These investments have been made because they are ostensibly cheaper than production expansions at Sudbury, and also because Inco has been unable to discover new nickel-sulfide deposits in Canada. Falconbridge has investments in nickel laterites in the Dominican Republic, but has recently been diversifying its investments away from nickel.[9] These investments, taken together, added another 87,000 metric tons of contained nickel capacity to Canada's control in 1979, which represented 35% of world capacity.[31]

Trade in Canadian nickel is relatively complex. As noted above, some nickel matte is exported to Wales and Norway for further processing. These Norwegian and Welsh refined products, and, in fact, the bulk of Canadian exports, ultimately are destined for the U.S. market. In 1979, for example, the United States imported 160,725 metric tons of nickel in all forms. Canada provided 72,672 metric tons, 45.2% of the total, for Canadian operations. However, imports of Canadian-based nickel from Norway and Wales accounted for another 17,466 metric tons. Thus Canada directly and indirectly accounted for 56.1% of total U.S. nickel imports in 1979.[35] Other major importers of Canadian nickel, especially in metal form, are the United Kingdom, the Netherlands, and Belgium-Luxembourg.

Canada's share of the world nickel export market is being eroded, just as is its production share. For one thing, consumption in the United States has slowed down somewhat during the latest recession. This is because over 35% of the nickel used in the United States is for consumer products; thus there is a recession-induced volatility in U.S. (and other nations') imports.[50] Second, Canada's traditional markets are now being threatened by new, aggressive producers. Some producers, such as those in the Dominican Republic, admittedly have high processing costs due to the lateritic nature of their ores. However, cheap wages and government policies often keep these producers going even during a recession. Other producers threatening some traditional Canadian markets are Cuba and the USSR. The USSR, in the first half of 1982, exported roughly 40,000 metric tons of nickel to the West, while it exported 32,000 metric tons in all of 1981. France and West Germany, in particular, increased their imports of Soviet nickel in 1982.[51]

The advent of major producers in developing and communist nations is a particularly threatening situation to Canada's producers because the former group rarely, if ever, follows the dictates of market forces. The impact has already been felt in Canada. Inco and Falconbridge closed mines, smelters, and refineries for extended periods of time in 1982 and early 1983.[52] The result was that Canada's output of nickel in 1982 was less than half the amount produced in 1981.

The outlook for Canadian nickel is cloudy for several reasons. First, it is unlikely that competing mines in developing and communist nations will cut back on their production. Second, there is an attempt in Canada to make pollution control laws even more stringent, especially with respect to emissions of sulfur dioxide (SO_2). Since nickel smelters produce most of Canada's SO_2 emissions, they have borne the brunt of past control regulations, and will probably continue to do so. For example, Inco's Copper Cliff (Sudbury) smelter released 7000 tons per day of SO_2 in the 1960s. Control regulations reduced output to 1950 tons per day in 1983, and there is a push to increase controls even more.[9] This is discussed in more detail in Chapter I-6.

A third influence will be the level of government taxation. As noted elsewhere, Canada's favorable tax laws were largely scrapped in the early 1970s, partly because of a view that mineral producers earned too high a profit on nonrenewable resources. Nickel, in particular, was, and still is, a very controversial mineral in this respect because of Inco's past as the dominant force in the nickel industry. Although some feel that Inco's nickel prices (once the world standard) have always been set just above normal costs,[50] others argue that Inco garnered a large monopoly profit. Moreover, even though Inco's monopoly profit has been dissipated because of new producers in the market, its Ricardian rents (based on low costs of production and high ore grades) are still thought to be high enough to withstand additional tax increases.[53] This issue is still being debated.

The last possible major influence on the future of Canada's nickel industry is the future exploitation of deep-sea manganese nodules. The high nickel content of these nodules could conceivably present an additional challenge to Canadian (and other) land-based producers. However, the technology is not developed, and the legal and political environment for extraction is still not fully settled. Furthermore, Inco and Falconbridge have joined nodule-mining consortia, presumably to keep tabs on the latest development, and make adjustments accordingly.

Platinum-Group Metals

Platinum-group metals (PGM) produced in Canada are derived from basic and ultrabasic rocks associated with nickel and copper sulfides. Their recovery as a by-product of nickel and copper mining makes production highly dependent on the state of the export markets for Canada's nickel and copper. For example, production of PGM in 1981 amounted to 5.7% of world output, which made Canada the third largest producer. The 53% drop in Canadian nickel production in 1982, a response to market oversupply conditions, caused PGM production to fall by 51% (Figure I-3-22).[16] As a result, Canada contributed only 3% of world PGM output in 1982.[16]

Canadian production of PGM first began in 1909 in the Sudbury nickel district. Until 1953, Canada and the USSR were the major producers worldwide, but South African production came fully on-stream at that point. Since then, Canada has generally ranked as the third largest PGM producer (behind South Africa and the USSR), with variations in output largely determined by nickel production and sales.

Figure I-3-17 shows that three mines currently produce PGM ores in Canada. Inco Limited recovers ores from its Sudbury, Ontario, and Thompson, Manitoba, nickel mines. It also recently extracted PGM ores from the Shebandowan district in northwestern Ontario. Falconbridge Nickel Mines recovers ores at its Fanconbridge, Ontario, operations. The composition of Sudbury district ores (the major source) is 43% platinum, 45% palladium, and 12% of the other four platinum-group metals.[31] The composition of all Canadian platinum-group ores averages 46% platinum, 40% palladium, and 14% of the minor platinum-group metals.[45] Total production capacity in 1979 was 500,000 troy ounces.[31]

Inco's nickel and copper refineries in Ontario produce platinum-bearing residues which are then shipped to Inco's Acton (London), England, plant for refining. Falconbridge ships nickel-copper matte to its Nikkelverk A/S refinery in Kristiansand, Norway. Platinum-group-rich sludge is then shipped to the United States, South Africa, and elsewhere in Norway for refining.[9] This scattered processing means that Canada's share of world PGM trade is somewhat hidden. This is especially true for U.S. imports of Canadian PGM. One estimate is that Canada's share of U.S. imports was recently as large as that held by South Africa.[45]

Canada's ability to remain an important producer and exporter of PGM is uncertain. Unless

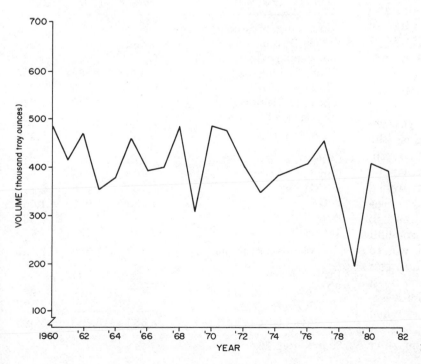

FIGURE I-3-22 Production of platinum-group metals, 1960–1982. (From Refs. 15, 16, and 35.)

nickel demand increases to a great extent, Canada's share of world production will fall in relation to Soviet and South African levels. With low nickel demand, Canadian producers may be content just to exploit already-recovered ores. Furthermore, labor problems, such as the eight-month-long strike in 1978–1979 at Inco's nickel plants, will always place some doubt on the stability of Canada's production.

Potash

Potassium salts (or potash) are normally found as evaporites in ancient lake or seabeds. Deposits in both Canada and the United States are normally in the form of sylvite (KCl, or muriate of potash), a mineral that is 52% potassium, or roughly 63% K_2O equivalent.[15] Potash was discovered in Saskatchewan in 1943, while drilling for oil. It did not take long to determine that the Saskatchewan beds were flat, thick, continuous, and of a very high grade. Canada has easily established itself as the premier producer of potash in the free world, and ranks second only to the USSR overall.

The potash industry has had an experience similar to that of Quebec's asbestos industry in that the provincial government acted to take over part of the industry. Until 1971, the potash industry enjoyed the same tax and regulatory environment of all other minerals. The 1971 provincial elections in Saskatchewan brought the New Democratic Party (NDP) to power, and things changed quickly. The new government felt that the province was not earning enough tax revenues from the potash industry, so taxes were increased dramatically. Industry taxes were C$5 million in 1972–1973, but increased to C$100 million in 1974.[54] The new taxes took approximately 90% of the industry's pretax profits.[9]

The industry, and the Canadian federal government, strongly protested the new provincial taxes. Among other things, the federal government joined an industry suit lodged against the Saskatchewan government. The federal government also removed the deduction of provincial income taxes from federal taxes, and threatened to tax provincial Crown corporations. The Saskatchewan government held firm, though, and in 1975 it declared its intention of acquiring one-half of the potash industry's assets. The motive was to ensure that the industry acted in the best interests of the province. This type of rhetoric scared most free-market supporters, including the U.S. government, but concerns were largely eliminated when the NDP government guaranteed a fair takeover price for those firms that consented to selling their assets.[54] Since then the NDP government has acquired roughly 40% of total industry assets and has placed them under the control of the Potash Corporation of Saskatchewan (PCS), a provincial Crown corporation. The final 10% of industry assets sought by the province is to be acquired through normal expansion of the industry.

Eleven potash mines were in operation in Saskatchewan in 1982. Most of the mines use a combination of room-and-pillar and stress-relief underground mining techniques, but one mine uses solution mining. Current recovery rates average about 40% of in-situ ore.[9] The production capacity of current and expected potash mines is illustrated in Table I-3-15, and their location is shown in Figure I-3-6. Current and projected capacity levels in Table I-3-15 account for roughly one-fourth of world capacity for the time period discussed.[31] Figure I-3-23 illustrates Canada's production and exports of potash since 1960.

Given potash's low unit value, it is only natural that most Canadian exports travel only as far as the United States. This, in fact, is the case. Over 68% of the potash used in North America is used for agricultural purposes,[9] and the U.S. agricultural market needs potash. Production of potash in the United States has been declining since 1966, due to

TABLE I-3-15 Production Capacity for Existing and Proposed Potash Mines in Canada (Thousand Metric Tons K_2O)

Company	Capacity		
	1982	1986	1990
Potash Corp. of Saskatchewan			
Allen Division	490	655	655
Bredenbury Division	—	655	1,960
Cory Division	830	1,090	1,090
Esterhazy Division	585	945	945
Lanigan Division	830	1,740	1,740
Rocanville Division	1,090	1,090	1,090
PCS subtotal	3,695	6,175	7,480
CCP	815	815	815
Cominco Ltd.	600	655	655
IMCC	1,750	2,110	2,110
PPG	1,055	1,055	1,055
PCA	440	635	635
Texasgulf	325	435	435
Total Saskatchewan	8,680	11,880	13,185
Denison Mines	—	765	765
PCA	—	545	545
Others	—	—	1,590
Total Canada	8,680	13,190	16,085

Source: Ref. 12.

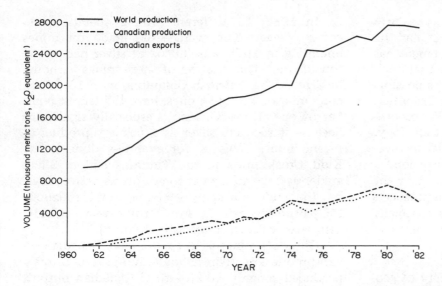

FIGURE I-3-23 Canadian production and exports of potash, 1960–1982. From Refs. 16 and 35.)

declining ore grades,[31] but the United States began importing potash long before then. In fact, by 1950, the United States was a net importer of potash. It now imports about one-half of its total needs. Of that amount, about 90% comes from Canada.[54] This represents about 70% of all Canada's potash exports.[55]

The North American potash market is basically split. Most U.S. production comes from mines in the Carlsbad, New Mexico, region, while some potash comes from mines in California and Utah. The Carlsbad potash is shipped to the southeastern United States, while the west coast is supplied by potash from California and Utah. All other U.S. markets (including the midwest grain belt) and Canada are supplied by the Saskatchewan mines. Saskatchewan potash is shipped by rail, truck, and Great Lakes freighters, but in the long run, transportation represents about one-half of total delivered cost; thus the split in market coverage.[54]

Canadian coverage of the U.S. market changed somewhat in late 1983 and early 1984, with the startup of two new potash mines in New Brunswick. The two mines, which are owned by the Potash Corporation of America and Denison Mines, are very close to the all-weather deep-water port at Saint John, New Brunswick. A C$40 million terminal and loading facility is under construction to facilitate the shipping of this potash. The two mines can get potash to dockside for $3.50 a metric ton, which is one-fourth the cost for moving potash from Saskatchewan to Vancouver.[54] As a result, these two mines are expected to take over some of the markets in the lower Great Lakes and southern U.S. regions. In addition, they are expected to tap markets in Central and South America, as well as Japan.[56]

Canada now also exports to the Pacific Rim nations and South America. These exports have lately accounted for 30% of total Canadian exports.[56] All told, Canada accounts for 37% of total world potash exports, which makes Canada the leading exporter worldwide.[31] In fact, Canada exports almost 95% of its potash production.[57] Overseas exports of potash are arranged by Canpotex, an umbrella organization representing all producers in Canada. As recently as mid-1982, over 60% of the exports sold through Canpotex were supplied by the Potash Corporation of Saskatchewan (PCS).[56]

The near-term prospects for Canada's potash producers appear unfavorable. Weak agricultural demand is projected to continue for some time. To make things worse, the recently established U.S. programs to pay farmers in-kind to keep land out of production have seen a response far beyond all expectations. Over 52% of all eligible farms have signed up for the programs, even though the programs have only been in effect since early 1983.[58] Officials of the U.S. Department of Agriculture have already predicted sharply declining sales of fertilizer as a result of these programs. This is likely to increase the number of shutdowns in the Canadian potash industry. For example, PCS has already shut down its mines to clear excess inventories.[55]

However, a political development may have the most significant short-term impact on the potash industry. The Crown corporation, PCS, had a hard time selling its potash during 1982, especially in the United States. This is thought to have been prompted partly by American desires to avoid involvement with the child of the New Democratic Party in Saskatchewan.[59] As a result, PCS announced its intentions to withdraw from Canpotex and to establish PCS International to market its own product.[60] The NDP government was recently defeated in a provincial election, however, and the new Progressive Conservative (PC) government has blocked the attempt to establish PCS International. This is presumably based on fears about a possible price war between PCS International and Canpotex, but some sources claim that the new PC government acted on the basis of its free-market ideology. Other recent moves to restrict provincial involvement in the market seem to verify this suspicion.[57]

With a burgeoning world population and the resultant demand for food, the long-term prospects for Canadian potash are good. This is especially true given the vast reserves Canada has to tap.

Silver

Silver is found in Canada in a variety of geological settings. It is recovered from many different mining operations, but most silver is recovered as a by-product of nickel, lead, zinc, and/or copper mining. As of the late 1970s, 82 percent of the silver produced in Canada was as a by-product of base-metal mining, and most of the remaining 18 percent was as primary silver.[15] Even as a by-product, silver's moderately high unit value makes its production somewhat less dependent on the output of associated base metals than is true of, say, cadmium. In fact, silver's profits often keep a base-metal mine producing that would otherwise be unprofitable.

Mine production capacity in 1978 was 62.1 million troy ounces, but production has usually been well below that level.[15] Canada produced 38.7 million troy ounces in 1981 and 41.0 million troy ounces in 1982, which accounted for 10.6 and 11.3% of world production, respectively.[16] Canada's 1982 output placed it as the fourth largest silver producer in the world behind the USSR, Mexico, and Peru.

In 1981, 51 different mining operations recovered silver. The average ore grade for mines operating in 1978 was 8.6 oz of silver per metric ton of ore.[9] The location of these mines is shown in Figure I-3-17. British Columbia, with 15 mines, and Ontario, with 13 mines, have the largest number of operations. Ontario is especially significant because it recovers silver as a primary product at several mines. This is, for example, done at the Kidd Creek mine in the Timmins region. Silver grades average 4.3 oz per ton of ore as native silver, but the grade is roughly 1.9 oz per ton throughout the entire deposit (a stratiform massive sulfide, associated with volcanoclastic rhyolite).[9]

Ontario's mines tend to dominate Canadian silver production. Throughout the 1970s, Ontario produced roughly 40% of total Canadian output. British Columbia accounted for 20%, New Brunswick produced roughly 13%, and Quebec accounted for 9%.[15] The three largest producers were Kidd Creek Mines, Brunswick Mining and Smelting (New Brunswick), and Cominco (primarily British Columbia). By 1981, the production of distribution changed somewhat, especially because of the startup of a number of silver-bearing copper mines in British Columbia. Out of a total of 1,203,000 kg of contained silver, British Columbia produced 32.3%, Ontario had 26.9%, New Brunswick had 16%, and the Yukon Territory produced 14.3%.[12]

Six different refineries produced primary silver in 1978. They are described in Table I-3-16, and their locations are plotted in Figure I-3-3. Capacity in 1978 was 53.5 million troy ounces.[12] The Cobalt, Ontario, refinery owned by Canadian Smelting & Refining Ltd. closed in February 1982, so refining capacity dropped to 1,533,660 kg per year.[12]

Figure I-3-24 indicates that Canada exports the bulk of its silver production. Since 1974, metal exports have exceeded ore exports (in contained

TABLE I-3-16   Silver Refineries in Canada

Company	Location	Annual capacity (kg)	Feedstock
Canadian Copper Refiners Ltd.	Montreal East, Que.	777,600	Blister copper, anodes
Cominco Ltd.	Trail, B.C.	373,200	Lead-zinc concentrate
Royal Canadian Mint	Ottawa, Ont.	217,705	Gold bullion
Canadian Smelting & Refining	Cobalt, Ont.	186,600	Silver-cobalt ore and concentrate
Brunswick Mining & Smelting	Belledune, N.B.	125,000	Lead concentrate
Inco Metals Co.	Copper Cliff, Ont.	40,250	Nickel-copper concentrates
		1,720,260	

Source: Ref. 15.

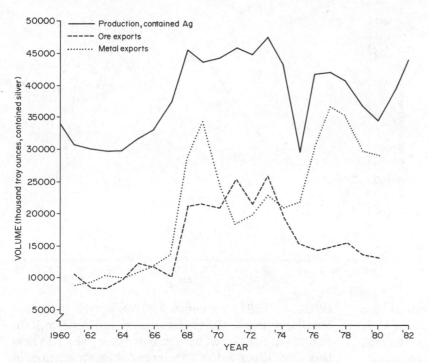

FIGURE I-3-24   Canadian production and exports of silver, 1960–1982. (From Refs. 15, 16, and 35.)

silver) by a factor of 2. Most exports have gone to the United States. From 1963 to 1980, between 70 and 92% of annual exports went to the United States.[35] This made Canada the major supply source for American imports of silver. From 1978 to 1981, Canada supplied 37% of American silver imports, while Mexico, the second-largest supplier, provided 24%.[9] Canadian producers expect a gradual rise in their production and export levels, but they face an increasing challenge in the U.S. market from Mexico and other Latin American suppliers.

Titanium

Titanium in Canada is derived from ilmenite $(FeTiO_3)$, whereas other producers (Australia, for example) derive their titanium from both ilmenite and rutile. Canada's production presently comes from one mine at the Lac Tio deposit in eastern Quebec (Figure I-3-12). The mine recovers an ilmenite-hematite ore from an anorthositic pluton. The deposit measures roughly 1200 by 740 ft in area, is 200 ft thick, and had 12,000,000 tons of reserves grading at 35% titanium dioxide (TiO_2).

Ore mined at Lac Tio is crushed and beneficiated by heavy media separation to yield a concentrate, grading 92.5% iron and TiO_2.[12] Concentrate is then carried by railway to Havre St. Pierre, a port on the north shore of the St. Lawrence River. From there, the concentrate is shipped to Sorel, where it is smelted in electric furnaces. Three products are derived from the smelting process: Sorelslag, a slag grading 70 to 72% TiO_2; Sorelmetal, a low-manganese pig iron; and Sorelflux, which is an ilmenite flux. Output in 1981 was 762,000 metric tons of Sorelslag, 538,000 metric tons of Sorelmetal and minor amounts of Sorelflux.[12]

The Lac Tio-Sorel complexes produced 653,000 metric tons of TiO_2 slag in 1982. This comprised roughly 16% of world output.[16] Canada accounted for even larger shares of world output—up to 25%—up through the 1970s, but production has fallen off.[31] Demand has decreased, and the Lac Tio mine had several strikes, most notably a four-month-long strike in 1979. Total production capacity, in terms of titanium content, is rated at 476,000 metric tons per year through 1985. This accounts for over 24% of world ilmenite-based capacity of 1,948,000 metric tons.[31] Figure I-3-25 shows production of Sorelslag in Canada from 1960 to 1982.

About 10% of the Sorelslag produced in Canada is used domestically. Its major use is titanium pigments for paints and paper. Pigments are produced by digesting the slag in sulfuric acid, then performing hydrolysis of the titanium sulfate. The resulting product, TiO_2, is then added to paints and other products. This process produces a great deal of waste acids and solids, which were once dumped into the St. Lawrence River. However, the Quebec government ordered pigment producers to change their practices by 1980. As a result, limestone is now added to the acids to produce gypsum. Unfortunately, demand for wallboard and other gypsum-based products has fallen, as has demand for paints, because of a slump in the construction industry.

The 90% of Sorelslag not consumed domesti-

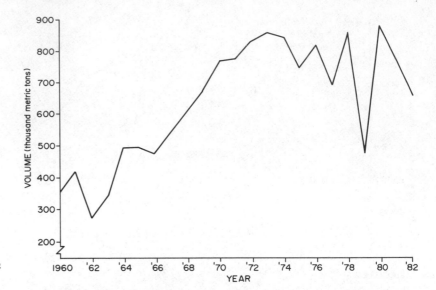

FIGURE I-3-25 Canadian production of TiO$_2$ slag, 1960–1982. (From Refs. 15, 16, and 35.)

cally is exported, with most of it going to the United States. These exports provided 34% of U.S. imports of ilmenite from 1978 to 1981.[16] The slag is used primarily for the production of pigments. Canada's demands for titanium metal are, in turn, met by U.S.-based producers and by some producers in the United Kingdom.

The Lac Tio-Sorel facilities are owned and operated by QIT-Fer et Titane, Inc., which itself is totally owned by Kennecott Copper of the United States. An important note is that QIT owns 32% of Richards Bay Minerals in South Africa. The latter facility produces rutile and an 85% TiO$_2$ slag. The point is that, if the demand were present, QIT's Canadian operations could conceivably also produce the 85% slag. This slag is of high enough grade to be used for titanium sponge production, as is apparently being done now in the USSR. Such a development, if it took place, would lessen U.S. dependence on Australian rutile as the feedstock for most of the sponge production in the United States. In fact, the Sorel facility is currently being modernized to yield an 80% Sorelslag.

Tungsten

The only operating tungsten mine in Canada is located at Flat River, Northwest Territories, about 3 km from the border with the Yukon Territory (Figure I-3-12). It had 3.04 million metric tons of minable reserves of contained tungsten and an ore grade of 1.32% in December 1982.[44] The mine, owned by Canada Tungsten Mining Corporation, began production in 1962 from a small open-pit mine. This operation was closed in 1974 and replaced by an underground (room and pillar) operation.[9] The mine and its on-site mill had a daily capacity of 1000 metric tons of tungsten trioxide

(WO$_3$) in 1981. Scheelite (CaWO$_4$) is the tungsten-bearing mineral. Current capacity is 7 to 8 million pounds of contained tungsten per year, which makes the Flat River mine the largest tungsten mine in the free world.[61] Production of 6,600,000 lb of contained tungsten in 1982 was the highest level of all noncommunist nations and placed Canada third behind China and the USSR as leading tungsten producers worldwide.[16]

The mine is reached only by an all-weather, 300-km road from Watson Lake in the Yukon Territory. Concentrates are carried by truck to Vancouver and then shipped elsewhere. Some of the concentrate reaches Fort Madison, Iowa, where AMAX, Inc., of New York, the parent of Canada Tungsten Mining Corporation, produces blue tungstic oxide and ammonium paratungstate.[61] Canada exports substantial proportions of its total production (all to the United States), and imports ferro-tungsten and metallic carbides. Production and export levels are shown in Figure I-3-26. This figure also shows the large share that Canadian exports make of total U.S. imports. From 1977 to 1981, that share was 21%.[16]

Canada's status as a world force in tungsten production was dramatically strengthened beginning in late 1983. The Sullivan Mining Group started up a new mine at Mount Pleasant, New Brunswick, in late 1983. The property, a multimetal porphyry, has 42 million metric tons of reserves, with an average grade of 0.2% WO$_3$. Expected production capacity will be 1800 metric tons of contained tungsten by mid-1984.[32] This raises Canada's production capacity to almost 12 million pounds of contained tungsten per year.

The second factor is the MacTung property which AMAX, Inc., is developing along the border

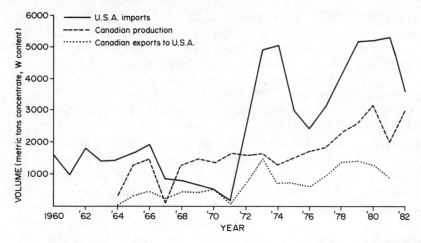

FIGURE I-3-26 Canadian production and exports to the United States of tungsten versus U.S. imports, 1960–1982. (From Refs. 16, and 35.)

of the Yukon and the Northwest Territories. Identified reserves are said to grade at 0.96% WO_3, and are roughly one-fourth the size of those in China. In other words, reserves are roughly 240 million metric tons.[62] This doubles the volume of Canada's existing tungsten reserves. Should this project continue on schedule, a tentative startup date of 1985 is expected.

All told, the future for Canadian tungsten looks fairly good. There are several stumbling blocks, though. Labor problems have already resulted in work stoppages at the Flat River plant. For example, a six-month strike occurred in 1981, resulting in a 30% drop in Canadian production in that year. Another obstacle to full development of new mines will be the pace at which economic growth picks up, especially in the United States. Low demand has reduced prices for tungsten to the point where the Flat River mine was closed in January 1983.[63]

Should this lull continue, the Mount Pleasant and MacTung projects will probably be delayed for an indefinite period.

Zinc

Canada has been the world leader in mine production of zinc since 1964, and has also been an important producer of zinc metal.[45] Canada's mine production of zinc in 1982 represented approximately 19% of the world's total.[16] Canada's production levels for zinc are shown in Figure I-3-27. Canadian zinc operations are strongly oriented toward export markets. Roughly 90% of mine output was exported annually throughout the 1970s, and the corresponding figure for 1981 was 86%.[15,35] Most of the demand for Canadian zinc exports (in any form) is ultimately based on the use of zinc for galvanizing and for die castings.

The dominant ore mineral, sphalerite, occurs in

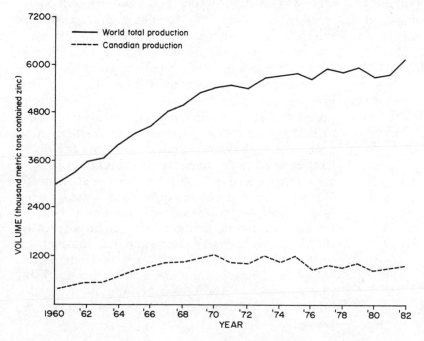

FIGURE I-3-27 Canada's share of world zinc production, 1960–1982. (From Refs. 15, 16, and 35.)

Canada in two major geological settings. The first, massive sulfides located in metamorphic rocks, is typified by the Sullivan deposit in British Columbia and the Kidd Creek Mine near Timmins, Ontario. This type of deposit contrasts with the second deposit: the irregular breccia or replacement, strata-bound deposits found in carbonates. This altered sedimentary occurrence is illustrated by the Polaris mine on Little Cornwallis Island in the Canadian Arctic. These two settings often contain up to 2% cadmium, and sometimes include germanium, gallium, and thallium. An important consideration is the trade-off between the high zinc ore grade in the sulfide deposit (often 10 to 20% zinc), and the simple and very cheap mining and processing of the sedimentary deposits (with ore grades of up to 5 or 6%).[31]

Figure I-3-9 shows the location of the 28 zinc mines currently operating in Canada. The location breakdown is shown further in Table I-3-17. These 28 mines had a nominal output capacity in 1980 of 1,362,000 metric tons per year. In decreasing order of capacity, the leading mine producers were Ontario, New Brunswick, the Northwest Territories, the Yukon Territory, Quebec, British Columbia, Manitoba, Newfoundland, and Saskatchewan.[12] Average ore grade is just under 6%, which is considerably higher than the 4% average of mines operating in the United States.[15]

Figure I-3-3 shows the location, and Table I-3-18 shows the annual capacity of the four metallurgical plants that process zinc in Canada. The rated annual capacity of 633,300 metric tons of zinc is in the process of being increased. By mid-1983, the Hoyle plant is due to increase its capacity by 19,000 metric tons. The Trail complex recently completed a 27,000-metric ton expansion, and the Valleyfield operation, by 1984, will have increased capacity by 9000 metric tons per year. In addition, Brunswick

Mining & Smelting has announced plans to construct a 100,000-metric ton per year reduction plant at Belledune, New Brunswick. The late 1984 startup target for this plant should bring total zinc refining capacity in Canada to 788,300 metric tons per year.[12] This should maintain Canada's status as the second largest zinc metal producer in the world, after Japan.

Canadian zinc refineries already hold a special status worldwide. The Trail and Valleyfield operations, even before their capacity expansions, are the two largest electrolytic zinc plants in the world.[15] Moreover, the Trail complex has started up a new zinc pressure leaching unit to replace its concentrate roaster. This highly acclaimed new concentration process produces elemental sulfur as a by-product instead of the roaster's sulfuric acid.[31]

The main destination of Canada's zinc exports, both directly and indirectly, is the United States. During the 1970s, Canadian zinc ores and concentrates made up 60% of U.S. imports of zinc.[31] This was about 14% of Canadian zinc exports. However, 55% of Canada's ore and concentrate exports went to Japan and Belgium-Luxembourg.[35] These exports made up one-half (or more) of the crude zinc imports of these nations. These countries then turned around and exported zinc metal to the United States. In 1980 the United States directly bought 59.5% of Canada's zinc metal exports.[35] All told, then, Canada's zinc metal exports to the United States were actually higher than the figures indicate.

Whether Canada can continue this level of exports will depend on a variety of factors. The dominant one is the strength of the global automobile industry. Exports to the United States should fall somewhat if many U.S.-based zinc smelters start up again in response to increased domestic demand. Another factor will involve Japan's barriers to imports of finished metals.

It appears that Canada will be able to at least maintain its current position, as long as it does not push too hard for more domestic processing of zinc. This is so because Japan and many Western European nations have very quickly become the dominant customers for Canadian ores and concentrates and will probably continue this dependence.[45] However, labor problems may alter Canada's competitiveness, both because of the image held by foreign customers, and because of the impact on the domestic industry. For example, contract disputes at the Bathurst, New Brunswick, zinc complex led to a four-month mine strike, and a 17-day smelter strike in 1980. This action cost Brunswick Mining & Smelting some $30 million due to lost sales.[9]

TABLE I-3-17 Distribution of Zinc-Mining Areas in Canada

Province	Number of mines
Northwest Territories	3
Yukon Territory	2
British Columbia	5
Saskatchewan	1
Manitoba	4
Ontario	5
Quebec	3
New Brunswick	2
Nova Scotia	1
Newfoundland	2

Sources: Refs. 15 and 98.

TABLE I-3-18   Zinc Metallurgical Plants in Canada, Year End 1981

Location	Company	Capacity (metric tons)
Hoyle, Ont.	Kidd Creek Mines	108,900
Flin Flon, Man.	Hudson Bay Mining & Smelting	74,400
Valleyfield, Que.	Canadian Electrolytic Zinc	205,000
Trail, B.C.	Cominco Ltd.	245,000
		633,300

Source: Ref. 12.

THE ENERGY INDUSTRIES

Coal

Canada is unusual in that it consumes, produces, and trades large volumes of coal. Few other countries, besides, say, the United States, Australia, and South Africa, are engaged in all aspects of the coal industry. Canada produces, exports, and imports both coking coal and bituminous coals.

Coking Coal. Like other steel-producing nations, Canada's steel industry needs large quantities of coke. This is produced by carbonizing bituminous coking coals. Since Canada has historically consumed, on a net basis, more coking coal than it produced, certain amounts were imported. The dominant supply source for Canada's imports has been the Appalachian region of the eastern United States. The Appalachian coals have qualities that allow for suitable blends with various coals produced in Canada. As an illustration, in 1978, Dominion Foundries and Steel Ltd. of Hamilton, Ontario, produced 1.16 million metric tons of coke from 1.46 million metric tons of coking coal. Of that amount, 85% came from the United States, and the remainder was a mixture of western and eastern Canadian coals.[15]

Figure I-3-28 indicates production, export, and import levels of Canadian coke. As noted above, coke imports come from the United States. Most coke exports also go to the United States, but the export destinations of Canadian coke have diversified somewhat lately. A large amount of coke is now exported to Japan and to Western Europe. By way of illustration, 92.9% of Canada's coke exports from 1963 to 1965 went to the United States. From 1970 to 1972, the figure was 56.7%. The figure for 1978–1980 was 59.1%.[35] One nation, Japan, received increasingly large amounts of Canada's coke exports. Canada provided 3% of Japan's coke imports in 1970; by 1979, the figure was 18%.[64]

Canada developed a number of coking coal mines in the 1970s, most of which were in British Columbia. (Physical details of the mines are presented in the following section.) Most of the mines, such as the Quintette project in northeast British Columbia, were developed to meet the growing import needs of a burgeoning Japanese steel industry. In fact, many of the mines were developed as joint ventures with a variety of Japanese firms. These agreements called for long-term purchases by Japan at fixed prices which were to be modified, if necessary, at predetermined times. The recent slump in world steel production has affected even Japan's industry, which saw production fall by some 5 to 10 million long tons in 1982. Thus, given

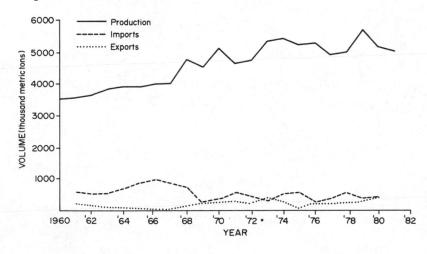

FIGURE I-3-28   Canadian production, exports, and imports of coke, 1960–1982. (From Refs. 15, 16, and 35.)

a reduced need for coke, Japanese firms are now trying to renegotiate prices with all their suppliers.

The United States recently granted a small price cut, but the Chinese, Australian, and Canadian coking coal suppliers are resisting. Japan has asked for a 22% cut in the price it pays for Australian coal and is now asking for at least a $3 per long ton reduction in the price of Canadian coal.[65] Existing contracts with Canadian suppliers call for sales of 100 million metric tons of coking coal over the next 15 years, at a price of roughly C$84 a long ton, but several large contracts are up for renewal in 1983.[66] Fording Coal, B. C. Coal, and Luscar Ltd., The Canadian firms scheduled for renewal of contracts, are fighting the price cut proposal on several grounds. They argue, first, that Japan led them to develop production capacity which may not have been needed. Second, they argue that their capacity, plus their long-term contracts, make them the base-case supply source for Japanese steel producers, and this capacity and dependability should be rewarded.

Thermal Coal. Coal deposits are widely scattered throughout Canada but tend to be restricted to five geologic provinces. The Cordilleran Orogen includes coal deposits in British Columbia and the Yukon Territory. The Interior Platform includes deposits in Alberta and Saskatchewan. Deposits in the Northwest Territories are situated in the In-

nuitian Orogen and the Arctic Platform. Maritime coal deposits are part of the Appalachian Orogen. Deposits in British Columbia and Alberta are usually of Cretaceous or Jurassic age. Saskatchewan coals are Paleocene in age, while Yukon and Northwest Territories coals are Cretaceous and Tertiary in age. Maritime provinces coal deposits are Pennsylvanian in age.[67]

Major coal deposits and their coal rank are shown in Figure I-3-29. Canadian coals range from low-volatile bituminous to lignites, with heat values ranging from 6000 Btu/lb for lignites, to 13,500 btu/lb for some bituminous coals. Minable coal reserves amount to some 16 billion metric tons. Lignites make up 3.2 billion, subbituminous coals comprise 7.3 billion, and bituminous coals account for 5.5 billion metric tons of the total.[67] Total coal resources, on the other hand, are estimated at over 360 billion metric tons, of which the vast majority are in British Columbia and Saskatchewan.[67]

Canada's consumption of thermal coal has followed a pattern, in the twentieth century, quite similar to that seen in many other developed nations. Although consumption levels increased in absolute terms in most years, coal accounted for a declining share of total energy consumption since the late 1940s, when oil and gas began to make their presence felt. Coal's share declined until, by 1979, it accounted for only 9.3% of Canada's primary energy consumption.[68] Since then, the share held by coal has increased somewhat. Coal is ex-

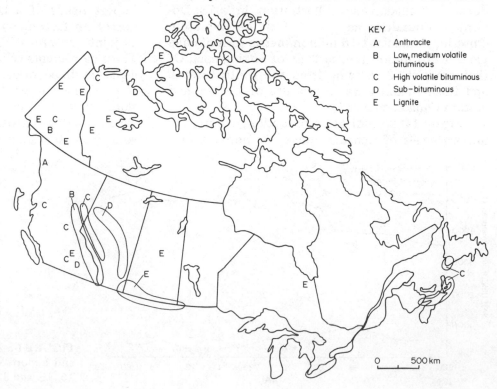

KEY

A   Anthracite
B   Low, medium volatile bituminous
C   High volatile bituminous
D   Sub-bituminous
E   Lignite

FIGURE 1-3-29 Location and rank of Canada's major coal deposits. (From Ref. 51.)

0   500 km

pected to account for 12 to 13% of Canada's primary consumption by the year 2000.

A second important note about Canadian coal is that production levels have only recently begun to approach consumption levels. In other words, Canada has had to import thermal coal. For example, consumption in 1968 was 24,782,275 metric tons, but production was 9,969,059 metric tons, or only 40.2% of consumption. By 1978, consumption was 31,738,310 metric tons and production was 30,476,855 metric tons, 96% of consumption.[15] Demand in 1980 was 30 million metric tons, and the 1990 level is expected to be from 64 to 83 million metric tons.[68] Recent increases in production capacity indicate a good possibility that domestic output can meet consumption demands. If this is not the case, Canada will continue to import coal.

Table I-3-19 shows coal mines currently operating, and soon to begin operations in Canada. It describes the type of coal produced, markets covered, and production capacity. Figures I-3-30A and I-3-30B indicate the location of these mines. Production in 1981 was dominated by Alberta, which produced 46.1% of total Canadian output. British Columbia accounted for 29.4%, Saskatchewan produced 17%, Nova Scotia produced 6.2%, and New Brunswick produced 1.3% of total output.[66] This distribution should remain roughly the same in the future, given expected provincial increases in production capacity. As Table I-3-19 illustrates, Alberta's production capacity in 1985 is expected to be 17.5 million metric tons per year, and capacity in British Columbia is expected to reach 35.2 million metric tons per year.

Canada will probably continue to import coal even if domestic production does exceed consumption levels. Because of the distances separating coal producers and coal consumers in Canada, and the usual dependence on rail transport for inland markets, it is in some instances cheaper to import coal from U.S. mines located closer to major Canadian demand centers. This has historically meant that the provinces of Ontario and Quebec have imported coal from the Appalachian region of the United States rather than from mines in western Canada. In turn, coal mined in the western part of Canada is usually exported overseas. This practice is quite sound economically, due to differences in transport costs and, therefore, the practice will almost certainly be continued.

Figure I-3-31 illustrates Canadian production, export, and import levels for bituminous coals from 1960 to 1982. It illustrates the gap between imports and exports steadily declined until exports exceeded imports in 1981. Canada has been involved in so much coal trade that, in the late 1970s, it was the third-largest coal importer in the free world behind Japan and France.[66] Canada was also the seventh-largest exporter. When measuring coal exports as a percentage of domestic production, Canada is the largest exporter of coal in the world. Exports made up 45% of production in 1981.[19]

Canada's imports of thermal coal all come from the United States, but Canada's export destinations are fairly diversified. Minor amounts go to the United States, and some goes to Brazil and other Latin American countries, but most exports go to the nations on the western Pacific Rim. Japan has traditionally been the primary destination for Canadian exports of thermal coal, but South Korea has lately imported a substantial amount. These two countries received over 80% of Canada's thermal coal exports in the late 1970s and the early 1980s.[35]

Future prospects for continued exports to Pacific Rim countries are such that producers in British Columbia and Alberta are expanding their facilities to meet this expected demand increase. British Columbia, for example, exported 11.5 million metric tons (mmt) of thermal coal and 7.4 mmt of coking coal to Japan in 1981. Producers in that province expect Japan to import, from British Columbia's mines alone, 25 mmt of thermal coal and 16 mmt of coking coal by 1985. Import levels could reach 40 mmt and 16 mmt, respectively, by 1990.[70] As a result, production capacity for both coking and thermal coals in British Columbia is expected to reach 27 mmt in 1985 and 74 mmt in 1990.[70]

Canada's desire to tap expanding export markets in the Pacific Rim is tempered with the knowledge that it is not the only producer capable of tapping the market. Japan's stated policy of supply diversification means that it is essentially shopping around for the best deal. Japan has invested in production facilities in Australia and other nations, all of which want to expand their exports to the Pacific Rim. The fact that Canada faces stiff competition is shown by the recent agreement by South Korea's largest electric utility (the Korea Electric Power corporation) to buy thermal coal from Alaska. Suneel Alaska Corporation is to provide 800,000 metric tons of thermal coal a year from 1984 to 1998.[71]

One factor that must be overcome if Canada is to compete economically for the Pacific Rim coal market is the high share of delivered cost made up by transportation. The major consideration here is inland transportation. Almost all Canadian coal transported inland is carried by economic unit

TABLE I-3-19 Coal Mining Operations in Canada[a]

Region	Mine name	Company name	Market, coal type	Coal rank	Annual capacity
Southeast British Columbia	Michel, Harmer Ridge	B.C. Coal Ltd.	Export, coking and steam	lvb	9.0 mmt by 1985
	Line Creek	Crows Nest Resources	Export, coking and steam	lvb, mvb	2.7 mmt by 1983
	Fording River	Fording Coal Ltd.	Export, coking and steam	mvb, hvb	5.5 mmt by 1983
	Coal Mountain	Byron Creek Collieries	Domestic and export, steam	mvb	3.6 mmt by 1985
	Greenhills	B.C. Coal and POSCO of Korea	Export, coking and steam	mvb	2.3 mmt by 1983
Northeast British Columbia	Quintette	Denison Mines	Export, coking and steam	lvb, mvb	5 mmt coking, 1.5 mmt steam by 1984
	Bullmoose	Teck Corporation	Export, coking and steam	lvb, mvb	1.7 mmt coking, 0.6 mmt steam by 1984
	Monkman	Petro-Canada and Canadian Superior Oil	Export, coking	lvb, mvb	3.3 mmt by 1984
Central Eastern Alberta	Highvale	Manalta Coal Ltd.	Domestic steam	sub B	9.7 mmt by 1984
	Whitewood	Manalta Coal Ltd.	Domestic steam	sub B	1.5 mmt by 1983
	Vesta	Manalta Coal Ltd.	Domestic steam	sub B	1 mmt by 1983
	Paintearth	Forestburg Collieries	Domestic steam	sub B	1.8 mmt by 1983
	Roselyn	Manalta Coal Ltd.	Domestic steam	sub B	0.8 mmt by 1983
	Genessee	Fording Coal Ltd.	Domestic steam	sub B	3 mmt by 1986
West Central Alberta	Cardinal River	Cardinal River Coals	Export, coking	mvb	2.9 mmt by 1983
	Smoky River	McIntyre Mines Ltd.	Export, coking and steam	lvb	3.5 mmt by 1983
	Coal Valley	Luscar-Sterco	Export and domestic, steam	hvb	5.0 mmt by 1983
	Obed-Marsh	Union Oil of Canada	Export, steam	sub A, hvb	3.0 mmt when contracts signed
	Mercoal	Manalta Coal Ltd.	Export, steam	hvb	2.0 mmt by 1985
	MacLeod River	Manalta Coal Ltd.	Export, steam	hvb	4.0 mmt by 1987

Province	Mine	Company	Rank	Use	Capacity
Saskatchewan[b]	Bienfait	Manitoba and Saskatchewan Coal Co.			
	Boundary Dam	Manitoba and Saskatchewan Coal Co.			
	Klimax	Manalta Coal Ltd.			
	Utility	Manalta Coal Ltd.			
	Souris Valley	Saskatchewan Power Corporation			
	Poplar River	Saskatchewan Power Corporation			
Nova Scotia[c]	Lingan	DEVCO			
	No. 26	DEVCO			
	Prince	DEVCO			
	Donkin	DEVCO			
New Brunswick	Minto	NB Coal Ltd.	hvb	Domestic, steam	0.5 mmt by 1983

Total Canadian capacity by 1987 = 81.876 mmt per year

[a] lvb, low-volatile bituminous; mvb, medium-volatile bituminous; hvb, high-volatile bituminous; sub B, subbituminous; sub A, subanthracite; mmt, million metric tons.
[b] All Saskatchewan coal mines recover lignite, which is used solely for provincial power plants. Capacity in 1981 was 6.8 mmt.
[c] All mines recover hvb coal from underground mines. Coking and steam coal is used domestically and exported. In 1981, capacity was 2.5 mmt.
Source: Ref. 69.

FIGURE I-3-30A Coal mining areas in Canada. (From Refs. 69 and 98.)

0   500km

FIGURE I-3-30B   Key to Figure I-3-30A: Coal Mining Areas in Canada

Region	Mine name	Company name
Southeast	1 Fording River	Fording Coal Ltd.
British Columbia	2 Coal Mountain	Byron Creek Collieries
	3 Line Creek	Crowsnest Resources Ltd.
	4 Michel	B.C. Coal Ltd.
	Harmer Ridge	B.C. Coal Ltd.
	5 Greenhills	B.C. Coal, POSCO
Northeast	6 Monkman	Petro-Canada, Canada Superior
British Columbia		Oil Ltd.
	7 Quintette	Denison Mines
	8 Bullmoose	Teck Corporation
West Central Alberta	9 Smoky River	McIntyre Mines Ltd.
	10 Cardinal River	Cardinal River Coals
	Coal Valley	Luscar-Sterco
	Mercoal	Manalta Coal Ltd.
	11 Obed-Marsh	Union Oil Co. of Canada
	MacLeod River	Manalta Coal Ltd.
Central Eastern Alberta	12 Whitewood	Manalta Coal Ltd.
	Highvale	Manalta Coal Ltd.
	Genesee	Fording Coal Ltd.
	13 Paintearth	Forestburg Collieries
	Vesta	Manalta Coal Ltd.
	14 Roselyn	Manalta Coal Ltd.
Saskatchewan	15 Souris Valley	Saskatchewan Power Corp.
	Utility	Manalta Coal Ltd.
	Poplar River	Saskatchewan Power Corp.
	16 Klimax	Manalta Coal Ltd.
	17 Bienfait	Man. & Sask. Coal Co.
	Boundary Dam	Man. & Sask. Coal Co.
New Brunswick	18 Minto	NB Coal Ltd.
Nova Scotia	19 Lingan	DEVCO
	No. 26	DEVCO
	Prince	DEVCO
	Donkin	DEVCO

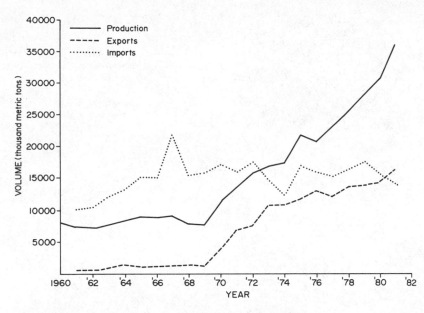

FIGURE 1-3-31   Canadian production, exports, and imports of thermal coal, 1960–1982. (From Refs. 19 and 50.)

trains, but the factor of importance is the long distances involved. Mines in British Columbia and Alberta, which produce for the export market, are an average of 700 miles (1120 km) from the west coast export terminals at Vancouver. This is the farthest distance coal must travel by rail to export ports in any of the major coal-exporting nations.[72]

Because of this distance, for example, 23% of the delivered price of C$82.62 a metric ton for B.C. Coal Company coal is based on the cost of rail transportation.[72] All in all, land-based transportation is more expensive in Canada than in all other major coal exporters, except the United States. This changed by the end of 1983, though, with the advent of revised Crow's Nest Pass grain rates. As noted above, minerals hauled on Canada's railways were charged high rates to subsidize the fixed low rates for grains. Now that this rate is being changed, haul rates for minerals should decrease somewhat, or at least not increase as rapidly as would otherwise be the case. In addition, long-overdue railway improvements should improve the efficiency of the lines and help hold costs down.[72]

Canadian coal exporters are also working to develop new and upgrade existing export terminals at the ports of Vancouver and Prince Rupert, British Columbia. These projects are explained in detail in the section on infrastructure. When these facilities are completed, Canada may be able to challenge Australia's low shipping costs to Japan. Australia now charges about C$8 per metric ton for ocean transport to Japan, whereas Canada's costs are somewhat higher.

Crude Petroleum

Given the vast geographic extent and geologic diversity of Canada's landmass, it is to be expected that there would be sedimentary basins suitable for the formation and entrapment of petroleum. The discovery, in 1949, of petroleum at Leduc, Alberta, proved this. Subsequent exploration and development work has since discovered petroleum deposits outside the major ones in Alberta (and to a lesser extent, Saskatchewan and British Columbia).

A number of promising discoveries have been made on Canada's outer continental shelf (OCS). The most significant discoveries to date have been in the arctic OCS (the Beaufort Sea and Sverdrup Basin discoveries), and in parts of the Atlantic OCS (especially the Hibernia field in Newfoundland's Grand Banks).[73] A third, and potentially significant addition to these petroleum resource areas is the massive oil sands found primarily in northern Alberta. Estimates of in-situ resources are huge. More conservative estimates put recoverable resources in the oil sands at 250 billion barrels, over 35 times the conventional oil reserves of Canada.[74] Figure I-3-32 shows the location of the oil sands, known offshore oil fields, and onshore producing areas.

Of all three areas, offshore fields are currently receiving the most attention from oil companies. This is due partly to a tax environment (described later) which reduced the incentive to work on-shore. This attention to OCS fields also indicates a very strong belief that total recoverable resources are very large. However, a number of obstacles make work on the OCS very difficult. Extreme cold and adverse weather and sea conditions are normal. Icebergs off Newfoundland are quite common and prompted the provincial government in early 1983 to request that Mobil withdraw its offshore exploration rigs until conditions improved. A storm in 1982 capsized such a rig and took all lives. The result of these adverse conditions is that oil exploration is quite expensive in these areas. Beau-

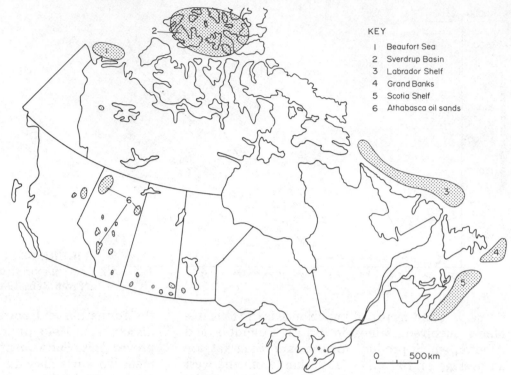

FIGURE I-3-32 Location of Canada's oil sands, offshore fields, and onshore producing areas. (From Refs. 73, 77, and 98.)

fort Sea wells typically cost up to $200 million each.[75]

Since oil sands are only lightly used now and offshore fields have not yet begun to produce commercially, Canada's current production of petroleum is dominated by the oil fields of Alberta, British Columbia, and Saskatchewan. The three accounted for 98% of total output in 1978, and this figure has remained relatively constant since then.[15] Figure I-3-33 shows total Canadian production of crude oil from 1960 to 1981. Total output peaked in 1973 and has been declining since then. Total production is the equivalent of about 75% of current domestic demand in Canada.[76]

Oil sands did receive a great deal of interest in the late 1970s and early 1980s. The intent behind development of the deposits was to help reduce Canada's dependence on imported oil. As noted by Bott,[77] at least eight different "megaprojects" were proposed to exploit the oil sands, especially those in the Athabasca region of northeastern Alberta. The eight projects described by Bott would have had a production capacity of almost 1 million barrels of oil per day.[77] This could have increased total Canadian production by some 20%. However, increasing costs, technical difficulties, and the inability of governments and oil companies to compromise combined to kill the megaprojects one by one.

Only two oil sands projects have ever been built. The 60,000-barrel per day (bpd) Suncor plant

started operation in 1967. The 80,000-bpd Syncrude Canada plant started up in 1978. However, both facilities have been plagued with operating

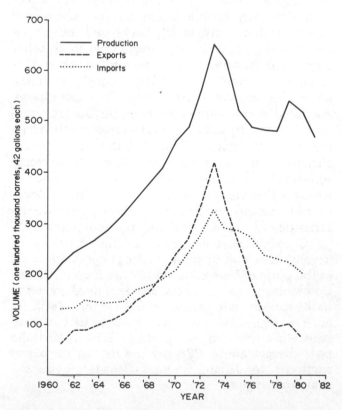

FIGURE I-3-33 Canadian production and trade of crude petroleum, 1960–1982. (From Refs. 15, 16, and 35.)

problems. These problems, and bad economics, recently killed a planned 60,000-bpd capacity increase for the Syncrude plant.[61]

As a result of the slow development of OCS and oil sand resources, Canada's production of crude petroleum still falls short of domestic consumption levels. Thus Canada is dependent on some imported oil. The export–import pattern seen with coal is also true with crude petroleum. That is, crude petroleum is often exported from Canada's western fields while petroleum is imported to supply markets in eastern Canada. Figure I-3-33 shows Canadian production, export, and import levels of petroleum. It shows that exports exceeded imports from 1969 to 1974. After 1974, Canada started to cut back on exports in an attempt to reduce its net dependence on imported oil. The United States has essentially been the only destination for exports of Canadian oil.

Canadian imports of oil come from a variety of suppliers. On the basis of long-term total volumes, oil producers in the Middle East are the major source of supply for Canada, with roughly 40% of Canada's imports. However, since 1960, Venezuela has almost always been Canada's primary supplier. Venezuela's relative share of yearly imports by Canada has been steadily dropping, yet Venezuela continues to be Canada's major source of oil imports. In 1964, Venezuela accounted for 71.3%, over 102 million barrels of the 147 million barrels of oil imported by Canada. In 1972, Venezuela provided 136 million barrels to Canada, 48.3% of the 281 million barrels imported that year. By 1980, Venezuela's exports to Canada were 63 million barrels, which was 30.6% of that year's total of over 205 million barrels of imported oil.[35]

The other supply sources that have taken away some of the Canadian import market from Venezuela are Iran, Saudi Arabia, and to a much lesser extent, the United States and Nigeria. Their relative shares of Canada's total imports have varied since 1960. An important note is that Iran, once the second largest source of Canadian imports, appears to have dropped out of the picture since 1979. This is apparently due to the effects of the Iran-Iraq war and, possibly, to Iran's anger over Canada's aiding the United States during the U.S. Embassy hostage crisis at Tehran. Also of importance is an apparent increase in U.S.-based exports to Canada. During the period 1978–1980, 21% of Canada's oil imports came from the United States.[35]

Increasing concern over Canadian dependence on imported oil ultimately resulted in a variety of attempts to correct the situation during the 1970s. One measure involved the gradual phase-out of oil exports to the United States. Conservation programs and the like were also developed. However, the most significant factor was the development and the release of the National Energy Program (NEP) in 1980.

One important feature of the NEP was the use of federal subsidies for oil imports and a tax on oil exports. The intent here was twofold: to protect domestic consumers from the rising world oil price (for imports), and to discourage exports. A second feature which pitted the federal and Alberta governments against each other was a split price for domestically produced oil. The compromise eventually reached favored the federal position. Oil produced from wells developed after 1981 ("new" oil) was to be priced at the going world rate. Oil from older wells, however, was to be priced at a maximum of 75% of the world rate, with scheduled adjustments according to market price variations.[76]

Implementation of the NEP, combined with other events, has caused a great deal of grief in Canada. The fall in world oil prices has meant very large decreases in tax revenues collected by Ottawa. One study indicates that there is a C$1 billion drop in federal tax revenues, spread over three years, for every $1 drop, from $32 per barrel, in the real price for world oil.[78] This revenue shortfall occurs because the NEP projected future world oil prices to continue to rise. This assumption led to the view that profits would be higher and, in turn, served as the foundation for the agreed-on two-thirds/one-third provincial–federal distribution of tax revenues. The falling world oil price has also meant harder times for the oil industry. As a result, Premier Peter Lougheed of Alberta cut provincial royalties on oil and natural gas until 1986. He also demanded that Ottawa not use this royalty decrease as an opportunity for raising federal taxes on the industry to make up tax revenue drops.[76]

A second result of the fall in oil prices has been that the level of shut-in oil production has increased dramatically in Canada. The daily average for shut-in production in all of 1982 was 58,000 barrels. By the end of January 1983, the average was 180,000 barrels a day. Moreover, the average expected to hold for all of 1983 is 75,000 to 100,000 barrels a day.[79] As a result, the oil industry is trying to remove the phase-out of exports to the United States. Canada now exports about 100,000 barrels a day of heavy oil to the United States, but producers now want to sell light oil to U.S. refineries to open up production.[79]

The third, and potentially most significant effect of the NEP's programs, in light of world events, is a decreased likelihood that oil sands or offshore resources will be developed soon. Oil from the Beaufort Sea is expected to cost C$30 to $40 a barrel (before taxes) to reach the market. It will cost C$20 to $30 per barrel for Hibernia oil.[80] At an exchange rate of U.S. 82 cents per C$1.00, the estimates run $24.30 to $32.40 per barrel for Beaufort Sea oil and $16.20 to $24.30 for Hibernia oil, before taxes. With an April 1983 benchmark price of $29 per barrel of Saudi crude, and spot prices well below this, there is some doubt as to when Canada's future resources will be economically developed. As a result, Canada can be expected to continue to be import dependent for petroleum.

Natural Gas

Figure I-3-32, presented in the preceding section, shows major natural gas deposits in Canada. The lone deposit that has yielded only natural gas is the Labrador Shelf area. Figure I-3-34 shows producing areas.

As with petroleum, production of natural gas is currently dominated by Alberta, with about 85% of total production to its credit. British Columbia produces roughly 10%, while Saskatchewan, Ontario, and the Northwest Territories produce most of the remainder.[15] Figure I-3-35 shows total Canadian production of natural gas from 1960 to 1981. It indicates that total output peaked in 1979 at approximately 3.7 trillion cubic feet (tcf) of gas.[35] Total Canadian demand for natural gas up through 1982 approached only 1.6 tcf per year,

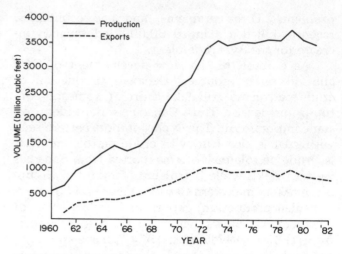

FIGURE I-3-35 Canadian production and exports of natural gas, 1960–1982. (From Refs. 15 and 35.)

and is expected to double to 3.1 tcf by the year 2000.[81] Should Canadian natural gas production not fall much below its current level, it seems likely to expect a continued production surplus through the remainder of the century.

This surplus has been in existence for quite some time, so Canada has long had a policy of exporting some of the annual excess to the United States. These exports are also shown on Figure I-3-35. As can be seen, exports have lately amounted to about 25% of total production. Canada, however, supplies only 4 to 5% of total U.S. natural gas

FIGURE I-3-34 Principal natural gas producing areas in Canada. (From Refs. 73, 77, and 98.)

demands.[82] These exports are occasionally increased over allotted levels, as was done during the 1976–1977 winter heating season when the United States desperately needed more natural gas. However, the disparity in the size of markets has had an adverse, if not predictable, impact on the Canadian suppliers. The reason is that a consumption shift in the United States which seems small to the United States is comparatively large to Canada.

What has happened is that U.S. consumption of natural gas has leveled off due to the recession. Demand in the United States for 1982 was 18.5 tcf, below domestic production of 18.7 tcf. Furthermore, 1983 demand is expected to be only 18 tcf, while deliverable natural gas supplies could reach 21 tcf.[82] This glut in the U.S. market has resulted in Canadian exports of only 40 to 55% of authorized volumes since 1980.[81] Moreover, the glut is not expected to clear up for several more years, which is bad news for Canada's exporters.

Partly in response to the demands of Canada's producers, Canada's National Energy Board recently increased the allowable volume of exports to the United States. Canadian producers are now allowed to export 9.3 tcf to the United States over the next 15 years.[83] Whether or not this will actually occur will depend on the extent of the U.S. economic recovery. However, another factor Canadian producers are also concerned about is what they consider to be an excessively high fixed price for Canadian gas exports to the United States. The current price is $4.94 per thousand cubic feet of gas. The National Energy Board is considering a flat price cut of up to $1 a thousand cubic feet, or an alternative variable price scheme based on the point of entry into the U.S. market. A price cut is almost certain, but even a $1 drop in the price may not be enough. With a $1 price cut, 1983 gas exports from Canada were less than the 1982 level of 780 billion cubic feet of gas.[84]

Given the expected slow upturn in the U.S. economy, Canada's National Energy Board also recently granted its approval to a proposal to sell liquefied natural gas (LNG) to Japan. The NEB approved a sale of 2.3 tcf over 15 years, starting in 1986.[83] Dome Petroleum, the major firm behind the proposal, is now spending $3.4 billion to build a gas liquefaction plant and other export facilities on Canada's west coast.[85]

Aside from the technical difficulties associated with LNG, Dome faces two barriers to success. The first barrier is price. The price now set for the exported LNG is $6.17 per thousand cubic feet.[85] The price is set equally on the basis of the posted price for Saudi and Indonesian oil, and on the price for

gas exports to the United States. Obviously, should the market glut continue, especially if the price of gas exported to the U.S. falls, the Dome LNG project will not be economically competitive. Dome argues that the price is reasonable for an expected project startup date of 1986, but only time will tell.

The other barrier to the success of Dome's project is competition from other natural gas producers. Canada is obviously not alone in its desire to pick up a larger share of a depressed market. Alaska, for example, seems a likely competitor. This is because the stalled Alaska Highway Natural Gas Pipeline, which was to carry Alaskan gas through Canada to major U.S. markets, is delayed indefinitely. There is only a small chance that the pipeline will be completed soon because of the current glut, among other reasons. As a result, Alaska in early 1983 proposed to sell LNG to Japan instead of gas to the lower 48 states.[86]

Canada's natural gas producers have also looked at possible LNG sales to Europe. An earlier attempt to move Arctic natural gas by ice-breaking LNG tankers to eastern Canada (and then the United States) fell through. However, the so-called Arctic Pilot Project is now considering Europe as an alternative market. France and West Germany have expressed an interest in the project, and Europe is the same distance from the Melville Island production site, as is eastern Canada and the United States.[87]

All told, Canada will probably remain a dependable supplier of small but important quantities of natural gas to the United States. Whether these exports will decrease because of increased domestic demand or export diversification is uncertain.

Uranium

Uranium deposits found in Canada are of three general types. The first, which accounts for three-fourths of Canada's existing reserves, is a Precambrian quartz-pebble conglomerate. This type of deposit is typified by the Elliot Lake ores on the north shore of Lake Huron. Major uranium minerals are brannerite and uraninite, and grades average 0.1 to 0.3% U_3O_8.[9] A second type of deposit, uranium, bearing pegmatites, is also found in Ontario. The third deposit, found normally in Saskatchewan, is uranium-bearing unconformities. These are typically quite rich, contain massive pitchblende, and are especially predominant along the edge of the Lake Athabasca sandstone formation.[88] Figure I-3-36 shows the major areas in Canada with geology favorable to the occurrence of uranium.

Canadian reserves of contained uranium are estimated at 230,000 metric tons, given a price of approximately C$135 per kilogram. Resources, at the same price, make up an additional 358,000

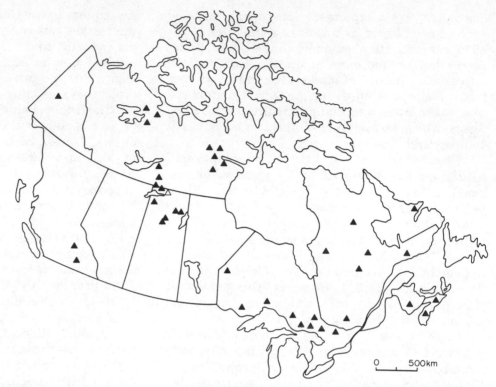

FIGURE I-3-36 Identified uranium deposits in Canada. (From Ref. 12.)

metric tons.[41] At a price of $200 per kilogram, reserves are 258,000 metric tons and resources are 760,000 metric tons. Using the $135 per kilogram price as a benchmark, Canada possesses almost 10% of the free world's uranium reserves.

Until 1981, uranium received more exploration attention and expenditures than practically any other mineral in Canada. The level has fallen somewhat since then, which is to be expected given the slump in uranium demand. Even so, uranium still receives a great deal of attention, especially in Saskatchewan. The Saskatchewan Mining Development Corporation, a provincial Crown corporation, focuses 90% of its exploration and development efforts on uranium. The corporation was involved in 275 uranium exploration projects in northern Saskatchewan alone in 1981.[9] One note of concern is that exploration in the Northwest Territories, which, according to Figure I-3-36 has many promising sites, has been blocked by native groups.[15]

Seven mines produced uranium in 1981. All mined uranium as a primary product. Of these seven mines, four were in Ontario and the rest were in Saskatchewan. The Ontario mines, which included the Elliot Lake mines, the largest in Canada, produced 60% of total output in 1981.[45] Saskatchewan should become the largest producer in Canada in late 1983 when the Key Lake mine starts up. The mine and mill operation taps the richest known uranium ores in the world and has a production capacity of 12 million pounds of U_3O_8 per year.[32] One Ontario mine closed in June 1982, so producing mines at the end of 1983 should be as described in Table I-3-20. Their location is plotted in Figure I-3-37. In addition, ESI Resources Limited of Calgary, Alberta, recovers uranium as a by-product of

TABLE I-3-20 Uranium Producing Mines in Canada, 1983

Company	Location	Notable feature
Agnew Lake Mines Ltd.	Agnew Lake, Ont.	Leach mining operation
Denison Mines Ltd.	Elliot Lake, Ont.	13,600 metric tons/day capacity
Rio Algom Ltd.	Elliot Lake, Ont.	9240 metric tons/day capacity
Madawaska Mines Ltd.	Bancroft, Ont.	1360 metric tons/day capacity
Gulf Minerals Canada	Rabbit Lake, Sask.	1500 metric tons/day capacity
Cluff Mining	Cluff Lake, Sask.	7–30% U_3O_8 ore grade
Key Lake Mines Ltd.	Key Lake, Sask.	12 million pounds U_3O_8 annual capacity

Sources: Refs. 9 and 12.

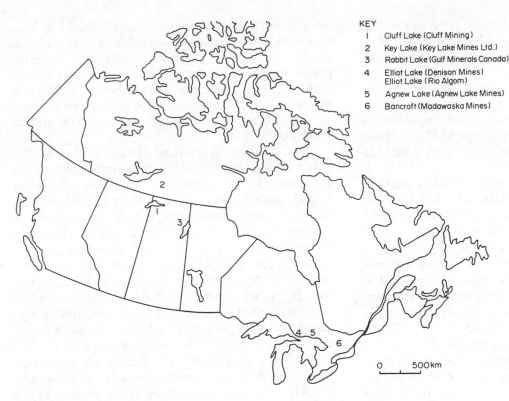

KEY
1   Cluff Lake (Cluff Mining)
2   Key Lake (Key Lake Mines Ltd.)
3   Rabbit Lake (Gulf Minerals Canada)
4   Elliot Lake (Denison Mines)
     Elliot Lake (Rio Algom)
5   Agnew Lake (Agnew Lake Mines)
6   Bancroft (Madawaska Mines)

0   500km

FIGURE I-3-37 Uranium mining areas in Canada. (From Refs. 15 and 98.)

phosphoric acid production. Production capacity is 40 metric tons of uranium per year. Phosphate rock for the plant is imported from Idaho and all recovered uranium is sold on contract to two U.S. utilities.[12]

Figure I-3-38 shows total production of uranium from 1960–1982. The massive swing in production levels during the 1960s can be largely attributed to the fact that most of the uranium produced is ex-

ported. What little uranium is consumed domestically is taken mainly by the 10 CANDU electric generating stations, which utilize natural uranium, containing 0.7% U-235. Total generating capacity in 1978 was 550 MWe and it is optimistically estimated that the capacity in 1988 will be 16,460 MWe.[88] Obviously, domestic demand for uranium is relatively low.

Uranium production in Canada became firmly

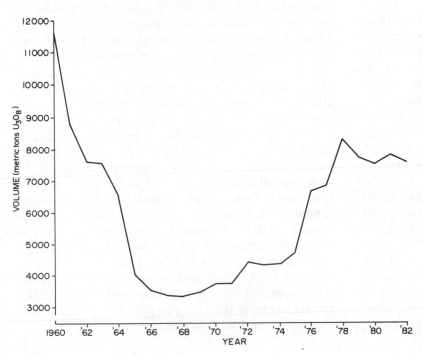

FIGURE I-3-38 Canadian production of uranium, 1960–1982. (From Refs. 15, 35, and 89.)

established when Canada provided uranium to the U.S. Manhattan Project during World War II. After the war's conclusion, especially after the mid-1950s, the Canadian uranium industry boomed in response to growing import demands by the United States. From 1948 to 1960, Canada sold over 90 percent of total uranium output under firm contracts to the U.S. Atomic Energy Commission (AEC).[90] However, in 1959 the AEC announced that contracts would not be renewed.

Figure I-3-38 indicates that Canada's production promptly slumped. Production fell by 71% from 1960 to 1968, and exports declined by 99%.[35] So many domestic producers were going out of business that the federal government established a stockpile to help absorb some of the excess production capacity. After 1960, the AEC banned the domestic enrichment of foreign-produced uranium for use in American nuclear reactors. This ban lasted for 10 years and effectively closed 70% of the world market to all non-U.S. producers.[88]

Another factor that adversely affected Canadian production and exports of uranium was the establishment of the U.S. uranium stockpile. In October 1971, the AEC announced that it was considering stockpile sales to U.S. and foreign enrichment plants.[88] This further threatened what little market there was for Canada and other producers. In June 1972, after previous discussions developed the foundation, Canada and other uranium producers barred from the U.S. market signed the Johannesburg Rules agreement. This established sales quotas and a price floor for uranium sales to customers other than the United States.

After 1973, nuclear power development projects started in many countries and so worldwide demand for uranium increased. Moreover, the United States began a gradual phase-out of its uranium import embargo in 1973. Thus world prices (and production) rose until, by late 1974, the Johannesburg Rules quota and price restrictions were withdrawn. The global expansion in uranium production and exports was reflected by activity in Canada. However, by the late 1970s, things tapered off again.

The United States, in the meantime, had taken out legal action against Canada and other parties to the Johannesburg Rules, arguing that the agreement was a cartel that violated the competitive nature of a free market. Canada retaliated by arguing two points. First, Canada would not have acted as it had if the United States had not prompted the initial state of overcapacity in the Canadian industry and then backed out of all contracts. Second,

even if the U.S. actions were justifiable, U.S. concern about the Rules agreement was not. The reason was that U.S. fears of unduly high uranium prices were unfounded, a point U.S. courts agreed with. In fact, Canada's price for uranium was below the monthly U.S. spot price for all but two months of the two and a half years the price floor was in existence.[88]

The burgeoning market of the 1970s, and the antitrust cases with the U.S. government, prompted the Canadian government to establish a uranium export policy. Policy goals were to ensure that domestic needs were always met, to maintain a healthy Canadian industry, to earn high economic returns on sales, and to ensure nuclear nonproliferation. The goal of high returns was furthered somewhat in 1977 when the Atomic Energy Control Board set price guidelines for uranium sales. Domestic producers and foreign consumers opposed the move, and the case was ultimately taken to a U.S. arbitrator. The decision was that the imposed price guidelines were fair to all concerned parties and were not an actual price-fixing scheme.[88]

With the turbulent experiences of the 1970s behind it, the Canadian uranium industry is concentrating on developing its export markets again. High ore grades and relatively low production costs give Canada's uranium a competitive advantage in the market, so it is thought that exports can increase. In fact, it has been argued that Canada could account for 20% of total world exports by the year 2000.[9]

However, several factors could prevent this from happening. The first possibility is that potential importers will refuse to buy Canada's uranium because of what they may see as excessively stringent contract safeguards against nuclear proliferation. Canada has indicated that it will not sell uranium to any country which does not agree to follow its safeguards. Furthermore, Canadian exports to nations that violated the agreement have been cut.

The second possible factor is that Canadian production may not rise to the point required to meet export demands. Lowered exploration levels for new deposits could result in this, but another problem exists. Since 1946, federal law has placed all uranium deposits and extraction facilities under federal control. Moreover, these facilities must be owned by companies incorporated in Canada. Foreign firms wishing to develop a prospect in Canada must go to the Foreign Investment Review Agency for permission. This requirement may lower foreign participation in the industry and could conceivably hurt future production and export levels.[88]

TRANSPORTATION INFRASTRUCTURE

The exploitation of mineral deposits goes hand-in-hand with the development of a diversified transportation network. This is because mining could not take place without the required facilities for transportation. Moreover, many potential transportation networks cannot be judged to be economically worthwhile unless these networks carry a certain level of traffic. The Canadian mineral industry's operations often provide just such a minimum required base of traffic for proposed new networks. All in all, four different transportation networks are important to the mineral industry: railways, roads, pipelines, and waterways.

Although these networks were developed at different periods of time in Canada's history, they all serve to reinforce a historical precedent. Early contacts between the United States and Canada were marked by American attempts to control Canada economically and politically. The Canadian government, to help resist American expansionism, decided to build a transcontinental railway, in the hope that such a railway would strengthen the loose interprovincial links that made up Canada. Thus Canada's transcontinental transportation network was given an east-west orientation instead of the north-south orientation which might have evolved given less resistance to American pressure.[91] This east-west orientation is strong even today.

Railways

Figure I-3-39 shows the extent of Canada's major railways as of 1981. The system currently has some 100,000 km of rail lines. It is immediately apparent from Figure I-3-39 that most development took place along an east-west axis located close to the border with the United States. The few lines extended north have all been developed to help tap rich mineral deposits. For example, the line from Sept-Iles to Schefferville allowed the recovery of iron ore from the Labrador Trough deposits.

There are only two Class I railways in Canada: Canadian National Railways (CN) and Canadian Pacific Railways (CP). These two railways had quite different origins. Canadian Pacific was the line built to link the Canadian provinces together in the nineteenth century so as to head off American expansionism. Canadian National, on the other hand, was formed by combining a number of smaller lines which failed due to the post–World War I drop in freight traffic in eastern Canada. Both CN and CP are quite similar in other respects, though. They both operate across almost all of Canada. Both own multimodal transportation firms and thereby operate efficiently. They also dominate in terms of freight carried, sharing some 90% of total Canadian railway traffic between themselves.

Canada is such a vast nation that CN and CP cannot possibly meet all railway expansion needs.

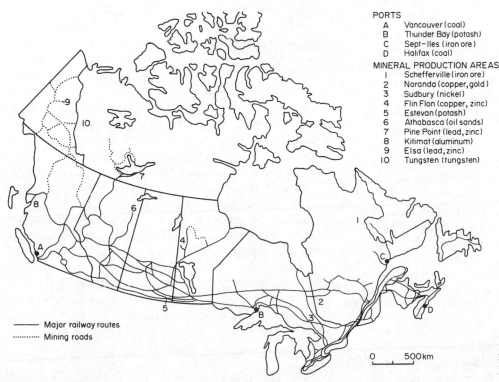

PORTS
A	Vancouver (coal)
B	Thunder Bay (potash)
C	Sept–Iles (iron ore)
D	Halifax (coal)

MINERAL PRODUCTION AREAS
1	Schefferville (iron ore)
2	Noranda (copper, gold)
3	Sudbury (nickel)
4	Flin Flon (copper, zinc)
5	Estevan (potash)
6	Athabasca (oil sands)
7	Pine Point (lead, zinc)
8	Kitimat (aluminum)
9	Elsa (lead, zinc)
10	Tungsten (tungsten)

——— Major railway routes
·········· Mining roads

0   500km

FIGURE I-3-39 Major railway routes and mining roads. (From Ref. 98.)

This is particularly important to the mineral industry. Several Class II railways (both public and private) operate in small geographic areas and often provide the crucial links with mining projects. These lines are usually dependent on traffic from the mines they serve. The Sept-Iles-to-Schefferville line mentioned above is such a case. The Quebec North Shore and Labrador Railroad was built and is run by the Iron Ore Company of Canada simply to tap the iron ore deposits of the Labrador Trough. The Faro–Watson Lake line in Yukon Territory got 75% of its traffic from the shipment of concentrates from the Cyprus Anvil lead-zinc mine in Faro. Now that the mine is closed, the rail line has been forced to close also.

However, distances are such that any railway expansion is an extremely costly venture. As a result, expansions are often undertaken only after government support is made available, for this assistance often determines the ultimate profitability of a line's expansion. This is shown by the current expansion of railways to carry coal produced by new mines in northeastern British Columbia. Both Canadian National and British Columbia Railway are upgrading and expanding their lines, with government assistance, to move coal to the Ridley Island export terminal at Prince Rupert. One specific example of government aid is the provision of C\$15 million for the electrification of the new 129-km Anzac-to-Tumbler Ridge spur being built by British Columbia Railway. This is the first major electrified rail system in all of Canada.[12]

Canadian National and Canadian Pacific have consistently lost money on western routes because of the low Crow's Nest Pass rates they are required to charge for shipment of grain. As a result, the coal industry, among others, must pay an artificially high rate to help CN and CP break even. The coal industry's competitiveness is hurt a second way because CN and CP have only reluctantly spent money on line improvements while the Crow's Nest Pass rates were effective.[12] However, the move in early 1983 to revise the rates has improved the outlook. The federal government is raising rates for grain and is going to spend roughly C\$1 billion on western rail lines from 1983 to 1986. This apparently has cleared the way for CN and CP to begin an estimated C\$16.5 billion of delayed system maintenance and expansion work.[92]

Roads

All-weather roads play an important role for the Canadian mineral industry because they are often cheaper to build than railways. Thus many small-scale mining projects or those located far from existing rail lines can still afford the expense of a road to link up with the railway system. Most roads of this type are built in the northern expanse of Canada, where population and economic activity is so low as to make a rail line uneconomic. These roads, then, as seen in Figure I-3-39, are literally an extension of the established network of highways and railways to the south. One example of a mine dependent on such a road is Canada Tungsten Mining Corporation's tungsten mine in the Northwest Territories.

Pipelines

For certain mineral commodities such as petroleum and natural gas, the most efficient and economic transport mode is the pipeline. The first major pipeline built in Canada was constructed in 1912. It was a 170-mile, 16-inch-diameter natural gas pipeline in Alberta. The first major oil pipeline was a 236-mile, 12-inch-diameter line built between Montreal and Portland, Maine, in 1941. Its purpose was to help ensure that supplies of oil to Montreal were not cut off by German submarines off the mouth of the St. Lawrence River.[91]

The boom in pipeline construction, especially for oil, hit with the Leduc basin oil discovery in 1947. In 1946, for example, there were 416 miles of oil pipelines. This increased to 5700 miles by 1956 and over 13,000 miles by 1966.[91] By 1981, some 20,100 miles of oil pipeline had been laid. By 1981, there were roughly 45,000 miles of natural gas pipelines. Figures I-3-40 and I-3-41 show the existing pipeline networks in Canada.

The pipeline collection system is largely focused on Alberta, British Columbia, and Saskatchewan. Due to geologic and market factors, the collection network has just as much of a north-south bias as it does east-west. However, the distribution system, once again, runs mainly east-west, for the eastern provinces are dependent on imports of oil and natural gas. The placement of distribution lines shows a continuing wariness of the United States. The first transcontinental oil pipeline in Canada was routed south of the Great Lakes, through the United States, because it was the most economic design possible to feed markets in southern Ontario. However, political forces prevented use of a similar route for the construction of the first transcontinental natural gas pipeline. The route chosen looped north over the Great Lakes before heading into southern Ontario, its major market destination. This was done partly to encourage the mineral industry of northern Ontario and partly out of fear of U.S. interference.[91]

Waterborne Transport

St. Lawrence Seaway. The dominant inland waterway network for Canada is the Great Lakes–St. Lawrence River route known widely as the St.

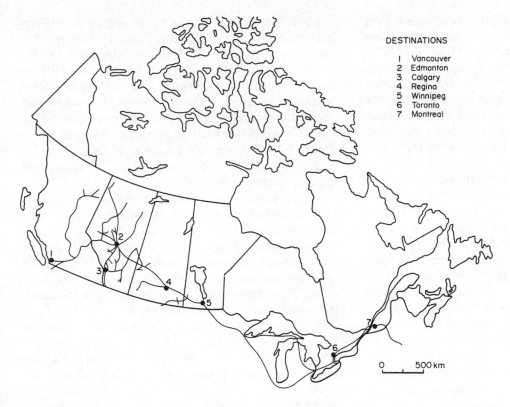

DESTINATIONS

1 Vancouver
2 Edmonton
3 Calgary
4 Regina
5 Winnipeg
6 Toronto
7 Montreal

0   500 km

FIGURE I-3-40 Major crude oil pipelines and destinations. (From Ref. 98.)

DESTINATIONS

1 Vancouver
2 Edmonton
3 Calgary
4 Regina
5 Winnipeg
6 Toronto
7 Ottawa
8 Montreal

0   500 km

FIGURE I-3-41 Major natural gas pipelines and destinations. (From Ref. 98.)

Lawrence Seaway. This network allows waterborne access to half of Canada's highly developed regions. Connections by rail with the western provinces result in another east-west-oriented transcontinental transportation network.

The St. Lawrence Seaway was the key to this new network. Most ships were unable to go farther upstream than Montreal because of shallow water and waterfalls in the St. Lawrence River. Major U.S. and Canadian firms, especially steel and iron ore companies, argued that a system of locks and dams to open up the entire Great Lakes–St. Lawrence River region would be economically advantageous.

The seaway was completed in 1959 and has

seen massive increases in traffic. Preconstruction traffic on the Montreal–Lake Ontario section was almost 12,000,000 tons of cargo in 1958. By 1971, the figure was 52,988,000 tons of cargo.[91] The same traffic increase holds true for the entire system. Figure I-3-42 shows the St. Lawrence Seaway, its locks and dams, and the predominant mineral traffic flow patterns. It shows that the United States and Canada have both benefited immensely from the Seaway.

However, the Seaway's usefulness is eroding in the face of the shift to large and more efficient ships. Upon its construction in 1959, the Seaway's 27-ft depth and 730-ft by 75-ft locks allowed 80% of the world's ships to utilize the Seaway.[93] This has since changed as shippers have gone to larger vessels to achieve greater economies of scale. This has presented problems with, for example, the export of coal. Small, 30,000 deadweight ton (dwt) coal vessels suitable for the Seaway are not efficient for ocean transport. Nor is it efficient to have to unload larger vessels partially at the mouth of the Seaway. This problem will not be resolved easily.

Ports. Canada has 25 major deep-water ports and 650 small ports and harbors.[94] These are administered by one of three different organizations. The most common type is the local port corporation. Other ports, especially east- and west-coast deep-water ports, are run by the National Harbors Board (NHB). Finally, facilities on the Great Lakes are run by private companies. All ports are ultimately subject to the control of the Canadian Marine Transportation Administration, a division of Transport Canada. The NHB is now a federal Crown corporation known as the Canada Ports Corporation (CPC).

The three leaders in tonnage shipped are Sept-Iles and Port-Cartier, Quebec, and Vancouver. The first two ports handle iron ore destined for the United States. Vancouver, on the other hand, has trade links with most Pacific Rim nations and other ports of the world. These three ports lead in tonnages shipped because they handle large volumes of bulk commodities such as iron ore, coal, and so on. Figure I-3-43 shows the location of 13 of Canada's largest ports, and notes some of the products handled by these ports.

Deep-water ports are probably receiving the most attention of any link in Canada's transportation networks at this moment. Increasing levels of mineral (and other) exports have called for the development of new facilities. One example is the new potash export terminal being built at Saint John, New Brunswick. It is being built to handle the output of the nearby mines being run by Denison Mines and the Potash Corporation of America.

The best example, though, of such attention is the expansion of the coal export facilities at Roberts Bank (Vancouver) and Ridley Island (Prince Rupert), both of which are on the west coast. For instance, the Roberts Bank facility will have the capacity to handle 22 million metric tons of coal per year, almost double its previous level. With low-tide water depths of at least 23 m, the port will easily handle ships of at least 250,000 dwt. Over C$1 billion is expected to be invested in expansion projects at 15 major ports through 1986.[95]

Overview

Several important points must be made about the transportation networks in Canada. The first is that the dominant east-west orientation of the

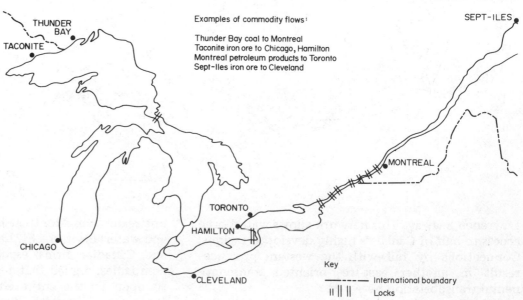

FIGURE I-3-42 St. Lawrence Seaway locks and select commodity flows. (Adapted from Refs. 15 and 91.)

Examples of commodity flows:

Thunder Bay coal to Montreal
Taconite iron ore to Chicago, Hamilton
Montreal petroleum products to Toronto
Sept-Iles iron ore to Cleveland

Key

------- International boundary
|| || || Locks

KEY
1 Vancouver (coal, ores)
2 Kitimat (aluminum)
3 Prince Rupert (coal)
4 Thunder Bay (iron ore, coal, potash)
5 Toronto (coal)
6 Montreal (petroleum)
7 Quebec (ores, coal)
8 Baie Comeau (aluminum)
9 Port-Cartier (iron ore)
10 Sept-Iles (iron ore)
11 Saint John (potash)
12 Halifax (ores, petroleum)
13 St. Johns

0   500 km

FIGURE I-3-43   Major Canadian ports. (From Refs. 94, 95, and 98.)

system is gradually shifting. This has already been seen in the case of natural gas and oil collection pipelines (Figures I-3-40 and I-3-41). However, it is likely that new extensions of road and railways into the northern reaches of the nation will occur because of the development of new mining ventures. Such extensions will be expensive because of the distances involved and the adverse climate, but they will eventually be built.

Even with the shift northward, the east-west axis will still remain a dominant one. Established settlement patterns dictate this. Moreover, expanded levels of trade with Europe and Asia will flow out from this east-west corridor. Finally, this multimodal transport corridor helps strengthen the ties between quite diverse Canadian provinces.[91]

The second point is that the transport network is partly geared to meet the need of export markets. Since the United States and Canada are each other's major trading partners, it is appropriate to ask how stable the transnational transport system is, for disruptions in the traffic flow could be quite costly and painful.

Given the time frame over which United States–Canada trade developed, it is not surprising to see that goods are carried along a variety of routes and by different modes. Many major rail systems in both nations are linked. Although this is partly due to different regulatory environments in the two nations, the point is that the links allow the easy movement of goods across borders. For example,

Canadian National has purchased five U.S. rail lines to expand its coverage into an uninterrupted loop around the Great Lakes and down through the U.S. industrial midwest.[96]

Trucking, should it become necessary for transportation of minerals across the border, would pose no problem given the extensive level of highway link-ups. Different regulatory environments could pose a problem, though. Pipelines for oil and natural gas pose no problem, except in the sense that several desirable ones have not been built for several reasons. The Alaska Highway Natural Gas Pipeline is such an example.

The situation is somewhat less optimistic for waterborne commerce. The reason is the St. Lawrence Seaway, specifically the portion running east from the Welland Canal between Lake Erie and Lake Ontario. Inadequate locks and dams are slowing down traffic and have prevented the upgrading of ship sizes to achieve greater economies of scale.

A possible problem that could occur is the stoppage of traffic through the narrow portion of the Seaway east of the Great Lakes. In fact, Garreau has suggested the possibility that a venting of political frustration in Quebec could result in an attempt to sabotage and sink a ship so as to block traffic in the Seaway.[97] Since this portion of the Seaway is entirely within Quebec's territory, it could be difficult to remedy the problem.

This act has not been tried in Canada, but has been elsewhere in the world. How likely it is in

Canada, especially in Quebec, depends on many factors. However, it seems fairly reasonable to assume that the chances are slim at best that such an act could occur soon. This is partly due to the nearly-universal shock felt in Canada after political terrorists became violent in Montreal in the early 1970s. All sides essentially agreed that violence, aimed at people or property, was not to be used again.

All in all, then, the reliability of the transport system between the two nations is good. Its cost-effectiveness may not be optimal, however, due to inadequate maintenance, problems with different regulatory approaches in the two nations, and the like. Still, it is reasonable to conclude that transportation poses little problem to continued mineral (and other) trade between the two nations.

REFERENCES

1. *What Mining Means to Canada*. Ottawa: The Mining Association of Canada, September 1981.
2. Toombs, R. B. "The Canadian Mineral Industry." In *Geology and Economic Minerals of Canada, Part A*. Edited by R. J. W. Douglas. Ottawa: Minister of Supply and Services Canada, 1976.
3. Anderson, Ian. "Metals That No Longer Glitter." *Maclean's*, 15 November 1982.
4. Energy, Mines and Resources Canada. *Mineral Policy Backgrounder*, No. 82/25(b) (8 March 1982).
5. *The Financial Post*, 12 February 1983.
6. Daly, D. J. "Mineral Resources in the Canadian Economy: Macro-economic Implications." In *Natural Resources in U.S.-Canadian Relations, Volume I: The Evolution of Policies and Issues*. Edited by Carl E. Beigie and Alfred O. Hero, Jr. Boulder, Colo.: Westview Press, Inc., 1980.
7. Govett, M. H., and G. J. S. Govett. "The Canadian Minerals Industry." *Resources Policy* 2 (March 1976).
8. *The Financial Post*, 31 July 1982.
9. "Spotlight on Canada's Resourceful Mining Industry." *Engineering and Mining Journal* 181 (November 1981).
10. Energy, Mines and Resources Canada. *Mineral Policy: A Discussion Paper*. Ottawa: Minister of Supply and Services Canada, 1982.
11. "Crumpled Maple Leaf." *The Economist*, 7 August 1982.
12. "Annual Mineral Review and Forecast." *Canadian Mining Journal* 103 (February 1982).
13. *The Wall Street Journal*, 3 March 1983.
14. Kaplan, Jacob. "U.S. Resource Policy: Canadian Connections." In *Natural Resources in U.S.-Canadian Relations, Volume I: The Evolution of Policies and Issues*. Edited by Carl E. Beigie and Alfred O. Hero, Jr. Boulder, Colo.: Westview Press, Inc., 1980.
15. Energy, Mines and Resources Canada. *Canadian Minerals Yearbook 1978*. Ottawa: Minister of Supply and Services Canada, October 1980.
16. U.S. Department of the Interior, Bureau of Mines. *Mineral Commodity Summaries 1983*. Washington, D.C.: U.S. Government Printing Office, 1983.
17. "IMF Sees 0.9% Growth in GNP during 1983." *The Globe and Mail*, 1 February 1983.
18. Kostuik, John. "Mining in the Next Decade." *CIM Bulletin* 73 (November 1980).
19. Freyman, A. J. *The Role of Smaller Enterprises in the Canadian Mineral Industry with a Focus on Ontario*. Toronto: Ministry of Natural Resources, 1978.
20. Wojciechowski, Margot J., Assistant Director, Centre for Resource Studies, Queen's University, Kingston, Ontario. Personal communication, 14 February 1983.
21. Dorr, André L., and John E. Tilton. "Bauxite and Aluminum." In *Natural Resources in U.S.-Canadian Relations, Volume II: Patterns and Trends in Resource Supplies and Policies*. Edited by Carl E. Beigie and Alfred O. Hero, Jr. Boulder, Colo.: Westview Press, Inc., 1980.
22. Sassos, Michael P. "Modernization Extends to Most Alcan Plants." *Engineering and Mining Journal* 182 (January 1982).
23. *The Financial Post*, 22 January 1983.
24. Russell, Allen S. "Pitfalls and Pleasures in New Aluminum Process Development." *Journal of Metals*, June 1981.
25. Skea, J. "Electricity Supplies for the Primary Aluminum Industry." *Resources Policy* 6 (March 1980).
26. Booth, Amy. "Hydro Sweetener Helps Giant Smelter Deal." *The Financial Post*, 22 January 1983.
27. Spector, Steward R. "Price and Availability of Energy in the Aluminum Industry." *Journal of Metals*, June 1981.
28. Dauphin, Roma. "Asbestos." In *Natural Resources in U.S.-Canadian Relations, Volume II: Patterns and Trends in Resource Supplies and Policies*. Edited by Carl E. Beigie and Alfred O. Hero, Jr. Boulder, Colo.: Westview Press, Inc., 1980.
29. Scales, Marilyn. "Canada Backs Asbestos Use, Third World Provides Markets." *Canadian Mining Journal* 103 (July 1982).
30. Freeman, Alan. "Canada Asbestos Firms Blame Problems Mainly on Recession, Rather Than Suits." *The Wall Street Journal*, 3 September 1982.
31. U.S. Department of the Interior, Bureau of Mines. *Mineral Facts and Problems, 1980 Edition*. Washington, D.C.: U.S. Government Printing Office, 1980.
32. *The Financial Post*, 12 March 1983.
33. National Research Council, Commission on Engineering and Technical Systems, National Material Advisory Board. *Tantalum and Columbium Supply and Demand Outlook*. Washington, D.C.: National Academy Press, 1982.
34. *The Wall Street Journal*, 14 March 1983.
35. U.S. Department of the Interior, Bureau of Mines. *Minerals Yearbook*. Washington, D.C.: U.S. Government Printing Office, various issues.
36. Whitney, John W. "Copper." In *Natural Resources in U.S.-Canadian Relations, Volume II: Patterns and Trends in Resource Supplies and Policies*. Edited by Carl E. Beigie and Alfred O. Hero, Jr. Boulder, Colo.: Westview Press, Inc., 1980.

37. Mackenzie, Brian W. "The Competitive Position of Canadian Copper Supply." *Resources Policy* 7 (December 1981).

38. *The Financial Post*, 19 March 1983.

39. Shao, Maria. "U.S. Copper Industry Is Being Devastated by Its High Costs As Well As by Recession." *The Wall Street Journal*, 16 September 1982.

40. *World Mine Production of Gold, 1981-1985*. Washington, D.C.: The Gold Institute, January 1983.

41. Soganich, John. "New Golden Era Could Be Coming." *The Financial Post*, 22 January 1983.

42. Hogan, William T. "Iron Ore." In *Natural Resources in U.S.-Canada Relations, Volume II: Patterns and Trends in Resource Supplies and Policies*. Edited by Carl E. Beigie and Alfred O. Hero, Jr. Boulder, Colo.: Westview Press, Inc., 1980.

43. "Quebec-Labrador Iron Ore/Steel Output Hurt by U.S. Steel Slump." *Engineering and Mining Journal* 182 (September 1982).

44. Cotter, Nicholas. "Anticipate Federal Clearance for Roche Bay Magnetite Mine." *The Northern Miner* 17 March 1983.

45. Hero, Alfred O., Jr., and Roderick M. Logan, "Other Minerals and Deep-Sea Nodules." In *Natural Resources in U.S.-Canada Relations, Volume II: Patterns and Trends in Resource Supplies and Policies*. Edited by Carl E. Beigie and Alfred O. Hero, Jr. Boulder, Colo.: Westview Press, Inc., 1980.

46. "Zinc-Lead Enigmas." *Mining Journal* 297 (9 October 1981).

47. Energy, Mines and Resources Canada. *Minerals*. SS 82-18. December 1982.

48. "Noranda Closes Molybdenum Mine." *The Globe and Mail*, 16 December 1982.

49. "Loss of By-product Support Is Critical to Copper and Base Metals, Study Shows." *Engineering and Mining Journal* 182 (April 1982).

50. Cameron, John I. "Nickel." In *Natural Resources in U.S.-Canada Relations, Volume II: Patterns and Trends in Resource Supplies and Policies*. Edited by Carl E. Beigie and Alfred O. Hero, Jr. Boulder, Colo.: Westview Press, Inc., 1980.

51. Welsh, Lawrence. "Inco, Falconbridge Facing Soviet Market Threat." *The Globe and Mail*, 2 October 1982.

52. "Nickel Operations Are in Trouble." *World Mining* 33 (December 1982).

53. Cairns, Robert D. "A Reconsideration of Ontario Nickel Policy." *Resources Policy* 7 (Autumn 1981).

54. Laux, Jeanne Kirk, and Maureen Appel Molot. "Asbestos." In *Natural Resources in U.S.-Canada Relations, Volume II: Patterns and Trends in Resource Supplies and Policies*. Edited by Carl E. Beigie and Alfred O. Hero, Jr. Boulder, Colo.: Westview Press, Inc., 1980.

55. "Canada's Potash Corp. to Resume Operations." *The Wall Street Journal*, 9 February 1983.

56. Branch, Stephen N. "Eastern Potash Industry Born As PCA Readies First Producer in NB." *Canadian Mining Journal* 103 (June 1982).

57. Greenspan, Edward. "Potash Producers Cast Worried Eyes South." *The Financial Post*, 15 January 1983.

58. Birnbaum, Jeffrey H., and Claudia Waterloo. "Price-

59. Support Program for Grain, Cotton Draws Extensive Response from Farmers." *The Wall Street Journal*, 23 March 1983.

59. Powers, Ned. "A Good Idea Gone Awry." *The Globe and Mail*, 30 October 1982.

60. *The Northern Miner*, 7 April 1983.

61. Dowsing, R. J. "Spotlight on Tungsten." *Metals and Materials*, November 1981.

62. Kundig, Konrad. "The Tungsten Market—From Chaos to Stability." *The Journal of Metals*, May 1981.

63. "Low Price, Write-Downs Hit Canada Tungsten." *The Northern Miner*, 7 April 1983.

64. Thomas, Alister. "Coal to Regain Prominence in Near Future." *The Financial Post*, 13 November 1982.

65. Bayless, Alan. "Price of Coal under Pressure." *The Globe and Mail*, 27 January 1983.

66. Mulvihill, R. P. *Coal: The Canadian Industry—1982*. Ottawa: Department of Industry, Trade and Commerce, Industrial Minerals Division, 1982.

67. Jeremic, M. L. "Coal Resources Are Plentiful but Unevenly Scattered." *World Coal* 7 (July-August 1981).

68. Energy, Mines and Resources Canada. *Discussion Paper on Coal 1980*. Resource EP 80-1E. Ottawa: Minister of Supply and Services Canada, 1980.

69. Schiller, E. A. "Coal in Canada." *Mining Magazine* July 1982.

70. Clancy, John, and Andrew J. Freyman. "The Pacific Perspective: Coal Development in British Columbia, 1981-2000." *CIM Bulletin* 75 (December 1982).

71. "Suneel, Korean Utility Set Contract for Coal." *The Wall Street Journal*, 21 January 1983.

72. Sigurdson, Albert. "B. C. Coal Chief Lauds Proposed Crow Solution." *The Globe and Mail*, 18 February 1983.

73. King, Robert E. "Canada: Offshore Search Continues." *World Oil*, May 1982.

74. Zehr, Leonard, et al. "Alsands Failure Is Major Loss for Canada, May Force It to Revise Economic Strategy." *The Wall Street Journal*, 3 May 1982.

75. Urqhart, John. "Canada Oil Exploration in the 'Frontier' to Double This Year despite High Costs." *The Wall Street Journal*, 25 January 1983.

76. Bayless, Alan. "Cheap Oil Is Causing Problems in Ottawa As Government Faces Decline in Revenue." *The Wall Street Journal*, 22 April 1982.

77. Bott, Robert. "The Megaproject Scorecard." *Canadian Business*, April 1981.

78. Cook, Peter. "OPEC Standoff Clouds Canadian Price Outlook." *The Globe and Mail*, 13 July 1982.

79. McCallum, Anthony. "Oil Exports Urged to Offset Slump." *The Globe and Mail*, 15 February 1983.

80. Anderson, Ian. "The World according to Yamani." *Maclean's*, 7 March 1983.

81. Lewington, Jennifer. "NEB Approves Record in Natural Gas Exports." *The Globe and Mail*, 28 January 1983.

82. Best, Dunnery, and Giles Gherson. "One Hurdle Keeps Gas Firms from Joy." *The Financial Post*, 5 February 1983.

83. "Gas Exports by Canada Seen Doubling." *The Wall Street Journal*, 28 January 1983.

84. Bayless, Alan. "Canada Likely to Cut Natural Gas Price

to U.S., but Rise in Canadian Exports Isn't Expected." *The Wall Street Journal*, 18 March 1983.

85. "Dome Petroleum Proceeds with Plan to Export LNG." *The Wall Street Journal*, 25 March 1983.

86. Harrison, Fred. "U.S. Clashes Threaten Natural Gas Markets." *The Financial Post*, 29 January 1983.

87. Urqhart, John. "Canada Firms Strive to Market in Europe Growing Surplus of Arctic Natural Gas." *The Wall Street Journal*, 4 August 1982.

88. Greenwood, Ted, and Alvin Streeter, Jr. "Uranium." In *Natural Resources in U.S.–Canada Relations, Volume II: Patterns and Trends in Resource Supplies and Policies*. Edited by Carl E. Beigie and Alfred O. Hero, Jr. Boulder, Colo.: Westview Press, Inc., 1980.

89. Welsh, Lawrence. "Value of Base Metals Mined Slides by $1.5 Billion." *The Globe and Mail*, 26 January 1983.

90. Salaff, Stephen, "Saskatchewan Uranium in the 1980's." *CRS Perspectives*, No. 12 (April 1982).

91. Schreiner, John. *Transportation: The Evolution of Canada's Networks*. Toronto: McGraw-Hill Ryerson Limited and Maclean Hunter Limited, 1972.

92. Eisler, Dale. "Bidding Farewell to the Old Crow." *Maclean's*, 14 February 1983.

93. U.S. Central Intelligence Agency, National Foreign Assessment Center. *The World Factbook—1981*. Washington, D.C.: U.S. Government Printing Office, April 1981.

94. Statistics Canada. *Canada Year Book 1980–81*. Ottawa: Minister of Supply and Services Canada, 1981.

95. "$1 Billion Slated for 15 Big Ports." *The Globe and Mail*, 12 July 1982.

96. Thomas, David. "Tough Enough to Hurt." *Canadian Business* 55 (November 1982).

97. Garreau, Joel. "Quebec." In *The Nine Nations of North America*. Boston: Houghton Mifflin Company, 1981.

98. Energy, Mines and Resources Canada. *Principal Mineral Areas of Canada*. Map 900A, 32nd ed. Ottawa: Surveys and Mapping Branch, 1982.

Chapter I-4

The Role of Foreign Investment

From its earliest days as a British and French trading settlement to the present, Canada has depended on inputs of foreign capital. This was especially true in the beginning because Canada's small economy produced just enough to meet domestic consumption needs. Thus there was little capital left over for investment purposes.[1] Canada, as with all other nations in the same position, had a choice. It could either accept slow economic development, based solely on investments from domestic savings, or it could bring in foreign capital to speed up growth. Rather than forgo the expected economic benefits of such growth, Canada chose to draw on available sources of foreign capital. It continues to do so today.

Continued acceptance of foreign investment has lately been tempered with concern over its negative effects on the Canadian economy. Two policies, the Foreign Investment Review Act and the National Energy Program, stand out as examples of the Canadian response. However, these policies may inadvertently subject the mining industry to serious problems.

HISTORICAL DEVELOPMENT

Forms of Investment

Capital invested in Canada, as in all other free-market nations, can be classified into two broad yet distinct categories, portfolio and direct. The distinction is critical. Portfolio investment usually entails the acquisition of bonds and the issuance of loans. It rarely amounts to anything else and, in any case, does not involve the legal, or voting, control of the assets of a commercial enterprise. Although direct investment is distinguished from portfolio investment mainly because the former entails legal control of a firm, there are other differences.

Globerman[2] noted three differences of particular concern: (1) whereas portfolio funds are usually contributed by an individual or an institutional investor, direct investment normally comes from a corporation; (2) direct investment is often a package of money, technology, and skills; and (3) direct investment is usually concentrated in certain industries such as manufacturing and the resource extraction firms. These characteristics of direct investment have well-known impacts on the Canadian economy. The impacts on the mining industry of an absence of direct investment is not as well understood. Thus this investigation focuses on the role of foreign direct investment in Canada.

Major Investors

United Kingdom. It is not surprising that England served as the primary source of investment capital during Canada's early years. British funds were first brought in to help establish trading companies such as the Hudson Bay Company when Canada was still a British colony. After an original emphasis on furs, British investment concerns fanned out into the resource sector, utilities, railroads, and manufacturing. By 1897, British direct investment in Canada amounted to $84.5 million. The mining sector received 54.1% of that total, or $45.7 million.[3] Table I-4-1 illustrates the growth of British direct investments in Canada from 1897 to 1967. It shows that as a proportion of total direct investments the British made in Canada, the mining sector received less and less emphasis with time. By 1967, the mining sector had received $281 million in British direct investments, or only 6.9% of the total direct investment stake in Canada.

The mining sector's relative decline in investment importance can be explained by a number of factors. An important one was the diversification of British mining investment into other Crown colonies and nations such as Australia. Another factor was the record of poor profitability of early British ventures in Canadian mines. Paterson attributed this record largely to two influences.[3] First, by normally investing only in producing properties, the British lost out on the potential windfalls often associated with bringing as-yet-undeveloped properties in from the outset. Second, the British engaged in poor mineral exploration practices. For example, they often brought in South African or Rhodesian mining engineers who knew little about the complex nature of Canadian ores. The outcome of these poor decisions was poor profitability, and that helped explain why British direct investment in the mining sector tended to increase at a declining rate.

TABLE I-4-1   Direct Investment In Canada by the United Kingdom and the United States, Selected Years (Million U.S. Dollars)

	U.S.			U.K.		
Year	Total invested	Mining investment	Mining percent of total	Total invested	Mining investment	Mining percent of total
1897	159.7	55.0	34.5	84.5	45.7	54.1
1908	405.4	133.0	33.5			
1909				117.3	45.9	39.1
1926	1,403	141	10.0	336	27	8.0
1933	1,933	211	10.9	376	26	6.9
1945	2,304	255	11.1	348	22	6.3
1960	11,198	1,329	11.9	2,639	236	8.9
1967	21,587	2,919	13.5	4,021	281	7.0
1970	22,801	3,014	13.2			
1980	44,640	2,997	6.7			
1981	46,950	3,360	7.1			

Sources: Refs. 3, 7, 14, and 15; Cleona Lewis, America's Stake in International Investments (Washington, D.C.: The Brookings Institution, 1938).

United States. American investment capital was initially not very welcome in Canada. American incursions into Canadian lands during the Revolutionary War and the War of 1812 had fostered Canadian fears of its more powerful neighbor in the south. One such fear concerned economic continentalism, or the dominance of the U.S. position in all North American economic affairs. However, the desire to protect Canada's economic and political sovereignty was undermined by both poorly coordinated Canadian policy and the desire of Canadians to improve their standard of living. In other words, Canadians recognized the United States as a vast source of the foreign investment capital so badly needed in Canada.

Canadian economic policy in the nineteenth century was centered on protectionist measures aimed at helping establish a young political and economic system in the North American continent.[4] However, the Canadian policies had loopholes that helped induce American investment in Canada. Scheinberg[5] pointed out three important Canadian policies that encouraged American investment: (1) a Canadian protective tariff system designed to discourage exports to Canada; (2) Canadian patent laws which required a foreign manufacturer to build a plant in Canada or license a Canadian manufacturer within a period of, normally, two years after the first export of that product, or else forgo the Canadian market; and (3) the 1897 imperial tariff preferences, which opened up the entire British empire to Canadian producers and exporters. All three of the policies above actually encouraged the establishment of U.S. branch plants, and other forms of direct investment, in Canada. As an example, Singer and Bell first set up branch plants in Canada in the 1870s and 1880s.

The desire for economic gain also motivated U.S. investment in Canada, and these investments were often favored by people in both the United States and Canada:[5]

> Any advantage sought by branch plant representatives would also accrue to some Canadian capitalists. Thus the agents of the branch plants became good members and even leaders of the Canadian Manufacturers Association, not because there was an American plot to subvert it but because they viewed their interests in the same manner as did Canadian industrialists. Canadian nationalism may have been exploited for economic ends, but the exploiters came from both sides of the border.

Canadian policymakers faced a tough dilemma. If they lowered protective tariffs or signed a reciprocity agreement with the United States, American exports would most likely flood the Canadian market and price out many domestic firms. However, by keeping tariffs high and maintaining patent laws to protect developing Canadian firms, American capital would be drawn into the market. This dilemma was not an easy one to resolve, so the easiest policy was to do nothing. At that time, there was little concern that the U.S. presence would grow to undesirable levels, so American investors were allowed to do as they pleased.

Although Americans invested primarily in Canada's manufacturing sector, investment was channeled to the mining sector and related infrastructure. Table I-4-1 illustrates the growth of American direct investment over the years. By 1897, the mining sector had received 34.5% (or $55 million) of the $160 million American direct investment stake in Canada.[2] By 1981, the United States had a book value of direct investment in Canada of $46.95 billion.[6] Of that amount, $3.36 billion was invested in the nonpetroleum mining sector.[7] Table I-4-2 shows that the United States invested far more than any other nation in the various relevant parts of the Canadian mining sector.

It was only a matter of time before Canada began to invest in the United States. Geographical proximity and cultural parallels certainly helped, but a major factor was the influence of bilateral business affairs. The result was that Canada and the United States have become each other's major destination for direct investments. Table I-4-3 shows that, by year end 1976, 53% of Canadian direct investment overseas was in the United States.[8] By 1981, some 20% of total U.S. foreign direct investment was in Canada, making Canada the largest recipient of American funds.[9]

Other Investors. Many other nations have invested in Canada, but perhaps the most notable one—especially for the western provinces now, and all of Canada in the future—is Japan. This is true for many reasons. First, although Japan's total (direct and portfolio) investment came to only $715 million in 1979,[10] of which $292 million was in the mineral resource sector, Japan's stated policy of mineral supply diversification bodes well for Canada. This is especially true now in the case of metallurgical and thermal coal. Japan has entered into several joint ventures in British Columbia and Alberta coal mines. These projects alone have increased the level of total Japanese investment to perhaps $1.3 billion.[11]

There is a second reason why Japan is important. It is that Japan traditionally depends less on direct equity investment than does the United States. As of 1979, 50% of the Japanese investment in Canada was in equity investment, yielding an equity-portfolio distribution similar to that of Japanese investment worldwide.[10] This 50% figure compares with a U.S. direct investment figure in Canada of over 60%. The high level of involvement in joint ventures also distinguishes Japanese investments from American investments.

CONCERN ABOUT DIRECT INVESTMENT

As noted above, a major distinction between direct investment and portfolio investment is that the former normally involves legal control over the business enterprise. This distinction was initially of no concern to most Canadians, but this soon changed.

By 1907, it was noted that most U.S. investment in Canada was centered in the resource and manufacturing industries, and was in the form of

TABLE I-4-2 Foreign Investment in Canada's Mining Sector, Year End 1976 (Million Canadian Dollars)

Investment sector	U.S.	U.K.	Other	Total
Iron and products	5,630	251	243	6,124
Nonferrous metals	2,014	146	222	2,382
Nonmetallic minerals	552	126	329	1,007
Oil and gas	9,355	1,042	1,380	11,777
Other mining and smelting	4,680	434	829	5,943
Railways	715	322	201	1,238
Utilities	2,188	99	208	2,495
	$25,134	2,420	3,412	30,966

Source: Ref. 8.

TABLE I-4-3 Canadian Direct Investment Abroad, Selected Years (Million Canadian Dollars)

Recipient	1951	1960	1970	1974	1976
United States	912	1,618	3,262	4,769	6,092
United Kingdom	74	257	586	904	1,037
Other nations	180	592	2,340	3,537	4,372
	$1,166	2,467	6,188	9,210	11,501

Source: Ref. 8.

direct investment "conferring managerial control."[5] By the beginning of World War I, U.S. direct investment in Canada had surpassed the corresponding British level (see Table I-4-2). After 1918, and especially after the end of World War II, British influence and direct investment declined worldwide. This left Canada in an awkward situation, for American investment in Canada not only increased rapidly through those years, it also came more and more in the form of direct investment. By 1978, total foreign direct investment in Canada was anywhere from 47 to 51% of the $102 billion then invested in Canada, depending on the definition used to indicate managerial control.[12] The vast bulk of that stake was held by the United States.

This high level of direct investment seemed indicative of an undue level of foreign, especially American, influence in the decision making of Canadian firms. True, the investment funds had been needed, as the 1957 Gordon Report noted:

> We found it difficult to escape the conclusion that economic advantage was enormously on Canada's side. In almost all instances, the Canadian economy could not have grown rapidly if not for United States investments, and in some cases nothing at all would have been accomplished.

Even so, by the late 1950s, many Canadians had become convinced that their high national standard of living was not enough; they wanted a greater sense of self-destiny. Throughout the entire twentieth century Canadians had watched the United States become the dominant power in the world, but Canadians also saw some of their own national identity and self-determination slip away with that increase in U.S. power. First, it was early U.S. attempts to annex Canada. Then came massive infusions of U.S. investment capital, mainly direct. By World War II, joint U.S.–Canadian defense agreements further linked the two nations. The NORAD agreement of the 1950s even further involved the United States in what once were strictly Canadian affairs. Finally, there was the American media and cultural presence in Canada.

However, two other factors seemed to clinch the issue of excessive U.S. influence. First, a number of critical sectors of the Canadian economy were not controlled by Canadians. For example, in 1978, 63.8% of the Canadian oil and gas industry's assets were controlled by foreigners.[13] Table I-4-2 shows that Americans held most of the foreign-controlled assets in that sector, as well as other resource sectors. This foreign control led to the second factor of concern. Many foreign-controlled firms were seen as not acting in the best interests of Canada.

Various reports were issued by Canadian government study groups from the late 1950s on, and all agreed with the observation that Canada was not gaining as much from the activity of these firms as would have been the case had the firms been controlled domestically.

A number of concerns were discussed by these studies. There was the concern that the foreign-controlled firms did not reinvest profits back into Canadian operations; that the firms did not hire enough Canadians, especially for management positions; and also that such firms adversely affected the competitiveness of domestic firms. Perhaps the most irritating aspect of the behavior of foreign-controlled firms, and this was and still is especially true of American-controlled firms, was the extraterritorial application of foreign laws to Canadian firms.

For example, the U.S. Trading with the Enemy Act of 1917 ostensibly prevented U.S.-based firms from engaging in business with such nations as China, Cuba, and North Korea. However, the act was also used to block American subsidiaries in Canada from engaging in business with these communist nations. As the Gray Report noted:[14]

> The Treasury Department authority under the Trading with the Enemy Act is not dependent on goods or technology having originated in the United States. The rules are applicable directly to exports made by foreign subsidiaries or affiliates of United States persons or firms. . . . It is not relevant legally to the United States authorities whether the goods involved are validly made under Canadian law and shipped abroad under a Canadian export permit.

All in all, Canadians were essentially being told by such actions that their nation was not an independent, sovereign entity economically or politically. This was bound to produce a reaction.

The Canadian Response

Political policies developed in democratic nations supposedly represent the wishes of the people. However, there is never any guarantee that the policies chosen are either the best of those available, or the ones most favored by the greatest number of people. The policies chosen seem rather to be the most practicable of the available alternatives. In any event, they are an initial attempt to deal with a situation seen to have political, economic, and cultural implications. The Canadian response to a situation of perceived excessive foreign influence in domestic economic affairs was based primarily on the development and implementation of two policies: the Foreign Investment Review Act of 1973 and the National Energy Program of 1980.

The Foreign Investment Review Act. The Act established the Foreign Investment Review Agency (FIRA) to monitor and review certain new foreign direct investment proposals, with the stated aim of ensuring that all allowed investments be of "significant benefit" to Canada. The Agency reviews two types of investment proposals: those to acquire control of an existing Canadian firm by a foreigner, and those to establish new businesses in Canada by a foreigner who either has no currently existing business activity, or who has an "unrelated" business activity, in Canada. A variety of standards has been developed to answer such questions as what a "related" business is, what constitutes foreign control, and how large a transaction can be (in the case of an acquisition) and still avoid the full review process. These standards have been and will continue to be the subject of much heated debate.

Perhaps the most important issue, however, is what exactly constitutes "significant benefit." The agency has issued a list of five criteria used to evaluate all reviewable investment proposals:[15]

1. The effect of the investment on the level and nature of economic activity in Canada, including the effect on employment, on resource processing, on the utilization of parts, components, and services produced in Canada, and on exports from Canada
2. The degree and significance of participation by Canadians in the business enterprise and in the industry sector to which the enterprise belongs
3. The effect on productivity, industrial efficiency, technological development, innovation, and product variety in Canada
4. The effect on competition within any industry or industries in Canada
5. The compatibility of the investment with national industrial and economic policies, taking into consideration industrial and economic policy objectives enunciated by a province likely to be significantly affected by the proposed investment

It should be noted that these five factors vary in their significance to FIRA depending on the nature of the proposal, the region it will affect, and so on. In other words, the factors are flexible enough to allow the agency to adapt its criteria to changing conditions in the federal and provincial economies. For example, several years ago FIRA would have favored an investment proposal which produced jobs in the Maritime Provinces more than a proposal which would produce those jobs in, say, Alberta. Now that even Alberta, with its once booming oil and gas sector, has felt the effects of the recession, namely unemployment, FIRA can change its evaluation criteria to suit the current situation.[16]

The variety of complaints made about the Foreign Investment Review Agency has been staggering. The complaints have all been united in the view that, at the very least, FIRA needs some basic revisions if it is to fulfill its objective of ensuring that Canada gains significant benefits from allowed investment.

One broad objection to FIRA was that initially correct decisions are sometimes not made. This has occurred for two reasons that revolve around the activities of FIRA's Assessment Bureau. The Bureau's function is to collect information pertinent to an investment proposal, consult with interested departments of involved provincial and federal governments, evaluate the proposal according to the five criteria, and then inform the Minister of Industry, Trade and Commerce of the proposal's likely benefits.[17] The Minister then forwards the assessment to the Cabinet for final action. In practice, the Cabinet rubber-stamps the reports from the Assessment Bureau.

It has been argued by some parties that the former Minister, Herbert Gray, modified certain assessments before presentation to the Cabinet. In other cases, the Assessment Bureau has tried to bargain with investors whose proposals were initially not acceptable. The intention was apparently to get the investor to up the ante, so to speak, in exchange for favorable assessment.[18]

Many other complaints have focused on the mechanics of the review process itself. Long processing times were a common complaint. Proposals were turned down with only the comment "not of significant benefit to Canada" given as an explanation. Finally, the cost of filing apparently convinced some potential investors not to submit a proposal. All told, many investors were turned off to the entire review process. They never submitted proposals in the first place, they withdrew applications during the actual review, or they perhaps decided never to try again after an earlier negative decision (or perhaps bad experience with FIRA). Thus is has been argued that FIRA's statistics on its approval rate are somewhat misleading and do not accurately represent the full extent of lost investment caused by the review process.

There is, in fact, some evidence supporting such a claim. A recent article noted that FIRA's approval rate for the fiscal year ending March 31, 1982, was 86% (out of 594 processed applications.[19] That figure did not include the 125 applications known to have been withdrawn from the review process before a final decision was made. The figure also did not count the hard-to-establish number of applications never even submitted because of the existence of FIRA. Thus the claim that the 86% approval rate is the second worst in FIRA's history is less than satisfying.

The situation may not be as bad as it has been made out to look. The Trudeau Administration recently made several shuffles in top positions, and two of those changes directly affect FIRA. Edward Lumley replaced Gray as Minister of Industry, Trade and Development, while Robert Richardson replaced Gorse Howarth as Commissioner of FIRA. Both Lumley and Richardson seem less influenced by strong nationalistic sentiments and more willing to allow foreign investment in Canada.

Furthermore, FIRA has revised some of its policies and standards for the review process. Now, explanations will be given as to why proposals were declined. Processing times will be further shortened. The standards set to determine "foreign control" have been increased, as have the transaction value limits, which means a larger number of proposals can skip the full review process.[17] In addition, FIRA now seems to be more willing to show potential investors how to file what will be a successful proposal. These changes have been welcomed by both private businesses and governments in the United States as in Canada.

Even with these changes, there is still some question as to whether FIRA performs a disservice to Canada's mining sector. Although it is commonly acknowledged that a minority of all proposed foreign direct investments are subject to review by FIRA—the estimate was 20% in 1978 and that was a high annual figure[20]—those investments subject to review are likely to involve large sums of capital. The mining sector by its very nature is often a recipient of these large infusions of direct investment. Furthermore, as noted earlier, direct investment often carries with it packages of technology and skills that are notable for their role in determining the success of a venture. It would be a mistake to assume a direct rate of substitution of portfolio investment for direct investment is always possible, especially for the mining sector. Thus, as Globerman[2] noted, the review process could actually involve wealth losses "if a sufficient number of acquisitions having 'spillover' economic benefits were discouraged by the review process." As noted in Chapter I-3, mining in Canada does produce such spillover economic benefits.

The National Energy Program. Although it deals with oil and natural gas, a resource sector principally outside the scope of this report, the National Energy Program (NEP) does need to be considered. This is because the NEP concerns itself with foreign investment and also because of the impact the NEP has had on the Canadian economy. A major objective of the NEP was to increase Canadian ownership of the domestic petroleum industry from a current level of roughly 30% to a minimum of 50% by 1990.

This attempt at Canadianization of the oil and gas industry was to be undertaken through a variety of measures, three of which are as follows. First, sliding-scale exploration incentives favor firms with high levels of Canadian ownership. Second, a 25% interest in all promising developments in the Canadian Arctic or offshore regions will be given to Petro-Canada, the public sector petroleum firm. Third, acquisitions of foreign-controlled firms are encouraged. The best example of this last measure is the $4.3 billion Dome Petroleum takeover of Hudson Bay Oil and Gas in 1981.

A variety of economic impacts caused by the NEP have been documented, such as the decrease in exploratory drilling done by foreign firms. However, perhaps the most significant impact of the NEP's policies was related to the acquisitions undertaken in the last few years. Canadian banks, which until recently were not subject to legal lending limits, had made large sums of finance easily available to firms about to engage in corporate takeovers. At that time, the security of the loans made to petroleum firms was judged to be quite high because of the expectation that real oil prices would increase. Real oil prices did not increase, but interest rates did. The result was that:[21]

> As the debt leverage of the acquiring Canadian companies increased, in an environment of high interest rates, the proportion of available cash flow of Canadian business consumed by interest expenses increased sharply. By 1981, that proportion was in excess of two-thirds for all Canadian business, and for those oil and gas companies active in acquisitions, it was frequently much higher.

In fact, Dome Petroleum's interest payments amounted to 107% of its cash flow.

The implications of the takeovers were staggering. In one year, 1981, a record of almost 21 years of positive direct investment inflows into Canada was reversed. In fact, there was a net outflow of direct investment of $5.3 billion caused by the takeovers.[21] Added to that, incidentally, was nearly $5 billion of actual Canadian direct investment abroad in 1981. These events combined to create great negative pressures on the Canadian dollar.[12]

Firms such as Dome Petroleum had to refinance their debts to survive. This often included short-term loans to cover interest payments. The worldwide oil glut only worsened matters by decreasing cash flows. Low profits, high debt–equity ratios, inflationary pressures, and the like brought Dome Petroleum to the brink of bankruptcy, forced then Finance Minister Allen MacEachen to ask banks to reduce their lending for takeover purposes, and probably helped along the downward slide in the Canadian dollar's exchange rate.

Many people questioned the actual gains from the foregoing events. True, Canadian control over the domestic oil and gas industry had been increased somewhat, but Canada had not only sustained strong blows to its dollar, banking institutions, and domestic firms, but had developed a bad reputation among foreigners. However, public and government support for Canadianization is still strong, so it appears likely that the push toward the goal of 50% Canadian ownership will continue. The pace, however, may be slow. The continued glut in the global oil market will not help, but an even more important obstacle exists. Simply put, it is unlikely that lenders, borrowers, and the Canadian central bank will soon repeat the unusual confluence of goals that led to the first round of corporate takeovers.[21]

LIKELY FUTURE EVENTS

Even considering the high savings rate of Canadians, it is quite likely that domestic funds will continue to fall short of the amount required to finance all worthwhile enterprises. Furthermore, even considering the scale of nationalistic sentiments in Canada, foreign investment will be accepted more frequently than not. This should be particularly true of the mineral industry, as well as other resource sectors, because of the contribution resource exports make to the overall strength of the Canadian economy.

The latest capital-spending report issued by the *Canadian Mining Journal* indicates nonpetroleum expenditures of C$12.984 billion through 1992.[22] Furthermore, the Mineral Policy Sector of Energy, Mines and Resources Canada has projected a capital requirement for the nonfuel mineral industry of almost C$44 billion (in constant 1979 dollars) through 1991.[23] Both figures would seem to indicate a continuing demand for foreign investment. How Canada will control foreign investment in the mining sector is not certain, except for one aspect. Since 66% of the mining industry is already Canadian-owned, there is no need for Canadianization measures similar to those embodied in the NEP.[24] Thus investment will probably be allowed within the existing FIRA framework. As noted above, this itself could create problems for the mining sector.

It seems obvious that Canadians will continue to grapple with the issue of foreign investment. Some will propose drastic alternatives to the current FIRA-NEP approach. The Mining Association of Canada has argued for the abolition of FIRA. Wonnacott has proposed replacing FIRA with an all-out welcome of all foreign direct investment, "with the only restriction being that the Canadian government would have the option to purchase the foreign firms' assets in, say, forty years at some reasonable prespecified guideline price."[25] Others will defend the present approach on the condition that further modifications be made when necessary. In any event, there will be a need to evaluate the impact of FIRA and the NEP, as the arguments above indicate.

Non-Canadians would do well to remember three things. First, even with FIRA and the NEP, Canada is still recognized as one of the best risks available for the foreign investor. For example, the semiannual country credit ratings of *Institutional Investor* recently gave Canada a fifth-place rating out of 105 nations.[26] Canada ranked below only the United States, Switzerland, Japan, and West Germany. This ranking is based on a survey of major international banks, which still gave a September 1982 ranking of fifth place to Canada, even after all the negative publicity about Canada's economic problems and especially the NEP.

Second, it is a mistake to single out FIRA and the NEP as the sole reasons behind Canada's economic problems of late. Canada is an interdependent part of the global trading community. Many of Canada's economic troubles are caused not so much by FIRA or the NEP but rather by the impact of the global recession, volatile interest rates, and exchange rates, and also by Canada's export orientation.

Third, and most important, it must be remembered that "cultural identity does not seem to be sufficient to satisfy most Canadians. The dominant desire is for control over national affairs."[27] This desire is hardly unique among nations, nor is it something Americans should be shocked about. President Nixon's address to the Canadian Parliament in April 1972 not only acknowledged Canada's desire to be economically and politically autonomous, it encouraged attempts to develop that autonomy.[28]

Canada will continue to experiment with its domestic economy. Fair and sincere advice and feedback from major trading partners would most likely be appreciated. Outright foreign hostility will not be tolerated if at all possible. Economic nationalism is by no means dead in Canada. Rather, its more extreme forms are being replaced by the recognition that Canada must survive first and maintain a distinct identity when and if possible. The recent changes made in FIRA indicate such an approach.

REFERENCES

1. Brecher, Irving, and S. S. Reismann. *Canada–United States Economic Relations*. Ottawa: Royal Commission on Canada's Economic Prospects, July 1957.
2. Globerman, Steven. *U.S. Ownership of Firms in Canada: Issues and Policy Approaches*. Montreal: C. D. Howe Research Institute, 1979.
3. Paterson, Donald G. *British Direct Investment in Canada 1890–1914*. Toronto: University of Toronto Press, 1976.
4. Creighton, Donald. *Towards the Discovery of Canada*. Toronto: Macmillan of Canada, 1972.
5. Scheinberg, Stephen. "Invitation to Empire: Tariffs and American Economic Expansion in Canada." In *Enterprise and National Development*. Edited by Glenn Porter and Robert D. Cuff. Toronto: A. M. Haekert Ltd., 1973.
6. King, John. "U.S. Investments' Book Value Up, but Equity Outlay Off." *The Globe and Mail*, 26 August 1982.
7. Whichard, Obie G. "U.S. Direct Investment Abroad in 1981." *Survey of Current Business* (U.S. Department of Commerce) 62 (August 1982).
8. Statistics Canada. *Canada Year Book 1980–81*. Ottawa: Minister of Supply and Services Canada, 1981.
9. American Embassy, Ottawa. "Canada." *Foreign Economic Trends and Their Implications for the United States*. Washington, D.C.: U.S. Department of Commerce, March 1982.
10. Gherson, Joan. "Japanese Investment in Canada." *Foreign Investment Review* 3 (Autumn 1979).
11. Sigurdson, Albert. "Coal Production, Projects Suffer As Exports Decline." *The Globe and Mail*, 17 September 1982; and Mulgrew, Ian. "B. C.'s Coal SuperProject: Dream or Nightmare?" *The Globe and Mail*, 4 September 1982.
12. Byleveld, Herbert C. "Foreign Investment in Canada: What's the Score?" *The Canadian Business Review* 9 (Summer 1982).
13. Energy, Mines and Resources Canada. *Mineral Policy: A Discussion Paper*. Ottawa: Minister of Supply and Services Canada, December 1981.
14. *Foreign Direct Investment in Canada*. Ottawa: The Government of Canada, 1972.
15. Foreign Investment Review Agency. *Foreign Investment Review Act, Businessman's Guide*. Ottawa: Minister of Supply and Services Canada, 1977.
16. Darisse, Alan. Chief, Communications Division, Foreign Investment Review Agency. Personal communication, 14 October 1982.
17. "How FIRA Works." *Foreign Investment Review* 6 (Autumn 1982).
18. Janigan, Mary. "Taking the Bite Out of FIRA." *Maclean's*, 27 September 1982.
19. "FIRA Approvals at 86% in Year." *The Globe and Mail*, 20 November 1982.
20. *Operating in a Changing Canada: A Rich Market, an Uncertain Future*. New York: Business International Corporation, 1978.
21. Sultan, Ralph G. M. "Canada's Recent Experiment in the Repatriation of American Capital." *Canadian Public Policy* 8 (October 1982).
22. Scales, Marilyn. "CMJ Capital Spending Report." *Canadian Mining Journal* 103 (October 1982).
23. Energy, Mines and Resources Canada, Mineral Policy Sector. *The Non-fuel Mineral Industry to 1991: A Quantitative Outlook*. Internal Report MRI 81/6. Ottawa, July 1982.
24. Erola, Judy. "Canada's Resource Policy." *American Mining Congress Journal* 69 (16 February 1983).
25. Wonnacott, Ronald J. "Controlling Trade and Foreign Investment in the Canadian Economy: Some Proposals." *Canadian Journal of Economics* 15 (November 1982).
26. "The 1982 Country Credit Ratings." *Institutional Investor* 16 (March 1982). See also the September 1982 issue.
27. Fayerweather, John. *Foreign Investment in Canada: Prospects for National Policy*. White Plains, N.Y.: International Arts and Sciences Press, Inc., 1973.
28. Barry, Donald. "Retrospective on Canada-U.S. Relations." *Foreign Investment Review* 6 (Autumn 1982).

Chapter I-5

Canadian Mineral Trade

Trade is of crucial importance to Canada because it helps overcome a major structural problem in the economy. Canada's small population base of 24 million (in 1981[1]) does not constitute a domestic market large enough to support the type of diversified industrial sector seen in, for example, the United States or Japan. To achieve such sectoral diversity requires a much larger market, either a domestic one or a foreign one. Since Canada's population is expected to reach only 28 million by 2000, Canada has exported to larger markets. As a result, trade with other nations now accounts for roughly 25% of Canada's gross national product (GNP).

Canada's major trading partner is the United States. This is to be expected, given such factors as geographical proximity, cultural similarity, and so on. The volume of annual trade between these two nations now exceeds that of any other bilateral trade flow in the world and has obviously benefited both nations. Since most Canadian products faced few obstacles to entering the U.S. market, Canada experienced an export boom while the American economy grew after World War II. In most respects, this has been good for Canada, as it dramatically improved the standard of living in Canada.

MINERAL TRADE WITH THE UNITED STATES

Canada and the United States each account for the largest share of mineral imports of its neighbor. Moreover, mineral trade between the countries is balanced. This would normally seem odd, given the different sizes of the two economies, yet this shows just how strong Canada's mineral industry is.

Canada's Imports from the United States

Table I-5-1 lists the crucial aspects of Canada's imports. The first and most important point is that Canada has a net import dependence for only a few minerals such as bauxite or phosphate. This stands in stark contrast to the condition of the United States and, to a greater extent, Western Europe or Japan. The second point is that the United States is the major source of many of Canada's mineral imports. The United States is either a domestic producer, as with phosphates, or it processes a mineral before export to Canada takes place.

TABLE I-5-1 Canada's Mineral Imports, Select Minerals

Item	Net import dependence, 1979	Total[a]	U.S. share (%)
Aluminum metal	0	158,785 mt	94.3
Copper, all forms	0	87,261 mt	81.2
Gold metal	0	2,249,000 to	82.4
Iron ore	0	5,875,000 mt	96.2
Lead metal	0	2,602 mt	99.7
Nickel, all forms	0	33,641 mt	49.6
Zinc, all forms	0	62,581 mt	87.3
Platinum group	0 (ores)		
	100 (metal)	34,227 to	68.7
Titanium metal	100	986 mt	94.5
Phosphate	100	4,207,000 mt	100.0
Bauxite	100	3,504,000 mt	1.0
Alumina	45	983,971 mt	28.1
Industrial diamonds	100	1,343,000 c	59.5
Chromium, all forms	100	30,090 mt	31.0
Manganese, all forms	100	128,822 mt	10.3

[a]mt, metric tons; to, troy ounces; c, carats.

Sources: Refs. 2 and 4.

The third point is that most of the minerals Canada imports from the United States are actually of Canadian origin. That is, some Canadian ores and concentrates are exported to the United States for processing, and are then imported from the United States. This is true of, for instance, zinc, copper, and platinum group metals. Most of the U.S.-based imports which did not initiate with Canadian ores are themselves items for which the United States itself has a net import dependence. Examples are chromium and industrial diamonds.

Energy, Mines and Resources Canada evaluated Canada's mineral import dependence and arrived at several important conclusions. First, after considering factors ranging from the diversity of supply sources to the ease of substitution, it was concluded that only chromium and manganese presented security of supply problems for Canada. Second, the United States plays a crucial role by processing many Canadian minerals. Although Canada could conceivably do some of this processing at home, the economics of such an act are questionable. Finally, it was recommended that both stockpiling and domestic production be considered. Canada has deposits of chromium and manganese that could conceivably be exploited with further advances in technology.[2]

Canada's Exports to the United States

The commodity descriptions in Chapter I-3 on Canada's mineral industry give some indication of specific examples of Canada's mineral trade with the United States. Table I-5-2 further illustrates Canada's importance as a supply source for the United States. Of the 15 minerals listed, Canada supplies over 50% of the U.S. imports for six minerals, over 30% for nine minerals, and over 20% for 12 minerals. As is readily apparent, many of these are items for which the United States is highly dependent on imports.

Canada's share of the value of U.S. mineral imports has increased with time. Canadian exports to the United States accounted for 30% of U.S. imports in 1965. By 1977, the Canadian share had reached almost 36%.[3] The converse of this trend is also true. Canada's exports have over time gone increasingly to the U.S. market. In 1960, 52% of Canada's mineral exports went to the United States. By 1978, this had increased to 70%, a figure that still holds relatively true.[4]

This increase in the U.S. share of total Canadian mineral exports (by value) can be explained by four factors. First, traditional European markets reduced their imports of Canadian minerals, as seen in Table I-5-3. Second, U.S. demand for imports increased significantly. Third, historically strong corporate linkages between the United States and Canada made it easy to export to the United States. Finally, the post-1973 increases in the price of crude oil significantly increased the American share of total value of Canada's mineral exports, since almost all of Canada's oil and gas exports go to the United States.

The increased mineral trade interdependence of

TABLE I-5-2 Canada's Share of U.S. Mineral Imports (Volumes of Selected Minerals)[a]

Item	Net dependence			Canada's share of imports			Alternative sources
	1980	1981	1982	1972-75	1975-78	1978-81	
Aluminum	E	E	4	72	60	62	Ghana, Norway
Asbestos	78	78	74	96	96	97	South Africa
Cadmium	54	65	69	29	22	27	Australia, Mexico
Columbium	100	100	100	NA	9	6	Brazil, Thailand
Copper	14	—	7	34	25	22	Chile, Peru
Gold	18	15	43	43	43	40	USSR, Switzerland
Ilmenite	32	54	72	77	42	34	Australia, South Africa
Iron ore	25	22	36	45	54	64	Venezuela, Brazil
Nickel[b]	71	68	75	72	63	55	Botswana, Australia
Potash	65	65	61	95	94	94	Israel
Silver	7	53	59	33	37	37	Mexico, United Kingdom
Tantalum	90	92	90	17	15	11	Thailand, Malaysia
Tungsten	53	50	48	24	23	21	Bolivia, China
Zinc	60	64	53	54	51	55	Spain, Mexico
Lead ore	E	1	1	30	17	19	Peru, Honduras

[a]E, net exporter; NA, not available; —, insignificant amounts.
[b]Includes Canada-based Norwegian exports to the United States.
Source: U.S. Department of the Interior, Bureau of Mines, *Mineral Commodity Summaries* (Washington, D.C.: U.S. Government Printing Office, various years).

TABLE I-5-3 Main Destinations for Canada's Mineral Exports, 1960-1978 (Million Canadian Dollars)

Year	Total exports	U.S. share (%)	U.K. share (%)	EEC share (%)	Japan share (%)
1960	1,683	52	21	11	3
1965	2,783	58	18	8	4
1970	5,202	55	14	11	8
1975	11,231	68	6	8	9
1976	12,075	68	5	8	9
1977	13,159	69	6	7	9
1978	14,503	70	4	6	9

Source: Ref. 4.

Canada and the United States initially seemed an ideal relationship. The vast U.S. market needed some amount of practically every mineral produced in Canada and allowed for significant increases in production capacity. The United States, on the other hand, had in Canada a source of most of the minerals it had to import. Canada fell short in its supplies of, say, cobalt and platinum-group metals, and could provide no chromium or manganese. Still, in Canada, the United States had an extremely wealthy and reliable source of minerals. In sum, the relationship seemed a good one.

The slow economic growth of the U.S. economy in the mid- to late 1970s led many in Canada to question the desirability of a further increase in export levels to the United States. Slow U.S. growth, plus relative prosperity in other nations, has meant that the United States now accounts for a smaller proportion of total world trade and is thereby holding down Canada's share. After World War II, the United States accounted for 25% of world trade, but this level fell to only 10% by 1982. This drop, combined with Canada's dependence on the U.S. market, meant that Canada's share of world trade fell from 5% in 1970 to 3.6% in 1982.[5]

Other major mineral-exporting nations did not suffer the same decline in their share of world trade that Canada did because they diversified their export destinations.[6] In other words, these nations attempted to market more of their minerals in those nations less affected by economic slowdowns than the United States and Western Europe. In particular, the Pacific Rim nations were chosen as good export markets. These markets have lately become increasingly attractive to the Canadians, given the continuing inability of the United States to solve its economic problems.

Canada is also trying to tap other export markets because recent U.S. trade policy has, in Canada's view, unfairly hurt Canada's exports. The first prominent incident involved the U.S. subsidization of Falconbridge and Hanna Mining during the 1950s

to reduce U.S. dependence on Inco as its primary source of nickel.[7] This act hastened Inco's decline as the world leader in nickel production.

In addition, Canada's uranium industry nearly collapsed during the 1960s after the United States cut off all imports in an attempt to protect its domestic industry. Finally, the threat of extraterritorial application of the U.S. Trading with the Enemy Act blocked Canada from selling 15,000 tractors to China in the early 1960s. This was a time when only Canada was considering the possibility of normalizing relations with China, and the United States was wholly against such a move.[8]

During the late 1970s, and especially during the early 1980s, the American government has threatened to use a variety of protectionist measures which could hurt Canadian exporters. Especially frustrating to Canada has been the mixed signals it receives from the United States. The U.S. Congress has been increasingly tempted to adopt protectionist measures, yet President Reagan has pushed hard for a free-trade approach. However, the Reagan Administration has clearly signaled its intention to oppose what it considers unfair Canadian export practices.[9] This included a protest of the Bombardier subway car sale, the threat of countervailing duties against lumber, and so on.

Such an American approach to trade, combined with Canada's desire to maintain a strong export industry, has lately caused Canada to modify its trade policies and goals. Should Canada be able to achieve its goals, its trade patterns will be quite different from those in existence today. This holds true for minerals as well as other trade items.

CANADA'S TRADE POLICY

There are four major elements in current Canadian trade policy: increased domestic processing, diversified export destinations, security of market access, and improved export competitiveness. Although these elements are not unique to mineral exports, they do influence mineral trade and will continue to do so.

Increased Domestic Processing

At present, one-half of Canada's mineral exports leave the country in relatively unprocessed forms.[2] Thus the first element of Canada's trade policy is to increase the level of domestic mineral processing before export. For example, Canada could export blister copper or refined copper instead of copper ores and concentrates. The intent behind increased domestic processing is twofold. On one hand, it allows the use of ores and concentrates grading below the levels acceptable to foreign buyers. This effectively increases the size of Canada's mineral reserves. On the other hand, increased domestic processing means more jobs, higher value added, greater tax revenues, and so on.

These gains have been noted and encouraged by official policy for many years. There are two primary government tools to encourage greater domestic processing. The first is government subsidization of required processing facilities so as to reduce the initial outlays made by private firms. Tax incentives, energy subsidies, and the like are quite helpful, yet they are also very expensive, given the capital-intensive nature of the processing industries. Given, for example, a government desire to boost employment, it would be far cheaper to use metal fabrication (end products) as opposed to processing.[2]

The second tool focuses on federal government export policies, four of which stand out. The 1946 Atomic Energy Control Act and subsequent policies required that all Canadian uranium exports be made in the most advanced level possible.[10] Canadian Mining Regulations require that all minerals produced in the Northwest Territories be processed to the refined or fabricated stage before export. The 1947 Export and Import Act, and its 1974 Amendment, allows the federal government to restrict exports of raw materials to encourage further processing. These restrictions are to be implemented only if multilateral trade negotiations and the like cannot expand exports of processed minerals.[11] Finally, one of the criteria used by the Foreign Investment Review Agency to determine the benefits from an investment proposal is the extent of mineral processing it calls for.[2]

In fact, trade with the United States has encouraged the development of additional mineral processing in Canada. Stringent environmental laws, high costs, and other factors have forced the shutdown of many U.S. mineral processing facilities. Thus the United States has allowed Canada and other major exporters to increase their level of processing before shipping minerals to the United States.[11]

Diversified Export Destinations

As noted above, the objective of diversifying exports is based largely on the slowdown in the U.S. economy. This objective initially envisioned a shift to other industrialized nations as a sufficient action, but the last few global recessions have proven this wrong. With even Japan experiencing unemployment, inflation, and slower economic activity, Canada had to broaden its export orientation. The next best market appears to be the newly industrializing countries (NIC's) such as Brazil and South Korea. From 1973 on, the NIC's experienced growth rates higher than even those of Japan. An increase in exports to these nations would appear to be called for.

However, Canada cannot ignore the United States and its other traditional export markets. This is due to two factors. First, the total level of economic activity in the developing world (including the NIC's) still does not approach the scale of the developed nations. Thus, even with far higher growth rates then the developed nations, the NIC's and others cannot immediately offer Canada the same volume of trade opportunities now established with developed nations. This could conceivably change in the medium- to long-term future, though. Second, many of the developing nations have adopted strong protectionist measures to restrict the level of imports and foster the development of domestic industries.[3] Thus, although these nations offer great long-term potential for Canada, the best course for now may be to continue to emphasize trade flows.

Secure Market Access

As noted by Hay,[12] an important corollary to access to a new market is the security of that access. It does Canada no good to commit resources over a period of time to an export scheme and then lose it for some reason. One reason, of course, could be an offer made by a competing exporter to the importing nation. In many cases, the assurance of secure market access will be provided only if the security is reciprocal. That is, Canada will usually have to guarantee access to its own domestic market in order to gain secure access to another market.

This will often entail a series of bilateral agreements, yet this should present little problem to Canada, as this type of arrangement has already been used. The most notable example of this is the Canada-U.S. Automotive Products Agreement of the 1960s. Other agreements include the broad trade compacts Japan and Canada signed in 1974, and the agreement to purchase Mexican oil in exchange for technical assistance.[12]

One potential problem, though, concerns the possible use of bilateral tariff reductions. The General Agreement on Trade and Tariffs (GATT) requires that any bilateral tariff concessions be extended multilaterally.[2] Such a requirement may kill certain otherwise desirable trade arrangements, given a desire not to open up a domestic market sector totally.

Canada could choose to explore multilateral trade arrangements. This was done when Canada signed broad trade agreements in 1974 with the European Community nations. In fact, Canadian policy explicitly states a preference for involvement in multilateral producer–consumer commodity arrangements.[11] This type of agreement meets Canada's objectives and helps ensure access to markets. Canada does not, however, support the use of cartels, partly because Canada imports such minerals as bauxite and manganese, and partly because a cartel runs contrary to the notion of a producer-consumer agreement. Canada's involvement in a uranium cartel during the 1970s was a reaction to U.S. protectionist measures. Moreover, the cartel was abandoned once the U.S. measures were ended.[11]

Improved Market Investigations

Obviously, Canada must not expect mineral importers to approach it with possible deals. Canada has had to actively market its minerals worldwide. The need to aggressively explore new market possibilities has recently increased with the advent of many new mines in developing nations. Although political instability often frightens prospective investors away from Third World producers, sometimes the deals are too good to pass up.[2]

Canada's federal government has been involved in export promotion and marketing activities since at least 1944. In that year, the Export Credits Insurance Corporation (ECIC) was formed to tap the expected increase in mineral demand in Europe after the end of World War II. Since wartime expenditures in Canada were falling, the attempt to expand mineral exports was doubly important to the task of reestablishing a peacetime economy. The ECIC was given C$750 million to lend for the purchase of Canadian goods, and to facilitate trade expansion.[10]

The ECIC was replaced in 1969 by the Export Development Corporation (EDC), a federal Crown corporation. The intent was to provide loan services, contract insurance, and other guarantees considered too risky for the private sector.[13] Since its establishment, the EDC has supported 40% of Canada's exports to countries other than the United States.[10]

Other federal and provincial activities emphasize initial trade promotion activities such as information exchanges and trade commissions. Ontario, British Columbia, Alberta, and Quebec have provincial trade offices in various cities around the world. The Canadian International Development Agency runs a Joint Venture Bureau to assist in the development of joint ventures between Canada and developing nations.[13] One example of federal support for trade involves the work done after a Japanese trade mission indicated in 1957 an interest in coking coals produced in western Canada.[10]

TRADE PROSPECTS WITH THE PACIFIC RIM NATIONS

The Pacific Rim nations present perhaps the best opportunity for a diversification of Canada's mineral (and other) exports. However, an expansion in trade with this region will most likely require that Canada compromise on one or more of its trade and economic goals. This is best seen by considering several markets Canada would like to further its involvement in.

China

Trade with China has been a great hope of Canada's for over 100 years.[14] Shortly after the Canadian Pacific transcontinental railway was built, Canadian Pacific Railways built its CP Empress ships to conduct trade with China and other Asian nations. Although trade with China eventually faltered for many reasons, Canada continued its involvement in China through the activities of government diplomats, missionaries, and various private citizens. This allowed a level of bilateral contact after the 1949 founding of the People's Republic of China unequaled by perhaps any other Western nation. Canada's record as the first major nation to reestablish formal diplomatic relations with China further improved Canada's image in the eyes of the Chinese.[8]

Trade between Canada and China is still small in scale, but is growing rapidly. Joint trade in 1970 was C$160 million and had exceeded C$1.2 billion by 1981.[8,14] Trade was lopsided in Canada's favor, though, with roughly $6 of Canadian exports to China to every $1 of imports.[14] Although wheat was the dominant Canadian export to China, such minerals as potash, sulfur, and nonferrous metals were also purchased. The bulk of Chinese exports to Canada was made up of textiles and clothing products.

Canada is expected to be able to increase its exports of minerals (especially potash and nonferrous metals) to China. However, to do so, Canada must do two things. It must first participate in joint ventures to help China develop, for instance, its coal and hydroelectric potential. Canada welcomes this, for it is an opportunity to display skills Canada has perfected. Second, Canada must try to even out the current imbalance in trade flows. Curtis feels that this is the main limit to any increase in trade levels.[14]

It would seem that such a balancing of trade flows would be perfectly acceptable to Canada given the prospect of increased exports to a market of over one billion people. However, China's competitive advantage lies in labor-intensive items such as textiles. To allow increased Chinese textile exports would threaten an already weak textile industry in Quebec. As a result, Canada refused to include textiles in the General Preferential Treatment agreement it signed with China in 1980.[14]

Australia

The Commonwealth of Australia represents an intermediate point between the two export market extremes represented by Japan and China. Australia's market is far smaller than that of Japan, yet its trade barriers are far lower. On the other hand, Australia's economy is more developed and diversified than is China's, and thereby presents a market for a wider range of Canada's mineral exports.

Australia and Canada are complementary trade partners because they have competitive advantages in different products. As a result, Canada's exports to Australia include asbestos, potash, sulfur, lumber, auto parts, mining equipment, and electronics. Canada imports meat, sugar, wool, and minerals like bauxite from Australia. One note of importance is that roughly 35% of Canada's exports to Australia are in the form of finished products.[15]

Trade in 1981 reached C$1.2 billion in value. Canada's exports accounted for nearly $800 million worth of the total, which continued a trend of an increasing Canadian trade surplus since 1978.[15] All told, Canada accounted for 3% of Australia's imports in 1981.[15]

Future prospects for increased trade look relatively good. Australia is perhaps the most stable and is undoubtedly the largest source of bauxite and alumina in the world, so Canada will continue to import from Australia. Canada's exports of minerals to Australia could increase, both in terms of crude industrial minerals and also finished products. This depends on the continued strength of Australia's economy.

Australia's growing mining industry provides one strong export opportunity for Canada's producers of mining equipment. Canada's ability to tap this market, though, will be constrained by two factors. The first is Canada's distance from Australia. Other producers of such equipment, most notably Japan, are closer and thereby enjoy a transportation cost advantage. Second, Canada's domestic mining equipment industry is still not well developed.[2] It may not be able to offer much to meet the specific needs of Australia's industry.

One situation could possibly hurt the prospects for expanded trade between Australia and Canada. Since both nations are major mineral exporters to Japan and other Pacific Rim nations, Canada, and Australia may be played off against each other by importing countries looking for the best bargain. The resulting competitive tensions could carry over into bilateral trade relations. Should this be the case, Canada could lose a wonderful opportunity to better develop an export market willing to accept finished goods from Canada.

Japan

Japan offers the most exciting prospects for expanded Canadian mineral exports. A Japan-Canada trade link appears to indicate the complementary needs of both countries. Japan's small domestic mineral production capacity forces it to import almost all of its mineral needs. Canada can supply many of the needed minerals and is one of the closest major mineral exporters to Japan.

In addition, Japan's economy is not only very large, it is perhaps the most stable of any in the world at this time. Japan had an annual real gross national product increase of 5.3% from 1971 to 1980 and even had a 3.3% real growth in 1982, the worst year of the last global recession.[16] Estimates are for at least 3% growth through 1985. This stable market translates to fairly stable mineral import demands, a factor of great importance to the mineral exporting nations like Canada. Finally, Japan's preference for joint ventures and minority equity positions is well suited to Canada's economic nationalism.[17]

Trade between the two nations has taken place for over 100 years, but did not achieve any truly significant levels until after World War II. One of the first Canadian products actively sought by the Japanese was coking coal. Other major minerals sought by Japan were (and are) lead, zinc, and copper. Table I-5-4 shows the level of trade between Japan and Canada for 1981. Several important points can be drawn from Table I-5-4. The first is that Canada had a trade surplus with Japan in 1981. In fact, this has historically been the norm.[17,18] Canada had an C$800 million trade sur-

TABLE I-5-4 Canadian Trade with Japan, 1981
(Million Canadian Dollars)

Item	Exports to Japan	Imports from Japan
Raw materials	1929.7	46.8
Fabricated materials	1380.7	630.2
End products	112.2	3286.7
Foods	1057.5	47.0
Other	5.2	28.3
	4485.3	4039.0

Source: Ref. 18.

plus with Japan up through September 1982.[19]

Second, trade is lopsided with respect to commodity type. Most of Canada's exports to Japan consist of raw materials. Only 2 to 3% of Canadian exports to Japan since 1960 have been in the form of manufactured products.[20] However, Japanese exports to Canada are mainly in the form of finished end products. This is an aggravating situation for Canada, given its desire to increase the level of processing its exports undergo.

Finally, Canada now accounts for only 4% of Japan's total annual imports (by value), but Canada's mineral exports often take a larger share of Japan's mineral imports.[20] In 1980, Canada accounted for 9.6% of Japanese imports of metallic ores, 15.1% of coal imports, and 5.4% of nonferrous metal imports.[17]

Prospects for increased trade are quite good. Several recent events point to a promising export market for Canada's minerals. First, as noted above, Japan's shift away from aluminum and other energy-intensive mineral industries should open up export gaps for Canada's products. This is particularly important because Canada's hydroelectric capacity gives it a strong competitive advantage over other producers.

Second, Japan has invested heavily in Canadian coal projects to diversify its supply sources somewhat. In fact, it is expected that Canada's steam coal exports to Japan will increase tenfold by 1995, and coking coal exports will double by 1985.[17] Third, Dome Petroleum is now pursuing a plan to export liquefied natural gas to Japan. Finally, many base metals and nonmetallic minerals such as potash are expected to continue to be imported by Japan.

However, there are tensions that underly this trade pattern. These tensions must be overcome if Canada is to exploit fully all potential opportunities for increased mineral exports to Japan. The first and most serious one is the dispute over the level of processing of each nation's exports. Canada feels that it is used largely as an exporter of raw materials by Japan's mineral processors, who then turn around and export finished products to Canada. Only 2% of the C$3.5 billion of Canadian goods imported by Japan in the first nine months of 1982 were in manufactured products.[19] Moreover, Canada's increased mineral trade with Japan from 1960 on was large enough to lower the overall proportion of minerals Canada exported in processed form.[11]

Japan counters this claim by noting that it has reduced its barriers to imports of finished goods. Moreover, the Japanese claim they are being blamed for problems that actually stem from the mismanagement of Canada's economy.[20,21] Should trade continue as it has, Canada's goal of increased domestic processing will be ignored. Yet, too stringent a push for increased processing of Canada's exports could eliminate an important part of Canada's second largest export market.

A second tension focuses on the danger an exporter like Canada faces when dealing with an importing nation determined to diversify its supply sources. Simply put, Canada's bargaining power is greatly reduced if it is not the major supplier to Japan. This situation came up late in 1982 with Japan's attempt to have coal producers in various nations lower their price and export levels. In addition to a slump in the steel market, Japan's need for Canadian coal fell because of overbuying from the United States in order to cover potential strike-induced shortfalls in Japan's supply from Australia.[20] However, the U.S. trade deficit with Japan caused the U.S. producers to turn down a proposal to lower exports and, in turn, put pressure on Canada's producers. Thus Canada's mineral producers will have to fight to stay ahead in an increasingly complex trade environment.

The third problem is that increased trade with Japan could exacerbate regional dissension in Canada. The bulk of Canada's exports to Japan comes from British Columbia and the Prairie provinces, which produce most of the raw materials sought by Japan. However, most Japanese exports to Canada are destined for Ontario and Quebec, two provinces that cannot beat the Japanese prices for many finished goods. Should trade also expand with the European Community, it is possible that Canada's internal problems could become greatly intensified.

OVERVIEW

Canada must aggressively market its mineral commodities if it wishes to retain its share of the world mineral trade. Otherwise, it will not be able to overcome two obstacles. The first is growing protectionism in almost all nations as governments try to solve economic problems.[8] The second is the advent of new producers in developing nations with no inclination to cut back their output.

Even should Canada be able to investigate new markets for its mineral exports, actual sales may not be guaranteed unless Canada's price is competitive. This factor, although constrained by geological factors out of human control, is determined largely by government mineral policy. The latter concern is discussed in the following chapter.

REFERENCES

1. U.S. Central Intelligence Agency, National Foreign Assessment Center. *The World Factbook—1981.* Washington, D.C.: U.S. Government Printing Office, April 1981.
2. Energy, Mines and Resources Canada. *Mineral Policy: A Discussion Paper.* Ottawa: Minister of Supply and Services, December 1981.
3. Drolet, Jean-Paul. "Strategic Importance of Canada as a Mineral Supplier to the World." *Mining Congress Journal,* February 1980.
4. Energy, Mines and Resources Canada. *Canadian Minerals Yearbook.* Ottawa: Minister of Supply and Services, various years.
5. *The Financial Post,* 26 February 1983.
6. Beckman, Christopher C. *Canada's International Trade Performance: A Survey of Recent Trends.* Executive Bulletin 23. Ottawa: The Conference Board of Canada, 1982.
7. Cameron, John I. "Nickel." In *Natural Resources in U.S.-Canadian Relations, Volume II: Patterns and Trends in Resource Supplies and Policies.* Edited by Carl E. Beigie and Alfred O. Hero, Jr. Boulder, Colo.: Westview Press, Inc., 1980.
8. Moumoff, S. J. "China: The Unexplored Prospects." *The Canadian Business Review,* Winter 1982.
9. Golt, Sidney. *Trade Issues in the Mid-1980's.* London: British-North American Committee, October 1982.
10. Wojciechowski, Margot J. *Federal Mineral Policies, 1945 to 1975: A Survey of Federal Activities That Affected the Canadian Mineral Industry.* Working Paper 8. Kingston, Ont.: Centre for Resource Studies, Queen's University, May 1979.
11. Patton, Donald J. "The Evolution of Canadian Federal Mineral Policies." In *Natural Resources in U.S.-Canadian Relations, Volume I: The Evolution of Policies and Issues.* Edited by Carl E. Beigie and Alfred O. Hero, Jr. Boulder, Colo.: Westview Press, Inc., 1980.
12. Hay, Keith A. J. "Canadian Trade Policy in the 1980's." *International Perspectives,* July-August 1982.
13. Wojciechowski, M. J., and C. E. McMurray. *Mineral Policy Update 1981.* Kingston, Ont.: Centre for Resource Studies, Queen's University, November 1982.
14. Curtis, John. "The China Trade." *Policy Options* 3 (January-February 1982).
15. Hatfield, Scott. "The Australian Parallel." *The Canadian Business Review,* Winter 1982.
16. Kanabayashi, Masayoshi. "Japan's Economy Expected to Stay Weak in 1983, with Some Recovery Late in Year." *The Wall Street Journal,* 1 February 1983.
17. External Affairs Canada. *Canada's Export Development Plan for Japan.* Ottawa: Government of Canada, August 1982.
18. "Report on Japan Business." *The Globe and Mail,* 10 December 1982.
19. *The Globe and Mail,* 17 January 1983.
20. Shuyama, Thomas K. "Canada-Japan/South Korea Trade Promises to Grow." *The Canadian Business Review,* Winter 1982.
21. Gilpin, Robert. "American Direct Investment and Canada's Two Nationalisms." In *The Influence of the United States on Canadian Development: Eleven Case Studies.* Edited by Richard A. Preston. Durham, N.C.: Duke University Press, 1972.

Chapter I-6

Canadian Mineral Policy

This chapter focuses on four elements of the system by which mineral policy is made in Canada: first, the nature of the domestic political environment, its actors, and their patterns of interaction; second, the influences that are external to the mineral industry yet have an impact on mineral policy; third, the nature of current mineral policy; and fourth, the likely direction of future policy in Canada.

DOMESTIC POLITICAL ENVIRONMENT

Policy regarding any sector of a nation's economy is invariably shaped by the structure of the policy-making environment. The environment in which Canadian mineral policy is developed is very complex and, therefore, warrants careful analysis. The first step is the identification of major actors in the policy process.

Major Actors

At first glance, it would seem there are only three actors in the policymaking environment as regards the mineral sector of Canada. These would be the Canadian federal government, the provincial governments, and the mining companies themselves. However, this simple description is highly inaccurate. A more accurate picture can be had if one envisions a hexagon with major role positions at each point. This model, shown in Figure I-6-1, is derived from the ideas of Smiley.[1] The major role positions in this model are the federal government (any of its relevant departments), mineral-consuming provinces, mineral-producing provinces, non-

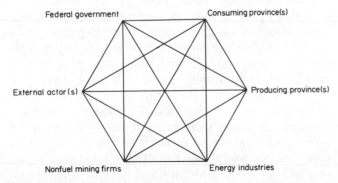

FIGURE I-6-1 Model of role positions in Canadian mineral policymaking.

fuel mining firms, energy companies, and relevant external parties.

Each of the six role positions can be represented by any of a number of actors. Stevenson[2] noted seven federal agencies of particular importance. The Department of Energy, Mines and Resources is primarily a technical support agency that provides geophysical surveys, base mapping services, and the like. This department has perhaps the closest relationship with private mining firms because of a high level of personnel exchanges between the two groups. The Department of Indian Affairs and Northern Development concerns itself with all matters involving the Yukon and Northwest Territories. Environment Canada has a task relatively similar to that of the Environmental Protection Agency in the United States. The Department of Regional Economic Expansion is involved in actions aimed at reducing regional economic disparities. It has lately taken over many of the funding activities formerly held down by the Department of Energy, Mines and Resources.

The federal Department of Finance has two sectors of importance to mineral policymaking. The Resource Programmes Division studies federal spending on resource development programs. The Tax Policy Branch has as its function the analysis and determination of federal taxation levels. The Department of Industry, Trade and Commerce has a self-explanatory name. One important function of this department is the administration of the Exports Permit Act. The final federal agency of importance is the Department of External Affairs, which is the rough equivalent of the U.S. Department of State.

The other major actor in the federal government's role position is the Parliament. Its importance lies in the distribution of power between major political parties, most notably the Liberal and Progressive Conservative parties. This point is discussed at greater length below.

The distinction between mineral-consuming and mineral-producing provinces is a variable one depending on the mineral in question. The only exception would perhaps be Prince Edward Island, which produces no minerals of significant value.

The importance of this distinction is simply that consumer and producer may not, and usually do not, agree on specific policy provisions. For example, the price of a commodity such as coal can be of crucial importance (for quite different reasons) to a consuming and producing province.

There is some degree of overlap between non-fuel mineral producing firms and the energy corporations, for many large corporations have diversified into both energy and nonfuel minerals. Still, the distinction holds for several reasons. First, energy firms (especially oil and gas) produce a commodity far less vulnerable to drastic swings in demand than is true of nonfuel minerals. Second, in many cases an energy production venture has a time horizon much shorter than the average nonfuel mineral operation. Third, in most cases the energy firm is thought (correctly or not) to have a higher profit margin than a nonfuel mining firm. Thus the two types of firms can have quite different concerns about mineral policy, as is seen later in this chapter.

Actors filling the role of relevant external parties to policymaking can be quite diverse. They can range from a community's citizen group to representatives of a foreign nation. Their concern, and role, in the policymaking structure is to ensure that their views are also considered.

This model provides a flexible yet relatively accurate representation of the policymaking organization. The number of actors can vary, as can the role players and their concerns, but the model offers an indication of the complexity of the policymaking environment. First, take the case of metallurgical coal. Table I-6-1 notes likely actors in the policymaking event. This case shows, among other things, the possible input of foreign parties.

An alternative illustration of the model's flexibility concerns the production of aluminum in Canada. Table I-6-2 shows the possible actors in this case.

TABLE I-6-1 Actors in a Hypothetical Coal Policy Case

Role position	Actors
1. Federal agency	Industry, Trade and Commerce; External Affairs
2. Producing province	Alberta
3. Consuming province	Ontario
4. Energy industry	McIntyre Mines Ltd.
5. Nonfuel industry	Algoma Steel Corporation
6. External actors	Coal Association of Canada; U.S. coal producers; Japanese steel firms

TABLE I-6-2 Actors in a Hypothetical Aluminum Policy Case

Role position	Actors
1. Federal agency	Department of Regional Economic Expansion
2. Producing province	Quebec
3. Consuming province	Ontario
4. Energy industry	Hydro-Quebec
5. Nonfuel industry	Canadian Reynolds
6. External actors	Foreign importers of aluminum

Rules of Interaction

There are four dominant rules that influence mineral policy formation in Canada. These lay out the groundwork for interaction between concerned political actors. The first and most important rule is the British North America Act (BNA) of 1867. Section 109 and subsequent legislative interpretations of the original BNA Act granted legal ownership to minerals to the governments of the provinces in which the minerals occurred. The federal government owned and controlled those minerals that occurred in territorial waters and in all land north of the 60th parallel. Other sections granted control of taxation and interprovincial and international trade to the federal government.

As LaForest noted, mineral production in 1867 was insignificant, so few people cared about the distribution of control over mineral resources.[3] For example, little attention was paid to the provisions for control of resources found under territorial waters. Unless otherwise claimed, a province effectively ceded control of these resources to the federal government upon its entry into the union. This provision, and others, has recently generated a great deal of conflict.

Canada patriated, or brought home from the United Kingdom, its constitution in 1982. A series of negotiations over proposed modifications to the constitution took place between the federal and provincial governments at that time. These negotiations resulted in three changes of importance to mineral policy.[4] First, the new constitution reaffirmed provincial control over resources within provincial boundaries. Thus provinces have the right to make laws concerning the exploration, exploitation, and conservation of those resources. Second, provinces were granted the power to make laws on the export of primary resources. However, this power is a qualified one. Federal jurisdiction in this area of pricing and export control is still paramount. Third, provinces are now allowed to impose direct taxes on mineral production (as only the federal government once could). This power is somewhat restricted by the slice already taken from taxable revenues by federal taxes.

The second rule of interaction is that government involvement in the economy is tolerated and even encouraged in Canada. An interesting example of this type of involvement is the establishment of federal and provincial Crown corporations. The intent behind such intervention in the market is to improve on, or speed up, the market process so as to, basically, build a nation.[5] This approach is strongly supported by the Canadian public because of vivid memories of the effects of the Great Depression. However, this approach can and has created conflicts with the business sector, as well as among the federal and provincial bodies.

The reason for this conflict is that federal and provincial governments have slightly different concerns. This diversity of concerns is the third general rule of interaction. The federal government focuses on such broad concerns as the national balance of trade. There is, then, a federal interest in the long-term strength of the domestic mining industry, because mineral exports contribute so much to the balance of trade. The federal government also concerns itself with developing stagnant regional economies so as to improve the overall quality of life. This often calls for extra sources of revenue to fund the grants and other support measures used in reducing regional economic disparity.

Provincial governments do not have exactly the same concerns, or at least not the same scale of concerns, as does the federal government. The concern of provincial governments is that resource exploitation serve as a springboard for the industrial diversification of the provincial economy. Ontario serves as the role model in this case. Losing access to the revenues of mineral exploitation is economically inefficient because the minerals are nonrenewable. Provinces want to garner the one-time benefits from these minerals when they can. Federal attempts to collect these revenues are viewed as a "deliberate attempt to forestall future industrialization."[6] In addition, provinces vary in their willingness to allow federal involvement of any sort. In general, the eastern provinces find federal involvement in the provincial economy to be more acceptable than do the western provinces.

The fourth rule of interaction concerns the shifting regional strengths of the major political parties in Canada. Although it once monopolized power at the federal level, the Liberal Party of Prime Minister Pierre Trudeau did not hold a single parliament seat from west of Manitoba. The Progressive Conservative Party is based in the western provinces and is the equivalent of the Republican Party in the United States ideologically. It has gained on the Liberal Party but could not settle on a party leader capable of challenging Trudeau in a national election. Joe Clark recently resigned his position as party leader after a vote failed to show enough support for him. The seeming groundswell of support for the pro-business views of the Progressive Conservatives proved able to overtake the base of Liberal support found in Ontario and Quebec. Any future mineral policy decisions must deal with this polarization of electoral strength. Otherwise, there is little chance that the policy will succeed in Parliament.

Bases of Intergovernmental Conflict

There are three basic types of intergovernmental conflict over mineral policy. The first involves conflict between the federal government and one or more provincial governments. The second involves conflict between various provinces. The third involves some factor unique to a province which can create tension with the rest of Canada.

Federal–Provincial Conflicts. This situation is perhaps the most widely publicized government conflict situation in Canada. It normally involves conflict over the extent to which either level of government can control the use of a natural resource. Although the BNA Act's power distribution seems clear enough, there have been problems. The struggle for control over oil and gas rents provides two good examples of this conflict.

The first example involved the struggle between the province of Alberta and the federal government in Ottawa over the production, pricing, and taxation levels for oil and gas produced in Alberta. Alberta wanted the freedom to allow prices to rise to world levels, and also to avoid what it considered excessively high federal taxes. Ottawa, on the other hand, claimed that world-level oil prices charged for Alberta's output would adversely affect the less economically well-off provinces and the nation as a whole. Furthermore, Ottawa's need for revenue encouraged it to tax oil and natural gas at a level many felt was excessively high. Alberta claimed that federal taxes raised the price of natural gas destined for the U.S. market to a level at which the United States refrained from purchasing agreed-on amounts.[7] This dispute was eventually resolved when Alberta and Ottawa agreed to a compromise policy. It contained a number of controversial provisions, such as a limit to 75% of world price for Alberta'a oil and gas. Most provinces, especially the western ones, were openly skeptical of the agreement.

A second example of this type of dispute concerned a conflict over control of the Hibernia oil and gas field in the territorial waters off Newfoundland. The British North America Act appeared to place this control with Ottawa on Newfoundland's decision to join the union in 1949. However, Newfoundland, as with other Maritime Provinces, has a history of legislative jurisdiction over its territorial waters which dates back to 1893.[3] As a result, Newfoundland argued that Ottawa had no right to control totally the development of the Hibernia resources.

Newfoundland fought long and hard to ensure its stake in the Hibernia project revenues because the province, the poorest in the nation, stood to gain a great deal of revenue. Ottawa's proposal to grant three-fourths of all revenues would have put an estimated $66 billion into Newfoundland's coffers over the next 35 years. This dwarfs Newfoundland's 1981 budget of $1.5 billion.[8] However, Newfoundland resisted, in an effort to gain even greater control over the development of the field. The dispute was referred to the Newfoundland Court of Appeal and the Supreme Court of Canada in the spring of 1982. Early in 1983, the Supreme Court ruled in favor of Ottawa. By not declaring its jurisdictional rights upon entry into the Canadian union, Newfoundland effectively ceded its rights to Ottawa.[9]

Interprovincial Conflicts. This situation involves conflicts over resource rents and the dependence of a provincial economy on those rents. One conflict involved Alberta and the manufacturing sectors of Ontario and Quebec. The latter group wanted affordable energy supplies to maintain their price competitiveness, while Alberta wanted the "maximum benefit" from its nonrenewable hydrocarbons.

A second example of interprovincial conflict involves the dispute Quebec and Newfoundland have over hydroelectric development in Labrador. Quebec receives most of the 5225 MW generated at the Churchill Falls plant under a 40-year contract, calling for a maximum price of 3.14 mills per kilowatt-hour.[10] This contract can also be extended another 25 years at rates even more favorable to Quebec. Quebec has refused Newfoundland's request to renegotiate the contract, yet Quebec exports some of the Churchill Falls electricity to the United States at a much higher price. Thus

Newfoundland subsidizes Quebec, especially Hydro-Quebec.

Newfoundland has also attempted to build transmission lines across Quebec to reach the American market. Quebec has blocked every attempt. By doing so, Quebec has destroyed the economics of proposed new hydroelectric plants elsewhere in Newfoundland. This act runs contrary to the intent of the British North American Act.[10] Furthermore, it runs contrary to Quebec's stated support for a province's right to control the use of its own resources when Alberta and Ottawa fought over oil and gas revenues. The lack of federal support for Newfoundland in this dispute may help explain why Newfoundland held out so long on its rights to the Hibernia field.

Conflicts Based on Unique Provincial Factors. There is little doubt that the best example of tension created by a factor unique to one province is the controversy over French-Canadian nationalism in the province of Quebec. Ever since the Quiet Revolution commenced in Quebec in the early 1960s, French Canadians have actively attempted to develop Quebec as a stronghold of cultural and economic identity. The seriousness of this intent was expressed in the form of the Parti Québécois, which advocated political separation from the rest of Canada. Quebec has since instituted a variety of policies that ostensibly serve to protect French Canada's cultural identity. However, these policies have had negative effects in both Quebec and other provinces. A first example, described above, concerned Quebec's conflict with Newfoundland over hydroelectric development.

A second example concerns more subtle policies with, perhaps, more far-reaching impacts. The required use of French as the official language in all businesses drove a number of corporations to relocate their headquarters outside Quebec.[11] A similar measure requiring French in all schools has apparently contributed to a high rate of English-Canadian emigration from Quebec.[12]

A final example concerns the use of pension fund investments to keep Quebec's companies under the control of Quebec interests. The Quebec government pension fund manager, the Caisse de Depot et Placement du Québec, justifies its actions by claiming that it seeks the best return available on its investments. However, other parties claim that the pension fund is acquiring an undue level of influence in the operations of businesses of significance to all Canada. A bill introduced before Parliament in November 1982 called for a 10% limit on the shares a provincial pension fund could acquire from an interprovincial transportation company.[13] This prevented the Quebec pension fund from acquiring more shares in the Montreal-based Canadian Pacific, the largest transportation firm in Canada and a firm with significant mineral investments (such as Algoma Steel and Cominco Ltd.).

EXTERNAL POLICY INFLUENCES

A variety of factors play an important role in defining the range of possible mineral policies adopted in Canada. Major ones are the following.

International Concerns

Canada has actively involved itself in attempts to aid developing countries around the world since at least 1951.[14] This involvement can be attributed to many factors, yet perhaps the most important one is Canada's own colonial history. Its experience while still a French and British colony, as well as its relationship with the United States, have sensitized Canadian political leaders and citizens to the plight of the developing nations. Furthermore, Canada's struggle to develop as a politically and economically sovereign nation is well known. Thus, many developing nations, especially in Anglophone and Francophone Africa, see Canada as capable of transferring skills and technology without the political and economic overtones of American or European aid.[15,16]

Canadian policy has been focused in two directions. The first has Canada serve as a mediator, or bridge, between the developed, industrialized nations and the developing nations. Canada's industrial development has not erased a history of economic problems, so Canada is in the somewhat unique position of a mediator truly aware of the concerns on both sides. However, success, especially in international forums such as the United Nations, has been rare.

The second direction of Canadian assistance has been the transfer of technology to developing countries. Not content to just distribute food handouts, Canada has attempted to assist others in areas in which Canada has expertise. So far, this assistance has been focused on agricultural, energy, and human resources development, but nonfuel minerals have also been touched. For example, Petro-Canada International was formed to aid in the development of hydrocarbon energy resources. Other programs have focused, for example, on hydroelectric development and the exploitation of nonfuel mineral resources.

A more specific example concerns Canadian involvement with Mexico. In May 1980, the two nations signed the Canada-Mexico Agreement on Industrial and Energy Co-operation. This pact provided for sales of Mexican oil to Canada in exchange for Canada's assistance in the development of Mexico's industrial base. Canada was sought out by Mexico as one of five nations with which it wanted to intensify relations, so as to reduce economic dependence on the United States.[17] As a result, trade between the two nations has increased dramatically for mineral and nonmineral products.

Government policy in this area affects the Canadian mineral industry in two ways. First, the government allows business managers access to information acquired by Embassy personnel and the Canadian International Development Agency. This can tip off businesspeople to prospects they may not have been aware of, with the likely result that Canadian investment spreads to areas once not considered. This can obviously yield benefits for both the developing nation and the Canadian investor. Trade is established, links become stronger, and the cycle repeats itself.

However, there is a second possible effect that this approach could have on the mineral industry. The development of infrastructure and mining prospects may ultimately facilitate a project quite competitive with Canadian firms back home. In other words, there is the possibility that a promising investment may come back to haunt those in Canada who made the investment. A case in point concerns the Latin American nickel investments made by Falconbridge and Inco. These projects had cost structures low enough that they were brought back on line during last year's nickel slump before mines at Sudbury were reopened. Such actions created a great deal of hostility, yet they are totally justifiable from the perspective of the companies involved. There is the possibility, however, that the continuation of such actions could result in a political backlash in Canada.

One major issue that directly relates to both Canada's mineral industry and the concern with North-South issues is the Law of the Sea Treaty (LOS). Canada signed the treaty upon its adoption on April 30, 1982, for three major reasons. The first was a general agreement with the provision to share profits with the developing nations. Second, Canada felt the need to support LOS provisions on pollution control, the Arctic oceans, continental shelf regions, and fisheries.[18] Third, Canada felt the LOS adequately met its concerns about the exploitation of deep-sea manganese nodules.

Canada believes that the LOS deep-sea mining

provisions adequately protect land-based Canadian mining interests from unfair competition based on nodule exploitation. This is because the exploitation of the nodules is not expected to be cost-efficient before the end of this century, as long as the exploitation is not subsidized.[19] Thus one provision which Canada pushed for—and got—was the antisubsidization and free-market access clause. Canada further feels that the likely haggling to take place over remaining LOS details will ensure that the nodules will not soon be a threat to its mineral industry.

Bilateral Concerns

Perhaps the most important foreign policy concern Canada has is its bilateral relationship with the United States. As described in earlier chapters on trade and investment, the two nations have developed a highly complex, interdependent relationship on all possible fronts. The intimate nature and long history of this bond is certain to produce disputes. For example, Canada and the United States have, since 1790, had conflicts over maritime boundaries and fisheries.[20,21] The variety of disputes is astounding and is well documented.

Disputes over mineral policy are generally based on two separate cases in which Canada and the United States have different perspectives. First, the two neighbors hold different views on what constitutes an acceptable level of government intervention in the economy. As noted above, Canada tends to be more tolerant of such practices. As a result, charges and countercharges of "unfair" subsidization have been made periodically, usually by the United States.[22] A good example of American distress involves the government-mandated aids Petro-Canada receives for its operations in oil and gas. The result of these aids, according to American, and some Canadian, oil firms is that Petro-Canada's subsidies grant it an unfair competitive advantage in the market.

The second difference in perspective concerns acceptable levels of foreign involvement in the domestic economy. In this case, Canadian frustrations are far stronger than American ones. This is partly because of the incredibly high level of American penetration of the Canadian economy, but it is also due to Canadian anger over U.S. actions to reverse a level of foreign control *far* less than is found in Canada. Canada, for example, has vehemently opposed the extraterritorial application of certain U.S. laws to Canadian subsidiaries of U.S. firms.[23] The United States has opposed programs such as the Foreign Investment Review Act which, directly or indirectly, tend to reduce U.S. influence in Canadian business. The United States argues that it agrees with the "Canadianization intent" of such programs but disagrees with the methods utilized.[24]

These types of disputes are not be to ignored. The bilateral ties are strong, but not invulnerable, and so lack of proper action could have severe economic and political consequences. Preventing this, or maintaining—if not improving—the bilateral ties, will require concerted, multichannel efforts on the part of both nations.[24]

Domestic Concerns

Canada's status as an industrialized nation in a world racked by recession means that Canada has strong concerns about inflation and sectoral unemployment. For example, unemployment in February 1983 was 12.5%, well above the 6 to 8% considered normal in Canada.[25] These and other problems are the long-term result of deficiencies in the economy which reduce competitiveness on the world market.

Economic Policy Concerns. The first policy concern of importance is the desire to reduce, or eliminate, the disparities in economic strength between provinces. Norrie[6] pointed out that all provinces strive to use the rents from mineral production to diversify and restructure their respective economies. One continuing point of conflict will be the desirable level of government taxation to fund welfare programs. The question is whether to allow growth as the market determines it, or to try to direct growth. In any case, the Canadian governments, especially the federal government, recently underscored their determination to maintain the welfare system, even in the face of declining resource revenues.[26]

A second policy concern deals with employment levels and labor productivity. Canada's recent record in both areas has been dismal, and implies further problems in the future. Annual growth in Canadian employment was 3.3 and 2.7%, respectively, for the periods 1963–1973 and 1973–1980. Annual growth in labor productivity, however, was 2.4 and 0.0%, respectively, for the same time periods.[26] Three factors in particular seem to have caused this drop in productivity. A major factor is the low level of research and development (R&D) activity in Canada. For example, in 1977 Canada spent only 0.92% of its gross national product (GNP) on R&D. This compared with the Japanese, West German, and American levels of 1.70, 2.00, and 2.39%, respectively. By 1982, Canada spent only 1.2% of its GNP on R&D.[27]

Labor problems make up the other two factors behind productivity. Even in the face of a worsening economy, Canadian unions have regularly pressed for wage increases. The recent action of the Canadian Chrysler union is an excellent example. Furthermore, unions have strongly resisted federal and provincial attempts to limit wage increases. Ottawa's "6 and 5" proposal (calling for 6 and 5% maximum wage increases for civil service workers in 1983 and 1984) of mid-1982 was soundly rebuked by unions and even by some businesses.[28] Quebec, with possibly the most strident union forces, recently saw the second example of union militancy—strikes—as a response to its own call for wage controls.

Canada, out of the entire Western world, has the highest per capita number of strikes. These strikes have occurred even during economic downturns, and have had significant impacts on the economy in general and on specific businesses in particular. For example, 10,000 United Steelworkers members went on strike at Inco's Sudbury nickel complex in May 1982 even though Inco recorded, in the previous quarter, the largest-ever corporate loss in Canadian history to that point (because of declining global nickel demand).[29]

The result of these factors is that Canada's productivity growth has decreased.[28] If Canada's competitive position in world trade is to be maintained, policies must be developed to deal with labor and R&D issues.

The third economic policy concern centers on the proper industrial development strategy for Canada. Canada wants to diversify its industrial structure to become less dependent on the vagaries of world mineral trade. However, it is faced with a difficult choice as to what direction its strategy should follow. Gilpin[30] identified three general approaches: (1) attempting to improve industrial efficiency along a broad front of different fields; (2) specializing in select industries; or (3) importing technology and industry, either by purchasing licenses or by allowing direct investment.

All three options suffer from serious problems. The first approach is constrained by problems of scale, since Canada's small domestic market will not support such undertakings. The approach is also constrained by the difficulties of raising investment capital and properly allocating resources. The second approach requires strong R&D efforts and the politically and socially painful rationalization of industry resources. The third approach has been followed widely. Japan buys licenses and uses its R&D capability to improve the technology. Canada allows direct investments to transfer the technology. However, Canadian problems with a small domestic market, low R&D expenditures, and concern over the side effects of direct investment may make the choice of a strategy a lengthy and difficult process.

The fact that a call for industrial rationalization may conflict with the goals of regional economic development only further delays the development of an acceptable strategy.

Regardless of the approach taken, it is argued that the continued production and export of minerals will serve as the financial foundation for Canada's industrial development.[31] Many parties have argued for further domestic processing of these minerals exports so as to capture a greater share of rent. The concern here is to find the proper balance between economically justifiable expansions in domestic processing and the import desires of major trading partners. Canada currently faces a variety of tariff and nontariff barriers to such an increase in domestic processing.

Indian Affairs. Indian affairs in Canada are very important for the mineral industry. One issue with an impact on mineral resource extraction concerns the disputed land claims of native peoples. These claims cover most of the Canadian territories north of the 60th parallel.[32] Even if these lands are not turned over to natives, they may be subject to increased levels of land-use control. The most important control relates to environmental protection, an issue of more importance to natives than any other group in Canada. Another important issue is the lack of infrastructure. As was done by Alaskan native groups, Canadian natives are trying to ensure that they benefit from any development and are not forced to bear all the costs.[33] Since it is thought that northern Canada holds a great deal of mineral deposits, any development must account for the demands of the native peoples.

Environmental Protection. With the exception of world mineral demand levels and taxes, concern about environmental protection is perhaps the most important recent issue to confront the Canadian mineral industry. This is because the Canadian people in general seem very determined to maintain a clean and safe environment. Whether this determination is actually the result of geographical and cultural proximity to the U.S.-based environmental movement is debatable.[16] The fact is that the concern exists both in the public and the government, and that this concern has been, and will continue to be, developed into policy.

By far the dominant issue in the field of environmental protection these days is that of acid rain in North America. Canada is a recipient of transboundary pollution from the United States to a greater extent than is true of the reverse. Most of Canada's emissions of sulfur dioxide (SO_2), the prime ingredient in acid rain, comes from nonferrous metal smelters, such as Inco's copper-nickel smelter at Sudbury. The bulk of U.S. emissions, however, comes from coal-fired power plants.[32]

Canada has accused the United States of dragging its feet on the issue, in terms of both studying the problem and acting on study conclusions. Moreover, Canada has acted unilaterally by already reducing its SO_2 emissions by one-fourth and offering to repeat this act if the United States would reciprocate. This has been explained by Rosencranz, who noted:[34]

> Ontario and Canadian federal officials made it clear that the Inco Control order was designed to strengthen Canada's position in the United States-Canada negotiations and to pressure the United States to take corresponding measures against U.S. sources of acid rain.

It has been estimated that a one-time investment of C$450 million (in 1980 dollars), and increased annual costs of C$100 million, would be required to remove 80% of the emissions of SO_2 from Inco's smelter at Sudbury.[32] This would result in an increase in the cost of nickel of some 20 to 30 cents per pound. Federal tax provisions, which reduce Inco's financial burden by over 50%, and some small cost increase passed on, would allow Sudbury nickel to remain price competitive on the global market. However, Noranda's nickel output from its Rouyn-Noranda smelter would not be as competitive because of different ores and operating characteristics. Obviously, an all-out attempt to control pollution, be it airborne or waterborne, has the potential to hurt the Canadian mineral industry seriously.

Quebec's Separatist Movement. The separatist movement faltered somewhat in 1980 when a referendum calling for political separation from Canada was defeated. However, the underlying current of frustration with English Canada still runs strong. Many have suggested that the shaky union between French and English Canada could be strengthened if Quebec could further develop its economy.[35] Quebec's approach to economic goals is no different than those of any other Canadian province. The need is for capital with which to build a diversified economy that is less dependent, in Quebec's case, on the production of items such as asbestos and textiles.

However, Quebec's desire for economic strength does not diminish its intention of developing a strong, and possibly politically independent, French Canada. Rather, the two goals go hand-in-hand. Thus policymakers have a dilemma. If English Canada funnels investment capital into Quebec, it may contribute to the eventual destruction of the Canadian union. However, if English Canada does not invest, Quebec will go elsewhere for capital. This will upset the English-Canadian nationalists who seek to reduce foreign control over Canada's economy. In fact, Quebec officials have noted that the availability of U.S. funds was what allowed them to nationalize Quebec's hydroelectric industry in the face of English-Canadian opposition.[30]

Neither option of the dilemma looks desirable, but something obviously must be done. Quebec's premier, René Levesque, announced in March 1983 that the Parti Québécois will soon run again on the separation issue.[36] Should his party gain 50% or more of the vote, Quebec will declare itself politically independent from the rest of Canada and will seek recognition from friendly nations and international organizations.

MAJOR POLICIES

The Goal of Mineral Policies

As is true with any other collection of policies, mineral policies have a central theme, or goal. Canada's goal for mineral policy has varied somewhat, but has always reflected an awareness of the importance of the industry to Canada. The goal outlined in 1973 by the Federal and Provincial Ministers responsible for mineral policy serves as a good example of that awareness. The goal developed in 1973 was to "obtain the optimum benefit for Canada from the present and future use of minerals."

The actual policy measures chosen were to achieve this goal by focusing on three areas of concern: quality of life, sovereignty and unity, and economic growth and development. The focal points of concern have also changed since 1973. The Mineral Policy Discussion Paper released by Energy, Mines and Resources Canada in 1981 suggested three new focal points: the strengthening of the resource base on which the industry rests, the adequacy of research activities needed for the development of the resources, and the further development of markets for these resources.[32]

One underlying theme was, and still is, the understanding that Canada will pursue a basically free-market approach to dealing with the mineral industry, but will allow and encourage government intervention when required. Although different actors in the policymaking environment have different views as to allowable levels of government involvement, they all admit the need for such involvement in some case or another.

Four major coherent blocks of mineral policy measures stand out from the complex maze of federal and provincial mineral policies. The first two, which concern Canada's investment climate and trade policy, are discussed in earlier chapters. The other two policy blocks concern government support for, and involvement in, mining and taxation. As noted above, the rationale for the establishment of government mineral policy is to maximize the economic returns and social benefits from mineral development in all its phases. As is shown below, this rationale also presumes an ability to collect some of these returns for uses not directly related to the mineral industry's needs.

Government Support for Mining

Government support for, and involvement in, mining can be roughly categorized into three major types: infrastructure development, basic support programs, and state intervention in the market.

Infrastructure Development. The provision of necessary social and physical infrastructure is often crucial to the success of a mining venture. It may be argued that mining in Canada would not have progressed as it has had it not been for the provision of infrastructure by government. With respect to social infrastructure, the provision of townsites and related facilities has been crucial to the success of mining operations in the far north. This was, for example, very important in the case of the Polaris mine on Little Cornwallis Island. Other programs have attempted to provide the necessary labor force. For example, Manitoba's Critical Trades Skills Training Program trains people to fill highly skilled positions where shortages exist.[37]

The fact that most of these mines, and many potential ones, are located in the Yukon and Northwest Territories has meant that the federal government normally is the source of funds for public infrastructure development. Federal expenditures for social infrastructure development amounted to $3.7 million in 1953. This had reached $162 million by 1975.[38] These funds came from a variety of federal departments, but perhaps the most important one was the Department of Regional Economic Expansion (DREE). Founded in 1969, DREE has spent over $3 billion, with at least $200 million going directly to the mineral industry's projects.[39]

Physical infrastructure development can focus on the provision of transport facilities, electric power supplies, and water. Perhaps the most important is the provision of transport facilities, the most significant of which has been the railway. From 1900 to 1950, most mineral development efforts were carried on along the expanded routes for railways. The Sudbury nickel, Cordilleran coal, and Cobalt, Ontario, silver deposits were identified during extension of railway lines.[38] This pattern changed after about 1945. Railways were then built into mineralized regions only after the commercial viability of a deposit was verified. Even so, by 1972 some 5600 km of new railway line had been built, most of which was carried out specifically for the mineral industry.

This expansion cost approximately $750 million, of which $200 million was provided by the federal government. For example, the $75 million, 690-km Great Slave Lake Railway connecting the Pine Point (Northwest Territories) mine with the Canadian National Railway (CNR) was financed totally by the federal government. CNR, in turn, operated most of these railway extensions.

As shown earlier in the section on infrastructure layout (see Figure I-3-39), no railways extend north of Pine Point and Hay River. As a result, the federal government instituted a series of road-building programs. The Northern Roads Program existed from 1966 to 1973 and spent some $100 million to construct 2061 km of permanent roads in territorial land. Its goal in constructing these permanent roads was to bring the roads to within roughly 300 km of mining projects.[38]

Another federal roadbuilding project was the widely publicized "Roads to Resources" program. It operated from 1945 to 1975 with the purpose of building roads to allow the delivery of minerals from a mine to the nearest railway or waterway. This program hinged on a trade-off. The federal government would build the roads, but only in exchange for a voice in the design and implementation of the entire project. Sometimes this was acceptable to a mining company (as with the Cyprus Anvil mine in Faro, Yukon Territory) and sometimes it was not (as with the Pine Point mine).

Waterborne transportation has not been ignored in Canada. The most obvious example of federal involvement is the joint Canada–United States St. Lawrence Seaway development project. This project involved the construction of locks and dams, the deepening of canals, and so on. It opened in 1959 at a cost of $1.054 billion. Of that total, the Canadian federal government paid $322 million.[38] The seaway is used mainly for the transportation of bulk commodities, especially those loaded at ports on the Great Lakes.

The National Harbors Board of Canada, a federal Crown corporation founded in 1936, provides important harbor facilities, such as wharves, piers, and onshore storage facilities. Its activities have focused primarily on the east and west coast ports, whereas ports along the lower St. Lawrence have mainly used private funding. For example, the National Harbors Board is currently involved in the development of the coal export terminal at Ridley Island, British Columbia.

The federal government's Northern Resources Airports Program builds airfields and airports for mining ventures. Started in 1965, the program pays 50% of the cost of construction for all facilities.

The final example of important physical infrastructure provided by governments in that of electric power. A federal Crown corporation founded in 1945, the Northern Canada Power Commission (NCPC) started out by building power plants for mines located in the Northwest Territories. Since then the NCPC has expanded its focus geographically and operationally. It now operates in the Yukon and Northwest Territories and also in the Maritime Provinces. Moreover, it is now involved in all phases of design, construction, operations, and ownership of a variety of power plants. By 1975, NCPC operated 56 diesel and 3 hydroelectric power plants.[38]

The provision of infrastructure is not an inexpensive task. However, the federal and provincial governments engage in this act because it is thought ultimately to yield benefits for the entire nation. Furthermore, the government expenditures are often not lost. Most transport facilities are provided on a cost-recovery basis whereby the application of user's fees will hopefully pay back the initial governmental expenditure. In addition, the governments normally provide only the funds needed for initial construction of facilities. Beyond that point, private operators are responsible for all additional expenditures required for the maintenance and expansion of the facilities. All in all, governments are likely to provide assistance if they are convinced that the social benefits, to the mining community and beyond, will exceed the costs of subsidization.

This subsidization, as noted above, is often crucial to the success of a venture. For example, 78 and 42% of the total costs of the Pine Point and Cyprus Anvil mines, respectively, were for infrastructure facilities. The federal government paid 100% of the infrastructure costs for the Pine Point mine and 55% for the Cyprus Anvil mine. One study has shown the impact of this subsidy. The expected lifetime before-tax, constant-dollar, internal rate of return on privately invested capital is estimated at 91% for the Pine Point mine and 15% for the Cyprus Anvil mine. This drops to 19 and 11.5%, respectively, when the rate of return is calculated on total (public and private) invested capital.[32]

Related Support Programs. These government programs expand knowledge about mining, exploration, processing, and so on. The intent is to use the knowledge to establish and maintain a competitive advantage with respect to other major mineral producers and exporters. This is particularly important to Canada because it helps to overcome the high labor costs and low ore grades that would otherwise tend to make Canadian minerals uncompetitive on the world market. These programs vary widely in their focus and in how much they are utilized by different provinces.

Ontario is the best example of provincial use of support programs. Ontario's programs have, as their general aim, the stimulation of further mineral exploration and development in the province. For example, in 1981 the $5 million Exploration Technology Development Fund was established to promote joint ventures with Ontario companies in geophysical and geochemical research.[37] The Ministry of Northern Affairs carries on a variety of activities which it funds on a joint basis with the federal government. The Ministry provides economic geologists for consultation with prospectors and it also funds airborne geophysical surveys.[40] The Ontario Mineral Resources Branch and Geological Survey carry on a variety of programs. They maintain an updated public-access data base which inventories and analyzes the geology and location of mineral deposits in Ontario. They have spent $5 million to establish a permanent drill-core storage facility. Finally, they have conducted studies to determine the feasibility of developing small, high-grade gold deposits in conjunction with the use of small on-site custom milling facilities.[37]

Federal support programs date back to at least the 1940s. After World War II, the federal government expanded its surveys and mapping program to assist the mineral industry in locating new deposits. It also began to conduct research on recovery techniques for low-grade, complex ores. Many of these activities are now conducted under the auspices of the Canada Centre for Mineral and Energy Technology (CANMET), a division of the Department of Energy, Mines and Resources. CANMET is currently attempting to improve the efficiencies of metallurgical processes (especially for aluminum) and is studying open-pit slope stability and underground rock mechanics problems.[41]

The Department of Energy, Mines and Resources is engaged in many other support activities. It is developing a computer data base listing of all the world's scientific literature on coal published after 1977. It also has entered into an agreement with the U.S. Department of the Interior to cooperate on the production of an aeromagnetic map of both countries and to share information on remote sensing, mineral technology, and earth sciences. Historically, these federal programs have yielded benefits. In addition, Canada has dominated the world market in the development of sophisticated geophysical survey equipment since the early 1970s. Moreover, nearly 85% of the airborne magnetic and electromagnetic surveying done worldwide is carried out by Canadian firms.[32]

State Market Intervention. As noted above, the Canadian government has not hesitated to become involved in the market. This usually involves the establishment of a Crown corporation designed to take risks that private firms would avoid. This pattern is true even of Quebec's Crown corporations, except that the Quebec corporations have extra motivations. As Bradbury pointed out, the extra incentive has been to liberate Quebec's economy from the control of Anglophones.[42] Among other things, this has involved the use of more French-Canadians in management positions.

One of the earliest examples of a federal Crown corporation involved in the minerals business was the Wartime Metals Corporation. Founded in March 1942, its objective was to explore for and develop deposits of minerals needed during World War II, including tungsten, magnesium, and molybdenum. Before its termination at the end of the war, the corporation had brought 14 projects on-line. Eight of these were joint ventures with the United States, in which American capital and Canadian management worked together to meet U.S. mineral needs.[39] Another Crown corporation, Eldorado Mining and Refining Ltd. (now Eldorado Nuclear), provided uranium to the Manhattan Project and continues to operate to this day.[38]

Some provincial Crown corporations have been the target of severe criticism because they were formed from the nationalized assets of private firms. The potash industry in Saskatchewan and Quebec's asbestos and hydroelectric industries are good examples. However, provincial desire to control foreign firms does not explain these actions. Saskatchewan nationalized the potash industry because it felt that it was not receiving adequate revenue from private firms. Quebec's asbestos actions occurred because private firms refused to increase the level of asbestos processing done in Quebec.[43] It now seems as though there is less governmental desire to intervene further in the market, however. In fact, in October 1983, the province of Saskatchewan repealed its requirement for compulsory Crown participation in all mining ventures in the province.[44]

Taxation

Perhaps no other type of government mineral policy has been as significant to the success of the mining sector as has taxation policy. Other policies are largely restricted to specific aspects of the industry, as, for example, policies concerning the development of infrastructure. Even trade and foreign investment policies tend to be subordinate to taxation. Trade and foreign investment take place only if a domestic mining venture is judged a safe risk, a probable success. However, success, or the likely profitability of a venture, often hinges on the tax structure.

Tax policy in Canada has a long and varied history. It has often included tax provisions of a positive nature, yet, as of late, the negative features have tended to stand out. The controversy over taxes developed for the nonpetroleum mineral industry should be considered, for these taxes have lately attracted the most attention.

Early Taxation Policy. Although some mineral taxes were in place by the first several decades of this century, it was not until the 1930s that important tax incentives were developed. The motivation to maintain a strong and productive domestic mining industry was based on the effect the Great Depression and World War II had on the economy and, in turn, the mining industry.[39] One of the most significant tax incentives developed at that time was a three-year tax holiday on income from new mines. The most important federal tax policies established by 1948 were as follows:

1. A $33\frac{1}{3}\%$ automatic depletion allowance for income earned up to the prime metal stage.
2. A 100% deduction for exploration and development expenses.
3. Provincial mining and income taxes were deductible from federal income taxes.
4. Three-year tax holiday for new mines, with deferred write-offs of development and capital costs.
5. Grubstakers and prospectors paid no income tax on gains from sale of mining properties.

The five tax provisions listed above, and most other tax policies of the time, were generally favorable in their treatment of the mining industry. Moreover, the interpretation of such policy was often even more generous. For example, the three-year tax holiday was applied not just to new mines, but also to newly discovered ore bodies on or beside producing properties. Furthermore, the deductions for exploration expenses were extended from mining companies alone to, first, firms converting ore to primary metals and, second, to firms that fabricated metals.[38] The overall result was that mining had one of the lowest tax levels of any sector in the Canadian economy. For this and other reasons, the Canadian mineral industry experienced a great deal of growth after World War II.

However, many people began to feel that the mining sector's tax incentives were excessively high. By the late 1950s, government reports stated that the industry was almost mature and stable enough to survive without the special tax provisions.[39] By the late 1960s, demand for a slower rate of resource exploitation pointed to tax incentives as one contributing factor to be modified. Canadian governments also felt that they had a right to tap some of the high profits being earned by the mineral resource sectors at that time.[45] Ultimately, the Benson White Paper of 1969 showed what was to come, when it agreed with the claim that mining was exceptionally risky yet enjoyed past tax benefits that were too high.

Policy Changes, 1969–1974. Subsequent revisions of the federal Income Tax Act took to heart the recommendations of the 1969 White Paper. New policies showed a continuing federal concern for the well-being of the industry, but they also showed the belief that successful mining operations needed to carry more of the tax load. A variety of tax changes were implemented during this time. The two most significant federal changes were the termination of the three-year tax holiday and the decision to no longer allow the deduction of provincial taxes from federal income taxes. Other changes included an earned depletion allowance of $1 for every $3 spent on exploration, new mines, and so on; and also an increase in the federal income tax rate from 25% to 36% and then to a nominal 46%.

Tax changes were not restricted to the federal level. During the early and mid-1970s, provincial tax and royalty levels were also increased. This was due partly to a desire to tap the high profits being earned by the mining industry. However, it was also due to the federal decision (by default) to continue the deduction of provincial mining duties and royalties from the federal income tax. Provincial royalty levels were then increased so much that the federal government soon had to declare the royalties nondeductible, or face losing its tax base from mining.[46]

It is obvious, in retrospect, that neither the federal government nor the provincial governments expected the joint impact of their tax-policy changes to be extremely adverse. However, that was essentially what happened. The mining industry's tax burden (federal and provincial income taxes, mining taxes, and royalties, as a percent of before-tax book income) increased from 21.5% in 1969 to 41.8% in 1975.[32] The combination of federal and provincial tax levies meant that, at least in British Columbia, mining firms paid taxes equal to over 100% of their net income. The overall impact was that the mining industry's ratio of after-tax to before-tax profits fell below the ratio for all other industries in 1974. Prior to that, the mining sector ratio had been an average 15 to 20% higher.[38]

Mineral exploration efforts suffered as a result of this decline in corporate net profits. As was explained in Chapter I-2, exploration expenditures are contingent on the availability of capital, and also on the belief that discovered properties can be profitably exploited. If taxes, or some other factor, lower the expected profitability of a venture, the incentive to explore decreases. DeYoung showed that real exploration expenditures in Canada declined at an annual average rate of 20% after 1971.[47] Over half the drop was accounted for by decreased exploration activity in British Columbia.

Figure I-2-3 illustrates the shift in exploration expenditures that occurred as a result of the tax changes from 1971 on. Provinces such as Quebec, which had lower provincial levies, tended to gain a larger share of the total domestic exploration outlays. The same was true for the Yukon and Northwest Territories, simply because mining firms there did not face the joint burden of federal and provincial taxes. In addition, Canadian mining firms spent much more of their capital on overseas exploration activities. In 1971, foreign outlays were roughly 20 percent of the total. By 1975, over 60% of total Canadian exploration outlays was spent abroad.[47]

Recent Tax Policy Changes. The drop in exploration activity by the mining sector was just the first of many indications that the combined federal and provincial tax burden may have been too high. The point became readily apparent during the recession of 1973–1974. As a result, both levels of government began in 1974 a long process of tax policy adjustment, which continues even to this day. Table I-6-3 summarizes some of the major federal and provincial tax provisions developed through 1982. Although it is by no means a definitive listing of all taxes and royalties, Table I-6-3 shows that many new tax policies have attempted to reduce their burden on the mining sector. This is illustrated, for example, by the reduced income tax rates for small businesses (those firms with annual incomes of $200,000 or less) and the incentives for research and development (R&D) activity.

However, the current tax framework for the mining sector is still strongly criticized. For example, certain specific provisions are particularly onerous. The Mining Association of Canada, for

TABLE I-6-3 Selected Federal and Provincial Tax Policies, 1945–1982

Government level	Policy
1. Federal	a. Investors in junior mining companies receive full write-offs of exploration and development expenditures
	b. Federal income tax is set at 46% with a 10% allowable abatement
	c. 33⅓% earned depletion allowance for social infrastructure and mineral processing equipment
	d. Iron ore mining now considered to conclude with pelletization and not pig iron formation
	e. Tax allowance for northern employees terminate as of December 31, 1982
	f. 100% deduction of R&D expenditures
	g. 50% deduction of increase in R&D expenditures over the average for the last three years
2. British Columbia[a]	a. 8% deduction of cost of processing equipment
	b. 15 to 20% tax exemption for mining firms that also do milling and concentrating
	c. Up to 20% exemption for mining firms that smelt or refine concentrate
3. Ontario	a. 100% deduction for all development expenses
	b. Tax exemption for first $250,000 of taxable profits from a new mine
	c. 25% provincial grant for new exploration efforts by firms not having conducted mining in province in last two years
4. Quebec	a. Corporate income tax lowered to 5.5% (3% for small businesses)
	b. 166.6% write-off for exploration expenses for limited-liability exploration firms

[a]British Columbia processing incentives not mutually exclusive.

Sources: Refs. 37, 41, and 48.

instance, strongly opposes the termination of the tax breaks for employees of firms operating in the far north. The argument is that the harsh climate and the isolation are hard enough on a person, so financial incentives are necessary to attract and keep employees in such areas.

On the other hand, the entire tax framework is often criticized. One major concern is that the system is uncoordinated and often capricious in its effects. All three levels of taxation (federal income, provincial income, and provincial mining duties or royalties) are quite complex and are revised frequently. Mining firms, faced with the task of complying with these laws, often find the financial cost of compliance to be unrealistically high. Moreover, the rate at which tax laws have changed lately has generated investment uncertainty among mining firms. Although not as severe as during the mid-1970s, this investment uncertainty has had an immeasurable impact on the industry itself and on the Canadian economy as a whole.[49]

The best example of this impact on the industry focuses on the junior mining companies. As noted in Chapter I-3, these firms account for a disproportionately high share of mineral discoveries relative to their exploration outlays. This is because they tend to take risks that larger firms typically avoid. However, the structure of current tax laws, in effect, discriminates against these firms.[32] For example, federal laws encourage the pooling of exploration ventures to share the risk. Most junior mining companies do not have enough capital to enter into such agreements. Furthermore, with no outside income, the companies often cannot utilize the 100% write-off on exploration expenditures until they discover and sell a minable deposit. Tax deductions for those firms with outside income are often deferred, which reduces the net present value of such deductions and thereby reduces their usefulness.

All told, the tax structure does not encourage activity on the part of junior mining companies. This is a cause for concern for three reasons. First, it is inequitable, which is all the more frustrating given the fact that current tax laws are complex partly because of a concern about equity. Second, the firms have an established record of successful exploration efforts. As noted above, Canada, to retain its position as a world leader in mineral production and exports, must discover and develop mines faster than it did during the 1970s. Third, the large firms that would possibly replace junior mining companies are controlled by foreigners to a far greater extent than is true for the junior firms.[32] This raises the prospect of decisions being made which are not in the best interests of Canada.

Future Prospects. Although it is impossible to predict the direction that future mineral tax policies in Canada will take, it is possible to identify the elements that will produce those changes. The first two elements are found primarily among federal and provincial government officials who establish tax policy. First, as noted by Parsons,[49] many lawmakers are tempted to use tax policy as a "fiscal tool to accomplish certain social, economic, and political objectives." The tax policies adopted by British Columbia's New Democratic Party government in the early to mid-1970s are a good example of this temptation.

Second, many lawmakers tend to think that nonfuel minerals enjoy the same monopoly and scarcity rents typically garnered by mineral fuels such as petroleum. The lawmakers further compound the problem by neglecting the cyclical nature of the nonfuel mineral industry. Nonfuel minerals may occasionally enjoy "windfall rents," but these are temporary. They are caused by supply-demand imbalances which eventually correct themselves and, unfortunately, often reverse themselves. Thus the occasional high profits do not serve as a legitimate, long-standing base of taxable income.[32] Chances are good, however, that the temporary gains will again seem large enough to warrant increased tax levels. This is to be expected given the government's desire to use tax revenues for a variety of purposes.

These two elements of government involvement are balanced by the various representatives of the mineral industry. These representatives will continue to bargain with government policymakers over the correct level of taxation. To do so, they will present evidence of the industry's dependence on unstable foreign markets, its problems with low ore grades and high labor costs, and many other items. All told, they will try to justify the continuation of tax breaks for an industry that is relatively strong, yet has serious weaknesses. Lately, this attempt to balance government desires has not been very successful. In fact, the industry's attempt to refute most of the conclusions of the 1969 White Paper was unsuccessful.

How this will be resolved is unknown except for one thing. The debate over mineral policy of all types, be it taxation or infrastructure development, will never be definitely concluded. There will always be a push for further adjustments. This will be of concern to those in the industry, and also to Canada's trading partners, yet it must be expected. Hopefully, however, those changes that do occur will not constitute extreme departures from past norms, for it seems that stability is industry's greatest concern about government mineral policy.

REFERENCES

1. Smiley, Donald V. "The Political Context of Resource Development in Canada." In *Natural Resource Revenues: A Test of Federalism.* Edited by Anthony Scott. Vancouver: University of British Columbia Press, 1976.

2. Stevenson, Garth. "The Process of Making Mineral Resource Policy." In *Natural Resources in U.S.-Canadian Relations, Volume I: The Evolution of Policies and Issues.* Edited by Carl E. Beigie and Alfred O. Hero, Jr. Boulder, Colo.: Westview Press, Inc., 1980.

3. LaForest, Gerard V. *Natural Resources and Public Property under the Canadian Constitution.* Toronto: University of Toronto Press, 1969.

4. Lucas, A. R. "Natural Resources and the New Constitution." *Resources,* No. 2 (September 1982).

5. Gaudet, Gerard. "Forces Underlying the Evolution of Natural Resource Policies in Quebec." In *Natural Resources in U.S.-Canadian Relations, Volume I: The Evolution of Policies and Issues.* Edited by Carl E. Beigie and Alfred O. Hero, Jr. Boulder, Colo.: Westview Press, Inc., 1980.

6. Norrie, Kenneth H. "Natural Resources Development and U.S.-Canadian Relations." In *Natural Resources in U.S.-Canadian Relations, Volume I: The Evolution of Policies and Issues.* Edited by Carl E. Beigie and Alfred O. Hero, Jr. Boulder, Colo.: Westview Press, Inc., 1980.

7. Goldstein, Walter. "Canada's Constitutional Crisis: The Uncertain Development of Alberta's Energy Resources." *Energy Policy* 9 (March 1981).

8. Gray, John. "Power and Price Peckford Foes." *The Globe and Mail,* 30 September 1982.

9. Joyce, Randolph. "Brian Peckford on the Rocks." *Maclean's,* 28 February 1983.

10. Crabb, Peter. "There Is More to Canada's Constitutional Problems Than Albertan Oil." *Energy Policy* 9 (December 1981).

11. "Balkanizing Canada: The Cost of Provincial Barriers." *Business Week,* 15 September 1980.

12. Watson, William. "We're Moving toward Two Nations in Quebec." *The Financial Post,* 2 October 1982.

13. Freeman, Alan. "Quebec Pension Agency Fights Bill to Stop Any Increase in Transportation Holdings." *The Wall Street Journal,* 18 November 1982.

14. External Affairs Canada, Domestic Information Programs Division. *Canada and the North-South Dialogue.* Canadian Foreign Policy Text 82/8 (September 1982).

15. Brault, M. A. "Canada's Advantage in Francophone Africa." *The Canadian Business Review,* Summer 1982.

16. Baker Fox, Annette. "The Range of Choice for Middle Powers: Australia and Canada Compared." *The Australian Journal of Politics and History* 26 (1980).

17. External Affairs Canada, Domestic Information Programs Division. *Relations between Canada and Mexico.* Canadian Foreign Policy Text 82/3 (September 1982).

18. Schabas, Bill. "Canada Backs Ocean Mining, U.S. Refuses to Sign Treaty." *Canadian Mining Journal* 103 (October 1982).

19. Hero, Alfred O., Jr., and Roderick M. Logan. "Other Minerals and Deep-Sea Nodules." In *Natural Resources in U.S.-Canadian Relations, Volume II: Patterns and Trends in Resource Supplies and Policies.* Edited by Carl E. Beigie and Alfred O. Hero, Jr. Boulder, Colo.: Westview Press, Inc., 1980.

20. U.S. Department of State, Bureau of Public Affairs. *The U.S.-Canada Maritime Boundary and Resource Agreements.* Current Policy 64 (April 1979).

21. External Affairs Canada, Bureau of United States Af-

fairs. *Notes on the Management of Canada/USA Relations*, June 1982.

22. See, for example, James A. Miller, "Ominous Trends in Canada's Resources Sector: Canada Development Corporation, Collapsing Megaprojects, the New Mineral Policy." *Alert Letter on the Availability of Raw Materials*, No. 13 (June 1982).

23. Valpy, Michael. "A Headache for Canada in the 80's?" *The Globe and Mail*, 25 October 1982.

24. U.S. Department of State, Bureau of Public Affairs. *U.S.-Canada Relations*. Current Policy 318 (October 1, 1981).

25. "Canada's Jobless Rate Rose to 12.5% in Month." *The Wall Street Journal*, 14 March 1983.

26. Rotstein, Abraham. "Independence When Times Are Hard." *Policy Options* 3 (September-October 1982).

27. "Curing Our Ills." *The Globe and Mail*, 19 August 1982.

28. Rose, Frederick. "Canada Gets Tougher on Wage and Price Increases." *The Wall Street Journal*, 19 August 1982.

29. "The Canadian Economy Is in Crisis." *Business Week*, 28 June 1982.

30. Gilpin, Robert. "American Direct Investment and Canada's Two Nationalisms." In *The Influence of the United States on Canadian Development: Eleven Case Studies*. Edited by Richard A. Preston. Durham, N.C.: Duke University Press, 1972.

31. Mulholland, W. D. "Toward a Mineral Industries Policy for Canada." *CIM Bulletin* 73 (October 1980).

32. Energy, Mines and Resources Canada. *Mineral Policy: A Discussion Paper*. Ottawa: Minister of Supply and Services Canada, December 1981.

33. Mitchell, Bruce. "The Natural Resources Development Debate in Canada." *Geoforum* 12 (1981).

34. Rosencranz, Armin. "The International Law and Politics of Acid Rain." *Denver Journal of International Law and Policy* 10 (Spring 1981).

35. Stanfield, Robert. "What to Do about Quebec's Isolation." *Policy Options* 3 (July-August 1982).

36. "Quebec Premier Plans Renewed Bid to Turn Province into Nation." *The Wall Street Journal*, 17 March 1983.

37. Wojciechowski, M. J., and C. E. McMurray. *Mineral Policy Update 1981*. Kingston, Ont.: Centre for Resource Studies, Queen's University, November 1982.

38. Wojciechowski, Margot J. *Federal Mineral Policies, 1945 to 1975: A Survey of Federal Activities That Affected the Canadian Mineral Industry*. Working Paper 8. Kingston, Ont.: Centre for Resource Studies, Queen's University, May 1979.

39. Patton, Donald J. "The Evolution of Canadian Federal Mineral Policies." In *Natural Resources in U.S.-Canadian Relations, Volume I: The Evolution of Policies and Issues*. Edited by Carl E. Beigie and Alfred O. Hero, Jr. Boulder, Colo.: Westview Press, Inc., 1980.

40. Beard, R. C. "Ontario Ministry of Natural Resources Incentive Programs to Encourage Exploration and to Facilitate the Development of Small Mines in Northwestern Ontario." *CIM Bulletin* 75 (September 1982).

41. "Canada—Crisis of Confidence." *Mining Journal* 12 (December 1982).

42. Bradbury, John H. "State Corporations and Resource Based Development in Quebec, Canada: 1960-1980." *Economic Geography* 58 (January 1982).

43. Von Riekhoff, Harald, John H. Sigler, and Brian W. Tomlin. *Canadian-U.S. Relations: Policy Environments, Issues, and Prospects*. Montreal: C. D. Howe Research Institute, 1979.

44. "Saskatchewan Requirement for Compulsory Crown Participation Repealed." *CIM Bulletin* 75 (December 1982).

45. Brown, Robert Douglas. "The Impact of Canadian Federal and Provincial Mining Taxation." Proceedings of the Council of Economics, AIME Annual Meeting, Denver, Colo., 26 February-2 March, 1978.

46. Hjorleifson, G. R. "A National Minerals Policy." *Resources Policy* 2 (March 1976).

47. DeYoung, John H., Jr., "Effect of Tax Laws on Mineral Exploration in Canada." *Resources Policy* 3 (June 1977).

48. Province of British Columbia, Ministry of Industry and Small Business Development. *Doing Business in British Columbia, No. 14: Tax Incentives* (November 1981).

49. Parsons, Robert B. "Time to Get Serious about Tax Simplification." *CIM Bulletin* 75 (December 1982).

Chapter I-7

Conclusions

It would be fair to say that the period of the late 1970s and early 1980s was one marked by national attempts to adjust to changing domestic and international conditions. Canada and the United States were by no means free from the need to make these adjustments. One such adjustment, involving mineral trade between the two nations, is still being made. Given its complex links with many other economic and political factors, the bilateral mineral trade issue is not likely to be resolved soon, or easily. As such, it pays to ask several important questions.

CANADA AS A SECURE SUPPLY SOURCE

With respect to Canada's mineral exports to the United States, the question to be asked is: Should Canada continue to be the major supply source for U.S. mineral imports? Several select issues bear heavily on Canada's decision.

Canada is an exporting nation and, as such, depends on trade with other nations. Decreased trade levels mean a slowdown in Canada's economic growth. Trade in minerals is particularly important in this respect for two reasons: (1) mineral demand, and therefore exports, is a cyclical event; and (2) Canada's exports of minerals contribute disproportionately to the overall balance of payments.

As a result, Canada needs trade links with economically stable nations capable of sustaining their imports from Canada. The slowdown in the U.S. economy and its vulnerability to external factors, such as the oil price hikes of the 1970s, indicate a need for Canada to diversify somewhat its exports away from the United States. This would have the effect of reducing Canada's vulnerability to slumps in the U.S. economy.

However, Canada cannot diversify its mineral (and other) trade patterns to a great extent. New markets, such as Japan and other Pacific Rim nations, are not willing to develop a large dependency on any one mineral supply source, a result of the problems of the 1970s. As a result, Canada's exports to nations other than the United States will be spread thinly among a large number of scattered nations. Moreover, there has lately been a trend toward economic regionalism, which will tend to block some of Canada's attempts to diversify its exports. The establishment of the European Community and the Japan-China-ASEAN trade linkages are just two examples.[1] As a result, Canada will have to continue to rely on the United States as its primary market for mineral exports.

This continued dependence on the U.S. market will naturally prompt Canadian concern over the strength of the U.S. economy. However, these concerns may be negated somewhat by other factors. One important one is that the United States will probably continue to import processed minerals from Canada, something not likely to occur should Canada be successful at penetrating the Pacific Rim markets. Another factor is that trade links between the United States and Canada are well established. Little effort will be required to sustain this trade compared with the requirements needed in the case of diversified export markets for Canada. Thus the savings in capital expenditures could be used elsewhere, a not undesirable situation given tight government and corporate budgets.

Finally, the breadth of the U.S. dependence on Canadian minerals will allow Canada the opportunity to use minerals as the "springboard" for a needed rationalization of Canada's economy.[2] Canada is justifiably concerned about being principally a supplier of raw materials to other nations, especially because Canada's industrial sector is not allowed to grow and mature as a truly diversified sector. The United States offers the market for, and the capacity to help develop, new expansion in Canada's industrial sector.

THE UNITED STATES AS AN IMPORTER

The next question refers to the other side of the coin: Should the United States continue to depend on Canada as a source of minerals? There is little doubt that the answer must be yes, and there are many reasons for this. One is that Canada provides a wide variety of needed minerals at a competitive price. Another is that Canada is a secure supply source. Although labor problems do occasionally occur, Canada is the most secure supplier for the

United States. Because they are geographical neighbors, transport between the two nations poses few problems compared to the vulnerability of imports of oil from the Mideast or of cobalt from Zaire.

An additional factor is that Canada's stability, although shaky at times, is still far better than that of any other supplier. The Mideast and African nations have already been mentioned. Latin American countries, such as Mexico and Peru, have a great deal to offer to the United States (such as oil, natural gas, lead, zinc, silver, and iron ore), but many are notorious for their developmental problems and political instability.[3] The situation does not look much better in Asia. Even Australia has problems. Labor strife is particularly bad, but Australia faces the added obstacle of being far away from the U.S. market.

BILATERAL OBSTACLES

It would appear, then, that the United States and Canada would find it in their own individual and joint best interests to maintain (and possibly expand) existing mineral trade levels. However, the ability to do so will be constrained by a variety of factors.

Economic Concerns

The United States and Canada are at different stages of economic development. The United States has a mature, well-developed industrial base which is now beginning to falter somewhat. Canada never fully developed its own secondary manufacturing sector, and what it has is not strong in most respects. Thus there are two inherent conflicts between the two nations.

The first conflict is based on the disparity in sizes of the national economies. The U.S. economy is over 10 times the size of the Canadian economy, so events thought to be relatively unimportant to the Americans can be overwhelming in Canada.[4] The second conflict is based on the different needs of the two economies. Many U.S. industries are just trying to save themselves from a further decline, while the Canadian industrial sector is trying to just become fully established. Each nation's policies could result in a bilateral conflict. For example, Canada's attempt to further process zinc domestically will certainly upset American workers put out of work because of a closed zinc refinery.

Since trade follows from, and is partly determined by, the nature of a country's industrial development, it is obvious that Canada has different concerns than the United States with respect to trade. Canada has less to offer the world than the United States in terms of a wide range of products,

so it must place special emphasis on its ability to market minerals and other items now exported. If trading partners fall short on their imports, Canada has to look to new markets until it can offer a new product of interest to old partners. The process is no different for the United States, but the constraints are. Given a stronger economic base and marketing sector, the United States can offer a larger variety of items to more countries. Thus the United States is less dependent on trade with any one nation than is the case with Canada.

Government Intervention

The two nations have different constitutions, so it is natural to expect different mandates regarding the "proper" level of government involvement in everyday affairs. Canada's constitution emphasizes "peace, order, good government" as a standard, whereas the United States was given "life, liberty, and the pursuit of happiness" as a goal. Time has shown that government involvement is now standard practice in almost every sector in Canada's life.[5] This was a necessity given the obstacles to be overcome at the time of Canada's founding and afterward. Such involvement has also been true in the United States, although to a much smaller extent.

Given the different perspectives, it is obvious that conflicts will result. The question is whether or not the conflicts are desirable. Canada obviously feels that, for example, government subsidies are required to bolster a developing industrial sector. The United States argues that this is unfair, yet has engaged in such acts itself. The United States has the right to react to damaging Canadian subsidies, but some of the policies undertaken by the United States have been not just reactive, but also hypocritical.

Historical Legacies

Aside from constitutional mandates, the two nations differ in other ways. The United States developed as a violent reaction against British colonial rule. However, by the mid-nineteenth century, relations between the two nations were fairly stable. This was partly because of the vast distance separating the two former antagonists.

Canada, on the other hand, formed partly as a result of U.S. imperialism, or rather a desire not to be American.[5] However, the United States and Canada are not separated as are the United States and Britain. Thus all U.S. activities, both good and, especially, bad, have been felt intensely and immediately in Canada. Moreover, the close proximity of the two nations has made it hard to prevent memories of past events from lingering. In Canada's case, it has been very hard to forget past U.S. attempts to control Canada economically and politically.

This is a particularly frustrating thing for Canada, because the desire to become more closely involved economically with the United States is countered by fears of cultural and political assimilation. As a result, Canada has over the years developed a series of protectionist policies aimed at fostering a sovereign and healthy Canada. Once adverse climate, tenuous links between quite different regions of the nation, and other factors are added to the fear of U.S. domination, it is easy to see why Canada's policies were adopted. One example of this is the Canadian attempt to control foreign investment.

The Need for Consultation

The history of U.S.-Canadian interaction is one of cooperation and controversy. The tension as of late has perhaps grown to new heights because so many formerly unrelated issues now are seen to be complexly linked. As a result, both Canada and the United States have a number of grievances with each other over a wide range of issues. These grievances cover, for example, trade and investment policy, environmental protection, nuclear weapons and defense policy, and so on. The danger of this is that conflicts over mineral trade issues may be lost in the struggle or, worse yet, linked with a totally unrelated issue.[5] Such an impasse could greatly harm both nations.

It would seem, then, that the two nations need actively to avoid unilateral actions that may be seen as a threat by the other nation. The recent past has seen a great many reactionary policies developed to counter such unilateral actions. The problem with such a reactive approach is that it does not pay attention to the foundations that led to the conflict, and unilateral actions, in the first place.[6] Such foundations were described briefly above.

Canada and the United States need each other. This is obviously true for minerals, for the United States needs Canada's supplies, and Canada needs the U.S. market. It is also true in terms of other trade, investment, cultural affairs, defense policies, and so on. It is even true in the sense that Canada needs the United States to help develop Canada's own sense of national identity.[4] One would hope, though, that the identity chosen is a reasonable one.

The bilateral relationship could perhaps be made stronger and less tumultuous if both sides were engaged in high-level consultations. The United States and Canada have a history of such talks, yet the intensity and frequency with which they have been held has declined since the early 1970s. Canada's Third Option and the U.S. attempt to regain control over its own affairs ultimately combined with the often burdensome nature of such talks to end the consultations.[7] It is unlikely that such talks will soon be resumed. Moreover, it is unclear what role minerals would play in a new round of talks.

However, replacing unilateral, reactive policies with joint anticipatory actions is ultimately a desirable process for both nations. It helps both countries deal better with each other, and in turn with the rest of the world. It does not mean loss of sovereignty but rather an effective adaptation to the problems inherent in any relationship. Given the intensity of their relationship, Canada and the United States should place special emphasis on using joint consultation to develop anticipatory policies. Joint trade in minerals, although immensely important, would be only one aspect of the relationship to gain from such an effort by the United States and Canada.

REFERENCES

1. Drolet, Jean-Paul. "Strategic Importance of Canada as a Mineral Supplier to the World." *Mining Congress Journal*, February 1980.
2. Mulholland, W. D. "Toward a Mineral Industries Policy for Canada." *CIM Bulletin* 73 (October 1980).
3. Hero, Alfred O., Jr. "Overview and Conclusions." In *Natural Resources in U.S.-Canadian Relations, Volume II: Patterns and Trends in Resource Supplies and Policies.* Edited by Carl E. Beigie and Alfred O. Hero, Jr. Boulder, Colo.: Westview Press, Inc., 1980.
4. Armstrong, Willis C., et al. "U.S. Policy towards Canada: The Neighbor We Cannot Take for Granted." *The Atlantic Community Quarterly* 19 (Fall 1981).
5. Drouin, Marie-Josée, and Harald B. Malmgren. "Canada, the United States and the World Economy." *Foreign Affairs* 60 (Winter 1981-82).
6. Volpe, John. *Industrial Incentive Policies and Programs in the Canadian-American Context.* Montreal: C. D. Howe Research Institute/Washington, D.C.: National Planning Association, January 1976.
7. Canadian-American Committee. *Improving Bilateral Consultation on Economic Issues.* Montreal: C. D. Howe Research Institute/Washington, D.C.: National Planning Association, October 1981.

PART II

SOUTHERN AFRICA

Chapter II-1

Introduction

In 1979, an Interagency Task Force issued the *Report on the Issues Identified in the Non-fuel Minerals Policy Review*[1] after a review of U.S. mineral policy. Not only did the report recognize the growing U.S. dependence on foreign sources of supply for certain nonfuel mineral commodities and the subsequent political consequences of this dependency; but the report went so far as to identify the "commodities of greatest concern." These minerals were chromium, cobalt, manganese, and platinum-group metals.[1]

The authors of this report chose these four minerals for several important reasons. Of course, the most obvious of these reasons was the lack of U.S. domestic production of these metals. Several reports in recent years by the U.S. Bureau of Mines have provided more information on possible U.S. domestic reserves and resources of chromium,[2] cobalt,[3] manganese,[4] and platinum-group metals,[5] including consideration of Alaska's[6] vast territory. In all cases, the United States does possess varying supplies of these key mineral commodities, but domestic production is generally dependent on government subsidies, higher prices for the metals, or unbuilt infrastructure.

The second major reason for choosing these metals as the most important had to do with concern about the continuing supply of these metals. Both production and reserves for these metals are highly concentrated in southern Africa (less so for

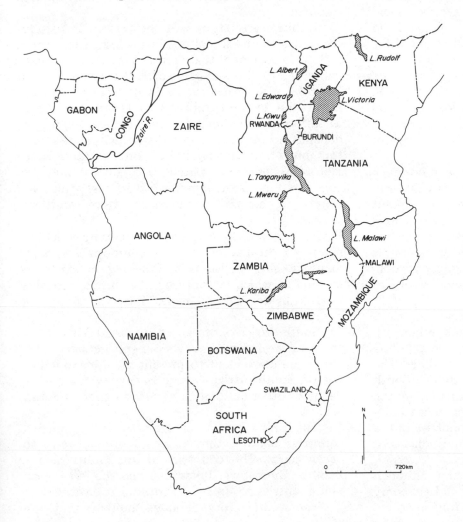

FIGURE II-1-1 Southern Africa, index map.

manganese). Concern about the supply of these metals is generated by both the economic and political climate in southern Africa.

The third most important factor generating concern for these metals is that alternative supplies of these metals are concentrated either in the USSR or in a country supportive of the USSR. Therefore, under strained conditions, especially war, these alternative supplies would be unavailable to the United States.

Another significant factor relating to the importance of these metals is their valuable end uses. They are crucial ingredients in chemical, industrial, and military applications. For many applications there are no substitute metals or efficiency is greatly reduced.

Fearing possible supply interruptions in nonfuel minerals from southern Africa, the U.S. government initiated a number of studies to identify possible factors that could have an effect on mineral supplies from the region. Of major concern were possible supply disruptions in South Africa,[7] Zaire,[8] Zambia,[9] and Zimbabwe.[10] Concentrating on South Africa, other government studies were prepared to determine the dependence of the United States and other OECD countries on mineral supplies from South Africa. In 1980, the U.S. Senate Committee on Foreign Relations released a report on *The Imports of Minerals from South Africa by the United States and the OECD Countries.*[11] In 1982, the U.S. Senate Committee on Foreign Relations again released a report concerning South Africa, which was entitled *U.S. Minerals Dependence on South Africa.*[12]

Although both reports establish the importance of South Africa as a mineral supplier, they differ in tone and conclusions. The reports reflect different approaches to South Africa, and tend to be somewhat biased toward their views of South Africa—one being predominantly concerned with human rights (Carter) and the other being predominantly concerned with strategic position and resources (Reagan).

Although South Africa is the most important country in the region (economically and militarily), it is crucial to realize several overlooked points about the southern Africa region as a whole. The principal fact to realize is that even though South Africa is politically isolated by countries in the region and world, no country in the region is totally independent of its neighbors. All countries in the region are affected by the actions and policies of their neighbors.

Second, it is essential to realize that although the South African mineral industry and economy dominate the region, mineral supplies (or the poten-

tial for them) from other countries in the region are also important contributors to future world mineral supplies.

The third salient phenomenon concerning the region is the propensity for political differences to translate into violence. Violence is not restricted to South Africa alone, nor to disagreements with its neighbors, but has also affected countries with black majority rule. This violence mainly threatens the mineral industries in the region, because it can effectively destroy the vulnerable infrastructure, such as transportation and power, that supports the mineral industries. Violence also tends to decrease capital investment from foreign sources in the region by making the return on investment less likely or more risky.

Political instability in the region is related to the major issues now affecting the region and include the following: colonialism versus black African independence; north-south dialogue; private versus state control and ownership of natural resources; apartheid in South Africa; independence for Namibia; various guerrilla wars in the region; and an East-West conflict for strategic position and resources.

Unlike many studies of the region, this study concentrates on all major mineral industries of the region. It is evident that factors that affect individual mineral industries often affect every country in the region, and that events in one country often affect the mineral industry of another country. Therefore, any competent policy toward the region depends not on one or two major independent factors, but on an understanding of many interrelated factors. These interrelated factors then may be condensed into several major factors that affect the supply of mineral commodities from southern Africa.

To accomplish this end, the countries in the region are divided into several subgroups. Although many possible divisions for the region exist, according to which factors are to be emphasized, this report divides the countries into the following groups.

South Africa is presented by itself. This is due to the size of the country's mineral industry. It is also more convenient to present the mineral industry of South Africa before reviewing the dependency of other national mineral industries on South Africa.

The first group is categorized as land-locked countries because they have no direct access to ocean ports. The countries in this group include Zambia, Zimbabwe, Botswana, and the less important countries of Malawi, Lesotho, and Swaziland.

The second group includes countries that possess ocean ports. Included in this group are Zaire, Angola, Mozambique, and Namibia/South-West Africa (SWA).

A recognized weakness in the report is an analysis of political conflicts and differences in the region. The report tends to limit the discussions of political factors by mainly listing the various political groups, their major differences, and the possible consequences to mineral supplies. This approach was chosen to preserve a more politically neutral approach to the region. Political judgments on events in southern Africa would tend to inflame readers who disagree, and cause them to dismiss the other useful aspects of the report.

The conclusion to the report presents important factors that may influence the supply of minerals from the region and mineral commodities affected by these possible supply disruptions.

REFERENCES

1. Interagency Task Force. *Report on the Issues Identified in the Nonfuel Minerals Policy Review*. Draft Report. Washington, D.C.: U.S. Government Printing Office, 1979.

2. U.S. Department of the Interior, Bureau of Mines. *Chromium Availability—Domestic*. IC 8895. Washington, D.C.: U.S. Government Printing Office, 1982.

3. U.S. Department of the Interior, Bureau of Mines. *Cobalt Availability—Domestic*. IC 8848. Washington, D.C.: U.S. Government Printing Office, 1981.

4. U.S. Department of the Interior, Bureau of Mines. *Manganese Availability—Domestic*. IC 8889. Washington, D.C.: U.S. Government Printing Office, 1982.

5. U.S. Department of the Interior, Bureau of Mines. *Platinum Availability—Market Economy Countries*. IC 8897. Washington, D.C.: U.S. Government Printing Office, 1982.

6. U.S. Department of the Interior, Bureau of Mines. *Critical and Strategic Minerals in Alaska*. IC 8869. Washington, D.C.: U.S. Government Printing Office, 1981.

7. U.S. Department of the Interior, Office of Minerals Policy and Research Analysis. *The Future Role of Central and Southern Africa in the Supply of Nonfuel Minerals to the U.S.—Qualitative Report on South Africa*, 20 June 1980.

8. U.S. Department of the Interior, Office of Minerals Policy and Research Analysis. *Implications of Current Events in Zaire for World and U.S. Cobalt Markets*, 27 April 1977.

9. U.S. Department of the Interior, Office of Minerals Policy and Research Analysis. *The Future Role of Central and Southern Africa in the Supply of Nonfuel Minerals to the U.S.—Qualitative Report on Zambia*, 11 July 1980.

10. U.S. Department of the Interior, Office of Minerals Policy and Research Analysis. *The Future Role of Central and Southern Africa in the Supply of Nonfuel Minerals to the U.S.—Qualitative Report on Zimbabwe*, 2 June 1980.

11. U.S. Congress, Senate, Committee on Foreign Relations. *Imports of Minerals from South Africa by the United States and the OECD Countries*. 96th Congress, 2nd Session, 1980. Washington, D.C.: U.S. Government Printing Office, September 1980.

12. U.S. Congress, Senate, Committee on Foreign Relations. *U.S. Minerals Dependence on South Africa*. 97th Congress, 2nd Session, 1980. Washington, D.C.: U.S. Government Printing Office, October 1982.

Chapter II-2

South Africa

The Republic of South Africa occupies some 1,222,480 km² of the southern portion of the African continent. Included in this area is the enclave of Walvis Bay, Namibia/South West Africa (1124 km²), Transkei (44,000 km²), and Bophuthatswana (38,000 km²).[1] With a coastline of 2881 km and excellent ports in the Atlantic and Indian Oceans, South Africa is strategically placed astride the major transportation routes for fuel and nonfuel minerals to Western Europe and the Americas. South Africa shares common borders with Namibia/South-West Africa, Botswana, Zimbabwe, Mozam-

bique, Swaziland, and completely surrounds Lesotho.

The population of approximately 30 million people is dominantly Africans (69.9%), whites (17.8%), coloureds (9.4%), and Asians (2.9%).[1] Except for native Africans, the population is quite urbanized, with 89% of whites, 77% of coloureds, 91% of Asians, and only 38% of Africans living in urban centers.[2]

The official languages are Afrikaans and English. However, there are several major language groups within each racial group. The multilingual nature

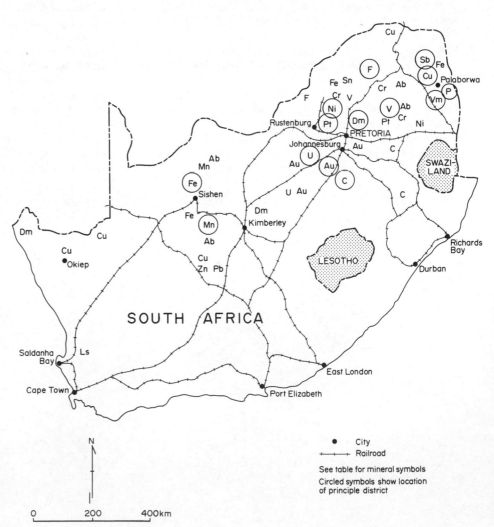

FIGURE II-2-1 Mines, processing facilities, and infrastructure of South Africa. (From U.S. Department of the Interior, Bureau of Mines. *Mineral Industries of Africa.*)

for the white population is presented in Table II-2-1, while that for the black population is presented in Table II-2-2.[2] Dominant languages within the Asian population are Engish, Tamil, and Hindi. Among the coloured population, Afrikaans is the dominant language.[2]

The present constitution for the Republic of South Africa came into being on 31 May 1961. Currently, the Administrative center for the government is in Pretoria, the legislative center is in Cape Town, and the judicial center is in Bloemfontein.

The head of government, Prime Minister Pieter W. Botha, is the leader of the country's politically dominant National Party. Other major white parties are the Progressive Federal Party, the New Republic Party, the Conservative Party, and the Herstigte Nationale Party. Internal dissension within the National Party has surfaced as a result of recent attempts to include some form of representation

and political power within the government for non-white racial groups.[3]

Despite these new efforts at increased representation in government for nonwhites, apartheid (racial segregation) remains institutionalized in South Africa. Apartheid in South Africa has been a major issue at the United Nations for a number of years. Domestic opposition among black Africans has given rise to several exiled political parties. The two most important black African parties are the African National Congress (ANC) and the Pan-Africanist Congress (PAC).

SOUTH AFRICAN ECONOMY

South Africa's economy is by far the dominant economy in southern Africa and is second only to Nigeria for the entire African continent. A comparison of the gross national product (GNP) of other southern African countries with that of South Africa is given in Table II-2-3.[4]

South Africa's economy is dominated by the manufacturing and mining industries, which were 27% and 21%, respectively, of the total GNP in 1979.[5] Table II-2-4 presents the contribution of individual economic sectors to the total GNP for recent years, and Table II-2-5 presents the GNP for South Africa in recent years.[5]

Manufacturing

The manufacturing sector grew from 21% of the total GNP in 1962 to 27% of the total GNP in 1979. Not only did the volume of manufactured goods increase during this period, but the diversity of goods increased in order to become more self-reliant. Table II-2-6 indicates the relative growth of various industries within the manufacturing sector.[5] As indicated, the total manufacturing sector has increased 30% during the period of 1970-1979.[5]

The two largest growth sectors were chemicals

TABLE II-2-1 Multilingual Nature of White South Africans

Linguistic group	1980 estimates[a]
Afrikaans	2,360,000
English	1,652,000
Afrikaans and English	47,200
Dutch	23,600
French	10,620
German	59,000
Greek	23,000
Italian	23,000
Portuguese	47,200
Yiddish	4,130
	4,249,750

[a]Estimates for 1980 based on an 18% population increase over a 10-year period.
Source: Ref. 2.

TABLE II-2-2 Ethnic Divisions of South Africa's Black Population

Ethnic group	1980 census
North Nguni	6,136,427
South Nguni	2,684,084
Transvaal Ndebele	617,165
Northern Sotho	2,264,544
Southern Sotho	1,792,687
Tswana	1,215,732
Uhavenda Lemba	185,058
Shangana-Tsonga	888,879
Other	185,443
	15,970,019

Source: Ref. 2.

TABLE II-2-3 Gross National Product for Selected Countries in Southern Africa, 1980 (Million U.S. Dollars)

Country	GNP 1960	GNP 1980
Angola	690	2,500
Lesotho	30	250
Malawi	170	1,420
Mozambique	830	2,360
South Africa	6,980	74,660
Zaire	130	6,160
Zambia	680	3,790
Zimbabwe	780	3,640

Source: Ref. 4.

TABLE II-2-4 South Africa's Economy: GNP by Sectors, 1962–1979, Selected Years (Million U.S. Dollars)

	1979[a]	1978[b]	1977[b]	1976[b]	1975[b]	1970[c]	1962[d]
Agriculture, forestry, fishing	3,700	3,198	3,013	2,656	2,899	1,362	954
Mining and quarrying	9,680	6,374	4,591	3,778	4,148	1,689	1,038
Manufacturing	12,066	9,447	8,595	8,136	7,715	3,914	1,566
Utilities	2,008	1,708	1,284	913	843	429	201
Construction	2,164	1,844	1,737	1,710	1,697	709	219
Transport, communications	4,718	4,137	3,777	3,152	3,288	1,615	775
Finance	5,841	4,478	4,077	3,701	3,964	2,063	827
General government service	4,844	4,034	3,660	3,237	3,340	1,584	676
Percent share of total							
Agriculture, forestry, fishing	8	8	8	8	8	8	10
Mining and quarrying	21	15	12	12	12	10	13
Manufacturing	27	22	22	23	22	23	21
Utilities	4	4	3	3	2	3	3
Construction	5	5	4	5	5	4	3
Transport, communications	10	10	11	10	10	10	10
Finance	13	10	11	11	11	12	11
General government service	11	10	10	10	10	10	9

[a]*South African Reserve Bank Quarterly Bulletin*, March 1980, SARB, Pretoria.
[b]*South African Reserve Bank Quarterly Bulletin*, December 1979, SARB, Pretoria.
[c]*South African Reserve Bank Quarterly Bulletin*, December 1978, SARB, Pretoria.
[d]*South African Statistics 1972*, Government of the Republic of South Africa, Pretoria.
Source: Ref. 5.

TABLE II-2-5 South African Gross National Product, 1970, 1973–1979 (Million U.S. Dollars)

	GNP at market prices	GNP at constant prices (1970)	Per capita real GNP (dollars)
1970	20,871	19,356	560
1973	28,621	20,239	580
1974	35,932	21,911	606
1975	37,181	22,549	607
1976	35,692	22,625	865
1977	40,361	22,845	847
1978	45,864	23,377	840
1979	57,785	20,838	742

Source: Ref. 5.

and chemical products (60% growth from 1970 to 1979), and basic metals and metal products (53% growth from 1970 to 1979).[5] Growth of these areas corresponds with the growth of the coal industry, especially Sasol, in the case of chemicals and chemical products; and a switch to downstream processing of mineral commodities in the case of basic metals and metal products. Another important growth area is South Africa's growing defense industry. Values for the various manufacturing industries are given in Table II-2-7.[5] The largest monetary gains were in the food and beverages, chemical and chemical products, and the basic metals.

Mining

The mining industry contributed 22.7% of the total South African GNP in 1980.[2] Of this amount, gold alone was responsible for 15.4% of the total GNP in 1980 and 68% of the mining sector.[2] In 1981, due to a drop in gold prices, gold accounted for only 12% of the GNP, or a total of $9.8 billion out of a total GNP of $81.8 billion. Total mineral sales for 1981 were $15.8 billion.[6]

While South Africa is the largest gold producer in the world (producing over half of the world's gold each year), the country is also an important leader in the production of other minerals vital to the economies of North America, Western Europe, and Japan. In particular, a few minerals have to be deemed strategic and/or critical to these countries because of their uniqueness and importance in military-related industries. Prominent among this type of minerals are asbestos, chrome, fluorspar, industrial diamonds, manganese, nickel, platinum-group metals, and vanadium. South Africa is also important for the production of several other basic minerals.

TABLE II-2-6   South African Manufacturing Production Index, 1968, 1970, 1974-1979 (Base 1970 = 100)

	1968	1970	1974	1975	1976	1977	1978	1979
Food	100.5	100.0	116.9	121.0	125.9	124.1	117.8	124.0
Textiles	86.9	100.0	111.8	105.5	115.0	106.5	108.8	113.5
Chemicals and chemical products	86.3	100.0	133.2	142.1	141.3	146.9	151.8	160.3
Nonmetallic mineral products	83.4	100.0	121.1	120.8	114.4	96.1	98.2	100.5
Basic metals and metal products	83.8	100.0	118.9	125.5	121.8	116.9	135.5	153.1
Machinery, including electrical machinery	93.8	100.0	109.5	113.0	108.4	89.7	88.5	91.4
Transport equipment	80.5	100.0	128.3	120.1	110.1	95.3	111.3	106.6
Other	80.4	100.0	126.0	129.1	129.9	117.7	123.0	132.2
	86.4	100.0	121.2	123.7	123.6	115.8	121.7	129.9

Source: Ref. 5.

TABLE II-2-7   Value of Manufacturing Sales, 1975-1979 (Thousand U.S. Dollars)

	1975	1976	1977	1978	1979
Food and beverages	4,542	4,212	4,824	5,125	6,337
Tobacco	422	420	464	492	594
Textiles	1,184	1,250	1,206	1,375	1,640
Clothing, footwear	987	926	888	991	1,155
Paper and printing	1,243	1,219	1,286	1,483	1,779
Chemical and chemical products	3,738	3,901	4,443	5,130	6,570
Nonmetallic mineral products	809	779	700	834	990
Basic metals	2,452	2,454	2,709	3,299	4,624
Fabricated metal products	2,038	1,969	1,666	1,865	2,320
Machinery	1,553	1,411	1,519	1,509	1,704
Electrical machinery	1,226	1,271	1,181	1,190	1,569
Transport equipment	550	473	485	455	498
	24,543	23,603	24,831	28,184	35,005

Source: Ref. 5.

The importance of South Africa's mineral production and mineral reserve base are given in Tables II-2-8 and II-2-9, respectively. Statistics were taken from information provided by the Bureau of Mines in *Mineral Commodity Summaries 1983*.[7] As can be seen from these tables, South Africa is a major contributor to world mineral production. More important, South Africa's mineral reserves indicate that South Africa has the potential to remain one of the world's most important mineral producers.

GOVERNMENT ROLE IN MINING INDUSTRY

It has been a long-held policy of successive South African governments to leave the exploitation of the country's mineral wealth to private enterprises, although the state is empowered to undertake mining operations itself. To encourage this policy, the country's mining laws have first been designed to encourage private enterprise by providing for security of prospecting and mining rights, and second, to provide a fair and attractive return on investment. To attract foreign investment, South Africa's mining laws have been designed not to differentiate between South African nationals and foreign investors in the acquisition of prospecting and mining rights. In recognizing the cyclic nature of the minerals industry and the large initial capital investment for a mineral venture, the South African government has adopted a taxation policy to encourage mineral ventures.

Mining Laws

Early mining laws in South Africa varied from colony to colony. These early laws were modernized and consolidated mainly in two mining laws. The Precious Stones Act of 1964 provides for the regulation and control of prospecting and mining

TABLE II-2-8   South Africa's Role in World Mineral Production, 1982

Mineral commodity	Production (metric tons)	World rank	Percent of world production
Antimony	8,616	2	15.2
Asbestos	200,000	2	5.0
Chromite[a]	2,630,000	1	29.9
Diamonds (industrial)[b]	5,500,000	3	19.6
Fluorspar	499	5	10.4
Gold[c]	21.20	1	51.7
Ilmenite[d]	380,940	5	9.3
Iron ore	26,416,000	8	3.4
Kyanite and related minerals	181,400	1	42.0[e]
Manganese[f]	3,990,000	2	19.1
Nickel	26,303	7	5.0
Phosphate	3,250,000	7	2.3
Platinum-group metals[g]	2,600	2	40.6
Rutile[h]	49,885	3	14.1[i]
Vanadium	11,340	1	34.4
Vermiculite	181,400	2	36.4
Zirconium	79,816	2	18.3[j]

[a] Gross weight.
[b] Million carats.
[c] Million troy ounces.
[d] Slag production (ilmenite production).
[e] In 1979 (Ref. 2) production (kyanite, etc.).
[f] Gross weight.
[g] 1000 troy ounces.
[h] Tons of concentrate.
[i] Excluding U.S. production.
[j] Excluding central planned economies.

Source: Ref. 7.

for and dealing in precious stones, which includes diamonds, rubies, and sapphires.[8] The Mining Rights Act, No. 20 of 1967, provides for the regulation of prospecting and mining for precious metals, base minerals, and natural oil. Precious metals are gold, silver, and platinum-group metals and ores. Base minerals are all minerals excluding precious metals, precious stones, and natural oil.[8] Natural oils have been defined as[8]

> any liquid or solid hydrocarbon or combustible existing in a natural condition in the earth's crust, but does not include coal or bituminous shales or other stratified deposits from which oil can be obtained by destructive distillation, or gas arising from marsh or other surface deposits.

The Atomic Energy Act, No. 90 of 1967, controls the prospecting and mining for and the processing, possession, and disposal of uranium and thorium or any material containing uranium or thorium above a specified concentration.[8]

With reference to The Mining Rights Act, No. 20 of 1967, all land is divided into three classes:

state land, alienated state land, and private land. Furthermore, distinct laws regulate the prospecting for and mining for precious stones, precious metals, base minerals, and natural oil on each.

State land is land owned by the state and for which the state is also the holder of the right to the precious or base minerals and natural oil. State lands may be open, in which case prospecting licenses may be issued to individual persons (up to 10 acres per magisterial district). Land that is not open to public prospecting requires the approval of the Minister of Mines, who determines that the applicant or tenderer has the necessary financial resources to carry out prospecting satisfactorily. A mining lease is granted only after the Minister of Mines and the Mining Leases Board have determined that the prospector's technical and financial resources are satisfactory to meet conditions for mining determined during negotiations.[9]

Alienated state land is land not owned by the state, or if state land, is held by a lessee. However, whether privately owned or owned by a lessee, the rights to precious or base minerals and natural oil

TABLE II-2-9 South Africa's Role in World Mineral Reserves, 1982

Mineral commodity	Reserves (metric tons)	World rank	Percent of world reserves
Antimony	326,520	2	7.2
Asbestos	22,000	2	21.2
Chromite[a]	2,267,500	1	67.6
Diamonds (industrial)[b]	50	4	5.7
Fluorspar	154,190	1	27.4
Gold[c]	700	1	52.6
Ilmenite	99,770	4	13.5
Iron ore[d]	3,719,000,000	7	3.8
Kyanite and related minerals	104,000,000[e]	1	35.0[e]
Manganese	1,995,400	2	40.7
Nickel[f]	5,830,000	5	6.0
Phosphate	1,800,000,000	4	4.5
Platinum-group metals[g]	970,000	1	80.8
Rutile[h]	5,442,000	4	4.6
Vanadium	7,802	1	47.1
Vermiculite	72,560,000	2	40.0
Zirconium	10,884,000,000	2	25.0

[a]Gross weight.
[b]Million carats.
[c]Million troy ounces.
[d]Recoverable (iron ore reserves).
[e]In 1979 (Ref. 2) reserves (kyanite, etc.).
[f]In 1979 (Ref. 2).
[g]1000 troy ounces.
[h]Tons of concentrate.
[i]Excluding U.S. production.
[j]Excluding central planned economies.
Source: Ref. 7.

are in the title deeds or lease. The owner of the surface or his nominee hold exclusive rights to prospecting on alienated state land. The prospecting license is obtained from the mining commissioner of the mining district concerned. A mining license is usually granted to the owner of surface rights or his nominee. However, the license may be refused to a nominee if the government mining engineer determines that the terms and conditions of the mining lease are detrimental to the state.[9]

Private land is land in which the state is not the holder of the right to precious or base minerals and natural oil. Most of the Bantu and tribal lands in South Africa come under this category. The holder of the mineral rights may grant permission to any qualified person to conduct prospecting on his land. Although no prospecting permit is required for base minerals, prospecting for precious minerals requires a permit. The holder of base mineral rights may grant permission to mine for base minerals. With regard to precious minerals, a mining permit can only be obtained from the Minister of Mines. The Minister of Bantu Administration and Devel-

opment is responsible for applications for prospecting and mining in Bantu areas and tribal lands.

In general, it can be concluded that the right for prospecting for natural oil is vested in the state, as well as the mining for and disposing of precious metals. In general, the rights for prospecting, mining, and disposing of base minerals is the province of the holder of the right to base minerals on that particular class of land.

Royalty

The royalty paid to the state is a deductible cost from the state income tax. The method for determining lease amounts depends on the type of mineral being mined and whether the mine is located on state land or on private land.

In the case of a gold mine, any income derived from the recovery of additional minerals (silver, osmiridium, uranium, pyrites, etc.) other than gold are also subject to the following formula:

$$y = a - \frac{ba}{x}$$

In this formula, y is the amount of tax paid to the

state and represents the percentage of profits after deductions of the capital redemption allowance and the capital allowance, and

$$x = \frac{P}{R} \times 100$$

where P represents profit less capital redemption allowance only and R represents mining revenue.[9]

The magnitude of a is dependent on factors such as the estimated grade, costs, the pioneering aspect, and the amount of risk involved. Historically, the amount has varied from approximately 10 to 30. Usually, b is 6 to 8. A return of 10 to 15% compound interest is considered adequate return on investment;[9] for example,

$$y = 12.5 - \frac{75}{x} \quad \text{or} \quad y = 12.5 - \frac{100}{x}$$

In the case of base minerals on state land, a lease consideration of 10% of profits before tax is deducted is the norm. On private land, the royalty is a matter of negotiation between the various parties involved.[9]

In the case of precious stones, particularly diamonds, royalty ranges from

$$y = 15 - \frac{120}{x}$$

for most alluvial or fissure workings

to

$$y = 20 - \frac{160}{x}$$

in the case of diamond mines (pipes)

In this instance, y and x have the same values as in the formulas for precious metals.[9] Royalty derived from oil and natural gas, which is payable to the state, is calculated on a sliding scale.[9]

In addition to the royalty payments, a tax equal to $1\frac{1}{4}$% of the state's share of profits is payable to the state in lieu of stamp and transfer duties. A tax of 30 cents per morgen per month is payable to the mining commissioner in accordance with the registration of the land in the deeds office.[9]

Taxation

Taxation of the mining industry is regulated by the Income Tax Act, 1962 (Act 58 of 1962, as amended). In the case of mining operations, specific allowances and sliding tax scales have been formulated which differ from the income tax laws of other industries.

The taxable income is determined to be the gross income reduced by the allowance for the redemption of capital expenditures and lease consideration or royalty payments.

Redeemable Capital Expenditure. Redeemable capital expenditure includes expenditure[8]

1. On shaft sinking and mine equipment and, in the case of a natural oil mine, the cost of laying pipelines from the mining block to the marine terminal or the local refinery, as the case may be, including any single renewal or replacement of such equipment or pipelines which, together with the accessories thereto exceed in cost forty thousand rand; and
2. On development, general administration and management (including interest and other charges payable on loans utilized for mining purposes) prior to the commencement of production or during any period of nonproduction.

Not allowable as capital expenditure are the cost of the mineral or surface rights or on legal and transfer fees connected with such rights. Surface surveys not connected with mining operations are not allowed as deductible capital expenditure. Financial charges for the formation of a company, underwriting commissions on a share issue, and so on, are also not permitted.[8]

Capital received from the sale of plant, machinery, equipment, and so on, are deducted from the capital expenditure of that year. If there is no unredeemed balance of capital expenditure after the sale of plant, machinery, equipment, and so on, the credit is added to the taxable income.

With respect to gold mines, a special capital allowance for further reduction of taxable mines is permitted to encourage deep-level gold mines and prolong the life of older gold mines. In this instance, a capital allowance is calculated as a percentage of a mine's redeemable capital expenditure according to Table II-2-10. This special allowance can be deducted only during a period in which the mine has an assessed loss or 10 years from the date in which a "deep-level mine" became an "other deep-level gold mine."[9]

Rates of Taxation. All nonmining taxable income is taxed at a flat rate of 40 cents for each rand (R1) of taxable income. There is a 5% surcharge and a loan levy of 15% are added. The effective rate of taxation is 48 cents per R1. Additionally, a simple interest of 5% per year is to be added to the loan levy, the interest being exempted from tax.[2]

In the case of base minerals and coal and platinum mines, the rate of tax, surcharge, and loan levy are the same as for both public and private

TABLE II-2-10 Special Capital Allowance for Gold Mines in South Africa

Type	Interest (%)[a]
New deep-level gold mines	5
(a) Lease granted after 30 June 1956 and	
(b) Principal object is to mine gold-bearing ores believed to exist at depths greater than 2286 m below surface, and	
(c) It will take at least seven years to reach production	
Other deep-level gold mines	
Any producing gold mine already mining deeper, or which will be mining deeper than 2286 m below surface within five years but excluding a new deep-level mine	5
Gold mines granted leases on or after 30 March 1963	6
Post-1966 gold mines: lease granted after 17 August 1966	8
Post-1973 gold mines: lease granted after 1 January 1974	10

[a]Annual compound interest rates.

Source: Ref. 9.

nonmining companies.[9] For precious stones, taxation is 45% of taxable income plus a 10% surcharge and a loan levy of 5% of the tax. There is an export duty of 15% of the value of all diamonds (whether cut or uncut).[9] The tax rate for natural oil is 40 cents for each R1 of taxable income plus a 5% surcharge and a 15% loan levy.[5]

The tax rate for gold mines is similar to the formula for determining royalty payments from gold mines to the state. To this amount is added a 5% surcharge and a 2½% loan levy. Mines using the assistance formula are exempt from this surcharge.[9]

The general formula for taxation of gold mines is

$$y = a - \frac{ab}{x}$$

As in the determination of royalty payments on a gold mine, y is the percentage of taxable income payable as tax and

$$x = \frac{P}{R} \times 100$$

where P is the taxable income and R the mining revenue. Application of this formula reveals that the lowering of the a factor would increase profits (i.e., an increase in a would decrease y). A raising of the b factor would increase the value of y. Therefore, tax on profits would be high and indirectly

encourage the mining of lower-grade ores (i.e., higher operating cost per ton would lower the profit per ton margin, and therefore the overall profit margin).

For gold mines before August 1966, the applicable formula is

$$y = 60 - \frac{360}{x} \quad \text{i.e., } a = 60, b = 6$$

To encourage marginal mines with profits less than R40,000 per year, the formula becomes

$$y = 20 - \frac{120}{x} \quad \text{i.e., } a = 20, b = 6$$

If taxable income exceeds R40,000 per year, the formula becomes

$$y = (20 + w)\left(1 - \frac{6}{x}\right)$$

where w equals completed multiples of R2500 over R40,000.[9]

The basic formula for post-1966 is the same except that b equals 8. Therefore, the basic formula becomes

$$y = 60 - \frac{480}{x} \quad \text{i.e., } a = 60, b = 8$$

The formula becomes

$$y = 20 - \frac{160}{x} \quad \text{i.e., } a = 20, b = 8$$

for mines whose taxable income is less than R40,000 per year. If taxable income exceeds R40,000, the formula becomes[9]

$$y = (20 + w)\left(1 - \frac{8}{x}\right)$$

To encourage marginal gold mines and provide for optimum recovery of gold reserves, a special tax rate applies to marginal gold mines. The formula is

$$y = 68 - \frac{601}{x}$$

or the standard formula, whichever results in a lower tax. In the case of "assisted" gold mines, there is no surcharge tax. A tax credit for the following year is given to companies where the assistance formula results in a negative tax.

To qualify as an "assisted" gold mine, a mine must have a mining life of eight years or less. Thus qualified, a mine may add any unredeemed balance to any accumulated assessed loss, and the resulting amount is allowed to reduce current taxable income to a level where no income tax is payable.[9]

Mining Organizations

South Africa is a world leader in shaft sinking, tunneling, and beneficiation processes for low-grade, complex base metals and uranium. It is also a world leader in the refining of platinum. Being a major exporter of mineral commodities, South Africa has also developed excellent abilities in the analysis of the world mineral industry. A number of public and private institutions have been developed in South Africa to support such technological and analytical capabilities.

State Organizations. The Government Mining Engineer's Office is mainly responsible for safety and health, and thus supervises all mines and works. As such, the office also recommends the terms of prospect and mining leases and on assistance to marginal gold mines.[10]

The Geological Survey of South Africa was established in 1912, with its primary function being the geological mapping of the country. The Survey initially performed early stratigraphic studies of South African mineral deposits. Since the mid-1930s, the Survey has shifted its attention more to applied geology and to geophysical and geochemical investigations.[10]

As South Africa became an increasingly important exporter in the world mineral market, the need for economic analysis of the market increased. In response to this need, the Minerals Bureau was established in 1975. Some basic functions of the Bureau are the following:[2]

1. Mineral and material commodity analysis to determine world supply and demand
2. Assessment of coal reserves and resources and formulation of a national energy policy
3. International mineral intelligence studies, including geopolitical studies, to determine mineral suppliers and consumers of South African imports and exports
4. Studies to determine mineral reserves and resources of South Africa and other countries in southern Africa
5. Assessment of physical infrastructure requirements for South Africa's growing mineral export development, with particular emphasis on harbors and the rail system
6. Development of a national mineral policy
7. The collection, compilation, and preparation of official statistics for South Africa's mineral industry
8. Forecasting of factors that might affect the mineral industry; such factors include production, local sales, imports, exports, services, and geopolitical constraints

Statutory Organizations. The Council On Mineral Technology (CMT) was established in 1966 and is located in Randburg. The CMT is headed by the Board of Control, which reports to the Minister of Mineral and Energy Affairs, and is comprised of members from the Department of Mineral and Energy Affairs and the mining and chemical industries.

The Institute is engaged primarily in applied research and development relating to mineralogy and mineral and process chemistry. Consequently, areas of study include ore dressing work, hydrometallurgy, electrometallurgy, and pyrometallurgy. The Institute has its own analytical facilities and supports research in affiliated universities.[2,10]

The CMT has contributed substantially to the mining industry in several areas, and recent areas of research and development include the following:[2]

1. Increases in the efficiency of the extraction of gold from ores, conventional plant residues, and waste solutions
2. Development of a continuous countercurrent ion-exchange process for the extraction of gold and uranium slurries without the need for clarification of the feed
3. Research into the development and use of plasma-arc furnaces in the production of ferroalloys as an alternative to the conventional submerged-arc furnace
4. Improvement and development of processes and reagents used in the extraction of base metals from the sulfidic ores of the northwestern Cape, with particular emphasis on flotation and other separation techniques
5. A new and patented process for the separation of the minor platinum-group metals, and a process for the extraction of platinum-group metals from the UG2 layer underlying the Merensky Reef in the Bushveld Complex

The South African Atomic Energy Board, founded in 1948, exercises statutory control over all nuclear materials and radioactive isotopes. At the National Nuclear Research Centre at Pelindaba, the Board conducts applied research on nuclear power, nuclear raw materials, and radioactive isotopes.[10]

The Fuel Research Institute in Pretoria is responsible for the issuance of a grading certificate on all exported coal. This statutory organization is under the Department of Industries. Its primary function is the investigation of South Africa's fossil-fuel resources and research relating to fuels and by-products.

The Council for Scientific and Industrial Research (CSIR) is involved in research concerning rock mechanics, mine ventilation, rope-guided conveyances in mine shafts, metallurgy, and physiological studies of mine workers.[10]

Research Units. A number of research units are attached to South African universities. Each unit specializes in a particular field within the mineral industry. Financing for these units originates with public and private funds, and denotes close cooperation among the state, private industry, and universities.

The Economic Geology Research Unit at the University of Witwatersrand carries out basic research-related problems of mineral exploration in southern Africa, as well as acting as a synthesizer of geological, geophysical, and geochemical information garnered from various prospecting and mining activities. The Precambrian Research Unit in the Department of Mineralogy and Geology at the University of Cape Town is engaged in studies of Precambrian geology and geochemistry in the western part of South Africa. The Institute for Geological Research on the Bushveld Complex was established in Pretoria in 1975 to conduct and coordinate research on the Complex, to collect and store information on the Complex, and to provide information to the mining and exploration companies engaged in the Complex.[10]

The Energy Research Institute at the University of Cape Town conducts research and does consulting work on the efficient use of energy. The Institute for Energy Studies at the Rand Afrikaans University is instrumental in acting as a forum for the formation of a national energy policy, provides information on all aspects of energy, and undertakes research on the economics of energy in South Africa.[10]

Chamber of Mines. The Chamber of Mines of South Africa, an employer's organization, is a coop of mining companies that donates money to other research organizations. In addition, the organization carries on its own research programs through the following laboratories:

1. The Mining Technology Laboratory is engaged in the development of new stoping methods and machinery in gold mines. It is also undertaking the development of communication telemetry and control systems applicable in the mining industry.
2. The Mining Operations Laboratory is concerned with the actual application of mining methods dealing with strata control in coal and gold mines.
3. The Metallurgy Laboratory is presently concerned with developing underground concentration methods so as to use the residue as backfill in stopes. Other areas of interest are the development of a portable gold analyzer, mill control, water purification, and firefighting.
4. The Environmental Engineering Laboratory is concerned with the improvement of the thermal environment in deep mines, reduction of fire hazards in mines, physiological studies of workers in deep mines, and ventilation of mines.
5. The Human Resource Laboratory provides information on demand for, supply of, and effective utilization of human resources, including factors that affect the labor force.
6. The Coal Mining Laboratory investigates the application of machinery, mining methods, and management in coal-mining operations.

INFRASTRUCTURE

The South African mineral economy enjoys several advantages that other countries in Africa do not. Among the most visible of these advantages is the development of the physical infrastructure to support a mining economy. Certainly, a major contributor to the successful growth of South Africa's mineral export market has been the development of South Africa's rail system and harbors. South Africa has the natural advantage of being able to ship its goods west into the Americas and Europe, as well as eastward into the Japanese and East Asian markets.

To supplement this natural advantage, the South African government has consistently maintained a national mineral policy that has advocated the development of an infrastructure to support a mineral export economy. As a result of this policy, South Africa has not only some of the finest export facilities for minerals in the world but also has an extensive rail system interconnected with the country's major mining centers. The country's road system reflects a similar degree of sophistication, and in some cases, offers a visible alternative to rail transport.

An additional advantage that the South African export market maintains is the availability of adequate, reliable, and relatively inexpensive electrical power. However, it appears certain that the demand for even more electric power will increase. This demand for additional power will undoubtedly affect the fuel commodities that can best provide this power in South Africa—coal and uranium.

A major disadvantage to the South African mining industry is the shortage and geographical location of water resources within the country. The geography and climate have combined to produce a relatively dry western half, and a relatively wet and centrally located mountainous zone. The result is a shortage of water. South Africa has undertaken numerous water-related projects to increase the efficient management of the country's water resources.

Transportation

Not only are South Africa's railroads strategic to South African interests, but other countries in southern Africa also rely on the system for imports and exports. A number of countries are dependent on South Africa's rail system and harbors. Railroads for southern Africa are presented in Figure II-2-2.

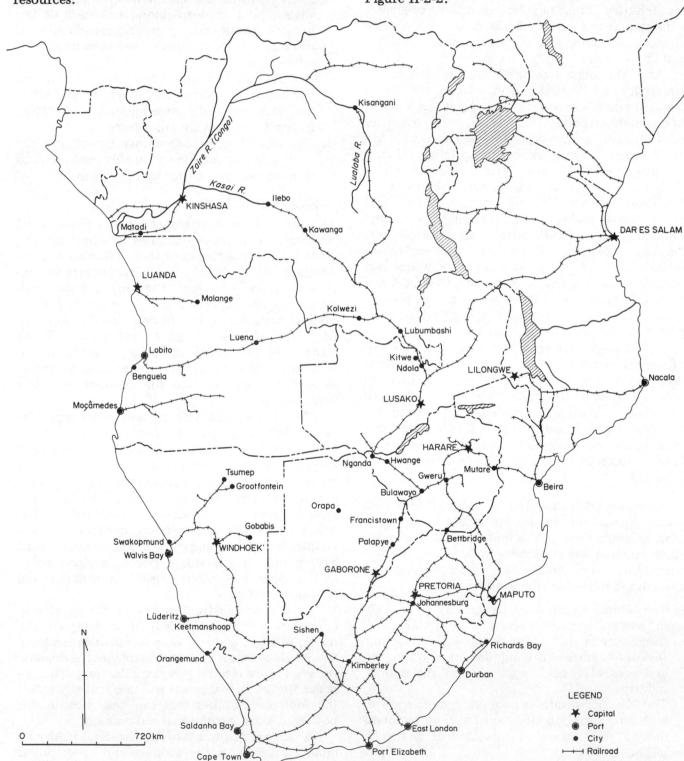

FIGURE II-2-2 Railroad system and major ports in southern Africa. (From National Geographic Society. Africa.)

Railroads. The contrast between the greater development of the South African rail system and other countries in the region can be seen in Table II-2-11. The total kilometers of track is nearly five times as large for South Africa's rail system as for that of its closest rival, Zaire. The amount of multiple track in South Africa is significantly greater than that of the other rail systems in South Africa.

The electrification of the rail line has proceeded to the extent that nearly 40% of the entire rail system is electrified.[2] Table II-2-12 lists the most important electrified lines.[2] There are advantages and disadvantages to the electrification of rail lines. It is true that electrification will lower transportation costs and mitigate South Africa's need for petroleum imports. However, the electrification of lines does increase their susceptibility to acts of sabotage. Electric, diesel, and steam lines are shown in Figure II-2-3.

A major portion of recent growth in the rail system is located in western South Africa. The rail system was expanded to accommodate the development of the base-mineral deposits in Northwestern Cape Province, and to facilitate the movement of raw materials for the steel industry. Additional work has been done to expand the rail system for the development of coal resources.

The rail and harbor system for South Africa (including the Namibia/South-West African rail and the harbors of Walvis Bay and Luderitz) is the responsibility of the South African Railways and Harbors (SAR&H) organization. The SAR&H is owned and controlled by the state under the Minister of Transport Affairs. The organization is controlled by a Railways and Harbors Board, with the Minister of Transport Affairs as chairman and three commissioners to offer advice.[2]

The railway network is divided into 10 geo-

TABLE II-2-11   Railroads in Southern Africa, 1981 (Kilometers)

Country	Total	1.067 gage	Other	Multiple track	Electrified
South Africa	22,560	21,854		5,292	5,000
Angola	3,189	2,879	310		
Botswana	726	726			
Lesotho	1.6	1.6			
Malawi	678	678			
Mozambique	3,436	3,288	148		
Namibia/South-West Africa	2,300	2,300			
Swaziland	292	292			
Zaire	4,859	3,573	1,286[a]		851
Zambia	2,014			13	
Zimbabwe	3,434	3,434		42	

[a]125 km, 1000 m; 136 km, 0.615 m; 1,025 km, 0.600 m.

Source: Ref. 1.

TABLE II-2-12   Major Electrified Lines of the South African Rail System

Province	In service as of 31 March 1980		Under construction or authorized	
	Route km[a]	Single track km[b]	Route km[a]	Single track km[b]
Cape	2,399.54	3,951.90	86.34	194.74
OFS	631.03	1,291.34	—	11.71
Natal	1,494.54	3,661.23	15.25	156.50
Transvaal	2,251.81	5,044.20	900.14	2,201.41
	6,776.92	13,948.67	1,001.73	2,564.36

[a]Route km, the distance from point A to point B.
[b]Track km, the total length of track between point A and point B (i.e., if the distance from A to B = 100 route km and there is a double railway line between these two points, the track km between A and B = 200 track km).

Source: Ref. 2.

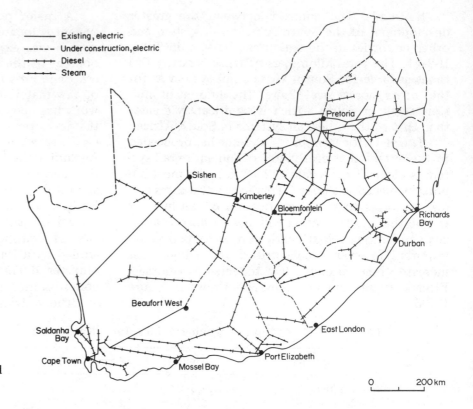

Existing, electric
Under construction, electric
Diesel
Steam

Pretoria
Sishen
Kimberley
Bloemfontein
Richards Bay
Durban
Beaufort West
Saldanha Bay
East London
Cape Town
Mossel Bay
Port Elizabeth

0   200 km

FIGURE II-2-3 South African rail and harbor system. (From Ref. 2.)

graphical areas or systems to facilitate local control under a system manager. Headquarters for these regions are in Johannesburg, Pretoria, Cape Town, Port Elizabeth, East London, Durban, Bloemfontein, Kimberley, Windhoek, and Saldanha.[2]

Harbors. There are six major harbors in South Africa. From east to west, these harbors are Richards Bay, Durban, East London, Port Elizabeth, Table Bay, and Saldanha. South Africa's proximity to one of the world's major shipping lanes and the country's mineral export trade have been major contributing factors to the excellence of the country's ports. Table II-2-13 indicates the number and type of ports in southern Africa.

Richards Bay, some 200 km north of Durban, is a recently built port and boasts the largest coal-exporting facility in the world. The Richards Bay Coal Terminal (RBCT) is the direct result of a

TABLE II-2-13 Ports in Southern Africa

Country	Number of ports	Type
South Africa	8	Major
Angola	3	Major
Mozambique	2	Significant
Tanzania	1	Minor
Namibia/ South-West Africa	2	Major
Zaire	1	Major

1970 contract between Japanese steelmakers and the Transvaal Coal Owners Association (TCOA) for an initial 27 million metric tons of blend coking coal.[11] While the TCOA built train-loading facilities in the Transvaal and ship-loading facilities at Richards Bay, the SAR&H administration built the port and railway infrastructure.

The coal terminal was commissioned 1 April 1976, with an initial capacity of 1 million metric tons of coal. A Phase II expansion project increased that capacity to 20 million metric tons/year. However, practical throughput is actually over 24 million metric tons/year.[11]

The RBCT is connected to 500 km of railway connecting the Natal and Transvaal coal fields to the terminal. As a result, coal from both the Transvaal and Natal coal fields has made the expansion of the RBCT necessary. Phase III is now in progress and will expand the capacity of the RBCT to 44 million metric tons in July 1985. Intermediate levels will be 28 million metric tons per year in 1982, 37 million in July 1983, and 40 million in January 1985.[11]

Presently, ship-loading capacity is 13,000 metric tons per hour with two ship loaders of 6500 metric tons per hour each. These will be upgraded to 8500 metric tons per hour and an additional ship loader with a 10,000-metric tons per hour rate will be added, bringing the capacity up to 27,000 metric tons per hour.[11] Richards Bay can accommodate vessels up to 150,000 dead weight tons (dwt). With dredging this can be increased to 350,000 dwt.

Another change in Phase III will affect the rail system. Presently coal is brought in unit trains of 90 cars with a payload of 5292 metric tons per train, with each car carrying 58.8 metric tons. The unit train is to be changed to 100 cars with each car carrying 84 metric tons. Before reaching Richards Bay, these trains will be doubled into 200 car units.

The expansion plan will also include facilities for the handling of steel, ferroalloys, and granite in Richards Bay Harbor. These facilities are not to be completed until 1985.

In addition to coal-loading facilities, there is a phosphoric acid pipeline with a loading capacity of 2000 tons per hour. The pipeline is owned by the fertilizer industry.

Durban, a land-locked bay, has an area of 1668 ha and a working depth of 12.8 m at Low Water Ordinary Spring Tide (LWOST), which means that any vessel in the South African trade can now enter and leave at any state of the tide. Total length of quayage available for commercial shipping is a little over 15,000 m.[2]

The facilities at Durban include a rail terminal equipped with two 35-ton rail/road transfer cranes, electrically driven appliances for the shipment of anthracite and coal, and privately owned bulk storage and handling facilities for manganese ore and alumina. There are also ship repair facilities, and a number of deep-water berths as well as berths for coastal vessels.[2]

East London is South Africa's only river port. The port has low-water depths ranging from 8.5 to 10.7 m and can accommodate tankers up to 204.2 m and a maximum loaded draft of 9.9 m.[2]

Port Elizabeth has a total of over 3400 m of quayage for commercial shipping, with a maximum depth of 12.2 m (LWOST). A 360-m wharf has been developed for the shipping of mineral ores. A mechanical ore-handling plant has a capacity of 1360 metric tons per hour.[2]

Table Bay is one of the most important ports in the southern hemisphere. It has commercial quayage of nearly 11,000 m and a maximum 14 m at low water. In addition, the port has extensive storage and handling facilities for grains and is the main export terminal for South Africa's fruit crops.

More important, the ship-repairing facilities at Sturrock graving dock are the largest in the southern hemisphere. With an extreme length of 360 m, width of 47.5 m, and a depth on the entrance sill of 13.7 m at high water, the facility is able to handle large oil tankers calling for repairs and bottom scraping.[2]

Saldanha Bay, some 110 km northwest of Cape Town, is the largest and deepest port on the west coast of Africa. With an area of nearly 5200 ha, the Bay is larger than the combined areas of South Africa's five other major ports. The present depth of 23.5 m (LWOST), with capacity for 250,000-ton ships, can be increased to 27.5 m to accommodate 350,000-ton ships.

Large-scale development of Saldanha Bay did not begin until 1973, at which time it was to be an ore export harbor for ISCOR's Sishen-Saldanha Iron Ore Export Project. Initial cost of the project was R600 million.[2] Soon after completion of the project in 1976, ISCOR began negotiations with SAR&H (now South African Transport Services) to take over the harbor area. However, on 1 July 1977, the SAR&H administration took complete control of the project, with the exception of the mine.[11]

The 861-km-long railroad line from Sishen in the northern Cape Province to Saldanha Bay is electrified at 50,000 V ac and uses 50-kV General Electric electric locomotives. The line is nearly straight for the whole distance and has 82 bridges and one 840-m-long tunnel near Elands Bay. The traffic is controlled by a single traffic controller in Saldanha using a computer-based system and microwave connections to the rest of the line.

Almost 14 million metric tons of iron ore was shipped through Saldanha Bay in 1981, and the amount for 1982 was over 15 million metric tons.[11] Of this amount, about 1 million metric tons per year originates from Associated Manganese, and the rest from ISCOR's mines.

The iron ore arrives in train units of 210 ore cars, with a total of about 18,000 metric tons per train. The trip from Sishen to Saldanha takes about 21 hours and the return trip about 18 hours. One of these trains arrives about every 8 hours, where it can be unloaded in about $3\frac{1}{2}$ hours, including shunting time.

Facilities for iron ore at the harbor are an area of 30 ha where about 3 million metric tons of iron ore can be stockpiled with a conveyor belt connecting the storage area with the ore dock, a computer-controlled ore-wagon tipper station with a capacity of 8000 metric tons per hour to empty train cars, and two shiploaders with tripper trailers at the two loading berths.[2,11]

Because of its deep-water capacity, ships carrying iron ore from Brazil to Japan and from Australia to Europe stop at Saldanha and have their cargoes topped off with South African iron ore.[11]

In addition to the export of iron ore, Saldanha Bay exports approximately 165,000 metric tons per year of lead, copper, and zinc concentrates from Black Mountain at Aggeneys. The concentrates arrive usually once a week in either 30 normal 39-metric ton capacity South African Railway cars (zinc), or in 60-car trains with 52 metric tons capacity (lead and copper) which have rubber seals and are closed to prevent contamination of grazing pastures along the right-of-way.

The export facilities for the concentrates are separate from the iron ore facilities. The berth and approach channel will accommodate vessels up to 30,000 dwt. However, dredging is expected to increase this capacity to 50,000 dwt.[2] A shed for storage has a 34,000-metric ton capacity. The concentrate loading dock is equipped with two 15-metric ton capacity wharf cranes with a loading rate of 270 metric tons per hour.

Roads. The road system in South Africa is more developed than that of any other country in southern Africa, as can be seen from the regional map in Figure II-2-4. Table II-2-14 clearly indicates the superior development of the South African highway system.[1] As indicated by Table II-2-14, the percentage of paved roads in South Africa (35%) is nearly twice the percentage of any other road system in southern Africa.

Paved roads are a more reliable alternative to a rail system than unimproved roads, because they are not as susceptible to closure under adverse weather conditions. The extensive road system also means that more major highways would have to be destroyed to interrupt the flow of road traffic.

Water Resources

South Africa's water resources are unevenly divided due to the topography of the country and its seasonal rainfall. The country is also within a drought belt.

It is estimated that only 3100 million cubic meters of water a year can be captured from storage dams from a flow of 52,000 million cubic meters of water for all of South Africa's rivers. An additional 1100 million cubic meters is presently recovered from groundwater. However, if the demand for water continues at its 7% annual increase, by the end of the century demand for water is estimated to be 29,300 million cubic meters a year.[2] Because of the uneven distribution of water supplies, it is expected that available water supplies will be insufficient and could possibly affect industrialization within the country.

As a result of this shortage, South Africa's water authorities have given priority to the following three solutions to solve the water shortage. These are development and utilization of natural water resources, improved efficiency in the use of developed supplies, and creation of new sources of supply (especially the desalination of seawater).[2]

The Directorate of Water Affairs, under the Department of Water Affairs, Forestry and Environmental Conservation, is responsible for all matters relating to water in South Africa. Major responsibilities of the Department include the following:[2]

1. Collecting and collating of hydrological data
2. Planning the best use of available water supplies
3. Providing detailed designs and working plans, specifications, contracts, and so on, necessary for construction of a project
4. Management, operation, and maintenance of government irrigation projects and regional water supply projects
5. Providing necessary engineering services for small water projects for farmers and local authorities
6. Drilling for underground water supplies for farms and small authorities

There are also a number of local water boards, which are corporate bodies that work in close cooperation with the Directorate of Water Affairs.

To enhance the country's water resources, several large water projects have been undertaken and are in various stages of development. The Orange River Project (ORP), in central South Africa's mountainous region, is the largest water project in South Africa. Exclusive of its main tributary (the Vaal River), the Orange River drains approximately 412,000 km[2] of South Africa. Its estimated flow of 7500 million cubic meters a year is approximately 15% of the country's water drainage. Some 7300 million cubic meters of this flow is captured by two reservoirs created by the Hendrick Verwoerd Dam and the P. K. le Roux Dam.[2]

The Verwoerd Dam was completed during 1971 with a gross storage capacity of 5958 million cubic meters and covers an area of 37,400 ha. The hydroelectric capacity of the dam is 320 MW. Water released from the dam travels 150 km downstream into the P. K. le Roux Dam, which opened in November 1977. The storage capacity of the reservoir is 3185 million cubic meters and has an installed hydroelectric capacity of 220 MW.

Another feature of the Hendrik Verwoerd Dam is an outlet from the base of the dam into the world's longest continuous water tunnel. The tunnel is 82.5 km long and has a diameter of 5.35 m. The tunnel transfers water from the reservoir to the Fish River valleys. Another project to be completed in the 1980s will transfer water from the Fish River valley to the Sundays River valley.

In another project being undertaken by Escom and the Directorate of Water Affairs, 196.7 mil-

FIGURE II-2-4 Road system in southern Africa. (From Michelin, Afrique Centrale et Sud-Madagascar.)

lion cubic meters of water will be transferred from the Tugela River to the Vaal River basin. The project is expected to produce 1000 MW of peak power.

The Boland project deals with the Riviersond-erend, Eerste, and Berg Rivers in western Cape. The Umgeni River supply project will furnish water to the industrial area along the Durban-Pietermaritz-burg axis and adjoining metropolitan areas.

TABLE II-2-14 Highways in Southern Africa, 1981 (Kilometers)

Country	Total	Paved	(Percent)	Crushed stone, gravel, or stabilized soil	(Percent)	Improved earth	(Percent)	Unimproved	(Percent)
South Africa	225,389	79,902	35.5	145,487	64.5				
Angola	73,828	8,577	11.6	28,723[a]				36,528	49.5
Botswana	10,784	800	7.4	1,540	14.3	5,407	50.0	3,037	28.2
Lesotho	3,916	250	6.4	1,200	30.6	946	24.2	1,520	38.8
Malawi	13,249	2,372	17.9	381	2.9	8,560	64.6	1,936	14.6
Mozambique	26,477	4,322	16.3			607	2.3	21,548	81.4
Namibia/South-West Africa	23,800	3,800	11.2	(Remainder gravel, earthrods, and tracks)					
Swaziland	2,805	390	13.9	1,158	41.3	1,257	44.8		
Zaire	168,979	2,654	1.6			140,979	83.4	25,346	15.0
Zambia	36,402	5,488	15.1	7,817	21.5	23,097[b]			
Zimbabwe	78,491	8,058	10.3	32,855[a]				37,578	47.9

[a] Also includes improved earth.
[b] Also includes unimproved earth.

Source: Ref. 1.

The Usutu River project in eastern South Africa will deliver water to the Camden power station (1600 MW), and from there to the Kriel power station (3000 MW). The building of Grootdraai Dam and pump stations near Standerton will deliver 44 million cubic meters of cooling water per year to Sasol II, and provide an additional 345 million cubic meters of cooling water per year in the coal-fields area of the eastern highveld of the Transvaal.[2]

Energy

The Department of Mineral and Energy Affairs was created in 1980, assuming the responsibility for all governmental activities relating to the acquisition, distribution, and policy formulations of the central government. Important components of the Department are Sasol, the Electric Supply Commission (Escom), the Atomic Energy Board (AEB), and the Uranium Enrichment Corporation (Ucor).

Present and Future Energy Consumption. South Africa consumes more than half of all the energy consumed in Africa. In 1980, the primary energy consumption was estimated at 2628 million gigajoules, the equivalent of 90 million tons of coal at 29.3 MJ/kg. After conversion losses, it is estimated that final energy consumption was 1540 million gigajoules. Consumption by sector in 1980 was as follows: industrial sector (49%), transport sector (30.3%), households (10.7%), agriculture (4%), and mining industry (6%).[2]

Electricity. Electricity is estimated to supply 20% of the total net energy for South Africa, with an expected increase to 40% by the end of the century. New electrical generating capacity has grown at an annual rate of 8% for the past decade.[10] To meet this anticipated demand, South Africa will be required to build at least one new 3600-MW power station every year by the 1990s.[12]

Currently, South Africa's electrical generating capacity is about 18,000 MW. A regional map of existing electrical generating facilities is given in Figure II-2-5. While most of the countries in southern Africa rely on hydroelectric power for the generation of electricity, the drier climate and terrain in southern Africa preclude this possibility. Figure II-2-6 indicates regional hydroelectric power facilities and rivers.

Electricity in Southern Africa. The Electricity Supply Commission (ESCOM) of South Africa supplies 93% of the electricity in the country, with the municipalities of Johannesburg, Cape Town, Pretoria, Port Elizabeth, and Bloemfontein providing the remainder. ESCOM was established by the Electricity Act, 1922 (Act 42 of 1922).[2] ESCOM is divided into six regional districts with head offices in Cape Town, East London, Durban, Witbank, Johannesburg, and Kimberly.

Electricity is transmitted over a national grid system as indicated in Figure II-2-7.[10] Major overhead transmission lines include 1030-km 553-kV (monopolar dc) lines; 965-km 220-kV, 12,421-km 132 to 165-kV, and 82,892-km 88-kV and lower-voltage lines.[2] In addition, some 5000 km of railroad lines is electrified. Major power stations in South Africa are listed in Table II-2-15.

With this national grid system, electricity can be supplied from the lowest-cost stations. Eventually, the thermal stations in the Cape Western distribution system will be used only for emergencies or peak-load periods. The added cost of shipping coal to these stations makes them higher-cost producers than stations located close to coal fields. The national control center is located at Simmerpan.

ENERGY MINERALS

The primary energy source for South Africa is coal, which provides roughly 80% of the country's primary energy needs. South Africa is also a major producer of uranium oxide (U_3O_8). The use of uranium as a major source of energy will depend on the successful growth of the country's nuclear industry and the relative cost between coal-produced electricity and nuclear-produced electricity. South Africa does not have much oil or natural gas, although some reserves have recently been discovered.

A comparison between South African energy minerals and the energy minerals of other countries in southern Africa can be seen in Figure II-2-8. Although a few of the other countries possess exploitable coal resources, none have developed these resources to the extent that South Africa has. However, other countries in southern Africa have the options of hydroelectric power and/or oil and natural gas. No other country in the region requires as much power as South Africa.

Coal

The successful exploitation of South Africa's coal is due to several contributing factors. The first factor is the lack of indigenous oil and natural gas, as well as the lack of substantial hydroelectric capacity. The second factor is the relative ease of mining coal (much of the coal can be strip-mined), and the fact that the country has substantial resources of exploitable coal. Over half of in-situ coal resources are at depths of less than 100 m.[2]

Coal plays a significant role in the economy of South Africa. Besides being the major source of

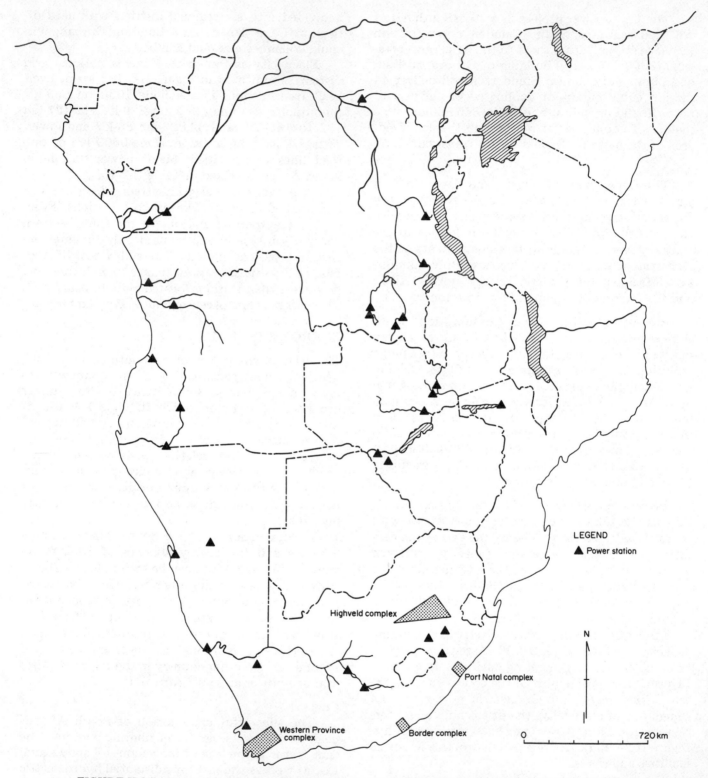

FIGURE II-2-5   Electrical generating facilities greater than 100 MW in southern Africa. (From Ref. 10.)

energy in South Africa, coal is the second largest revenue earner of all South African minerals. This dual role may mark coal as South Africa's most important mineral commodity, because without

energy the mining and transportation of other mineral commodities would be very difficult.

Coal supplies nearly 80% of South Africa's energy requirements.[2] In 1981, 60% of all domes-

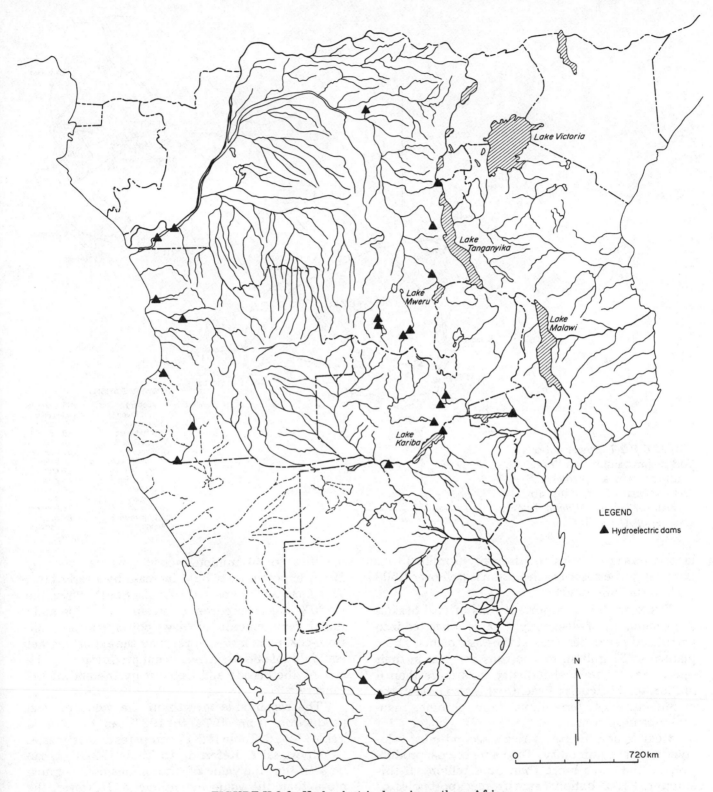

FIGURE II-2-6 Hydroelectric dams in southern Africa.

tic coal sales were for coal-fired stations.[6] In the same year, 40% of all domestic coal sales were for pit-head power stations, or those with captive collieries.[2] This demand will probably increase, as plans to expand and create new power stations are well under way. In fact, there are four power stations under construction that will require 10 to 13 million metric tons per year of coal. South Africa's

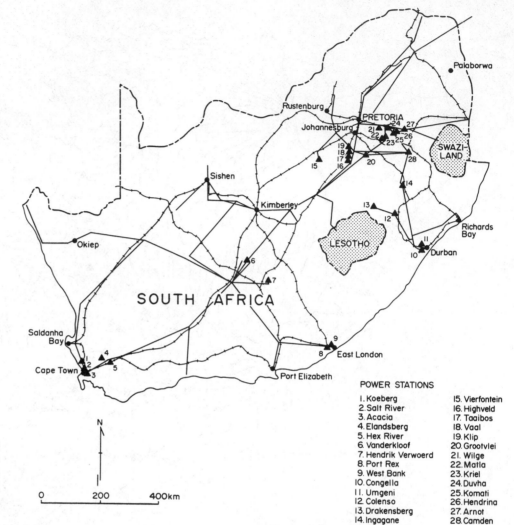

POWER STATIONS

1. Koeberg	15. Vierfontein
2. Salt River	16. Highveld
3. Acacia	17. Taaibos
4. Elandsberg	18. Vaal
5. Hex River	19. Klip
6. Vanderkloof	20. Grootvlei
7. Hendrik Verwoerd	21. Wilge
8. Port Rex	22. Matla
9. West Bank	23. Kriel
10. Congella	24. Duvha
11. Umgeni	25. Komati
12. Colenso	26. Hendrina
13. Drakensberg	27. Arnot
14. Ingagane	28. Camden

FIGURE II-2-7 Power stations and major transmission lines in southern Africa. (From U.S. Department of the Interior, Bureau of Mines. *Mineral Industries of Africa*; Ref. 10.)

largest power stations require only 6 million tons per year.[2] The cost of each station is well over R500 million (in 1980 rand).

There are other important uses of coal besides the generation of electricity. Coal is converted into petrol and petrochemicals at the Sasol plants. These plants use 32 million metric tons per year in their operations.[2] The steel industry uses more than 6 million tons of coal a year. Coal is also used in the manufacture of ferroalloys from chrome, manganese, and silicon.

Coal is now South Africa's second most valuable mineral commodity. The value of coal production in 1981 was more than $2.4 billion. Of this amount, $1.12 billion came from export sales of 30 million metric tons, and $1.30 billion from domestic sales of 98 million metric tons. The average price for exported steam coal was R32 per ton in 1981. The average price for domestic steam coal was R11.6 per ton in 1981.

With an increase in Richards Bay coal export

facilities to 44 million metric tons per year by 1986, export sales should increase by nearly 50%. This increase is expected to make South Africa the world's largest exporter of steam coal.[2] The additional requirements of new power stations, oil-from-coal plants, steelworks, and the export market could double South African coal production by the end of the decade and treble it by the end of the century.[2]

The remarkable growth in the value of coal production from 1974 to 1981 can be seen in Table II-2-16. Table II-2-17 compares domestic sales to export sales. Referring to Table II-2-16, it can be seen that the value of bituminous coal is much more than the value of anthracite. However, the value of both has grown at a rapid rate. In 1970, the value of coal exports was only 9%, whereas in 1981, the value of coal exports was 46%. The rapid increase in export sales from 1976 onward resulted in the opening and continued expansion of the Richards Bay coal terminal. If coal production

TABLE II-2-15 Major Power Stations in South Africa

Type	Complex	No.	Plant	Capacity (MW)
Coal fired		1	Drakensburg	100
		2	Ingagane	500
		3	Colenso	100
			Tutuka	3600
	(Planning stages)		Lethabo	3600
			Matimba	3600
	Highveld	4	Vierfontein	360
		5	Highveld	480
		6	Klip	420
		7	Wilge	240
		8	Matla	3600
		9	Kriel	3000
		10	Camdem	1600
		11	Taaibus	480
		12	Vaal	310
		13	Grootvlei	1200
		14	Duvha	3600
		15	Arnot	2100
		16	Komati	1000
		17	Hendrina	2000
	Port Natal	18	Umgeni	240
		19	Congella	100
	Border	20	West Bank	100
	Western Province	21	Elandsberg	1000
		22	Hex River	120
		23	Salt River	300
Gas turbine	Border	24	Port Rex	170
	Western Province	25	Acacia	170
Hydroelectric		26	Vanderkloof	220
		27	Hendrick Verwoerd	320
	Port Natal	28	Drakensburg Project[a,b]	1000
Nuclear	Western Province	29	Koeburg[a]	1800

[a]Under construction.
[b]Located in the Tugela River Basin above Durban.
Source: Ref. 10.

does treble by the end of the century, coal revenues will be an even more significant part of South Africa's mineral economy.

Geology. Coal formation in South Africa occurred during the Permian period. Coal was preserved in the intercratonic Karoo Basin. During the period preceding the Karoo system, South Africa was part of Gondwanaland, which had migrated across the South Pole. As the ice sheets melted, the vast shallow Ecca Sea was established, accumulating sediments from Early to Middle Permian time in the Karoo Basin. Orogenic activity during the Permo-Triassic Beaufort period created a series of fresh-water lakes of diminishing size. During another northward fluvial influx, orogenic activity char-acterized deposition of the Triassic Molteno Formation.[13] Table II-2-18 presents a simplified stratigraphy of the important Karoo coal-bearing seams.[8]

Coal Provinces and Coal Fields. The traditional coal fields are in northern Natal, the northern Orange Free State, the eastern Transvaal highveld, and the South Rand. Scattered coal fields are also located in the northeastern Cape Province and in the central and northern Transvaal.

Generally, the rank (measure of the degree of maturity or metamorphism) of coal increases from west to east. Coal in the west is higher in moisture and volatile content. Taken as a whole, South African coal is higher in ash, relatively low in

FIGURE II-2-8 Energy minerals in southern Africa. (From U.S. Department of the Interior, Bureau of Mines, Africa, August 1976.)

TABLE II-2-16 Value of Coal Sales in South Africa, 1974–1981 (Thousand U.S. Dollars)

| | Anthracite | | | | Bituminous | | | | Grand total | |
Year	Domestic	Export	Total	Percent growth per year	Domestic	Export	Total	Percent growth per year	Anthracite and bituminous	Percent growth per year
1974	7,448	16,452	23,900		254,886	15,385	270,271		294,171	
1975	13,223	24,846	38,069	59.3	367,767	26,186	393,953	45.7	432,022	46.9
1976	12,854	34,395	47,249	24.1	463,658	84,546	548,204	39.2	595,453	37.8
1977	13,207	56,691	69,898	47.9	569,676	229,143	798,819	45.7	868,717	45.9
1978	14,648	63,733	78,381	12.1	617,144	310,090	927,234	16.0	1,005,620	15.8
1979	25,599	93,736	117,335	49.7	131,202	512,077	1,243,279	34.1	1,360,614	35.3
1980	29,664	129,343	159,007	35.5	1,007,560	755,126	1,762,686	41.8	1,921,693	41.2
1981	31,691	194,137	225,828	42.0	1,273,053	928,419	2,201,472	24.9	2,427,300	26.3

Source: U.S. Bureau of Mines, Minerals Yearbook, 1974–1981.

TABLE II-2-17   Domestic Sales versus Export Sales for Coal, 1970–1981
(Thousand U.S. Dollars)

Year	Domestic	Percent of total	Export	Percent of total	Total
1970	139,544	91	14,335	9	153,889
1971	153,340	92	13,788	8	167,128
1972	153,559	94	10,370	6	163,929
1973	199,640	91	20,153	9	219,793
1974	262,334	89	31,837	11	294,171
1975	380,990	88	51,032	12	432,022
1976	476,512	80	118,941	20	595,453
1977	582,883	67	285,834	33	868,717
1978	631,792	63	373,828	37	1,005,620
1979	754,801	55	605,813	45	1,360,614
1980	1,037,224	54	884,469	46	1,921,693
1981	1,304,744	54	1,122,556	46	2,427,300

Source: U.S. Bureau of Mines, Minerals Yearbook, 1970–1981.

TABLE II-2-18   Karoo Stratigraphy in
South African Coal Fields

Stormberg Series	Drakensburg Stage
	Cave Sandstone Stage
	Red Beds Stage
	Molteno Stage
Beauford Series	Upper Stage
	Middle Stage
	Lower Stage
Ecca Series	Upper Stage
	Middle Stage
	Lower Stage
Dwyka Series	Upper Stage
	Tillite Stage

Source: Ref. 8.

calorific value, and physically harder than coal in the northern hemisphere.[11] The coal is also lower in sulfur content than are American and European coals.

The major coal fields of South Africa are grouped in Table II-2-19 according to their stratigraphic sequence. The map in Figure II-2-9 shows the geographical location of these various coal fields.

South African coking coals for the metallurgical industries are medium to below medium in quality. Coking coals for blending are found in the Vryheid, Klip River, Waterberg, Soutpansberg, and Limpopo coal fields.[8]

Production. Over 73% of coal in South Africa is mined by the board-and-pillar method. Unfortunately, much of the coal is left in pillars to support the roof. Stooping (extraction of pillars) and long-wall mining are increasingly used as the value

of coal increases. Approximately 30% of the coal is mined by open-cast (pit) mining using giant draglines.[2] By the mid-1980s nearly 35% of the coal will be mined by this method.[11]

Production figures for coal and coke are given in Table II-2-20. For the period 1970–1981, production of anthracite, bituminous coal, and coking coal have increased 139%, 139%, and 71%, respectively.

There are three significant coal-producing companies in South Africa. The AMCOAL group, owned by the giant Anglo-American Corporation, produces some 34 million metric tons per year (26% of the nation's coal output) from 14 operating collieries.[12] AMCOAL's total resources are estimated at 11 to 12 billion metric tons of run-of-mine coal. AMCOAL's coal-export commitments were 7.8 million metric tons per year in 1981. This amount is expected to rise to 12.8 million metric tons per year by 1986.

The General Mining Union Corporation (GENCOR) is South Africa's second-largest coal producer. Its subsidiary, the Trans-Natal Coal Corporation Ltd., owns nine collieries in the Transvaal and three in Natal. Another subsidiary, Clydesdale (Tvl) Collieries Ltd., owns an additional colliery in the Transvaal and Orange Free State.[11] GENCOR exports 1.25 million metric tons per year. Projects are under way to increase this amount by 4.75 million metric tons per year of bituminous coal and 1.5 million metric tons per year of anthracite. These increased exports are with Phase III of the Richards Bay coal terminal.

A few of the major collieries are listed in Table II-2-21. The Bosjesspruit colliery at Secunda is the largest colliery in South Africa, supporting

TABLE II-2-19   Coal Fields in South Africa

Stage	Series	Province	Coal field
Ecca	Upper Ecca	Transvaal	Springbok Flats
			Waterburg
			Soutpansberg
			Limpopo
			Tshipise-Pafuri
			Lebombo Range
	Middle Ecca	Transvaal	Witbank
			Highveld
		Eastern	Carolina-Wakerstroom
		Transvaal-Natal	Utrecht
			Vryheid-Paulpietersberg
			Newcastle-Dundee
			Colenso-Ladysmith
		Orange Free State	Odendaalsrus-Virginia
			Vierfontein
		Orange Free State-	Vereeniging-Sasolburg
		Transvaal	South Rand
		Northern Transvaal	Waterburg
			Soutpansburg
			Swaziland
			Nongoma-Lowesburg-
			Maguda Basin
Stormberg	Molteno	Northeastern Cape	Molteno-Indwe

Source: Ref. 8.

both the Sasol II and Sasol III facilities. The list of collieries in South Africa is incomplete, and production figures may not be accurate. The Bosjesspruit colliery at Secunda in the Transvaal Province is the largest underground coal mine in the world, having achieved a production rate of 1 million metric tons per month in July 1981.[12]

Other significant developments in production of coal in 1981 occurred at the Kriel, Matla, and Sigma collieries. Kriel produced a record 207,393 metric tons in August for a conventional mechanized mine. Matla produced a record 118,794 metric tons in May 1981 for a conventional mechanized drum miner. The Sigma colliery achieved a world record for long-wall mining of 210,000 metric tons in October 1981.[12]

Sasol. The South African Coal, Oil and Gas Corporation was established in 1950, with shareholding by the government via the Industrial Development Corporation. Construction of Sasol I started in 1952, and the first synthetics products were produced in 1955.[2] Sasol I is located some 40 km south of Johannesburg. After the oil crisis in 1973, the government requested a feasibility study on the building of an additional Sasol plant. The approval for an additional plant was given in December 1974, and the site at Secunda was chosen

in May 1975. Site preparation for Sasol II started in July 1976, with production commencing in 1980.[14] Full production was achieved in 1982.[2] Following the Iranian oil crisis in 1979, the South African government decided to build Sasol III. Sasol III reached full production in 1984.[11] The cost of Sasol II was R2503 million,[2] and the estimated cost of Sasol III is R3276 million, excluding infrastructure. Sasol II and III are approximately 1 km apart and are located in Secunda, 150 km southeast of Johannesburg in the highveld.

A schematic for the production of petro and petrochemical products is presented in Figure II-2-10. Using high-pressure Lurgi gasifiers, a gas, composed of hydrogen and carbon monoxide, is produced. At full production Sasol II and III will produce over 2 billion cubic feet of gas daily in 72 Lurgi Mark 4 gasifiers.[11] Water, heavy hydrocarbons, and other components are then precipitated by cooling. The purified raw gas is introduced into the Fischer-Tropsch units with a gas consisting of hydrocarbons and oxygenated chemicals. The hydrocarbon liquid is then refined using conventional methods. The other products are chemically treated to produce alcohols and ketones. The uncondensed methane is recovered as a substitute natural gas or reintroduced into the synthol unit as hydrogen and carbon monoxide. Tar and petrol units are utilized to up-

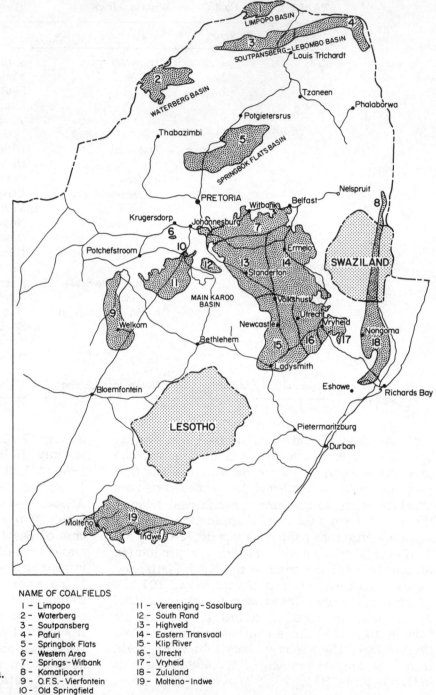

NAME OF COALFIELDS

1 – Limpopo	11 – Vereeniging - Sasolburg
2 – Waterberg	12 – South Rand
3 – Soutpansberg	13 – Highveld
4 – Pafuri	14 – Eastern Transvaal
5 – Springbok Flats	15 – Klip River
6 – Western Area	16 – Utrecht
7 – Springs - Witbank	17 – Vryheid
8 – Komatipoort	18 – Zululand
9 – O.F.S. - Vierfontein	19 – Molteno - Indwe
10 – Old Springfield	

FIGURE II-2-9 Coal fields in South Africa. (From Ref. 11.)

grade petrol and diesel components.[11] Altogether, more than 27 different fuel and chemical products are produced. The most important of these are gasoline, diesel fuel, jet fuel, motor oil, and fertilizer products. Completion of Sasol III will make South Africa nearly self-sufficient with regard to petro products. A decision will be reached as to whether other Sasol units will be required for future oil-from-coal demands.

Reserves. South African coal reserves were estimated at approximately 82,000 million metric tons in 1975, with only 744 million metric tons being anthracite and the rest bituminous coal. Of total reserves only 25,000 million metric tons were considered economically extractable.[2] However, new estimates by the Minerals Bureau have restated in-situ coal reserves at 110 billion metric tons and commercially extractable reserves at 51 billion

TABLE II-2-20 Coal and Coke Production in South Africa, 1970–1981 (Thousand Metric Tons)

	Coal			Coke		
Year	Anthracite	Bituminous	Total	Coke oven and beehive	Gashouse: low and medium temperatures[a]	Total
1970	1,678	52,934	54,612	3,184	100e	3,284
1971	1,841	56,825	58,666	3,592	100e	3,692
1972	1,336	57,104	58,440	3,583	101	3,684
1973	1,408	60,944	62,352	3,593	99	3,692
1974	1,435	64,621	66,056	3,600	100	3,700
1975	1,591	67,849	69,440	4,443	105	4,548
1976	2,459	74,600	77,059	4,608	110e	4,718
1977	2,559	82,852	85,411	5,201	110e	5,311
1978	2,150	88,208	90,358	4,869	110e	4,979
1979	3,309	100,459	103,768	4,951	100e	5,061
1980	3,895	111,225	115,120	5,377	94	5,471
1981	4,017	126,362	130,379	5,500	100	5,600

[a]e, estimated.

Source: U.S. Bureau of Mines, Minerals Yearbook, 1970–1981.

TABLE II-2-21 Major Collieries in South Africa

Company	Colliery	Province — magisterial district	Production (million metric tons/year)
AMCOAL	Goedehoop	Transvaal — Witbank	3
	Kleinkopje	Transvaal — Witbank	2.4 (coke)
	Kriel	Transvaal — Bethal	8.5
	New Denmark	OFS — Sasolburg	10.0 (mid-1980s)
	New Vaal	Transvaal	6.5 (1985)
GENCOR	Coalbrook	OFS	4.4
	Optimum	Transvaal — Middelburg	3.5
	Ermelo	Transvaal — Ermelo	3.0 (for export)
ISCOR	Durnacol	Natal — Dannhauser	1.3
	Grootegeluk	Transvaal — Waterburg	1.8 (coking)
			8.1 (steam) (1985)
ESCOM	Matla	Transvaal — Bethal	9.5
Rand Mines	Duhua	Transvaal — Middelburg	10.4
	Rietspruit	Transvaal — Witbank	9.0
Sasol I	Sigma	OFS — Sasolburg	6.0
Sasol II and III	Bosjesspruit	Transvaal — Secunda	30.0 (1985)

Sources: Refs. 6, 15, and 26.

metric tons.[6] It is expected that as the value of coal increases, these reserve figures may also increase.

Exports. Exports of South African coal have increased dramatically in the last decade. Figures for the export of anthracite and bituminous coal are given in Table II-2-22. The rapid increase in coal exports since 1976 can be attributed to the development of the Richards Bay coal terminal, which by 1986 will have an annual export capacity of 44 million metric tons of coal.

South Africa's main customers are European countries (especially France and Italy) and Japan. In February 1982, South Africa became Japan's main supplier of thermal coal. At 12,141 yen per ton in 1981, South African coal was well below 14,386 yen per ton for Canadian coal and 17,235 yen per ton for Australian coal.

Shell, British Petroleum, and Total (the local

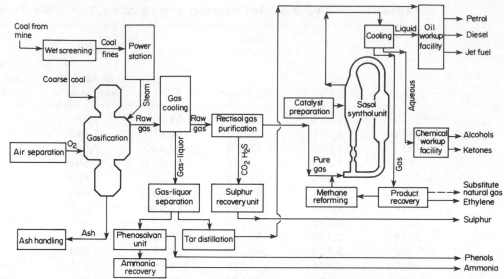

FIGURE II-2-10 Schematic for Sasol Synthol oil-from-coal process. (From Ref. 11.)

TABLE II-2-22 Exports of South African Coal, 1970–1981 (Thousand Metric Tons)

Year	Anthracite	Bituminous	Total
1970	1,203	659	1,862
1971	1,167	764	1,931
1972	1,168	323	1,419
1973	905	1,039	1,944
1974	1,035	1,245	2,280
1975	1,122	1,565	2,687
1976	1,530	4,431	5,961
1977	2,224	10,478	12,702
1978	2,528	12,866	15,389
1979			23,341
1980	3,409	25,741	29,150
1981	4,289	25,592	29,881

Source: U.S. Bureau of Mines, *Minerals Yearbook*, 1970–1981.

subsidiary of France's CFP) have been given export allocations of up to 13.5 million metric tons per year in exchange for agreeing to furnish South Africa with adequate supplies of crude petroleum. These companies have established partnerships with recently established mines and are actively prospecting in the eastern Transvaal and northern Orange Free State.[11]

Outlook. For the long term, the demand for South African coal looks very promising. A strong demand for domestic coal supplies is assured by the continued expansion of coal-fired power stations. Coal exports are expected to continue to grow as more countries switch to coal-fired power stations and away from expensive and unpredictable oil supplies. Looking to the future to assure an adequate domestic supply of coal, the Minister of Mineral and Energy Affairs has announced an 80-million metric ton per year export ceiling for coal.[6]

Uranium

The great majority of uranium produced in South Africa is as a by-product of gold producing mines in the Witwatersrand Basin. There are very few mines with uranium as the primary commodity, although this could change with a substantial increase in demand for uranium with a concurrent rise in price. However, the production of uranium remains very dependent on the price of gold.

South Africa is a major exporter of uranium to the world market, and exports account for the lion's share of revenues. Figures for domestic sales are not available, and given the possible military uses of uranium coupled with South African politics and world opinion, they are not likely to become available.

Other than possible military applications, domestic use of uranium for nuclear power plants will likely dominate future uranium use. Currently, the only nuclear power plant being constructed in South Africa is located near the west coast, north of Cape Town, in Duynefontein. The Koeberg power plant will consist of two 922-MW pressurized-water reactor units. Both plants were to be completed at the end of 1982 and early 1983.[2] However, various delays, including sabotage of the plant site, have delayed completion of the units. By the end of the century, South Africa is expected to generate some 15,000 MW of electricity from nuclear power.[10] South Africa's only uranium enrichment plant is located at Valindaba, near Pretoria.

Geology. The main uranium-bearing mineral in South Africa is uraninite. Both gold and uraninite are thought to have been mechanically deposited with pebbles as detrital particles. During the time of deposition, the atmosphere is believed to have been an oxygen-poor mixture, which explains why uranium oxide minerals do not dominate in South Africa.

Although the greatest reserves of uranium (along with gold) are located in the quartz-pebble conglomerates of the Dominion Reef, Witwatersrand, and Ventersdorp Systems of the Witwatersrand Basin, there are several other sources for uranium production or possible uranium production within South Africa. One of these other sources is the uranothorianite found in the phoscorite and carbonatite of the Palabora Complex. A possibility for future exploitation of uranium exists from the sandstone of the Lower Beaufort Stage of the Karoo System[8] and uraniferous coal in the Springbok Flats of the northern Transvaal.[2]

Production. Production of uranium oxide (U_3O_8) has increased steadily since 1970, although production has fallen off since the peak in 1980. Production figures are given in Table II-2-23, as well as export sales for 1975–1981.

Major uranium producers for 1981 are given in Table II-2-24. The Beisa mine, opened in 1981, is the only mine that produces uranium as the primary commodity and gold as the by-product. Ergo and the OFS Joint Metallurgical Scheme are recovering gold, uranium, and sulfuric acid from the old slime dams left from previous gold production.

Currently, South Africa is stockpiling uranium-bearing slimes for future treatment. The storage of the slimes is considered cheaper than treating the slimes and then storing the concentrate produced.[12]

The National Institute of Metallurgy has recently developed improved contractors for use in the extraction of uranium by the continuous-ion-exchange (CIX) method. In the last four years six new plants have been commissioned using this method.[2]

Almost all of the country's uranium is processed and marketed by the Nuclear Fuels Corporation of South Africa (Pty.) Ltd. (Nufcor). Nufcor is a privately owned company whose shares are owned by the various gold mines that produce uranium.

Reserves. Since uranium is a by-product of gold, reserves of uranium depend largely on the price of gold. The higher the price of gold, the longer the gold mines will remain open, and thus the amount of uranium likely to be recovered increases.

Uranium reserves for major producers are given in Table II-2-25. Additional resources do exist but will depend on the future demand and price of uranium.

Outlook. Future demand for uranium will depend primarily on the future of foreign nuclear power generating stations. Presently, South Africa exports to Japan, Taiwan, and European countries. However, because uranium is recovered as a by-product of gold mining, production should remain steady.

Production of uranium was expected to be as follows: 16,500,000 lb of U_3O_8 in 1982; 17,600,000 lb in 1983; 18,400,000 lb in 1984; 18,300,000 lb in 1985; 18,100,000 in 1987; and 17,800,000 lb in 1988.[12] If U.S. uranium production continues to decline at its present rate, South Africa could become the Western world's second-largest producer by the end of the decade.[12]

Petroleum and Natural Gas

South Africa has no significant land-based deposits of petroleum or natural gas and imports whatever quantity of products cannot be met by Sasol's collective production. There are a few producing wells offshore, and in recent years, exploration efforts have increased offshore. In 1979, after years of exploration, the land search for economic deposits of petroleum and/or natural gas was terminated by the Southern Oil Exploration Corporation (Pty.) Ltd. (SOEKOR), a state-financed organization established in 1965 to undertake, encourage, and coordinate the search for oil in South Africa.[2]

TABLE II-2-23 Uranium Oxide Production in South Africa, 1970–1981

Year	U_3O_8 (metric tons)	Export sales (thousand U.S. dollars)[a]
1970	3,737	
1971	3,800	
1972	3,629	
1973	3,094	
1974	3,198	
1975	2,934	52,281
1976	3,254	70,315
1977	3,962	104,424
1978	4,672	244,953
1979	5,637	303,057
1980	7,295	415,000e
1981	7,234	300,000e

[a]e, estimated.

Source: U.S. Bureau of Mines, Mineral Yearbook, 1970–1981.

TABLE II-2-24   Major South African Uranium Producers, 1981

Finance group[a]	Mine	Ore milled (thousand metric tons)	Production (kg)	Grade (kg/metric ton milled)	Profit (thousand rand)[b]
RM	Blyvooruitzicht	1,933	315,502	0.163	5,754
GMU	Buffelsfontein	3,135	631,750	0.202	19,554
GF	West Driefontein	1,309	242,327	0.185	8,620
ATC (UP)	Hartebeestfontein	3,185	478,663	0.150	12,900
JCI (UP)	Randfontein	3,351	591,774	0.177	12,034
AAC	Vaal Reefs	8,501	1,693,569	0.199	42,523
GMU (UP)	Stilfontein				
AAC	Western Deep Levels	2,331	212,484	0.091	5,072
JCI	Western Areas				
RM (UP)	Harmony	6,944	580,428	0.084	NA
GMU	West Rand Cons.	673	190,258	0.283	15,868
AAC[c] (UP)	Anglo-American OFS Joint Metallurgical Production Scheme		1,093,416		
GMU	Beisa	1,200		0.6	
Rio Tinto Zinc Corp. (British)[d]	Palabora Copper		234,206		
AAC[d]	East Rand Gold and Uranium Ltd. (Ergo)		302,194		

[a]AAC, Anglo-American Corp. of South Africa Ltd.; ATC, Anglo-Transvaal Consolidated Investment Co. Ltd.; GF, Gold Fields of South Africa Ltd.; GMU, General Mining Union Corp. Ltd.; JCI, Johannesburg Consolidated Investment Co. Ltd.; RM, Rand Mines Ltd.

[b]NA, not available.

[c]Consist of the following mines: Free State Geduld, Free State Saaiplaas, President Brand, President Steyn, Welkom Western Holdings.

[d]Ref. 6.

Source: Ref. 11.

Recent offshore exploration has been concentrated along the west and south coasts. There are two rigs operating in each area. In 1981, SOEKOR spent some $75 million in exploration cost. Much of this amount was for offshore drilling of 13 drill holes with a combined depth of 43,988 m.[6] Four drill holes were along the west coast and 10 along the south coast. Exploration activity along the southern coast has been concentrated some 90 km south of the Mossel Bay area, where gas and oil have been discovered, and south of the Plettenburg Bay area, where gas has been discovered. Exploration along the west coast has centered around the Orange River area.

Pipelines. South Africa has some 3000 km of pipeline, of which 836 km are for crude oil, 1748 km for refined products, and 322 km for natural gas. Significant developments in the expansion of the pipelines were the completion of the pipeline from Durban to Johannesburg in 1965 and to the inland refineries in Sasolburg in 1969; the expansion of lines from Johannesburg to Pretoria, Potchefstroom, and Klerksdorp in 1972; and the completion

of the line from Sasol II and III to the Witwatersrand in 1979.[2] Table II-2-26 presents the pipeline systems of other countries in southern Africa.

Refineries. Major refineries in South Africa are located in the port of Durban and adjacent to the Sasol I plant. Production from refineries is given in Table II-2-27, and imports of petroleum products are given in Table II-2-28.

The successful completion and uninterrupted operation of Sasol II and Sasol III oil from coal

TABLE II-2-26   Pipelines in Southern Africa (Kilometers)

Country	Length	Type
South Africa	836	Crude oil
	1748	Refined products
	322	Natural gas
Angola	179	Crude oil
Mozambique	306	Crude oil
Zaire	390	Refined products
Zambia	724	Crude oil
Zimbabwe	8	Crude oil

Source: Ref. 1.

TABLE II-2-25 Uranium Reserves of Major South African Uranium-Producing Mines

Mine	Ore reserves (thousand metric tons)	Uranium oxide grade (kg/metric tons)	Uranium oxide (total metric tons)	Uranium content (metric tons)
Blyvooruitzicht	5,338	0.221	1,179.7	
Buffelsfontein	11,626	0.312	3,627.3	
Free State Geduld	8,744	0.09	787.0	
	10,176	0.09	915.8	
	11,580	0.09	1,042.2	
Harmony	30,607	0.156	4,774.7	
Hartesbeestfontein	14,508	0.23	3,336.8	
President Brand	9,092	0.09	818.3	
	10,226	0.09	920.3	
President Steyn	14,490	0.11	1,593.9	
	16,272	0.11	1,789.9	
	17,563	0.11	1,931.9	
Randfontein				
Cooke No. 1 Shaft	3,671	0.165	605.7	
Cooke No. 2 Shaft	2,517	0.340	855.8	
Vaal Reefs				
Vaal Reef	25,487	0.41	10,449.67	
	26,534	0.40	10,613.6	
VCR and Elsburg Reef	4,263	0.10	426.3	
	6,350	0.10	635.0	
Dominion Reef	67	1.54	103.2	
(Afrikander lease area)	70	1.52	106.4	
Western Areas	1,891	0.705	1,333.2	
Middle Elsburg Reefs				
Western Deep Levels	3,022	0.24	725.3	
Carbon Leader				
Western Holdings	18,559	0.10	1,355.9	
	22,074	0.10	2,207.4	
	24,341	0.10	2,434.1	
TOTAL			55,802.37	47,319.3

Source: Ref. 12.

TABLE II-2-27 Refinery Products in South Africa, 1970-1979

Year	Gasoline	Jet fuel	Kerosene	Distillate fuel oil	Residual fuel oil	Lubricants	Other	Refinery fuel and losses	Total
1970	19,962	792	2,266	14,551	13,975	782	3,154	5,181	60,663
1971	22,778	623	2,991	15,828	15,298	710	3,438	5,127	66,793
1972	24,589	1,002	3,315	20,493	21,727	1,292	4,810	7,606	84,834
1973	25,487	2,589	2,479	21,301	20,427	1,680	4,941	5,313	84,217
1974	26,039	2,297	2,339	20,923	21,440	1,812	2,741	5,123	82,714
1975	31,939	3,144	3,400	30,220	29,454	2,171	7,715	6,651	114,694
1976	29,783	2,196	3,132	31,279	22,027	2,228	8,581	5,561	104,787
1977	30,083	2,349	3,338	31,918	22,036	2,262	8,607	5,624	106,217
1978	30,090	2,440	3,333	32,973	22,178	2,240	7,372	5,700	106,326
1979	30,660	2,550	3,285	33,580	22,630	2,555	6,205	4,015	105,485

Source: U.S. Bureau of Mines, Minerals Yearbook, 1970-1981.

projects will lessen South Africa's dependence on imported petroleum crude and refined products. The concentration of refinery capacity, together with power generation, makes refineries a particularly high priority for sabotage. In early 1980, approximately $5.5 million worth of liquid fuels was destroyed in Sasolburg (Sasol I) and Secunda (Sasol II and III).[15]

TABLE II-2-28   Imports of Refinery Products to South Africa, 1970–1980
(42-Gallon Barrels)

Year	Mineral jelly and wax	Pitch and pitch coke	Petroleum coke	Bitumen and other residues	Bituminous mixtures
1970	237,000	4,983	33,468	17,720	
1971	251,000	4,461	23,788	16,344	
1972	182,000	2,221	38,815	15,968	4,733
1973	260,000	8,500	66,099	18,896	5,964
1974	384,000	11,061	163,843	11,146	9,102
1975	237,000	14,740	162,371	6,393	5,502
1976	327,000	3,069	87,093	4,181	14,126
1977	351,000	165	187,380	3,806	3,194
1978	354,303	970	223,461	3,171	3,049
1979	388,315	6,450	311,583	1,594	2,211
1980	407,985	997	448,189	1,595	2,595

Source: U.S. Bureau of Mines, Minerals Yearbook, 1970–1981.

MINERAL INDUSTRY

The mineral industry of South Africa is the fourth largest in the world, with only the United States, the USSR, and Canada having larger mineral industries. The country is nearly self-sufficient in minerals, bauxite and petroleum being the significant exceptions.

Although South Africa is known mostly for its gold and diamonds, the country's mineral endowment is very important for other minerals as well. Chief among these are minerals considered indispensable to a modern industrial society, which include the following: for metals—platinum group, chromium, iron ore, manganese, nickel, vanadium, antimony, copper, lead, zinc, titanium, zirconium; and for nonmetals—asbestos, fluorspar, phosphate, and vermiculite. Fuel minerals have already been discussed, and included coal, uranium, petroleum, and natural gas. South Africa's mineral endowment is related to crustal evolution in the subcontinent. A brief look at some unique geologic features in South Africa follows.

Unique Geology of South Africa

Geological time in South Africa can be divided into three main eras. Each era represents a unique developmental stage in the crustal evolution of South Africa. These three eras, from oldest to youngest, are the Archeozoic (before 3200 to 2400 million years ago), the Proterozoic (from 3200–2400 to 700–500 million years ago), and the Phanerozoic (700 to 500 million years ago to the present).[10] Both the Archeozoic and Proterozoic are often referred to as the Precambrian era. Variations in these years are due to different stages of development in different parts of the world.

South Africa's mineral uniqueness is due to the age of its deposits and the resulting nature of its ores. During the Archeozoic, South Africa's crust became stable around 3000 million years ago, much earlier than other parts of the world, except perhaps eastern Siberia.[10] This early stabilization allowed for the preservation of depositional processes from the Archeozoic basement rocks.

This early Archeozoic crust was dominated by siderophile elements. Siderophile elements have a relatively weak affinity for oxygen and sulfur, and tend to combine readily with iron. Since the early Archeozoic crust had a high concentration of iron, minerals found within the crust formed iron compounds. These siderophiles were preserved because they formed before an oxygenated atmosphere had developed. Other areas of the world did not preserve these types of minerals because of a somewhat later crustal development in an oxygenated atmosphere.

The other categories for minerals include biophiles formed by decaying plants and animals (coal, oil, etc.); lithophiles formed in stony matter or slag as oxides (especially silicates); and chalcophiles formed in combination with sulfur, and commonly forming the sulfide ores of base metals. This sulfur usually originates from decaying organic matter or from deep gas emanations.

The relative economic importance of these different elements is reflected in the value of mineral products extracted from them. In the period 1852–1975, siderophile elements accounted for 78% of the total value of mineral production, while chalcophiles had produced only 5% of the total.[10]

Preservation of these early-formed siderophiles in the Early Proterozoic mineralization period, between 3000 and 1500 million years ago, has made the Early Proterozoic the Golden Age of South African mineralization. Through 1975, 88% of the total value of all South Africa's mineral production was derived from this geologic era. During the same period, this era also accounted for 98% of the value of precious metals mines, 77% of the value of base metals, 15% of the value of gemstones, 67% of the value of nonmetallics, 38% of the value of sources of energy, 92% of the value of siderophiles, 60% of the value of chalcophiles elements, and 86% of the value of lithophile elements.[10] In contrast, the next most important era, the Phanerozoic, contributed only 12% of the total value of mineral production in this same period.[10]

Crustal Evolution. Figure II-2-11 depicts the four main structural provinces in South Africa, which are generally covered by various thicknesses of the Phanerozoic deposition.[1] The Kaapvaal fragment, in northeastern South Africa, represents the earliest crustal formation, which stabilized some 3000 million years ago. Although this structure

covers only 50% of the land area, it accounts for 94% of the value of mineral production.[10] The Limpopo Province, north of Kaapvaal, stabilized around 2000 million years ago. It accounts for only 2.28% of total land area and 1.19% of its mineral wealth. The Sanama Province, south of Kaapvaal, stabilized 1000 million years ago, accounting for 40% of land area and 4.23% of the country's mineral wealth. The Cabo Province in the southern portion of the country and along the western coast, stabilized only 500 million years ago. It accounts for 7.56% of the land area and only 0.45% of the total mineral wealth.[10]

Another distinguishing geological feature of South Africa is the Brakbos lineament. This geological fracture runs southeastward from the common border of Namibia/south-West Africa–Botswana–South Africa through Upington to the coast between East London and Port Elizabeth. Southwest of this line are located only 14 mineral districts and deposits, whereas northeast of this line there are more than 66 major mining localities.[10]

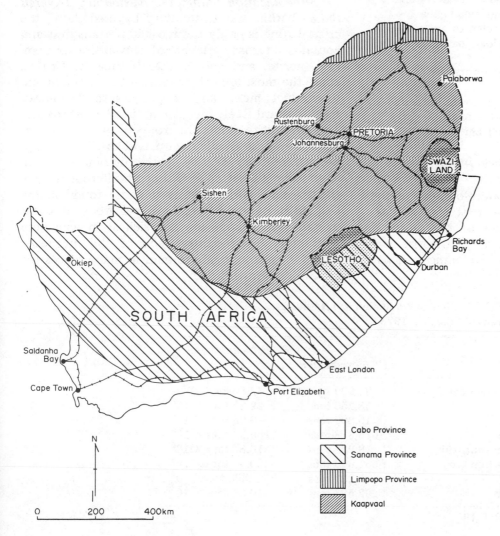

FIGURE II-2-11 Four main structural provinces in South Africa. (From U.S. Department of the Interior, Bureau of Mines, *Mineral Industries of Africa*; Ref. 10.)

Northeast South Africa (roughly the Kaapvaal) contains the country's most valuable and most strategic mineral deposits. The two dominant mineralized provinces in this region are the Bushveld Igneous Complex and the Witwatersrand. A brief geological description of each follows.

Bushveld (Igneous) Complex. The Bushveld Complex is the largest known intrusive of its kind in the world, covering some 65,000 km^2.[13] The complex is known for the persistence and regularity of the layering in the ultrabasic and basic rocks. Within this complex are the world's largest reserves of platinum-group metals, chromite, and vanadium. The complex also produces gold, nickel, copper, magnetite iron ore, tin, and fluorspar. Table II-2-29 indicates the mineral production for 1975 from the Bushveld Complex, as well as posited resources for these minerals.

The Bushveld Complex is not a single giant lopolith but consists of four separate lobes.[13] These lobes are:[11]

1. A western lobe extending from near Pretoria westward to Rustenburg and around the 1.4-Ga Pilanesburg alkaline complex to and along the southern flank of the Mokoppa Dome Archean granitoids and gneisses.
2. A southeastern lobe, largely covered by Mesozoic rocks.
3. An eastern lobe that continues northward from the southeastern lobe, extending to and beneath the Mesozoic cover some 50 km east of Potgietersrus
4. A northern lobe that extends from south of Potgietersrus to Villa Nova

The cause and historical development of the Bushveld Complex are still a matter of debate. D. R. Hunter's *Some Enigmas of the Bushveld Complex* reviews some of the major theories concerning both the origin and development of the Bushveld Complex.[16]

Stratigraphy of the Bushveld (Igneous) Complex. Hall (1932) divided the Bushveld Complex into the following five distinct phases:[16]

1. First or earlier basic volcanic phase
2. Second or later acid volcanic phase
3. Norite lopolith or earlier plutonic phase
4. Sill phase of the norite lopolith
5. Granite or later plutonic phase

To conform with recommendations laid down by the South African Stratigraphic Code and to develop similar general terminology for other layered intrusions around the world, Tankard et al. have recently developed a new stratigraphic nomenclature for the Bushveld Complex.[3] Using this nomenclature, Hall's phases 2 to 5 have been renamed as follows:[16] (2) Rooiberg Group, (3) Rustenburg Layered Suite, (4) Rashoop Granophyre Suite, and (5) Lebowa Granite Suite. The most important stratigraphic zone is the Rustenburg Layered Suite, which contains chromite, platinum, vanadium, and by-product nickel and copper.

Mineralization within the Rustenburg Layered Suite. Within the Rustenburg Layered Suite, the critical zone is easily the most important unit and contains alternating layers of chromitite, pyroxenite, norite, and anorthosite. Within the Critical zone, the most important sources for platinum and by-product nickel and copper are the Merensky Reef, Bastard Reef, and giant mottled anorthosite.[8] The chromitite seams are grouped into a Lower or Main Group, a Middle Group, and an Upper Group because of their relative spatial distribution. Vanadium mineralization is within titaniferous magnetitite seams. While these seams are found in the Critical and Main zones, the lower group of seams in the Upper zone are the richest in vanadium pentoxide. The country rock of the magnetite seams is generally gabbro, and in many places the footwall is anorthosite.[8]

TABLE II-2-29 **Mineral Production and Resources of the Bushveld Complex, 1975**

	1975 production	Estimated total resource
Platinoid metals and gold	85,000 kg	62.89 kg × 10^6
Nickel	18,250 tons	22.80 tons × 10^6
Copper	10,800 tons	9.95 tons × 10^6
Chromite	2,075,378 tons	156.0 tons × 10^6
Vanadium pentoxide	19,002 tons	16.8 tons × 10^6
Magnetite iron ore	1,561,670 tons	1030.5 tons × 10^6
Tin	5,232 tons	0.05 ton × 10^6
Fluorspar	202,583 tons	40.0 tons × 10^6

Source: Ref. 13.

Fluorspar within the Bushveld Complex is located within the granites of the Lebowa Granite Suite.[13] Tin mineralization is generally associated with the acid phase of the Bushveld Complex. Tin mineralization is both syngenetic and epigenetic within the granites of the Lebowa Granite Suite. Within the Granophyre, deposits include zonal ore bodies, breccia ore bodies, and replacement ore bodies.[8] Cassiterite is also found as replacement lodes in felsites.[8]

Witwatersrand. The Witwatersrand is noted for its gold and uranium deposits. Indeed, the Witwatersrand accounts for 55% of all the gold ever mined[13] and continues to produce around 50% of the world's annual gold production. Some 300,000 workers are engaged in gold mining, and they develop 1000 km of underground tunnels at an average depth of 1650 m every year.[13]

Most gold mined in the world is derived from Phanerozoic host rocks. However, the age of the Witwatersrand has been determined at between 2.8 and 2.3 Ga,[13] firmly within the Proterozoic era. The Witwatersrand basin generally reflects a general regressive trend through time. The older deposition indicates a marine environment, while younger deposition near the top indicates fluvial deposition.

Stratigraphy of the Witwatersrand. D. A. Pretorius, a recognized expert on the Witwatersrand, describes the Witwatersrand as "an intermontane, intercratonic, yoked basin with a fault-bounded northwestern edge and a gently downwarping more passive southeastern boundary."[17] The basin was either a shallow water lake or inland sea. Furthermore, as the basin became more unstable, interference folding occurred, and gold fields developed in downwarps between basement domes. The deposits were laid down in a shrinking basin, resulting in two distinctive stratigraphic layers. Conditions in the Lower Witwatersrand Division reflect a period of transgression, while those in the Upper Division reflect regressive conditions, as the basin edge moved southeastward by 60 km and the depositional axis by about 10 km.[17]

The Witwatersrand Supergroup is divided into the younger West Rand Group and the Central Rand Group. Subgroups of the West Rand Group (from oldest to youngest) are Hospital Hill, Government, and Jeppestown. The Central Rand Group has two subgroups, the younger Johannesburg and Turffontein subgroups.

The West Rand Group is approximately 42,000 km^2 in extent.[10] The thickness averages 4650 m but varies from 830 m in the Evander area to 7500 m northwest of Krugersdorp. In general, the sequence contains shales and sandstones in equal proportions and 250 m of volcanic rocks. Deposition occurred predominantly in marine shelf and tidal environments with minor episodes of fluvial accumulation.[13]

The Central Rand Group is approximately 9750 km^2 and lies above the West Rand Group. The Central Rand Group is predominantly made up of coarse-grained subgraywacke with less than 10% conglomerate and quartz arenite, subordinate lacustrine or shallow marine shale and silt, and minor lava.[13] Deposition was dominated by alluvial-fan environments.

Cycles of sedimentation within the Central Rand Group vary in thickness between 30 and 600 m, averaging 250 m. The composition of these cycles is essentially minor amounts of gravel and major developments of sand, with or without sericite.[18] The cycle is composed of a base of scour surfaces, on which pebble lags or gravel bars form. On top of these quartzites form trough-cross beds. These quartzites in turn are followed by subgraywackes and siltstones.

Three separate facies can be recognized within the alluvial fan deposits. The fanhead facies is characterized by the coarsest clastics and lower gold and uranium concentrations and represents a high-energy environment. The midfan facies is distinguished by a medium-small-pebble conglomerate with relatively high concentrations of heavy minerals. Gold is concentrated in this area. The fanbase is characterized by fine-grained sediment and a low percentage of conglomerates, in which uranium is more concentrated than gold.[7] However, gold is concentrated here within the kerogin of fossil algal mats.[13]

Mineralization within the Witwatersrand. Of the more than 70 ore minerals recognized in Witwatersrand placers, over 40 are detrital. Important minerals within the Witwatersrand reefs include arsenopyrite, brannerite, cobaltite, chromite, gold, platinoids, pyrite, uraninite, and zircon.[13] Of these minerals, only the following are currently of economic importance: gold, uraninite, pyrite, and brannerite.

The source area for the Witwatersrand basin deposits is believed to be a granite-greenstone terrain. Gold originated from quartz veins, while uranium and zircon originated from granitic rocks. Chromite and platinoids originated from mafic and ultramafic rocks.[8]

Deposition of minerals varies according to grain size, relative availability of mineral, and the local hydrolic conditions. For gold and uranium, deposition was in both the initial and terminal stages of successive depositional cycles. In the initial stages, gold and uranium can be found in the matrices of the conglomerates and in heavily pyritic sands filling erosion channels. Deposition in the terminal stages can be seen on quartzites along a plane of unconformity, on shales along planes of unconformity, and in carbon seams on or adjacent to planes of unconformity.[17] Usually, gold can be found concentrated in the midfan area, and uraninite farther down the paleoslope.[18]

Mineral Economy

Although gold sales still dominate the mineral economy and national economy of South Africa, there are some important developments within the industry. To investigate recent trends, several tables are given which contrast domestic and export sales, as well as the role of metals, nonmetals, and energy minerals within each category.

It can easily be seen in Table II-2-30 that export sales constitute the major portion of mineral revenues. For the period 1970–1977, the percentage of the total for domestic sales was increasing slightly. However, the rise in gold prices offset this trend after 1977.

The contribution of metals, nonmetals, and energy minerals for domestic and export sales are given in Tables II-2-31 and II-2-32, respectively. For the period 1970–1974, in exports sales the percentages for the metal category are actually higher. During these years, the category "other minerals" was predominantly platinum and uranium sales.

In domestic sales, the percentage for metals has steadily declined, while that for nonmetals has risen slowly. The percentage for energy minerals has grown from 37.8% in 1970 to 53.3% in 1981.

Because of gold exports, metals dominate export sales and have for a number of years. Table II-2-33 shows a decline in the percentage for nonmetals, while showing an increase for energy minerals, especially with the opening of the Richards Bay Coal Terminal.

Table II-2-33 also presents the percentages for metals, nonmetals, and energy minerals of the total domestic and export sales. The percentage value for metals has remained approximately the same from 1970 to 1981. The percentage value for nonmetals has declined somewhat, while that for energy minerals has more than doubled, from 7.0% in 1970 to 15.9% in 1981.

These tables illustrate several basic facts about the present South African mineral economy. First, the mineral economy is export oriented. Second, the value of metals has remained proportionally constant from 1970 to 1981, and the value for nonmetals has decreased somewhat. Finally, the value for coal, both for domestic uses and exports, has increased dramatically in the past decade.

PRECIOUS MINERALS

Gold

Gold was discovered in the eastern Transvaal in 1868. However, it was not until 1886 that gold was discovered in the quartzite hills of the Witwatersrand by George Harrison on the farm Langlaagte, near Johannesburg.[2] Today, South Africa is the world's largest producer of gold, accounting for just over 50% of gold production in 1982. This gold comes from the deepest mines in the

TABLE II-2-30   South African Domestic Mineral Sales versus Export Mineral Sales, 1970–1981 (Thousand U.S. Dollars)

Year	Domestic	(Percent)	Export	(Percent)	Total
1970	369,145	16.7	1,835,490	83.3	2,204,635
1971	366,546	16.7	1,831,513	83.3	2,198,059
1972	365,704	14.6	2,144,847	85.6	2,510,551
1973	518,840	12.6	3,591,434	87.4	4,110,274
1974	662,813	11.5	5,115,957	88.5	5,778,770
1975	934,523	15.7	5,031,394	84.3	5,965,917
1976	1,001,992	18.6	4,377,385	81.4	5,379,377
1977	1,225,240	19.9	4,924,070	80.1	6,149,310
1978	1,349,997	17.0	6,584,282	83.0	7,934,279
1979	1,635,375	14.4	9,718,275	85.6	11,353,650
1980	2,190,135	11.7	16,546,641	88.3	18,736,776
1981	2,448,541	16.0	12,842,194	84.0	15,290,735

Source: U.S. Bureau of Mines, Minerals Yearbook, 1970–1981.

TABLE II-2-31   South African Domestic Mineral Sales, 1970-1981 (Thousand U.S. Dollars)

Year	Metal	(Percent)	Other[a]	(Percent)	Nonmetals	(Percent)	Energy	(Percent)	Total
1970	108,532	29.4	41,027	11.1	80,042	21.7	139,554	37.8	369,145
1971	86,407	23.6	57,594	15.7	69,205	18.9	153,340	41.8	366,546
1972	84,995	23.2	58,579	16.0	68,571	18.7	153,559	42.0	365,704
1973	146,733	28.3	83,218	16.0	89,249	17.2	199,640	38.5	518,840
1974	204,975	30.9	89,366	13.5	106,138	16.0	262,334	39.6	662,813
1975	247,245	26.5			306,288	32.8	380,990	40.7	934,523
1976	250,605	25.0			274,875	27.4	476,512	47.6	1,001,992
1977	307,565	25.1			334,792	27.3	582,883	47.6	1,225,240
1978	341,784	25.3			376,421	27.9	631,792	46.8	1,349,997
1979	360,086	22.0			520,488	31.8	754,801	46.2	1,635,375
1980	508,409	23.2			644,502	29.4	1,037,224	47.4	2,190,135
1981	494,505	20.2			649,292	26.5	1,304,744	53.3	2,448,541

[a]"Other" is comprised primarily of platinum and uranium, and therefore would increase the metal percentile.

Source: U.S. Bureau of Mines, Minerals Yearbook, 1970-1981.

TABLE II-2-32   South African Export Mineral Sales, 1970-1981 (Thousand U.S. Dollars)

Year	Metal	(Percent)	Other[a]	(Percent)	Nonmetal	(Percent)	Energy	(Percent)	Total
1970	1,439,750	78.4	204,471	11.2	176,664	9.6	14,335	0.8	1,835,490
1971	1,503,035	82.1	147,008	8.0	167,682	9.2	13,788	0.8	1,831,513
1972	1,757,032	81.9	189,965	8.9	187,480	8.7	10,370	0.9	2,144,847
1973	2,961,813	82.5	275,032	7.7	334,436	9.3	20,153	0.6	3,591,434
1974	4,355,596	85.1	400,840	7.8	327,684	6.4	31,837	0.6	5,115,957
1975	4,574,186	90.9			406,176	8.1	51,032	1.0	5,031,394
1976	3,831,622	87.5			426,822	9.8	118,941	2.7	4,377,385
1977	4,114,014	83.5			524,222	10.7	285,834	5.8	4,924,070
1978	5,469,470	83.1			740,989	11.2	373,823	5.7	6,584,282
1979	8,219,331	84.6			893,131	9.2	605,813	6.2	9,718,275
1980	14,683,155	88.7			979,016	5.9	884,469	5.3	16,546,641
1981	11,052,250	86.1			667,388	5.2	1,122,556	8.7	12,842,194

[a]"Other" is comprised primarily of platinum and uranium, and therefore would increase the metal percentage.

Source: U.S. Bureau of Mines, Minerals Yearbook, 1970-1981.

TABLE II-2-33   Value of Metal, Nonmetal, and Energy Minerals in South Africa, 1970-1981 (Thousand U.S. Dollars)

Year	Metal	(Percent)	Nonmetal	(Percent)	Energy	(Percent)	Total
1970[a]	1,548,282	70.2	256,706	11.6	153,889	7.0	2,204,635
1971[a]	1,589,442	72.3	236,887	10.8	167,128	7.6	2,198,059
1972[a]	1,842,027	73.4	256,051	10.2	173,712	6.9	2,510,551
1973[a]	3,108,546	75.6	423,685	10.3	219,793	5.3	4,110,274
1974[a]	4,560,571	78.9	433,822	7.5	294,171	5.1	5,778,770
1975	4,821,431	80.8	712,464	11.9	432,022	7.3	5,965,917
1976	4,082,227	75.9	701,697	13.0	595,453	11.1	5,379,377
1977	4,421,579	71.9	859,014	14.0	868,717	14.1	6,149,310
1978	5,811,254	73.2	1,117,410	14.1	1,005,615	12.7	7,934,279
1979	8,939,503	78.7	1,413,619	12.4	1,360,614	11.9	11,353,650
1980	15,191,565	81.0	1,623,518	8.7	1,921,693	10.3	18,736,776
1981	11,546,755	75.5	1,316,680	8.6	2,427,300	15.9	15,290,735

[a]Percentage values for 1970-1974 are less than 100%. Missing percentage values were under "Other minerals." Percentage values for metals are greater than given.

Source: U.S. Bureau of Mines, Minerals Yearbook, 1970-1981.

world, which are located along the rim of the old Witwatersrand Basin.

Gold remains South Africa's single most valuable commodity. As illustrated in Table II-2-34, gold sales form a significant portion of total mineral revenues and the total GNP for the country. Table II-2-35 presents gold production, revenues, and average price of gold for 1970–1981. Gold revenues for Tables II-2-34 and II-2-35 were taken from export sales only. In 1980, gold reached a record high of over $800 per troy ounce. Gold prices have since fallen, and have oscillated roughly between $300 and $500, with an average of over $400 per troy ounce for the last several years.

Geology. As noted previously, the Witwatersrand Basin extends some 320 km on a northeasterly axis, with a northwesterly width of about 150 km. The economic gold zone is predominantly within the Witwatersrand Supergroup. Within this supergroup, the Central (Upper) Rand Group is the most economically important ore zone. The relevant stratigraphy is presented in Table II-2-36.

The gold fields have developed around the rim of the Witwatersrand Basin and are usually referred to as the Evander, West Rand, Central Rand, Eastern Rand-Heidelberg, Far West Rand, Klerksdorp, and Orange Free State (OFS) gold fields. For the location of these fields, see Figure II-2-12. Of these

TABLE II-2-34   Value of Gold Sales in South Africa, 1970–1981 (Million U.S. Dollars)

Year	Gold	Total mineral value	Percent gold	GDP	Percent gold
1970	1,162.5	2,204.6	52.7	16,700	7.0
1971	1,249.9	2,198.1	56.9	18,400	6.8
1972	1,499.8	2,510.6	59.7	21,800	6.9
1973	2,585.5	4,110.3	62.9	27,200	9.5
1974	3,853.7	5,778.8	66.7	32,900	11.7
1975	3,498.3	5,965.9	58.6	34,500	10.1
1976	2,737.2	5,379.4	50.9	33,370	8.2
1977	3,237.2	6,149.3	52.6	38,230	8.4
1978	4,485.0	7,934.3	56.5	46,000	9.7
1979	6,951.9	11,353.6	61.2	57,000	12.2
1980	13,329.1	18,736.8	71.1	80,210	16.6
1981	9,831.5	15,290.8	64.3	81,800	12.0

Source: U.S. Bureau of Mines, Minerals Yearbook, 1970–1981.

TABLE II-2-35   Average Price of Gold, 1970–1981

Year	Gold production (thousand troy ounces)	Value of gold (thousand U.S. dollars)	Average price (dollars)
1970	32,164	1,162,470	35
1971	31,389	1,249,963	40
1972	29,245	1,499,771	51
1973	27,495	2,585,524	94
1974	24,388	3,853,745	158
1975	22,938	3,498,268	152
1976	22,936	2,737,196	119
1977	22,502	3,237,240	144
1978	22,649	4,485,038	198
1979	22,617	6,951,983	307
1980	21,669	13,329,098	615
1981	21,121 *651 m tons*	9,831,548	465
1982[a]	21,200		

[a]Ref. 7.

Source: U.S. Bureau of Mines, Minerals Yearbook, 1970–1981.

TABLE II-2-36 Stratigraphy of Witwatersrand Basin

Supergroup	Group	Series
Transvaal	Wolkberg	
	Chuniespoort	
	Wolkberg	Black Reef
Ventersdorp	Pniel	
	Platberg	
	Klipriviersberg	
Witwatersrand	Center Rand (Upper)	Turffontein
		(Kimberley — Elsburg)
	West Rand (Lower)	Johannesburg (Main — Bird)
		Jeppestown
		Government
		Hospital Hill
	Dominion	

Source: Ref. 13.

GOLD FIELDS

1. Evander
2. West Rand
3. Central Rand
4. Eastern Rand-Heidelberg
5. Far West Rand
6. Klerksdorp
7. Orange Free State

FIGURE II-2-12 Gold fields of South Africa. (From U.S. Department of the Interior, Bureau of Mines, *Mineral Industries of Africa*; Ref. 10.)

fields, the OFS, Far West, and Klerksdorp have produced some 70% of the production to date.[11] Economic ore zones vary among the different gold fields. Important ore zones for the various gold fields are given in Table II-2-37.

Production. Gold production for 1982 was an estimated 21,200,000 troy ounces of gold (52% of the world production), down from the high of 32,164,000 troy ounces in 1970. The gold fields of the Witwatersrand employed an average of 44,818 whites and 434,120 nonwhites during 1981.[11]

Production figures for individual mines for 1981 are given in Table II-2-38. Mines are grouped with respect to respective gold fields. Production figures for individual mining houses are given in Table II-2-39.

The major mines produce a gold bullion that is composed of 88% gold, 10% silver, and 2% base-metal impurities in 26-kg ingots. These ingots are delivered and assayed at the Rand Refinery Ltd., at Germiston. In 1982, the Chamber of Mines announced that it would market 99.9% pure gold

bars overseas. The bars will be marketed in a limited number of 400-oz bars and a new 1-kg bar.[12]

South Africa will continue to sell its Krugerrand coins at the current standard of 95.5% gold (the rest being silver). In 1981, Intergold, which markets Krugerrands overseas, reported sales of 3,128,727 of the 1-oz coins, 178,014 of the $\frac{1}{2}$-oz coins, 725,428 of the $\frac{1}{4}$-oz coins, and 1,321,022 of the $\frac{1}{10}$-oz coins. In addition, 18,538 of the 2-rand coin and 19,274 of the 1-rand coin were sold. The total weight of the South African coins sold was 3.54 million troy ounces.[6]

South Africa, and in particular the Chamber of Mines, have contributed greatly to the production of gold by establishing and financing a number of research projects which have resulted in improved technologies to mine gold. Contributions to this technology have included advances in the metallurgy to extract the gold from ores, air-refrigeration techniques and clothing for deep-level mining, lighter and more efficient drills for blasting, and new portable gold analyzers.[11]

TABLE II-2-37   Important Reefs of the Witwatersrand

Gold field	Supergroup	Group	Subgroup	Reefs
1. Evander	Witwatersrand	Central Rand	Turffontein	Kimberley
2. East Rand	Witwatersrand	Central Rand	Turffontein	Kimberley
3. Central Rand	Witwatersrand		Johannesburg	South
				Main Leader
				Main
4. West Rand	Transvaal	Wolkberg		Black
	Ventersdorp	Klipriviersberg		Ventersdorp
				Contact
	Witwatersrand	Central Rand	Turffontein	Kimberley
			Johannesburg	Bird
				Livingston
				South
5. Far West Rand	Transvaal	Wolkberg		Black
	Ventersdorp	Klipriviersberg		Ventersdorp
				Contact
	Witwatersrand	Central Rand	Turffontein	Elsberg
			Johannesburg	Main
				Carbon Leader
6. Klerksdorp	Transvaal	Wolkberg		Black
	Witwatersrand	Central Rand	Turffontein	Gold Estates
			Johannesburg	Vaal
				Commonage
		West Rand	Government	Government
			Dominion	Dominion
7. OFS	Transvaal	Central Rand	Turffontein	A and B
			Johannesburg	Leader
				Basal

Sources: Refs. 8 and 13.

TABLE II-2-38   South African Gold Production by Mine, 1981

Gold field	Mine	Finance group[a]	Ore milled (thousand metric tons)	Production (kg fine)	Rank	Grade (g/ton milled)
Evander	Winkelhaak	GMU	2,135	13,932.3	18	6.53
	Kinross	GMU	1,650	9,654.5	22	5.85
	Bracken	GMU	995	3,485.6	33	3.50
	Leslie	GMU	1,235	3,994.6	31	3.23
Hiedelburg	Witwatersrand Nigel	AEC	308	1,124.6	36	3.65
East Rand	Marievale	GMU	819	1,216.9	35	1.49
	Grootvlei	GMU	1,745	6,453.3	26	3.75
	East Rand Property Mines	RM	2,624	11,632.2	20	4.43
Central Rand	ERGO[b]	AAC	19,200	6,300.0	27	0.3
West Rand	Durban Deep	RM	2,245	8,248.0	23	3.67
	West Rand Cons.	GMU	2,110	2,829.1	34	1.34
	Randfontein	JCI	4,525	23,679.0	13	5.23
Far West Rand	Randfontein (Cooke section)					
	Western Areas	JCI	4,291	17,706.0	15	4.13
	Venterspost	GF	1,414	5,638.8	28	3.99
	Libanon	GF	1,680	10,127.2	21	6.03
	Kloof	GF	2,035	29,767.1	8	14.63
	East Driefontein	GF	2,760	36,346.2	5	13.17
	West Driefontein	GF	2,785	39,667.3	3	14.24
	Western Deep Levels	AAC	3,150	39,012.9	4	12.36
	Blyvooruitzicht	RM	2,124	18,794.4	14	8.85
	Doornfontein	GF	1,446	11,889.5	19	8.22
	Elandsrand	AAC	1,215	5,294.3	30	4.28
	Deelkraal	GF	1,211	5,335.2	29	4.41
Klerksdorp	Stilfontein	GMU	1,912	14,891.3	17	7.79
	Hartebeestfontein	ATC	3,004	30,533.1	7	10.16
	Buffelsfontein	GMU	3,345	28,133.3	9	8.41
	Vaal Reefs (North and South)	AAC	8,481	73,502.0	1	8.67
	Vaal Reefs (Africander)	AAC	21	5.3	37	0.26
OFS	Loraine	ATC	1,636	6,576.7	25	4.02
	Free State Geduld	AAC	2,967	27,490.3	10	9.27
	St. Helena	AAC	7,856	41,667.7	2	5.31
	Western Holdings	GMU	2,140	15,569.0	16	7.28
	Welkom	AAC	Merged with Western Holding			
	Free State Saaiplaas	AAC	Merged with Western Holding			
	President Steyn	AAC	3,794	24,501.7	12	6.46
	President Brand	AAC	3,323	26,729.7	11	8.04
	Unisel	GMU	1,071	7,330.5	24	6.84
	Harmony	RM	7,641	31,946.0	6	4.18
	Anglo-American OFS Metallurgical Scheme	AAC		3,496.2	32	
	Other			5,898.3		
			Total	650,409.8		

[a]AAC, Anglo-American Corp. of South Africa Ltd.; AEC, African Exploration Co. Ltd.; ATC, Anglo-Transvaal Consolidated Investment Co. Ltd.; GMU, General Mining Union Corp. Ltd.; GF, Gold Fields of South Africa Ltd.; JCI, Johannesburg Consolidated Investment Co. Ltd.; RM, Rand Mines Ltd.

[b]ERGO, East Rand Gold Operation.

Sources: Refs. 11 and 12.

TABLE II-2-39   Production of Gold by Mining House, 1981

Mining house	Production (kg fine)	Percent of total
AAC	247,920.1	38.1
AEC	1,124.6	0.1
ATC	37,109.8	5.7
GF	138,771.3	21.3
GMU	107,580.2	16.5
JCI	41,385.0	6.4
RM	70,620.6	10.9
Other	5,898.3	0.9

Sources: Refs. 11 and 12.

Reserves. South Africa is estimated to have the largest gold reserves in the world. Expected reserves for individual mines are presented in alphabetical order in Table II-2-40. Of course, possible gold reserves will vary depending on the price of gold and the cost of production.

Outlook. Prospects for future gold production in South Africa continue to be good. A number of expansion projects are under way.[11] A lower price for gold will probably cause smaller operations to amalgamate to survive lower gold prices.[12] A stronger demand and higher price for uranium would contribute to the continued operation of marginal mines, as will the state income tax for marginal mines. With gold being so important to the South African economy, it is unlikely that there will be anything but a gradual decline in South African gold production.

Platinum-Group Metals

The platinum-group metals are composed of six metals, divided into two groups of three metals each. The first group [ruthenium (Ru), rhodium

TABLE II-2-40   South African Gold Reserves by Mine

Mine	Reserves (thousand metric tons)	Grade (g/ton gold)
Blyvooruitzicht	5,338	20.9
Bracken	2,900	5.3
Buffelsfontein	11,626	10.57
Deelkraal		
Ventersdorp CR	1,079	6.2
Deelkraal Reef	597	4.6
Doornfontein		
Carbon leader	3,585	11.6
Main Reef	1,290	8.9
Driefontein Cons.		
East Driefontein		
Ventersdorp CR	9,461	19.6
Carbon leader	1,491	8.9
Main Reef	2,007	6.8
West Driefontein		
Carbon leader	3,942	23.9
Ventersdorp CR	2,515	13.4
Main Reef	1,329	7.7
Durban Deep	6,117	5.3
E.R.P.M.	7,724	7.3
Elandsrand	854	7.68
	910	7.44
Free State Geduld	8,744	13.60
	10,176	12.37
	11,580	11.38
Grootvlei		
Kimberley Reef	5,500	4.2
Main Reef	3,200	2.7
Harmony	30,607	6.2
Hartebeestfontein	14,508	11.9
Kinross	7,700	6.4
Kloof		
Ventersdorp CR	4,162	20.0
Kloof Reef	491	9.2

TABLE II-2-40 South African Gold Reserves by Mine (Cont'd.)

Mine	Reserves (thousand metric tons)	Grade (g/ton gold)
Leslie	3,500	4.9
Libanon		
Ventersdorp CR	3,703	11.5
Main Reef	6,120	5.4
Elsburg Reef	781	6.3
Kimberley Reef	175	8.1
Loraine	5,827	8.1
(additional, not included above)	391	7.8
Marievale		
Main Reef	20	4.3
Kimberley Reef	380	4.6
President Brand	9,092	12.00
	10,266	11.10
	11,392	10.28
President Steyn	14,490	10.15
	16,272	9.43
	17,563	8.95
Randfontein		
Cooke No. 1 shaft	3,671	11.4
Cooke No. 2 shaft	2,517	8.6
St. Helena		
Basal	11,400	10.0
Leader	300	4.1
Stilfontein	7,061	8.76
Unisel		
Basal	3,200	7.2
Leader	200	4.4
Middle	100	11.0
Vaal Reefs		
Vaal Reefs	25,487	13.43
	26,534	13.08
VCR and Elsburg Reef	4,263	4.29
	6,350	3.80
Dominion Reef	67	1.12
(Afrikander lease area)	70	1.12
Venterspost		
Main Reef	6,938	5.2
Ventersdorp CR	1,719	7.4
Western Areas		
Ventersdorp CR and Upper Elsburg Reefs	8,595	6.1
Middle Elsburg Reefs	1,891	2.8
Western Deep Levels		
Carbon leader	3,022	24.03
	3,054	23.84
VCR	2,949	11.94
	3,370	11.03
Western Holdings	18,559	9.81
	22,074	8.74
	24,341	8.17
West Rand Cons.	4,305	7.13
Winkelhaak	11,400	7.1
Wit. Nigel	976	6.11

Source: Ref. 12.

(Rh), palladium (Pd)] is of lower atomic mass, atomic number, and relative density. The second group [osmium (Os,) iridium (Ir), platinum (Pt)] is of greater atomic mass, atomic number, and relative density. As a group, these metals are indispensable to the electrical, chemical, and petroleum industries, where they are used as catalysts and corrosive-resistant materials.

South Africa produces all six metals and is estimated to have the world's largest reserves of these metals located within the Bushveld Igneous Complex.[19] South Africa is the world's largest producer of platinum and the world's largest exporter of platinum-group metals. South African platinum-group metal production is unique in that platinum-group metals are the primary products mined. In the USSR and Canada, platinum-group metals are by-products of nickel. As a consequence, South African producers can more easily regulate platinum-group metal production.

Geology. The main ore zones for platinum-group metals are located within the Critical zone of the Rustenberg Layered Suite.[13] Like gold, the main ore zones are called "reefs." Within the Critical zone, the most important ore zone, from which most of South Africa's platinum-group metals are produced, is the Merensky Reef. The reef is composed mostly of pyroxenite, containing a thin (2.5 cm) chromite marker band. The reef is mined by stoping methods to a thickness of 75 to 100 cm. The platinum-group metal content is 4 to 15 g per metric ton. The ore also contains 0.1 to 0.2% nickel and 0.1 to 0.15% copper, both as sulfides.[11]

Another reef, the Upper Seam of the upper group of chromite seams (UG2), located some 330 to 1155 ft below the Merensky Reef, is being mined by Western Platinum Ltd.[20] Another possible source of platinum-group metals is the Plat Reef, located at the contact between the Bushveld Complex and underlying rocks.[20]

Another important source for platinum-group metals, especially osmiridium, is within the Proterozoic conglomerate reefs of the Witwatersrand. The platinum-group metals are a by-product of gold.[21]

Production. Data on platinum-group production, sales and exports are not made public by the South African government or commercial sources. Production figures are given for all platinum-group metals in Table II-2-41 for 1970–1981. Of the platinum-group metals, platinum is the dominant metal, followed by palladium. The drop in production between 1980 and 1982 reflects a general

TABLE II-2-41 Platinum-Group Metal Production in South Africa, 1970–1981

Year	Metal content of concentrate, matte, and refinery products (thousand troy ounces)
1970	1503
1971	1253
1972	1453
1973	2363
1974	2830
1975	2600
1976	2700
1977	2870
1978	2860
1979	3017
1980	3100
1981[a]	3000
1982[a]	2600

[a]Ref. 7.
Source: U.S. Bureau of Mines, Minerals Yearbook, 1970–1981.

decline in the price for platinum-group metals rather than a drop in platinum-group metal-mining capacity.

There are three principal platinum-group metal producers in South Africa, with all three producing in the Merensky Reef within the Bushveld Igneous Complex. The largest producer of platinum-group metals in South Africa are the Rustenburg platinum mines, which are composed of four sections: the Rustenburg, Amandelbult, Atok, and Union, which are located in the Bophuthatswana Homeland. The mines are part of the Rustenburg Platinum Holding Company and the Matthey Rustenburg Refiners (Pty.) Ltd. of the United Kingdom. The management company is the Johannesburg Consolidated Investment Co. Ltd.[22] The refinery is located in Wadesville, Transvaal. Platinum-group metal production was estimated at 1.3 million troy ounces per year in 1981,[6] with an estimated capacity of 1.4 million troy ounces per year.[11] By-products include copper, gold, nickel, and silver.

The second largest producers are the combined mines of the Impala Platinum Holding Company. These mines are the Bafokeng North and South mines and the Wildebeestfontein North and South mines.[22] The mines are managed by the Union Corporation Ltd.[22] Production was an estimated 945,000 troy ounces per year in 1981,[6] with a capacity of 1 million troy ounces per year.[11] The refinery for the mines is located in Springs, Transvaal. By-products include copper, gold, nickel, and silver.

Lonrho S. A. Ltd., managed by Lonrho Ltd., is

the third largest platinum-group metals producer. Its Western Platinum Ltd. (Wesplat) mine produced some 132,000 troy ounces in 1981,[6] from the Merensky Reef. However, the company has plans to exploit the UG2 reef, which contains 27 to 37% chromium and 4.8 g per ton of platinum-group metals.[11] A new 60,000-ton per month concentrator is being built to produce a metals concentrate, a chromite-rich (35%) reject to be stockpiled, and tailings.

Of the 122,000 tons of ore per month being produced from the Merensky Reef, some 12,000 tons is being stockpiled to be used in the new concentrator.[6] The matte from the concentrator is sent to the Falconbridge refinery in Norway. From Norway the precious metals are sent back to Lonrho's refinery at Brakpan, Transvaal.

Another platinum-group metal mine, Der Brochen, located south of the town of Lydenburg, may be developed in the late 1980s. The property has been explored and mineral rights are owned by Platinum Property Ltd. of Lydenburg.[20] Location of the various mines are depicted in Figure II-2-13.

Previously, the conventional method of extracting platinum-group metals was costly and took from four to six months. A new process of solvent extraction for the extraction of gold and platinum-group metals has been developed by the South African National Institute of Metallurgy (NIM). The OPNIM process is now being used at a commercial plant at the Lonrho platinum refinery at Brakpan, South Africa, for Rh, Ru, Ir, and Os. Impala Platinum is also using the process. Matthey

Rustenburg Refiners is also conducting research on solvent-extraction techniques.[20] The OPNIM process reduces the refining time to 20 days and achieves 99.95% purity for products.[20] The solvent process also uses 20% less labor, and the equipment involved is 50% less expensive than that used in the conventional process.[20] Flowsheets for the conventional refining process and the new OPNIM refining process are given in Figures II-2-14 and II-2-15, respectively.

Platinum Sales. Almost all of the revenue from platinum-group metals is derived from export sales. These figures are not public information. However, average producer and dealer prices for 1977–1982 are provided in Table II-2-42. Prices are based on constant 1981 U.S. dollars.[23] Rhodium is the most expensive of the platinum-group metals, followed by iridium, platinum, osmium, palladium, and ruthenium.

Producer prices are set by the major producing countries: South Africa, Canada, and the USSR. These prices are usually applied to industrial accounts and long-term purchases. In contrast, dealer prices are determined by spot purchases, exchange prices, hedging, and speculative accounts.[23]

The value of export sales for platinum-group metals for 1975–1980 are reported to be the following by the U.S. Bureau of Mines *Minerals Yearbooks:* in 1975 ($529 million), in 1976 ($460 million), in 1977 ($465 million), in 1978 ($462 million), in 1979 ($793 million), and in 1980 ($1100 million).

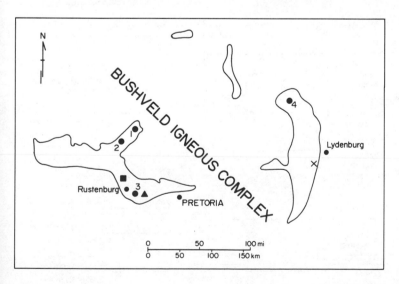

● Rustenburg Sections – J.C.I.
  1. Amandelbuit
  2. Union
  3. Rustenburg
  4. Atok

■ Impala Sections – Union Corp.
  Bafokeng North & South
  Wildebeestfontein

▲ Western Platinum Ltd.– Lonrho

✕ Der Brochen (undeveloped)

FIGURE II-2-13 Platinum mines in South Africa. (From Ref. 24.)

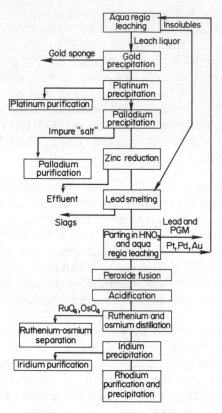

FIGURE II-2-14 Conventional platinum refining process. (From Ref. 20.)

Reserves. Reserves for platinum-group metals in South Africa vary considerably. Much of these discrepancies can be attributed to the different schemes used to define reserves and resources of these metals. In all analyses, South Africa is estimated to have the world's largest reserve base of platinum-group metals. Table II-2-43 compares the various proportions of platinum-group metals in the Merensky Reef, UG2, and the Platreef.[22]

Recent estimates of platinum-group metal reserves and reserve base by the U.S. Bureau of Mines are given in Table II-2-44.[23] Estimates were based on reserves to a depth of 2000 ft and do not include possible reserves from the Platreef.

Outlook. Future demand for platinum-group metals will be affected by the performance of the automotive, electrical, chemical, and petroleum industries. For platinum, the biggest single end use is for jewelry in Japan, followed by the demand for emission converters in automobiles in the United States.

The small number of producers in South Africa and the world make it possible to maintain a steady price more easily by corresponding cuts in production and availability.[12] South Africa enjoys a definite advantage in the production cost of platinum-group metals, as indicated by Table II-2-45.[20] The lower production cost and ready availability of

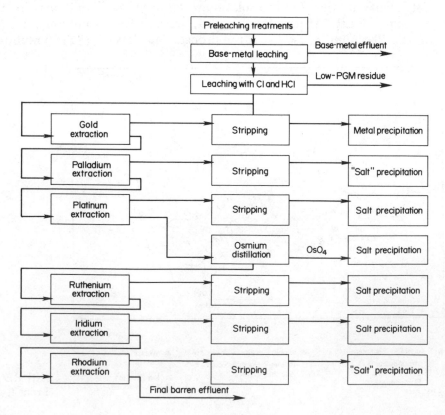

FIGURE II-2-15 New OPNIM refining process. (From Ref. 20.)

TABLE II-2-42   Yearly Average Production and Dealer Prices for Platinum-Group Metals, 1977–1982 (1981 Constant Dollars per Troy Ounce)[a]

	Platinum		Palladium		Rhodium		Iridium		Ruthenium		Osmium	
Year	P	D	P	D	P	D	P	D	P	D	P	D
1977	226	157	83	49	611	409	357	258	83	35	242	130
1978	306	261	92	63	658	524	387	240	72	33	194	130
1979	419	445	134	120	872	770	306	280	54	32	179	130
1980	479	677	234	201	837	729	552	666	49	35	164	130
1981	475	446	130	95	641	498	600	529	45	32	150	130
1982	448	327	104	67	566	323	566	359	42	26	142	130

[a]P, producer price; D, dealer price.

Source: Ref. 23.

TABLE II-2-43   Relative Proportions of Platinum-Group Metals in the Merensky Reef, UG2, and Platreef

	Merensky Reef	UG2	Platreef
Pt	59	42	42
Pd	25	35	48
Ru	8	12	4
Rh	3	8	3
Ir	1	2.3	0.8
Os	0.8		0.6
Au	3.2	0.7	3.4

Source: Ref. 22.

TABLE II-2-44   South African and World Reserves of Platinum-Group Metals[a] (Million Troy Ounces)

	South Africa	Percent of world total	World total
Platinum	385	85.6	450
Palladium	253	66.6	380
Rhodium	50	100	50
Platinum group	790	79.0	1000
Total platinum group	970	80.8	1200

[a]To a depth of 2000 ft; does not include the Platreef.

Source: Ref. 23.

future supplies do much to prohibit the entry of new producers outside South Africa.

Silver

South Africa is not a major world producer of silver. Until recently, the majority of silver produced in South Africa was a by-product of the gold fields of the Witwatersrand Basin. In mid-1980, the country's first major lead-silver deposit at Broken Hill in north-central Cape Province, near Namibia/South-West Africa, was developed.

Production. Silver production has remained steady, with a slight decline until 1981, as indicated in Table II-2-46. In 1981 there was a dramatic increase in silver production as the Black Mountain Mineral Development Co. (Pty.) Ltd., managed and owned (51%) by Gold Fields of South Africa Ltd. (GFSA) and the U.S.-based Phelps-Dodge (49%), came into full production.

Production was from the upper high-grade lens of the Broken Hill ore body, which averaged some 5.4 troy ounces of silver per ton.[15] In 1981, the mine milled some 1.15 million tons of complex lead-silver-copper-zinc ore.[6] The ore was concentrated in a computer-controlled concentrator using selective flotation methods.[15] The concentrates

TABLE II-2-45   Estimated Typical Operating Costs of World Platinum Mines

Mine type	Ore treated (million metric tons/year)	Weighted-average dollars per troy ounce refined platinum				
		Mine	Mill	Misc.[a]	Refining	Total
Underground						
South Africa	1.3–15.8	151	33	28	16	228
Zimbabwe	0.36	796	107	175	16	1094
Placer: U.S. and and Colombia	12.0–13.0	468	6	0	27	501

[a]Includes costs related to town-site facilities.

Source: Ref. 20.

TABLE II-2-46 Primary Silver Production in South Africa, 1970–1981 (Thousand Troy Ounces)

Year	Production
1970	3527
1971	3378
1972	3294
1973	3652
1974	2699
1975	3084
1976	2821
1977	3130
1978	3104
1979	3236
1980	3125
1981	7568

Source: U.S. Bureau of Mines, *Minerals Yearbook*, 1970–1981.

were nearly 100,000 tons of lead, with more than 3 million troy ounces of silver, and a copper concentrate with 4570 tons of copper and nearly 1 million troy ounces of silver.[6] A zinc concentrate of nearly 21,000 tons of zinc and about 45,000 troy ounces of silver was also produced.

The concentrates were trucked over a 166-km road constructed by the Provincial administration to the mine's loading and storage facilities on the Sishen-Saldanha Railroad. The lead and copper concentrates were taken by rail to Saldanha Bay, where they are shipped overseas. The zinc concentrates were shipped to the Zinc Corp. of South Africa Ltd. (ZINCOR) refinery at Springs in the Transvaal Province. ZINCOR is a subsidiary of GFSA.

Reserves. The reserves of silver of the Broken Hill ore body are estimated to be 2.7 troy ounces of silver per ton from an estimated 38 million troy tons of ore available.[15] Silver is recovered from the gold fields in an average of 1 part silver to 10 parts gold. However, the silver recovered from the mines is alloyed in the 26-kg gold bullion ingots produced (88% gold, 10% silver, 2% base-metal impurities). The silver is delivered to Rand Refining Ltd. at Germiston.[6] Total silver reserves are estimated to be 8700 metric tons.[2]

Diamonds

The first diamond find in South Africa was near Hopetown on the Orange River in 1866. However, it was not until a few years later in 1870 that South Africa's first mineral "rush" resulted from diamond finds in the Kimberley area from what came to be known as kimberlite pipes.[8] Soon after, South Africa became the world's largest producer of gem-quality diamonds—a position it still enjoys.

Geology. There are three source types of diamondiferous deposits in South Africa: (1) the ancient conglomerates of the Witwatersrand (a very unimportant source), (2) the kimberlite pipes and fissures, and (3) alluvial gravel and boulder beds of fluvial and marine origin.

The kimberlite pipes are volcanic intrusions whose diameters vary from less than 30 m to some 1.5 km in diameter. The pipes are usually oval in shape and possess nearly vertical sides. The kimberlite material is usually classified as either basaltic or micaceous kimberlite. The basaltic kimberlite "consists of phenocrysts of olivine, with a small amount of phlogopite and ilmenite, embedded in a groundmass of secondary minerals rich in perovskite, apatite, iron ores and chromite."[8] Micaceous kimberlite "consists of phenocrysts of olivine and abundant phlogopite in a dark fine-textured matrix of phlogopite, olivine, apatite, ilmenite, perovskite, and chromite, with serpentine and calcite."[8]

The kimberlite body itself usually consists of (1) yellow ground, representing hydrated and oxidized kimberlite nearest the surface; (2) blue ground, which does not crumble as readily; and (3) a "hardebank," which is a solid variety of kimberlite that does not disintegrate with exposure to weather.[8]

Diamonds in the pipes are found in well-formed crystals, broken crystals, and cleavage fragments, which are the most abundant. Color and size of the diamonds varies greatly.

Both the fluvial and marine diamond deposits are believed to have originated from the weathering of kimberlite pipes and fissures. Fluvial deposits consist of (1) shallow deposits of high-level gravel over floodplains, (2) well-defined gravel terraces bordering the present river beds, (3) deep gravel or deep placers representing ancient gravel-filled river channels, and (4) accumulations of gravel in deep pools and hollows along river beds.[8]

Marine (coastal) deposits occur both on raised beach deposits and offshore. These terrace deposits consist of marine sand, grit, gravel, and boulder beds. The distribution and concentration of diamonds in beach gravels are controlled by factors of time, turbulence, bedrock topography, and bedrock lithology.[8]

Production. South Africa is the largest gem diamond producer in the world and produces a significant amount of industrial diamonds as well. Diamond production has increased slightly over the past decade, as indicated by Table II-2-47. The drop in production from 1980 to 1981 reflects a deliberate policy to cut production due to the uncertain and flooded diamond market.

TABLE II-2-47 Diamond Production in South Africa, 1970–1981

Year	Gem	Industrial (thousand carats)	Total	Export sales (thousand U.S. dollars)
1970	3,758	4,354	8,112	105,734
1971	3,169	3,862	7,031	90,558
1972	3,370	4,025	7,395	116,407
1973	3,448	4,117	7,565	234,620
1974	3,440	4,070	7,510	210,444
1975	3,435	3,860	7,295	238,039
1976	3,340	3,683	7,023	247,334
1977	3,099	4,544	7,643	296,232
1978	3,078	4,649	7,727	512,656
1979	3,539	4,845	8,384	651,373
1980	3,403	5,117	8,520	710,882
1981	3,429	6,097	9,526	390,562

Source: U.S. Bureau of Mines, Minerals Yearbook, 1970–1981.

Production from individual diamond mines is given in Table II-2-48. Of the various diamond divisions, the diamond pipes of the Kimberley division are located north and south of Kimberely and west of Bloemfontein, the Premier mine is northeast of Pretoria, and the Namaqualand marine alluvial diggings are on the west coast north of Lambert's Bay and up into Namibia/Southwest Africa, as indicated in Figure II-2-16.

De Beers Consolidated Mines Ltd. owns or leases from the government most of the diamond mines in South Africa. In the Kimberley Division, De Beers owns the De Beers Mine, Kimberley, Kamfersdam, Koffiefontein, and Wesselton, and leases the Bultfontein and Du Toits mines from its subsidiaries, the Consolidated Company Bultfontein Mine Ltd. and the Griqualand West Diamond Mining Company Dutsitspan Mine Ltd. The Finsch mine is owned 30% by De Beers, with the other 70% being leased by De Beers from the government.[12]

The Premier Mine in the Transvaal is owned 40% by De Beers, with the other 60% being leased by De Beers from the government.[12] In the Namaqualand area along the western seacoast, De Beers owns 75% of mining operations at Dreyers Pan, Koingness, Langhoogte, Mitchell's Bay, Sankop, and Tweepan, and leases the other 25% from the government. The government's 30.7% interest in the Annex Kleinzeee is also leased by De Beers, which owns the other 69.3%.[12]

TABLE II-2-48 Diamond Mines in South Africa, 1981

Division	Mine	Total tonnage treated	Carats recovered	Carats per 100 metric tons
Kimberley		12,530,900	5,923,267	
	De Beers mine	663,300	127,989	19.30
	Dutoitspan mine	586,800	98,758	16.83
	Bultfontein	663,200	268,032	40.41
	Bultfontein dumps	1,131,400	367,510	32.48
	Wesselton	1,377,200	274,399	19.92
	Finsch	4,845,500	4,463,944	92.13
	Koffeefontein	3,263,500	322,635	9.89
Namaqualand		6,537,000	1,214,077	18.57
	Koingnass area	1,400,000	550,119	32.81
	Buffels complex	4,500,000	622,486	13.72
	Langhoogte mine	323,500	41,472	12.82
	Alexander Bay State Alluvial diamond diggings		125,405	
Premier Mine		5,200,000	1,501,157	28.81
	Premier retreatment plant	1,600,000	538,654	33.64

Source: Ref. 6.

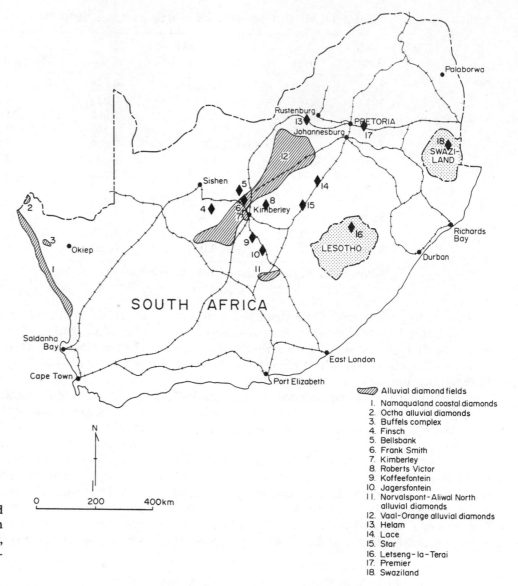

Alluvial diamond fields
1. Namaqualand coastal diamonds
2. Octha alluvial diamonds
3. Buffels complex
4. Finsch
5. Bellsbank
6. Frank Smith
7. Kimberley
8. Roberts Victor
9. Koffeefontein
10. Jagersfontein
11. Norvalspont–Aliwal North alluvial diamonds
12. Vaal-Orange alluvial diamonds
13. Helam
14. Lace
15. Star
16. Letseng–la–Terai
17. Premier
18. Swaziland

FIGURE II-2-16 Diamond mines in South Africa. (From U.S. Department of the Interior, Bureau of Mines, *Mineral Industries of Africa*; Ref. 10.)

Export Sales. South Africa's diamonds are marketed through the De Beers Central Selling Organization. Export sales for 1970–1981 are given together with production figures in Table II-2-47. Revenue for 1981 was down sharply from the high of nearly $711 million in 1980. The contribution of diamond sales to the overall value of total mineral sales has declined steadily over the years.

Reserves. No official figures for diamond reserves are given.

Outlook. Future diamond production will depend greatly on the demand for jewelry and industrial diamonds. The recent major discovery of diamonds in Western Australia by the Ashton Joint Venture will affect sales of diamonds from South Africa. The company is expected to produce between 2 and 3 million carats annually by 1990.

Gem diamonds are estimated to make up 10% of this production. De Beers has acquired rights to sell the gem-quality diamonds from the Ashton mines, Northern Mining, and the Australian government.[12]

Other significant actions likely to affect diamond sales are the selling of diamonds by the governments of the USSR and Zaire (which recently broke away from the De Beers Central Selling Organization).[12]

FERROUS MINERAL COMMODITIES

Ferrous minerals are minerals that can be combined with iron to form steel and special alloy steels. The major ferrous metals produced by South Africa are chromium, iron ore, manganese, nickel, and vanadium.

Chromium

South Africa is the world's largest producer of both chrome ore and ferrochromium, followed by the USSR. South Africa also possesses the largest reserves of chromium in the world.

There are three distinct classifications of chrome ore, depending on the ratio of chromium to iron. For the metallurigcal grade, the Cr/Fe ratio is >2:1 (>45% Cr_2O_3), chemical-grade ore has a Cr/Fe ratio of <2.0:1 (40 to 45% Cr_2O_3), and refractory-grade ore, which requires a consistency of composition, contains more than 20% Al_2O_3 and more than 60% Al_2O_3 and Cr_2O_3 combined. Chrome ore is further classified as either lumpy (remains coherent) or friable (breaks down to individual grains).[8]

Geology. South African chromite production is from the mostly stratiform deposits found in the Bushveld Igneous Complex in the Transvaal. There are some high-grade podiform deposits within the Bushveld Igneous Complex, but production from these sources is small. South African chromite from the Bushveld is generally friable in composition and is predominantly of chemical grade.

There are three main groupings of the stratiform chromite seams within the Bushveld Igneous Complex, with an increasing Cr/Fe ratio downward. These divisions have come to be known as the Lower (or Main) Group, Middle Group, and Upper Group. Mining operations are usually confined to the Lower Group layers, because of the higher Cr/Fe ratio. It is the usual practice to mine only the thickest seams of chromite.[24] The chromite is present as seams in the pyroxenite, norite, and anorthosite units of the complex.[8]

In the Western part of the Complex, the main seam is referred to as LGG (also known as the Main or Magazine seam) with a strike length of some 70 km. In the eastern part of the complex, this seam is known as the Steelpoort seam and has a strike length of 90 km. The Steelpoort seam is worked only north of Steelpoort. South of Steelpoort, the main seam is the F seam, and has a strike length of 35 km. The seams of Potgietersrus and Zeerust areas are much lower in the petrographic sequence than the LGG or the F seams and have a higher Cr/Fe ratio.[24]

Recently, Western Platinum Ltd. has begun to mine the Upper Group 2 (UG2) seam found below the Merensky Reef. The UG2 contains both chromium and platinum-group metals. The chromium can be beneficiated up to 40 to 42% Cr_2O_3.[8]

Production. South Africa's chromite industry has been greatly affected by the world recession and the subsequent drop in world steel and ferro-alloy production. Production figures for the three grades of chromite are given in Table II-2-49. Growth rates have slowed considerably since 1977, and both 1981 and 1982 showed major declines in production.

Major chrome producers are given in Table II-2-50. In 1981, several of the large producers shut down some of their mines, because of the oversupply of chrome in the last several years. Among those mines affected by shutdowns were the large Winterveld mine near Steelpoort, which shut down

TABLE II-2-49   Chromite Production in South Africa, 1970-1981 (Thousand Metric Tons)

Year	>48% Cr_2O_3	44-48% Cr_2O_3	<44% Cr_2O_3	Total	Percent growth
1970	97.1	934.9	395.2	1427.2	
1971	100.8	1031.3	512.1	1644.2	+15.2
1972	48.4	977.8	456.9	1483.1	-9.7
1973	30.1	957.8	661.7	1649.6	+10.1
1974	20.5	938.6	917.8	1876.9	+13.8
1975	15.2	1127.5	932.6	2075.3	+10.6
1976	25.6	1311.9	1071.7	2409.2	+16.1
1977	53.0	1607.0	1399.9	3059.0	+27.0
1978	33.0	1524.0	1588.0	3145.0	+2.8
1979	33.0	1633.0	1631.0	3297.0	+4.8
1980	20.0	1989.0	1405.0	3414.0	+3.5
1981	36.0	1561.0	1273.0	2870.0	-15.9
1982[a]				2630.0	-8.4

[a]Ref. 7.

Source: U.S. Bureau of Mines, *Minerals Yearbook*, 1970-1981.

TABLE II-2-50   Major Chrome Producers in South Africa

Management company	Mining company	Operation active
Anglo-American Corp. Ltd. Associated Ore & Metal Corp. Ltd.	Vereeniging Refractories Ltd.	Marico Mineral Co. (Pty.) Ltd. Zeerust Chrome Mines Ltd.
Barlow Rand Ltd. (Rand Mines Ltd.)	Transvaal Consolidated Land & Exploration Co. Ltd.	Henry Gould (Pty.) Ltd. Millsell Chrome Mines (Pty.) Ltd. Winterveld (T.C.L.) Chrome Ltd.
BAYER AG	Chrome Chemicals (Pty.) Ltd.	Rustenburg Chrome Mines (Pty.) Ltd.
General Mining Union Corp. Group	Transvaal Mining & Finance Co. Ltd.	Groothoek Chrome Mine Kroondal Chrome Mine Montrose Chrome Mine Zwartkop Chrome Mine
International Mineral & Chemical Corp.	Lavino, S.A. Pty. Ltd.	Grootboom (Lavino) Chrome Mine
Metallurgy S.A. (Pty.) Ltd.		Waterkloof Chrome Mines (Pty.) Ltd.
Mining Corp. Ltd.	Bantu Mining Corp.	Dilokong Chrome Mine (Pty.) Ltd.
S.A. Manganese Amcor Ltd.	Bathlako Mining Ltd. Cromore Pty. Ltd.	Ruighoek Chrome Mine Grasvally Chrome Mine Mooinooi Chrome Mine
Union Carbide Corp.	Union Carbide Africa & Middle East Inc.	UCAR Chrome Co. (S.A.) (Pty.) Ltd.

Source: Ref. 24.

three sections of the mine.[6] GENCOR closed its mines in the Rustenberg area.[6] The location of the major mines is given in Figure II-2-17.

The U.S. Bureau of Mines estimates the mine capacity of chromium as 1,500,000 short tons of contained chromium in 1981 and 1982, which will

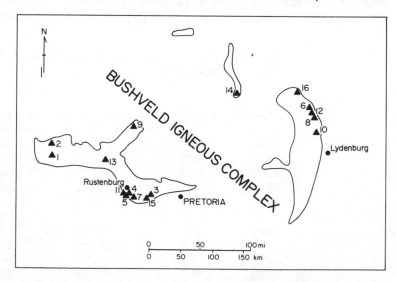

1. Marico Mineral Co. (Pty.) Ltd.
2. Zeerust Chrome Mines Ltd.
3. Henry Gould (Pty.) Ltd.
4. Millsell Chrome Mines (Pty.) Ltd.
5. Rustenburg Chrome Mines (Pty.) Ltd.
6. Groothoek Chrome Mine
7. Kroondal Chrome Mine
8. Montrose Chrome Mine
9. Zwartkop Chrome Mine
10. Grootboom Mine
11. Waterkloof Chrome Mines (Pty.) Ltd.
12. Dilokong Chrome Mine (Pty.) Ltd.
13. Ruighoek Chrome Mine
14. Grasvally Chrome Mine
15. Mooinooi Chrome Mine
16. UCAR Chrome Co. (S.A.)(Pty.) Ltd.

FIGURE II-2-17   Major chrome mines in Bushveld Igneous Complex. (From Ref. 24.)

increase to 1,800,000 short tons of capacity in 1985.[25] In 1981, South African mines operated at only 63.1% of capacity.

Ferrochromium Production. South Africa was the world's largest producer of ferrochrome in 1981, having produced some 490,000 short tons of contained chromium. In 1981, ferrochrome production was 90.7% of the rated capacity of 540,000 short tons of contained chromium for ferrochrome producers. This capacity is expected to fall to 520,000 tons for 1982 to 1985.[25] Ferrochrome production for 1975–1980 is given in Table II-2-51.

Rapid expansion of South Africa's ferrochromium producing capacity is due mainly to the development of the argon–oxygen decarburization (AOD) process, which allowed the use of South Africa's charge chrome in the manufacture of stainless steel. The building of ferrochrome facilities within South Africa also ensured that more profits would be realized from chromite ore production. Major ferrochrome producers are given in Table II-2-52.

South Africa ferrochromium producers enjoy several advantages in ferrochromium production. First, the AOD process can use the cheaper South African chromite (chemical grade). Second, South

African producers are closer to the source of ore. Finally, South Africa has lower cost structures.[24]

Reserves. Reserve figures for chromite vary depending on the classification system used and the depth to which reserves are estimated. The U.S. Bureau of Mines estimates world chromite reserves at 1000 million tons shipping ore, gross weight, and a reserve base of some 2000 million short tons. Of this amount, South African reserves are put at 910 million tons (91%) of reserves and 910 million short tons (45.5%) of reserve base.[25]

Chromium Sales. The value of chromium has been increasing steadily since 1970, with domestic sales capturing a growing share of the market in the period 1970–1981. The value of chromite sales is given in Table II-2-53. The average annual prices for a ton of chromite and ferrochromium are given in Table II-2-54. By adding ferrochromium production within South Africa, the chromite industry gained substantial additional revenue.

Exports. South Africa exports a large percentage of both its chromite ores and ferrochromium. Exports of chromite and ferrochromium are given in Table II-2-55 for 1973–1980. The nearest seaport is Maputo, Mozambique, some 55 km from the eastern belt and up to 790 km from the farthest portion of the western belt.[8]

Outlook. Long-term prospects for South African chromite and ferrochrome are considered good. Many of the most recent world ferrochrome producers are using the AOD process, thereby establishing the demand for lower-grade South African chromite. It is also apparent that with the cost of production advantageous in producing ferrochrome and other ferro-alloys, South Africa has a strong competitive edge in the world market.

Competition from Zimbabwe will depend largely on the internal policies and politics the

TABLE II-2-51 Ferrochrome Production in South Africa, 1975–1980 (Thousand Metric Tons)

Year	Production	Percent growth
1975	299	
1976	350	+17.1
1977	350	0.0
1978	660	+88.6
1979	780	+18.2
1980	800	+2.6

Source: U.S. Bureau of Mines, *Minerals Yearbook*, 1975–1981.

TABLE II-2-52 Major Ferrochrome Producers in South Africa

Smelter	Shareholders	Locality
Ferroalloys Ltd.	Associated Manganese U.S. Steel Corp.	Machadodorp
Ferrometals Ltd.	South Africa Manganese Corp. (SAMANCOR)	Witbank
Palmiet Chrome Corp. (Pty.) Ltd. RMB Alloys (Pty.) Ltd. Tubatse Ferrochrome (Pty.) Ltd.	General Mining Union Carbide	Steelpoort
Consolidated Metallurgical Industries Ltd.	Johannesburg Consolidated Inv. Showa Denko British Steel	Lyndenburg

Sources: Refs. 6 and 24.

TABLE II-2-53   Value of Chromite Sales in South Africa, 1970–1981 (Thousand U.S. Dollars)

Year	Domestic	Percent total	Export	Percent total	Total	Percent growth
1970	2,833	19.2	11,945	90.8	14,778	
1971	3,761	18.5	16,520	81.5	20,281	+37.2
1972	2,919	17.7	13,527	82.3	16,446	-18.9
1973	4,967	20.3	19,513	79.7	24,480	+48.8
1974	7,089	24.5	21,890	75.5	28,979	+18.4
1975	16,701	29.9	39,090	71.1	55,791	+92.5
1976	25,537	26.5	44,430	63.5	69,967	+25.4
1977	41,361	36.8	71,103	63.2	112,464	+60.7
1978	31,357	30.8	70,354	69.2	101,711	-9.6
1979	44,181	41.6	61,957	58.4	106,138	+4.4
1980	51,717	44.3	65,095	55.7	116,812	+10.1
1981	38,357	38.7	60,807	61.3	99,164	-15.1

Source: U.S. Bureau of Mines, Minerals Yearbook, 1970–1981.

TABLE II-2-54   Time–Price Relationships for Chromium
(Average Annual Price, Dollars per Short Ton[a])

Year	Chromite		Ferrochromium	
	Annual prices	Based on constant 1981 dollars	Actual prices	Based on constant 1981 dollars
1960	62	175	419	1181
1961	56	156	407	1137
1962	56	154	437	1199
1963	49	132	341	922
1964	51	136	327	870
1965	54	141	366	953
1966	53	134	334	843
1967	56	137	358	877
1968	53	124	346	812
1969	58	129	336	750
1970	72	153	364	771
1971	79	159	421	849
1972	81	160	383	742
1973	74	137	355	651
1974	87	147	544	917
1975	159	245	963	1486
1976	192	283	831	1218
1977	187	259	749	1038
1978	181	234	623	804
1979	195	232	858	1021
1980	201	220	882	963
1981	198	198	863	863
1982e[b]	207	195	914	862

[a]Custom value per ton of chromium contained in imported material.
[b]e, estimated.
Source: Ref. 25.

indigenous government assumes with respect to Zimbabwe's chromite producers. With over 90% of the world's chromite reserves located within South Africa, the country's chromite industry's long-term prospects are good.

The recovery of the chromite industry in the short-term depends largely on the world economy. Recessions generally are followed by lower steel and ferroalloy production, which greatly affect the principal demand for chromium.

TABLE II-2-55   Exports of South African Chromite and Ferrochrome, 1973-1980 (Thousand Metric Tons)

Year	Chromite ore and concentrate, gross weight	Ferrochrome
1973	948.3	175.2
1974	780.8	194.8
1975	907.3	283.7
1976	597.5	340.6
1977	1085.3	330.3
1978	1129.6	674.8
1979	1269.0	599.3
1980	1376.0	691.4

Source: U.S. Bureau of Mines, Minerals Yearbook, 1973-1981.

Manganese

South Africa is the second largest producer of manganese ore in the world, producing over 5 million metric tons in 1981. South Africa also contains the largest known deposits of manganese ore in the world.[2] South Africa is the world's largest producer of electrolytic manganese. It is the world's largest exporter of ferromanganese and electrolytic manganese metal and a major supplier of metallurgical ore and manganese chemicals.

Manganese is essential in the production of steel and certain steel alloys. It is used in making dry-cell batteries and is important in the chemical industry.

Geology. High-grade manganese ore in South Africa is found within the Transvaal System of the Postmasburg and Kalahari fields in northwest Cape Province. Ore grades range from 28 to 50% Mn and 4.5 to 26% Fe.[15] Chemical-grade ore is obtained from shallow deposits in a belt extending from Kurgersdorp, Transvaal, northwest to the Botswana border.

The Postmasburg field, with an Eastern Belt and a Western Belt, extends from Postmasburg 65 km northward to Sishen. The field was discovered in 1922.[8] Most mining occurs in the Western Belt. Manganese ore is found as replacement bodies in the Gamagara shale or as bodies of ore that have slumped into the underlying Dolomite Series.[8]

The Kalahari field extends from some 45 km north of Sishen eastward to the Kuruman River. The Kalahari sand covers this field, and prospecting did not occur until the early 1950s.[8] The manganese mineralization is thought to have originated syngenetically as chemical precipitates, followed by a subsequent supergene enrichment and replacement. These deposits are located within the banded iron formations with three horizons of manganese ore near the base of the Daspoort Stage, with cal-careous dolomite higher up in the succession. The three ore bodies are known as the Bottom, Middle, and Top bodies.[8] The manganese ore in the western Transvaal is located in the dolomites of the Dolomite Series. The deposits are not large and are usually rather shallow.

Production. South Africa produces both metallurgical- and chemical-grade manganese ore. The great majority of ore produced is of the metallurgical grade. Within both the metallurgical and chemical grades of ore, various concentrations of manganese are produced. Figures for manganese ore and concentrate production for 1970-1981 are given in Table II-2-56.

South African Manganese Amcor Ltd. (SAMANCOR) continues to be the largest manganese ore producer in South Africa. The government's share in SAMANCOR is 45%.[6] With the purchase of the Middelplaats underground mine from Anglo-American Corporation in March 1981, SAMANCOR has a mining capacity of well over 4 million metric tons per year.[26] Except for its Lohatla Manganese Mine, SAMANCOR's mines are located in the Kalahari field.

The Associated Manganese Mines of South Africa Ltd. is the second largest producer of manganese ore in South Africa, with a mining capacity of over 2 million metric tons per year in 1978.[26] Mines of the Postmasburg and Kalahari fields are given in Table II-2-57.

There are various mines operating in the Transvaal Province. Chief among these companies is the Rand London Manganese Mines (Pty.) Ltd., which operates eight small mines in the Transvaal Province. Of these, the Ryedale mine and the Brandvlei mine in the Ventersdorp district each produce from 3000 to 4000 metric tons per month of 35 to 40% MnO_2. These two mines are the chief suppliers of MnO_2 for the extraction of uranium from uranium ores. In addition to the Transvaal Province mines, Rand London Manganese Mines (Pty.) Ltd. also owns the Gopane mine, located within the Boputhatswana Homeland in northern Cape Province. This mine produces 4000 metric tons per year of battery-grade ore (76.5% MnO_2) and is the only battery-grade ore producer in South Africa. Total annual production for the combined mines is 400,000 metric tons per year of uranium-grade ore, and 50,000 metric tons per year of chemical ores other than uranium grade.

Ferromanganese production in South Africa has grown rapidly in recent years. Until recently, South Africa exported most of its ferromanganese. However, with the expansion of the steel industry, more ferromanganese is being purchased for domes-

TABLE II-2-56   Manganese Production in South Africa, 1970-1981 (Thousand Metric Tons)

(a) Metallurgical Grade

Year	48% Mn	45–48% Mn	40–45% Mn	30–40% Mn	Total metallurgical	Percent of total
1970	664	113	109	1666	2552	95.3
1971	689	87	239	2140	3155	97.5
1972	743	116	175	2143	3177	97.1
1973	935	278	246	2629	4088	97.9
1974	1138	218	264	3026	4647	97.9
1975	199	1379	232	3872	5683	98.5
1976	270	1517	209	3358	5354	98.2
1977	263	1198	577	2839	4877	96.6
1978	262	1131	430	2357	4180	96.8
1979	296	998	763	2897	4954	95.6
1980	290	942	997	3099	5328	93.6
1981	368	1226	676	2429	4699	93.2

(b) Chemical Grade

Year	65% MnO_2	35–65% MnO_2	35% MnO_2	Total chemical	Percent of total	Grand total	Percent growth
1970	11	116		127	4.7	2679	
1971	7	74		81	2.5	3236	+20.8
1972	8	86		94	2.9	3271	+1.1
1973	8	79		87	2.1	4175	+27.6
1974	5	93		98	1.5	4745	+13.7
1975	7	79		86	1.5	5769	+21.6
1976	3	95		98	1.8	5452	-5.5
1977	—	171		171	3.4	5048	-7.4
1978	—	118	19	137	3.2	4317	-14.5
1979	—	153	76	229	4.4	5183	+20.0
1980	—	166	201	367	6.4	5695	+9.9
1981	—	45	296	341	6.8	5040	-11.5

Source: U.S. Bureau of Mines, Minerals Yearbook, 1970-1981.

tic use, then exported in various steel and ferro-alloy products. Production of ferromanganese from 1975 to 1980 is given in Table II-2-58. Major producers are given in Table II-2-59, with major mines located in Figure II-2-18.

Manganese Sales. The value of manganese exports continues to be much greater than domestic sales. Both domestic and export sales have grown steadily in the last decade, as indicated in Table II-2-60.

Reserves. Reserves of South Africa's manganese ores have grown dramatically in the last decade. South African sources provide the following figures for proved and probable ore in-situ:[26]

More than 44% Mn	310,800,000
40 to 44% Mn	30,000,000
30 to 40% Mn	7,181,000,000
20 to 30 Mn	4,613,000,000
Total	12,139,800,000

The U.S. Bureau of Mines estimates South Africa's reserves at 407,000,000 short tons of a world total of 1,000,000,000 short tons; and a reserve base of 2,900,000,000 short tons of a world total of 4,000,000,000.[27] In both instances, South Africa possesses the world's largest land-based reserves of manganese ore.

Exports. South Africa is the world's largest exporter of ferromanganese and electrolytic manganese, and a major supplier of metallurgical ores and manganese chemicals. Exports of manganese ores and ferromanganese are given in Table II-2-61.

Manganese ore exports are delivered to the 6.5-million ton capacity ore outloading facilities at Port Elizabeth. The overall distance from the mines in the Kalahari and Postmanburg fields is about 1000 km.[15]

Outlook. The demand for manganese is tied directly to the production of steel. Therefore, the

TABLE II-2-57   Manganese Mines in the Postmasburg and Kalahari Fields, South Africa

Field	Producer	Mine	Type of mine[a]	Grade of ore
Kalahari	SAMANCOR	Wessels	UG	44–65% Mn, 6–18% Fe
		Mamatwan	OP	40% Mn, 7% Fe
		Hotazel	OP	40% Mn, 7% Fe
		Middelplaats	UG	
	AMMOSAL	Adams	OP	
		Belgravia	OP	
		Black Rock	UG	40–48% Mn, 10–20% Fe
		Devon	OP	
		Gloria	OP	
		Perth	OP	
		Santoy	OP	
		N'Chwaning	UG	
	National Manganese Mines (Pty.) Ltd.	Annex Langdon		
Postmasburg	SAMANCOR	Lohatla	OP	
	AMMOSAL	Southern Farms		
	Consolidated African Mines Ltd.	Pensfontein and Rooinekke		35% Mn, 25–27% Fe

[a]UG, underground; OP, open pit.

Sources: Refs. 6, 8, and 15.

TABLE II-2-58   Ferromanganese Production in South Africa, 1975–1980 (Thousand Metric Tons)

Year	Production
1975	337
1976	350
1977	310
1978	330
1979	560
1980	520

Source: U.S. Bureau of Mines, Minerals Yearbook, 1975–1981.

TABLE II-2-59   Ferromanganese Producers in South Africa

Company	Location
SAMANCOR	Kookfontein (Transvaal)
	Newcastle (Natal)
	Mayertown (Transvaal)
AMMOSAL	Cato Ridge (Natal)
AAC (Highveld Steel & Vanadium Corp.)	Witbank (Transvaal)
ISCOR	Vanderbijlpark (Transvaal)
	Pretoria (Transvaal)
	Newcastle (Natal)
Delta Manganese (Pty.) Ltd.	Nelspruit (Transvaal)

Source: U.S. Bureau of Mines, Minerals Yearbook, 1978–1981.

recovery of the manganese industry will depend on an end to the world recession.

Vanadium

In 1980, the South African government classified all statistics relating to the production, sales, and exports of vanadium ores or products.[15] Vanadium, like platinum and nickel, is regarded as a critical commodity. However, reasonable guesses can be made from recent statistics and prevailing market conditions. South Africa remains the world's largest producer of vanadium.

Vanadium is primarily used as an alloying agent for iron and steel, and especially for high-strength low-alloy steels. It is also used as an alloying agent with nonferrous metals, especially with titanium in aerospace applications. It is also used as a catalyst in the chemical industries.

Geology. The largest source of vanadium in South Africa is within the titaniferous magnetite seams of the Upper Zone of the Bushveld Igneous Complex. The lowest seams in the Upper Zone contain ores of vanadium pentoxide (V_2O_5), in excess of 1.5%, while the amount of V_2O_5 diminishes to 0.3% in the upper seams. Of this lower group of seams, the richest is referred to as the Main magnetite seam. The Main seam is the primary source of V_2O_5 in the eastern Transvaal, and there are two to three of the lower seams in excess of 1.5% V_2O_5 being exploited in the western Transvaal.

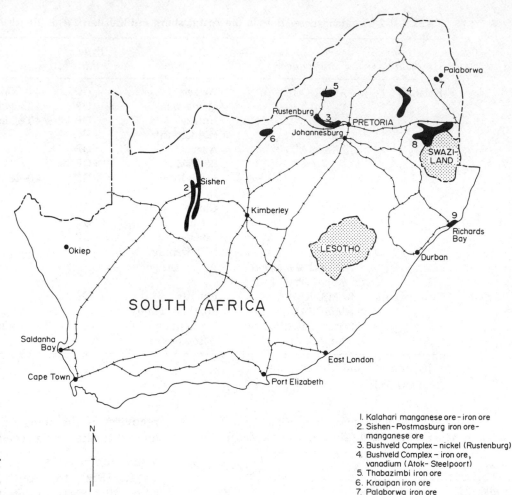

1. Kalahari manganese ore – iron ore
2. Sishen – Postmasburg iron ore – manganese ore
3. Bushveld Complex – nickel (Rustenburg)
4. Bushveld Complex – iron ore, vanadium (Atok – Steelpoort)
5. Thabazimbi iron ore
6. Kraaipan iron ore
7. Palaborwa iron ore
8. Barberton Mountains – iron ore
9. Richards Bay iron ore

FIGURE II-2-18 Ferrous metal mines in South Africa. (From U.S. Department of the Interior, Bureau of Mines, *Mineral Industries of Africa*; Ref. 10.)

0 200 400km

TABLE II-2-60 Value of Manganese Sales in South Africa, 1970–1981 (Thousand U.S. Dollars)

Year	Domestic	Percent total	Export	Percent total	Total	Percent growth
1970	7,024	18.9	30,187	81.1	37,211	
1971	8,090	15.8	43,243	84.2	51,333	+38.0
1972	8,626	17.9	39,599	82.1	48,225	-6.1
1973	9,446	12.3	67,125	87.7	76,571	+58.8
1974	16,427	14.3	98,707	85.7	115,134	+50.4
1975	22,990	16.5	116,434	83.5	139,424	+21.1
1976	24,099	15.9	126,906	84.1	151,005	+8.3
1977	24,837	18.1	112,690	81.9	137,527	-8.9
1978	31,644	24.0	100,377	76.0	132,021	-4.0
1979	54,124	26.0	153,715	74.0	207,839	+57.4
1980	46,983	25.1	140,025	74.9	187,008	-10.0
1981	55,625	29.2	134,701	70.8	190,326	+1.8

Source: U.S. Bureau of Mines, *Minerals Yearbook*, 1970–1981.

Production. There are three main producers of vanadium in South Africa. The largest producer of V_2O_5 in the world is Highveld Steel and Vanadium Corp. Ltd., a subsidiary of the Anglo-American Corporation of South Africa Ltd. The ore is ex-

tracted at the open-pit Mapochs mine in the Roosenekal area, 90 km north-northeast of Middleburg, in the eastern Transvaal. The V_2O_5 is recovered from a vanadium-rich slag after the recovery of iron for steel production from the magnetite

TABLE II-2-61   Manganese Exports from South Africa, 1970-1980

Year	Metallurgical ore and concentrate (thousand metric tons)	Oxides and hydroxides (metric tons)	Metal (electrolytic) (metric tons)	Ferromanganese (thousand metric tons)
1970	2,005	27	8,857	213
1971	2,623	16	9,782	170
1972	2,504	476	11,896	230
1973	3,509	320	—	276
1974	3,021	710	—	347
1975	3,401	1,327	—	276
1976	3,047	8,247	—	330
1977	2,896	4,986	51,984	310
1978	2,532	128	17,434	372
1979	3,055	69	19,282	445
1980	3,428	500	21,797	288

Source: U.S. Bureau of Mines, Minerals Yearbook, 1970-1981.

ore. The slag (25% V_2O_5) is then roasted in one of the eight roasting facilities at Highveld's facilities near Witbank. Production capacity of the Highveld is 12,000 tons per year of V_2O_5 and other vanadium-bearing products.[15]

Ucar Minerals Corp. Ltd., a subsidiary of Union Carbide S. A. Ltd., is the second largest producer of V_2O_5, with a capacity of 2000 tons per year from production facilities at Brits and Bon Accord, northwest of Pretoria in the western Transvaal.[15] Three seams are exploited, with V_2O_5 contents of 1.8%, 1.55%, and 1.44%.[21]

The third producer of V_2O_5 in South Africa is Transvaal Alloys Pty. Ltd., a subsidiary of the Otavi Mining Co. Ltd. The capacity of its plant

near Wapadskloof in the eastern Transvaal is 1500 tons per year of V_2O_5.[15]

Like other alloying agents for steel, vanadium production declined in 1982 by 11%.[7] Vanadium production for 1970-1982 is given in Table II-2-62.

Reserves. South Africa has the largest reserves of vanadium in the world. Reserves are estimated at 7.7 million tons of metal—some 47% of world reserves.[7]

Exports. Nearly all of the country's vanadium products are exported. As a result, export sales account for nearly all of the revenue for vanadium. The value of domestic and export sales is given in Table II-2-63.

TABLE II-2-62   Vanadium Production in South Africa, 1970-1982 (Metric Tons)

Year	Vanadiferous slag, gross weight	Vanadium content of: Vanadiferous slag	V_2O_5 and vanadate products	Total
1970	—	4,400	2,418	6,818
1971	26,286	3,680	2,240	5,920
1972	31,519	4,413	2,905	7,318
1973	34,074	4,770	3,437	8,207
1974	34,522	4,833	3,318	8,151
1975	41,690	5,837	4,808	10,645
1976	50,000	7,000	2,875	9,875
1977	53,969	7,556	3,682	11,238
1978	54,381	7,600	3,650	11,250
1979	55,500	8,400	3,900	12,300
1980	60,000	8,620	4,080	12,700
1981	62,000	8,980	3,820	12,800
1982				11,360[a]

[a]Ref. 7.

Source: U.S. Bureau of Mines, Minerals Yearbook, 1970-1981.

TABLE II-2-63   Value of Vanadium Sales in South Africa, 1970-1981[a]

Year	Domestic sales	Export sales
1970		26,084
1971		23,645
1972		27,523
1973		33,139
1974	398	39,532
1975	1,048	51,921
1976	8	57,195
1977	22	60,524
1978	39	61,180
1979	47	77,369
1980	50e	80,000e
1981	50e	65,000e

[a]e, estimated.

Source: U.S. Bureau of Mines, Minerals Yearbook, 1970-1981.

The main export items are V_2O_5 and vanadium slag. Export figures for vanadium for 1970-1980 are presented in Table II-2-64.

Outlook. The future for vanadium from South Africa is good, although the demand for vanadium is tied directly to the world steel industry. A growing amount of steel production is toward high-performance materials, especially high-strength low-alloy steels.

Of some concern to South African producers is the growing vanadium trade between China and

TABLE II-2-64   Vanadium Exports from South Africa, 1970-1980 (Metric Tons)

Year	
V_2O_5 (fused)	
1970	5,392
1971	3,619
1972	4,692
Vanadium materials V_2O_5 equivalent	
1973	11,000
1974	11,600
1975	9,905
1976	12,885
1977	13,635
1978	15,000
Oxides and hydroxides	
1979	4,406
1980	3,949

Source: U.S. Bureau of Mines, Minerals Yearbook, 1970-1981.

Japan. Japan is South Africa's leading purchaser of exported vanadium products. Chinese producer's price was $2.80 to $2.85 per pound compared to Highveld's $3.16 per pound.[12]

Nickel

South Africa was the world's eighth largest producer of nickel in 1982, producing a little over 5% of the world's nickel output.[7] Nickel is a by-product of platinum-group metals. Nickel production, sales, and export figures are no longer made public.

Geology. Nickel is a by-product of platinum-group metals mined from the Merensky Reef located within the Critical Zone (see the section on the platinum-group metals). Both nickel and copper are present within the Merensky Reef as sulfide ores, with a mill grade of 0.0019% for Ni. The relative proportion of the nickel-bearing ores is as follows: 44.3% pyrrhotite (Fe, Ni)S, 34.1% pentlandite (Ni, Fe)$_2$S$_8$, 17.7% chalcopyrite, and 3.4% pyrite.[8]

Production. Two major platinum-group metal mining companies account for South Africa's nickel production. The Rustenburg platinum mines are the largest producer and are believed to have produced some 17,000 tons of electrolytic nickel in 1982.[12] The new $15.9 million nickel/copper refinery, inaugurated in October 1981, has a planned capacity of 19,000 tons per year of refined nickel. The plant is owned by Matthey Nickel Ltd., a subsidiary of Johnson Matthey of the United Kingdom.[6]

Impala Platinum Holding Company is estimated to have produced some 12,000 tons of nickel in 1982.[12] Matte produced in the smelters near the mine is refined at Impala's refineries in Springs, Transvaal.

Total nickel production has fallen in recent years because of the planned lower production of platinum-group metals. Production figures are given in Table II-2-65.

Reserves. Reserve figures for nickel vary considerably. The U.S. Bureau of Mines estimated South African nickel reserves at 1,000,000 short tons.[7] However, South African sources (Minerals Bureau) in 1979 estimated nickel reserves as 5,830,000 metric tons in the Bushveld Complex as follows: Merensky Reef, 3,074,000 metric tons; UG2, 901,000 metric tons; and Platreef, 1,040,000 metric tons. Estimates were made to a depth of 600 m.[21] In 1976, Von Gruenewald estimated these resources at 23,800,000 metric tons at a depth of 1200 m.[21]

TABLE II-2-65 Nickel Production in South Africa, 1970-1981 (Metric Tons)

Year	Mine output, metal content	Metal, electrolytic
1970	11,557	9,000
1971	12,757	10,000
1972	11,656	10,000
1973	19,426	15,000
1974	22,100	17,000
1975	20,754	14,000
1976	27,000	22,371
1977	22,760	21,955
1978	28,700	22,500
1979	30,290	8,040
1980	25,700	18,100
1981	26,400	17,960
1982[a]	26,300	

[a]Ref. 7.

Source: U.S. Bureau of Mines, Minerals Yearbook, 1970-1981.

Nickel Sales. Nickel revenue is dominated by export sales as indicated in Table II-2-66. Export sales have declined considerably since 1977-1978, due to less production and lower nickel prices from a world surplus of nickel.

Exports. Most of South Africa's nickel production is exported. However, in the future, an increasing amount of nickel may be used in South Africa's expanding steel and ferroalloy industry. Nickel export statistics from the U.S. Bureau of Mines are inconsistent in coverage. These figures are given in Table II-2-67.

TABLE II-2-66 Value of Nickel Sales in South Africa, 1970-1981 (Thousand U.S. Dollars)

Year	Domestic	Export	Total	Percent growth
1970	21,450	16,110	37,560	
1971	9,265	21,087	30,352	-19.2
1972	7,721	25,751	33,472	+10.3
1973	13,559	30,769	44,328	+32.4
1974	19,041	50,682	69,723	+57.3
1975	12,204	71,991	84,195	+20.8
1976	14,830	77,883	92,713	+10.1
1977	13,983	86,410	100,393	+8.3
1978	15,199	87,630	102,559	+2.3
1979	11,265	39,234	50,499	-50.8
1980	14,949	68,981	83,930	+66.2
1981	12,084	66,194	78,278	-6.7

Source: U.S. Bureau of Mines, Minerals Yearbook, 1970-1981.

Outlook. Nickel's future is dependent in some degree to the demand for platinum-group metals and the recovery of the world's steel industries.

Iron Ore

South Africa was the eighth largest producer of iron ore in the world in 1982, with a production of over 28 million tons. This represented only over 3% of world production. South Africa's potential iron ore reserves are considered very large.

Geology. Iron ore is produced from both hematite and magnetite in South Africa. There are four main iron-ore-producing areas in South Africa. For hematite ores, the production areas are by Postmasburg and Sishen in Cape Province and the Thabazimbi area, approximately 150 km northwest of Pretoria. The magnetite ores are produced from the titaniferous magnetite ores of the Bushveld Igneous Complex and from the Palabora Igneous Complex in northeastern Transvaal.[8]

In the Sishen-Postmasburg area, iron ore is recovered from both the lower Dolomite Series and the higher Pretoria Series of the Transvaal System. The iron ore from the Dolomite series is referred to as Thabazimbi type, and occurs in the banded iron formation. The ore is found in lenticular and irregular bodies up to 25 m thick and up to 460 m long. The ore has massive, laminated, or brecciated texture.[8] Within the Pretoria Series, two types of ore bodies are found within the Timeball Hill Stage. The first is a laminated Gamagara type that averages 15 to 30 m in thickness and was formed by ferruginization of the basal shale of the Timeball Hill Stage.[8] The second and higher layer is a conglomeratic (gritty) Gamagara type, with layers averaging 5 to 15 m thick. The ore was formed with the ferruginization of conglomerate and/or grit.[8] The iron ore in the Thabazimbi area in central Transvaal also occurs within the banded iron formation of the Dolomite Series and averages 18 to 25 m in thickness.

Ore grades are as follows: The higher-grade ores average 69.9 to 66% Fe, 0.8 to 4.5% SiO_2 + insolubles; medium grade 65.9 to 63% Fe, 4.6 to 8.5% SiO_2 + insolubles; and the lower grade ores average 62.9 to 60% Fe, 8.6 to 12.5% SiO_2 + insolubles.

Magnetite ores occur as titaniferous magnetite in the Main Magnetite Seam of the Upper Zone of the eastern Bushveld Igneous Complex near Roossenekal (see the section on vanadium). Grade of ore averages 55 to 60% Fe, 8 to 20% TiO_2, and 0.3% V_2O_5.[8]

Magnetite ores also occur in the carbonatites and phoscorite, averaging 27% magnetite. However, the central carbonatites average only 15 to 30%

TABLE II-2-67   Exports of Nickel from South Africa, 1970–1980

Year	Ore and concentrate	Matte and speiss	Metal including alloy		
			Scrap	Unwrought	Semimanufactures
1970	16	—	41	3,454	185
1971	—	349	12	4,801	187
1972	9,984	207	157	10,193	245
1973				10,300	
1974				14,434	
1975				17,895	
1976				19,200	
1977				19,694	
1978	30,038	3,947	570	25,321	503
1979	41,432	3,910	1,092	26,591	1,058
1980	48,319	3,998	1,648	21,815	1,081

Source: U.S. Bureau of Mines, Minerals Yearbook, 1970–1981.

magnetite, while the phoscorites can average up to 50% magnetite by mass. The TiO_2 content averages 1% in the carbonatite and increases to 5% in the phoscorite.[8]

Production. The production of hematite ores was more than 24 million tons (85% of iron ore production), while magnetite ores accounted for nearly 4.25 million tons (15% of production).

ISCOR's open-cast Sishen mine, 64 km north of Postmasburg, is the largest producer of iron ore in South Africa. The mine produces nearly 20 million tons of iron ore a year,[15] and has an estimated capacity of 27 million tons per year.[26] In 1981, the Sishen mine provided 7.3 million tons per year of ore to ISCOR's steelworks at Vanderbijlpark, southeast of Johannesburg and Newcastle, Natal. Over 12 million tons per year is exported by rail, 861 km to Saldanha.[15]

Other iron ore mines in the Postmasburg area include the Beeshoek iron mine (AMMOSAL), Kapstewel iron ore mine, Omnia iron ore mine, and the Springbok iron mine.

ISCOR's Thabazimbi mine in northern Transvaal produced 2.1 million tons of ore in 1981. The mine has a capacity of 3 million tons per year.[26] The iron ore is shipped to ISCOR's steelworks in Pretoria.

Vanadium-bearing titaniferous magnetite ores are produced by Highveld's Mapochs mine, located near Roosenekal, in the eastern Transvaal. Production was 1.9 million tons in 1981.[6] The ore is shipped to the steelworks in Witbank.

Both the Palabora Mining Company and the Phosphate Development Corp. Ltd. (FOSKOR) produce magnetite ore from the Palabora Complex. Production was 120,000 tons of finished magnetite concentration during 1981.[6]

Overall iron ore production has fallen from a high of 31,565,000 tons in 1979. Production figures for 1970–1981 are given in Table II-2-68.

Reserves. South Africa's iron ore reserves are estimated at 9,500,000,000 metric tons at a depth of 30 m by South African sources.[2] The U.S. Bureau of Mines estimated reserves of 4,100,000,000 tons of recoverable iron. Reserve figures for the Sishen area are from 1000 million to 3000 million tons.[8] For the Bushveld Complex, some 4000 to 5000 million tons of reserves are estimated.[8]

Iron Ore Sales. Both domestic and export sales of iron ore have grown dramatically in the last decade. Relevant statistics for iron ore sales are given in Table II-2-69.

TABLE II-2-68   Iron Ore Production in South Africa, 1970–1982 (Thousand Metric Tons)

Year	Ore and concentrate, gross weight	Iron content	Pig iron[a]
1970	7,354		3,924
1971	10,496		4,004
1972	11,223		4,409
1973	10,955		4,331
1974	11,553		4,621
1975	12,298		5,177
1976	15,663		5,795
1977	26,481	9,789	6,114
1978	24,206	16,948	5,910
1979	31,565	15,492	7,031
1980	26,312	20,202	7,515
1981	28,319	16,840	NA
1982[b]	26,416	18,124	

[a]NA, not available.
[b]Ref. 7.

Source: U.S. Bureau of Mines, Minerals Yearbook, 1970–1981.

TABLE II-2-69 Value of Iron Ore Sales in South Africa, 1970–1981 (Thousand U.S. Dollars)

Year	Domestic	Percent total	Percent growth	Export	Percent total	Percent growth	Total	Percent growth
1970	16,331	40.3		24,214	59.7		40,545	
1971	17,271	39.4	+5.8	26,514	60.6	+9.5	43,785	+8.0
1972	18,200	46.4	+5.4	20,996	53.6	-20.8	39,196	-10.5
1973	20,562	40.3	+13.0	30,489	59.7	+45.2	51,051	+30.2
1974	22,769	44.1	+10.7	28,909	55.9	-5.2	51,678	+1.2
1975	25,848	44.9	+13.5	31,693	55.1	+9.6	57,541	+11.3
1976	31,050	33.8	+20.1	60,853	66.2	+92.0	91,903	+59.7
1977	41,588	17.6	+33.9	194,741	82.4	+220.0	236,329	+157.2
1978	62,893	24.9	+51.2	189,802	75.1	-2.5	252,695	+6.9
1979	86,662	24.8	+37.8	262,995	75.2	+38.6	349,657	+38.4
1980	136,668	25.9	+57.7	234,992	64.1	-7.2	380,660	+8.9
1981	166,437	40.1	+21.8	248,538	59.9	+1.9	414,975	0.0

Source: U.S. Bureau of Mines, Minerals Yearbook, 1970–1981.

Exports. Almost all iron ore exports are through the Saldanha iron ore loading facilities (see the section on infrastructure, Saldanha Bay). Almost all the ore is hematite from the Sishen mine and Associated manganese mines (2 million tons per year[12]) in the Postmasburg-Sishen area. The distance from mines to Saldanha is some 850 km. Saldanha has an 18-million ton per year iron ore loading capacity[15] and can accommodate vessels over 250,000 dwt.[2] Three trains arrive daily, carrying some 18,000 tons of ore per train. The trains are nearly 2 km long. The trip takes about 19 hours, with trains averaging 45 to 50 km per hour.

Exports of iron ore and iron ore products are given in Table II-2-70. Exports of iron ore have more than doubled since 1970, owing mainly to the opening of Saldanha export facilities.

Outlook. The future of iron ore depends on the world steel industry. Exports are favorable because of the excellent ship-loading facilities and ship-tonnage capacity at Saldanha. Domestic sales are increasing because of South Africa's growing steel industry. Higher producer prices would help alleviate increasing production and transportation costs.[12]

Steel

Steel production in South Africa has expanded considerably since 1970 for both domestic use and exports. South Africa has several advantages, both planned and natural, that contribute to the present and future steel industry.

First, South Africa has ample natural resources of iron ore, chrome, manganese, vanadium, and nickel, used in the making of steel and steel alloys.

TABLE II-2-70   Exports of Iron Ore Products from South Africa, 1970–1980

Year	Ore and concentrate (thousand metric tons)	Pig iron, cast-iron powder shot (metric tons)[a]
1970	5,527	455
1971	5,545	372
1972	5,120	495
1973	3,419	442
1974	2,849	198
1975	3,337	17
1976	4,213	32
1977	11,440	NA
1978	14,328	18,776
1979	17,079	102,208
1980	14,286	92,088

[a]NA, not available.

Source: U.S. Bureau of Mines, Minerals Yearbook, 1970–1981.

The only major steelmaking components not present in significant quantities are molybdenum and coking coal.

Second, South Africa has developed a cheap source of energy from its coal-fired electrical generating plants by concentrating on economies of scale. The major steel producing plants are all located near South Africa's major coal fields. This close proximity to the country's major coal fields has also reduced the cost for transporting coal to the steelworks.

Production. Due to a lack of large coking-coal reserves, some steel production in South Africa will be by the "direct reduction process," and treated iron ore is referred to as "direct reduced iron" or sponge iron. Using several different direct reduction processes, oxygen is removed from the iron ore, either hematite (Fe_2O_3) or magnetite (Fe_3O_4), leaving a pure iron.

Direct reduction processes operate well below the melting point of iron. As a result of this lower temperature, (1) the silica present in the iron ore is not reduced to silicon, and (2) less carbon is picked up by the iron—less than 1.5% compared to 4 to 4.5% for pig iron from a blast furnace.[28]

The major steel producer in South Africa is the government-owned South African Iron and Steel Corporation (ISCOR). In 1981, ISCOR produced 6.86 million tons of steel, which was 52% ingots and 48% continuously cast blooms and slabs. Production by individual steelworks was 3.5 million tons at Vanderbijlpark, 1.5 million tons at Pretoria, and 2.0 million tons at Newcastle.[6] In 1981, Lurgi Chemie und Huttentechnik GmbH of West Germany was awarded a contract for a four-kiln, 600,000-ton per year direct-reduction plant at Vanderbijlpark by 1984.[6]

Another ISCOR subsidiary, Cape Town Iron and Steel Works (Pty.) Ltd., authorized Danieli Engineering S.P.A. to plan a new electric steelworks of 75,000 tons per year for the production of billets. Present capacity is 52,000 tons per year.[15]

Scaw Metals Ltd. was constructing a 75,000-ton per year plant at Gemiston in 1981, using a process developed by Direct Reduction Corp. of New York.[6]

Middelburg Steel and Alloys Ltd., a subsidiary of Barlow Rand (71%) and AAC (18%), was expanding its stainless steel capacity at its Southern Cross Steel Co. (Pty.) Ltd. plant at Middelburg from 23,000 tons per year to 63,000 tons per year in 1981. Southern Cross Steel produces all of South Africa's stainless steel. Cost of the project was $193 million.

Of significance to South Africa's steel and chromium industry was the development of low-cost corrosion-resistant 3CR12 steel. The steel alloy (0.03% carbon, 0.6% nickel, and 12% chromium) is intended to replace coated carbon steel.[29] The 3CR12 is currently used in the domestic market in such uses as for ore chute liners, agitation leach tanks, and other installations requiring resistance to wet abrasion and corrosion.[6] The use of 3CR12 by foreign steel producers would greatly increase world consumption of chromium.

Anglo-American Corporation's Highveld Steel and Vanadium Corp. continued as the country's leading producer of specialty steels and products in Pretoria. Its Vantra Division at Witbank in Transvaal was to expand its capacity by one-third in 1983.[15]

Production of crude steel in South Africa for 1970–1980 is presented in Table II-2-71. The production of crude steel has nearly doubled in the last decade.

Exports. As with production, exports of steel products has grown tremendously in the last decade. The export of steel products is given in Table II-2-72.

Outlook. The future of steel in South Africa is considered good. Significant domestic use of steel over the last decade has been for the development of South Africa's large coal-fired generating plants and the expansion and upgrading of South Africa's rail system. Many of these large projects will be completed by the early 1990s.

The export of steel products will depend on major foreign steel producers. South Africa's main advantage lies with the fact that it has large reserves of material needed for the production of steel and

TABLE II-2-71 Crude Steel Production in South Africa, 1970–1980 (Thousand Metric Tons)

Year	Ingots	Castings	Total
1970	4757		
1971	4920		
1972	5340		
1973	5628		
1974	5832		
1975	6367	213	6580
1976	6926	230	7156
1977	7175	201	7376
1978	7735	167	7902
1979	8667	201	8868
1980	8863	205	9068

Source: U.S. Bureau of Mines, *Minerals Yearbook*, 1970–1981.

TABLE II-2-72 Exports of Steel from South Africa, 1970–1980 (Metric Tons)

Year	Primary forms	Semimanufactures
1970	21,503	370,794
1971	801	280,022
1972	8,087	516,535
1973	90,800	479,300
1974	100,400	511,900
1975	31,300	292,400
1976	182,800	752,800
1977	500,400	1,414,000
1978	110,629	800,870
1979	173,193	1,074,757
1980	203,840	1,107,593

Source: U.S. Bureau of Mines, *Minerals Yearbook*, 1970–1981.

steel alloys, whereas other major world producers do not. South Africa also has cost advantages in labor and energy.

NONFERROUS MINERAL COMMODITIES

Nonferrous minerals included in this report are antimony, copper, lead, zinc, tin, titanium, and zircon. Nonferrous metal mines are located in Figure II-2-19.

Antimony

South Africa is the second largest producer of antimony (Sb) in the world. Production in 1982 was 8617 metric tons, 15% of the total world production of 56,597 metric tons.[7]

The primary consumption of antimony is in the form of oxides used in the manufacture of a flame retardant for plastics. Other important uses include antimonial lead in electric batteries, and various uses in the ceramic and glass and chemical industries.

Geology. The only production of antimony in South Africa is from the Murchison Range in northeastern Transvaal. Antimony mineralization occurs with dolomitic lenses, which are surrounded by talc carbonate and talc schists, within the sedimentary division of the Murchison syncline.[8] The antimony, usually stibnite, occurs as vein-type deposits, disseminations, and stockworks in the dolomitic lenses along the "antimony line." Small amounts of gold are also present in the mineralized ore.[8]

Production. All production of antimony is from a series of mines from the Murchison Range. All mines are owned and managed by Consolidated Murchison Ltd. (CML), owned 25% by Jonannes-

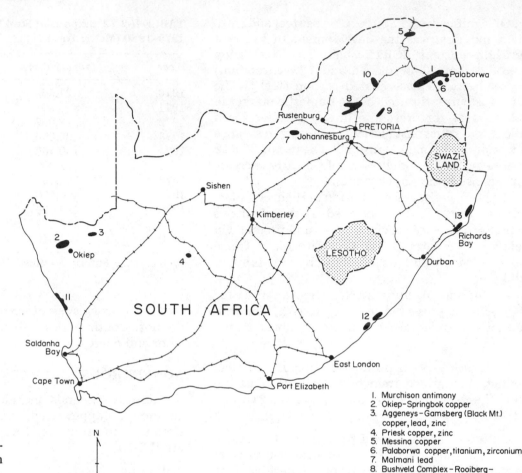

FIGURE II-2-19 Non-ferrous metal mines in South Africa. (From U.S. Department of the Interior, Bureau of Mines, *Mineral Industries of Africa*; Ref. 10.)

1. Murchison antimony
2. Okiep-Springbok copper
3. Aggeneys-Gamsberg (Black Mt.) copper, lead, zinc
4. Priesk copper, zinc
5. Messina copper
6. Palaborwa copper, titanium, zirconium
7. Malmani lead
8. Bushveld Complex - Rooiberg - Nylstroom tin
9. Bushveld Complex - Mutue Fides tin
10. Bushveld Complex - Zaaiplaats - Potgietersrus tin
11. West Coast black sands titanium, zirconium
12. South Coast black sands titanium, zirconium
13. Richards Bay titanium, zirconium

burg Consolidated Investment Co. Ltd. (JCI). Production was reduced in 1981 and 1982 due to a world surplus of antimony. Production figures for the years 1970–1982 are given in Table II-2-73.

Reserves. The U.S. Bureau of Mines estimates South Africa's antimony reserves at a little over 325,000 metric tons,[8] approximately 7.2% of world reserves.

Antimony Sales. Domestic and export sales are given in Table II-2-74. Only in recent years have the domestic sales of antimony been greater than export sales. Antimony Products (Pty.) Ltd. (APL) is a major domestic purchaser of antimony. APL manufactures an antimony oxide fire retardant at its factory on CML property in Gravelotte in Transvaal. The factory treats about 80% of CML's annual production of concentrate. Gold is produced as a by-product of the slag discharged by the APL factory.[6]

TABLE II-2-73 Antimony Concentrate Production and Exports from South Africa, 1970–1982

Year	Production gross weight	Metal content	Exports ore and concentrate
1970	28,759	17,370	27,522
1971	23,996	14,246	10,704
1972	24,109	14,571	35,267
1973	25,870	15,705	29,794
1974	25,212	15,170	22,530
1975	26,160	15,924	12,247
1976	18,341	10,698	5,958
1977	20,053	11,535	8,452
1978	16,395	9,094	3,173
1979	20,086	11,657	11,379
1980	22,372	13,067	
1981	16,599	9,810	
1982		8,617	

Source: U.S. Bureau of Mines, *Minerals Yearbook*, 1970–1981.

TABLE II-2-74   Value of Antimony Sales in South Africa, 1970–1981 (Thousand U.S. Dollars)

Year	Domestic	Export	Total
1970		20,542	20,542
1971		7,092	7,092
1972		14,303	14,303
1973		17,985	17,985
1974	161	28,417	28,578
1975	15,226	15,208	30,434
1976	5,752	19,926	25,678
1977	9,941	10,375	20,316
1978	9,269	3,244	12,513
1979	11,601	14,146	25,747
1980	13,534	4,119	17,653
1981	10,490	7,158	17,648

Source: U.S. Bureau of Mines, Minerals Yearbook, 1970–1981.

Exports. Exports of antimony were dominated by ore and concentrates. However, since the production of antimony oxides by APL, oxides constitute the main component of export material.

In 1980, 6393 tons of oxides and hydroxides was exported compared to a metal content of exported ores and concentrates of 1937 metric tons.[6] Exports are given in Table II-2-73.

Copper

South Africa produces about 2.5 to 3% of the world's copper supply. Although South Africa is not a major world producer of copper, revenue from copper exports has grown steadily in the last decade. Revenue from domestic sales has nearly tripled in the same period.

There are two main copper-producing areas in South Africa, and one minor area. The main sources for copper are from the northwest Cape Province and the northeast Transvaal. Minor amounts of copper are produced as a by-product in the mining for platinum in the Bushveld Igneous Complex.

Production. Copper production increased in 1980, and again in 1981 from a low in 1979. However, several major copper producers continue to experience financial problems, due to the world surplus of copper and consequent depressed price for copper.

There are two major copper producers in the northwest Transvaal. The Palabora Mining Company, a subsidiary of Rio Tinto-ZINC Corporation, is the nation's largest copper producer, with an open-pit mine in the Palabora Complex. The Palabora pipe is a carbonatite intrusive. The copper sulfide ore is found as chalcopyrite in the inner transgressive carbonatite, and as bornite in the outer phoscorite rock.

The Palabora Complex has a number of other minerals recovered from the carbonatite and phoscorite. In 1981, other minerals produced were as follows: magnetite, 120,000 tons; U_3O_8 concentrate, 234.2 tons; zirconium oxide (from baddeleyite), 4941 tons; precious metals (mostly silver), 15.5 tons; and vermiculite, 175,125 tons.[6]

While other major mines are high-cost producers, Palabora remains one of the world's largest low-cost copper producers. In 1981, Palabora Mining Company milled 100.7 million tons of ore, producing 121,000 tons of electrolytic copper. An additional 6359 tons was refined for other producers.[6]

Messina-Transvaal Development Company's Messina Mine continued to experience financial difficulties. Production cost ($1742 per ton) continued to exceed the average price of copper ($1644 per ton).[6] Copper mineralization occurs as hydrothermal veins or replacement in brecciated rocks along the Messina fault within the various metamorphosed gneisses that occur in the Limpopo metamorphic belt.[8] In 1981, Messina produced 7489 tons of copper in concentrate from 600,000 tons of 1.3% copper ore. The concentrate and an additional 45,000 tons of low-grade concentrate were smelted to produce 8590 tons of copper ingots.

There are three major copper producers in Cape Province. Prieska Copper Mines (Pty.) Ltd. (PEM), jointly owned by Anglo-Transvaal Consolidated Investment Co. Ltd. (Anglovaal) and the United States Steel Corp., continues to produce copper from its Prieska mine, near Copperton, southeast of Prieska. Prieska mined and milled 2.76 million tons of ore in 1981, from which 103,424 tons of copper (as well as zinc, lead, and iron pyrite) concentrate was recovered and smelted by the facilities at O'okiep.[6] The copper ore, chalcopyrite, is recovered from a massive sulfide ore deposit within highly metamorphized volcanic rocks.[8] Production costs per ton are approaching selling price per ton, and reserves are small. With the current price for copper, the Prieska mine will remain open for only several more years.[12]

In 1981, the O'okiep Copper Company Ltd., owned 57.5% by Newmont Mining Corp. and 14.4% by AMAX, Inc., produced 1.7 million tons of 1.45% copper ore. From this, 68,000 tons of 32% copper concentrate was extracted, yielding 22,334 tons of blister copper. Another 16,460 tons of blister copper from toll concentrates were produced.[6] The O'okiep copper mine is near Springbok, and its smelter is located in Nababeep. There are a number of mines in the O'okiep district (1500 km^2), which include mining centers at Nababeep, O'oKiep, and Carolusberg.

Copper is found in a number of relatively small transgressive basic rock structures, called noritoids. Mineralization is as sulfides occurring in finely disseminated form, in coarser aggregates with interconnecting veinlets, and in massive form.[8]

The Black Mountain Mineral Development Company, managed and owned by Gold Fields of South Africa and partly owned by Phelps Dodge Corp. of the United States, recovered 4517 tons of copper in argentiferous (silver) concentrate from 1.15 million tons of ore.[16] The ore is in the metasediments of the Kheis System.[8]

Copper is also recovered as a by-product of platinum from the Merensky Reef. Although production is not large, reserves are substantial.

Total copper production for 1970–1981 is given in Table II-2-75. Copper-producing areas are located in Figure II-2-19.

Reserves. Reserve figures for 1978 are presented in Table II-2-76.

Copper Sales. Copper revenue for domestic and export sales were nearly equal in 1980 and 1981. However, at the beginning of the 1970s, export sales were much larger than domestic sales, as indicated by Table II-2-77.

Exports. The export of refined copper increased dramatically in the last few years, while exports of blister and anode declined. Export figures for 1978–1980 are given in Table II-2-78.

TABLE II-2-75 Copper Production in South Africa, 1970–1981 (Metric Tons)

Year	Mine output, metal content	Smelter	Refined
1970	149,205	144,700	75,348
1971	157,470	152,300	79,220
1972	161,927	167,800	79,300
1973	175,797	150,400	90,600
1974	179,111	147,800	88,500
1975	178,927	166,800	92,400
1976	196,880	168,000	95,600
1977	208,287	188,400	145,900
1978	205,745	191,400	149,100
1979	190,591	178,000	150,757
1980	200,683	180,819	140,887
1981	208,700	199,424	144,100

Source: U.S. Bureau of Mines, *Minerals Yearbook*, 1970–1981.

Outlook. The future prospects for low-cost producer Palabora are considered excellent because of world-competitive production costs. However, unless copper prices rise, the future of Messina mines and Prieska mines is not promising.

Lead

The only lead produced in South Africa was by the Black Mountain Mineral Development Co. (Pty.) Ltd., near Aggeneys in northwestern Cape Province. The underground operation produced nearly 100,000 metric tons of lead in 1981, from its complex lead-silver-copper-zinc ore. The lead ore was exported through Saldanha Bay. Exports earned $41,493,000 in 1980 and $72,396,000 in 1981.[6]

Tin

South Africa's tin production was only 1.1% of total world tin production in 1981.[7] The growing domestic demand for tin has decreased the amount of tin available for export.

There are three principal tin producers in South Africa. The largest tin producer is the Rooiberg Minerals Development Co. Ltd., managed by Gold Fields of South Africa Ltd. The Rooiberg tin field is some 60 km west of Warmbaths and includes the Hartebeestfontein, Nieupoort, and Vellefontein mines.[12] With the completion of the new 2000-ton per year smelter at Rooiberg, Rooiberg now produces tin metal as its primary tin commodity, with less than 10% of its tin being sold in concentrate form.[12] Rooiberg produced 2005 tons of refined tin in 1981.

Union Tin Mines Ltd., another subsidiary of GFSA, produced only 401 tons of tin in concentrates for 1981. Zaaiplaats Tin Mining Co. Ltd., controlled by the AAC managerial subsidiary ZIMRO (Pty.) Ltd., produced only 201 tons of tin in 1981.

All tin production is presently from the Bushveld Igneous Complex. The main deposits occur either as "syngenetic pegmatite-pneumatolitic pipes and impregnation in the granite or as epigenetic hydrothermal veins in impervious traps, or as replacement deposits within quartzitic and arkosic sediments."[21] Tin production and sales for 1970–1981 are given in Table II-2-79.

Titanium-Zirconium

In 1982, South Africa produced 380,940 metric tons of ilmenite slag, which was 9.3% of world ilmenite production, and was the fifth largest producer of ilmenite ($FeO\text{-}TiO_2$) in the world. In 1982, South Africa was the third largest producer of rutile, having produced 49,885 metric tons of rutile concentrate. In 1982, South Africa was the

TABLE II-2-76   Copper Reserves for South Africa, 1978

Area	Million metric tons	Percent copper	Copper metal (metric tons)
Cape Province			
O'oKiep	27,198	1.67	454,200
Prieska	48,000	1.74	835,200
Black Mountain	80,000	0.80	640,000
Broken Hill	60,000	0.40	240,000
Transvaal Province			
Messina	4,300	1.37	59,000
Palabora	509,091	0.55	2,800,000
Bushveld Complex			
Merensky Reef	1,712.5	0.08	1,370,000
		Total	5,028,400

Source: Ref. 26.

TABLE II-2-77   Value of Copper Sales in South Africa, 1970–1981 (Thousand U.S. Dollars)

Year	Domestic	(Percent)	Export	(Percent)	Total	Percent growth
1970	58,638	30.1	136,153	69.9	194,791	
1971	46,076	30.4	105,483	69.6	151,559	-22.2
1972	44,943	29.8	105,809	70.2	150,752	-1.0
1973	92,210	37.3	155,140	62.7	247,350	+64.1
1974	126,167	38.4	202,243	61.6	328,410	+32.8
1975	75,319	37.7	124,581	62.3	199,900	-19.2
1976	64,933	32.9	132,630	67.1	197,563	-1.2
1977	60,619	25.2	179,995	74.8	240,614	+21.8
1978	72,398	30.0	169,442	70.0	241,840	+0.5
1979	123,328	35.2	227,097	64.8	350,425	+44.9
1980	193,877	50.3	191,235	49.7	385,112	+9.9
1981	165,705	52.0	153,262	48.0	318,967	-17.2

Source: U.S. Bureau of Mines, Minerals Yearbook, 1970–1981.

TABLE II-2-78   Copper Exports from South Africa, 1978–1980 (Metric Tons)

Year	Ore and concentrate	Matte and speiss	Sulfate	Metal, including alloys		
				Scrap	Unwrought	Semimanufactures
1978	48,837	—	597	1,498	235,553	2,958
1979	72,001	214	342	1,088	181,601	6,859
1980	84,909	88	—	1,068	180,423	2,519

Source: U.S. Bureau of Mines, Minerals Yearbook, 1978–1981.

second largest producer of zirconium in the world, having produced nearly 80,000 metric tons.[7]

Production. Although there are enormous reserves of titaniferous magnetite in the Bushveld Igneous Complex, the ilmenite and magnetite are so intimately intergrown that the titanium content is not currently utilized. At present, Highveld discards its titaniferous slag from the production of vanadium from its Mapochs mine in the Steelpoort-Roosenekal area of Transvaal.

The main producing area for titanium and by-product zirconium is from the $300 million Richard's Bay project, approximately 160 km north of Durban on the east coast. Both producers, Tisand (Pty.) Ltd. and Richards Bay Iron and Titanium Ltd. (RBIT), are owned by Kennecott Copper Company's Quebec Iron and Titanium Inc. of Canada

TABLE II-2-79   Tin Production and Sales in South Africa, 1970–1981

Year	Production (metric tons)			Sales		
	Gross weight	Metal content	Metal primary	Domestic	Export	Total
1970	3,299	2,011	613	2,256	4,651	6,907
1971	3,418	2,029	713	1,944	4,523	6,467
1972	3,682	2,159	779	2,425	4,433	6,858
1973	5,056	2,677	874	3,960	7,030	10,990
1974	5,149	2,542	854	7,030	10,481	17,511
1975	5,652	2,643	780	5,999	11,373	17,372
1976	5,625	2,799	683	5,806	10,960	16,766
1977	6,139	2,864	582	6,640	21,096	27,736
1978	6,120	2,886	637	8,740	25,729	34,469
1979	5,706	2,697	819	10,958	23,983	34,941
1980	6,160	2,913	1,100	26,712	22,678	49,390
1981	6,950	2,811	2,056	10,490	16,042	26,532

Source: U.S. Bureau of Mines, Minerals Yearbook, 1970–1981.

(40%), Gencor Ltd. (30%), IDC (20%), and the South African Mutual Life Assurance Society and the Southern Association (10%). Kennecott also manages the RBIT.[15]

In 1980, RBIT produced 360,000 tons of 85% titania slag from Tisand's ilmenite concentrate, with 195,000 tons of low-manganese pig iron as a by-product. Rutile concentrate was 50,400 metric tons and 103,500 metric tons of zirconium silicate was produced.[15] The deposit included some 700 million tons of heavy mineral sands, with an average composition of 6% ilmenite, 0.3% rutile, and 0.6% zircon. A land-based electromagnetic-electrostatic plant, 8 km from the dredge site, separated rutile, zircon, and monazite from the black sands.[6] The ilmenite concentrate was smelted in RBIT's electric furnace with Natal anthracite, reducing the iron to molten metal, leaving the titania slag. Note that this is not the "direct reduction" route used by other steelmakers in South Africa. The slag was exported mostly to countries in the northern hemisphere.[15]

Zirconium (baddeleyite) was also produced as a by-product of the phosphate and copper mining at Palabora. In 1981, the Palabora Complex produced 4941 metric tons of zirconium oxide.

Production figures for titanium and zirconium are given in Table II-2-80. Sales for the two metals are given in Table II-2-81. Exports for 1973–1980 are presented in Table II-2-82.

Reserves. South African sources estimate titanium (metal) reserves at 33,256,000 metric tons, 15% of world reserves.[2] Titanium-bearing sands are also found along the western coast, but are not presently exploited.

TABLE II-2-80   Titanium and Zirconium Production in South Africa, 1973–1981 (Metric Tons)

Year	Titanium		Zirconium concentrate (baddeleyite)
	Rutile concentrate	Slag	
1973			4,956
1974			11,978
1975			11,594
1976			11,252
1977	4,500	—	16,825
1978	18,100	90,700	36,000
1979	41,740	286,700	82,000
1980	48,000	344,000	80,000
1981	49,900	370,000	80,000

Source: U.S. Bureau of Mines, Minerals Yearbook, 1973–1981.

Zirconium (metal) reserves are estimated at 4,000,000 metric tons, 10% of world resources.[2] However, the U.S. Bureau of Mines estimates South African zirconium reserves at 10,884,000 metric tons, 25% of world reserves.[7]

Zinc

In 1981, South Africa produced only 1.5% of the world's zinc.[7] Although South Africa does export zinc in concentrates, the domestic demand for zinc requires imports of zinc. ZINCOR (Zinc Corporation of South Africa Ltd.) sources report that 75% of domestic zinc use is for galvanizing.[12]

Production. There are only two zinc producers in South Africa. Anglovaal's Prieska mine in northern Cape Province is South Africa's largest zinc producer. Zinc is secondary to copper production.

TABLE II-2-81 Value of Titanium and Zirconium Sales in South Africa, 1974-1981[a] (Thousand U.S. Dollars)

	Titanium			Zirconium		
Year	D	E	T	D	E	T
1974				589	2,110	2,699
1975				326	1,577	1,903
1976				0	1,084	1,084
1977				1	3,884	3,885
1978				No longer given		
1979	3,844	8,359	12,203	Classified data		
1980	5,739	11,973	17,712			
1981	5,276	13,117	18,393			

[a]D, domestic; E, export; T, total.

Source: U.S. Bureau of Mines, Minerals Yearbook, 1974-1981.

TABLE II-2-82 Exports of Titanium and Zirconium from South Africa, 1973-1980 (Metric Tons)

	Titanium				Zirconium
Year	Ore and concentrate	Slag and dross containing Ti	Oxides	Metal, including alloys, all forms	ore and concentrate
1973					4,883
1974					6,256
1975					4,102
1976					–
1977					–
1978	7,321	–	–	–	15,719
1979	42,845	27,144	892	115	48,108
1980	54,367	–	1,142	9	83,258

Source: U.S. Bureau of Mines, Minerals Yearbook, 1973-1981.

Other by-products include lead and iron pyrite. Zinc (sphalerite) is more concentrated near the outer boundaries of the massive sulfide ores.[8]

In 1981, Prieska produced 120,872 tons of zinc concentrate from some 2,857,000 tons of milled ore. The average grade of milled ore was 2.67%, and the average grade of concentrate was 52.8%.[12] The concentrate is shipped by rail to GFSA's ZINCOR smelter-refinery (hydrometallurgical)[15] near Springs in Transvaal Province.

South Africa's other zinc producer was GFSA's Black Mountain operation, near Aggeneys in northwestern Cape Province. Other by-products from mining operations include copper, lead, and silver. Black Mountain produced 41,137 tons of zinc concentrate from the milling of some 1,152,000 tons of ore.[12] Zinc concentrates were shipped to the ZINCOR smelter-refinery.

Production of zinc in South Africa is given in Table II-2-83. Zinc production and refining have increased steadily since the opening of Prieska in 1973.

Reserves. Zinc reserves are directly affected by the mining of copper and lead. Reserves of zinc

TABLE II-2-83 Zinc Production in South Africa, 1970-1981 (Metric Tons)

Year	Concentrate, gross weight	Metal content	Metal, smelter
1970	–	–	26,900
1971	315	158	43,400
1972	4,017	2,009	47,200
1973	34,091	17,016	58,100
1974	67,993	33,995	65,400
1975	127,624	63,812	63,700
1976	149,922	74,961	66,200
1977	139,262	69,631	76,000
1978	130,318	65,159	79,100
1979	107,646	53,823	75,400
1980	158,137	79,068	81,400
1981	174,377	87,172	87,200

Source: U.S. Bureau of Mines, Minerals Yearbook, 1970-1981.

(metal) are given as 12,067,000 metric tons,[2] derived from ores of Prieska (11,293,000 metric tons of ore at 2.36% Zn), Black Mountain (38,000,000 metric tons of ore), and Gamsberg (143,000,000 metric tons of ore).

Zinc Sales. In general, the value of zinc sales has increased steadily since the early 1970s. Since 1977, domestic sales have dominated total revenues, as indicated by Table II-2-84.

Exports–Imports. Although South Africa exports zinc concentrates, domestic demand requires the imports of zinc ores and products. Since the completion and expansion of ZINCOR's smelter-refinery, the imports for unwrought and semi-manufactured zinc and zinc products have dropped considerably. The production of zinc concentrates from Black Mountain, starting in 1978, reduced the imports of zinc concentrates. Exports and imports of zinc are given in Table II-2-85.

Outlook. Although future demand for domestic zinc appears to be strong, the production of zinc will depend largely upon the demand for co-products copper and lead. The exploitation of the Gamsberg deposit would significantly increase South Africa's zinc production.

NONMETALS

In 1980 and 1981, the value of nonmetals was 8.7% and 8.6%, respectively, of the total value of mineral production in South Africa. Export sales of nonmetal minerals amounted to 5.9% of total export sales in 1981. However, the percentage of nonmetals for domestic sales was considerably higher in 1980 (29.4%) and 1981 (26.5%).

Of the nonmetals produced in South Africa, only the Kyanite-related minerals (andalusite, sillimanite, kyanite), asbestos, and fluorspar are discussed. Nonmetal mines are located in Figure II-2-20.

TABLE II-2-84 Value of Zinc Sales in South Africa, 1972–1981 (Thousand U.S. Dollars)

Year	Domestic	Exports	Total
1972	161	—	161
1973	2,029	2,197	4,226
1974	5,309	6,806	12,115
1975	7,654	12,685	20,339
1976	10,552	15,143	25,695
1977	13,143	9,454	22,597
1978	12,346	4,811	17,157
1979	14,076	5,768	19,844
1980	18,180	3,747	21,927
1981	25,911	6,973	32,884

Source: U.S. Bureau of Mines, *Minerals Yearbook*, 1972–1981.

TABLE II-2-85 Zinc Exports–Imports in South Africa, 1970–1980 (Metric Tons)

Year	Ore and concentrates, gross weight		Metal, including alloys unwrought	
	Exports	Imports	Exports	Imports
1970	26,957	11,538		24,894
1971	22,271	34,851		17,598
1972	16,679	21,599		6,683
1973	12,631	16,036		918
1974	39,422	38,077		4,691
1975	66,706	20,704	3,000	1,961
1976	86,048	22,095	7,000	—
1977	59,569	1	16,600	466
1978	41,600	4,196	8,160	11
1979	37,367	2	33	<1
1980	52,846	<1	100	1,825

Source: U.S. Bureau of Mines, *Minerals Yearbook*, 1970–1981.

Andalusite-Sillimanite

South Africa is the world's largest producer and exporter of the anhydrous aluminum silicates of aluminum ($Al_2O_3 \cdot SiO_2$) used to make high-grade refractory materials and ceramics. Andalusite and sillimanite (as well as kyanite) are polymorphs of $Al_2O_3 \cdot SiO_2$. All three minerals are formed as a result of the metamorphism of aluminum-rich sediments, and therefore are found in most thermally and regionally metamorphosed sedimentary rocks.

Andalusite. South Africa's andalusite production is concentrated in the Transvaal Province. The main deposits are located between Pietersburg and Lydenburg in northeastern Transvaal, Thabazimbi area, and the Zeerust-Groot Marico area in western Transvaal. Mines located in these areas are given in Table II-2-86. The largest producer of andalusite is the Weedon's Minerals (Pty.) Ltd., operating near Thabazimbi. Production, export, and sales of andalusite are given in Table II-2-87.

Sillimanite. Sillimanite deposits are scattered throughout the intensely folded and metamorphosed rocks of the Kheis System in Namaqualand. The deposits are either sillimanite-corundum deposits or sillimanite-schists deposits. The main deposits are located around the Pella Mission Farm near Aggeneys in northwestern Cape Province.[8] Pella Refractory Ores S. A. (Pty.) Ltd. and R. G. Niemoller are the chief producers of sillimanite from this area. Production, export, and sales of sillimanite are given in Table II-2-88. Major production areas are located in Figure II-2-20.

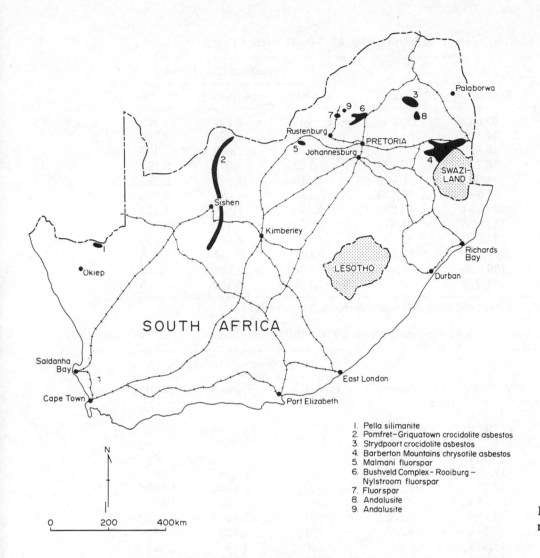

1. Pella silimanite
2. Pomfret–Griquatown crocidolite asbestos
3. Strydpoort crocidolite asbestos
4. Barberton Mountains chrysotile asbestos
5. Malmani fluorspar
6. Bushveld Complex– Rooiburg –
   Nylstroom fluorspar
7. Fluorspar
8. Andalusite
9. Andalusite

FIGURE II-2-20 Nonmetal mines in South Africa.

TABLE II-2-86   Andalusite Mines in South Africa

Area	Mine	Geology
Zeerust-Marico	Andafrax	Metamorphosed shale of the
	Exandaly	Daspoort Stage and alluvial
	General Overseas	deposits from Daspoort shale
Thabazimbi	Timeball (Weedons)	Timeball Hill shale of Pretoria Group
Pietersburg	Hoogenoeg	Timeball Hill and Daspoort shales
	Lager (Lebowa Homeland)	and alluvial deposits from both
Lyndenburg	Havercroft (Lebowa Homeland)	
	Hudson (Lebowa Homeland)	
	Krugerspost	
	Marico Minerals	

Source: Ref. 30.

Reserves. Reserves of andalusite and sillimanite are estimated at 104,000,000 metric tons and account for 34% of world reserves.[2]

Asbestos

South Africa is the third largest producer of asbestos fibers in the world. South Africa is the fourth largest producer of chrysotile, the world's largest producer of crocidolite, and the world's only producer of amosite.

Chrysotile. Chrysotile deposits are scattered throughout Transvaal and Natal Provinces. There are two major producing areas of chrysotile. The first is in the Barberton area. In the Transvaal, the chrysotile ore bodies occur as serpentine intrusives

TABLE II-2-87   Andalusite in South Africa, 1970–1981 (Metric Tons)

Year	Production	Exports	Sales (thousand U.S. dollars)		
			Domestic	Export	Total
1970	42,522		507	2,916	3,423
1971	44,471		572	2,224	2,796
1972	45,857		714	1,041	1,755
1973	60,702	16,637	1,105	2,070	3,175
1974	64,008	18,759	1,445	1,149	2,594
1975	77,149	23,773	2,482	1,696	4,178
1976	77,464	16,357	2,741	1,230	3,971
1977	113,076	71,379	3,780	7,144	10,924
1978	112,040	54,457	4,444	5,852	10,296
1979	134,177	69,810	6,492	7,953	14,445
1980	196,516		8,872	13,888	22,760
1981	181,272		11,555	7,100	18,655

Source: U.S. Bureau of Mines, Minerals Yearbook, 1970–1981.

TABLE II-2-88   Sillimanite in South Africa, 1970–1981 (Metric Tons)

Year	Production	Export	Sales (thousand U.S. dollars)		
			Domestic	Export	Total
1970	31,916				
1971	17,460				
1972	9,476				
1973	19,317	15,363			
1974	13,087	18,737	1	1,592	1,593
1975	16,911	16,524	39	1,445	1,484
1976	25,733	12,960	91	1,414	1,505
1977	15,455	14,235	181	1,717	1,898
1978	9,540	14,195	236	1,726	1,962
1979	19,574		808	2,537	3,345
1980	16,194		826	2,989	3,812
1981	15,504		784	2,251	3,035

Source: U.S. Bureau of Mines, Minerals Yearbook, 1970–1981.

in the Archean System, and subsequent metamorphism. In the Carolina, Lyndenburg, Pilgrim's Rest and Groblersdal districts of the Transvaal, chrysotile occurs in the serpentinites of the Dolomite Series.[8]

Crocidolite-Amosite. Crocidolite and amosite are produced in the Pietersburg and Letaba Districts, between Chuniespoort in the west and Steelport River in the west. The asbestos occurs in the banded Ironstone Stage of the Dolomite Series, in the Lower, Main, Short Fibre, and Upper Zones.[8]

The Penge mine is the only amosite mine and produces only amosite. Most of the amosite is exported. GENCOR owns all of the crocidolite-producing mines in South Africa.

Similar to the Transvaal Blue crocidolite, Cape Blue crocidolite is situated in Griqualand West, in a continuous line from Prieska northward to the Botswana border. Most production is concentrated along the line from east of Prieska to northwest of Kuruman. The crocidolite has also developed in the Banded Ironstone and Upper Dolomite Stages of the Dolomite Series.[8]

Production of asbestos in South Africa from 1970 to 1981 is presented in Table II-2-89. Production has declined gradually over the past decade due to the recession and declining popularity of asbestos fibers in some uses.

Most asbestos produced in South Africa is exported. Export figures for asbestos are given in Table II-2-90 for 1970–1981. Export sales far exceed domestic sales and have almost trebled since the early 1970s. Sales of asbestos are given in Table II-2-91. South African sources place reserves of asbestos at 8,500,000 metric tons, which is about 5% of world asbestos reserves.[2]

TABLE II-2-89   Asbestos Production in South Africa, 1970–1981 (Metric Tons)

Year	Amosite	Chrysotile	Crocidolite	Total
1970	97,380	52,801	137,235	287,416
1971	98,464	61,161	159,671	319,296
1972	102,278	58,512	159,838	320,628
1973	106,477	69,807	155,941	332,225
1974	94,543	82,430	156,299	332,450
1975	88,411	99,660	164,727	352,798
1976	78,898	111,025	178,411	368,334
1977	66,983	111,575	201,056	379,614
1978	40,526	79,511	137,288	257,325
1979	39,058	91,828	118,301	249,187
1980	51,646	106,940	118,148	276,734
1981	56,834	76,772	102,337	235,943

Source: U.S. Bureau of Mines, Minerals Yearbook, 1970–1981.

TABLE II-2-90   Exports of Asbestos from South Africa, 1970–1980 (Metric Tons)

Year	Amosite	Chrysotile	Cape Blue	Transvaal Blue	Total
1970					290,380
1971					299,802
1972					295,846
1973	101,361	57,737	154,902	2,347	316,347
1974	90,109	71,669	159,029	610	321,417
1975	173,165	112,415	257,273	—	542,853
1976	66,978	81,056	170,412	—	318,446
1977	67,027	93,631	172,750	90	333,498
1978	64,134	71,235	131,841	117	267,327
1979					389,470
1980					330,730

Source: U.S. Bureau of Mines, Minerals Yearbook, 1970–1981.

TABLE II-2-91   Value of Asbestos Sales in South Africa, 1970–1981 (Thousand U.S. Dollars)

Year	Domestic	Export	Total
1970	4,331	44,064	48,395
1971	4,710	48,789	53,499
1972	3,721	45,453	49,174
1973	3,355	64,506	67,861
1974	4,026	75,720	79,746
1975	6,471	118,603	125,074
1976	6,581	128,705	135,286
1977	5,041	153,404	158,445
1978	6,570	123,595	130,165
1979	6,861	120,526	127,387
1980	11,718	119,582	131,300
1981	9,999	123,330	133,329

Source: U.S. Bureau of Mines, Minerals Yearbook, 1970–1981.

Fluorspar

South Africa is the second largest producer of fluorspar in the Western world, producing a little over 10% of world fluorspar production in 1982.[7] There are two major producing areas in South Africa. In the western Transvaal, fluorspar is produced south of Zeerust in the Marico District. The fluorspar occurs as metasomatic replacement deposits associated with lead and zinc ores.[8] The second producing area is from the Bushveld acid rocks between Thabazimbi and Warmbad.[8] Major fluorspar producers are given in Table II-2-92.

Three grades of fluorspar are produced in South Africa: acid, ceramic, and metallurgical. Acid-grade dominates fluorspar production in South Africa. Fluorspar production figures for South Africa from 1970–1981 are given in Table II-2-93.

Metallurgical-grade fluorspar is used as a flux in

TABLE II-2-92   Major Fluorspar Producers in South Africa

Magisterial district	Mine	Manager	Capacity (MT/yr)	Type
Potgietersrust	Buffalo	GENCOR	240,000	Acid grade
Marico	Marico	RAND	180,000	Acid grade
	Ottoshoop	Phelps Dodge	100,000	Metallurgical grade
Rustenburg	Slipfontein			
Marico	Van Den Heever	Derby Metals & Minerals S.A.		
Bronkhorstspruit	Vergenoeg	Bayer A.G.	115,000	

Sources: Refs. 6 and 15.

TABLE II-2-93   Fluorspar Production in South Africa, 1970–1981 (Metric Tons)

Year	Acid grade	Ceramic grade	Metallurgical grade	Total
1970	71,199	5,257	96,538	172,994
1971	141,022	13,848	84,171	239,041
1972	142,883	17,861	50,062	210,806
1973	185,304	4,475	20,545	210,324
1974	193,565	4,989	9,379	207,933
1975	172,270	10,294	20,019	202,583
1976	210,874	39,502	40,342	290,718
1977	234,649	65,660	50,370	350,679
1978	297,591	14,907	80,778	393,276
1979	387,305	8,477	55,330	451,112
1980	448,783	9,823	64,112	522,718
1981	451,614	6,118	38,789	496,512

Source: U.S. Bureau of Mines, Minerals Yearbook, 1970–1981.

the iron and steel industries. Acid-grade fluorspar is used to produce hydrofluoric acid (HF), which is used extensively in the chemical industry and in the production and refining of aluminum. Ceramic-grade fluorspar is used in the ceramic, glass, and enamel industries.[8]

Fluorspar revenue is dominated by exports, which have nearly quadrupled in the last decade. Export figures for fluorspar are given in Table II-2-94. The value of fluorspar sales, domestic and export, are given in Table II-2-95.

South African reserves of fluorspar are the largest in the world. According to South African sources, total recoverable reserves of 15 to 30% CaF_2 are estimated at 190,000,000 metric tons.[12] It is expected that South Africa will become an increasingly important source of fluorspar to the Western world. It should also be noted that the

TABLE II-2-94   Exports of Fluorspar from South Africa, 1970–1979 (Metric Tons)

Year	Acid grade	Ceramic grade	Metallurgical grade	Total
1970				120,381
1971				180,301
1972				125,051
1973	97,308	—	17,228	114,536
1974	111,382	—	12,206	123,588
1975	126,061	39	454	126,554
1976	178,479	28,845	8,591	215,915
1977	246,586	9,167	32,873	288,626
1978	309,168	10,640	64,579	384,387
1979				471,498

Source: U.S. Bureau of Mines, Minerals Yearbook, 1970–1981.

TABLE II-2-95   Value of Fluorspar Sales in South Africa, 1970–1981 (Thousand U.S. Dollars)

Year	Domestic	Export	Total
1970	605	4,050	4,645
1971	748	6,639	7,387
1972	746	5,003	5,749
1973	828	5,937	6,765
1974	868	6,816	7,684
1975	1,481	9,194	10,675
1976	1,616	15,954	17,570
1977	1,835	18,542	20,377
1978	2,020	24,789	26,809
1979	2,559	33,402	35,961
1980	3,284	43,855	47,139
1981	3,207	52,029	55,236

Source: U.S. Bureau of Mines, Minerals Yearbook, 1970–1981.

USSR does not produce enough fluorspar for domestic use and does not have significant reserves.

CONSTRAINTS TO MINERAL PRODUCTION

Those factors likely to affect future mineral production within South Africa and the subsequent future availability of South African mineral exports can be divided (for convenience sake) into internal and external factors. It should be recognized that some of these factors are certainly interrelated, and are the result of distinct cause-and-effect relationships. The clearest example of this is the structure of the political system within South Africa, and the world response to this political system.

Internal Factors

Internal constraints are those factors that are indigenous to South Africa, are either natural or political in nature, and will have a bearing on the future of South Africa's mineral industry.

Infrastructure. There are three factors relating to infrastructure that will have a future influence on the mineral industry of South Africa: the future availability of water resources, energy supplies, and skilled workers (both black and white).

The future availability and distribution of water resources in South Africa will affect the mineral industry. The mining and benefication of some minerals in South Africa requires large volumes of water. Unfortunately, much of South Africa is an arid region and subject to seasonal rainfall. Water resources are not evenly distributed, and many of the mining districts are not within easy reach of water supplies. Many water projects have been completed, and more are being either built or planned. However, the volume of water in the rivers of South Africa will depend on seasonal rains. As the South African economy diversifies, the mining industry will have competition for water resources.

The mining industry and related industries will require vast amounts of power for the mining, benefication, and manufacturing of minerals and mineral products. As the dominant source of energy in South Africa is derived from either coal or coal products, South Africa's future will in the short to medium term depend on the coal industry, especially the large coal-fired electric-generating plants and the Sasol projects for synthetic fuels.

Although South Africa's coal reserves and resources are large, in the long run South Africa will have to rely on energy derived from nuclear power. South Africa certainly has enough domestic reserves of uranium to supplement a nuclear power industry.

The third factor relating to infrastructure that will affect the future of South Africa's mining industry is the future availability of skilled workers, both white and nonwhite. The vast majority of mine workers are black. Black workers are either from South Africa (including the black homelands) or from neighboring black nations. In recent years, a growing percentage of black workers are from South Africa or the black homelands.[2]

Recent legislation and governmental commissions have improved the relative wages between whites and nonwhites, and an increasing number (although a still relatively small percentage) of nonwhites are moving into managerial positions. A growing number of nonwhites have formed or joined labor unions.

The extent of progress in labor reforms depends on the willingness of white labor unions and white management to accommodate these reforms. Strikes by white unions dissatisfied with labor reforms are equally as disruptive to the mining industry, as strikes by nonwhites disenchanted with the present system.

Political. It is no secret that political power, and the means to enforce it, resides with the white population in South Africa. However, the white population is not politically homogeneous in character. The whites are divided on major political decisions relating to the sharing of political power with nonwhites, especially the large majority black population.

Government efforts are concentrated on establishing black homelands for the various tribal groups. Within each homeland, the various black tribal groups would supposedly have complete political power. However, black leaders realize that real and significant economic power would still reside with the white population. Therefore, the white national government will still have considerable potential to influence internal decisions and policies of black homelands.

The white-controlled government, on the other hand, is adamantly opposed to a "one man, one vote" political system, where the white population would be outnumbered four to one. In such a system, the whites believe they would lose both political and economic control,[2] which is, of course, unacceptable to them.

The resolution of the political, social, and economic status of nonwhites in South Africa is of tantamount importance to the future mineral industry of South Africa. Unsatisfactory progress in these areas through legitimate means is an open invitation for external intervention.

External Factors

External factors that can potentially affect the future of the South African mineral industry are political or economic in nature, and can be implemented either nonviolently or violently.

Political. Opposition to the current government of South Africa has taken many forms. Major goals of this opposition seem to fall into two general categories. While almost all nations want major political and economic advancements for nonwhites in South Africa, there is disagreement over the degree of political control that should be exercised by whites and nonwhites. A number of nations consider black majority rule as the only lasting political solution, whereas others envision some degree of political power sharing between whites and nonwhites. Black nations, the United Nations, and Eastern Block countries are the most adamant in opposing the current political system in South Africa.

Economic. An economic boycott of South African exports and limited imports into South Africa has been a major cornerstone of United Nations policy toward South Africa for a number of years. Despite U.N. sanctions toward South Africa, the South African economy has grown and diversified over the last two decades. As long as South Africa remains a major exporter of minerals critical to industrialized societies, the chance for the success of implementing an economic boycott are slim. South Africa can always retaliate by withholding key mineral supplies critical to industrialized countries, thus hurting the economies of those countries in return.

Violent Actions. For some opposition groups, the failure of nonviolent means (political and economic) to change the current government of South Africa justifies the use of violent means for change. This form of action can affect the performance of the mining industry in several ways: by sabotage of the mining infrastructure and mines, terrorizing or killing workers in the mining industry, or direct intervention by outside forces in a revolutionary movement.

Although South Africa has perhaps the most technologically advanced mining industry in the world, the industry is susceptible to disruption by sabotage. South Africa's electricity-producing facilities are built for economy of scale, which has tended to restrict the number of electrical-generating facilities. Because of the national grid system, any serious disruption of electrical power would have to be coordinated in order that more than several electrical-generating facilities could be disrupted at once. Of course, power lines are an easier target, and continuous destruction of power lines would also disrupt the economy. Since a significant part of the rail system is electrified, disruption of electricity would also affect these lines. Canals, pipelines, and dams are also susceptible to sabotage.

The labor force employed in mining and related industries could be affected by acts of terrorism, thus interrupting mining operations. Given an increasing level of violence, it is not improbable that skilled personnel might wish to leave South Africa.

In a worst-case scenario, it is possible that under extremely violent conditions (revolution, anarchy, etc.) intervention by foreign powers could take place. Under these circumstances, it is impossible to predict the damage to mining infrastructure and employees.

CONCLUSIONS

South Africa is amply endowed with mineral resources and is a major world producer of several key minerals which are indispensable to an industrialized society. Furthermore, South Africa possesses significant reserves of these important minerals.

The mining industry is the backbone of the South African economy. Cooperation between the public and private sectors has encouraged the exploitation of mineral resources through tax incentives and a free-market system. Cooperation has benefited research efforts and mining technology. Due to ample mineral reserves, low labor cost, advanced technology, relatively inexpensive energy, and excellent exporting facilities, South Africa is a major world producer and exporter of mineral commodities.

However, due to the current political system in South Africa, and worldwide opposition to this system, the future availability of minerals from South Africa over the long run is uncertain. A change in the political system does not guarantee continued mineral availability. Nor does black majority rule guarantee a strifeless society of equality under the law, or automatic political alignment with the West.

It can be stated with certainty that the South African social and political structure is under stress, from within and without. Furthermore, major changes in this structure are likely to occur at some point, and they will undoubtedly affect the mining industry.

REFERENCES

1. U.S. Central Intelligence Agency, National Foreign Assessment Center. *The World Factbook—1981*. Washington, D.C.: U.S. Government Printing Office, 1982.
2. Government of South Africa, Department of Foreign Affairs and Information. *Official Yearbook of the Republic of South Africa*, 8th ed. Johannesburg, South Africa: Chris van Rensburg Publications, 1982.
3. Stocks, Kevin. "Botha Regime Facing a Test in South Africa." *The Wall Street Journal*, 2 May 1982, p. 32.
4. World Bank, International Bank for Reconstruction and Development. *World Development Report 1982*. New York: Oxford University Press, Inc., 1982.
5. U.S. Department of Commerce, International Trade Administration. "Marketing in South Africa." *Overseas Business Reports*. OBR 81-03. Washington, D.C.: U.S. Government Printing Office, March 1981.
6. U.S. Department of the Interior, Bureau of Mines. "The Mineral Industry of the Republic of South Africa." *Minerals Yearbook 1981*. Washington, D.C.: U.S. Government Printing Office, 1983.
7. U.S. Department of the Interior, Bureau of Mines. *Mineral Commodity Summaries 1983*. Washington, D.C.: U.S. Government Printing Office, 1983.
8. Coetzee, C. B., ed. *Mineral Resources of the Republic of South Africa*, 5th ed. Pretoria, South Africa; The Government Printer, 1976.
9. Storrar, C. D., ed. *South African Mine Valuation*. Johannesburg, South Africa: Chamber of Mines of South Africa, 1977.
10. van Rensburg, W. C. J., and D. A. Pretorius. *South Africa's Strategic Minerals—Pieces on a Continental Chessboard*. Johannesburg, South Africa; Valiant Publishers 1977.
11. Dayton, Stanley H., and John R. Burger. "Mining in South Africa." *Engineering and Mining Journal*, November 1982, pp. 52-139.
12. U.S. Department of State. *Industrial Outlook Report: Minerals South Africa—1981*. Prepared by the American Embassy in Pretoria, 11 June 1982.
13. Tankard, A. J., et al. *The Crustal Evolution of Southern Africa*. New York: Springer-Verlag New York, Inc., 1982.
14. Geertsema, A. "SASOL's Expansion Programme—Economic Implications for South Africa." Paper presented at the 13th AISEC Annual Economic Congress, Cape Town, 7 July 1981.
15. U.S. Department of the Interior, Bureau of Mines. "The Mineral Industry of the Republic of South Africa." *Minerals Yearbook 1980*. Washington, D.C.: U.S. Government Printing Office, 1982.
16. Hunter, D. R. *Some Enigmas of the Bushveld Complex*. Economic Geology Research Unit. Information Circular 92. Johannesburg, South Africa: University of the Witwatersrand, 1974.
17. Pretorius, D. A. *The Depositional Environment of the Witwatersrand Goldfields: A Chronological Review of Speculations and Observations*. Economic Geology Research Unit. Information Circular 95. Johannesburg, South Africa: University of the Witwatersrand, March 1975.
18. Pretorius, D. A. *Gold and Uranium in Quartz-Pebble Conglomerates*. Economic Geology Research Unit. Information Circular 151. Johannesburg, South Africa: University of the Witwatersrand, January 1981.
19. U.S. Department of the Interior, Bureau of Mines. "Platinum-Group Metals." *Mineral Facts and Problems 1980*. Washington, D.C.: U.S. Government Printing Office, 1982.
20. U.S. Department of the Interior, Bureau of Mines. *Platinum Availability—Market Economy Countries*. IC 8897. Washington, D.C.: U.S. Government Printing Office, 1982.
21. Vermaak, C. F. *The Global Status of the South African Minerals Economy and Data Summaries of Its Key Commodities*. Review Paper 1. Johannesburg, South Africa: The Geological Society of South Africa, 1979.
22. Buchanan, D. L. "Platinum—Great Importance of Bushveld Complex." *World Mining*, August 1980, pp. 56–59.
23. U.S. Department of the Interior, Bureau of Mines. "Platinum-Group Metals." *Mineral Commodity Profiles 1983*. Washington, D.C.: U.S. Government Printing Office, 1983.
24. Buchanan, D. L. Bureau for Mineral Studies. *Chromite Production from the Bushveld Complex and Its Relationship to the Ferrochromium Industry*. Report 3. Johannesburg, South Africa: University of the Witwatersrand, October 1978.
25. U.S. Department of the Interior, Bureau of Mines. "Chromium." *Mineral Commodity Profiles 1983*. Washington, D.C.: U.S. Government Printing Office, 1983.
26. U.S. Department of the Interior, Bureau of Mines. "The Mineral Industry of the Republic of South Africa." *Minerals Yearbook 1978/79*. Washington, D.C.: U.S. Government Printing Office, 1981.
27. U.S. Department of the Interior, Bureau of Mines. "Manganese." *Mineral Commodity Profiles 1983*. Washington, D.C.: U.S. Government Printing Office, 1983.
28. Government of South Africa, Department of Mines, Minerals Bureau. *Alternative Routes to Steelmaking in South Africa*. Preliminary Report 2175. Braamfontein, South Africa, 1975.
29. U.S. Department of the Interior, Bureau of Mines. *Minerals Yearbook 1982*. Washington, D.C.: U.S. Government Printing Office, 1984.
30. Government of South Africa, Department of Mines, Minerals Bureau. *Reserve and Resource Analysis Nonmetallic Minerals Andalusite, Sillimanite, Kyanite*. Internal Report 6. Braamfontein, South Africa, 7 February 1977.

Chapter II-3

Land-Locked Countries

There are six land-locked countries in this study: Lesotho, Swaziland, Malawi (all of minor importance), Botswana, Zambia, and Zimbabwe. Having no direct access to ocean transportation, these countries must rely on neighboring countries with ports and harbors for exports of their goods and imports of necessary supplies. Therefore, not only do land-locked countries have their own problems, but they must contend with the problems of their neighbors.

Regional maps for railways and roads were presented in Figures II-2-2 and II-2-4. Land-locked countries must depend either on Tanzania, Zaire, Angola, Mozambique, Namibia/South-West Africa, and South Africa for ocean-related commerce.

LESOTHO

The Kingdom of Lesotho is a land-locked nation of some 30,303 km². The country is located within the borders of South Africa, due south of Johannesburg about 350 km. It is predominantly a mountainous region surrounded by South Africa. Its natural resources are confined to limited diamonds, water, and agricultural and grazing land. The popu-

FIGURE II-3-1 Lesotho: Mines, processing facilities, and infrastructure. (From U.S. Department of the Interior, Bureau of Mines. *Mineral Industries of Africa*, 1976.)

lation is approximately 1,350,000 and is 99.7% Sotho, with very few Europeans and Asians.[1]

Lesotho was granted independence by the United Kingdom on October 4, 1966, and remains a member of the Commonwealth. The capital is Maseru, located along the northwestern border. Before 1970, Lesotho was a constitutional monarchy. It remains a monarchy under King Moshoeshoe II. However, Chief Leabua Jonathan, the prime minister, remains the central authority after the suspension of the constitution. Jonathan's Basutoland National Party (BNP) has remained in political control, although there has been some conflict with the Basutoland Congress Party (BCP) led by Ntsu Mokhehle, who was forced to flee Lesotho. The judicial system consists of lower courts, which administer customary law for Africans, a High Court, and a Court of Appeal at Maseru, which has appellate jurisdiction.[1]

The economy is primarily subsistence agriculture, livestock raising, and some diamond mining. Key economic indicators for Lesotho are given in Table II-3-1. By far the largest share of the economy is repatriated funds and earnings from mine workers in South Africa, as indicated in Table II-3-1 under "Basotho earnings abroad."

Thus Lesotho is dependent on South Africa for a substantial portion of its income. Additional revenue accrues from the fact that Lesotho is a member of the Southern African Customs Union. Receipts from the Customs Union have provided nearly two-thirds of current government revenues.[2] The budget continues to expand. In 1970–1971 it was $15.7 million, $92.5 million in 1977–1978, and $141 million in 1979.[2]

Mineral Industry

The mineral industry of Lesotho consists of the mining of diamond and stone. The results of uranium exploration in Lesotho's Maluti Mountains have been withheld.[2]

The diamond industry is dominated by the Letseng-la-terai mine of the De Beers Lesotho Mining Corp. (Pty.) Ltd., a subsidiary of the De Beers Consolidated Mines Ltd. (South Africa). The mine closed operation in 1982. During 1980, the Lesotho government received approximately $2.4 million from a 10% diamond export levy; the government also owned 25% equity in the diamond mine in 1980.[2] Workers also worked a diamond pipe mine at Kao, just west of Letseng-la-terai mine. The value of the diamond exports was approximately $29.0 million (8.5%) of the GDP of $339.5 million in 1979–1980; Lesotho's diamond production for 1976–1980 is given in Table II-3-2.

Mineral Potential. Exploration for minerals in Lesotho is limited. The Lesotho National Development Corporation (LNDC), a government-owned corporation, is responsible for promoting tourism, mining, and commercial projects in Lesotho.[3] So far there has been little response for mineral ventures.

Lesotho has the potential for hydropower. The Lesotho Highlands Water Project will involve damming the Orange River and some of its tributaries in the Lesotho Highlands. The water will be diverted to South Africa. The project is estimated to cost between $1 and $4 billion.[3] The first stages of the project are not to be completed until the early 1990s.[3] This may prove to be Lesotho's biggest source of income in the future.

Certainly among Lesotho's some 200,000 migrant workers in South African mines, Lesotho will have at least a semiskilled labor force for potential mining projects.

Lesotho's infrastructure remains poor in this mountainous region. There is only one short rail line connecting Maseru to Marseilles on the Bloemfontein-Natal main line.[4] In general, goods are transported by road to the nearest rail station in the Orange Free State.

Currently, Lesotho's power comes from South Africa's ESCOM power grid.[2] Other sources include small diesel-powered generators. This will change dramatically with the completion of the Highlands Water Project.

REFERENCES

1. U.S. Central Intelligence Agency, National Foreign Assessment Center. *The World Factbook—1981.* Washington, D.C.: U.S. Government Printing Office, 1982.
2. Weisfelder, Richard. "Lesotho: Changing Patterns of Dependence." In *Southern Africa: The Continuing Crisis,* 2nd ed. Edited by Gwendolyn M. Carter and Patrick O'Meara. Bloomington, Ind.: Indiana University Press, 1982.
3. U.S. Department of the Interior, Bureau of Mines. "Lesotho." *Minerals Yearbook 1980.* Washington, D.C.: U.S. Government Printing Office, 1982.
4. U.S. Department of Commerce. "Kingdom of Lesotho." *Foreign Economic Trends and Their Implications for the United States.* FET 81-086. Washington, D.C.: U.S. Government Printing Office, August 1981.
5. U.S. Department of State. "Kingdom of Lesotho." *Background Notes.* Washington, D.C.: U.S. Government Printing Office, December 1981.
6. U.S. Department of the Interior, Bureau of Mines. "Lesotho." *Minerals Yearbook 1981.* Washington, D.C.: U.S. Government Printing Office, 1983.

TABLE II-3-1 Key Economic Indicators for Lesotho
(Million Dollars Unless Otherwise Noted)[a]

Indicator	1977–1978	1978–1979	1979–1980
Gross national income	380.6	472.1	574.1
Gross national product	213.9	286.3	339.5
Population	1,239,400	1,261,800	1,289,900
Per capital income	328.9	373.7	445.1
Per capital GNP	173.6	227.3	263.2
Imports	202.6	224.5	368.7
Exports	14.0	36.2	43.2
Basotho earnings abroad	168.1	186.8	235.9
Foreign assistance (est.)	80.5	119.0	147.6
Principal exports			
Diamonds	1.4	21.2	29.0
Mohair	2.2	5.8	5.7
Wool	3.0	3.5	3.2

	1975–1976	1976–1977	1977–1978
Principal imports			
Food and livestock	29.8	45.1	51.3
Beverages and tobacco	6.4	9.4	10.9
Fuels and lubricants	9.5	15.2	20.9
Chemicals	8.2	11.3	12.5
Machinery and transport equipment	17.6	26.7	29.2
Manufactured goods	43.7	55.9	61.9

	1976–1977	1978–1979	1979–1980
Government accounts			
Receipts	60.3	97.5	126.4
Capital expenditures	35.1	52.8	66.9
Current expenditures (do not include all foreign assistance)	59.1	51.9	57.5
Real GNI	196.1	217.3	228.8
Real GNP	109.7	130.9	132.6
Real earnings abroad	86.4	86.4	96.2
Price indices (1970–1971 base)	1.94	2.17	2.50

[a]Figures are based on Lesotho government, World Bank, and International Monetary Fund data. The price index is based on changes in both domestic and South African prices, a mix of which is available to Basotho consumers. From April 1976 until March 1979, M1.00 (or R1.00) = $1.1522. From March 1979 the maloti steadily appreciated to a high M1.00 = $1.35 in October 1980. Since then the maloti depreciated to M1.00 = $1.19 in June 1981.

Source: Refs. 4 and 5.

TABLE II-3-2 Lesotho Mineral Production, 1976–1981[a] (Metric Tons)

	1976	1977	1978	1979[p]	1980[e]	1981
Diamond						
Gem (carats)	891	7,576	14,333	10,484	10,743	10,921
Industrial (carats)	4,125	34,514	57,332	41,937	42,971	42,000
	5,016	42,090	71,665	52,421	53,714	52,921

[a]p, preliminary; e, estimated.
Source: Ref. 6.

SWAZILAND

The Kingdom of Swaziland is a land-locked nation of some 17,364 km². It is located within the borders of South Africa (northeastern section), but shares a common border with Mozambique in the east. Its natural resources are asbestos and coal, agriculture, and forest products. Its population of 564,000 is 96% African, 3% European, and 1% mulatto.[1]

Swaziland was granted independence by the United Kingdom on September 6, 1968, and remains a member of the Commonwealth. The administrative capital is Mbabane, located in western Swaziland.

After independence, Swaziland was governed

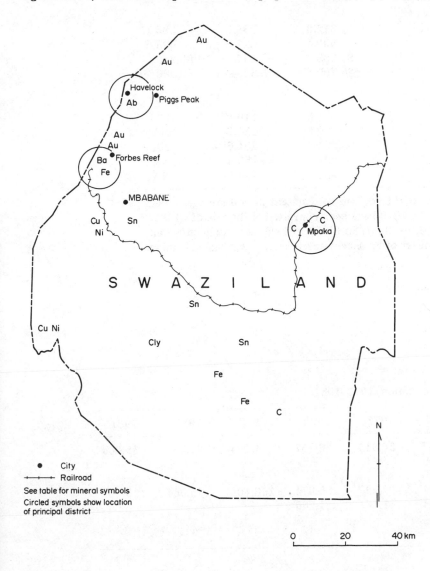

FIGURE II-3-2 Swaziland: Mines, processing facilities, and infrastructure. (From U.S. Department of the Interior, Bureau of Mines. *Mineral Industries of Africa*, 1976.)

by a two-party system and a two-house parliament, with the king as constitutional head of state. On April 16, 1973, the constitution was suspended by King Sobhuza II.[2] In 1979 a new bicameral parliament opened. Fifty members of the parliament were chosen by an 80-member electoral college and 20 members by the king. The king's approval is required before parliamentary acts become law. The judiciary is part of the Ministry of Justice but is independent of the executive and legislative branches.[1]

The Swazi economy's most important sector is agriculture, accounting for approximately 20% of the GDP and 75% of the work force.[3] Sugar, followed by forest products, are the two highest export earners. Mining is third, with asbestos and coal being the primary export-earning commodities. Key economic indicators for Swaziland are given in Table II-3-3. The figures indicate a slowly expanding economy.

Mineral Industry

Mining's contribution to total export values has declined steadily from the early 1970s. In the early 1970s, mining's contribution was nearly 40% of the GNP; this contribution has declined to less than 15% in 1980.[4] Mineral production for Swaziland is presented in Table II-3-4.

A regional geochemical sampling program was completed in 1980 and a report was issued indicating potential mineral occurrences of copper, lead,

TABLE II-3-3   Key Economic Indicators for Swaziland[a]
(Money Values in Million U.S. Dollars Except Where Noted)

	1978	1979	1980[b]
Income, production, employment			
GDP at constant (1976) prices[c,d]	312.0	338.5	385.1
Per capita GNP (1976 prices)[c,d]	592.4	622.8	686.6
Employment[e]	80,883	83,567	85,242
Population[c]	526,700	543,500	560,900
Balance of payments and trade[e]			
Gross international reserves	121.6	116.8	153.5
External debt	50.5	114.2	NA
Exports (f.o.b.)	200.0	234.6[f]	NA
Imports (f.o.b.)	224.4	259.4	NA
Trade balance	-24.4	-24.8	NA

[a]Data have been converted to U.S. dollars to represent period averages unless otherwise indicated. Data for 1978 have been converted at the rate of E1.00 = $1.15, for 1979 E1.00 = $1.20, for 1980 E1.00 = $1.29. Sources include Central Bank of Swaziland, Ministry of Finance, Department of Economic Planning, *Annual Statistical Bulletin.*
[b]NA, not available.
[c]Embassy projection based on 1976 data.
[d]Fiscal year.
[e]End of period.
[f]Provisional.
Source: Ref. 3.

TABLE II-3-4   Swaziland Mineral Production, 1976–1981

	1976	1977	1978	1979	1980	1981
Asbestos: chrysotile	41,847	38,046	36,957	34,294	32,833	35,264
Barite	369	—	—	—	—	—
Clays: kaolin	989	—	—	—	—	—
Coal: anthracite	154,524	128,990	165,874	168,409	175,984	163,780
Iron ore, direct-shipping grade, gross weight (thousand metric tons)	1,744	1,441	1,266	—	—	—

Sources: Refs. 5 and 7.

zinc, manganese, cobalt, nickel, molybdenum, and tin.[5] Several foreign firms have indicated interest in prospecting for tin, gold, diamonds, kaolin, and coal.[5]

Mining of iron ore at Ngwenya mine, in western Swaziland, ceased in 1977, due to declining grade of ore. The mine began production in 1964.

Chrysotile asbestos, from the Havelock mine, was the largest mineral export earner. Exports from the mine were valued at $19.5 million in 1978 and an estimated $20.2 million in 1979.[4] The mine is owned 60% by Turner and Newall Ltd. of the United Kingdom and 40% by the Swaziland government.[4] A larger portion of short asbestos fibers are being encountered with increasing depth.

The prospects for continuing and expanding coal production in Swaziland are good. The Mpake mine, the Swaziland Coal Corp.'s one existing mine, produces anthracite, a high-energy coal with a low sulfur content. In general, Swaziland coal is expected to have high calorific content and low sulfur content, making it a desirable source of energy.

Approximately 200 million tons of coal reserves have been identified.[5] In 1980, two new coal mines for eastern Swaziland were proposed. The Japanese and Swaziland governments were negotiating over details for a deep coal exploration and drilling project at Libhuku, within the Lowveld.[5] Shell Coal Swaziland (Pty.) Ltd. submitted a mining rights application for anthracite coal at Mhlume. A 115,000-ha area in the southern Lowveld which had been granted under a new Special Exclusive Prospecting License for coal exploration to Inter-Coal Trading A. G. Ltd., was transferred to Inter-Coal Mining and Trading Ltd. of Switzerland.[5]

West Germany continued to assist in evaluation of residual tin deposits in the Makwanekop-Motshane and Sinceni areas. A small-scale treatment plant operated at Sinceni in 1980.[5]

Foreign Investment Incentives. The Swazi government continues to encourage foreign development of its mineral industry. Mining companies are taxed at a variable rate of 27% on the first E28,000 and 37½% on the remainder of taxable income.[6] The normal depreciation allowance consists of "the difference between the original cost and the sum of (1) the accumulated depreciation allowance, and (2) the amount realized when the item was scrapped."[6]

In addition, 30% of the cost of new industrial buildings, plants, and machinery can be deducted in the first year.[2] Losses may be carried forward indefinitely, to offset profits in succeeding years.

Mineral Potential. The climate for future mineral projects in Swaziland is good and the government is actively encouraging foreign investment in the mineral industry. The rail line, which originally went to the port of Maputo, was operated by the Mozambique Railway until 1975, when the Swaziland Railway took control. To facilitate future coal exports the railroad has been expanded south to Lavumisa and on to the coal terminal at Richard Bay and to Durban. Construction for another line to extend northward to Komatipoort in eastern Transvaal has started.

Although there is some hydroelectric power in Swaziland, the mountainous terrain and water resources could be developed for more hydroelectric power. In fact, future hydroelectric generation is called for.

Although Swaziland is significantly dependent on South Africa, it is less so that Lesotho. The United Kingdom receives about 33% of Swaziland's exports, South Africa 20%, and the United States 10%. However, 90% of Swaziland's imports are from South Africa.[3] The number of mine workers in South African mines continues to decline. As with Lesotho, receipts from the Southern Africa Customs Union continues to be the single most important source of revenue for Swaziland. These revenues have been approximately 60% of the total revenue.[3] In 1980–1981 these receipts amounted to $123 million, but are expected to fall to $100 million in fiscal year 1981–1982.[3]

The future of Swaziland remains uncertain. The new king will face several major problems. Involvement with guerrilla groups against South Africa could bring swift reprisals against Swaziland, either by economic intimidation or by threatened destruction of Swaziland's infrastructure. On the other hand, noninvolvement may bring reprisals by anti-South African guerrilla groups. Future development will depend on maintaining foreign investment and some political distance from the political struggle for South Africa.

REFERENCES

1. U.S. Central Intelligence Agency, National Foreign Assessment Center. *The World Factbook—1981*. Washington, D.C.: U.S. Government Printing Office, 1982.
2. Vilakazi, Absolom. "Swaziland: From Tradition to Modernity." In *Southern Africa: The Continuing Crisis*, 2nd ed. Edited by Gwendolyn M. Carter and Patrick O'Meara. Bloomington, Ind.: Indiana University Press, 1982.
3. U.S. Department of Commerce. "Kingdom of Swaziland." *Foreign Economic Trends and Their Implications for the United States*. FET 81-113. Washington, D.C.: U.S. Government Printing Office, September 1981.
4. U.S. Department of the Interior, Bureau of Mines. "Swaziland." *Minerals Yearbook 1978/79*. Washington, D.C.: U.S. Government Printing Office, 1981.
5. U.S. Department of the Interior, Bureau of Mines.

"Swaziland." *Minerals Yearbook 1980*. Washington D.C.: U.S. Government Printing Office, 1982.

6. U.S. Department of Commerce. "Marketing in Swaziland." *Overseas Business Reports*. OBR 77-36. July 1977.

7. U.S. Department of the Interior, Bureau of Mines. "Swaziland." *Minerals Yearbook 1981*. Washington, D.C.: U.S. Government Printing Office, 1983.

MALAWI

The Republic of Malawi is a land-locked nation sharing borders with Mozambique, Zambia, and Tanzania. Malawi is approximately 95,053 km^2 in area and has a population of some 6,130,000 people. There are over 99% native Africans and less than 1% European and Asians.[1]

Malawi is a one-party state under the leadership of President Dr. H. Kamuzu Banda of the Malawi Congress Party (MCP). The cabinet is appointed by the president. There is a unicameral National Assembly, and a High Court with a chief justice and at least two other justices.[1]

The economy is primarily agricultural. In 1980, the agricultural sector was nearly 40% of the GDP ($700 million in constant 1973 prices) and 90% of export earnings.[2] Agriculture also accounted for nearly 50% of employment in 1980. Malawi's mining industry is dominated by cement and industrial minerals production.

Mineral Industry

The mining of limestone for cement production remains Malawi's biggest mining industry. Deposits of asbestos, chromite, corundum, iron ore, kyanite, magnesite, nickel laterite, bauxite, uranium minerals, and zircon have been reported, but none have been exploited.[3] Mineral production for Malawi is given in Table II-3-5.

The Ngana coal field in northern Malawi reportedly contains 14 million tons of reserves and has been considered for development by the government. It is unlikely that Malawi will have a mineral industry of any significance in the foreseeable future. However, Malawi does have enormous water

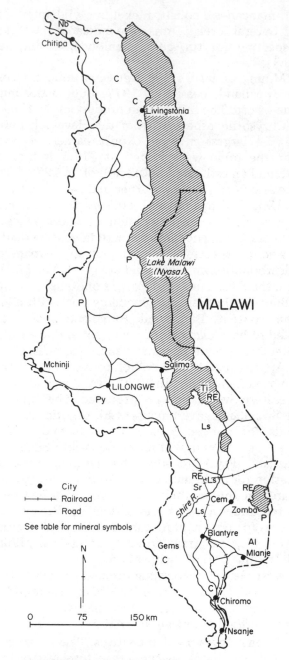

FIGURE II-3-3 Malawi: Mines, processing facilities, and infrastructure. (From U.S. Department of the Interior, Bureau of Mines. *Mineral Industries of Africa*, 1976.)

TABLE II-3-5   Malawi Mineral Production, 1976–1981 (Metric Tons)

	1976	1977	1978	1979	1980	1981
Cement, hydraulic (1000 tons)	85	94	103	113	92	78
Gem and ornamental stone: agate	4	4	4	6	7	7
Kyanite	90	250	100	—	—	—
Lime	250	—	—	—	—	—
Stone: limestone	149,254	116,653	155,229	168,604	122,814	116,118

Sources: Refs. 2 and 4.

resources and a railway running into Mozambique in southern Malawi to the port at Beira and eastward to the port of Nacala.

REFERENCES

1. U.S. Central Intelligence Agency, National Foreign Assessment Center. *The World Factbook—1981*. Washington, D.C.: U.S. Government Printing Office, 1982.
2. U.S. Department of the Interior, Bureau of Mines. "Malawi." *Minerals Yearbook 1980*. Washington, D.C.: U.S. Government Printing Office, 1982.
3. U.S. Department of the Interior, Bureau of Mines. "Malawi." *Minerals Yearbook 1978/79*. Washington, D.C.: U.S. Government Printing Office, 1981.
4. U.S. Department of the Interior, Bureau of Mines. "Malawi." *Minerals Yearbook 1981*. Washington, D.C.: U.S. Government Printing Office, 1983.

BOTSWANA

The Republic of Botswana is a land-locked nation of approximately 600,000 km[2]. It shares a common border with Namibia/South-West Africa the west and north, Zimbabwe on the northeast, and South Africa in the south and southeast. Botswana is characteristically a desert region (75%) with the population, industries, and agriculture located in the eastern part of the country. The population of approximately 800,000 is composed of 94% Tswana, 5% Bushmen, and 1% European.[1]

Before independence, Botswana was a British Protectorate. In 1966, committed to a new constitution and a parliamentary government, Bechuanaland became the Republic of Botswana. It has remained a member of the Commonwealth. The capital is Gaborone, located in the south, adjacent to the South African border.

The government, headed by President Quett K. J. Masire, consists of an executive branch with an appointed cabinet. The legislative branch consists of a Legislative Assembly and a House of Chiefs with deliberative powers only. The judicial branch is made up of a High Court and subordinate courts which have criminal jurisdiction over all residents, with the Court of Appeal having appellate jurisdiction. Local courts administer customary laws.[1]

The economy of Botswana, although slowly

FIGURE II-3-4 Botswana: Mines, processing facilities, and infrastructure. (From U.S. Department of the Interior, Bureau of Mines. *Mineral Industries of Africa*, 1976.)

diversifying, is dependent mainly on the raising and processing of livestock and the mining of diamonds, copper, nickel, and coal. Indirectly, wages earned by migrant workers in South African mines have also been an important contributor to the Botswana economy, as indicated by Table II-3-6. General features of the Botswana economy are presented in Table II-3-7.

Government Role in Mineral Industry

The government owns all the mineral rights in Botswana, but encourages the development of its mineral resources by foreign sectors. Before independence, mineral rights were controlled by various tribal groups and private interests. In 1967, all mineral rights, except building materials, were assumed by the government with the consent of the various tribes. After a substantial tax increase on privately held lands in 1972, the mineral rights of large areas of land reverted to the government.

The government believes that the development of its mineral resources is an essential element to Botswana's modernization. This development is expected to create employment and income in the extractive and processing industries, to increase revenues to government and foreign currency earnings, and to create indirect benefits from the development of secondary and tertiary industries and infrastructure facilities.[2]

Sustained mineral development will depend on continuing geological investigations, the encouraging of prospecting by the private sector, creating a skilled labor force, and protecting the environment.

Mineral Administration and Regulation. The Ministry of Mineral Resources and Water Affairs is the main governmental body overseeing the development and exploitation of natural resources. Under its jurisdiction are the Geological Survey Department, the Department of Mines, the Department of Water Affairs, the Department of Electrical Engineers, the Botswana Power Corporation, and the Water Utilities Corporation.[3]

The Geological Survey Department is responsible for the preparation of geological maps, monitoring the exploration activities of the private sector, and undertaking its own mineral investigations.[3]

The Department of Mines evaluates all proposed mineral development and monitors the extractive and processing industries to ensure compliance with governmental regulations. By routine sampling and analysis of ores, the Department of Mines ensures that standards are maintained and appropriate revenues on mineral products are collected by the government.[3]

Three major types of mineral concessions exist in Botswana: a reconnaissance permit, a prospecting license, and a mining lease. A reconnaissance permit is valid for one year and allows the exploration of a wide area without financial obligation, with a view to finding a smaller area for actual prospecting under license. The prospecting license is restricted to an area of 1000 km^2 to a particular mineral commodity. The license requires the submission of a work program and estimated expenditures to the Department of Geological Survey. The license is legal for three years with an additional two renewals of two years each. The Minister of Mineral Resources and Water Affairs must be notified of mineral discoveries within 30 days. A mining lease, valid for 25 years, is issued by the Department of Mines to those holding a prospecting license only.[4]

Mineral Agreements. The government grants mineral licenses only to those companies that have the technical, financial, and marketing capabilities to discover and develop those minerals requested in a prospecting license. The interested company must also submit a plan of development to the government for consideration. The government reviews this plan for possible economic and infrastructure benefits to Botswana.

To attract foreign investment, the government recognizes that foreign companies must realize a

TABLE II-3-6 Employment in South African Mines

Year and quarter	Number of employees	Number of recruits	Earnings of deferred pay (thousand pula)	Remittances (thousand pula)
1973		28,446		
1974		27,104	4,102	640
1975	24,076	33,337	6,972	1,098
1976	25,456	40,390	9,964	1,834
1977	25,297	38,564	10,695	2,571
1978	20,804	23,360	9,706	2,895
1979	20,307	19,523	9,120	4,333

Source: Ref. 8.

TABLE II-3-7 Gross National Product for Botswana, 1968–1979 (Million Pula at Current Prices)

GNP by output	1968–1969	1971–1972	1973–1974	1974–1975	1975–1976	1976–1977	1977–1978	1978–1979
Agriculture	23.2	33.1	62.4	61.2	65.7	74.4	71.7	78.1
Mining and quarrying	0.2	11.2	16.0	18.0	33.6	42.0	55.8	118.2
Manufacturing	2.8	5.1	10.1	15.5	20.9	25.3	24.4	41.4
Electricity and water	0.3	1.3	3.3	6.9	11.1	9.2	10.0	11.5
Construction	1.9	10.0	20.1	20.1	18.8	15.3	17.1	25.3
Wholesale and retail trade	5.1	17.5	27.9	34.3	42.5	55.5	71.9	101.7
Transport and communications	3.4	3.8	5.3	5.5	7.6	1.5	7.3	7.6
Financial institutions	3.5	6.1	13.3	14.6	18.2	24.6	29.9	46.8
General government	9.5	11.8	18.2	24.9	36.1	48.0	52.5	73.5
Household, social, and community services	1.3	3.5	6.8	8.8	13.1	12.5	15.2	18.9
Dummy sector	—	-0.8	-0.07	-3.3	-2.7	-7.9	-9.3	-17.6
GNP at market prices	51.2	102.6	182.7	206.5	264.9	300.4	346.5	503.4

Source: Ref. 8.

fair rate of return on investment. Therefore, the government has demonstrated a flexible bargaining position in mineral agreement negotiations.[5]

The government generally collects revenues from mineral development in four ways: royalty on sales revenue, free government equity of 15 to 25%, application of normal income tax (presently 35%), and a withholding tax of 15% on repatriated dividends.[6]

Royalties are an allowable tax deduction and are based on the gross market value of minerals (sales revenue) less transportation and insurance costs prior to sale. The following royalties are generally excepted, but subject to government change, depending on the circumstances.

	Royalties (%)
Precious stones	10
Petroleum and natural gas	10
Coal	5
Radioactive minerals	5
Semiprecious stones	5
Precious metals	5
Oil shale	3
Building and industrial materials	3
Other minerals and mineral products	3

Source: Ministry of Mineral Resources and Water Affairs, Republic of Botswana.

The government believes that equity (usually 15 to 25%) participation together with membership on the boards of directors will enhance communication between the government and the private company. This equity is considered to compensate for infrastructure and geological information provided for by the government. Mining ventures face a 35% tax rate, as does any other commercial project in Botswana. The 15% withholding tax on dividends repatriated to foreign countries was instigated to tax early windfall profits due to Botswana's generous depreciation laws. Table II-3-8 indicates revenue collected by the Department of Mines in recent years.

Infrastructure

Botswana remains a largely undeveloped country. Development in the eastern part of Botswana far exceeds that in the west. There are few roads, one railway, and limited amounts of water in the interior. Providing infrastructure to mining ventures is a cornerstone of Botswana's mineral policy. In return, the government is compensated by 15 to 25% equity in the company. The government has placed a high priority on the development and expansion of its infrastructure in order to attract new mineral ventures.

Transportation

Railroads.   There is only one main railroad line in Botswana. It enters Botswana from Zimbabwe in the northeast above Francistown, and runs southward through the capital Gaborone into South Africa. The rail line is operated and owned by the government of Zimbabwe, although Botswana plans to nationalize the railroad by 1986.[7]

Because of the civil war in Zimbabwe, Botswana has been very dependent on South Africa's railroads for its exports and imports. To alleviate this dependency and the dependency of other southern African countries on South Africa's rail system, Botswana has initiated planning on a proposed $800 million trans-Kalahari Desert railroad line from Palapye or Francistown, wesward to the railhead at Gobabis, Namibia, into Walvis Bay Port.[7] This would require about 850 km of track in Botswana, and an additional 110 km from the western border to Gobabis. A 1979 study indicated that the project would become economically feasible if Botswana could export 10 million tons of coal per year along this route.[7]

Roads.   Botswana's major highway parallels the rail line. In 1981, the final unpaved portion, between Lobatse and Ramatlabama, south of Gaborone, was completed.[7] Another highway connects Francistown with Nata to the northwest. From Nata, an all-weather road extends northward to Kazungula, where Namibia, Zambia, and Zimbabwe intersect. By ferrying across the Zambezi River, there is a major highway to Lusaka, the capital of Zambia. Another paved, all-weather road extends westward from Nata to Maun, near the Okavango swamp. From there a partially improved road extends to the southwest on to Ghanzi and then on to Gobabis, Namibia. From Gobabis, a highway extends all the way through Windhoek to Walvis Bay. Another partially improved road extends from Ghanzi to Lobatse in the southeast of Botswana.

There are also paved roads connecting the major mines to the infrastructure in the east, to facilitate the movement of machinery and supplies.

Water. Botswana is mostly a desert region. The Department of Water Affairs under the Ministry of Mineral Resources and Water Affairs is responsible for the water in Botswana.

There are three major rivers in Botswana. The Cubango River flows from central Angola, across Namibia, and empties into the Okavango swamps in northwest Botswana. The Molopo River parallels the border between South Africa and Botswana in the southwest. The Limpopo River parallels the border between South Africa and Botswana in the southeast. Development of any of these rivers will require some form of international agreement.

TABLE II-3-8   Revenue Collected by the Department of Mines, Botswana

	1981–1982	1980–1981	1979–1980
Diamonds			
Royalties and dividends	14,797,385	50,550,355	50,112,041
Leases	11,472	11,472	11,472
	14,808,857	50,561,827	50,123,513
Nickel-copper			
Royalty[a]	0	375,000	1,787,684
Leases	12,576	12,576	12,576
	12,576	387,576	1,800,260
Coal			
Royalty	245,665	148,721	143,408
Leases	1,104	1,104	1,104
	246,769	149,825	144,512
Talc[b]			
Royalty	—	562	829
Leases	—	48	48
Semiprecious stones: royalty	—	126	0
Blasting licenses	1,029	1,024	846
Sundries	256,878	160,790	27,902
	15,326,109	51,261,778	52,097,910

[a]Revenue is accounted for on a cash basis only. All base metal royalties were deferred in 1981 except for P375,000, which was credited directly to the Ministry of Finance and Development Planning.
[b]Talc operations ceased during 1979, and the mining lease was formally surrendered November 1981.

Source: Republic of Botswana, Department of Mines, *Annual Report*, 1981.

Rainfall in Botswana is seasonal. There are various small scale projects to capture this rainfall. Most notable of these was a project to capture the runoff from the Okavango swamps through the seasonal Boteti River southward to a natural pan, some 60 km from the Orapa diamond mine. Water is pumped from this reservoir by pipeline to the Orapa diamond mine.

There is a dam on the Sheshe River which provides water to the Selebi-Pikwe mine in eastern Botswana. A dam also supplies water to Gaborone.

Energy. The Botswana Power Corporation (BPC) was established in 1970 and is responsible for the generation, transmission, supply, and distribution of electricity in areas approved by the Minister responsible for power.[3] Power generation has increased from 38,781,000 kwh in 1971 to 473,408,000 kwh in 1980.[8]

There are two main operating divisions of the BPC. The southern division is located in Gaborone, where a 25-MW station is located. An additional 6-MW extension for this station is under construction and is due to be completed in late 1982.[7]

The Northern division is a 60-MW station at Selebi-Pikwe which supplies electricity to the mine and Francistown. This station was initially run on diesel fuel but has since converted to coal from the Morupule colliery.

To meet Botswana's longer-term power supply needs, a 180-MW power plant is to be built near Palapye-Morupule, taking advantage of the large coal reserves nearby. The plant will have an initial output of 90 MW by 1986 and an additional 90 MW by 1989. A high-voltage transmission network will connect this plant with most of the nation. The initial portion will be built to Gabarone. Cost of the project is estimated to be at $200 million.[7]

Mineral Industry

The mineral industry of Botswana is dominated by diamond, copper-nickel, and coal production. Government revenues from these mineral commodities form a substantial percentage of government revenue through equity on the companies, taxes, and royalties. The value for Botswana's mineral production are given in Table II-3-9. Conversion factors for the Pula are given in Table II-3-10. The

TABLE II-3-9 Botswana Mineral Production, 1973–1981

Year	Copper-nickel matte				Coal		Diamonds	
	Total matte produced (metric tons)	Copper content (metric tons)	Nickel content (metric tons)	Estimated value of production (thousand pula)	Production (metric tons)	Estimated value of production (thousand pula)	Total production (thousand carats)	Estimated value of production (thousand pula)
1973	—	—	—	—	20,299	130	2,453	23,229
1974	6,663	2,752	2,577	7,857	32,732	310	2,718	30,135
1975	16,513	6,504	6,447	22,185	71,248	641	2,414	29,583
1976	32,506	12,473	12,581	51,586	224,099	1,719	2,361	33,829
1977	30,776	11,783	12,099	43,000	294,039	2,170	2,691	48,567
1978	39,516	14,614	16,049	51,315	314,486	2,322	2,799	78,972
1979	39,823	14,563	16,173	74,068	355,115	2,696	4,369	165,719
1980	40,099	15,553	15,442	83,258	371,395	3,461	5,101	214,293
1981	46,565	17,819	18,278	79,439	380,698	4,237	4,960	202,082

Source: Republic of Botswana, Department of Mines, Annual Report, 1981.

TABLE II-3-10   Rate of Exchange
Between Pula and Dollar[a]

Date	1 pula =	U.S. dollar
Dec. 1970		1.3943
Dec. 1971		1.3068
Dec. 1972		1.2274
Dec. 1973		1.4900
Dec. 1974		1.4501
Dec. 1975		1.1500
Dec. 1976		1.1500
Dec. 1977		1.2075
Dec. 1978		1.2075
Dec. 1979		1.2679
Dec. 1980		1.3473

[a]1970–1975 figures are rates against the South African rand;
1976–1980 figures are rates against the pula.

Source: Ref. 8.

general trend for the value of mineral production
can be seen in Figure II-3-5. A general geologic description of Botswana is given in Figure II-3-6.

Diamonds.[9] Diamond exploration in Botswana
began in 1955 by De Beers to discover the source
of three small diamonds found along the banks of
the Motboutse River in eastern Botswana. In 1967,
Orapa, some 240 km west of Francistown, was discovered. It turned out to be the second largest
diamond pipe in the world. The De Beers Botswana
Mining Company (Debswana) was formed on
June 23, 1969, to undertake the development of
the mine. Initially, the government's equity in the
project was 15% but increased to 50% in 1975 for
all diamond ventures.[5] The mine came into production on July 1, 1971.

Initially, the mine was designed to produce 2.5
million carats per year. Current production is nearly
4.5 million carats per year after recent expansions
in the mine. Gem-quality diamonds are about 10%
of production.

Some 40 km southeast of Orapa, two smaller
pipes were discovered in 1969. The Letlhakane
mine came into production in 1977. Initial production was about 300,000 carats and increased to
400,000 in 1980 after further expansion of the
mine.

In 1973, De Beers' geologists discovered another diamond pipe at Jwaneng, 125 km west of
Gaborone, discovered under 30 to 50 m of Kalahari
sediment—the first time a payable pipe has been
discovered beneath a sand overburden of this magnitude. In April 1978, De Beers and the Botswana
government reached agreement on the development of this mine. The Jwaneng mine was officially
opened in August 1982. The mine is expected to
produce 20% gemstones. Mine output was expected
to be 3 million carats in 1982, increasing to 4.5
million carats by 1985.[10] By 1985 it is estimated
that Debswana's three mines will be producing
about 9.5 million carats per year.[10]

In 1978 and 1979, the total value for diamond
production was $92 million and $208 million,
respectively. This amounted to 59% and 68% of
the total value of mineral production in 1978 and
1979, as well as 41% and 46% of total export earnings.[11] In 1980, the value of diamond production
increased to $302 million, 62% of total exports.[12]

In 1981, revenue from diamonds decreased
$110 million, and the forecast for 1982 was even
bleaker because of the fall in diamond prices and
world overproduction. In 1981, De Beers sales

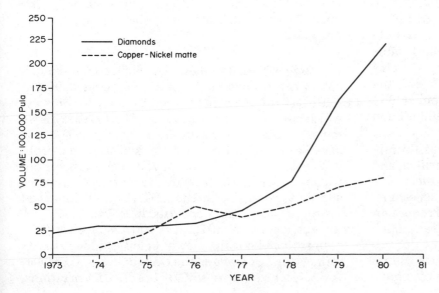

FIGURE II-3-5 Value of mineral production in Botswana, 1973–1981. (From
Ref. 8.)

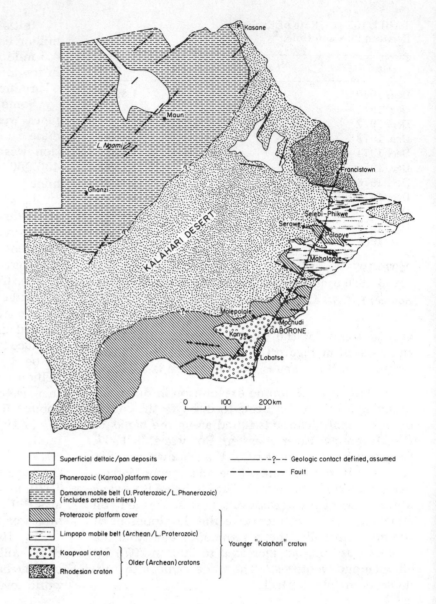

FIGURE II-3-6 Geological map of Botswana. (From Botswana Geological Survey. Mineral Resources Report No. 4, 1977.)

plummeted to $1.4 billion, a decrease of 46%. The value of their stockpiles doubled to over $1 billion dollars.[7] As a result, Debswana was forced to hold back nearly $150 million worth of marketable diamonds. Consequently, Botswana is experiencing its worst financial crisis since the 1930s, with the country's balance of payment moving from an $81 million surplus in 1980 to a deficit of $68 million in 1981.[7]

Diamond resources were estimated as 60 million carats proven, 25 million carats identified, and 25 million carats undiscovered.[11] However, in a country covered 75% by desert sands, such estimations may prove to be meaningless. Production figures for the Orapa mine and Letlhakane mine are given in Table II-3-11.

Diamonds are marketed through De Beer's Central Selling Organization. At one time diamonds were sent to London, but now Botswana has developed its own facilities for polishing and grading diamonds.

Copper-Nickel-Cobalt. The main copper-nickel-cobalt ore bodies in Botswana are located 90 km southeast of Francistown in Selebi and Phikwe, which are about 14 km apart. The Selebi-Phikwe mines began production in February 1974 and have been plagued by financial and technical problems ever since. BCL Ltd. operated the concession, which was owned 15% by the Botswana government and 85% by the Botswana RST (BRST). Financial restructuring of BRST occurred in 1978 and 1979.[11] An agreement in 1978 had AMAX receiving the copper-nickel-cobalt matte being shipped to its refinery in Louisiana in the United States. However, AMAX, which owns 25% of the BRST operations,

TABLE II-3-11   Diamond Mining Orapa Mine and Letlhakane Mine: Production Statistics

Year	Area mined (1000 m³)	Amount treated (thousand metric tons)	Grade (carats/metric ton)	Diamonds recovered (thousand carats)
Orapa Mine				
1970	—	—	—	538[a]
1971	—	695	0.89	913[a]
1972	1438	2701	0.89	2446
1973	1499	2815	0.86	2453
1974	1570	2954	0.92	2718
1975	1448	3048	0.79	2414
1976	1570	3429	0.69	2361
1977	1579	3450	0.68	2345
1978	1640	3515	0.70	2454
1979	2903	6140	0.67	4091
1980	3061	7330	0.64	4700
1981	3048	7401	0.61	4531
Letlhakane Mine				
1977	523	1087	0.32	346
1978	586	1111	0.30	331
1979	654	1304	0.23	303
1980	945	2113	0.19	401
1981	965	2229	0.19	430

[a]Includes prospecting diamonds.

Source: Republic of Botswana, Department of Mines, *Annual Report*, 1981.

has requested a 25% reduction in the copper-nickel matte shipped to its refinery because of a 40% reduction in earnings for 1981. There was speculation that AMAX had offered to sell its share of BLC to Anglo-American Corporation of South Africa Ltd. in 1971.[7] Normally, AMAX receives approximately 40,000 tons of matte per year for processing. To meet this 25% reduction of approximately 10,000 tons/year, negotiations are under way with both RTMZ and Falconbridge Nickel Mines of Canada to refine this tonnage in their own smelters.[7]

Average content of the matte produced by the mine is 40% copper, 40% nickel, and 0.6% cobalt. The production of the copper-nickel-cobalt matte was valued at $62 million in 1978, $94 million in 1979,[11] and $112 million in 1980.[12] Production figures for the Selebi-Phikwe mines are given in Table II-3-12.

Additional copper deposits are known to exist in northeastern Botswana, as well as copper deposits in northwestern Botswana, which are similar to the copper deposits of Zaire and Zambia.[9]

Coal. The Morupule Colliery in eastern Botswana is the only major coal-producing site in Botswana today, although recent studies have indicated that coal resources for Botswana may be near 40 billion tons.[11] Other possible coal fields are located in western Botswana. Coal production was valued at $2.8 million in 1978, $3.4 million in 1979,[11] and $5.5 million in 1980.[12]

The Morupule Colliery currently produces around 380,000 tons of coal per year,[7] although the government had hoped to produce 1 million tons of coal per year by 1980.[12] Recent exploration by several companies has indicated that reserves of this coal field are 3.5 billion tons of proven reserves and 13 billion tons of indicated reserves.[7] These reserves are in two main seams—18 ft and 8 ft thick—separated by about 70 ft of sandstone. The coal was located in Karoo sediments of the Late Carboniferous to Jurassic Age.[11] Botswana coal is classified as a steam coal having a medium heat content, a medium ash content, and a low sulfur content.[11]

This coal is currently used to produce electricity for the Selebi-Phikwe power station and will be the source of fuel for the government-planned 180-MW power station near the mine. Government plans call for the startup of a new, larger mine between the towns of Serowe and Palapye.

The government hopes to greatly expand coal production in order to export this coal. The low

TABLE II-3-12 BCL Ltd. Production Statistics, 1975–1981

	1975	1976	1977	1978	1979	1980	1981
Ore mined (tons)							
Phikwe mine							
Underground	509,163	1,121,328	1,152,393	1,476,717	1,531,312	1,568,149	1,629,107
Open pit	620,889	1,001,451	748,149	627,488	596,437	635,814	—
Selebi mine	—	—	—	—	98,814	344,773	826,683
	1,130,052	2,122,779	1,900,542	2,104,205	2,226,563	2,548,736	2,445,790
Metric tons milled	1,149,868	2,079,855	1,946,966	2,101,801	2,233,249	2,541,589	2,463,476
Head grade (%)							
Ni	1.15	1.07	1.02	1.04	0.99	0.93	0.89
Cu	0.94	0.86	0.83	0.82	0.78	0.82	0.81
Concentrate (tons)	262,678	402,605	375,850	429,511	517,535	625,571	712,135
Recoveries mill (%)							
Ni	70.4	70.3	73.5	75.9	79.1	80.6	86.8
Cu	91.2	94.2	94.1	94.9	95.2	94.7	96.3
Recoveries smelter (%)							
Ni	70.0	79.0	78.0	91.5	91.5	88.6	90.0
Cu	71.0	74.0	73.3	82.4	84.3	83.7	85.2
Co	15.0	21.0	16.6	23.5	27.2	22.1	22.0
Matte produced (tons)	16,514	32,506	30,772	39,517	39,823	40,099	46,565
Contained metal (tons)							
Ni	6,447	12,581	12,094	16,049	16,173	15,442	18,278
Cu	6,504	12,473	11,788	14,615	14,563	15,553	17,819
Co	92	198	165	261	294	226	254
Grade (%)							
Ni	39.04	38.71	39.31	41.80	40.62	38.51	39.25
Cu	39.38	38.37	38.31	36.98	36.57	38.79	38.26
Co	0.56	0.61	0.53	0.66	0.74	0.56	0.55
Estimated value of production (thousand pula)	22,185	51,586	43,000	51,315	74,068	83,258	79,439

Source: Republic of Botswana, Department of Mines, Annual Report, 1981.

sulfur content makes Botswana steam coal desirable for environmentally concerned countries. During 1979, a mission from the Japanese Ministry of International Trade and Industry visited both the Morupule and Mmamabula coal fields with a view to developing these fields for possible development and exportation of coal to Japan.

In July 1982 the government of Botswana signed agreements with Shell Coal International to set up a coal development company. In this arrangement the government receives 15% equity and an option to purchase an additional 10% at a later date. The feasibility study is expected to take two years. The coal will be shipped via Ellisras, South Africa, to Richards Bay. Production could begin by 1985, with up to 5 million tons being produced per year.[13]

Although the government has long-range expectations for the exploitation of coal in Botswana, there are several constraints to consider. The first constraint is transportation, and the second constraint is water availability.

Presently, Botswana's two main export routes are through Zimbabwe to the Mozambique port of Maputo or through South Africa to Richards Bay port. Both routes are sensitive to political pressures and possible disruptions. A government feasibility study has indicated that 10 million tons of export coal a year would finance a rail line from Palapye to Gobabis, Namibia. However, the problems of water availability must first be overcome for this to be possible.

Because of its medium-ash content, Botswana coal would need large amounts of water to process the coal for exportation. Botswana is largely a desert region, with water located on its borders. Coal development would require use of this water, which would in turn require international agreements on water use. Whether South Africa would encourage competition in coal exports is questionable. At any rate, extensive coal development in Botswana would require capturing more of the seasonal rainfall, use of underground water reservoirs, and construction of water pipelines.

Mineral Potential

An airborne magnetic survey conducted by the Canadian International Development Agency (CIDA) and the Botswana Geological Survey Department in the mid 1970s was released in 1978. The survey, combined with a computer-assisted interpretation of the possible subsurface geology, indicated potential areas of mineralization. These areas were to be investigated further by ground survey teams.[6] Other exploration programs are also

under way or planned through bilateral technical assistance arrangements with other governments. These include West Germany, France, Japan, and the United Kingdom.[6] A list of companies holding prospecting licenses and mining leases is given in Table II-3-13.

Chances for further mineral discoveries are good. Given the degree of mineralization in surrounding countries of Namibia, South Africa, and Zimbabwe it seems likely that mineral-producing environments also exist in Botswana. Advances in

TABLE II-3-13   List of Companies Holding Prospective Licenses and Mining Leases in Botswana

Company	Minerals
Shell Coal (Botswana) (Pty.) Ltd.	Coal, oil shale, and radioactive minerals
Falconbridge Explorations (Botswana) (Pty.) Ltd.	Cu, Pb, Zn, Co, Ni, Sb, Au, Ag, W; radioactive minerals; precious stones
BCL Ltd.	Cu, Ni, Co, Ag, Au, Pt, Pd
Billiton Botswana (Pty.) Ltd.	Cu, Pb, Zn, Ni, Au, Ag, Co, Mo, Ba, Cd
N.M. Michaelides & Partners	Au, Ag
AMAX Exploration (Botswana), Inc.	Coal, radioactive minerals
BP Coal Ltd.	Coal, radioactive minerals
C.D.F. Botswana Pty. Ltd.	$NaCo_3$, $NaHCO_3$, NaCl, $NaSo_4$, Br, Li, Cu, Ni, Co, Pb, Zn, Ag, Au, Pd, Pt; precious stones
Masego Mining Co. (Pty.) Ltd.	Au, Ag
Michaelides, Edds & Partners	Au, Ag
Golden Sands (Pty.) Ltd.	Au, Ag
Gauta Goldfields (Pty.) Ltd.	Au, Ag
Total Coal Botswana (Pty.) Ltd.	Coal, radioactive minerals
De Beers Prospecting Botswana (Pty.) Ltd.	Precious stones
De Beers Botswana Mining Co. (Pty.) Ltd.	Precious stones (Orapa) (Letlhakane) (Jwaneng)
Bamangwato Concessions Ltd.	Cu-Ni ore (Selebi-Phikwe)
Morupule Colliery (Pty.) Ltd. Coal	Coal (Morupule)
Shamrock (Pty.) Ltd.	Au, Ag in mine dumps
Rainbow Mine Sands (Pty.) Ltd.	Au, Ag in mine dumps
Masego Mining Co. (Pty.) Ltd.	Au, Ag
Golden Sands (Pty.) Ltd.	Au, Ag in mine dumps

Source: Republic of Botswana, Geological Survey Department, "Areas Held under Prospecting Licenses and Mining Leases as at 1st July 1982" (map), Lobatse, Botswana.

technological exploration techniques of discovering mineralization beneath the sand blanket of Botswana and Namibia will be forthcoming. Yet discovery does not guarantee exploitation. The economic feasibility of these deposits depends on several factors.

Constraints to Mineral Development

Of the major countries in southern Africa, Botswana remains one of the few to actively encourage foreign investment. The encouragement of foreign capital, coupled with a realization that investment by companies requires a good rate of return on investment, should be viewed as an inducement to investment in the Botswana mineral industry. However, potential investors will undoubtedly recognize that there remain several major constraints to mineral development in Botswana.

The lack of infrastructure, especially transportation and water, have already been discussed. The government will have to develop at least a skeletal network of infrastructure across the country. Otherwise, the cost of infrastructure to private companies may be so high in most instances as to make the cost of a project economically unfeasible.

Being a land-locked nation, Botswana is also vulnerable to political developments on its borders. Zimbabwe (formerly Rhodesia) has just emerged from years of civil war in its country. Ownership and control of Namibia has been a major issue in world politics for some time. Pressure on Botswana to oppose South Africa will undoubtedly increase as black Africa presses for majority rule within South Africa. The presence of guerrilla troops opposed to South Africa in Botswana could bring retaliation against Botswana.

South African influence in Botswana's economy is significant, especially for rural workers. "For most rural districts, between 20 and 30 percent of the males in the working age-group are in South Africa at any one time, and in the southeast of the country the proportion of male absentees is even higher than this."[14] For many rural men who do not own cattle, migration into South Africa forms a significant share of their monetary resources. As much of Botswana's development is aimed at urban development, loss of South African jobs could be significant to an already poor rural population.

The majority of trade, both imports and exports, is also within the Customs Union (Botswana, South Africa, Lesotho, Swaziland, and Namibia/South-West Africa). This 1969 Agreement, which contained provisions for the distribution of income, free interchange of goods, consultation, freedom of transit, trade with nonmembers of the Union, and protection of infant industries in the less-developed regions, is dominated by South Africa. Interruption in trade between South Africa and Botswana would be significant.[15] The importance of the common customs area can be seen from Table II-3-14 for exports and Table II-3-15 for imports. Figure II-3-7 indicates in graphic form the importance of revenue from the Custom Area to the total imports of Botswana.

Low commodity prices for mineral commodities, especially diamonds, would significantly alter Botswana's balance of trade. The percentages of diamond sales for total exports were 41% in 1978, 46% in 1979,[11] 62% in 1980,[12] and down to 42% in 1981.[7] The percentage is not expected to rise soon because of low worldwide diamond prices and oversupply. Just how the tremendous increase in diamond production from the newly opened Jaweng diamond mine affects revenues remains to be seen. Present copper and nickel prices remain depressed also.

TABLE II-3-14   Exports from Botswana, 1973–1979 (F.O.B. Value, Thousand UA)

Year and quarter	Common customs area[a]	Other Africa	Europe United Kingdom	Other	North and South America	All other countries	Total imports
1973	11,071	4,338	40,154	3,013		591	59,200
1974	30,771	3,436	35,450	3,200	8,617	516	81,990
1975	24,772	4,704	49,727	2,186	22,644	1,077	105,040
1976	23,188	11,494	63,319	2,708	52,067	396	153,172
1977	18,131	13,111	61,515	21,336	42,014	546	156,653
1978	26,244	14,281	9,051	88,862	53,094	1,144	192,676
1979	25,631	30,972	27,765	197,433	63,609	1,843	367,253

[a]The common customs area comprises Lesotha, South Africa, and Swaziland.

Source: Ref. 8.

TABLE II-3-15   Imports into Botswana, 1973–1979 (Duty Inclusive, C.I.F. Value, Thousand UA)

| Year and quarter | Common customs area[a] | Other Africa | Europe | | North and South America | All other countries | Total imports |
			United Kingdom	Other			
1973	79,530	12,438	6,358	2,214		1,860	114,964
1974	94,393	17,265	4,324	2,498	5,064	1,874	125,418
1975	127,109	20,310	3,875	3,404	3,831	759	159,288
1976	147,574	22,136	3,014	3,048	3,678	1,935	181,385
1977	205,480	23,818	3,916	1,377	4,068	946	239,605
1978	260,036	30,536	5,096	3,021	6,172	2,229	307,090
1979	384,261	30,263	9,903	4,130	5,855	3,877	438,289

[a]The common customs area comprises Lesotha, South Africa, and Swaziland.

Source: Ref. 8.

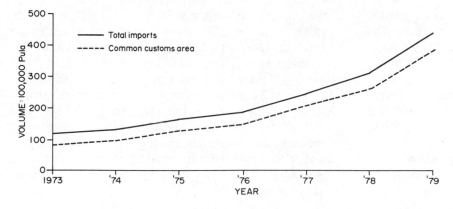

Note: Common customs area comprises Lesotho, South Africa, Swaziland, and South-West Africa

FIGURE II-3-7 Imports to Botswana, 1973–1979. (From Ref. 8.)

Government revenues from the mineral industry are necessary for plans to expand and modernize the infrastructure and create more jobs. The loss of this revenue, without aid from various world organizations, will directly inhibit future growth in Botswana.

REFERENCES

1. U.S. Central Intelligence Agency, National Foreign Assessment Center. *The World Factbook—1981.* Washington, D.C.: U.S. Government Printing Office, 1982.
2. Government of Botswana, Government Information Services. *Mining in Botswana.* Gaborone, Botswana: Government Printer.
3. Government of Botswana, Ministry of Mineral Resources and Water Affairs. *Mirwana News.* Gaborone, Botswana: Government Printer.
4. Government of Botswana, Mines and Mineral Act, 1976. *Botswana Government Gazette*, Supplement A, 31 December 1976.
5. Cobbe, James H. *Government and Mining Companies in Developing Countries.* Boulder, Colo.: Westview Press, Inc., 1979.
6. Johnson, Charles J. "Mineral Objectives, Policies, and Strategies in Botswana—Analysis and Lessons." *Natural Resources Forum* 5 (1981), pp. 347–367.
7. U.S. Department of the Interior, Bureau of Mines. "The Mineral Industry of Botswana." *Minerals Yearbook 1981.* Washington, D.C.: U.S. Government Printing Office, 1983.
8. Government of Botswana, Ministry of Finance and Development Planning. *Statistical Bulletin* 5, No. 4. Gaborone, Botswana: Government Printer, December 1980.
9. Information provided by the Embassy of Botswana, Washington, D.C.
10. "Jwaneng Diamond Mine, Botswana Officially Opened." *Mining Magazine*, October 1980, pp. 263–265.
11. U.S. Department of the Interior, Bureau of Mines. "The Minerals Industry of Botswana." *Minerals Yearbook 1978/79.* Washington, D.C.: U.S. Government Printing Office, 1981.
12. U.S. Department of the Interior, Bureau of Mines. "The Mineral Industry of Botswana." *Minerals Yearbook 1980.* Washington, D.C.: U.S. Government Printing Office, 1982.
13. U.S. Department of Commerce. *Botswana.* Prepared by American Embassy, Gaborone, December 1982.
14. Colclough, Cristopher, and Stephen McCarthy. *The*

Political Economy of Botswana: A Study of Growth and Distribution. New York: Oxford University Press, Inc., 1980.

15. Kowet, Donald Kalinde. *Land, Labour Migration and Politics in Southern Africa.* New York: Africana Publishing Company, 1978.

ZAMBIA

The Republic of Zambia is a land-locked country, sharing a common border with a number of countries. Zambia is surrounded by Angola, Zaire, Tanzania, Malawi, Mozambique, Zimbabwe, and Namibia/South-West Africa. Its land area is some 745,920 km², with a population of about 6 million people. The percentage of non-Africans is less than 2%.[1]

Zambia, formerly Northern Rhodesia, became an independent country on 24 October 1964. The capital, Lusaka, is located in south-central Zambia. It has remained a member of the British Commonwealth. President Kenneth Kaunda has remained head of state since independence. Zambia became a one-party state in December 1972. The ruling party is the United National Independence Party (UNIP).[1]

The economy of Zambia, although diversified, depends on the export of copper and cobalt for over 90% of its export earnings. The labor force is much more diversified. The work force is divided as follows: 15% mining, 9% agriculture, 9% domestic service, 19% construction, 9% commerce, 10% manufacturing, 23% government and miscellaneous services, and 6% transportation.[1] Key economic indicators are given for Zambia in Table II-3-16.

Government Role in Mineral Industry

Control of Zambia's mining industry in 1964 was primarily in the hands of foreign companies. This control extended to marketing, management, and ownership. Since copper sales were by far the largest export earner, control of the copper industry became a prime objective of the new government.

In April 1968, President Kaunda made his Mulungushi Declaration in which he announced that the government would take over 51% of the equity of some 27 companies in manufacturing, wholesale, and retail trade.[2] In August 1969, an invitation to the copper companies to sell 51% of their equity to the government was extended. In 1970 the Zambian government became majority holder of the copper industry. In 1973, Zambia paid off the bonds supporting the government's debt for the 51% assumed equity. The government also canceled all existing management and marketing contracts by assuming these functions

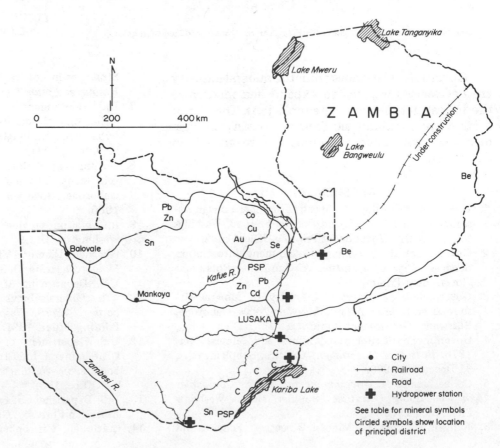

FIGURE II-3-8 Zambia: Mines, processing facilities, and infrastructure. (From Ref. 4.)

TABLE II-3-16   Key Economic Indicators for Zambia, 1978–1980[a]

	1978	1979	1980
Income, production, and employment			
GNP at current prices	2201	2623	3038
GNP at constant 1970 prices	1459	1342	1354
Per capital GNP at current prices (kwacha)	402	464	521
Per capital GNP 1970 prices (kwacha)	267	238	232
GNP composition at current prices			
Agriculture, forestry, fishing	358	375	434
Mining	287	484	535
Manufacturing	384	433	538
Gross fixed capital formation	555	472	594
Gross fixed capital formation at 1970 prices	213	154	162
Index of industrial product (1973 = 100)	103	105	112
Copper production (1000 tons[b])	656	579	610
Cobalt production (tons[c])	1560	3271	3309
Marketed Corn Production (1000 tons[b])	657	331	381
Employment (excluding traditional agriculture; thousand)	368	372	384
Population (million)	5.5	5.7	5.8
Balance of payments			
Merchandise exports[d]	674	1091	985
Copper	597	910	808
Cobalt	30	139	79
Merchandise imports (f.o.b.)	496	598	894
Oil	68	97	154
Imports from the U.S.	39	55	64
Services (net)	-187	-288	-259
Investment income (net)	-110	-99	-162
Current account (including transfers)	-184	46	-400
Capital account (net plus errors and omissions and SDRs)	-74	112	229
Overall balance	-259	158	-171
Payments arrears[e]	145 (508)	-149 (359)	133 (472)
Other financing (net, including IMF, reserve movements, and currency realignments)	114	-9	58

[a]All figures in millions of kwacha unless otherwise indicated. Average exchange rates are as follows: 1978, K1 = $1.23; 1979, K1 = $1.26; 1980, K1 = $1.27.

[b]1 ton = 2240 lb.

[c]Probably underestimates.

[d]Estimate. Export figure for 1980 may be revised downward 5 to 8%.

[e]Began in 1975; number in parentheses is cumulative arrears.

Source: U.S. Department of Commerce, International Trade Administration, "Zambia," *Foreign Economic Trends and Their Implications for the United States*, FET 82-023, March 1982.

itself.[3] Subsequently, control of other mining operations was also taken over by the government of Zambia.

Government Organization of Mining Industry. Soon after gaining control of the copper industry, the government began to reorganize the industry. Operations previously under the control of Anglo-American Corp. Group (Anglo) were consolidated into Nchanga Consolidated Copper Mines Ltd. (NCCM). Those operations under the Roan Selec-

tion Trust Group became known as Roan Consolidated Mines Ltd. (RCM). On 1 August 1974, Anglo (NCCM) relinquished its management agreements to the government for $51.5 million. In February of the following year, Amax (RCM) also gave up its managerial position for $34.3 million.[4] Metal-marketing activities were taken over by the Metal Marketing Corp. of Zambia Ltd., a government-owned company.

On 1 April 1982, both the NCCM and RCM divisions were reorganized into a new copper company, the Zambia Consolidated Copper Mines Ltd. (ZCCM). This new company is now the world's second largest copper mining group. In May 1981, President Kaunda presented arguments for the merger in a press conference.[5]

> Both companies are seeking to combat the effects of falling ore grades in the older mines and a general depletion of ore reserves. Based on current commercial assessment, it is expected that the RCM ore reserves mined as at present will be exhausted well before those of NCCM. While both companies are involved in mining for copper and cobalt, each has different problems and advantages, some of which are complementary. It is vital to the industry as a whole that the various operations of the mining businesses be closely coordinated and that the necessary priorities are given to their activities.

Benefits to be gained from the merger include the following:[6]

1. Coordinated planning and direction for the capital expenditure, extraction, and production of both companies
2. Creation of a more financially secure unit for raising money in international money markets
3. Better deployment of scarce technical skills and financial resources
4. Combined efforts on research and development, prospecting, and exploration and mine development
5. Important economies of scale in marketing, purchasing, and utilization of stores and other specialist services

Presently, majority state-owned enterprises and parastatals comprise 60 to 70% of the Zambian economy. Most of these are under the authority of the Zambia Industrial and Mining Corporation Ltd. (ZIMCO). ZIMCO includes the ZCCM and other mineral industry firms, most energy industries, the transport industry, and other major industries.[3]

Foreign Investment. Although government ownership of industries in Zambia is significant, Zambia also recognizes that private enterprise is important to the Zambian economy. The Industrial Development Act of 1977 defines current government policy regarding the private sector and foreign capital investment. Priority enterprises are those that use domestic raw materials, produce intermediate goods used by other industries, create substantial employment, improve domestic skills and technology, and promote industries in rural areas. Such enterprises may receive incentives in the form of preferential treatment for import licenses, rebates on imported goods, relief from sales and employment taxes, and some exemption from in-

come tax. These incentives may also be available for companies that export a large amount of its products.[3]

Corporations are taxed at 50% of profits. There is a mineral tax of 51% on net profits for copper, and varying amounts for other minerals. There is also a 45% income tax levied on profits net of the mineral tax.[3] A withholding tax of 15% on dividends and management fees and 10% on interest and royalty income was introduced in 1972.[3] In 1974, the special depreciation allowance for mines was canceled.[3]

Infrastructure

Zambia's main industrial zone is located in central Zambia within the copper-belt region, along the railways, south to the capital Lusaka. Western and northeastern Zambia remains largely undeveloped. Zambia has several potential routes to ocean ports. However, due to various political crises in surrounding countries, Zambia's access to ocean ports remains tenuous at best. Zambia's water resources are excellent, having been endowed by nature with numerous rivers. There are also several large lakes with hydroelectric capacity.

Transportation. Because Zambia is a landlocked nation, transportation to ocean ports is of critical importance to the Zambian economy. Routes do exist to both western and eastern ports in the Atlantic and Indian Oceans. However, political turmoil in neighboring countries has often made use of these various routes impossible. It is possible that the resolution of the conflict in Zimbabwe will stabilize Zambia's eastern routes to Mozambique's ports, as well as the southern routes to South Africa's ports.

Railroads. The Benguela route connects the Zambian copper region with the port of Lobito, Angola. The route extends from the copper belt, northward through the Zairian copperbelt to Lubumbashi, and then westward through Zaire into Angola at Luau. From Luau, the railway travels westward to Benguela and the port of Lobito. Unfortunately for both Zambia and Angola, the Benguela Railroad has remained closed due to the continuing conflict in Angola. The FLNA and UNITA parties control a large section of eastern Angola and have virtually stopped rail traffic from Zaire along the Benguela Railroad to Lobito.

The TAZARA rail route travels northeast from the copper belt through Tanzania to the port of Dar es Salaam. The TAZARA Railroad was completed in 1975. The railroad was engineered and built by the Chinese. However, this route has also had its problems. Heavy rains in April and May of

1979 washed out approximately 100 km of track within Tanzania.[7] The vulnerability of this route was demonstrated by Rhodesian troops when in 1979 they blew up the railroad bridge spanning the Chambishi River in northeast Zambia.[7]

The southern rail route from the copper belt travels south through Lusaka and Livingston into Zimbabwe. From Zimbabwe, rail traffic can proceed to either Beira or Maputo ports in Mozambique, or into South Africa to the port of East London. Zambia would prefer not to use the South African rail system. Although Zambia prefers to use facilities in countries other than South Africa for export, many imports into Zambia continue to arrive via South Africa. Freight charges are also lower on the South African route.[8] South African ports are superior to those of Dar es Salaam and Beira.

There are two developments that may significantly affect future rail transportation in Zambia. The first is the possibility of constructing a railway from the existing line in eastern Botswana, westward to Walvis Bay, Namibia. It is hoped that such a line will provide a means to avoid using the rail system in South Africa. For more detail, see the section on Botswana.

The other development concerns the possible electrification of Zambia Railways. A feasibility study for electrification of the lines began in 1981 and was to be completed in 1982. Electrification of lines would save a substantial amount of money on diesel used in locomotives. The study is being funded by the World Bank Railway Project.[8]

Roads. Due to various interruptions in the rail transport system, the use of roads in Zambia has occasionally played a major part in flow of goods to and from Zambia. The Tanzam Highway generally follows the TAZARA highway to the port of Dar es Salaam in Mozambique. Occasionally, this route has also been damaged by washouts caused by seasonal rains.

A highway extends from Lusaka, eastward to Malawi. From here it is possible to reach the port of Beira and also Nacala, a port in northern Mozambique. A highway also extends northward into Zaire's Shaba province—Zaire's copper-belt region.

From Kafue, just south of Lusaka, a highway runs into Zimbabwe to Harare (formerly Salisbury), with connections to Beira. Another highway runs through Kazungula into Botswana. There is a pontoon ferry across the Zambeshi River at Kazungula, making this route vulnerable to damage to the ferry, as Rhodesian troops demonstrated in 1979.[7] However, traffic resumed in 1980 when the ferry was restored to service.[9]

Water. Zambia's natural water resources are abundant. Major rivers, as well as a number of large lakes, are found throughout Zambia. The Zambezi River in southern Zambia forms Lake Kariba. The Kafue River runs through central Zambia and forms a large reservoir below Lusaka. Lake Mweru and Lake Bangweulu are large lakes in northeastern Zambia. The southernmost part of Lake Tanganyika is also in northeast Zambia.

Energy. The major source for energy in Zambia is electricity from hydroelectric facilities. The second most important source of energy is from imported oil, which forms a substantial part of Zambia's import bill. Coal is Zambia's third most important source of energy. Total installed generating capacity was 1708 MW by 1977.[3]

Less than 10% of electric sales were used by household and commercial users. The copper belt accounts for approximately 76% of electric consumption, yet covers only approximately 4% of the land area.[3] Excess electricity is exported to the copper region in Zaire.

There are three main hydroelectric installations in Zambia. The Victoria Falls facilities in southern Zambia on the Zambezi River have two facilities producing 60 MW and 40 MW. The Kariba I facility, built in 1960, has a rated capacity of 600 MW. The Kariba II facility was finished in 1977 with an additional 600 MW of capacity. The Kariba facilities are shared equally with the Zimbabwe government. The Kafue I Hydroelectric station generated 600 MW on its completion in 1972. In 1977, two additional facilities of 150 MW were added, bringing the total to 900 MW. Plans for additional hydroelectric capacity are under study.

In addition to hydroelectric generation, 100 MW are generated by gas turbine and thermal installations in the copper belt. In 1968, a 1719-km pipeline from Dar es Salaam to Ndola was completed. Initial capacity was 760,000 tons per year. Soon after 1968, extra pipe was added and the capacity of the line was expanded to over 1 million tons per year. The pipeline transported crude petroleum to the Indeni Refinery at Ndola, which had a 1.1-million ton per year refining capacity.

Electricity runs north from the hydroelectric station through Lusaka, north to the copper belt. As a consequence, most industrial and commercial users have located along this electrical network.

The electrification of Zambia's rail system would considerably lower the cost of transportation in Zambia. It would also reduce Zambia's petroleum import bill as locomotives are currently powered by diesel. However, transportation cost beyond Zambia's borders would remain an important variable in total transportation cost.

Mineral Industry

Although the mineral industry is only 20 to 25% of the Zambian GNP, it remains the major contributor to export earnings. In general, since 1964, revenue from copper exports has not fallen below 90% of total export earnings. Revenue from cobalt exports has increased steadily over the years as the price of cobalt has soared. Zambia is the world's fifth-largest copper producer and the world's second-largest cobalt producer. The other import minerals mined in Zambia are lead and zinc from the Broken Hill mine. Unlike several other countries in southern Africa, neither diamonds nor gold contributes much to the Zambian economy. Mineral production for Zambia is given in Table II-3-17, and the role of copper exports is given in Table II-3-18.

TABLE II-3-17   Zambia Mineral Production, 1969–1982 (Metric Tons)

| | Copper | | Cobalt | | Lead | Zinc |
	Mine	Refine	Mine	Metal	Mine	Mine
1969	719,467	603,199	1,812	1,798	22,900	68,200
1970	684,064	580,722	2,400	2,052	32,900	65,800
1971	651,396	534,339	2,080	2,079	33,900	68,900
1972	717,700	615,222	4,300	2,664	31,400	70,500
1973	706,574	638,509	3,840	2,379	37,000	73,200
1974	697,956	676,854	2,846	1,962	35,776	80,704
1975	676,921	629,150	3,164	1,843	26,082	67,298
1976	708,867	694,157	3,262	2,175	15,549	48,777
1977	656,000	648,043	3,677	1,704	13,542	45,018
1978	642,972	627,744	3,741	2,063	15,853	50,000
1979	588,334	561,940	4,280	3,176	17,640	46,600
1980	595,757	607,592	4,400	3,310	13,900	31,985
1981	588,000	560,446	4,960		17,152	40,557
1982[a]	540,000		5,000			

[a]Figures for 1982, *Mineral Commodities Summaries 1983*.

Source: U.S. Bureau of Mines, *Minerals Yearbook*, 1969–1981.

TABLE II-3-18   Zambian Revenues[a] from Copper Exports, 1968–1981

	Total exports	Copper exports	Percent copper of exports	Total imports	Exports minus imports
1968	762.2	744.2	97.6	455.3	+306.9
1969	1073.1	1046	97.5	436.5	+636.6
1970	1000.7	984.6	98.4	477.0	+523.7
1971	679.2	658.1	96.9	559.3	+119.9
1972	845.1	813.5	96.3	630.1	+215.0
1973	1157.4	1131.2	97.7	541.1	+616.3
1974	1406.6	1303.1	92.6	985	+421.6
1975	809.8	733.5	90.6	1137.7	−327.9
1976	1050.3	965.3	96.5	791.9	+258.4
1977	893.8	818.7	91.5	797.9	+ 95.9
1978	818.4	760.3	92.9	718.8	+ 99.6
1979	1412.0	1167.8	82.7	904.3	+507.7
1980		1279			
1981		1002			

[a]Based on Handy and Harman, "U.S. Producer Copper Refinery Cents per Pound (Wire Bars)," *Metals Week*.

Source: U.S. Bureau of Mines, *Minerals Yearbook*, 1968–1981.

Copper-Cobalt. Since cobalt production is a by-product of copper mining, aspects of the cobalt industry are included in the discussion of the copper industry.

Geology. The copper belt is a region in north-central Zambia and the Shaba (formerly Katanga) Province located in southern Zaire. It is a belt some 500 km long and 30 km wide located on the high lands between the Zaire River to the north and the Zambezi River to the east and south.[10]

The economic copper deposits are stratiform deposits, Proterozoic (Precambrian) in age, and are located in the Lower Roan group of the Katanga system. Several of the Zambian deposits have been uplifted and folded, resulting in gentle to overturned dips.[10] Usually, the ore occurs to depths of many hundred meters and must be mined by underground methods. Due to deformation of the original deposit, occasionally ore bodies approach the surface and are mined by open-pit methods.

The Katanga sediments and metasediments rest unconformably on a granite-schist-quartzite basement.[11] The Ore Formation is generally overlain by an alternating series of arenites and argillites.

The host rock is usually shale or dolomitic shale to the southwest of the Kafue Anticline and arkose-arenite to the northeast of the Kafue Anticline.[11]

The copper, with minor traces of iron and cobalt, occurs predominantly in disseminated bornite, chalcopyrite, and chalcocite. The lower part of the Ore Formation is more mineralized than the upper part.[11]

Table II-3-19 lists the major copper mines, with percent content of copper and cobalt, as well as the lastest production and reserve figures.

Organization. With the merging of the former NCCM and RCM companies into the Zambian Consolidated Copper Mines Ltd. (ZCCM), ZCCM became the second-largest copper mining company in the world after Chile's Corporación Nacional del Cobre de Chile (Codelco). ZCCM operates eight copper mines at Mufilira, Luanshya, Baluba, Chibuluma, Chambishi, Rokana, Chingola, and Kondola.

As of 31 March 1981 it was estimated that ZCCM had fully and partially developed reserves of 93,600,000 tons of ore at 3.45% copper. Of this amount, there are an estimated 28,400,000 tons of ore with an average grade of 0.17% cobalt.[6] Accordingly, Zambia has estimated reserves of 3,229,200

TABLE II-3-19   Copper Production and Ore Reserves, by Company and Mine

Company and mine	Ore mined and treated			Ore reserves		
	Gross weight (thousand metric tons)	Copper (%)	Copper content (metric tons)	Gross weight (thousand metric tons)	Copper (%)	Cobalt (%)
Roan Consolidated Mines[a]						
Mufulira (underground)	5,459	2.10	114,639	109,000	3.10	—
Luanshya (underground)	4,045	1.30	52,585	55,000	2.51	—
Baluba (underground)	1,691	1.70	28,747	59,000	2.54	0.16
Chambishi (underground)	2,080	1.56	32,448	31,000	2.84	—
Chibuluma (underground)	484	2.41	11,664	8,000	3.66	0.20
Kalengwa (open pit)[b]	215	2.47	5,311	—	—	—
Total or average	13,974	1.76	245,394	262,000	2.84	0.16
Nchanga Consolidated Copper Mines[c]						
Nkana (underground)	4,757	1.60	76,112	113,433	2.35	0.08
Bwana Mkubwa (open pit)	757	2.26	17,108	452,000	3.50	—
Chingola (underground and open pit)	9,016	3.34	301,134	281,089	3.16	—
Konkola (underground)	1,687	2.85	48,080	193,710	3.72	—
Kansanshi (open pit)	307	3.21	9,855	16,658	2.57	—
Total or average	16,524	2.74	452,289	1,056,890	3.31	0.01
Grand total or average	30,498	2.29	697,683	1,318,890	3.22	0.04

[a]Roan Consolidated Mines Ltd., Annual Report, 1981.
[b]Stockpiled ore was used for Kalengwa mill feed.
[c]Nchanga Consolidated Copper Mines Ltd., Annual Report, 1981.
Source: Ref. 8.

tons of copper and 48,280 tons of cobalt as of 31 March 1981. Table II-3-19 provides information from the RCM and NCCM *Annual Reports* of 1981.[8] RCM estimated the life of its mines at another 25 years, while NCCM estimated a life of 40 years for its reserves.[6] Processing facilities for both copper and cobalt are located at the various mines, and ore (at various processed stages) from one mine is often shipped to processing facilities at another.[6,10]

Within the old RCM division, Mufulira is the most important center, possessing milling and concentrating, smelting, and refining facilities.[8] There is also a copper refinery at Ndola.[6] A new cobalt vacuum refining plant at Chamblishi will make cobalt metal suitable for use in superalloy production.

Within the old NCCM, Rokana has become an important processing center. In addition to copper smelting and refining facilities, Rokana was the only producer of sulfuric acid in Zambia as of May 1982.[6] Rokana has also become the major center for processing cobalt. At Rokana, the old cobalt plant was extended and new roasting, leaching, and electrowinning facilities for cobalt were instigated.[6] Chingola has constructed new leaching facilities to be completed in 1985 for an estimated 15 years of reserves from old tailings at the mine site. The plant is expected to produce 524,000 tons of copper over a 15-year period.[5]

Exports. The dominant buyers of Zambian copper exports are Japan, France, Italy, the United States, West Germany, the United Kingdom, and India.[8]

Lead-Zinc. The main producer of lead-zinc in Zambia is the Broken Hill Division, a division of the old NCCM company. The ores are found in dolomite. The ore bodies are generally of three types: (1) lead-zinc lodes; (2) oxidized iron-zinc lodes; and (3) single pure, massive zinc lodes. The main sulfide minerals are sphalerite, galena, pyrite, and a little chalcopyrite.

Lead and zinc reserves were estimated at just under 1,200,000 tons of ore at average grades of 11.1% lead and 23.1% zinc.[6] Exploration at the mines has indicated only low-grade mineralization to date.[8]

Phosphate. ZIMCO has discovered large phosphate deposits at Kaluwe, 220 km east of Lusaka. The carbonatite ore contains an estimated 200 million tons of low-grade ore. At Petauke district in the Eastern Province, 2 million tons of high-grade apatite ore reserves have been discovered.[5] In the past, all phosphate (about 50,000 tons per year) has been imported for the manufacture of compound nitrate fertilizers at Nitrogen Chemicals of Zambia at Kafue. Development of these phosphate reserves can be an important addition to the Zambia agriculture sector, replacing foreign imports of this commodity.

Mineral Fuels. Coal production at Maamba collieries in the Southern Province should increase due to installation of new machinery financed by the African Development Bank and by West Germany.[8] The coal mine produces bituminous coal. Production at the mine was 708,000 metric tons in 1977, 1,169,000 tons in 1978, and down to 599,000 tons in 1979, 569,000 tons in 1980, and 527,000 tons in 1981.[8]

The Zambian Geological Survey has discovered a major coal belt near the Kafue National Park, located several hundred kilometers west of Lusaka.[8]

In 1982, a Canadian consortium began searching for oil in the Zambezi River basin area of Western Province and in the Lunaga River valley in eastern Zambia. The four-year prospecting venture will cost approximately $30 million.[5] A team from the World Bank began exploring for oil in 1982.[5]

Uranium exploration in Zambia continues. Companies involved in uranium exploration include Agip of Italy, Cogenna of France, Saarberg Interplan of West Germany, and Power Reactor Nuclear Fuel Development of Japan.[5] Little information of exploration results has been released at the request of the Zambian government.

Constraints to Mineral Development

The Zambian mineral industry, especially copper and cobalt production, is adversely affected by low commodity prices, transportation problems, skilled worker shortages, and lack of spare parts for mining-related machinery. Political unrest in neighboring countries threatens Zambia's carefully maintained neutrality.

The cessation of political turmoil in neighboring countries would do much for Zambia's mineral economy. First, it would provide greater access to ocean ports and help ease Zambia's import and export problems. Second, transportation remains a major cost in copper production. Any lowering of transportation cost would increase revenues from copper production.

Zambia is an outspoken critic of South Africa's apartheid policies and advocates majority rule in South Africa. Zambia is dependent on imports of food and machinery from South Africa, as well as on its transportation system. Zambia's significant involvement in effective action to promote majority rule in South Africa will probably result in some form of reprisal from South Africa. Active involvement in armed struggle would undoubtedly result in some form of physical reprisal. Raids on Zambia's transportation system or hydroelectric facilities could seriously harm Zambia's copper production. The results would be disastrous to Zambia's economy and export earnings. Table II-3-20 indicates Zambia's export and import trade routes for 1980.

Hydroelectric power capacity already exceeds demand, and Zambia exports power to both Zimbabwe and Zaire. Electrification of Zambia's railroads could lower transportation cost, generate more revenue from copper exports, and save on Zambia's fuel-import cost. Increased hydroelectric facilities could generate additional revenue. The discovery of oil from exploration activities could further reduce Zambia's import bill, as well as add security to Zambia's petroleum demands.

REFERENCES

1. U.S. Central Intelligence Agency, National Foreign Assessment Center. *The World Factbook—1981.* Washington, D.C.: U.S. Government Printing Office, 1982.

2. Cobbe, James H. *Governments and Mining Companies in Developing Countries.* Boulder, Colo.: Westview Press, Inc., 1979.

3. U.S. Department of Commerce, Industry and Trade Administration. "Marketing in Zambia." *Overseas Business Reports.* OBR 78-24. June 1978.

4. U.S. Department of the Interior, Bureau of Mines. *Mineral Industries of Africa.* Washington, D.C.: U.S. Government Printing Office, 1976.

5. "Zambia." *Mining Annual Review*, June 1982, pp. 437-441.

6. "Zambia: World's Second Largest Copper Producer Is Born." *World Mining*, May 1982, pp. 71-76.

7. U.S. Department of the Interior, Bureau of Mines. "The Mineral Industry of Zambia." *Minerals Yearbook 1978/79.* Washington, D.C.: U.S. Government Printing Office, 1981.

8. U.S. Department of the Interior, Bureau of Mines. "The Mineral Industry of Zambia." *Minerals Yearbook 1981.* Washington, D.C.: U.S. Government Printing Office, 1983.

9. U.S. Department of the Interior, Bureau of Mines. "The Mineral Industry of Zambia." *Minerals Yearbook 1980.* Washington, D.C.: U.S. Government Printing Office, 1982.

10. White, Lane. "Zambia." *Engineering and Mining Journal*, November 1982, pp. 146-183.

11. Evans, Anthony M. *An Introduction to Ore Geology.* New York: Elsevier North-Holland, Inc., 1980.

TABLE II-3-20   Zambia: Exports and Imports by Trade Routes, 1980 (Metric Tons)

Route	Quarter 1	Quarter 2	Quarter 3	Quarter 4	Total, 1980
			Exports		
Exports to Zaire	7,661	17,301	13,882	7,067	45,911
Dar es Salaam (road)	45,310	55,862	48,337	46,807	196,316
Dar es Salaam (rail)	24,702	88,697	85,052	64,419	262,870
Malawi-Mozambique	2,281	1,890	155	100	4,426
Zimbabwe-South (road)	—	—	1,183	4,867	6,050
Zimbabwe-South (rail)	127,516	51,996	41,028	46,298	266,838
Air freight[a]	1,235	693	797	772	3,497
Total Zambian exports	208,705	216,439	190,434	170,330	785,908
Zaire's exports in transit[b,c]	83,507	90,588	82,280	101,932	358,307
Total export traffic	292,212	307,027	272,714	272,262	1,144,215
			Imports		
Imports from Zaire	110	395	21	—	526
Dar es Salaam (road)	42,436	44,454	42,031	40,206	169,127
Dar es Salaam (rail)	56,395	58,104	69,371	59,940	243,720
Malawi-Mozambique	1,480	777	685	312	3,254
Zimbabwe-South (road)	12,549	27,649	25,011	8,046	73,255
Zimbabwe-South (rail)	88,622	92,572	121,403	68,258	370,855
Air freight	5,818	4,939	4,690	4,199	19,646
Total Zambian imports	207,410	228,800	263,212	180,961	880,383[d]
Zaire's imports in transit[c]	98,537	109,505	94,563	102,030	404,635
Total import traffic	305,947	338,305	357,775	282,991	1,285,018[d]

[a]Includes cobalt metal.

[b]Includes an average of 9480 tons per month of concentrates to port of East London for shipment to Japan.

[c]Does not include any significant amount of traffic with Dar es Salaam.

[d]Does not include 744,289 tons of crude oil imported by Tazama pipeline.

Source: Ref. 9.

ZIMBABWE

The Republic of Zimbabwe (formerly Southern Rhodesia) is a land-locked nation of approximately 391,090 km[2]. It shares common borders with the countries of South Africa, Botswana, Zambia, and Mozambique. The current population is over 7.5 million people.[1] Population growth for Zimbabwe during the period 1980–2000 is projected to be 4.3%, the highest for any country reported in the *World Development Report, 1982*.[2]

Of the current population, over 70% are members of the Shona subtribes, 20 to 25% are members of the Ndebele tribe, approximately 3% are European, and less than 0.5% are of mixed races.[1]

On 11 November 1965, Southern Rhodesia declared its independence from Great Britain in a dispute over the issue of majority rule. This Unilateral Declaration of Independence (UDI) was immediately followed by U.N. sanctions of imports into Rhodesia and exports of Rhodesian products. This

was the beginning of a long and costly struggle over majority rule in Rhodesia. On 1 March 1970, Rhodesia declared itself a republic. Political maneuvering by world governments, and an increasingly costly civil war, led to British supervision of elections held on 27, 28, and 29 February 1980. As a result of this election, Robert Mugabe, head of the Zimbabwe African National Union (ZANU), was elected Prime Minister of the Republic of Zimbabwe, creating another black-ruled African nation.[3] Zimbabwe became an independent nation on 18 April 1980.

The capital of Zimbabwe remains at Harare (formerly Salisbury). The government is a British-style parliamentary democracy. The parliament consists of a 100-member House of Assembly and a 40-member Senate. Although the new government is governed by majority rule, whites continue to hold some power in the government. Twenty seats are reserved for whites in the House of Assembly, and 10 of the seats in the Senate are elected by white

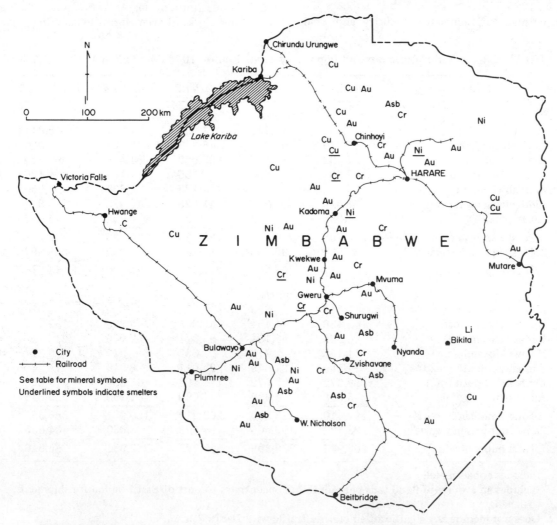

FIGURE II-3-9   Zimbabwe: Mines, processing facilities, and infrastructure. (From Ref. 8.)

members of the House of Assembly. Whites also occupy other important posts in the government.

A split in the ZANU-PF party and the Zimbabwe African People's Union (ZAPU), headed by Joshua Nkomo, could develop into a civil war between the two predominant tribal groups in Zimbabwe: the Shona and the Ndebele. Nkomo was dismissed from his cabinet post over the discovery of buried arms—linked to a possible coup attempt. The dispute eventually lead to Nkomo fleeing the country in early 1983.

Despite U.N. economic sanctions, the economy of Zimbabwe is the second largest in southern Africa. The economy is well diversified and is enhanced by a national infrastructure second only to South Africa's. In recent years, agriculture has contributed 16% to the GDP and has accounted for 37 to 45% of the country's exports. Agriculture accounts for the major portion of income for over 70% of the population.[4] Of this amount, the nearly 5000 commercial estates (mostly owned by whites) accounted for 89% of sales.[5]

In the same period, mining has accounted for only about 7.5% of the GNP and employs only 6% of the labor force. However, mining and mineral products combined are the leading foreign exchange earners.[4]

The manufacturing sector accounted for approximately 25% of the GNP.[4] The manufacturing industry is closely linked to the agricultural and mining industries. Zimbabwe has the potential to become an important supplier in some areas for neighboring countries trying to shift dependency on supplies from South Africa to other regional countries.

Government Role in Mineral Industry

Historically, the mineral industry of Zimbabwe has been controlled by private interests. The government sought to improve the infrastructure and encourage private investment. After UDI, U.N. sanctions were designed to halt imports to Zimbabwe and exports to other countries. However, U.N. sanctions did not bring down the white Rhodesian government, but instead, did much to encourage the diversification of the economy. As in the case of South Africa, Southern Rhodesia was forced to manufacture much of its own needs.

It is difficult to assess the ultimate effect of the new regime's eventual goals for the mineral industry. Mugube realizes that Zimbabwe's mining industry will flounder without foreign investment. However, his socialist leanings have led to some speculation that the government's role in the mineral industry may increase dramatically. The effect may be to reduce foreign investment further, owing to fears concerning security of investment.

Mineral Marketing Corporation. One indication of future government participation in the minerals industry was the formation of the Minerals Marketing Corporation (MMC) in April 1982.

The MMC is controlled by a nine-member board appointed by the Minister of Mines. There are three representatives from industry, five from government, and one from labor. The Chamber of Mines offers advice on the selection of these representatives.[6]

The MMC has the authority to decide when, and to whom, mineral products will be sold. It can also negotiate prices and approve or disapprove existing sales contracts. It determines stockpile levels and can purchase minerals for its own account. It can also sell or otherwise dispose of minerals. As a state agency it is exempt from any liabilities in carrying out its function.[6] In short, the MMC has assumed a greater position of power in the marketing of mineral products than that of the private mining companies. The effect has been to impose a certain amount of insecurity regarding contractual agreements made by private companies. However, the MMC has been in existence for only a short period of time, and it remains to be seen what the long-range effect will be on the mineral industry. It can be said that the powers of the MMC will do little to encourage foreign investment in the mining industry.

Another government institution—the Mineral Development Corporation (MDC)—is the government's newest attempt to gain more control over the mining industry. As of this writing, the composition, powers, and ultimate purpose of the MDC have not been determined. However, it is assumed that the new agency will involve state participation in existing and new mineral ventures. The vehicle for this participation will probably involve some sort of equity percentage in the range of 20 to 40%.[6]

It would be premature to speculate on the effects of government participation in the mineral industry. It must be remembered that other countries, notably Botswana, also have equity participation in mining ventures, and have representation on these companies' boards of directors. The amount of equity is however, the major issue, rather than whether there will be government equity in the future. These recent moves have tended to create an atmosphere of uneasiness over the government's ultimate goals. This uneasiness will in turn limit future investment—an unfortunate development, since Zimbabwe needs more capital than it can raise locally to provide growth, jobs, and export earnings.[5]

Taxation. A deficit of $433.8 million was estimated for 1981–1982.[7] An increase in goverment spending was to be offset by a 30% increase in tax revenue. As a result, several of the tax laws have undergone some fundamental changes.

The percentage deduction of initial capital expenditure has been reduced from 100% to 30%. A capital gains tax of 30% on immovable property and marketable securities has been initiated. A 20% tax on foreign dividends has been extended to include local dividends. There is now a 5% surcharge on customs duties and an increase in sales taxes. There was also some discussion of removing the 5% depletion allowance.[7] The overall effect will likely be a decrease in new mineral ventures in the country. However, it must be noted that Zimbabwe still maintains relatively favorable tax treatment for the mining industry.

The corporate tax rate, effective 1 April 1981, is 45% plus a surcharge of 15%, for a total of 51.75%. In addition, branches of foreign companies also pay a profits tax equivalent to 15% of 56% of taxable income.[4] Although company gains are not taxed, neither are capital losses tax deductible. Depreciation allowance for purchase of buildings and equipment ranges from 2 to 20% per year.[4]

A prospecting and exploration allowance permits deduction of all prospecting and exploration expenses during a year from gross income. Expenses can be carried forward from previous years.[8] Profits from the sale of a claim may be spread equally over four years.[8]

Capital Regulation. New venture capital may be repatriated after two years, minus any amount that may have already been transferred out of country. Capital in excess of the net balance can be transferred through six-year, 4% tax-free bonds. Interest on a special issue of government bonds can be transferred in foreign currency biannually. These bonds can be redeemed in six installments beginning one year from date of purchase. This amount can also be remitted in foreign currency. The sale or liquidation of investments, other than new investments, requires the remittance of proceeds through this six-year bond procedure.[8] These procedures are to prevent the flight of capital from Zimbabwe.

Mining Rights. The Mines and Minerals Act of 1961 remains the basis for governing mining in Zimbabwe. All rights for searching and disposing of both hard minerals and petroleum rest with the president. Administration of mining laws is the province of the Ministry of Mines and Energy Resources.[8]

A Prospecting License entitles a prospector to search for precious metal, precious stone, or base metals on any land open to prospecting for two years. However, claims must be for small sized deposits only.[8]

An Exclusive Prospecting Order is issued on an individual basis. It specifies the rights, terms, and conditions of an exclusive prospecting order. The Exclusive Prospecting Order also specifies the duration and the minerals covered. In addition, six- to 12-month prospecting plans must be submitted from time to time. A special grant can also be issued for working coal, mineral oils, or natural gas within the holder's reservations.[8]

A Mining Lease entitles holders of registered mining locations to mine any ore or deposit, other than coal or petroleum, within the leased area. The discovery of another mineral, not mentioned in the lease, can be mined only if the Mining Commissioner is notified of the discovery. The leaseholder may apply for an order forcing the owner to sell any land on which the lease is located.[8]

The right to mine coal, oil, or gas requires a Special Grant. The company involved must submit company financial information and development plans.

Possession of precious stones or gold in commercial quantities requires a license; special regulations apply to precious stone mining, and to possession and dealing in gold.[8]

Infrastructure

Zimbabwe's infrastructure is second only to South Africa's. It has an excellent transportation system, abundant energy resources for both the present and future, and accessibility to water resources. It also has a sophisticated manufacturing sector to complement its mineral resources.

Transportation. Although Zimbabwe has an excellent transportation system, it is a land-locked nation. Its mineral exports are either through the ports of Beira and Maputo in Mozambique or ports in South Africa. Zimbabwe also has access to ports in Tanzania and Angola, by using its rail line into Zambia. However, at present, this route carries little export or import trade.

Railroads. With the cessation of civil war in Zimbabwe and the emergence of the new government of Prime Minister Robert Mugabe, all borders with Zimbabwe are again open to rail traffic. Before Zimbabwe's border with Mozambique was closed in 1976, nearly two-thirds of its rail traffic went to Beira and Maputo.[8] Since 1976, a growing percentage of traffic has been routed through South Africa.

Attempts to reverse this trend of rising South African routes has met with little success due to intermittent sabotage of Mozambique railroads by guerrillas opposed to FRELIMO, Mozambique's ruling party. Mozambique's ports at Beira and Maputo are inferior to ports in South Africa. Beira and Maputo have inadequate loading and unloading facilities, causing a heavy backlog of agricultural and iron and steel products.[6] A decline in the use of Maputo by all of southern African countries can be seen. In 1969 Maputo received more than 2000 vessels and handled more than 13 million tons of cargo. By 1979, it received fewer than 200 vessels with only 1.5 million tons of cargo.[8] Another drawback to the port at Beira is that it can only handle vessels up to 25,000 tons, and the port must be dredged continuously.[4] The capacity of Maputo is 65,000 tons.[8] In 1981, only about 10% of Zimbabwe's trade passed through Mozambique ports.[7]

In contrast to the ports in Mozambique, South African ports offer excellent rail routes and ports. Zimbabwe has two major rail connections with South Africa. The first route is through southern Zimbabwe to Beitbridge, South Africa. The connection with Beitbridge was not completed until 1978. The second and older route was from the rail center at Bulawayo, southwest through Botswana and into South Africa. The link with South Africa's rail system enables Zimbabwe to use the ports of East London, Port Elizabeth, Durban, and Richards Bay. Approximately 90% of Zimbabwe's export-import traffic makes use of South Africa's rail system and ports.[8]

It is estimated that Zimbabwe was losing $7 million per week because of rail transport inefficiencies[7] at midyear 1981. In response, the new government created the Rail Priorities Committee. This body's main responsibility was to determine both the number and destination of rail wagons used for export. Priority was given to exports earning the highest foreign exchange.[6,7] The government also banned all employees of the National Railway of Zimbabwe (NRZ) from strikes or trade union actions for better working conditions or higher wages. In 1981, to alleviate the shortage of workers, the government invoked the Emergency Powers Act compelling all workers to be at the disposal of the NRZ full time.[6]

In 1982, 60 new GM locomotives arrived from the United States and Canada. Twenty-two locomotives in the existing fleet were also given new engines. This should help improve the rail system.

Zimbabwe's new development plans call for an estimated $350 million for the reequipment and development of the railroads. Of this amount $280 million is designated for the electrification of the rail system from Harare to Maputo, Beitbridge, and later Bulawayo to the southwest.[8] The first stage of this project involves the electrification of rail lines between Harare and Gwelo, and includes approximately 335 km of main line and 130 km of tracks for subsidiary stations. The completion of the first stage was set for the end of 1982.

Roads. Major highways in Zimbabwe and to foreign ports parallel the railroads. The exception is a lack of a direct major highway to the port at Maputo, although a partially improved road does exist. This road is paved most of the way, but portions of the road are still unpaved and susceptible to closure due to heavy rains. The highway system within Zimbabwe is good and accounts for a significant portion of trade traffic within the country.

Water. Zimbabwe's water resources are considered excellent. Part of the northern border is formed by the Zambezi River and Lake Kariba. The Sabi River runs through the southeastern part of the country. Numerous other rivers run throughout Zimbabwe. The hydroelectric potential of these rivers is not mentioned.

There are over 7000 dams in Zimbabwe. Of these, 45 are major dams. The water is used mainly for municipalities and irrigation. It is estimated that only 10 to 15% of water storage capacity has been utilized.[9]

Energy. Zimbabwe's electricity demands are approximately 1000 MW annually,[4] while electricity consumption has increased an average of 10% for the last 10 years.[8] Zimbabwe produces only 65% of its electricity demand. Approximately 30% of Zimbabwe's electricity was imported from the hydroelectric facility on Zambia's side of Lake Kariba, at a cost of $13 million in 1980.[4]

The mining industry uses some 20% of the electricity in Zimbabwe. Ferrochrome production alone accounts for nearly 10% of total electricity consumption within the country.[8]

More than 90% of Zimbabwe's electricity is generated at the two hydroelectric facilities at Lake Kariba on the Zambezi River. One of the facilities is in Zambia and the other is in Zimbabwe. Total capacity of the two facilities is around 1200 MW, and the output is divided equally between the two countries. The rest of the electricity is provided by thermal-power stations in Harare, Umnyati, and Bulawayo.[8]

Future increased electrical output will come from two sources. Investigations are proceeding on the location of new hydroelectric facilities along the Zambezi River and expansion of the facilities at Lake Kariba. The other source is the completion of a two-stage project involving coal-fired generating plants at the Wankie coal mines. Wankie I, started in the early 1970s (but interrupted), was to come on-line sometime in 1983. The plant will have a 480-MW capacity.[8] Wankie II is expected to come on-line in 1986 with an additional 800 MW, for a total capacity of nearly 1300 MW.[8]

Even with this additional capacity, Zimbabwe may not meet its demand for electricity. If demand for electricity continues at a 10% increase every year, by 1990 the demand will be nearly 18 billion kilowatt-hours. This is nearly a 300% increase over the 1979–1980 demand of 6.7 billion kilowatt-hours.[8] Zambian exports of electricity are expected to decrease as demand increases in that country. To counter this, Zambia and Zimbabwe are considering joint hydroelectric projects at Zambia's Mpata and Devil's Gorges.[8]

All of Zimbabwe's petroleum needs are met by imports. A petroleum pipeline, built by Lonrho, was to be completed between Beira and Umtali in eastern Zimbabwe in 1982.[10] There are also plans to produce ethanol from sugarcane. Of course, the electrification of Zimbabwe's rail system will reduce the demand for petroleum.

Labor. Like South Africa, many of the skilled jobs in Zimbabwe were reserved only for white workers. The flight of skilled white workers has left many vacancies in areas requiring skilled workers. As a result, delays in repairs, together with shortages of parts, have contributed to transportation and management problems. The greatest losses due to emigration were among engineers, mechanics, and administrative and clerical staffs.[8]

About 60,000 people were employed in mining in 1980. Of these around 60% were black Zimbabweans, but none were above the level of foreman.[8] Training programs will have to be expanded greatly to meet vacancies left by whites. A number of private companies offer training programs and sponsor university scholarships in technical disciplines. Two-year learnership contracts are also available to train miners and operators.

The government is encouraging skilled and educated Zimbabweans to return to the country. The government is also offering incentives for white skilled workers to remain. Plans are also under way to increase vocational training programs.[8]

Mineral Industry

Unlike its close neighbors, Botswana and Zambia, Zimbabwe's mineral industry is fairly diversified. The minerals of major importance in Zimbabwe are asbestos, gold, nickel, copper, coal, and chromite.

In recent years, asbestos has been the most valuable mineral export, although in 1980, gold's export value rose from $98.6 million for 1979 to $170.5 million. This was a reflection of the high price of gold in 1980 rather than a surge in the amount of gold exported.

However, taken altogether, Zimbabwe's mineral exports represent the single most valuable group of exports. As can be seen from Table II-3-21, mineral exports as a percentage of total exports fell from a high of 53.8% in 1980 to only 40.2% in 1981.[8] This drop can be attributed largely to a decrease in gold prices and the general decrease in demand for exports due to the world recession. It is impossible to assess to what extent the new government's mineral policies have contributed to this drop. In Table II-3-21 the value of chromite is the dominant portion of ferroalloy exports.

Although in 1980, the mineral industry contributed only 7.5% to the GDP, and only employed 6% of the labor force,[4] the contribution of mineral products to foreign exchange will remain significant.

Despite U.N. sanctions of Rhodesian mineral products after UDI, the value of mineral commodities rose steadily from 1965 to 1979. Figure II-3-10 clearly shows the dramatic increases in the value of mineral products during these years. The value of Zimbabwe's mineral production for 1979–1980 is given in Table II-3-22.

Geology. Except for coal, Zimbabwe's most valuable minerals are located within the greenstone belts of Zimbabwe. These metamorphosed igneous and sedimentary rocks are located in central and eastern Zimbabwe. The western part of Zimbabwe belongs to tribal groups, and very little exploration has been done. Within the greenstone belt are located all the chromite, and most of the asbestos, nickel, copper, iron ore, and gold deposits.[8]

Great Dyke. The most prominent geological feature of Zimbabwe is the Great Dyke. The Great Dyke begins about 110 km north-northwest of Harare and continues south-southwest for approximately 530 km. It passes some 55 km west of Harare, 30 km east of Gwelo, and ends approximately 65 km east-northeast of Gwanda. The width of the Great Dyke varies from 5 to 11 km, it covers nearly 7500 km^2, and extends to a depth of

TABLE II-3-21   Value of Exports of Principal Mineral Commodities from Zimbabwe and Their Share of Total Exports, 1978-1981 (Million Current U.S. Dollars)

Commodity	1978	1979	1980	1981
Asbestos	82.6	104.4	118.5	105.6
Cement	1.3	1.6	2.8	2.1
Coal	4.5	4.6	5.8	3.5
Cobalt oxide	—	3.4	4.2	1.1
Coke	6.8	9.5	9.5	10.8
Copper, metal	38.1	46.3	36.4	25.5
Gemstones	1.1	1.9	2.2	3.3
Gold	66.4	98.6	170.5	88.8
Iron and Steel				
Ferroalloys[a]	45.5	68.6	130.3	111.7
Ingots and billets	32.8	38.4	50.7	33.6
Bars, rods, sections	24.5	47.3	48.9	24.5
Wire	4.9	5.7	8.1	8.2
Other	15.3	19.1	30.8	26.4
Lithium ores	1.4	1.5	2.6	3.0
Nickel, metal	52.6	56.0	78.1	65.0
Precious metal waste	1.2	2.7	5.2	7.5
Tantalum ores	0.5	2.8	4.2	2.8
Tin, metal	9.0	12.6	12.7	11.5
Tungsten ore and concentrate	2.3	1.9	1.9	1.1
Principal mineral export earnings	390.8	526.9	723.4	536.0
All export earnings	877.4	1059.2	1345.6	1333.4
Principal mineral share of all export earnings (%)	44.5	49.7	53.8	40.2

[a]Chrome exports are included in the ferroalloys.

Source: Ref. 6.

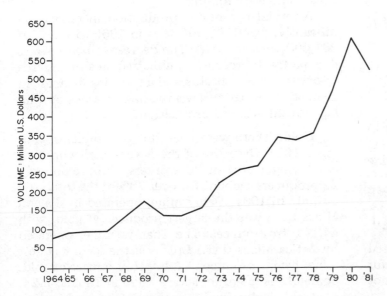

FIGURE II-3-10   Value of mineral production in Zimbabwe, 1964-1981. (From U.S. Department of the Interior, Bureau of Mines. *Minerals Yearbook*, 1964-1981.)

over 3000 m.[11] The age of the dyke is estimated to be late in the early Precambrian time.[11]

The Great Dyke is known for its unique stratiform type chromite seams of high-grade metallurgical chromite. The dyke is thought to have formed from four separate feeders, creating a nearly continuous body of four segments. The segments are known as the Musengezi Complex, Hartley Com-

TABLE II-3-22 Value of Zimbabwe Mineral Production, 1979–1980
(Thousand Metric Tons and Million U.S. Dollars)

	Volume		Value		Percent of total production (by value)	
Mineral	1979	1980	1979	1980	1979	1980
Gold (thousand troy ounces)	386	367	120.1	229	25.7	25.0
Asbestos	259	250.9	97.8	111	20.9	16.9
Nickel	14.6	15.1	66.9	88	14.3	13.3
Copper	29.6	27.0	52.2	56	11.2	8.5
Coal	3188	3134	38.4	44	8.2	6.7
Chromite	541.8	553.5	24.0	29	5.1	4.4

Sources: Refs. 5 and 8.

plex, Selukwe Complex, and the Wedza Complex. Starting from north to south the complexes are described as follows.

1. The Musengezi Complex is 44 km long, at least 3660 m thick, and contains at least seven chromite seams.[8] It is the northernmost segment of the dyke. None of the chromite seams are of minable width or grade.[11]
2. The Hartley Complex is 314 km long and greater than 3100 m thick, and contains 11 chromite seams.[8] Seams 5 to 11 are all of metallurgical-grade chromite, are generally thicker than seams in the other complexes, and occur over nearly 200 km of the complex.[11]
3. The Selukwe Complex is 97 km long and has been explored only to a depth of 1800 m.[8] It contains seven chromite seams, of which 1 and 2 are the best from a mining standpoint.[11]
4. The Wedza Complex is 80 km long, 1800 m thick, and contains six chromite seams.[8] The number 2 seam varies from 45 to 60 cm in thickness.[11]

The various rock types grade into each other in the general sequence (from the bottom up) (1) serpentine and dunite; (2) Harzburgite, olivine pyroxenite, and picrite; and (3) gabbro, gabbroic anorthosite, norite, and quartz gabbro. In general, the gabbroic rock types are lenses which are centrally located in the serpentinites and pyroxenites.[11]

Asbestos. Zimbabwe's asbestos is considered excellent by world standards,[4] and prospects for growth are good. Reserves and resources are not measured but are considered large.

Most of the chrysotile is of low-iron, spinning-grade quality, with low conductivity, and can be used for some defense purposes. Because of this, Zimbabwean asbestos is considered a "strategic" commodity.[8]

Almost all of Zimbabwe's asbestos is mined underground. The Shabani mine is the largest asbestos mining operation in Africa, where some 30 irregular pod-shaped ore bodies are in various stages of development.

The asbestos mines operated by Shabani and Mashaba Mines (Pvt.) Ltd., and owned by Turner and Newall, account for nearly 90% of Zimbabwe's asbestos output.[8] (See Table II-3-23 for asbestos mines in the country.)

Of the mines operated by Pangani Asbestos Mines (Pvt.) Ltd., the Pangani, Vanguard, and Boss mines were closed in May 1981. Pangani depleted its ores, and production costs at the Vanguard and Boss mines were too high.[7]

After UDI, asbestos production increased significantly, from 163,000 tons in 1965 to a high of 281,000 tons in 1976. The decrease since 1976 is due partly to recently publicized health hazards associated with asbestos and a resulting decrease in demand. New safety laws relating to asbestos mining have also slowed production.

Gold. There were over 280 gold mines operating in 1981. Over 250 of these were small workings, employing less than 20 workers.[7] Approximately 22 producers account for over 78% of the total gold output. In 1981, Dalny mine, operated by Falcon Mines Ltd., was the largest producer of gold, with 44,111 troy ounces. The Shamva mine, operated by Attica Mines (Pvt.) Ltd., was the second largest mine, having produced 17,400 troy ounces of gold. Some of the major mines are listed in Table II-3-24.

Gold production has declined considerably since UDI, due to security at the mines. During World Wars I and II, gold production reached a high of approximately 900,000 troy ounces.[8]

Gold mineralization occurs over a wide region in Zimbabwe's greenstone belts. There is some allu-

TABLE II-3-23   Asbestos Production in Zimbabwe, 1975–1981[a]

Annual output (tons)	
1965	160,000
1970	188,000
1975	262,000
1976	281,000
1977	273,000
1978	244,000
1979	260,000
1980p	251,000
1981e	248,000

Asbestos mines	Owner	Operator
Gaths, King, Shabani, Temeraire	Turner and Newall	Shabani and Mashaba Mines (Pvt.) Ltd.
Boss, Rex, Pangani, Vanguard	Asbestos Investments Pty. Ltd.	Pangani Asbestos Mines (Pvt.) Ltd.
Kudu	Essexvale Asbestos (Pvt.) Ltd.	Kudu Mines (Pvt.) Ltd.
Thornwood	Thornwood Asbestos Mines (Pvt.) Ltd.; A.D. Theron & Sons (Pvt.) Ltd.	A.D. Theron & Sons (Pvt.) Ltd.
D.S.O.	D.S.O. Asbestos (Pvt.) Ltd.	D.S.O. Asbestos (Pvt.)
Ethel	NA	NA

Asbestos mills	Throughput (tons/year)	Source of materials processed
Kudo	360,000	Open-pit mine
Gaths, King	2,160,000	Underground mines
Shabani		
New ore	2,400,000	Open-pit and underground mines
Reclaimed ore	720,000	Waste dumps
Temeraire	480,000	Waste dumps

[a]p, preliminary; e, estimated; NA, not available.
Source: Ref. 8.

vial gold in the north and northeast, but production from this source is insignificant. By-products of gold production in Zimbabwe include tungsten, copper, antimony, arsenic, and silver. Almost 30% of Zimbabwe's silver is a by-product of gold conduction.[8]

In recent years, the value of gold exports has increased significantly with the rise of gold prices, surpassing the value of asbestos exports in 1980, when the price of gold was at an all-time high.

Gold production is likely to increase if proposed base-metal projects of copper and nickel come into production. Gold is a by-product of copper and nickel in Zimbabwe.

Although no large high-grade gold deposits have been discovered recently, a rise in gold prices will increase the number of mines as lower-grade de-

posits become economical to mine. Gold production in Zimbabwe is given in Table II-3-24.

Nickel. Most nickel deposits in Zimbabwe are parallel to the Great Dyke. Nickel deposits are found in five different geological environments: hydrothermal shearzones, lateritic ore deposits, ultrabasic lava deposits, unlayered mafic intrusions, and layered mafic intrusions.[12]

The Noel Mine southeast of Bulawayo was the only producer of nickel from hydrothermal shear zones until its closure in 1957. As of 1980, lateritic deposits (Musengezi Complex) and layered mafic intrusions all along the dyke were under investigation and had not been mined.[12] Nickel in ultrabasic lava flows is mined at the Trojan Mine from talc and serpentinite host rocks. The rocks are easy to

TABLE II-3-24   Gold Production in Zimbabwe, 1975-1981[a]

Annual output (troy ounces)	
1965	550,000
1970	438,000
1975	354,000
1976	387,000
1977	402,000
1978	399,000
1979	386,000
1980p[b]	368,000
1981e[b]	371,000

Gold mines	Owner	Operator
Arcturus, Mazoe, Muriel, Inyati	Lonrho-Investment Group	Corsyn Consolidated Mines
Shamva, How	Lonrho-Investment Group	Attica Mines Pvt.
Old West, Redwing	Lonrho-Investment Group	Independence Mining Co.
Commoner, Old Nic, Dawn	Falcon Mines Ltd.	Olympus Consolidated Mines Ltd.
Golden Oriole, Arlandzer, Turkois, Venice, Mascot, Dalny	Falcon Mines Ltd.	Falcon Mines Ltd.
Blanket	Falconbridge Nickel Corp.	Blanket Mines (Pvt.) Ltd.
Athens, Falcon	Homestake Mines (Pvt.) Ltd.	Homestake Mines (Pvt.) Ltd.
Brompton, Patchway, Cam & Motor, Renco	Rio Tinto Zimbabwe Ltd.	Rio Tinto Zimbabwe Ltd.

Mill	1979 throughput (tons/year)	1979 production (troy ounces)	Mine reserves gross weight (tons)	Grade (troy ounces/ton)
Arcturus	103,000	24,274	446,000	0.30
Blanket	150,000	15,000	NA	0.13
Dalny	244,000	52,888	1,600,000	0.33
Mazoe	105,000	18,037	245,000	0.24
Muriel	66,000	22,505	235,000	0.52

[a]p, preliminary; e, estimated; NA, not available.
[b]1980-1981 figures from Ref. 7.
Source: Ref. 8.

mine, but beneficiation of the ores is expensive.[12] Nickel in unlayered mafic intrusions is produced at the Madziwa Mine from gabbroic or dioritic rock. However, the rock was difficult to drill, and mineralization was unevenly distributed.[12]

Ore reserves for producing mines are estimated at 60 million tons, ranging in grade from 0.6 to 1.2% nickel.[8] Additional reserves, in the Hartley and Wedza Complexes, are 2000 million tons at 0.25% nickel, 0.25% copper, and 3 to 5 g of combined platinum and palladium.[8] Another possible source of nickel is at Hunters Road near Que Que. These reserves are estimated at 15 to 20 million tons of ore at 0.75% nickel.

Although nickel mining began in 1928, it was not until the late 1960s that nickel production began to increase steadily. Generally, current nickel mines do not have large reserves. As a result, new mines will have to be developed to keep production at current levels. However, lower nickel prices could impede such growth. Nickel production, mines, and mills are given in Table II-3-25.

There are two nickel smelters in Zimbabwe. The two smelters are at Rio Tinto Zimbabwe Ltd.'s Eiffel Flats and the Bindura Nickel Corp.'s (BNC) at Bindura. The smelters produce nickel cathode. Rio Tinto's capacity is 9000 tons/year and Bindura's capacity is 16,000 tons per year, for a total capacity of 25,000 tons per year. Nickel production in 1981 was estimated at 13,018 tons. Thus only about 52% of smelter capacity was used in 1981.

The value of nickel production in 1979 was $67 million[8] and increased to $87 million in 1980.[4] The value of nickel exports were $56.0 million in 1979, $78.1 million in 1980, and only $65.0 million in 1981. The drop in 1981 reflects a decrease both in production and exports, due to the world recession.

By-products of nickel include cobalt, silver, gold, and platinum-group metals. Cobalt brought

TABLE II-3-25 Nickel Production in Zimbabwe, 1975–1981[a]

Annual output (tons)	
1965e	470
1970	8,585
1975	9,121
1976	14,604
1977	16,671
1978	15,701
1979	14,591
1980p	15,075
1981e	13,018

Primary nickel mines	Startup date	Owner	Operator
Trojan	1964	Bindura Nickel Corp.	Trojan Nickel Ltd.
Madziwa	1969	Bindura Nickel Corp.	Madziwa Mines Ltd.
Empress	1971	Rio Tinto Zimbabwe Ltd.	Empress Nickel Mining Co.
Perserverance	1972	Rio Tinto Zimbabwe Ltd.	Empress Nickel Mining Co.
Shangani	1975	Johannesburg Consolidated Investment Co., Ltd., 49%; Bindura Nickel Corp., 37%	Shangani Mining Corp. Ltd.
Epoch	1976	Bindura Nickel Corp.	Trojan Nickel Ltd.

Mill	Throughput (tons/year)	Grade (% Ni)
Epoch	420,000e	0.75
Madziwa	500,000	0.60
Trojan	850,000	0.68
Empress	880,000	1.3[c]
Perserverance	33,000	
Shangani	900,000	0.45–0.78

[a] e, estimated; p, preliminary.

[b] Although most nickel mines were opened only within the last 10 years, several are not expected to last more than 10 years at current rates of production. The Perserverance mine was to have exhausted its original reserves of 2 million tons of ore by October 1980. In addition, because of problems stemming from rock characteristics and insufficient skilled mining personnel, production losses have occurred, as at the Empress, Perserverance, and Shangani mines. Mining operations at Shangani have been highly sensitive to metal prices over the period of operation (since 1976), and financial losses have been significant. The fiscal year ending June 1980 was Shangani's first profitable year. Underground mining development at Shabani has been under way for three years, and underground ore is now about one-sixth of mill throughput.

[c] Also copper.

Source: Ref. 8.

in nearly $8 million in 1979 and 1980 when world cobalt prices were high.

Copper. The majority of copper mines and reserves are located in northern Zimbabwe, near Sinoia. Copper is a by-product of a number of gold mines, and nearly 3000 tons are produced as a by-product of nickel mining.[8] Copper production, mines, and mills are given in Table II-3-26.

Copper mining was first recorded in 1904, but it was not until the mid-1950s that copper mining increased. After 1955, copper production increased each year until its peak of 51,900 tons in 1973.[8]

Copper production has decreased steadily since then, reaching an estimated output of only 24,583 tons in 1981. The value of copper exports has also decreased in recent years—from $46.3 million in 1979, $36.4 million in 1980, to only $25.5 million in 1981.[6] Currently, copper is approximately 9% of the total value of mineral production. It seems that low copper prices in 1973, coupled with rising energy cost, was a major factor in the demise of the copper industry. Before the decline, copper was the most valuable mineral commodity in Zimbabwe from 1969 to 1974 (except for 1971).[8]

Copper resources are not available. However,

TABLE II-3-26 Copper Production in Zimbabwe, 1975–1981[a]

			Annual output (tons)		
		1965		18,000	
		1970		30,000	
		1975		47,600	
		1976		41,300	
		1977		34,800	
		1978		33,800	
		1979		29,600	
		1980p		26,921	
		1981e		24,583	

Mine	Status	Reserves (1000 tons)	Grade (% Cu)	Owner	Operator
Miriam	Active	14,000	1.27	Messina (Transvaal Development Company	M.T.D. (Mangula) Ltd.
Norah	Active	3,000	1.35	Messina (Transvaal Development Company	M.T.D. (Mangula) Ltd.
Copper Queen	Care and maintenance	NA	NA	Messina (Transvaal Development Company	M.T.D. (Sanyati) Ltd.
Shackleton	Active	837	1.26	Messina (Transvaal Development Company	Lomagundi Smelting & Mining (Pvt.) Ltd.
Avondale	Under development	320	1.14	Messina (Transvaal Development Company	M.T.D. (Mangula) Ltd.
Alaska	Care and maintenance	579	1.11	Messina (Transvaal Development Company	M.T.D. (Mangula) Ltd.
Inyati	Care and maintenance	870	1.2e	Lonrho Investment Group	Corsyn Consolidated Mines

Mill	Throughput (1000 tons)	Grade (% Cu)	Cu in concentrates (tons)
Inyati	193	NA	2,964
Miriam	1,203	0.86	9,641
Muriel	63	NA	348
Norah	534	1.07	5,332
Shackleton	466	1.05	4,800e

[a]e, estimated; p, preliminary; NA, not available.
Source: Ref. 8.

deposits are generally small and exhibit complex geologic structures. Ore grade is also declining. Production from current mines is estimated to be no more than 10 years. New exploration for deposits where copper is the main product is currently under way. Several mines are currently under care and maintenance because of high production cost and depressed world prices. Copper production in Zimbabwe is presented in Table II-3-26.

Western Europe has been the traditional market for Zimbabwe's copper exports. The trend has been toward producing a more upgraded product to maximize revenues. As a result, a new $7 million refinery has been constructed next to the copper smelter of the Alaska mine.[8]

Chrome. Zimbabwe's Great Dyke is unique for its high-grade metallurgical chromite seams. How-

ever, the majority of chromite in Zimbabwe is produced from podiform deposits. Currently, 75% of Zimbabwe's annual chromite production is converted into ferrochrome.[8]

Figures for Zimbabwe's chromite reserves vary considerably. Estimates range from 500 to 564 million metric tons.[8] Resources are estimated to be from 3.2 billion tons to more than 10 billion tons.[8] Resources of 10 billion tons would equal one-third of the world's resources of chromite at 33 billion tons. South Africa has the remaining two-thirds of the world's resources for chromite. This overwhelming concentration of chromite in such a concentrated area makes chromite one of the most strategically placed of all mineral commodities in the world. Table II-3-27 indicates the reserves and resources for Zimbabwe's chromite.

It can be seen from the map in Figure II-3-11

that chromite production in Zimbabwe parallels the Great Dyke. Chromite production peaked in 1975 and 1976, at 876,000 and 864,000 tons per

TABLE II-3-27 Estimates of Zimbabwe Chromite Reserves and Resources (Million Metric Tons)

	Reserves[a]	Resources
Podiform[b]		
Selukwe	1	76
Mashaba	1	8
Belingwe	1	13
Total podiform	3	97
Stratiform[b]	448	3,137
Eluvial	54	—
Total Zimbabwe	500[b]	3,234[b]
	564[c]	10,000[c]
Total world	NA	32,700[c]
Total South Africa	NA	22,700[c]

[a]NA, not available.
[b]Von Gruenewaldt, as quoted in Ref. 8.
[c]U.S. Bureau of Mines.
Source: Ref. 8.

year.[8] Chrome ore production for 1964–1982 is given in Table II-3-28. The boost in production between 1970 and 1971 of nearly 83,000 tons may have been due to the passage of the Byrd Amendment (Public Law 92-156) in 1971. The Amendment allowed for the lifting of U.S. sanctions for certain minerals considered "strategic" to the United States. However, in 1977, U.S. legislation once again embargoed Southern Rhodesian products, contributing to a decline in production of some 187,000 tons from 1976 to 1977.[8] U.N. sanctions against Zimbabwe were lifted 16 December 1979. In September 1981, the government of Zimbabwe requested duty-free status for shipments of low-carbon ferrochrome and ferrosilicon chrome to the United States.[7]

Two Selukwe mines, Selukwe Peak and Railway Block (podiform deposits), account for nearly 70% of Zimbabwe's chrome. The mines are owned by Union Carbide. The mines are now operated under the name Zimbabwe Mining and Smelting Co., and previously went under the name of Rhodesia Chrome Mines. Together the mines are reported to have mined 10 million tons of chromite, but re-

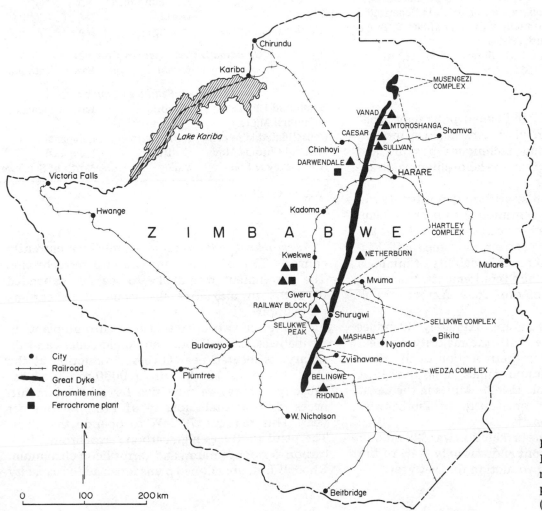

FIGURE II-3-11 Great Dyke, major chromite mines, and ferrochrome plants in Zimbabwe. (From Ref. 8.)

TABLE II-3-28   Chrome Ore Production in Zimbabwe, 1964–1980

Year[a]	Zimbabwe[b]	South Africa[c]	USSR[a]
1964	448	850	1300
1965	586	940	1420
1966	586	1060	1500
1967	558	1150	1570
1968	433	1150	1650
1969	371	1200	1700
1970	504	1430	1750
1971	672	1650	1800
1972	755	1480	1850
1973	790	1650	1900
1974	703	1880	1950
1975	876	2080	2080
1976	864	2410	2120
1977	677	3060	2180
1978	478	3150	2300
1979	542	3300	2400
1980e	554	3416	2450
1981[d]	526	2866	
1982e[d]	499	2630	

[a]e, estimate by U.S. Bureau of Mines. All USSR figures are also estimated by the U.S. Bureau of Mines.
[b]Zimbabwe *Monthly Digest of Statistics*, September 1980.
[c]Rounded figures based on production totals in Department of Mines Quarterly Information Circular *Minerals*, Bramfontein, Republic of South Africa.
[d]1981–1982 figures from U.S. Bureau of Mines, *Mineral Commodity Summaries 1982*.
Source: Ref. 8.

sources are still large. Individual pods may range from 15 to 400 m in diameter,[8] making modern large-scale underground techniques possible and less expensive to mine than the labor-intensive stratiform deposits.

The chrome ore in stratiform deposits varies in thickness from a few centimeters to nearly 40 cm. However, the distance between seams precludes the mining of multiple seams from the same shaft. Factors determining the economic viability of mining a particular seam are labor requirements, thickness, friability, and grade of ore. These factors vary all along the Great Dyke.[8]

In 1979, there were approximately 30 mines along the Great Dyke, with the majority of them being located in the northern section of the Hartley Complex.[8] Chromium mines along the Great Dyke are given in Table II-3-29. Mines in the Great Dyke produce about one-fourth of Zimbabwe's chrome ore production.[12]

The value of chrome production was $29 million in 1980 for 553,500 tons and was only 4.4% of the total value of mineral production for that year.[4]

TABLE II-3-29   Chromium Mines on the Great Dyke[a]

Owner	Name of Mine	Location
Salisbury District, Hartley Complex		
Rhodesian Alloys (Anglo-American)	Noro	East of Sinoia
	Vanad A	East of Sinoia
	Vanad B	East of Sinoia
	Feoch	East of Sinoia
	Poort	East of Sinoia
	Caesar A	East of Sinoia
	Caesar B	East of Sinoia
	Sutton A	East of Sinoia
	Sutton B	East of Sinoia
	Darwendale	Darwendale
	McGowan	Darwendale
	Divide	Darwendale
	Greenvale	Darwendale
	Hay	Darwendale
	Noro	Darwendale
Zimbabwe Mining (Union Carbide)	Umvukwes	Umvukwes
F.A. Gerber	Arthur's Luck	Matoroshonga
	Mema	Matoroshonga
	Royal	Matoroshonga
	Shiel's Luck	Matoroshonga
Ore Recovery Co.	Evol	NA
	Humming Bird	NA
	Oofbird	NA
Rhoshwa Mining Co.	Impinge	NA
Gatooma District, Hartley Complex		
NA	Negesi	East of Gatooma
Gwelo District, Selukwe Complex		
Hammond-Rhodesian Cambrai Mines	Cambrai	East of Lalapanzi
Consolidated Minerals	Netherburn	Lalapanzi
Midland Chrome Mines	Central	Lalapanzi
H.H. Davey & Son	Valley	North of Selukwe

[a]NA, not available.
Source: Ref. 8.

Ferrochrome Production. Zimbabwe currently converts 75% of its chrome ore into ferrochrome. The government would like to see this expanded to 100% to maximize the value of its chrome production.

There are two operating ferrochrome plants in Zimbabwe. Rhodall Ltd., an Anglo-American subsidiary, is located near Gwelo. Production at the plant began in 1953, with a 9000-ton per year capacity. There are now five ferrochromium furnaces, with a total capacity of 113,000 tons per year. This requires 70 MW to operate the plant. The plant produces high-carbon ferrochrome, low-carbon ferrochromium, and ferrosilicon chromium. Rhodall Ltd. abandoned plans for an added capacity of 30 MW, which was to produce 50,000 tons per year of high-carbon ferrochromium.[7,12] Reasons cited for abandoning the expanded capacity were high production costs, mainly from high wages.[12]

The other ferrochrome plant in Zimbabwe is the Zimbabwe Mining and Smelting Co., a Union Carbide subsidiary, located at Que Que. Capacity at the plant was expanded by 90,000 tons per year with the recent addition of two new 24-MW furnaces which produce 45,000 tons per year each.[7] The plant already had a capacity of near 50 MW with four furnaces ranging from 7.5 to 15 MW, with a total capacity of 90,000 tons. With an annual capacity of 180,000 tons per year, the Zimbabwe Mining and Smelting Co. is now Zimbabwe's largest ferrochromium facility.

Together both plants have a capacity of 293,000 tons per year. Presently, both plants use 10% of Zimbabwe's electricity consumption.[8] Significant expansion of ferrochromium facilities will require new electric generating capacity and higher ferrochromium prices.

Coal. The two main coal deposits in Zimbabwe are located in the Zambezi Valley in western Zimbabwe near Wankie and in the Sabi Limpopo Valley in the southeast. Estimates of resources and reserves for coal vary. Recent announcements posit a possible resource base of 30 billion tons and reserves of 2.5 billion tons, of which 750 million tons are extractable.[12] Of this extractable amount, 66% is thought to be coking coal of low ash content, and the other 34% is thought to contain a high ash content, suitable for steam generation.[8] The sulfur content is high—from 1.5 to 3%—a factor limiting possible coal exports of its coking coal.

The Wankie colliery in western Zimbabwe has the only producing coal mine in Zimbabwe. The mines have been producing for the last 58 years and were acquired by Anglo-American in 1953. The government is a minor shareholder in Wankie colliery but owns all other coal deposits in Zimbabwe.[7] Coal production for Zimbabwe is given in Table II-3-30.

As mentioned previously, a new coal-fired power plant is being built at Wankie. Wankie I, which was to have been completed in late 1982, is to have a capacity of 480 MW. Wankie II, to be completed in 1986, will add another 800-MW power plant. Wankie coal will be used for the new power plants. Consequently, coal production will have to increase to an estimated 4.4 million tons per year to provide fuel for the power plant and some exports.[8]

Although the labor force at Wankie is sufficient, it is believed that there is a shortage of trained personnel to open a comparable mining operation in other parts of the country. Hence, other coal fields, such as Bubi and Sengwa, are expected to be open pit whenever possible.[8]

Coking coal is exported to several surrounding countries, but its use is somewhat limited because of its high sulfur content. Primary destinations are copper-smelting plants in Zambia and Zaire.[8] Otherwise, most of Zimbabwe's coal is used in its own thermal power plants.

By-products recovered from the coke plant in 1981 included 8600 tons of crude tar, 1107 tons of ammonia, and 7914 gallons of crude benzol.[7]

Constraints to Mineral Development

The new government recognizes the importance of mineral exports, as indicated by plans for increased electrical generating capacity (Wankie I and II and increased hydroelectric capacity on the Zambezi River) and the eventual electrification of the railroad system.

Zimbabwe's new three-year development plan, announced in July 1981, calls for expenditures of $6 billion and an annual growth rate of 8%. The plan divides the money roughly in half between the private and public sectors.

Public sector spending includes $121 million for a refugee program, $250 million for reconstruction, $363 million for training programs, and $1.2 billion for agricultural development (including irrigation and credit to small farmers).[8]

Private-sector spending would include power development of $1.7 billion, $676 million for transport and communications, and $930 million for urban housing.[8] Investment from foreign sources for the private sector is expected to be $689.7 million during this period. Table II-3-31 gives a breakdown of the proposed development plan for 1981–1984.[4] It is not clear whether the new government

TABLE II-3-30 Coal Production in Zimbabwe, 1975–1981[a]

	Annual output	Bituminous	Coke metallurgical
1965	3,609,000	3,509,000	100,000e
1970	3,765,000	3,520,000	245,000e
1975	3,570,000	3,300,000	270,000e
1976	3,820,000	3,593,000	227,000
1977	3,223,000	3,029,000	194,000
1978	3,244,000	3,065,000	179,000
1979	3,389,000	3,188,000	201,000
1980p	3,369,000	3,134,000	235,000
1981e	3,067,000	2,867,000	200,000

[a]e, estimated; p, preliminary.
Sources: Refs. 7 and 12.

TABLE II-3-31 Cost Breakdown of Proposed Development Plan, 1981-1984
(Value in Thousand U.S. Dollars)

	1981-1982	1982-1983	1983-1984	Total
Commercial, agriculture	54,685	60,257	81,750	196,693
Water development	30,269	38,822	16,059	54,937
Tourism	607	607	579	1,794
Energy	465,124	620,000	620,000	1,705,124
Transport and communications	214,416	231,535	230,615	676,567
Health	16,667	9,357	18,750	44,774
Information	12,403	10,707	5,659	28,769
Urban development	248,373	326,540	357,395	932,309
Government services	56,692	56,430	56,080	169,204
	1,099,239	1,345,259	1,386,890	3,840,390

Source: Ref. 4.

understands the extent of alarm with which its new policies toward foreign-owned mining companies is being viewed.

A report prepared by the American Embassy in Harare listed as one of the constraints to mineral development the lack of foreign investment in Zimbabwe since independence.[5] Chester Crocker, the assistant secretary of state for African affairs, reflected on this problem when he stated:[13]

Americans lack the climate [in Zimbabwe] that makes them feel secure. They need predictability. [It is] disturbing to businessmen that Zimbabwe refuses to sign an accord with the Overseas Private Investment Corp., a U.S. agency that grants political-risk insurance to U.S. firms.

Another constraint on foreign investment in Zimbabwe, besides uncertainty over security of investment, is Zimbabwe's future relationship with South Africa.[5] A significant portion of Zimbabwe's trade (both imports and exports) utilizes South Africa's rail system and ports. Although Zimbabwe plans to use Mozambique's transportation network and ports more extensively in the future, these facilities are inferior to South Africa's. Also, the infrastructure of Mozambique has been the target of sabotage by guerrilla forces operating within Mozambique that are opposed to FRELIMO, the ruling party in Mozambique.

Adding further to an uncertain investment climate is the growing political friction between Robert Muzabe's ZANU-PF party and Joshua Nkomo's ZAPU party. The split in parties began with the arrest of some prominent members of ZAPU after arms were discovered hidden on land belonging to ZAPU members. Mugabe dismissed Nkomo from the Ministry of Home Affairs (which included control of the police force in Zimbabwe). The situation has deteriorated further and Nkomo

was placed under house arrest, and Nkomo finally fled the country in 1983. It has been reported that elements of the Fifth Brigade of the army sent to Matabeleland to subdue Ndebele guerrillas have killed between 1000 and 3000 villagers in efforts to destroy rebel sanctuaries.[14] The importance of this growing violence cannot be overestimated. If the conflict widens, it will turn into a war between Shona and Ndebele tribes and could lead to yet another civil war in Zimbabwe.

Another period of armed struggle in Zimbabwe could seriously harm the mineral industry. It could lead to the flight of more skilled whites, thereby increasing the shortage of skilled workers in the mining industry. Another armed conflict would undoubtedly decrease foreign investment or, more likely, halt it altogether. The country's infrastructure could become a major target of guerrilla action.

Perhaps the greatest danger resulting from conflict in Zimbabwe is the possible involvement of other foreign governments in the struggle, most notably South Africa and/or the USSR. Nkomo and his ZIPRA guerrillas were supported by the USSR during the struggle for majority rule. It is unlikely that South Africa would not become involved to some degree to protect its infrastructure along the borders and to prevent an openly hostile government from assuming power. Depending on the degree of involvement of other governments (if any), there is the danger of a conflict spilling over Zimbabwe's borders and turning into a regional conflict.

Just as the political climate for Zimbabwe has been transformed, so too have the names of former Rhodesian cities. The new official names for major principalities within Zimbabwe reflect the older tribal linguistic heritage of black Africans. A list of name changes is provided in Table II-3-32.

TABLE II-3-32   Name Changes in Zimbabwe

Present name	New name
Salisbury	Harare
Fort Victoria	Nyanda
Gwelo	Gweru
Que Que	Kwekwe
Gatooma	Kadoma
Marandellas	Marondera
Hartley	Chegutu
Sinoia	Chinhoyi
Umtali	Mutare
Shabani	Zvishvane
Selukwe	Shurugwi
Umbuma	Mvuma
Enkeldoorn	Chivhu
Essexvale	Esigodini
Melsetter	Mandidzudzure
Vila Salazar	Sango
Belingwe	Mberengwa
Wankie	Hwange
Somabula	Somabhula
Mashaba	Mashava
Mangula	Mhangura
Sipolilo	Chipuriro
Mtorashanga	Mutorashanga
Inyazura	Nyazura
Dett	Dete
Balla Balla	Mbalabala
Nuanetsi	Mwenezi
Chipinga	Chipinge
Mtoko	Mutoko
Mrewa	Murewa
Tjolotjo	Tsholotsho
Nkai	Nkayi

REFERENCES

1. U.S. Central Intelligence Agency, National Foreign Assessment Center. *The World Factbook—1981.* Washington, D.C.: U.S. Government Printing Office, 1982.

2. The World Bank, International Bank for Reconstruction and Development. *World Development Report 1982.* New York: Oxford University Press, Inc., 1982.

3. O'Meara, Patrick. "Zimbabwe: The Politics of Independence." In *Southern Africa: The Continuing Crisis,* 2nd ed. Edited by Gwendolyn M. Carter and Patrick O'Meara. Bloomington, Ind.: Indiana University Press, 1982.

4. U.S. Department of Commerce, International Trade Administration. "Marketing in Zimbabwe." *Overseas Business Reports.* OBR 81-21. August 1981.

5. U.S. Department of Commerce, International Trade Administration. "Zimbabwe." *Foreign Economic Trends and Their Implications for the United States.* FET 82-041. June 1982.

6. U.S. Department of the Interior, Bureau of Mines. "Recent Government Actions Affecting the Minerals Industry of Zimbabwe." *Minerals and Materials*, August–September 1982, pp. 42-53.

7. U.S. Department of the Interior, Bureau of Mines. "The Mineral Industry of Zimbabwe." *Minerals Yearbook 1981.* Washington, D.C.: U.S. Government Printing Office, 1983.

8. U.S. Department of the Interior, Bureau of Mines. "Zimbabwe." *Mineral Perspectives*, August 1981.

9. Stoneman, Colin, ed. *Zimbabwe's Inheritance.* New York: St. Martin's Press, Inc., 1981.

10. "Zimbabwe: 18 Months After." *Mining Journal*, 30 October 1981, pp. 329-330.

11. Ridge, John Drew. "Zimbabwe." In *Annotated Bibliographies of Mineral Deposits in Africa, Asia (exclusive of the USSR) and Australia.* Elmsford, N.Y.: Pergamon Press, Inc., 1976, pp. 21-46.

12. U.S. Department of the Interior, Bureau of Mines. "The Mineral Industry of Zimbabwe." *Minerals Yearbook 1980.* Washington, D.C.: U.S. Government Printing Office, 1982.

13. Mufson, Steve. "Investors Unconvinced on Zimbabwe." *The Wall Street Journal*, 5 March 1982, p. 25.

14. "Zimbabwe Leader Tries to Ignore Them, but Massacre Charges Keep Multiplying." *The Wall Street Journal*, 28 February 1982, p. 25.

Chapter II-4

Ocean-Bordering Countries

Of the four ocean-bordering countries in this section—Mozambique, Angola, Namibia/South-West Africa, and Zaire—both Mozambique and Angola now have Marxist-type regimes, after liberation from Portuguese rule. Namibia/South-West Africa's future government is uncertain between Marxist-backed SWAPO guerillas operating out of Angola and the local administration controlled by South Africa. In Zaire there have been recent attempts from secessionists to form a separate country for the mineral-rich Shaba (Katanga) Province—Zaire's principal source of income.

The turmoil in these individual countries has led to disruptions and general disrepair of the critical link between land-locked countries and ocean-bordering countries. In effect, localized problems have become regional problems.

MOZAMBIQUE

The People's Republic of Mozambique (GPRM) is located in southeastern Africa, adjacent to the Indian Ocean, across from the island of Madagascar. Mozambique shares common borders with Tanzania to the north, Malawi, Zambia, Zimbabwe, South Africa, and Swaziland. Mozambique has an area of approximately 786,762 km^2, possessing a 2470-km coastline.

Its population of 10,500,000 is 99% native African and less than 1% European and Asian. The official language remains Portuguese, and there are many tribal dialects of native African languages.[1]

Mozambique became independent from Portugal on 25 June 1975, after 11 years of guerrilla warfare organized by the Frente de Libertação de Mozambique (FRELIMO). Samora Moises Machel has remained President since independence, and his FRELIMO party is now the only legal party in Mozambique.[1] The capital is Maputo, which is also Mozambique's major port and industrial center.

The major industries of Mozambique are food processing, chemicals, petroleum products, and nonmetallic mineral products.[1] Mozambique generates a substantial portion of its revenues from the transit of goods to and from Zambia, Zimbabwe, and South Africa. The sale of electricity to South Africa from its hydroelectric facility at Cabora Bassa is also an important source of revenue.

Mining does not presently play a dominant role in Mozambique's economy. However, Mozambique is believed to be well endowed with mineral resources. There are several large-scale mineral exploration programs under way, and numerous smaller-scale exploration programs.[2]

Columbium-tantalum and coal are the country's most important minerals. However, the country is believed to possess exploitable deposits of diamonds, diatomite, gold, mica, fluorite (600,000 tons at 65% fluorite, and 60 million tons of lower 16 to 26% grade ore), iron (49 million tons reserves and 245 million tons resources), and titanium and zircon minerals (35 million tons).[3] In addition to these minerals, natural gas has been found, and exploration for petroleum, both onshore and offshore, is under way by major international oil companies.[2]

Government Role in Mineral Industry

After independence, the government of Mozambique claimed full ownership of the land and subsoil resources, which includes all mineral rights.[3] Since independence, the great majority of the country's 250,000 Portuguese have left the country, taking with them the technical expertise to run the country's industries. As a result, the new government asked for, and received, technical aid and experts from Eastern European countries.

The loss of technical expertise and a stagnating economy were countered by new economic directives in early 1977 by the FRELIMO Party Congress. Property left for 90 days reverted to the state without compensation. Businesses were taken over by the workers or operated by state-appointed administrative committees. The distribution of commodities was often assumed by state agencies. But the most important government moves were the nationalization of most banks, larger mines, and the country's only oil refinery.[4]

On 12 May 1978, the government of the People's Republic of Mozambique nationalized the country's only colliery at Moatize, because of a lack of safety precautions and to gain control over a key sector of the economy.[4] The owner, Com-

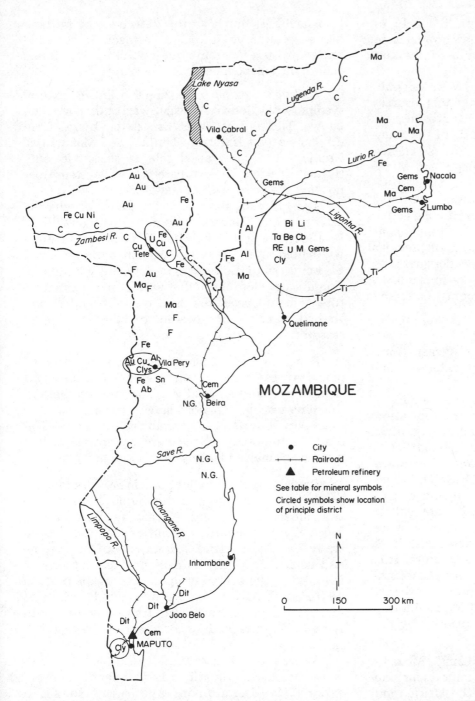

FIGURE II-4-1 Mozambique: Mines, processing facilities, and infrastructure. (From Ref. 3.)

panhia Carbonifera de Mozambique (CCM), was denied compensation on grounds that mineral rights and subsoil were state property.

Shortly thereafter, a delegation led by the Minister of Coal and Energy from East Germany (GDR) and the GPRM signed an agreement whereby the GDR would provide workers and equipment to operate the Moatize colliery with the stipulation that some coal exports be reserved for the GDR.[4]

The lack of indigenous mining expertise led to other mineral agreements with the GDR. All exploration and exploitation of the tantalum-rich pegmatite deposits of Muiane, Marropino, and Morrua (200 km northeast of Quelimane) were reserved for the GDR in 1980.[5] It is believed that all microlite and tantalite produced are exported to the GDR.[5] Previous to that, an accord signed by the Secretary of the GDR State Ministry of Geology on 22 June 1978 provided GDR technicians to operate copper and gold mines in the Manica district, the site of Mozambique's only copper mine.[4]

The lack of trained personnel has forced the government to rely on foreign expertise for the country's growing mineral exploration program.

The Mozambique government, with the help of other Eastern European nations, has programs to develop an indigenous technical mining force, but progress has been slow.

In late July 1981, the Council of Ministers published two new decrees relating to mineral resources in Mozambique.[2] Decree No. 10/81 provides for GPRM regulation of trade and exports of the country's natural resources. Decree No. 11/81 established the National Directorate for Geology and Mines and Protection of the Subsoil as the agency responsible for precious metals. The agency was responsible for the regulation and supervision of mining activities relating to precious metals. To complement this agency and increase state control of precious metals, the Bank of Mozambique was designated as the sole agent for buying or selling precious metals, as well as the sole custodian of coins, bars, and ingots.[2]

Infrastructure

Mozambique's infrastructure is facing some major problems brought about by external conflict, internal strife, and past neglect.

During the last few years of the Rhodesia/Zimbabwe conflict, Mozambique's transportation facilities were a major target of Rhodesian punitive action for harboring Zimbabwean guerrillas.[6] Recently, the Mozambique National Resistance Movement, a group opposed to FRELIMO, has inflicted damage to the transportation and energy sectors.[5] These destructive actions, in conjunction with scarcities of a skilled maintenance force, spare parts, and routine maintenance schedules, have contributed to a deteriorating infrastructure. It is also a deterrent to other countries using Mozambique facilities for imports and exports. However, the general upgrading and repair of Mozambique's infrastructure is considered a high priority by the government.

Transportation. Countries utilizing Mozambique's rail system for exports and imports include Malawi, Zambia, Zimbabwe, South Africa, and Swaziland. Of these countries, only Zambia does not have direct access to Mozambique ports via its rail system, and must traverse either through Malawi, Zimbabwe, or South Africa. Mozambique's ports at Maputo, Beira, and Nacala play a major role in the trade of southern Africa. Hence the transportation system to these ports assumes an importance disproportionate to Mozambique's own mineral industry.

The civil war in Zimbabwe, and growing internal guerrilla activity, have tended to increase the use of superior South African ports and transportation facilities, much to the detriment of Mozambique's economy. However, the black states would prefer to use Mozambique's facilities to lessen economic dependence on South Africa.

Railroads. Maputo is directly linked by rail system with South Africa, Swaziland, and Zimbabwe. The port at Beira is accessible by rail from Zimbabwe (via Harare to Umtali) and Malawi (via Nsanje). The port at Nacala is accessible only through Malawi. Unfortunately, a rail system does not link the ports with one another.

The railroads have been the target of guerrilla activity within Mozambique in the past few years. In 1981, rail traffic between Zimbabwe and Beira was halted after the bridge over the Pungwe River, 65 km west of Beira, was destroyed by an explosion in early October of that year. A pontoon structure that replaced the bridge in December was washed out by floodwaters of the early rainy season.[2]

Roads. A major highway connects Maputo with South Africa. A paved road also links Maputo and Zambabwe, but the road is sometimes unusable in the rainy season. A major highway parallels the railroad between Harare, Zimbabwe, and Beira. The roads to the port at Nacala are susceptible to bad weather. A major highway links Beira and Maputo.

Ports. The three major ports in Mozambique are at Maputo in southern Mozambique, Beira in central Mozambique, and Nacala in northeastern Mozambique. Maputo's facilities are used extensively by South Africa for exports of its chrome and coal. South African personnel are extensively involved in the everyday operations of the port. In 1978, South Africa's exports accounted for 85% of Maputo's shipping traffic, and transit fees for South Africa's imports and exports were averaging $12 million a year.

Swaziland also uses the Maputo facilities. This route was used primarily for Swaziland's iron ore exports. However, iron ore exports from Swaziland have ceased, but may be replaced by coal exports. In 1969, Maputo was receiving more than 2000 vessels and handling close to 13 million tons of cargo. By 1979, Maputo received less than 200 vessels and handled only 1.5 million tons of cargo. Maputo can berth vessels only up to 65,000 tons.[7]

The port at Beira is used extensively by Zimbabwe and is important to Malawi and Zambia. However, use of Beira has decreased since 1975, being replaced by South African ports. The reason for this decrease has been attributed to the civil war in Zimbabwe and to Beira's inferior harbor facilities. Beira can only berth vessels up to 25,000 tons.[7] Another drawback is that Beira's harbor must be continuously dredged.[7]

The port at Nacala is one of Africa's best natural Indian Ocean harbors. The facilities at Nacala are especially equipped for handling roll-on, roll-off container freight.[5] The GPRM plans to upgrade the rail between Malawi and Nacala and further upgrade Nacala's freight facilities.[5]

In 1980, plans were announced for a $515 million extension to Maputo harbor.[5] South African contractors were engaged to add an additional 2.2 million tons of coal export capacity per year to the existing facilities. Overseas contractors were being sought to construct a new port to export 6.5 million tons of coal per year.[5] This expansion program was intended to handle increased coal exports from South Africa, Mozambique, and possibly Swaziland.

Water. Mozambique's water resources are extensive, with major rivers located throughout the country. The Zambezi River, flowing through Angola, Zambia, and Zimbabwe, empties into the Indian Ocean in central Mozambique. The Cabora Bassa Dam forms a lake some 250 km long by 30 km at its widest point and contains approximately 60 billion cubic meters.[5] Lake Malawi forms part of the northwest border of Mozambique.

Energy. The main source of electricity in Mozambique is from the Cabora Bassa Dam on the Zambezi River in northwestern Mozambique. The current generating capacity of 2075 MW is produced from five generating groups of 415 MW each.[5] Future plans call for the construction of an additional 1750 MW of capacity to be built on the north bank of the Zambezi River and the extension of power transmission lines to all sections of the country.[5] Through the export of power into neigh-

boring countries, this additional capacity will become a major source of revenue for the country.

A major portion of the electricity from the Cabora Bassa Dam continues to be exported via a high-tension, direct-current power line to the major transformer near Pretoria, some 1450 km to the south. Other areas supplied by power from Cabora Bassa are the Maputo area, Tete, and the Moatize coal mining center.[5] Revenue from South Africa's use of this electricity was about $4 million per month in 1978.[4]

Disruption of these power lines has been a main target of the Mozambique National Resistance Movement, in order to disrupt the Mozambique economy. As a result, the power line was inoperable 60% of the time in 1981.[2]

Mozambique has only one petroleum refinery, which is located in Maputo. Refining capacity for the plant is given in Table II-4-1.

Petroleum is exported through Mozambique facilities to Malawi, Zimbabwe, South Africa, and Swaziland. Transit of petroleum through these facilities earned between $20 and $25 million in 1980, while refined petroleum products (from Mozambique's refinery) accounted for $37 million.[5] Petroleum supplies to Zimbabwe are through the pipeline connecting Beira to Umtali. This pipeline has also been a target of guerrilla saboteurs. A section of the line was destroyed, together with the bridge and railway, when the bridge over the Pungwe River was destroyed in 1981.[2]

Mozambique's own petroleum needs are provided chiefly by imports of crude from Iraq. In 1979, imports of crude and refined petroleum products amounted to $116 million.[5] In 1980, petroleum products from Iraq remained the number one import commodity and were 17.7% of the country's total import costs.[5]

Due to Mozambique's hydroelectric capacity

TABLE II-4-1 Petroleum Refinery Products in Mozambique, 1977-1981 (Thousand 42-Gallon Barrels)

	1977	1978	1979	1980p	1981e
Gasoline	621	510	336	510	NA
Kerosine	178 ⎫	296	282	330	NA
Jet fuel	128 ⎭				
Distillate fuel oil	865	634	1668	746	NA
Residual fuel oil	1438	1132	236	1332	NA
Other	63	85	227	323	NA
Refinery fuel and losses	229	270	128	130	NA
	3592	2927	2877	3371	NA

[a]p, preliminary; e, estimated; NA, not available.
Source: Ref. 2.

and lack of extensive demand for coking coal for the mineral industry, coal contributes a relatively minor role in the country's power needs. However, there are plans to greatly increase coal production for exports.

Mineral Industry

Although the mineral potential of Mozambique is believed to be much larger than the current level of mineral production indicates, the mineral industry in Mozambique at present comprises only a small percentage of the total GDP.

The minerals of greatest importance in Mozambique continue to be columbium-tantalum and coal. As indicated previously, the economic potential for other mineral deposits is high. The figures for production of coal, columbium-tantalum, and copper are given in Table II-4-2.

Columbium-Tantalum. The main source of columbium-tantalum in Mozambique is the pegmatite deposits of Zambezia Province. Of these deposits, the Muiane, Marropino, and Morrua deposits are located 200 km northeast of Quelimane.[2] All exploration and exploitation of these deposits are reserved for the German Democratic Republic, and it is believed that all microlite and tantalite produced is shipped to the GDR.

Production of columbite, microlite, and tantalite has decreased since independence. Figures for production since 1980 are unavailable, but production of tantalum for 1981 was believed to be between 29,000 and 32,000 kg, nearly half of preindependence production.[2]

Other minerals produced from Zambezia Province pegmatites include beryl, gemstones, bismuthinite, feldspar, and mica. These pegmatite deposits accounted for nearly one-half of the value for mineral production, with tantalite being the most valuable commodity.[2]

Copper. Production of copper resumed in 1978 with the reactivation of Lonrho's Edmundian copper mine near Manica, near Umtali.[4] Technicians from the GDR help to operate copper and gold mines in this Manica district. Reserves are believed to be small, and production is only one-fourth of the 1974–1975 production levels. Other copper deposits are near the Zambia-Mozambique border area in northwestern Mozambique.

Coal. The only coal production is at the Moatize colliery in northwestern Mozambique. Coal deposits of economical potential also exist in northern Mozambique, near Tanzania, and are parallel to the Lugenda River and Lake Malawi.[3] The coal industry was nationalized in May 1978 and is now referred to as the Carbonifera de Mozambique (Carbomoc). The Moatize colliery consists of four mines.

Future plans call for increased coal production of up to 6 million tons of high-quality coal per year. The government of Brazil is involved in these plans and hopes to receive coal for its steel industry.[2] In conjunction with this increased coal production, the government of Mozambique is also expanding its coal-exporting facilities.

It must be noted that an increase of coal production to 6 million tons per year will be quite an

TABLE II-4-2   Mozambique Mineral Production, 1970–1981[a]

Year	Coal (bituminous, thousand metric tons)	Columbium and tantalum ores and concentrates[b] (kg)			Copper (metal content, metric tons)
		Columbite	Microlite	Tantalite	
1970			63,000		130
1971	329		15,000		414
1972	336		61,000	42,000	176
1973	394		54,000	31,000	407
1974	426		62,000	46,000	805
1975	575		44,000	46,000	803
1976	553	17,000	55,921	28,000	700e
1977	310	2,300	39,866	36,300	—
1978	118	2,300	39,866	36,300	130
1979	320	2,300	31,750e	31,750	225
1980	408	NA	NA	NA	200
1981	600	NA	NA	NA	200

[a]e, estimated; NA, not available.
[b]Figures for columbium-tantalum for 1970–1975 are rounded to the nearest 1000 kg.
Source: U.S. Bureau of Mines, Minerals Yearbook, 1970–1981.

achievement. Although coal production has finally reached pre-independence levels, projected levels of output have fallen far short of government-determined goals.

Constraints to such an enormous increase in production (nearly 10 times the current level) include inadequate rail facilities from Moatize to Beira, severe shortages of trained workers and technical staff, critical lack of spare parts for Mozambique's aging equipment, and inadequate coal export facilities at Beira.[5]

Natural Gas and Petroleum. Mozambique's only refinery, in Maputo, is owned by Emprêsa National Petróleo de Mozambique (Petromac). The company is an amalgamation of two earlier companies that were nationalized in 1977.[4] Refer to Table II-4-1 for the plant's refining output.

Gas fields are in the Pande, Buzi, and Temane areas. Pande natural gas has been estimated at 18 billion cubic meters of proven and recoverable gas, and possible recoverable reserves at 60 billion cubic meters.[2] These gas fields are located south of Beira. Before independence, the Pande field has been proposed as a possible energy source for an industrial complex to produce ammonia fertilizers, urea, and sponge iron. However, no progress had been made since independence toward this goal until a French company, Compagnie Général de Géophysique (CGS), was awarded a $6 million contract to further evaluate the gas resources of Pande, Buzi, and adjacent areas.[2]

Although Mozambique does not currently produce its own petroleum, exploration for petroleum, onshore and offshore, is in progress. In June 1981, contracts for onshore and offshore seismic surveys were awarded to Western Geophysical Co. (a U.S. company), and the Geophysical Co. of Norway.

The surveys were to extend up to 2000 m, with more extensive work up to 500 m.[2] The survey included the coastline from Maputo to Beira and from Beira to Rovuma.[2] The Secretariat of Coal and Hydrocarbons announced that these companies would bear the cost of the surveys in exchange for part of the profits from any development of their discoveries.[2] The survey results were to be sold openly to interested oil companies. By the end of 1981, over 30 oil companies had shown interest in the results.

The value of petroleum imports are a significant portion of Mozambique's total import bill. However, the total demand for petroleum is not large. Therefore, any significant find, depending on the size, could make Mozambique as free from oil dependency as Angola.

Other Minerals of Importance. Titanium (ilmenite) and zirconium minerals are present in beach deposits between Quelimane and Mozambique. Resources are estimated at 35 million tons.[3]

Iron deposits are scattered throughout Mozambique. Reserves were estimated at 49 million tons of reserves and 245 million tons of resources. Exploitation of Mozambique's iron ore, and the development of potential gas fields could lead to an expanded steel industry within the country. However, exploitation of iron ore will depend not only on the grade of deposits, but will depend greatly upon the country's ability to transport the ore. The current rail system would need to be improved substantially.

Mozambique also has deposits of fluorite estimated at 600,000 tons of 65% fluorite and 60 million tons of lower 16 to 26% grade ore.[3] These deposits are located in western Mozambique, between Tete and Beira.

Constraints to Mineral Development

Major constraints to future development of Mozambique's mineral resources include improvement of the infrastructure, development of a technically qualified work force, capital investment, avoidance of internal and external strife, and its relationship with South Africa.

Even though Mozambique is not a land-locked nation, it is still dependent on its neighbors and affected by their actions. This dependency arises from the revenue generated by trade through Mozambique's ports, and dependency on South Africa's technical assistance in running key facilities at the Maputo port.

Revenue from exports of electricity to South Africa and revenue from Mozambique's workers in South African mines are important sources of income. However, the number of workers from Mozambique has decreased in recent years. South Africa has also become an important source of imports.[6] In 1980, imports from South Africa were $103 million (14.4% of total imports), while exports to South Africa were valued at only $12 million.[5]

The extent and future effects of the growing unrest and political repression in Mozambique are difficult to judge at this time. Certainly, guerrilla activity has affected the economy by disrupting key transportation and energy facilities. However, this type of guerrilla warfare cannot be used to determine accurately the number and strength of those opposed to the present government. An expansion of the internal conflict could possibly lead to a request for foreign assistance, and intervention from foreign powers, most notably Eastern Block involvement. However, South Africa would certainly be apprehensive of such an arrangement. Indeed, such an arrangement would likely be perceived as intolerable from South Africa's viewpoint and result in some form of intervention. South African military ventures into Angola support the validity of such a scenario, for Mozambique, unlike Angola, shares a common border with South Africa.

Lack of capital is a major constraint to mineral development. Recently, the government has admitted that the nationalization program has not produced the desired results. In May 1980, President Machel invited Portuguese nationals back into the country and promised to restore nationalized industries and businesses to their former owners. Delegates to a Round Table on Western Investment in Mozambique, sponsored by the Bank of Mozambique and arranged by Business International (BI), were impressed by the country's economic potential. However, they noted that a lack of hard economic data and trained personnel to support new ventures would limit future investments.[5]

The future of the mineral industry will also depend on the results of current mineral exploration programs. Historically, Mozambique's importance to Portugal was as a food producer. The mineral industry was neglected, and consequently, detailed geological exploration and economic evaluation of Mozambique's mineral wealth were limited.

Exploration activity in Mozambique is now escalating and includes a number of important projects. The USSR is active in geophysical studies and prospecting operations for coal, gas, pegmatites, and gemstones.[2] In early 1981, Hunting Geology and Geophysics Ltd. of Great Britain negotiated a contract for the mineral inventory of 275,000 km^2, which included the use of airborne magnetometers and gamma-ray spectrometers.[2] Goezavod of Belgrade, a Yugoslavian state enterprise, is prospecting in six of Mozambique's provinces.[2] Czechoslovakia has been contracted to explore for iron, feldspar, gold, and tin in the provinces of Nampula, Zambezia, and Niassa.[2] The United Nations is involved in detailed mineral exploration in Tete Province (15,000 km^2), Zambezia Province (6000 km^2), Manica Province (700 km^2).[2] This detailed study will contribute to an inventory of mineral wealth and provide more hard geological data to access the economic feasibility of exploiting Mozambique's mineral wealth.

REFERENCES

1. U.S. Central Intelligence Agency, National Foreign Assessment Center. *The World Factbook—1981*. Washington, D.C.: U.S. Government Printing Office, 1982.

2. U.S. Department of the Interior, Bureau of Mines. "The Mineral Industry of Mozambique." *Minerals Yearbook 1981*. Washington, D.C.: U.S. Government Printing Office, 1983.

3. U.S. Department of the Interior, Bureau of Mines. *Mineral Industries of Africa*. Washington, D.C.: U.S. Government Printing Office, 1976.

4. U.S. Department of the Interior, Bureau of Mines. "The Mineral Industry of Mozambique." *Minerals Yearbook 1978/79*. Washington, D.C.: U.S. Government Printing Office, 1981.

5. U.S. Department of the Interior, Bureau of Mines. "The Mineral Industry of Mozambique." *Minerals Yearbook 1980*. Washington, D.C.: U.S. Government Printing Office, 1982.

6. Hodges, Tony. "Mozambique: The Politics of Liberation." In *Southern Africa: The Continuing Crisis*, 2nd ed. Edited by Gwendolyn M. Carter and Patrick O'Meara. Bloomington, Ind.: Indiana University Press, 1982.

7. U.S. Department of the Interior, Bureau of Mines. "Zimbabwe." *Mineral Perspectives*, August 1981.

ANGOLA

The People's Republic of Angola, which includes the enclave of Cabinda, is adjacent to the Atlantic Ocean, with a total land area of 1,245,790 km^2 and a coastline of 1600 km.[1] Angola shares a common border with Zaire, Zambia, and Namibia/South-West Africa; Cabinda is surrounded by Zaire and Congo.[1]

The population of Angola is approximately 7 million, and the population of Cabinda is a little over 100,000. The population is ethnically divided into 93% Africans, 5% European, and 1% mixed.[1]

Angola is a former colony of Portugal, and became an independent country on November 11, 1975. The official language of Angola remains Portuguese. The capital is located at Luanda, adjacent to the Atlantic coast in northern Angola.

Independence was followed by civil war, the introduction of Cuban forces, and the eventual emergence of Agostinho Neto as the country's new leader.[2] The current President of Angola is José Edwardo dos Santos.[1] The Popular Movement for the Liberation of Angola (MPLA) is the only legal party in Angola. However, the two other political parties contending for control of Angola are still conducting guerrilla warfare against the present government. These two political parties are the National Front for the Liberation of Angola (FNLA), operating in northeastern Angola; and the National Union for the Total Independence of Angola (UNITA), operating in southeastern Angola and led by Jonas Savimbi.[2]

The economy of Angola is dominated by the mineral and agricultural sectors. The mineral economy of Angola is dominated by petroleum and

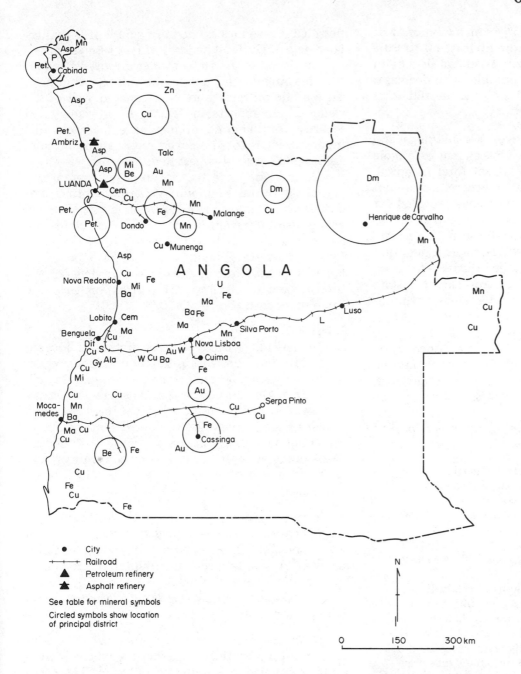

FIGURE II-4-2 Angola: Mines, processing facilities, and infrastructure. (From Ref. 3.)

diamonds, which constitute the main portion of foreign exchange earnings. Other potentially economic mineral deposits in Angola (although not currently produced) are iron ore, copper, manganese, phosphates, and uranium.[3,4]

Government Role in Mineral Industry

General Economy. Prior to independence in 1975, there was little state participation in the general economy, and the mineral industry was privately owned. Exports were dominated by crude petroleum (30%), coffee (30%), diamonds (10%), and iron ore (6%).[5]

After independence in 1975, the economy contracted sharply. Contributing to the collapse of the economy were several factors. Independence and the ensuing civil war were responsible for the mass exodus of 250,000[2] to 350,000[4] skilled Portuguese workers and countless blacks, which resulted in the collapse of the infrastructure. Other factors were the continuing guerrilla insurgency in northeastern and southeastern Angola, and lack of foreign investment caused by the civil strife and Angola's nationalization and state operation of major economic sectors (80% of Angola's enterprises are state-run).[4]

Government control of the economy has not performed as anticipated and foreign investment has decreased drastically. Angola moved from the world's fourth-largest coffee producer (growing 243,000 metric tons in 1973) to its current level of

21,500 metric tons in 1981.[6] From an exporter of food products, Angola is now at least 50 to 90% dependent on food imports.[6] A general decline in the mineral industry is also notable, with decreases in petroleum and diamond production and total cessation of iron ore production.[8]

New Government Incentives. Given this general decline in the economy, the government of Angola instigated new measures to attract foreign capital. In 1979, Angola produced new general investment laws for the country.[5] The new laws provided for repatriation of profits and provided guarantees of compensation in case of nationalization. For certain industries, the government will consider exemption from certain taxes and customs duties. However, certain strategic industries are reserved for state control. These industries included such categories as defense, banking, insurance, communications, electricity, and water supply. A satisfactory time scale for new ventures has been determined to be between 10 and 15 years. Joint projects would be controlled by the government (51%), although wholly owned private enterprises were to be allowed.[5]

In May 1979, a new mining law was approved with the following provisions:[5]

1. Government ownership of all mineral resources.
2. Government control of geological mapping.
3. Government control of resource evaluation.
4. Utilization of mineral resources as required by the planned economy.
5. Concession of exploration rights exclusively to Government mining enterprises, with some justifiable exceptions.
6. Acceptability of operating contracts with certain approved foreign agencies with technical capacity to develop the extraction industry.
7. Foreign companies were obligated to abide by Angolan law and comply with needs of the Angolan people, but at the same time were allowed the right to investment recovery and profits.

The government has increased foreign investment in offshore oil ventures by dividing up the coast into 14 offshore blocks (including Cabinda), and by mid-1982, over half of the blocks had been sold.[6] The government has also managed to sign new exploration and production agreements with at least 11 Western oil companies.[6] Angola expects investments of over $1 billion dollars in the Cabinda and Soyo oil-producing areas by 1985. The government expects production to rise from 130,000 barrels per day to 250,000 to 300,000 barrels per day.[6] In 1978–1979, Sonangol also increased its share of petroleum revenues by acquiring 51% of the Angolan assets of Gulf Oil and Petrofina of Belgium.[6]

An increase in the production of oil would brighten Angola's economic outlook tremendously, since crude oil sales were an estimated 78% of the value of all exports in 1981.[7] Improvements in security for diamond mining have increased diamond revenues for the government. Diamond sales are Angola's second-largest export earner.

Despite the expected rise in governmental revenues, the overall economic situation continued to deteriorate. Almost 50% of Angola's foreign exchange earnings were used for national defense,[6] and the external debt had increased to $2.3 billion. Even though this amount of foreign debt is not large by Third World standards, the servicing of the foreign debt in 1981 was 15% of export earnings, three times the ratio for 1979.[6]

Infrastructure

In general, Angola's infrastructure is located along the coastline, within the mountainous region in western Angola, and along three separate railroads that are not connected. Angola's infrastructure was a major casualty of the civil war and remains a frequent target of guerrilla insurgency in the country. As a result, only one of its three railroads operates with any degree of continuity,[6] and frequent power shortages are reported in Luanda and other major cities.[7]

Transportation. The transportation system in Angola consists of three separate, unconnected railways, traveling in an east-west direction. A major highway system services the mountainous region of western Angola and travels in a generally north-south direction, connecting Angola's major ports with one another.

Railroads and Ports. The map of Angola in Figure II-4-2 reveals the fact that the railroads were built to service the minerals industry. There are three railroads in Angola; one is located in northern Angola, and another in southern Angola. The third railroad (Benguela Railroad) in central Angola is the only rail system to extend beyond Angola's border.

The northern railroad extends from the port at Luanda some 512 km eastward to Malange.[3] The route traverses through iron ore and manganese districts.

The southern railroad extends from the port at Mocamedes some 880 km to Menongue,[3] with a spur to the former Cassinga iron ore mines and another spur extends to berryllium deposits.

The central Benguela Railway connects neighboring Zaire with the port at Lobito and is some 1414 km long.[3] This route was a major export route for Zairian copper and manganese exports, as well as Zambian copper exports. As a result of guerrilla activity, the Benguela Railway has remained all but closed to major export traffic from Zaire and Zambia since the civil war in 1975. Other problems along the railway include a deteriorating signal system and maintenance and repair of rolling stock. In 1978 it was estimated that upgrading of the railroad would require $106 million.[5]

Water. Angola has many rivers, but also has several arid and semiarid regions. Unlike most other major southern African countries, Angola does not possess a major interior body of water.

Energy. Total electricity capacity was 525 MW in 1977.[1] Of that amount, 343 MW was hydroelectric, and 135 MW was from thermal power plants in 1974.[3] A project on the Cuenene River in southern Angola was to produce an eventual 2975 MW of electricity.[3] South Africa was the dominant force behind the project, and some electricity was to be diverted to the mineral industry in Namibia/South-West Africa. However, South Africa withdrew from the project during Angola's civil war. South Africa has now completed the hydroelectric facilities on the Namibian side and now generates electricity from the facility. Major cities in Angola continue to experience power failures, due in part to intermittent rainfall,[8] old equipment, and management inefficiencies.[7]

Angola's petroleum demand is roughly 10 to 15% of its production. The country's only refinery is located at Luanda and has a capacity of 33,000 barrels per day. In 1981, the refinery was damaged by sabotage, and several storage tanks and one of the distillation units were destroyed. Imports from Gabon and the Canary Islands were required.[7] The refinery, which is majority-owned by Angola's Petrangol, is to be rebuilt with double capacity at a cost of $200 to $300 million.[6] Current production at the refinery is given in Table II-4-3.[7] Figures for 1980 and 1981 are not available.

Mineral Industry

Before independence, revenue from petroleum and diamond exports were each 30% of total export value. By 1978, the value of petroleum exports accounted for about 70% of the total export value.[5] In 1980, world price increases in crude petroleum contributed to a 138% increase in petroleum revenues in Angola and generated $2 billion from petroleum exports.[8] By the end of 1981, crude petroleum exports amounted to 78% of the value of all exports.[7] During this period of time, diamonds remained the second most important source of foreign exchange, followed by coffee exports.

Fuel Minerals

Petroleum. The importance of crude petroleum exports to the Angolan economy is recognized by the government and has been instrumental in opening Angola to foreign investment. A number of foreign countries are now operating in Angola.

Angola has recently divided its coastline into 14 offshore blocks, to be acquired for exploration of teams of foreign oil companies. By mid-1982 over half the blocks had been sold.[6]

Production. Currently, Angola is the fifth-largest oil producer in Africa[5] and the third-largest producer of oil among black African nations. Production levels for natural gas and petroleum are given in Table II-4-4.

In 1980, Angola produced about 164,000 barrels per day (bpd) of crude petroleum.[8] The government of Angola expected an 8% increase to 178,000 bpd for 1981.[8] However, actual production of oil

TABLE II-4-3 Petroleum Refinery Products in Angola, 1970-1981[a]
(Thousand 42-Gallon Barrels)

	1977	1978	1979	1980p	1981e
Gasoline	526	510	500e		
Jet fuel	441	480	450e		
Kerosine	163	160	160e		
Distillate fuel oil	1469	1567	1500e	NA	NA
Residual fuel oil	3725	3796	3700e		
Other	143	189	150		
Refinery fuel and losses	229	368	300e		
	6696	7070	6760e	NA	NA

[a]e, estimated; p, preliminary; NA, not available.
Source: Ref. 7.

TABLE II-4-4 Petroleum and Natural Gas Production in Angola, 1970–1981

	Crude petroleum (thousand 42-gallon barrels)	Natural gas (million cubic feet)	
		Gross	Marketable
1970	36,499	340,000	1,500
1971	33,922	27,000	1,500
1972	50,932	31,393	2,000
1973	58,852	36,000	2,300
1974	61,392	37,500	2,400
1975	57,943	35,000	2,300
1976	36,700	22,000	2,000
1977	62,437	42,400	2,500
1978	47,450	46,500	2,500
1979	49,640	48,600	2,500
1980	55,034	58,000	2,500
1981	52,000	55,000	2,500

Source: U.S. Bureau of Mines, Minerals Yearbook, 1970–1981.

for 1981 turned out to be an estimated 140,000 bpd. As a result, the 1982 development plan was revised, and the government reduced government-directed investment by 40% from the 1980 level.[7] The drop in production was attributed to technical difficulties, an exhaustion of reserves in the oldest oil fields, and a drop in world oil demand and prices.[8] New development of oil fields by foreign companies is expected to increase production from 1981's 140,000 bpd to an estimated 250,000 to 300,000 bpd.

Roughly two-thirds of Angola's petroleum production comes from Cabinda (Block 1), produced by Gulf Oil Company and Sociedade Angolane de Combustíveis de Angola (Sonangol). Production is expected to increase from 90,000 bpd for 1981 to 200,000 bpd by 1985.[6] This increase is to be obtained in two ways. The first is to inject 78 million cubic feet of natural gas into two wells in 1982.[7] The total cost of gas injection to increase reservoir pressure and production is an estimated $168 million and will increase production by 15,000 to 20,000 bpd.[6] Second, development of the new Takula field, at an estimated cost of $600 million, is expected to increase production by another 70,000 bpd by 1985.[6] Additional economic benefits to be derived from the area were to be achieved by extracting natural gas liquids. Fractionators were also to be installed to produce propane, butane, and natural gasoline.[7] Facilities were to be installed to transfer liquefied petroleum gas from a refrigerated storage unit via pipeline to waiting vessels. Natural gasoline would be mixed with crude oil and sent to shore via pipeline to storage units.[7] Sonangol is also developing a major supply and service center in the region on Kwanda Island, 5 km from Soyo in Zaire Province.[7]

Production at the Soyo field near the Congo River in northern Angola is about 32,000 to 35,000 bpd.[6] Texaco and Sonangol are involved in the field. Texaco already produces 12,000 to 13,000 bpd and expects to produce more soon.[6] Texaco and Sonangol are building an 800,000-barrel new terminal and base camp in the area.[6] Proven reserves for the country are estimated at 1.5 billion barrels.[6] However, current exploration is expected to increase total reserve figures.

Foreign Oil Companies. Various foreign companies have begun petroleum exploration and development in recent years. In 1979, Texaco Petroleos de Angola signed an agreement with Angola to develop three more fields (Escungo, Etele, and Cuntala) at $250 million, and an eventual 50,000-bpd increase in production.[5] The agreement also included expenditures of $110 million on exploring Block 2, an area of 4000 km^2 in the Congo River basin.[5]

Texaco (the operator) will have a 40% share in the project and Sonangol reserves the right to assign part of its 60% to another company. The profit oil will be taxed on a sliding scale of 70 to 95%. An additional tax of 50% on the profit oil will be levied. If Sonangol retains its 60% share of the project, its realized revenues from profit oil will vary from 88% (minimum) to 98% (maximum).[5]

Gulf Oil and Sonangol signed a new contract in which Angola assumed 51% of the ownership and title to Cabinda's offshore concessions. Gulf Oil Company's share of 49% was to be taxed at a rate of 16% for royalty and a 65.75% income tax.[5]

Elf-Aquitaine, a French company, was granted the exploration rights to a 4000-km^2 area in Block 3 (Soyo area). Elf would own 50% interest and would be the operator. Sonangol's share will initially be 15% but will rise to 95% later. Elf is to invest $40 million in a three-year period.[5] Elf believes it has found a potential 100,000-bpd field in Block 3.[6]

Total Compagnie Africaine des Pétroles (Total-CAP) and Sonangol were to invest $30 million in exploration cost over a five-year period on a 4000-km^2 block near Luanda. Petrofina was nationalized by the government on 1 January 1979. Production from its holdings were 7.7 million barrels in 1978.[5] Petrobas, Brazil's national petroleum company, was involved with Angola by providing oil equipment to Angola and training of Angolan personnel. Petrobas has also signed a contract for offshore oil exploration.[5] Brazil was to increase oil purchases 25% from its estimated 2.5 million barrels of crude oil in 1980.[6] In 1981, Sonangol announced that whenever possible, it would market its oil directly.

Natural Gas. Only a small portion (4.5%) of Angola's natural gas was marketed in 1981.[6] Because of a lack of infrastructure for gas production, most of Angola's natural gas is flared. The government has initiated efforts to increase marketable gas production.[6]

Coal. Coal has been produced only sporadically in Angola. Thermal power plants in Angola commonly use gas or oil rather than coal to generate electricity.

Reserves of 500 million tons of lignite have been identified near Lueno in east-central Angola. Deposits are near the Benguelo Railway. However, a lack of demand for coal will delay development of this deposit. Lignite is unsuitable for metallurgical processing.

Uranium. Uranium deposits have been identified in Angola,[3] but little information about resources is available.

Nonfuel Minerals

Diamonds. Diamonds continue to be Angola's second most important foreign exchange earner. Angola's main diamond deposits are located in Luanda Norte Province in northeastern Angola. Smaller deposits are also located between Quela and Caungula in north-central Angola. Angola's diamond deposits are alluvial deposits.[6]

Diamond production in Angola is given in Table II-4-5. Gem diamond production is far more valuable than industrial diamonds, being 75% of

total diamond production in 1981.[7] Diamond production is still below 1971's high of over 2 million carats, but has increased since a post-independence low of only 340,000 carats in 1976. An increase in gravel ore treatment and tighter security contributed to the increase.

The nationalized Companhia de Diamantes de Angola (Diamang) increased its share in the diamond industry to 77.21% in 1979, by the nationalization of 283,039 shares of stock belonging to Portuguese entities.[5] Belgium Societe Generale has 15% of the remaining stocks, with Swiss and British banks owning the remainder.[6] Diamang is managed by Mining and Technical Services (MATS), a De Beers Company. Marketing is through De Beers' Central Selling Organization (CSO).[6]

Copper. The government of Angola is considering opening the Movoio Mine, 28 km from Maquelo do Zombo in northwestern Angola. The mine was worked in the 1940s and 1950s by Empress do Cobre Angola (ECA). Gold and silver were also produced as by-products.[7] Drilling in the vicinity of Titelo near Aigl resulted in the discovery of 10 million tons of ore grading 3 to 4% copper. The deposit was reported to be at considerable depth.[7]

Iron Ore. Iron ore production was halted in 1976, as indicated by Table II-4-6. Iron ore deposits at Cassinga in southern Angola were estimated at 100 million tons of alluvial hematite, grading 62 to 64% iron, and 2 billion tons of itabirite-type ore (taconite-hematite and silica).[8] The area reportedly had 21.7 million tons of detrital iron ore that could be mined by using scrapers in open-cast mining.[7] Iron ore from the mine would be taken to the port of Mocamedes, where a 600-m-long pier can load a 100,000-dead-weight-ton vessel at 5000 tons per hour.[7]

Austromineral, a subsidiary of the Austrian company Voest-Alpine AG, was drawing up mining

TABLE II-4-5   Diamond Production in Angola, 1970–1981 (Thousand Carats)

	Industrial	Gem
1970	599	1797
1971	603	1810
1972	539	1616
1973	531	1594
1974	485	1455
1975	135	400
1976	85	255
1977	88	265
1978	175	525
1979	210	630
1980	375	1125
1981	400	1200

Source: U.S. Bureau of Mines, *Minerals Yearbook,* 1970–1981.

TABLE II-4-6   Iron Ore Concentrate Production in Angola,[a] 1970–1976 (Gross Weight, Thousand Metric Tons)

1970	6052
1971	6165
1972	4831
1973	6052
1974	5600
1975	2600
1976	—

[a]No iron ore production after 1975.

Source: U.S. Bureau of Mines, *Minerals Yearbook,* 1970–1976.

plans for the Cassala-Quitungo deposit. The mine is slated to produce approximately 2 million tons of iron ore pellets per year.[5] Cassanga-Quitungo magnetite deposits were estimated at 92 million tons of ore grading 32% iron.[8]

In 1979, all assets and rights of Companhia Mineira do Lobito, S.A.R.L. (an iron ore mine operator at Cassinga Mine), and in 1980, the Companhia do Manganese de Angola (CMA) (an operator and owner at Cassala) were nationalized.[8]

Manganese. The Companhia do Manganese de Angola (CMA) was nationalized. Manganese is no longer mined in Angola. Important deposits exist near Malange in northern Angola and near the Zairian border in eastern Angola.

Phosphate. Prior to 1976, 27 million tons of phosphate resources had been identified in Angola.[3] In late 1981, production of phosphate rock began at Quindonacache in Zaire Province. The open-pit operation produced 15,000 tons per year of phosphate containing more than 30% P_2O_5.[7] Bulgareomin of Bulgaria had surveyed 720 km^2 in Zaire Province prior to their design and construction of the phosphate mine, plant, and infrastructure.[7]

Constraints to Mineral Development

Future development of Angola's mineral industry will depend mainly on the resolution of the continuing guerrilla insurgency in Angola. This resolution, together with recent government incentives, will create a more favorable climate for foreign investment. It would also enable the government to invest more funds in mineral ventures and infrastructure rather than in defense (estimated at 50% of government expenditures in 1981).[7]

Angola will have to work toward a resolution of the SWAPO–Namibia/South-West African conflict. South Africa has invaded Angola in past years in pursuit of SWAPO military facilities in Angola.

The presence of Cuban troops in Angola is a point of dissention with South Africa, the United States, and other Western allies. A report issued in November 1982 by the chairman of the Subcommittee on Security and Terrorism (Senator Jeremiah Denton of Alabama) entitled *Soviet, East German and Cuban Involvement in Fomenting Terrorism in Southern Africa*[9] considers the alliance of SWAPO and communist goals a threat to American interest in southern Africa. Angola must develop an indigenous, technically oriented work force to run its own mineral industry and maintain repairs on the country's infrastructure.

REFERENCES

1. U.S. Central Intelligence Agency, National Foreign Assessment Center. *The World Factbook—1981.* Washington, D.C.: U.S. Government Printing Office, 1982.
2. Marcum, John. "Angola: Perilous Transition to Independence." In *Southern Africa: The Continuing Crisis,* 2nd ed. Edited by Gwendolyn M. Carter and Patrick O'Meara. Bloomington, Ind.: Indiana University Press, 1982.
3. U.S. Department of the Interior, Bureau of Mines. *Mineral Industries of Africa.* Washington, D.C.: U.S. Government Printing Office, 1976.
4. Savastuk, David J. "Angola: U.S. Business Involvement Is Significant, Expanding." *Business America,* 9 August 1982, p. 25.
5. U.S. Department of the Interior, Bureau of Mines. "The Mineral Industry of Angola." *Minerals Yearbook 1978/79.* Washington, D.C.: U.S. Government Printing Office, 1981.
6. "Wanted: Capitalist Comrades." *The Economist,* 24 July 1982, pp. 65–67.
7. U.S. Department of the Interior, Bureau of Mines. "The Mineral Industry of Angola." *Minerals Yearbook 1981.* Washington, D.C.: U.S. Government Printing Office, 1983.
8. U.S. Department of the Interior, Bureau of Mines. "The Mineral Industry of Angola." *Minerals Yearbook 1980.* Washington, D.C.: U.S. Government Printing Office, 1982.
9. U.S. Congress, Senate, Committee on the Judiciary, Subcommittee on Security and Terrorism. *Soviet, East German and Cuban Involvement in Fomenting Terrorism in Southern Africa.* 97th Congress, 2nd Session. Washington, D.C.: U.S. Government Printing Office, 1982.

NAMIBIA/SOUTH-WEST AFRICA

Namibia/South-West Africa is adjacent to the Atlantic Ocean and shares a common border with Angola, Zambia, Botswana, and South Africa. The country has an area of 832,620 km^2 with a coastline of some 1489 km.[1] The land is predominately desert region except for an area along the northern border and an interior plateau.

The population is a little over 1 million people. Ethnic divisions within the country are 12% white, 6% mulatto, and 82% African. Almost half of the African population consists of members of the Ovambo tribe in the northern section of the country. Afrikaans is spoken by about 70% of the white population; German (22%) and English (8%) are the other European languages. There are several African languages.[1]

The political history of the country is rather complicated and has been a major point of debate in the U.N. for some time. The country was formerly the German colony, South-West Africa. After World

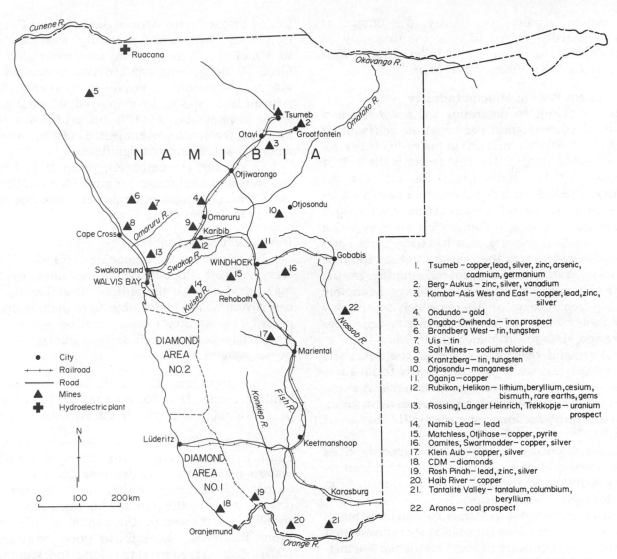

1. Tsumeb – copper, lead, silver, zinc, arsenic, cadmium, germanium
2. Berg-Aukus – zinc, silver, vanadium
3. Kombat-Asis West and East – copper, lead, zinc, silver
4. Ondundo – gold
5. Ongaba-Owihenda – iron prospect
6. Brandberg West – tin, tungsten
7. Uis – tin
8. Salt Mines – sodium chloride
9. Krantzberg – tin, tungsten
10. Otjosondu – manganese
11. Oganja – copper
12. Rubikon, Helikon – lithium, beryllium, cesium, bismuth, rare earths, gems
13. Rossing, Langer Heinrich, Trekkopje – uranium prospect
14. Namib Lead – lead
15. Matchless, Otjihase – copper, pyrite
16. Oamites, Swartmodder – copper, silver
17. Klein Aub – copper, silver
18. CDM – diamonds
19. Rosh Pinah – lead, zinc, silver
20. Haib River – copper
21. Tantalite Valley – tantalum, columbium, beryllium
22. Aranos – coal prospect

• City
⊢━━⊣ Railroad
──── Road
▲ Mines
✚ Hydroelectric plant

FIGURE II-4-3 Namibia/South-West Africa: Mines, processing facilities, and infrastructure. (From U.S. Department of the Interior, Bureau of Mines. *Mineral Perspectives*, August 1983.)

War I, the League of Nations mandated administrative control of South-West Africa to South Africa. On 27 October 1966 the United Nations formally ended South Africa's mandate and has since recognized the predominately Ovambo-controlled South-West Africa People's Organization (SWAPO), led by Sam Nujoma, as the country's legal representative.[2] South Africa does not recognize the new U.N. mandate. The current government is controlled by the Democratic Turnhalle Alliance (DTA), a political party composed of traditional tribal chiefs and the white Republican Party.[2]

The country, like South Africa, is divided into zones of white settlement and tribal homelands. Traditional chiefs and representative bodies exercise limited autonomy in their respective areas. An Administrative-General is appointed by the South African government and has veto powers over the Namibian National Assembly's legislative powers as well as the Ministers' Council, an executive body made up of members of the National Assembly.[3] Police, defense, and customs relations powers are South African, although a Southwest African Territory Force was established on 1 August 1980.[1] The capital is Windhoek, located in the central part of the country.

Namibia/South-West Africa's economy is dominated by mining, ranching, and fishing. Approximately 60% of the labor force is in agricultural sector and 10% in the mining sector.[3] Important minerals are diamonds, uranium, copper, lead, and zinc. Namibia is also a member of a common customs union, which includes South Africa, Botswana, Lesotho, and Swaziland.[3] In a reflection of current

U.S. policy toward the country, the Bureau of Mines officially refers to the country only as Namibia, and has dropped the co-term South-West Africa in its 1981 *Minerals Yearbook*.[4]

Government Role in Mineral Industry

It is difficult to determine the exact role and extent of governmental participation in Namibia/South-West Africa's mineral industry. This is a result of several factors. The first factor is the lack of separate, official, and current statistics for the country. Trade figures for the nation have been included in the statistics for South Africa since 1966.[3]

Control of mining in Namibia/South-West Africa was centered in Pretoria, South Africa, from 1968 to 1978. Control was then shifted to the capital of Namibia, Windhoek. Control of mining in the country is now with the Directorate of Economic Affairs, Department of Mines, which administers the country's mining laws according to the Mines Works and Mineral Ordinance of 1968.[4]

All mineral rights are vested in the state. The government can grant exclusive rights to prospect in any defined area for any or all minerals. A stipulation to this prospecting grant requires a minimum expenditure obligation that varies with the size of grant.[4]

Upon discovery of minerals, prospecting titles are converted to mining titles. The government requires a minimum of 25% local participation in large foreign mineral ventures.[4]

However, in order to attract foreign investment, there are no provisions for obligatory reinvestment of profits or recruiting of local personnel. Furthermore, the processing of minerals to a more refined stage in order to capture added value to mineral products is not obligatory, nor is there any restriction on the output volume.[4]

Taxes on mineral commodities vary with the mineral. Diamond mines are taxed at an effective rate of 60%. Uranium is taxed on a sliding scale or formula basis, from zero to over 70%. This tax enables a company to economically vary the grade of ore mineral based on the price of uranium. Other minerals were taxed at the normal 40% corporate rate.

To increase foreign investment in mining, there are liberal write-offs on capital expenditure that postpone tax payments until capital investment has been fully recovered. This law has enabled Rossing Uranium Ltd. (RUL) to defer taxes until 1983. Production at the uranium mine commenced in 1976.[3]

Mining on former black homelands was controlled by the South African Bantu Mining Corporation until 1978, when authority was transferred to Windhoek. Mining in an area controlled by a black ethnic government had two important provisions to provide revenue to the area. Local African labor was to be employed, where possible, and a revenue tax of 10% on profit was to be awarded the local government. However, revenues from this source are not significant.[4]

To encourage companies, prospecting licenses are issued to companies for any place in Namibia. However, individuals could prospect and peg claims only in their own ethnic areas.[4]

Infrastructure

The expansion and upgrading of Namibia/South-West Africa's infrastructure continues to be a major goal of both the indigenous and South African governments. Major objectives include upgrading of the country's transportation systems and Walvis Bay port, as well as the country's electric power systems.

Transportation. The transportation system is built to service the country's mining industry and provide a major link with South Africa.

Railroads. The railway connects Namibia/South-West Africa to South Africa in the southeastern part of the country near Nakop. The rail divides at Seeheim, in south-central Namibia, with one line going to the port at Luderitz and the other traveling northward to the capital at Windhoek. From Windhoek the railroad goes westward to Walvis Bay. At Kranzberg, along the Windhoek-Walvis Bay route, the railway proceeds northward to the mines near Tsumeb. In all, the Tsumeb-Walvis Bay railroad is about 2350 km long.[1]

There have been some discussions and feasibility studies concerning the extension of the railroad from Gobabis in east-central Namibia/South-West Africa, eastward into Botswana, where the two countries' rail systems will be linked. The project was estimated to cost around $1 billion, to be financed by extra traffic to Walvis Bay, and the development of coal deposits in the Morupule area of southern Botswana or the newly discovered fields in western Botswana. It was estimated that completion of the new railroad would take nearly 10 years.[4]

Roads. Of the estimated 33,800 km of roads in the country, only some 3800 km is paved.[1] One of the country's important roads connects Ghanzi in western Botswana and Gobabis to Walvis Bay.

Ports. There are two major ports in the country. One is located at Walvis Bay, in central Namibia/South-West Africa, and the other is at Luderitz in the southern part of the country.

Deepening the harbor at Walvis Bay is intended to increase the facility's capacity to handle cargo ships up to 70,000 tons.[4] Walvis Bay is estimated to handle over 80% of the country's exports.[5] South Africa maintains that Walvis Bay belongs to it.[2] An independent Namibia/South-West Africa, without Walvis Bay, would have insufficient facilities for exports. The nation would be forced to build a new port, expand capacity at Luderitz, or export through South African facilities. Thus South Africa would be able to exercise considerable influence over the country's economy.

Water. Namibia/South-West Africa is a semiarid and arid land. Rains are seasonal. Development of mining in the country will require the construction of pipelines to carry water. There is a large river (the Orange River) that forms the southern border with South Africa, and another river (the Kunene River) along the northern border with Angola.

Energy. The average demand for electricity was estimated to be about 140 MW in 1981.[6] Electricity was supplied by the 120-MW thermal station at Windhoek and the 45-MW thermal emergency generator at Walvis Bay.[4] A total of 100 MW was available to the country through the South African power network, with an additional 100 MW to be available by 1982.[6]

The Consolidated Diamond Mines (CDM) in the southern part of the country was connected to South Africa's power system (ESCOM), north of the Orange River's mouth, where CDM now has an 18-MW standby power station.[6]

In a major power project for the country, the Ruacana Falls hydroelectric power station is to be tied to South Africa's power grid system. In January 1980 the station generated 240 MW of power to a power-switching station at Omaruru, 160 km northwest of Windhoek. Work continues on the 320-MW power line[5] from Windhoek to South Africa.[6] Namibia/South-West African energy sources are presented in Table II-4-7.

The country does not currently produce petroleum, although exploration programs off the coast and in the northern section of the country may identify potential oil deposits. The Government-owned Southern Oil Exploration Corporation (South-West Africa) (Pty.) Ltd. is responsible for granting oil exploration concessions. The parent company, Southern Oil Exploration Corporation (Pty.) Ltd. (SOEKOR), holds offshore concessions south of Walvis Bay.[4]

The Etosha Petroleum Company, owned by Briland Mines Ltd., holds a concession over an area in the northern section of the country. In 1980 the company signed a letter of intent giving Superior Oil Co. a contract for oil-drilling exploration over a portion of the Etosha area.[6]

There is no coal production in Namibia. However, estimated resources of 3 billion tons have been reported by the Anglo-American Corporation Ltd. and its affiliate CDM, along the Nossob River in the Aranos area. However, excessive overburden and an underground lake over portions of the

TABLE II-4-7 Namibia Energy Sources, 1982

Plant and transmission line location	Annual capacity (MW)	Energy type	Startup date[a]
Ruacana Project Angola-Namibia	240[b]	Hydroelectric	1977
Van Eck Windhoek, Namibia	120	Coal-fired thermal	1972
Oranjemund Namibia	100	Coal-fired thermal[c]	1978
Aggeneys South Africa	100	Coal-fired thermal[c]	1982
Walvis Bay South Africa	45	Diesel	NA

[a]NA, not available.
[b]Wet season.
[c]Received through Electricity Supply Commission, Republic of South Africa transmission lines.
Source: U.S. Bureau of Mines, "Namibia," *Mineral Perspectives,* August 1983.

deposit will hamper exploitation of the deposit.[4] Exploration continues along the Kaokoveld coast in the northwestern section of the country and in the Etosha pan area in the north-central section of the country.[4]

Mineral Industry

The mineral industry's contribution toward the country's GNP in recent years is approximately one-half of the total GNP. Namibia/South-West Africa's nonfuel mineral industry is the fourth most valuable of all African countries. In 1980, the GNP was estimated at $1.7 billion, while in 1981, the GNP was only $1.5 billion. The drop in GNP from 1980 to 1981 can be attributed to the general decline in the value of diamond exports, which was due to a general decline in the value of gem diamonds. Gem diamonds have accounted for 95% of diamond production.[4] The government had expected $142 million in revenues from diamond production in 1981. Instead, only $44 million was generated. This drop of approximately $100 million in revenue was about 15% of the GNP for 1981.[4] Diamond production was estimated to be 30% of the GNP in 1980, or roughly $500 million.[6] Namibia was the world's third-largest producer of gem diamonds in 1982.[7]

Uranium is the country's second-most-valuable mineral commodity, although Rossing Uranium Ltd. still pays no taxes on profits until investment has been recovered.[4] Namibia is one of the world's largest uranium producers and has the world's largest single open-pit mine.[4]

Other important minerals include arsenic, copper, lead, silver, vanadium, tin, and zinc. Namibia/South-West Africa is the sixth- or seventh-largest producer of arsenic in the world.[7]

Diamonds. Diamonds are mined predominantly along the country's southern coast, using a bucket wheel excavator. Diamonds occur in marine gravels, and mining is to within 100 m seaward of the high-water mark.[4] The main area of mining extends along the coastline approximately 100 km northward of the Orange River.[5] Stripped overburden, amount of gravel treated, ore grade, and diamond production are given in Table II-4-8. At the end of 1981, stripped reserves were six months ahead of mining requirements.[4] The diamonds are mined by Consolidated Diamond Mines of South-West Africa (Pty.) Ltd. (CDM), a wholly owned subsidiary of De Beers Consolidated Mines Ltd. The company holds the mineral rights in its concession areas until December 31, 2010.[4]

Reconnaissance prospecting is being conducted over large areas of the country. However, the only encouraging site was located along the north bank of the Orange River within Diamond Area No. 1.[4]

The country's diamond production represented 22% of De Beers' after-tax profits in 1978.[5] In 1981, after-tax profits to De Beers were $40 million.[4]

Uranium. Uranium production in Namibia/South-West Africa is based on the world's largest open-pit uranium mine, 65 km northeast of Swakopmund. Production did not begin until 1977.

Reserves of the mine were determined to be 375 million tons of 0.4% U_3O_8.[5] The principal ore is uraninite. Steps in the preparation of yellow cake (U_3O_8) first include the grinding of uraninite to a pulp. The pulp is then treated by a sulfuric acid leach followed by ion exchange, solvent extraction, and an ammonia precipitation of the uranium oxide (U_3O_8).[4] The Rossing Uranium Ltd. (RUL) operation is owned by Rio Tinto Zinc Corporation Ltd. (RTZ) of the United Kingdom (46.5%); General Mining and Finance Corporation Ltd.; IDC of the Republic of South Africa; Total Compagnie Minière et Nucléaire of France; and Rio Algom Mines Ltd. of Canada.[6] RTZ has controlling interest in the mine. Major purchasers of the uranium oxide were the United Kingdom Atomic Energy Commission, Kansai Electric Power Company of Japan, Total (France), and Urangesellschaft (West Germany).[5]

TABLE II-4-8   Diamond Production in Namibia/South-West Africa

Year	Quantity of overburden (million tons)	Quantity of gravel treated (million tons)	Ore grade (carats/ton)	Diamonds (thousand carats) Gem	Diamonds (thousand carats) Industrial
1977	59	14.90	0.1343	1901	100
1978	55	15.14	—	1803	95
1979	54.9	16.03	0.1031	1570	83
1980	56.6	16.82	0.0927	1482	78
1981	37.7	12.54	0.0995	1186	62

Source: U.S. Bureau of Mines, Minerals Yearbook, 1977–1981.

While annual design capacity of the mine is 5000 tons of U_3O_8,[4] The most U_3O_8 produced in one year was 4767 tons in 1980.[6] A series of design and technical difficulties have prevented the plant from reaching full capacity.[5] Annual production of U_3O_8 was 700 tons in 1976, 2760 tons in 1977, 3175 tons in 1978, 4518 tons in 1979, 4767 tons in 1980, and 4681 tons in 1981.[4]

A number of other uranium deposits were undergoing exploration activity and economic evaluation. However, the country's political climate and depressed uranium prices have so far stifled development.[6]

General Mining was considering a $350 million operation near Tinkas, on the Langer Heinrich uranium deposit. GFSA was evaluating the Trekkopje uranium deposit 20 km northeast of Rossing.[5] There were also two deposits at Tubas and Assinanis being considered by the French Elf-Aquitaine oil company and the Anglo-American Corporation.[6]

Nonferrous Metals. Tsumeb is the country's largest producer of nonferrous metals. The company produces copper, lead, zinc, and silver. It is also the country's only producer of arsenic and cadmium.[6] In 1981, Tsumeb produced the following mineral commodities from four mines and three mills: 37,800 tons of contained lead, 30,300 tons of copper, 6950 tons of zinc, 1.89 million troy ounces of silver, 3800 tons of sulfur, and 1370 tons of arsenic. The cadmium plant was shut down in 1981, and the arsenic plant operated only part time.

The company's copper and lead smelters produced 39,719 tons of blister copper and 41,729 tons of refined lead. However, only 66% of the copper and 75% of the lead originated from Tsumeb's own operations.

Tsumeb was owned and managed by Newmont Mining Corporation of the United States (29.6%), AMAX, Inc. (30.2%), Selection Trust Ltd. of Britain (14.2%), O'okiep Copper Company Ltd. (9.5%), South West Africa Company Ltd. (SWACO) (2.4%), and the other 14.1% owned by other companies.[4] The following are mines operated by Tsumeb: Tsumeb mine, Matchless mine, Kombat mine, Asis West mine, and Otjihase mine.

Copper. Copper production in Namibia/South-West Africa for 1981 was 46,185 metric tons, up from 1980's 42,300 metric tons. However, Table II-4-9 reveals that production for both copper and blister copper peaked in 1977. Copper production by individual mines is given in Table II-4-10. The Tsumeb mine, north of Windhoek, remains the country's largest copper mine. In December 1980, Tsumeb signed a joint-venture agreement with

TABLE II-4-9 Copper Production in Namibia, 1970–1981: Mine Output, Metal Content of Concentrate (Metric Tons)

Year	Copper	Blister copper
1970	31,393	28,593
1971	32,039	26,922
1972	26,186	26,119
1973	34,168	36,049
1974	32,478	46,612
1975	39,034	36,410
1976	43,500	36,570
1977	51,200	53,371
1978	39,000	45,919
1979	44,800	42,707
1980	42,300	40,004
1981	46,185	39,719

Source: U.S. Bureau of Mines, *Minerals Yearbook*, 1970–1981.

Otjihase Mining Company Ltd., acquiring 70% interest in Otjihase for $41 million.[4]

The second-largest producer of copper in Namibia is the Oamites Mining Company (Pty.) Ltd., owned by Falconbridge Nickel Mines Ltd. of Canada (74.9%) and the Industrial Development Corporation (IDC) (25.1%). In 1981, the mine produced 4827 tons of recoverable copper, down 17% from 1980.

Lead and Zinc. The Tsumeb mine, besides being the dominant copper producer, is also Namibia's largest lead producer. The Tsumeb Mine was also one of only two Namibian zinc-producing mines. Lead is also produced at Tsumeb's Kombat and Asis West mines.

The country's largest zinc producer is the Rosh Pinah mine (open-pit) in the Namib desert, 27 km north of the Orange River. Rosh Pinah is operated by and for Imcor Zinc (Pty.) Ltd., a subsidiary of the South African Iron and Steel Industrial Corporation Ltd. (Iscor). In the 1980–1981 business year, all of the mine's 44,000 tons of 40% zinc concentrate, and 17,900 tons of 45% of lead concentrates were trucked 180 km south to South Africa's railhead at Aus. From there, the concentrates were taken to Iscor's steelworks at Vanderbijlpark in the Transvaal Province of South Africa.[4] The producers of zinc and lead concentrates is given in Table II-4-11. The production of lead and zinc have gradually declined since 1970, as indicated in Table II-4-12.

Other Minerals. Arsenic, cadmium, and silver are all by-products of Namibia's copper-lead-zinc production. Namibia is the sixth- or seventh-largest producer of arsenic in the world. However, the pro-

TABLE II-4-10   Copper Producers in Namibia, 1981[a]

Mine	Quantity of ore (tons)	Grade of ore	Concentrates	Grade of concentrates	Reserves	Grade of reserves
Tsumeb	493,708	3.60% Cu 7.32% Pb 2.08% Zn	23,429	35.49% Cu 33.3 oz Ag/ton	3,528,000	4.27% Cu 5.99% Pb 1.77% Zn
Matchless	111,247	2.37% Cu 11.47% S	12,313	20.46% Cu 30.55% S 2.5 oz Ag/ton	371,000	2.22% Cu
Kombat	285,411	2.80% Cu 1.90% Pb	20,317	29.76% Cu NA Pb 9.3 oz Ag/ton	2,442,000	1.83% Cu 2.07% Pb
Asis West[b]	12,522[b]	7.54% Cu 2.31% Zn	2,551	33.12% Cu NA Pb 8.0 oz Ag/ton	1,580,000	5.54% Cu 2.77% Pb
Otjihase	206,900	1.54% Cu 16.60% S	14,242	20.2% Cu 2.2 oz Ag/ton	10,117,000	2.23% Cu
Oamites	530,000	1.10% Cu	NA	NA	2,027,000	1.13% Cu 0.5 oz Ag/ton

[a]NA, not available.
[b]Asis West Mine production is milled and concentrated at Kombat facilities.
Source: U.S. Bureau of Mines, Minerals Yearbook, 1978-1981.

TABLE II-4-11   Lead and Zinc Producers in Namibia, 1981[a]

Mine	Quantity of ore (tons)	Grade of ore	Concentrates	Grade of concentrates	Reserves	Grade of reserves
Tsumeb	493,708	7.32% Pb 2.08% Zn	97,970 Pb	31.1% Pb 8.1% Cu 6.46% Zn 7.9 oz Ag/ton	3,528,000	4.27% Cu 5.99% Pb 1.77% Zn
			1,196 Zn	51.94% Zn 3.02% Cu 9.67% Pb 10.4 oz Ag/ton		
Kombat	285,411	2.80% Cu 1.90% Pb	6,825 Pb	30.5% Pb 14.0% Cu 4.0 oz Ag/ton	2,442,000	1.83% Cu 2.07% Pb
Asis West	12,522	7.54% Cu 2.31% Zn	177 Pb	34.8% Pb 11.5% Cu	1,580,000	5.54% Cu 2.77% Pb
Rosh Pinah			47,200 Zn 17,900 Pb	40.0% Zn 45.0% Pb	NA	NA

[a]NA, not available.
Source: U.S. Bureau of Mines, Minerals Yearbook, 1978-1981.

duction of arsenic has declined from a high of 8147 metric tons in 1973 to only 1370 metric tons in 1981.[4] Arsenic production for 1970–1981 is given in Table II-4-13. In 1981, silver production of 3258 metric tons nearly doubled the 1977 level of 1758 metric tons.[4]

Tin. Tin production came mainly from Iscor's Uis mine near Brandberg, northeast of Swakopmund. The tin concentrates, like Rosh Pinah's

lead and zinc concentrates, were sent to Iscor's Vanderbijlpark Steelworks in South Africa.[4] Recent drilling has expanded mine reserves significantly.[4]

Tantalum. In 1981, Southern Mining and Development Company Ltd., a subsidiary of Utah International Inc. of the United States, located a deposit with a potential of 18,000 kg of tantalite per year.

TABLE II-4-12 Lead and Zinc Production in Namibia, 1970-1981 (Metric Tons)

Year	Lead mine output, metal content of concentrate	Refined lead	Zinc mine output, metal content
1970	73,119	70,129	46,685
1971	71,498	58,820	43,696
1972	59,990	63,961	34,742
1973	61,694	63,592	37,919
1974	56,761	64,342	47,186
1975	48,300	44,300	43,251
1976	46,400	35,598	41,308
1977	38,000	42,723	38,300
1978	34,800	39,512	36,600
1979	44,200	41,695	23,300
1980	50,200	42,654	31,908
1981	59,121	48,479	39,600

Source: U.S. Bureau of Mines, Minerals Yearbook, 1970-1981.

TABLE II-4-13 Arsenic Production in Namibia, 1970-1981 (Metric Tons)

Year	Production
1970	4478
1971	3701
1972	2370
1973	8147
1974	6640
1975	6663
1976	5122
1977	2615
1978	2401
1979	2221
1980	1288
1981	1370

Source: U.S. Bureau of Mines, Minerals Yearbook, 1970-1981.

Constraints to Mineral Development

Namibia/South-West Africa's future mineral industry is, at best, uncertain. The political uncertainty created by the government of South Africa and the U.N. over the legal government of Namibia/South-West Africa is the most critical factor affecting Namibia's investment climate. If foreign investors fear that a transition in government to a SWAPO-dominated black government will bring about nationalization of mining companies and state ownership of natural resources, foreign investment is likely to slow down or disappear altogether.

However, South Africa does not seem anxious to give up the country. South Africa's position is that the U.N. has no authority to rescind the League of Nation's mandate.[2] South Africa continues to improve Namibia's infrastructure. South Africa also has several large military bases in the country.[2]

South Africa's position is also based on arguments of national security. The government of South Africa believes that a government of Namibia under SWAPO control would provide yet another base in the region for guerrilla insurgency against the present white-minority government of South Africa.

The lack of water and a large sand overburden covering much of the country may also serve as deterrents to mineral development in Namibia. Large investments in infrastructure will require either unusually rich or large ore reserves and/or cheap labor and/or cheap and available power.

REFERENCES

1. U.S. Central Intelligence Agency, National Foreign Assessment Center. The World Factbook—1981. Washington, D.C.: U.S. Government Printing Office, 1982.
2. Landis, Elizabeth S., and Michael I. Davis. "Namibia: Impending Independence." In Southern Africa: The Continuing Crisis, 2nd ed. Edited by Gwendolyn M. Carter and Patrick O'Meara. Bloomington, Ind.: Indiana University Press, 1982.
3. U.S. Department of State, Bureau of Public Affairs. Namibia/South-West Africa. Background Notes, September 1980.
4. U.S. Department of the Interior, Bureau of Mines. "The Mineral Industry of Namibia." Minerals Yearbook 1981. Washington, D.C.: U.S. Government Printing Office, 1983.
5. U.S. Department of the Interior, Bureau of Mines. "The Mineral Industry of the Territory of South-West Africa." Minerals Yearbook 1978/79. Washington, D.C.: U.S. Government Printing Office, 1981.
6. U.S. Department of the Interior, Bureau of Mines. "The Mineral Industry of Namibia (Territory of South-West Africa)." Minerals Yearbook 1980. Washington, D.C.: U.S. Government Printing Office, 1982.
7. U.S. Department of the Interior, Bureau of Mines. Mineral Commodity Summaries 1983. Washington, D.C.: U.S. Government Printing Office, 1983.

ZAIRE

The Republic of Zaire is the third largest country in Africa, with a total land area of 2,343,950 km^2 and access to the Atlantic Ocean along a coastline of 37 km, which is located between Angola and its enclave of Cabinda. Located in south-central Africa, Zaire shares a common border with the following countries: Congo, Central African Republic, Sudan, Uganda, Rwanda, Burundi, Tanzania, Zambia, and Angola. The country is covered by a large tropical rain forest (45% of land area) and numerous rivers. Potential agricultural land (22% of land area) is also enormous, but only an estimated 1% is used.[1]

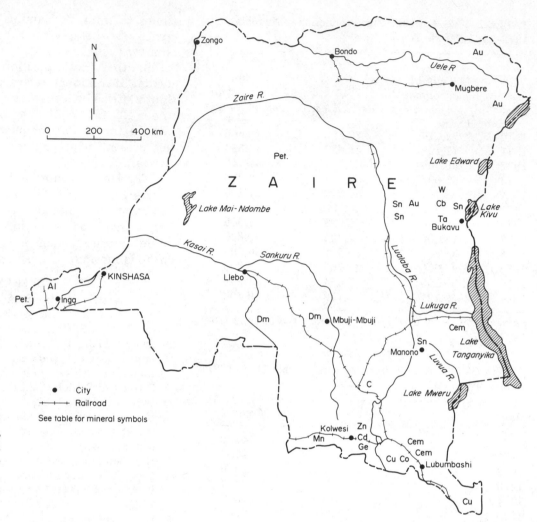

FIGURE II-4-4 Zaire: Mines, processing facilities, and infrastructure. (From U.S. Department of the Interior, Bureau of Mines. *Mineral Industries of Africa*, 1976.)

The population is the fourth largest in Africa, but is not homogeneous, being comprised of over 200 African ethnic groups. However, three Bantu tribes (Mongo, Luba, Kongo) and one Hamitic tribe (Mangbetu-Azande) account for roughly 45% of the population. As a result of this diversification, French, English, Lingala, Swahili, Kikongo, and Chiluba are all considered official languages.[1]

Zaire is a former colony of Belgium, from which it gained independence on 30 June 1960. Prime Minister Patrice Lumumba, leader of the popular Mouvement National Congolais (MNC), was assassinated in January 1961.[2]

Lumumba's assassination, coupled with the seccession of the Katanga Province (Zaire's copper and cobalt region), prompted the introduction of U.N. troops until the Katanga rebellion ended in 1963.[2]

The current President of Zaire, Mobutu Sese Seko, was chief of staff of the national army under Lumumba and seized control of the country in 1965. In May 1965, the Mouvement Populaire de la Révolution (MPR) was created by Mobutu, and now is the country's only legal political party.[2] The capital, Kinshasa, is located on the Congo River in western Zaire.

The Zairian economy is dominated by the mineral industry. While the potential for agriculture and timber is considered good, government neglect of these sectors has considerably impeded their development. Key economic indicators for Zaire are given in Table II-4-14.

Zaire is the world's largest producer of cobalt and industrial diamonds. Until recently, Zaire was also the world's largest producer of germanium, but the government no longer releases production figures for germanium. Zaire is the world's sixth-largest producer of copper. Other important minerals include gold, manganese, and tin.

Government Role in Mineral Industry

On 31 December 1966 the government of Zaire announced its intentions to expropriate Société Générale de Belgique, which then controlled nearly 70% of the Congolese economy.[2] Through control of the Union Minière du Haut-

TABLE II-4-14   Key Economic Indicators for Zaire, 1974-1980[a]

	1974	1975	1976	1977	1978	1979	1980
GNP (constant 1970 prices)[b]	2340	2210	2115	2074	1920	1901	1935
Real annual growth (%)	5.4	0.6	-0.3	-1.9	-1.0	NA	NA
Money supply[b]	779	908	815	1066	1307	NA	NA
Total exports[b]	1283	799	1063	1239	1399	1860	2089
Total imports	911	974	1432	1258	1410	1450	1469
Trade balance[b]	372	-175	-369	-19	-11	+410	+620
Budget balance[b]	-413	-206	-310	-313	NA	NA	NA
Balance of payments[b]	-237	-508	-384	-508	-524	-324	-288
Total debt[b]	NA	2561	3088	3306	35377	3952	4370
Total debt payment due[b]	NA	196	176	317	476	560	527
Debt-service ratio (%)	NA	24.5	16.6	25.6	34.0	30.1	24.5
Consumer price index[c]	200	273	446	738	1439	2374	3490
All minerals index[c]	120	115	101	106	101.8	100	114
Agricultural index[c]	112	106	118	107	100	100	100
Manufacturing index[c]	130	119	109	101	102.3	100.3	100
Copper production[d]	498	496	444	490	420	370	4257

[a]NA, not available.
[b]Value in millions of U.S. dollars.
[c]1970 = 100.
[d]Thousand metric tons.

Source: Ref. 5.

Katanga, the Société Générale de Belgique also held exclusive mining rights through 1990 over most of today's Shaba Province.[2] This nationalized Belgium company has eventually come to be known as La Générale des Carrières et des Mines du Zaire (GECAMINES). The company is the country's major mineral producer and revenue earner.

The government has also nationalized a number of other industries, especially under its "Zairianization" efforts of the early 1970s.[2] Under this program some industries not only were absorbed by government, but management responsibilities were given to Zairians, whether or not they were qualified.

In late 1976, President Mobutu announced that his "Zairianization" program had been a failure, and that Zairianized businesses would be returned to their former owners with the stipulation that 40% of each business would be "available" for Zairian participation. As of the end of 1977 most Zairianized businesses had been returned to their former owners.[4] The government of Zaire, however, retained the exclusive right to bargain and to become partners with foreign firms in the areas of energy, mining, forestry, and transportation.[4]

Since 1969, the government has allowed the formation of two nongovernment mining companies in Zaire. The Société de Développement Industriel et Minier du Zaire (SODIMIZA) concession was granted in 1970, with 85% holdings by Japanese firms. Also in 1970, Société Minière de Tenke-

Fungurume (SMFT) was granted a concession with only a 30% government share.[2] Both companies are producers of copper.

The government also controls diamond production through the Société Minière de Bakwanga (MIBA), managed and 20% owned by the Société Générale de Belgique and the Oppenheimer Group.[3] In May 1981, President Mobutu authorized the Société Zairoise de Commercialisation des Minerais (SOZACOM) to commence marketing diamonds, and closed the Meltax and Britmond/Central Selling Organization (CSO). The firm had received commissions of 20% including cost during 1980.[3]

Other state-owned mining enterprises include: Offices des Mines d'Or de Kilo-Mots (OKIMO), a gold producer in the northeast Haute-Zaire region; Société Minière de Kisenge (SMK), a manganese producer west of Kolwezi; SOMINKI (28% government-owned), the leading producer of tin and tungsten concentrates; and the Société Zaire-tain (50% state owned), a producer of cassiterite oxides, tin metal, and titaniferous slag.[3] The government's share of mineral production is sold by the Société Zairoise de Commercialisation des Minerais (SOZACOM), a state-owned agency.

Mineral Rights. In Zaire, under the 1967 Mining Code (Law No. 67/614), ownership of land does not include mineral rights, except for "produits de carrier" (quarry products).[4] Otherwise, minerals are considered "substances concessibles" (concessionable substances) and require permission from the government for exploration and exploitation. In certain circumstances, the government may class any mineral under a special category referred to as "substances réservés" (reserved substances), and then proceed to establish special rules to govern the exploitation of these minerals.[4]

Concern with state participation in mineral ventures prompted the passage of Legislation Minière National Zairoise and Ordanice-Law No. 67/416 on 11 May 1967. Under these provisions, all mining ventures must have state participation. The president of Zaire has the authority to grant exceptions to this rule, as well as being the sole authority to create mining companies through the State Commissioner for Mines and Energy.[4] Apparently, the percentage of state participation may vary from venture to venture.

Investment and Repatriation of Capital. General investment law in Zaire is described in "Marketing in Zaire," an *Overseas Business Report* (OBR 79-44) published in December 1979.[4] For investment of over $250,000 the following laws generally apply:

1. Profits of newly established enterprises are exempt from the company profits tax for five years.
2. Enterprises that extend or modernize their productive capacity are exempt, for a period of up to five years, from company income tax upon that of their taxable profit exceeding the average taxable profit declared in the past three years.
3. New enterprises are exempt from the special tax on expatriates' salaries until the commencement of production.
4. Dividends distributed to subscribers of new shares, issued to finance an extension of a firm's capital stock, are exempt from the tax on dividends and interest for up to five years.
5. New enterprises are exempt from property tax for a period of five years after the purchase of property, as are existing firms upon real estate acquired for purposes of expansion or modernization.
6. Approved enterprises are exempt from import duties and the import turnover tax on machines, equipment, and related materials, as long as products of equivalent quality and price cannot be obtained domestically.
7. Undisturbed profits that are applied to a reserve fund with the purpose of being reinvested within Zaire are relieved by half the company income tax due upon them.

All proposed investments are reviewed by an investment commission comprised of representatives from various departments and the Bank of Zaire. The chairman of the commission is appointed by the president of Zaire.[4]

The repatriation of the capital is governed by the Bank of Zaire's Circular 121, issued in December 1968.[4] Under these provisions, foreign companies are permitted to transfer to foreign stockholders their share of profits earned by Zairian companies, as well as the transfer of funds to cover certain foreign expenses.[4] However, transfers require the special permission of the Central Bank. The Central Bank, in turn, requires proof of the amount of foreign ownership and proof that all taxes have been paid.[4] In practice, however, foreign firms have generally not been able to transfer money. The Bank of Zaire has turned applications down, citing a lack of foreign exchange within the country.[4]

The U.S. Department of Commerce's International Trade Administration issued a report on Zaire in December 1981 advising that[5]

> Zaire remains less than ideal at this time for foreign direct investment. Investment is hindered by companies' inability to repatriate earnings, interference from an unskilled but powerful public sector, and a general depression in the manufacturing sector, with most factories running about one-third of capacity.

Infrastructure

Zaire's infrastructure is in a general state of decline. Important factors contributing to this decline have been lack of trained maintenance personnel, spare parts, interruptions in fuel supplies, and general government neglect.[6] Recently, the government (with IMF guidelines and requirements) has given top priority to rehabilitation of the country's basic infrastructure.[8]

Transportation. Zaire's transportation system is dominated by the country's two largest rivers—the Zaire (Congo) River and the Kasai River—and their tributaries. Both rivers account for some 10,000 miles (16,100 km) of navigable waterways.[4] Due to the size and distribution of the river system, both railroads and roads were designed to complement the river system. As a result, the country has less than 3000 miles (4800 km) of railroads.[4] The road network of 90,000 miles (144,840 km) was designed to provide local access roads or feeder roads to the rail/river system.[4]

Railroads. The country's only continuous rail system is found in Shaba (Katanga) Province, and was built to provide transportation to the mining industry within the country. Within the southern mining Province of Shaba, the railroad crosses into Zambia at several points—one crosses near Mufulira and the other near Ndola. The routes allow Zaire access to the port at Dar es Salaam or southward to South African ports. The Zambia-Zimbabwe border is now open, making shipments to Mozambique's ports also possible.

From the copper-cobalt region in southern Shaba Province the rail line proceeds northward to Tenke, where the railway continues northward or westward. The western route proceeds from Tenke to Kolwezi to Dilolo and then crosses into northeastern Angola on the Benguela Railway to the Angolan port of Lobito. However, the Benguela Railway has remained all but closed since the Angolan civil war in 1975, due to guerrilla insurgency in Angola.

The northern route from Tenke proceeds northwest to Llebo, a railhead on the Kesai River. Several hundred miles from Tenke, along the northern route, the railway divides at Kamina and a line proceeds to the northeast. The line divides again at Kabalo, with one line proceeding eastward to Lake Tanganyika (ending at Kalemie), and the other line proceeding northward along the Zaire River. The Mayumbe railroad in northern Zaire connects Bumba on the Zaire River, with Bondo on the Vele River and Isiro in northeastern Zaire. An important line connects the capital of Kinshasa with the port at Matadi in western Zaire. Electrification of the Matadi Kinshasa railroad was slated for 1981.[6]

Roads. The road system in Zaire is large but is built to complement the rail/river system. The majority of unimproved roads are susceptible to interruption during Zaire's rainy season.

Rivers. The Zaire (Congo) River is Zaire's main river. The river forms a border with Congo. The Zaire River originates from drainage in eastern Zaire, travels westward, and then southwestward into the Atlantic Ocean, passing the capital of Kinshasa and the port at Matadi. Matadi is over 100 km inland from the mouth of the Congo River and Atlantic Ocean.

The Kasai River is formed by tributaries and drainage from northeastern Angola and southcentral Zaire. The river, in conjunction with the railroad from Shaba Province to Llebo, forms the country's most important export–import route.

Export-Import Routes. Zaire can transport exports and imports along several routes. The Benguela Railway through Angola and southern routes

through Zambia have already been mentioned. Zaire does conduct some trade through the city of Kalamie on Lake Tanganyika. Goods are taken by barge across Lake Tanganyika to the railhead at Kagoma, Tanzania. From Kagoma, the railway extends to the port at Dar es Salaam. In anticipation of increased traffic to Dar es Salaam, Zairian and Tanzanian officials have discussed improvements to the TAZARA railway and warehouse facilities in Dar es Salaam.[7] Current inadequacies of port facilities at Dar es Salaam and the TAZARA railway have been discussed in the section on Zambia.

Zaire's main export/import route is along the railway from Shaba Province to the railhead at Llebo. From Llebo, material is loaded onto either barges or trucks and taken to Kinshasa. From Kinshasa, material on barges must be loaded onto the railroad or truck to the port at Matadi, where it must once again be transferred—this time to ocean-going vessels. Trucks from Llebo to Kinshasa can proceed to Matadi. A direct rail route from Llebo to Kinshasa is under study.[8]

In 1979, 53% of the export traffic was carried on the river barge and rail Voie Nationale and shipped from Matadi. Another 39% was shipped through South African ports, and 8% went to Dar es Salaam by either the lake-barge route or the TAZARA railway.[6]

In 1979, virtually all cobalt was shipped by air, which was economical due to cobalt's high price. Very low prices would tend to make air transportation less economical. South African ports transported 51% of the copper, 46% went to Matadi, and only 3% went to Dar es Salaam.[6] More than 90% of the tin shipments went to Matadi, and less than 10% went to Dar es Salaam.[6]

During the first eight months of 1981, Zaire exported 139,375 tons of copper, 27,753 tons of zinc, and 65,881 tons of copper concentrate through South African ports. Returning rail cars carried 40,394 tons of coke (mostly from Zimbabwe's Wankie colliery), 7178 tons of sulfur for the manufacturing of sulfuric acid used in extractive metallurgical processes, 32,705 tons of coal, and 82,585 tons of maize and wheat.[7]

Water. Zaire's water resources are enormous. The country is covered with major rivers and lakes. Eastern Zaire's tin and gold industry is near Lake Mweru in the southeast and Lake Tanganyika, Lake Kivu, Lake Edward, and Lake Albert in the northeast. The Lualaba (Congo) (Zaire) River and its tributaries also run through this area, and the river originates near GECAMINES' Kolwezi district. Another large river and lake in the copper-cobalt Shaba Province is the Lufira River and the Lac de Retenue.

The Kesai and Lulua Rivers are near Zaire's diamond area. The Kesai River originates near Angola's alluvial diamond deposits in northeastern Angola.

Power. Zaire's potential for hydroelectric power is one of the largest in the world. However, Zaire has yet to fully develop this potential. The largest hydroelectric facility in Zaire is the Inga II plant on the Zaire River, less than 50 km northeast of Matadi. The installed capacity of 1272 MW is generated by eight 159-MW turbines.[7] Power is to be transferred along a 1700-km HVDC (high-voltage, direct-current) power line to Shaba Province's copper-cobalt mining districts. In August 1981 the last pylon was completed in Kolwezi. However, the transmission of electricity had to wait the completion of converter terminals at the Shaban mining centers in 1983.[7] The first pylon was erected in the Lower Zaire Province in 1974.[7]

Previous to the Inga II power station, electricity came from four hydroelectric stations built by GECAMINES' precedessor, Union Minière du Haut-Katanga. In 1930, Franqui was installed on the Lufira River, northeast of Likasi, with a capacity of 47 MW. This was followed by the 77-MW Bia generator in 1945. In 1953, the Delcommune dam and a 120-MW plant north of Kolwezi on the Lualaba River were completed. The 276-MW Le Marinel station was completed in 1956, just 35 km downstream from the 120-MW plant.[7]

Zaire has the natural resources to become a major exporter of hydroelectric power in central and southern Africa. Major hydroelectric projects could help stimulate the diversification of the Zairian economy and provide cheap electricity for a domestic manufacturing industry.

Coal production in Zaire was an estimated 280,000 metric tons of low-grade coal in 1981.[7] Zaire's two producers of coal are the Société des Charbonnages de la Luena north of Kolwezi, and the Makala coal mine near Kalemie on Lake Tanganyika. Makala coal was used in nearby cement plants. Luena coal was used in GECAMINES' nearby cement and metallurgical plants. However, the bulk of metallurgical coal originates from the Wankie Colliery, Zimbabwe, and South African sources, as well as from several European countries.[7]

Zaire's petroleum production continues to increase. In 1981, Zaire produced 7.4 million barrels compared to only 6.5 million barrels in 1980.[7] The Zairian-Italian Refining Company has completed

its new crude petroleum desalinization plant. The plant will enable the oil from offshore wells to be blended with lighter imported crudes at the refinery at Banana.[7] Saudi Arabia supplied $23,549,000 of petroleum crude to Zaire in 1979.[7]

Most of Zaire's petroleum production comes from Gulf Oil's offshore wells. The wells are owned 50% by U.S. Gulf Oil, 32% by a subsidiary of Teikoku Oil Company of Japan, and 18% by Soliza, a Zairian firm. Moanada Oil Company, a subsidiary of Cometra Oil Company, is producing an increasing percentage of oil from the Mibale structure. Esso-Zaire, an affiliate of the U.S. Exxon Oil Corporation, has a well at Mbandaka in the Equator region and one near Lokolama in the Bandunda region.[7]

Production of bituminous coal and crude petroleum are given in Table II-4-15, and refinery production is given in Table II-4-16. Statistics for 1980 and 1981 production are unavailable.

Mineral Industry

The mineral industry continues to be Zaire's most important foreign exchange earner. Recently, the percentage of foreign exchange attributed to the exports of minerals has increased. In 1979, mineral exports were estimated at $1646.3 million and were 60% of foreign exchange earnings.[4] In 1980, the value of total exports was $2209.5 million, and mineral exports amounted to $1897.3 million, 85% of total exports.[6]

Although the total value of mineral exports decreased to only $1500 million in 1981, the percen-

TABLE II-4-15   Coal and Petroleum Production in Zaire, 1970-1981[a]

Year	Bituminous coal (thousand tons)	Crude petroleum[b] (thousand 42-gallon barrels)
1970	102	
1971	114	
1972	128	
1973	115	
1974	95	
1975	89	25
1976	100	9075
1977	239	8255
1978	215	6604
1979	225	7535
1980p	287	7500
1981e	280	7500

[a]p, preliminary; e, estimated.
[b]No petroleum production before 1975.
Source: U.S. Bureau of Mines, *Minerals Yearbook*, 1970-1981.

TABLE II-4-16   Refinery Production in Zaire, 1976-1981 (Thousand 42-Gallon Barrels)[a]

Products	1976	1977	1978	1979	1980	1981
Gasoline	579	275	279	483	NA	NA
Kerosene and jet fuel	318	142	231	319	NA	NA
Distillate fuel oil	698	320	289	682	NA	NA
Residual fuel oil	878	351	529	1252	NA	NA
Liquefied petroleum gas	12	15	—	21	NA	NA
Refinery fuel and losses	245	110	125	173	NA	NA
	2730	1213	1453	2930	NA	NA

[a]NA, not available.

Source: U.S. Bureau of Mines, Minerals Yearbook, 1976-1981.

tage of the value of mineral exports to the value of total exports increased to 86%. The value of total exports for 1981 fell to $1750 million.

The growing importance of mineral exports as a foreign exchange earner reveals several important trends in the Zairian economy and mineral industry. First, the growth in value of mineral exports from 1979 to 1980 reflects higher mineral prices, especially for cobalt and gold. The lower value for mineral exports in 1981 is related to the subsequent drop in the prices of metals, especially copper, cobalt, and gold. Second, the rise in importance of mineral exports reflects indirectly the poor performance of the agricultural and manufacturing sectors. Table II-4-14 indicates that the agriculture and manufacturing indices remain at 100.0 for 1980; the mining sector has increased to 114.0 for 1980.

The drop in the value of mineral exports from 1980's $1897.3 million to 1981's $1500 million (a drop of nearly $400 million) can be attributed mainly to the drop in the value of cobalt and industrial diamonds. Table II-4-17 indicates the value of mineral exports for 1973-1981 (except for 1978). Table II-4-18 indicates the percentage of the value of mineral exports compared to the value of all exports for 1973-1981 (except for 1978).

Zaire's mineral industry is also an important contributor to the world mineral industry. Zaire's copper industry is the sixth largest in the world and is Zaire's most valuable industry. Cobalt, a by-product of copper, is Zaire's second-most-valuable mineral commodity. However, strategically speaking, cobalt is considered Zaire's most important mineral for two reasons: first, because of Zaire's dominance as the number one producer in the world; and second, due to the importance of cobalt

TABLE II-4-17   Value of Mineral Exports, 1973-1981[a,b] (Million U.S. Dollars)

Product	1973	1974	1975	1976	1977	1979[c]	1980	1981
Copper	617.9	859.1	433.0	380.4	427.2	630.0	898.1	825
Cobalt	77.3	71.4	92.9	120.0	119.2	655.0	529.0	200
Zinc	0.2	52.0	30.2	49.0	38.4	50.0	13.0	65
Germanium	0.7	2.3	2.0	NA	NA	NA	NA	NA
Diamonds	57.4	62.3	50.2	57.0	64.8	202.1	256.3	112
Silver	5.2	7.4	6.2	7.2	6.8	17.5	55.5	<20
Gold	9.2	15.9	12.5	9.4	6.0	15.4	32.6	33
Cassiterite	13.4	21.7	18.8	20.0	24.0	34.9	42.8	NA
Manganese	2.8	5.6	2.0	25.2	23.8	—	—	—
Wolfram-tungsten	0.8	0.7	1.3	2.0	3.4	NA	NA	NA
Lead	3.6	3.4	2.4	3.6	3.9	NA	NA	NA
Other	54.5	0.8	4.0	17.6	19.1	NA	NA	NA
Petroleum	NA	NA	NA	NA	NA	26.0	60.0	120
	843.3	1101.2	655.6	688.8	732.0	1646.3	1897.3	1500

[a]Excluding 1978; 1973-1979 from OBR 77-22; 1980-1981 from Refs. 3 and 6.
[b]NA, not available.
[c]Estimated.

Sources: U.S. Department of Commerce, Industry and Trade Administration, OBR 77-22; Refs. 3 and 6.

TABLE II-4-18   Value of Mineral Exports versus Total Exports, 1973–1981[a] (Million U.S. Dollars)

	1973	1974	1975	1976	1977	1980	1981
Total exports	1004.0	1295.0	826.0	898.0	958.0	2209.5	1750.3
Mineral exports	843.3	1101.2	655.6	688.8	732.0	1897.3	1500.9
Percentage of mineral exports of total exports	84.0	85.0	79.4	76.7	76.4	85.8	85.7

[a]U.S. Bureau of Mines, *Minerals Yearbook*, 1973–1981.
Sources: U.S. Department of Commerce, Industry and Trade Administration, OBR 77-22; Refs. 3 and 6.

in industrial and military applications. Zaire is also the world's largest producer of industrial diamonds. Other important minerals produced by Zaire include tin and its by-products tungsten and columbium-tantalum; gold, silver, zinc, germanium, and manganese.

Copper/Cobalt. As mentioned previously, Zaire is the world's largest producer of cobalt and the world's sixth-largest producer of copper. In 1980 and 1981, Zaire produced 50% and 51.3%, respectively, of the world's cobalt. For these same years, Zaire produced only 6.1% and 6.2%, respectively, of the world's copper. Since cobalt is a by-product of copper, Zaire's production of cobalt is dependent on copper production, which, in turn, is dependent on world demand for copper. However, several major copper mines in Zaire are only marginally profitable, and their future is dependent not only on the price of copper, but also the price of cobalt.

The price of cobalt has undergone tremendous fluctuations in recent years due to anxieties over the security of supplies from Zaire. This anxiety was created by the second invasion of the Shaba Province in less than two years. On 13 May 1978 armed invaders struck the Kolwezi copper operations. By about 14 June, operations had returned to normal.[8]

There were several significant effects from this invasion. First, it appears that European workers bore the brunt of both casualties and looting. As a result, many of the European workers left Zaire, and there was a dearth of skilled personnel to run mining operations. This has also affected the recruitment of European workers since the invasion. Financial losses from this invasion to GECAMINES and its personnel were estimated at $49 million. The 16-day interruption in production was estimated at $28 million, for a total of $77 million.[2] However, this loss in revenue was offset by a subsequent increase in the price of cobalt following the Shaba II invasion in 1978–1979, as indicated in Table II-4-19. It should also be noted that cobalt production increased in 1978, by some 3000 metric

TABLE II-4-19   Price Rise of Cobalt, 1978–1979

Dates	Producer price	Spot price
December 21, 1977	$ 6.40	
February 24, 1978	6.85	
May 24, 1978	8.50	
July 20, 1978	12.50	
September 14, 1978	18.00	
October 27, 1978	20.00	$39.00
November 10, 1978	20.00	49.00
February 1, 1979	25.00	39.00
May 23, 1979	25.00	42.00
July 27, 1979	25.00	30.00
August 24, 1979	25.00	29.00

Source: Ref. 8.

tons from the previous year. It therefore appears that the "Shaba invasions" had mixed results. By 1982, Zaire had a surplus of cobalt and the price had fallen to below $6 on the free-market exchange.[9] Zaire's surplus of cobalt in 1982 was greatly reduced when the General Service Administration of the United States purchased 5.2 million pounds of cobalt for the National Defense Stockpile at $15 a pound, for a total of $78 million.[3]

Revenue from copper and cobalt exports constitutes the major portion of Zaire's foreign exchange. Thus prices for these commodities can greatly affect the Zairian economy. In 1980, revenue from exports of copper and cobalt were $898.1 million and $529.0 million, respectively. This amounted to 75% of the total value of all mineral exports and 65% of the total value of all exports.[6]

In 1981, falling prices for both copper and cobalt generated $825 million for copper and only $200 million for cobalt. This amounted to only 68% of the value of all mineral exports and only 59% of the value of all exports.[3] The loss in revenue from both minerals was a little over $400 million from 1980 to 1981. This $400 million loss in revenue represents 23% of the total exports of $1750 million in 1981.[3]

Geology. The copper deposits in Zaire are a continuation of the Zambian copper belt. The main differences in the Zairian and Zambian copper deposits are (1) a difference in age; (2) the ore-bearing zone in Zaire underwent more complex folding; and (3) cobalt mineralization is greater in the Zairian deposits.[10]

While copper deposits in Zambia are located within the Lower Roan Group of the Katangan Supersystem, copper deposits in Zaire are located in the Upper Roan Group of the Katangan Supersystem. In Zaire, the ore-bearing zone was deposited from the uplift and erosion of underlying Kibaran rocks (during the Kibaran orogenic cycle) and as a platform deposit as the orogeny died out.[10]

The principal ore for copper is bornite, while the principal ores for cobalt-bearing minerals are the more abundant carrollite ($CuCO_2 \cdot S_4$) and linnaeite ($CoCO_2 \cdot S_4$). There is some controversy as to whether simple copper sulfides are primary or secondary.[10]

Production. Copper production has increased steadily in recent years, as indicated in Table II-4-20. Zaire exports copper ore, blister copper, leach cathodes, and refined copper.

The government-owned firm GECAMINES produces most of Zaire's copper and cobalt, from three operating centers located in Shaba Province. These three centers are often referred to as the Western Group at Kolwezi, the Central Group at Likasi, and the Southern Group at Lubumbashi.

TABLE II-4-20   Copper Production in Zaire, 1970-1982 (Metric Tons)

Year[a]	Copper (metal content, mine output)	Blister and leach cathodes	Refined
1970	387,116	386,900	189,000
1971	407,064	404,600	207,842
1972	435,741	428,260	216,200
1973	488,567	460,479	223,557
1974	494,428	462,000	254,552
1975	496,331	462,600	225,900
1976	444,432	413,000	66,018
1977	481,550	443,000	98,708
1978	423,800	390,700	102,797
1979	399,584	370,100	103,214
1980	459,392	425,745	144,161
1981p	497,000	470,000	165,000
1982e[b]	480,000		

[a]p, preliminary; e, estimated.
[b]1982 estimate from Ref. 12.
Source: U.S. Bureau of Mines, *Minerals Yearbook,* 1970-1981.

The Western Group at Kolwezi consists of open-pit and underground mines at Kamoto. Grades of ore in 1979 were 4.3% Cu in the open-pit mines and 4.2% Cu and 0.35% Co in the underground mine.[11] Facilities in the Kolwezi area include flotation plants at Kolwezi and Kamoto, a wash plant at Mutoshi, a new concentrator at Dima, and a copper-cobalt refinery in Luila and a new flash smelter.[3,11] In addition, there is an electrolytic zinc plant at Kolwezi, with a capacity of 6000 metric tons per year of zinc and 300 metric tons per year of cadmium.[11]

The Central Group at Likasi consists of the Kakanda open-pit mine, the Kamboue open-pit and underground mines, and the Shituru and Panda copper-cobalt refineries.[3,11] Ore grade in 1979 was 3% Cu at the Kamboue mines and 5 to 6% Cu in the Kakanda mines.

In 1979, the hydrometallurgical plants at Shituru and Luilu had the following capacities. Shituru's capacity was 130,000 metric tons per year of electrolytic copper and 7000 metric tons per year of cobalt. Luilu's capacity was 175,000 metric tons per year of electrolytic copper and 9000 metric tons per year of cobalt. Combined capacity for the two plants were 305,000 metric tons per year of electrolytic copper and 16,000 metric tons per year of cobalt.[11]

The Southern Group at Lubumbashi consists of the underground mine at Kipushi, concentrating facilities, and a smelter. The Kipushi mine is important not only for its copper production, but also for its zinc production. There are two principal ores. One is a high-copper ore (7% Cu-10% Zn) and the other a high-zinc ore (1% Cu-40% Zn).[11] Copper comes primarily from chalcopyrite and bornite, while zinc comes primarily from sphalerite. The concentrator has a capacity of 75,000 metric tons per year of zinc contained in concentrate, and a capacity of 60,000 metric tons per year of copper contained in concentrate. The smelter produces blister copper.[11]

In addition to copper and zinc, the Kipushi mine also produces by-product germanium from a germanium-rich bornite mineral, renierite, by selective mining and magnetic separation.[11] Metallurgie Hoboken-Overpelt S.A. of Belgium has acknowledged that its output of 30 tons of germanium per year was from Zairian raw material.[3] The mine also produces some lead, silver, and gold.

The other major copper producer in Zaire is Société de Développement Industriel et Miniere du Zaire (SODIMIZA), owned 80% by a Japanese consortium and 20% by the government of Zaire. SODIMIZA operates the Kinsenda and Mushoshi mines, as well as a concentrator at Mushoshi.

SODIMIZA produced 87,377 tons of concentrate containing 40.35% Cu metal in 1981. The copper concentrates are shipped south to the South African port at East London, and from there to Japan.[3]

Another potential source of copper and cobalt is from the Société de Tenke-Fungurume (SMTF), operated by the Compagnie Générale des Matières Nucléaires (Cogema, a subsidiary of BRGM). The mine was bought in 1975 from GECAMINES, which had halted work because of the closure of the Benguela Railroad. The present ownership of the mine includes the government of Zaire (20%), Cogema (26.5%, which it purchased from the U.S. firm Amoco Minerals Co. in 1979), French Bureau de Recherches Geologique et Minieres (BRGM) (8.5%), and Chartered Consolidated Ltd. (28%, which was the former operator). Reserves at the mine are an estimated 55 million tons of ore at 5.7% Cu and 0.45% Co.[3]

Table II-4-21 depicts copper production by area in 1980, and Table II-4-22 indicates cobalt

TABLE II-4-21   Details of 1980 Copper Production in Zaire by Area (Thousand Metric Tons)

Area[a]	Ore	Concentrate	Copper content
Mine[b]			
Western group			
Open pits	7,482	—	352.1
Kamoto UG	3,050	—	127.7
Central group			
Open pits	1,129	—	40.7
Kambove UG	1,559	—	47.9
Kipushi UG[c]	1,461	—	66.5
SODIMIZA UG	1,305	—	36.0
	15,986	—	670.9
Concentrator			
Western group			
Mutoshi	2,536	50	14.6
Kamoto	4,217	439	162.2
Dima	1,325	218	62.2
Kolwezi	4,201	655	147.6
Central group			
Kambove	1,511	92	41.0
Kakanda	786	100	23.6
Kipushi[c]	1,456	189	54.0
SODIZIMA	1,305	93	34.3
	17,337	1,836	539.5

[a]UG, underground.
[b]Gross weight.
[c]The Kipushi ore also contained 103,500 tons of zinc, and the Kipushi concentrator produced 123,000 tons of zinc concentrate containing 67,000 tons of zinc.

Source: Ref. 7.

TABLE II-4-22   Cobalt Production in Zaire, 1970-1982 (Metric Tons)

Year[a]	Cobalt mine output, metal content	Refined
1970	13,958	13,374
1971	14,518	14,518
1972	13,112	13,043
1973	15,052	15,052
1974	17,632	17,565
1975	14,000	13,638
1976	11,000	10,696
1977	10,200	10,215
1978	13,300	13,125
1979	15,000	14,100
1980	15,500	14,700
1981p	15,500	13,000
1982e [b]	15,410	

[a]p, preliminary; e, estimated.
[b]1982 estimate from Ref. 12.
Source: U.S. Bureau of Mines, *Minerals Yearbook,* 1970-1981.

production in Zaire from 1970 to 1982. In 1982, Zaire produced 51.3% of the world's cobalt production.[12]

Diamonds. Zaire is the largest producer of industrial diamonds in the world. In 1982, Zaire produced approximately 27% of the world's industrial diamond output, and its reserve base of 200 million carats was 23% of the world's resource base.[12] Table II-4-23 indicates diamond production in Zaire from 1970 to 1982. The percentage of industrial diamonds has increased steadily since 1970, leveling off to around 96% of total production since 1979. Diamonds were Zaire's third-largest export earner until 1981, when revenue from petroleum exports was worth $8 million more than diamonds—$112 million for diamonds versus $120 million for petroleum.[7]

Zaire's chief producer of industrial diamonds is the Société Minière de Bakwange (MIBA). The company is managed by Société Générale de Belgique and the Oppenheimer Group, which own only 20% of the company. The government of Zaire owns 80% of the company.[7]

In May 1981, President Mobutu authorized SOZACOM to market Zaire's diamonds, instead of Meltax and Britmond/Central Selling Organization.[7] This move ended the monopoly that Britmond-Zaire enjoyed since 1967. The company formerly received a commission of 20% on its sales of diamonds.

Although Zaire's marketing of its own diamonds has met with initial success, it is no longer

TABLE II-4-23   Diamond Production in Zaire, 1970-1982[a]
(Thousand Carats)

Year	Gem	Industrial	Total	Percent of industrial
1970	1649	12,438	14,087	88.3
1971	1274	11,468	12,742	90.0
1972	1339	12,051	13,390	90.0
1973	1294	11,646	12,940	90.0
1974	620	12,991	13,611	95.4
1975	395	12,415	12,810	96.9
1976	498	11,323	11,821	95.8
1977	533	10,681	11,214	95.2
1978	640	10,603	11,243	94.3
1979	294	8,440	8,734	96.6
1980	345	9,890	10,235	96.6
1981p	350	9,650 (7200)[b]	10,000	96.5
1982e	NA	(7,500)		

[a]p, preliminary; e, estimated; NA, not available.
[b]From Ref. 12.
Source: U.S. Bureau of Mines, Minerals Yearbook, 1970-1981.

guaranteed a market—an arrangement under the previous Central Selling Organization. Australia's entry into the diamond market in 1983, and an increased expansion in 1985 to 15 to 20 million carats per year, will increase world production some 26 to 31%.[12] Australia's diamond deposits produce approximately 75% industrial diamonds, and Australian reserves are 46% of world reserves.[12] This increased production could affect SOZACOM's ability to market its diamonds.

Tin. In 1982, Zaire was the tenth-largest producer of tin in the world. However, its output was less than 1.0% of the world total, and Zaire's reserves are only 2% of the world total.[12] Columbium, tantalum, and tungsten are by-products of tin mining in Zaire. In 1982, Zaire was the fifth-largest producer of columbium and tantalum, with less than 1.0% of the world's production and reserves.[12] Zaire's tin, columbium, and tantalum are given in Table II-4-24. There has been a general decline in

TABLE II-4-24   Tin, Columbium, and Tantalum Production in Zaire, 1970-1982[a]

Year	Mine output, metal content	Smelter primary	Columbium (1000 lb metal content)	Tantalum (1000 lb metal content)
1970	6458	1396		
1971	6455	1350		
1972	5960	1915		
1973	5442	969		
1974	4675	571	29	25
1975	4562	647	43	43
1976	3776	478	48	46
1977	5073	765	41	56
1978	4390	496	10	9
1979	3879	458	20	15
1980	3000	458	20	19
1981p	2200	550	45	46
1982e	2000	NA	40	40

[a]p, preliminary; e, estimated; NA, not available.
Source: Tin: U.S. Bureau of Mines, Minerals Yearbook, 1970-1981; Mineral Commodity Summaries 1982. Columbium, tantalum: Mineral Commodity Summaries, 1974-1982.

tin ore production since 1970. However, recovery of columbium and tantalum have increased in 1981 and 1982 from the previous years of 1978–1980.

Geology.[10] The tin belt in Zaire is located in eastern Zaire. The belt runs from Punia in the north to Mitwaba in the south, a distance of approximately 1300 km. The deposits are of middle Precambrian Age.

Tin production is from tin-bearing pegmatites and eluvial and alluvial placers from source pegmatities and quartz veins. The main tin deposits are the Manono-Kitotolo pegmatite (the largest ore-bearing pegmatite in the world) and the Kalima-Punia area.

The pegmatites in the Manono area contain tin, tungsten, columbium, and tantalum. The principal minerals found in the pegmatites are feldspar, quartz, spodumene, and mica. The cassiterite is uniformly spread throughout the pegmatite in grains 0.4 to 1 cm in diameter. The Maniema district is known for its eluvial and alluvial detrital tin deposits due to the region's "quiet and steady stream erosion."

Producers. The Société Zairetain produces tin and tantalum near Manono (400 km north of Lubumbashi) in northeastern Shaba Province from several open-pit mines. The company is 50% state owned and 50% owned by Géomines Cie, a privately owned Belgian company. Concentrates were smelted into tin metal and tantaliferous slag at a Manono foundry belonging to the company. There is some discussion of producing lithium from spodumene, which occurs with the cassiterite.[7]

The Entreprises Minières du Zaire (EMZ), successor to the Sermikat Pilolet Group, has hired the British engineering firm of Guest, Keen, and Nettlefolds Ltd., to rehabilitate Mitwaba, Bukena, Kibambo, and Mkoy tin mines south of Manono. The EMZ properties contain wolframite (tungsten) in addition to cassiterite.[7]

The French government mining company, owned 69% by BRGM, the Compagnie Française des Mines, is reopening the Kania open-pit tin mine south-southwest of Manono in the eastern Shaba region.[7]

A U.S. company, Metallurg (a major consumer of columbium), agreed to participate in a new venture that could prove to be one of the world's top three sources of columbium. Other participants include Somikivu (a French-Belgian-Zairian consortium) with a 70% share in the project; Sominki, a 10% share (28% of which is owned by the Zairian government); and the Zairian government, a free

20% of the share. The Lueshe deposit contains large reserves of pyrochlore.[7]

SOMINKI is owned 28% by the government and 72% by Cogemin of the Belgo-French Empain Group. SOMINKI is Zaire's largest producer of tin and tungsten concentrates, and produces columbium, tantalum, gold, and monazite as well. The company is an amalgamation of many smaller operations in the Kivu region, with the largest mine and headquarters at Kalima. Other mines were nearby, while some were clustered near Buni in the north and Kamituga to the east.[6,7]

Gold-Silver. Zaire's chief producer of gold is the state agency, Office des Mines d'Or de Kilo-Moto (OKIMA). Other producers are SOMINKI and GECAMINES.[6] Kilo-Moto's mines are located in northeastern Zaire near the Ugandan and Sudanese borders in the Haute-Zaire region. Kilo-Moto plans to increase production from 50,000 troy ounces to 160,000 troy ounces of gold per year. Reserves are estimated at 2.9 million troy ounces.[6] All exports are flown to Kinshasha for export.[7]

GECAMINES produces 3500 troy ounces per year from tankhouse slimes derived from Likasi and Kilwezi cobalt-copper refineries.[7] SOMINKI produces some 15,000 troy ounces per year from its alluvial tin, tungsten, and tantalum deposits in the Kiva region.[7]

At one time, Zaire was one of the world's largest gold producers. However, gold production has fallen substantially since 1970, as indicated by Table II-4-25. GECAMINES is the major producer of silver as a by-product of its mining operations.

TABLE II-4-25   Gold and Silver Production in Zaire, 1970–1981 (Troy Ounces)

Year	Gold (troy ounces)	Silver (thousand troy ounces)
1970	177,128	1,709
1971	174,513	1,534
1972	140,724	2,078
1973	133,650	1,995
1974	130,603	1,694
1975	103,217	2,291
1976	91,093	2,472
1977	80,418	2,730
1978	76,077	4,391
1979	69,992	3,892
1980p	39,963	2,733
1981e	70,000	3,000

Source: U.S. Bureau of Mines, *Minerals Yearbook,* 1970–1981.

As indicated in the table, silver production has increased steadily since 1970, but has declined from a high of 4,391,000 ounces in 1978.

Manganese. Zaire produced manganese from several mines west of Kolwezi until 1975, when the Angolan civil war closed the Benguela Railway. The mines were run by the Société Minière de Kisenge (SMK). In 1981, there were 500,000 tons of crushed carbonate ore and 700,000 uncrushed tons of carbonate ore at Kisenge. SMK had planned to build a dry-cell battery factory, by building a refinery to convert the manganese carbonate into electrolytic manganese oxide for battery manufacture.[7] In 1976 and 1977 manganese exports were worth $25.2 million and $23.8 million, respectively. Table II-4-26 indicates Zaire's manganese production since 1970.

Zinc-Germanium. Zinc production in Zaire is primarily from GECAMINES' Kipushi mine, as is most of Zaire's germanium production. Production of both zinc and germanium is given in Table II-4-26. Zaire has been the world's largest producer of germanium in the past. However, Zaire no longer releases production figures for germanium. Reserves for germanium are considered to be very large.[12]

Constraints to Mineral Industry

The future availability of mineral supplies from Zaire to the West should not be taken for granted. Current events in Zaire that can have an effect on the mineral industry are the political climate and management of the economy.

Although the MPR is the only legal political party in Zaire, there are other political parties opposed to Mobutu's rule. Prominent among these are the Front for the National Liberation of the Congo (FLNC), headed by Nathaniel Mbumba. The FLNC was responsible for the Shaba I and Shaba II invasions in 1977 and 1978. However, Mobutu's reconciliation with the MPLA in Angola has led the Angolan government to withdraw support of the FLNC.[2] The Katanganese People's Movement (KPM), led by Gerard Moke, is derived from the secessionist regime of Moise Tshombe in the early 1960s.[2] The People's Revolutionary Party, headed by Laurent Kabila, is the longest-surviving opposition group and operates in the Kivu region near Lake Tanganyika. Kabila was in the Stanleyville (Kisangani) rebellion, which nearly seized power in the mid-1960s.[2] The Party for the National Conscience has support from the African churches in the Shaba region.[2] Committee for the Liberation of the Congo (CLC) is headed by Mungul-Diaka and serves as an umbrella group for various other opposition groups.[2] Finally, the Comite Zaire, led by Philippe Borel, is composed of Belgians opposed to Mobutu's rule.[2]

Political opposition to Mobutu is largely due to the present government's alleged corruption, which includes appointment of family members to high government office and the displacement of large

TABLE II-4-26   Manganese, Zinc, and Germanium Production in Zaire, 1970-1981[a]

| Year | Manganese[b] ore and concentrate gross weight (metric tons) | Zinc[b] | | Germanium[c] (kg of metal) |
		Mine output metal content (metric tons)	Metal primary electrolytic	
1970	346,940	105,082	63,750	20,000
1971	329,066	109,200	62,673	50,000
1972	369,481	99,252	66,652	55,000
1973	333,963	87,559	66,026	61,652
1974	308,775	84,464	66,182	26,000
1975	308,525	79,300	65,588	50,000
1976	182,184	67,800	61,677	23,000
1977	41,019	73,000	51,049	NA
1978	—	73,700	43,500	NA
1979	—	68,000	43,508	NA
1980p	16,586	67,000	43,800	NA
1981e	10,000	69,500	58,000	NA

[a]p, preliminary; e, estimated; NA, not available.
[b]U.S. Bureau of Mines, Minerals Yearbook, 1970-1981.
[c]U.S. Bureau of Mines, Mineral Commodity Summaries, 1971-1983.

sums of public money into private hands (led by Mobutu himself).

Galen Spencer Hull declares that "although the MPR is allegedly a mass movement, relatively few benefit from participation in the party."[2] According to Hull there are three distinct classes within Zaire. These classes are:[2]

1. The bureaucratic bourgeoisie, composed of 50-odd kinsmen of President Mobutu and the presidential brotherhood of several hundred elite representing major ethnic groups
2. The petite bourgeoisie, composed of several thousand middle-level officials, university teachers, and graduates
3. The masses

This particular political arrangement is not much different from that of other developing nations. However, for the present government to remain in power, the masses will have to derive some economic benefit.

Zaire's current economic problems stem from a lack of economic planning. In the 1970s, imports rose twice as fast as foreign exchange from exports. As a result, external debt increased, while the price for copper fell. In addition, international inflation and increases in imported fuels increased domestic inflation.

All of these factors combined to force Zaire to seek large loans from the IMF. These loans were granted on the condition that certain reforms be initiated in the Zairian economy. Key reforms in 1978-1979 included: (1) key expatriate advisors were appointed to key posts in the Bank of Zaire and the Ministry of Finance; (2) the local currency (zaire) was devalued, followed by a drastic demonitization program; (3) the Customs Service and other public corporations were reorganized; (4) the tax collecting system was reorganized; and (5) Zaire signed standby agreements with the IMF, with certain conditions.[2]

In June 1981 the government of Zaire signed an agreement with the IMF for an Extended Fund Facility (EFF).[5] The agreement runs from mid-1981 through 1983 and extends $1 billion in several installments to the Zairian government. This agreement again called for the devaluation of the Zairian currency and strict credit ceilings on external borrowing and internal lending.

Of particular interest to the availability of future mineral products, the EFF called for an investment of almost 7 billion zaires in the country's infrastructure. Of this amount, 34% will go to the mining sector. Funds are also mandated for investment in the transportation and energy sectors to upgrade roads and ports and to reduce energy costs.[5] The EFF plan also calls for a reduction in state participation in Zaire's economy, with a greater role for private enterprise.[5]

It is unlikely that Zaire will increase foreign investment within the country until companies are able to repatriate earnings. Current laws dealing with repatriation of money exist, but little money is ever allowed to leave.

REFERENCES

1. U.S. Central Intelligence Agency, National Foreign Assessment Center. *The World Factbook—1981.* Washington, D.C.: U.S. Government Printing Office, 1982.
2. Hull, Galen Spencer. *Pawns on a Chessboard: The Resource War in Southern Africa.* Washington, D.C.: University Press of America, Inc., 1981.
3. U.S. Department of the Interior, Bureau of Mines. "The Mineral Industry of Zaire." *Minerals Yearbook 1981.* Washington, D.C.: U.S. Government Printing Office, 1983.
4. U.S. Department of Commerce, Industry and Trade Administration. "Marketing in Zaire." *Overseas Business Reports.* OBR 79-44, December 1979.
5. U.S. Department of Commerce, International Trade Administration. "Zaire." *Foreign Economic Trends and Their Implications for the United States.* FET 81-143. Washington, D.C.: U.S. Government Printing Office, December 1981.
6. U.S. Department of the Interior, Bureau of Mines. "The Mineral Industry of Zaire." *Minerals Yearbook 1980.* Washington, D.C.: U.S. Government Printing Office, 1982.
7. U.S. Department of the Interior, Bureau of Mines. "The Mineral Industry of Zaire." *Minerals Yearbook 1981.* Washington, D.C.: U.S. Government Printing Office, 1983.
8. U.S. Department of the Interior, Bureau of Mines. "The Mineral Industry of Zaire." *Minerals Yearbook 1978/79.* Washington, D.C.: U.S. Government Printing Office, 1981.
9. "Zaire: Problems Mount." *Mining Journal,* 22 October 1982.
10. Ridge, John Drew. "Zaire." In *Annotated Bibliographies of Mineral Deposits in Africa, Asia (exclusive of the USSR) and Australia.* Elmsford, N.Y.: Pergamon Press, Inc., 1976, pp. 177-195.
11. White, Lane. "Zaire." *Engineering and Mining Journal,* November 1979, pp. 188-206.
12. U.S. Department of the Interior, Bureau of Mines. *Mineral Commodity Summaries 1983.* Washington, D.C.: U.S. Government Printing Office, 1983.

Chapter II-5

Conclusions

Countries in southern Africa, especially South Africa, supply a significant percentage of the world's mineral supplies. Some of the minerals most important to industrialized nations are supplied from this region. Of particular importance are chrome, cobalt, fluorspar, manganese, platinum-group metals, and vanadium. Other important minerals include antimony, copper, gold, industrial diamonds, titanium, uranium, and zirconium. For major uses of some of these minerals there are no substitutes, or where substitution is possible, there is a significant loss of efficiency in use. The importance of the region for mineral production and reserves can be seen from Table II-5-1.

In general, the region is in a state of economic decline due to civil strife, reliance on one-crop economies, declining food production, increasing foreign debt, deteriorating infrastructure, unemployment, and lack of skilled labor. The exception to this rule is South Africa. Problems facing countries in southern Africa are presented in Table II-5-2.

Although many critics argue against any theories of "resource war" being waged in southern Africa or elsewhere between the East and West, there certainly exist incentives for both sides to engage in such a war. Critics for and against the resource war are given in Table II-5-3. A few strategic advantages for possessing or controlling southern Africa are given below, together with some critical unanswered questions.

A major oversight in many discussions of the resource war is the role of time and mineral reserves. With the passage of time, those countries with large reserves of a particular mineral commodity will become increasingly important for that mineral. In southern Africa, particularly important minerals for reserves are chromium, cobalt, industrial diamonds, fluorspar, gold, manganese, nickel, platinum-group metals, titanium, vanadium, and zirconium. It is interesting to note how the addition of these mineral commodities to the USSR's already impressive mineral resources would certainly put the USSR in a dominant world position for many of these minerals. These figures are presented in Table II-5-4.

Another misconception or preconception is to believe that the USSR cannot eventually secure the

southern Africa. The USSR need only secure the ocean-bordering countries in the region to have an economic stranglehold on the land-locked countries in southern Africa. Angola, Mozambique, and Zimbabwe are already Marxist states. If the U.N. prevails, Namibia will be controlled by Marxist-oriented SWAPO. If racial policies in South Africa are not improved, eventually there may be large-scale disorders in South Africa itself. Given the current state of affairs in the U.N., Soviet intervention would not receive worldwide condemnation. Of course, direct intervention would be costly, with the risk of all-out confrontation between the East and West.

However, the USSR need not enter into direct intervention in South Africa to gain major economic and strategic advantages. Since the mineral resources of the USSR and South Africa are similar, any major disruptions in the availability of South African minerals would increase the economic and strategic importance of the USSR's mineral production. Disruptions in South African mineral supplies, together with Soviet unwillingness to sell certain strategic minerals to the West, would seriously affect the economies and military industries of the West.

Furthermore, with South Africa's gold production and reserves, European dependency on Soviet energy supplies, and cooperation with Arab oil producers, how strong would the ruble appear at some future date? What if, under such circumstances, the world reverts to a gold standard and the United States and Western Europe are importing more than they export? How strong would the dollar be then?

Finally, what would be the relative positions of the East and West in a worst-case scenario in which the USSR and the United States have a military confrontation? Short of total conflagration with nuclear weapons, the West imports many minerals from southern Africa that the USSR produces to a degree of self-sufficiency. With future air or naval bases in just Angola and Mozambique, the USSR could possibly stop shipping around southern Africa. Thus both mineral and oil supplies would be in jeopardy.

Although these scenarios may seem a little far-

TABLE II-5-1 Ranking of Mineral Production and Reserves
in Southern Africa, 1981[a]

Mineral	Country	Mineral production ranking (1981)	Mineral reserve ranking
Antimony	South Africa	2	2
Asbestos	South Africa	4	2
	Zimbabwe	3	12
Beryllium	South Africa[b]	3	NA
Chromium	South Africa	1	1
	Zimbabwe	4	2
Cobalt	Zaire	1	1
	Zambia[b]	2	2
Columbium	Zaire[b]		4
Copper	Zaire	6	6
	Zambia[b]	5	4
Diamonds	Botswana	3	3
(industrial)	South Africa	2	4
	Zaire	1	2
Fluorspar	South Africa[b]	3	1
Germanium	Zaire		Large
Gold	South Africa	1	1
	Zimbabwe	11	10
Iron ore	South Africa	8	7
Kyanite, etc.	South Africa	1	1
Lithium	Zaire		3
	Zimbabwe	2	5
Manganese	South Africa	2	2
Nickel	Botswana	10	10
	Zimbabwe	11	7
	South Africa[c]	7	3
Platinum group	South Africa	2	1
Platinum	South Africa	1	1
Palladium	South Africa	2	1
Rhodium	South Africa	1	1
Tantalum	Zaire	6	4
	Mozambique	NA	NA
Thorium	Zaire	5	
Tin	South Africa	10	
	Zaire	12	12
Titanium			
Ilmenite	South Africa	5	4
Rutile	South Africa	3	4
Vanadium	South Africa	1	1
Vermiculite	South Africa	2	2
Zirconium	South Africa	2	2

[a]NA, not available.
[b]Noncommunist world.
[c]South African estimates from Bushveld Igneous Complex.

Sources: U.S. Bureau of Mines, Minerals Yearbook, 1981; Mineral Commodity Summaries 1983.

fetched, they cannot be totally overlooked. Many complacent authorities believe that no matter who rules in southern Africa, they will have to sell mineral exports to the West. However, the question arises whether it will be through a free-market system or a Soviet-controlled marketplace where supplies can be withheld or mineral cartels are formed with substantially higher prices.

The fate of southern Africa depends largely on future events in South Africa. The clamoring of the

TABLE II-5-2   Major Problems Affecting Southern Africa

	Land-locked	Ocean-bordering	Marxist	Foreign troops	Guerilla activity	Declining Infrastructure	Limited water resources	Limited Mineral resources	Dependent on mineral industry	Decreasing foreign investment	Civil war in last 10 years
Angola		X	X	X	X	X			X	X	X
Botswana	X						X		X		
Lesotho	X							X	X		
Malawi	X							X			
Mozambique		X	X	X	X	X				X	X
Namibia/ South-West Africa		X		X	X		X		X		X
South Africa		X							X		
Swaziland	X							X	X		
Zaire		X		X	X				X	X	X
Zambia	X						X		X	X	
Zimbabwe	X		X	X	X	X			X	X	X

TABLE II-5-3   Survey of Literature Relating to Resource War in Southern Africa

Author	Resources-related	Not resources-related
Strauss[a]	X	
Szuprowicz (21st Century Research)[b]	X	
Fine[c]	X	
Shabad[d]		X
Vogely[e]		X
Strishkov (BOM)[f,t]		X
Bowman[g]		X
Shafer[h]		X
Rees (Institute for the Study of Conflict)[i]	X	
Council of Economics and National Security[j,k]	X	
Resources for the Future[l]		X
Soviet Analyst[m]	X	
Meyer[n]	X	
Tilton and Landsberg[o]		X
Foreign Affairs Research Institute[p]	X	
ALARM (Miller)[q]	X	
South African Embassy[r]	X	
Slay[s]	X	

[a]Simon D. Strauss, "Mineral Self-Sufficiency — The Contrast between the Soviet Union and the United States." *Mining Congress Journal*, November 1979.

[b]Bohdan Szuprowicz, *How to Avoid Strategic Materials Shortages*. New York: John Wiley & Sons, Inc., 1981.

[c]Daniel I. Fine, "Mineral Resource Dependency Crisis: Soviet Union and United States." In *The Resource War in 3-D: Dependency, Diplomacy, Defense*. Edited by James A. Miller, Daniel I. Fine, and R. Daniel McMichael. Pittsburgh, Pa.: World Affairs Council of Pittsburgh, 1980.

[d]Theodore Shabad, "The Soviet Mineral Potential and Environmental Constraints." Mineral Economics Symposium, Washington, D.C., 8 November 1982.

(continued)

[e]William A. Vogely, "Resource War?" *Materials and Society* 6, No. 1 (1982).

[f]U.S. Department of the Interior, Bureau of Mines. "The Mineral Industry of the U.S.S.R." By V. V. Strishkov. Preprint from the *1980 Minerals Yearbook*. Washington, D.C.: U.S. Government Printing Office, 1980.

[g]Larry W. Bowman, "The Strategic Importance of South Africa to the United States: An Appraisal and Policy Analysis." *African Affairs* 81, No. 323 (April 1982).

[h]Michael Shafer, "Mineral Myths." *Foreign Policy*, No. 47, (Summer 1982).

[i]David Rees, *Soviety Strategic Penetration in Africa*. London: Institute for the Study of Conflict, 1976.

[j]Council on Economics and National Security. *The "Resource War" and the U.S. Business Community: The Case for a Council on Economics and National Security*. Washington, D.C.: Council on Economics and National Security, 1980.

[k]Council on Economics and National Security. *Strategic Minerals: A Resource Crisis*. Washington, D.C.: Council on Economics and National Security, 1981.

[l]"What Next for U.S. Minerals Policy?" *Resources*, No. 71 (October 1982).

[m]Brian Crozier, "Strategic Decisions for the West." *Soviet Analyst* 10, No. 5 (4 March 1981).

[n]Herbert E. Meyer, "Russia's Sudden Reach for Raw Materials." Reprinted from *Fortune*, 28 July 1980.

[o]John E. Tilton and Hans H. Landsberg, "Nonfuel Minerals: The Fear of Shortages and the Search for Policies." Paper presented at the RFF Forum, Washington, D.C., 21 October 1982.

[p]Foreign Affairs Research Institute, "The Need to Safeguard NATO's Strategic Raw Materials from Africa." No. 13 (1977).

[q]" 'Resource War' Targeted as a Major National Problem." *ALARM*. Edited by James A. Miller. No. 1 (May 1981).

[r]"South Africa: Persian Gulf of Minerals. Strategic Minerals Threat." Backgrounder issued by the Information Counselor. South African Embassy. No. 9 (December 1980).

[s]General Alton D. Slay, "Minerals and National Defense." *The Mines Magazine*, December 1980.

[t]Bohdan Szuprowicz, "Critical Materials Cut-off Feared." *High Technology*, November–December 1981.

Source: Philip R. Ballinger. "An Analysis of the Energy and Nonfuel Minerals of the U.S.S.R." Master's thesis. University of Texas at Austin, 1983.

TABLE II-5-4 Combined Mineral Production of South Africa and the USSR, 1981

Mineral	Percent combined production	Percent combined reserves
Chromium	87.9	68.2
Industrial diamonds	50.3	11.5
Fluorspar	20.2	31.3
Gold	72.5	72.5
Manganese	61.2	85.2
Platinum group	93.2	97.5
Vanadium	63.2	71.8
Zirconium	33.8	37.5

Sources: U.S. Bureau of Mines, *Minerals Yearbook*, 1981; *Mineral Commodity Summaries 1983*.

U.N. for "one man, one vote" and black majority rule in South Africa does not guarantee a strifeless society with equality for all. There are several large black tribes within South Africa. With black majority rule, would a new struggle ensue with tribal divisions as the deciding factor? It has happened in neighboring countries.

There are many uncertainties regarding the future of southern Africa and hence the future availability of minerals from the region. However, the importance of the region for world mineral supplies is conclusive—southern Africa is a dominant supplier of several essential minerals and has the potential to continue as such.

PART III **AUSTRALIA**

Chapter III-1

Introduction

WHY AUSTRALIA?

Australia is among the five countries—the United States, the USSR, Canada, Australia, and South Africa—that produce the majority of the world's minerals. Because of its relatively small domestic market, Australia exports a large percentage of its output, making the nation one of the most vital sources of a wide range of minerals.

Australia is the world's leading producer of alumina, rutile, and zircon, and is among the top five producers of iron ore, lead, zinc, nickel, tin, manganese, and ilmenite (Table III-1-1). As an exporter, Australia is also highly involved in coal, copper, and uranium. In the near future Australia will become increasingly important in the diamonds, gold, and liquefied natural gas markets.

Being a virtual mineral treasure house and having a stable democratic government have made Australia an irreplaceable supplier to most Western nations, especially Japan, who is Australia's most important trading partner, accounting for over 40% of her mineral exports.

Australia has a long history of mining. Within five years after the continent was settled, coal was discovered (1791) in New South Wales. Traces of

TABLE III-1-1 Australian Percentage of World Mineral Production and Resources, 1982

Commodity	World production	World rank	World resources	World rank
Rutile	64	1	7.7	2
Bauxite	30	1	20.7	2
Ilmenite	30	1	6.4	6
Nickel	12	3	9.1	5
Lead	13	3	15.8	2
Iron ore	11	3	10.9	3
Zinc	10	3	9.9	3
Manganese	7	7	1.8	3
Tin	5	7	3.5	9
Tungsten	5	5	3.8	7
Copper	3	10	3.1	10
Black coal	3	8	5.0	7
Diamonds	1	8	46.0	1

Source: Ref. 1.

gold were first reported in 1823, and occurrences of other minerals were found in the early part of the nineteenth century. Because of the nature of the colony and its small population, the first metal mines were not opened until the 1840s.[2] Later in the century tin, copper, and zinc deposits were developed. However, it was the gold rushes in the 1850s that put Australia on the map. Peaking in the 1890s with discoveries in Western Australia, the rushes took prospectors all over the continent.

Mining remained dormant during the early decades of the twentieth century, with only the discovery at Mt. Isa and a gold revival in the 1930s to stir interest. With the expansion of Japanese industry in the 1960s, Australian mining found new life as iron ore and coking coal led an impressive growth surge in mineral production (Figure III-1-1). The 1970s brought international interest in Australia's energy resources, and a new "resource boom" revolving around steam coal and uranium was heralded. Before the boom could get under way, the world recession of 1981–1983 put a damper on hopes of increased production.

Australia has been dependent on exports in one form or another for the last 40 years, beginning with the wool boom of the 1940s, a beef boom in the 1950s, and the iron ore and coking coal boom in the 1960s and 1970s. Today farming and minerals comprise three-fourths of Australia's exports (Table III-1-2). Although these industries contribute only one-eighth to gross national product (GNP) and 7% to employment, Australia has to export a large percentage of its production to keep its domestic industry healthy. Roughly 50% of Australia's mineral production and 70% of its food and fiber production is for the export market. Black coal and iron ore are still Australia's most important minerals, totaling some 50% of mineral exports by value (Figure III-1-2).

Australia is a country of vast area and small population. Being roughly the size of the continental United States, it is the home of fewer than 15 million people, of which some 90% live in the fertile rim between Brisbane and Melbourne. This has meant that Australia has had to overcome problems of insufficient capital and huge infra-

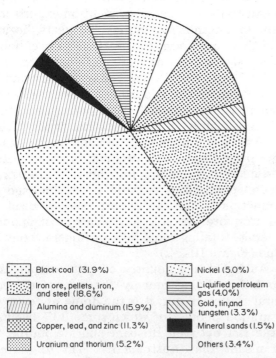

FIGURE III-1-1 Growth in the Australian mineral industry, 1952–1978. (From Ref. 2.)

TABLE III-1-2 Australian Exports, 1981–1982 (Million Dollars)

Metalliferous ores and scrap	3,372.1
Cereals and cereal preparations	2,503.7
Coal, coke, and briquettes	2,295.0
Textile fibers and their wastes	1,888.9
Manufactures of metal	1,629.3
Meat and meat preparations	1,381.0
Machinery and transport equipment	1,102.3
Sugar, sugar preparations, and honey	793.7
Petroleum, petroleum products, etc.	557.4
	19,585.9

Source: Ref. 3.

Black coal (31.9%)
Iron ore, pellets, iron, and steel (18.6%)
Alumina and aluminum (15.9%)
Copper, lead, and zinc (11.3%)
Uranium and thorium (5.2%)
Nickel (5.0%)
Liquified petroleum gas (4.0%)
Gold, tin, and tungsten (3.3%)
Mineral sands (1.5%)
Others (3.4%)

FIGURE III-1-2 Australian mineral exports by value, 1982. (From Ref. 4.)

structure requirements. Australia's small but skilled population, with its English influences, has been the source of labor problems. Trade partners have complained that this has made Australia an unreliable supplier.

The object of this chapter is to analyze the production trends of Australian minerals, as well as observing what trends in trade patterns will mean to future exports. In addition, the policies of the Australian government toward mineral production and exports, and the problems that the mineral industry faces, are considered in detail. From this study it is hoped that key deficiencies in production and policy can be isolated and that feasible solutions to industry problems may be suggested.

REFERENCES

1. U.S. Department of the Interior, Bureau of Mines. *Mineral Commodity Summaries 1983*. Washington, D.C.: U.S. Government Printing Office, 1982.
2. Govett, G. J. S., and M. H. Govett. "The Role of the Australian Mineral Industry in World Mineral Supplies." *Resources Policy*, June 1980.
3. Australian Department of Trade and Resources. *Exports Australia*. Canberra: Australian Bureau of Statistics, 1982.
4. Govett, G. J. S., and M. H. Govett. "Australian Minerals." *Resources Policy*, December 1982.

Chapter III-2

Nonfuel Minerals

Gold and the traditional base metals were the key to Australia's past mineral success. Iron ore, bauxite, nickel, and mineral sands led the nation's mineral resurgence in the 1960s and are likely to continue to be high revenue earners. Bauxite could be Australia's most important nonfuel mineral in the future as the nation attempts to increase her level of processing. In this chapter the Australian nonfuel mineral industry is discussed, highlighting production, reserves, exports, and the status of each of the major nonfuel mineral commodities.

ALUMINUM

Production, Resources, and Reserves

Australia is the world's largest producer of bauxite and alumina, with about 30 and 25% of global annual output. Aluminum production is surprisingly low, being only 2% of annual world output.[1] This should change, as production levels of alumina and aluminum are expected to rise as a consequence of new refineries and smelters to be built before 1990. With these new facilities Australia could possibly become the world's largest aluminum exporter.[1]

The total bauxite resources of Australia are 6179 million metric tons, of which 4443 million metric tons are economic reserves.[1] Guinea has the world's largest reserve base of bauxite, totaling 5900 million metric tons as of 1981.[2] Table III-2-1 presents Australian bauxite resources as of 1980.

TABLE III-2-1   Australian Bauxite Resources, 1980 (Million Metric Tons of Ore)

Economic resources	
Demonstrated	2703
Inferred	1740
	4443
Subeconomic resources	
Para-marginal	1599
Submarginal	137
	1736
	6179

Source: Ref. 3.

Table III-2-2 details production statistics for bauxite, alumina, and aluminum since 1975.

Exports and Trade

Australian consumption of aluminum has increased from 0.5 to 1.4 kg per capita between 1945 and 1980, with most metal used in the building and packaging industries and in the construction of transportation equipment.[4] In 1980 domestic consumption was 84.7% of Australian-produced aluminum.[5] The Australian Department of Trade and Resources expects that Australian consumption of aluminum will increase at 5% per annum, and will reach 260,000 metric tons by 1985.[3]

Australia is a significant exporter of bauxite, ranking third, behind Guinea and Jamaica, with about 20% of the world's exports. Japan (around 40%) and Western Europe are the two largest importers of Australian bauxite.[3] The USSR imported 175,000 metric tons of bauxite from Gove, Northern Territory, in 1981 to supplement its growing need for bauxite. Australia leads the world in alumina exports with about 50% of the market. High-volume destinations of Australian alumina exports are North America, Japan, and the Middle East.[3] Australia is only a minor aluminum exporter (less than 2% of the world total), with its largest markets in Japan, China, and Southeast Asia.[3] Table III-2-3 provides import and export statistics of bauxite, alumina, and aluminum since 1976.

Uses

Over 90% of the world's production of bauxite is used in making alumina. The remainder is consumed by the abrasive, refractory, and chemical industries. Aluminous chemicals, made from bauxite, are used for water and sewage treatment, as well as dyeing, tanning leather, and sizing paper. Other uses for bauxite include high-alumina cement, as an absorbent or catalyst in the oil industry, in welding rod coating and fluxes, and as a flux in making steel and ferroalloys.[5] About 6% of alumina production is used in abrasives, refractories, chemical filters, and in the production of artificial sapphires; the other 94% is smelted into aluminum.

Aluminum can be fabricated in a variety of

TABLE III-2-2   World Bauxite, Alumina, and Aluminum Production, 1976–1982

	1976	1977	1978	1979	1980	1981	1982
Bauxite (million metric tons)							
World	80.7	84.6	83.9	88.0	89.9	85.7	75.8
Australia	24.1	26.1	24.3	27.6	27.2	26.8	23.0
Jamaica	10.3	11.4	11.7	11.5	12.1	11.7	8.0
Guinea	11.3	10.8	12.1	12.2	13.3	12.1	10.2
USSR(e)[a]	6.7	6.7	6.7	6.5	4.6	4.6	4.6
Alumina (million metric tons)							
World	27.98	30.51	30.83	32.30	33.20	32.30	
Australia	6.20	6.66	6.78	7.42	7.25	7.08	6.63
U.S.	5.81	6.03	6.00	6.45	6.81	5.96	
USSR(e)[a]	3.20	3.25	3.30	3.20			
Jamaica	1.64	2.05	2.14	2.07	2.48	2.55	
Japan	1.66	2.05	1.77	1.82	1.94	1.34	
Aluminum (thousand metric tons)							
World	13,212	14,339	14,754	15,217	15,365	15,068	12,879
Australia	232	248	263	270	304	371	381
U.S.	3,519	4,118	4,358	4,557	4,653	4,488	3,265
USSR(e)[a]	2,220	2,220	2,300	2,400	1,787	1,790	905
Japan	919	1,188	1,058	1,010	1,091	770	363

[a]e, estimated.

Sources: Refs. 1 and 2.

ways. It is used in the transportation industry to save weight in autos, ships, rail car, and aircraft parts. Other uses include packaging (cans, foil) building construction, electrical equipment, and consumer goods (Table III-2-4).

Aluminum use is expected to rise by 4 to 5% per year over the next 20 years,[6] with the largest expansion of aluminum use in containers, construction, and transportation. For example, aluminum use in automobiles is expected to increase by 33% between 1980 and 1990.[7] One industry that has the potential to become a huge consumer of aluminum is electric cars. Research continues on the aluminum-air battery. Such a system would use aluminum as a fuel to produce electricity to power cars. The battery is only in the prototype stage, but if the aluminum-powered auto becomes a reality, it would greatly increase the demand for aluminum metal.

Aluminum recycling has grown substantially over the past 10 years. In the United States, recycling is equal, on a tonnage basis, to nearly 17% of domestic primary capacity. In Japan, production from secondary sources now exceeds primary production. This is important in that smelting scrap aluminum uses less than 20% of the energy required to smelt primary aluminum.[6] It is estimated that 20% of total world supplies of aluminum metal will come from secondary aluminum by the year 2000.[6]

History of the Industry

Difficulties in importing aluminum metal during World War II led to the passage of the Aluminum Industry Act of 1944.[8] Tasmania was chosen as the site of the first smelter due the low hydroelectric rates available there. Construction of the 25,000-metric ton per year refinery and 13,000-metric ton per year smelter at Bell Bay began in 1948. By 1961 both facilities had reached full production.[8] The development of alumina and aluminum productive capacity, between 1960 and 1980, is shown in Figures III-2-1 and III-2-2.

The reddish cliffs above the Gulf of Carpentaria were first described by a British mariner named Matthew Flinders in 1802, but it was not until 1955 that the Weipa area was recognized as one of the largest high-grade, low-cost bauxite deposits in the world. Mining was initiated by Comalco in 1963.

The other major bauxite mining area, the Darling Range, 45 km southeast of Perth, was originally thought to have grades too low for eco-

TABLE III-2-3   World Bauxite, Alumina, and Aluminum Trade, 1976–1982 (Thousand Metric Tons)

	1976	1977	1978	1979	1980	1981	1982
Bauxite							
Exports							
World	32,016	34,884	35,298	34,997			
Australia	6,904	7,314	6,563	6,854			
Guinea	8,576	9,146	10,225	9,338			
Jamaica	7,166	7,447	7,591	7,831			
Imports							
U.S.	13,111	13,616	15,333	14,506			
Japan	4,275	5,318	4,743	4,596			
West Germany	4,087	4,091	3,614	3,694			
USSR	4,140	4,140	3,610	3,612			
Alumina							
Exports							
World	11,683	13,029	12,947	13,409			
Australia	5,442	6,153	6,085	6,809	6,938	6,509	5,977
Jamaica	1,787	1,935	2,148	1,977			
Surinam	811	907	1,159	1,010			
Imports							
U.S.	3,288	3,759	3,967	3,767			
Norway	1,312	1,312	998	1,174			
Canada	909	822	1,057	953			
Japan	627	940	757	759			
Aluminum (unwrought)							
Exports							
World	3,797	4,011	4,406	4,185	3,695		
Australia	65	78	80	76	46	79	156
Canada	508	655	863	551	785		
Norway	562	555	630	565	521		
U.S.	139	89	102	164	608		
USSR(e)[a]	520	520	580	560			
Imports							
Japan	430	534	740	748	910		
U.S.	517	611	686	517	527		
West Germany	417	440	434	511	589		

[a]e, estimated.

Sources: Ref. 1; Australian Bureau of Mineral Resources.

nomic recovery. While the Weipa bauxite grades at over 50% alumina, the Darling Range ore averages only 32 to 35% alumina. What made the Darling Range deposits economical to mine was that they were close to established infrastructure, necessary for development. In addition, the bauxite is composed entirely of gibbsite, which can be treated at a lower temperature than boehmite, therefore requiring less caustic soda, which is a major cost item in alumina refining. With these factors to offset the low grade of the Darling Range desposits, Western Aluminum began mining at Jarrahdale in 1963.[8] Figure III-2-3 shows the development of bauxite mining in Australia.

TABLE III-2-4   Consumption of Aluminum by Application, 1980 (Kilograms per Capita)

Application	U.S.	Japan	West Germany
Transportation	5.9	4.5	5.9
Construction	6.7	6.7	4.1
Packaging	6.5	1.3	2.1
Electrical	3.0	2.1	1.4
Consumer durables	2.3	1.2	1.8
Machinery and equipment	1.9	1.2	2.0
Other	1.9	2.0	1.7
	28.2	19.0	19.0

Source: Ref. 6.

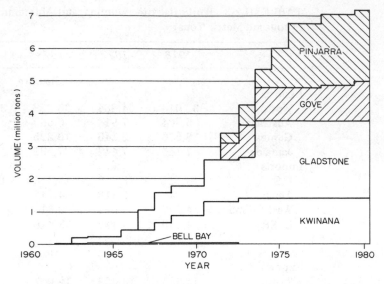

FIGURE III-2-1 Development of Australian alumina refining, 1960–1980. (From Ref. 3.)

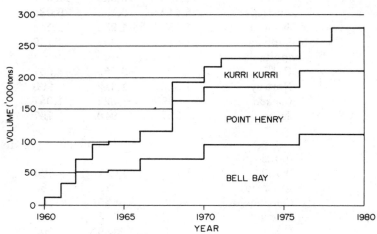

FIGURE III-2-2 Development of Australian aluminum smelting, 1960–1980. (From Ref. 3.)

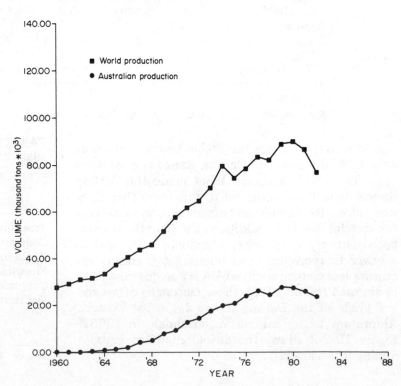

FIGURE III-2-3 World and Australian bauxite production, 1960–1982. (From Australian Bureau of Statistics.)

Geology of the Major Deposits

The overburden of Weipa averages about 1 m in thickness. The bauxite is flat lying, averaging 2.4 m thick, and is composed of 20% boehmite and 80% gibbsite. The ore is unconsolidated and can be mined with a front-end loader or hydraulic loader without blasting.[8]

The Gove deposit is across the Bay of Carpentaria in the extreme northeast of Arnhem land, Northern Territory. The ore at Gove averages 3.5 m in thickness and is near or at the surface. The bauxite is 97% gibbsite and only 3% boehmite.[8]

The Darling Range deposits, Western Australia, are overlain by loose sandy gravel 0.3 to 3 m thick, which usually overlies a hard cap, with a maximum thickness of about 1.5 m. The average ore thickness is 4 m.[9] Figure III-2-4 is a map of important locations in the Australian bauxite/aluminum industry.

Mining Facilities

Of the five major bauxite mines in Australia, three are located in the Darling Range. The other two, Weipa and Gove, are on the Bay of Carpentaria. Reserve figures of individual mines are given in Table III-2-5, and Table III-2-6 is a summary of current mining operations.

The Weipa mine is the world's largest bauxite mine, supplying 12 to 15% of global annual requirements. Over half the bauxite produced at Weipa is shipped to the alumina refinery at Gladstone, Queensland. The rest is exported to Japan and Western Europe. About half of the bauxite mined at Gove is exported; the rest is refined on-site.

Both the Weipa and Gove deposits required massive amounts of new infrastructure before they could be productive. The bauxite is transported by road, rail, and conveyor, then shipped out of the ports via Weipa and Gove. The refinery at Gove is situated next to the port facilities.

These are remote locations; access by road is possible only during the dry season, and even then the roads are difficult to traverse. Personal and materials transport is accomplished primarily by sea and air routes. The sea link from Brisbane takes 14 days. The cost and difficulty of building the required port and harbor facilities, beneficiation plants, and township and housing complexes, was great, even higher than the development cost of the mine sites. Such infrastructure was financed by the operating companies, as well as the state and commonwealth governments.

Little additional infrastructure was required to develop the Darling Range deposits since they occur near Perth. All of the bauxite from Alcoa's three existing mines in the area (Jarrahdale, Del

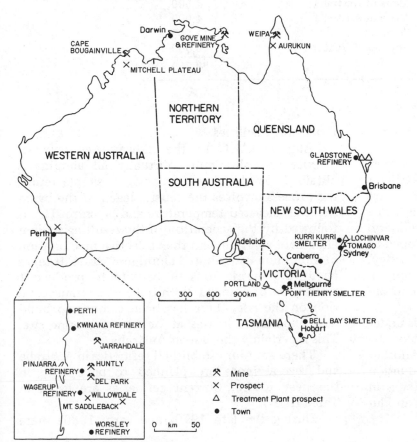

FIGURE III-2-4 Australian bauxite/aluminum prospects, mines, and treatment facilities, 1980. (From Ref. 3.)

TABLE III-2-5   Reserves of Individual Bauxite Mines in Australia
(Million Metric Tons)

Location	Measured	Indicated	Inferred	Total
Cape York Pen. (Queens.)				
Weipa	599	794	1740	3133
Aurukun	370			370
Other	75	137		212
Darling Ranges (W. Austr.)				
Alcoa leases		500[a]		500
Mt. Saddleback		250		250
Chittering	200			200
Other	34			34
Kimberley area (W. Austr.)				
Mitchell Plateau	235			235
Cape Bougainville		995		995
Gove (N. Terr.)	250			250
	1763	2676	1740	6179

[a]Includes measured.
Source: Ref. 3.

TABLE III-2-6   Current Bauxite Mining Operations in Australia

Facility	Ownership	Capacity (thousand metric tons/year)
Jarrahdale (W. Austr.)	Alcoa of Australia	5,000
Del Park (W. Austr.)	Alcoa of Australia	4,500
Huntly (W. Austr.)	Alcoa of Australia	4,000
Weipa (Queens.)	Comalco	11,250
Gove (N. Terr.)	Gove Joint Venture	5,000
		29,750

Source: Ref. 3.

Park, and Huntly) is refined at the nearby refineries at Pinjarra and Kwinana.

There are a number of bauxite deposits in various stages of development. Alcoa has two projects in the Darling Range nearing completion. Both are being built to supply new alumina refineries. The Mt. William mine will supply the Wagerup refinery, and the Mt. Saddleback mine is being developed to provide bauxite for the new refinery at Worsley. The Wagerup project should begin operations in 1983, and the Worsley refinery should come onstream in 1984.

A multimillion dollar feasibility study is under way to look into the possible development of the Mitchell Plateau area of Western Australia. Another area being studied for possible bauxite production is the Aurukun deposit just south of Weipa in Queensland. This may also be the site for an alumina refinery.

Processing Facilities

Alumina (Al_2O_3) is the midway stage between bauxite and aluminum. Virtually all alumina is obtained via the Bayer process. In simple terms, the process involves the caustic leach of the bauxite at an elevated temperature and pressure. This is followed by the separation of the resulting sodium aluminate solution, and then selective precipitation of aluminum as hydrated aluminum oxide. It takes about 2.5 metric tons of bauxite to produce 1 metric ton of alumina. With the cost of transportation on the rise, there has been a move to build new alumina refineries at or near the mine site. This is certainly the case in Australia.

There are four established refineries in Australia and several more are planned to be opened in the near future. Current refineries are listed in Table III-2-7.

During the late 1970s and early 1980s there

TABLE III-2-7 Capacities of Established Australian Alumina Refineries

Facility	Ownership	Capacity (thousand metric tons/year)
Kwinana (W. Austr.)	Alcoa of Australia	1400
Pinjarra (W. Austr.)	Alcoa of Australia	2500
Gladstone (Queens.)	Queensland Alumina	2400
Gove (N. Terr.)	Gove Joint Venture	1200
		7500

Source: Ref. 3.

were many plans for the expansion of existing alumina plants and the construction of new ones. Several of these plans have been altered due to the changing aluminum market.

Alcoa of Australia, which runs the Kwinana and Pinjarra refineries, has had to cancel its expansion plans for Western Australia due to the depressed aluminum market of 1982. Alcoa, which had previously increased the capacity of its Pinjarra refinery from 2.2 to 2.5 million metric tons per year (mtpy), had to delay the startup of the Wagerup complex, in which the first 500,000-ton per year stage was completed in 1981. When fully operational, the Wagerup refinery will have a 2 mtpy capacity.[10] The Gladstone refinery plans to expand its capacity to 2.75 mtpy by 1984. These plans could be delayed, since in 1982 Gladstone had to cut back production to 70% of capacity.[11] There are no plans to increase the capacity of the Gove refinery; in fact, production was cut to 50% of capacity in October 1981.

The refinery planned for Worsley is expected to begin operations in 1984 with a capacity of 1 mtpy. Later expansions to 2 mtpy are possible. On a longer time horizon, the Mitchell Plateau and Cape Bougainville deposits could supply bauxite to a planned 0.7- to 1.0-mtpy refinery at Mitchell Plateau. If feasible, construction could begin in 1984 and alumina production in 1987.

The smelting of alumina is the final stage in the processing of aluminum. The Hall-Heroult process is the most common method of smelting alumina. In this method the aluminum is dissolved in molten sodium aluminum fluoride (synthetic cryolite) and aluminum fluoride in large electrolytic furnaces known as pots or cells. The carbon cell wall acts as a cathode, and a carbon anode is suspended in the liquid. An electric current passed through the liquid causes the breakdown of the aluminum oxide. The aluminum metal migrates toward the cathode, collects at the bottom of the cell, and is then drained off at intervals. The molten aluminum is mixed with alloying materials and cast into ingots, ready for use in manufacturing.[5]

As the price of oil rose during the 1970s, there was a great deal of interest in building smelters in areas with inexpensive energy supplies. Australia, with its vast coal fields, is one such area. Electricity requirements of an average smelter are 8 kWh/lb, and 6 kWh/lb for the newest smelters. Energy costs account for 25 to 30% of the total aluminum production costs.[12] Countries with high electricity costs, such as Japan, are at a competitive disadvantage with nations with low-cost energy sources, such as Brazil with its hydropower or Australia with its coal. Also affected by rising fuel costs are transportation expenses. Since it takes 1.9 units of alumina to produce 1 unit of aluminum metal, aluminum companies have an incentive to locate new smelters near existing bauxite and alumina facilities.

There are three existing aluminum smelters in Australia and a host of new facilities in various stages of construction and planning. Table III-2-8 indicates the status of the established smelters.

In the late 1970s there were many ambitious plans to increase Australian aluminum output. From less than 300,000 metric tons, it was hoped that capacity would reach 1.7 million metric tons by 1985. Plans are still going ahead, but by 1982 the capacity figure planned for 1985 had dropped to 800,000 tons.

All the existing smelters have expanded capacity, or plan to do so, in the early 1980s. The Point Henry smelter increased its capacity from 100,000 to 165,000 metric tons per year (tpy).[13] Alcoa's new project, a 132,000-tpy smelter at Portland, Victoria, had been expected to start production by 1983.[13] Delays caused by the drop in demand for aluminum and escalating electricity rates have pushed the completion date to mid-1985. At one point in 1981, the State Energy Commission of Victoria boosted the price of power to be supplied to the new plant by 25%. This was on top of a 15%

TABLE III-2-8   Capacities of Established Australian Aluminum Smelters

Facility	Ownership	Capacity (thousand metric tons/year)
Point Henry (Vict.)	Alcoa of Australia	165
Bell Bay (Tas.)	Comalco	117
Kurri Kurri (N.S.W.)	Alcan Australia	90
		372

Source: Ref. 3.

increase announced earlier in the year.[14] Alcoa's reaction was to suspend construction for five months until the Commission agreed to review its pricing policy. Currently, Alcoa is looking for partners for the project to aid its insufficient cash flow.

Bell Bay, which was the oldest smelter in the nation, increased its capacity to 117,000 tpy in 1981. Plans are to raise the capacity again to 129,000 tpy.[15] The smelter is owned by Comalco, which also has a 30% interest in the Boyne smelter near Gladstone, Queensland. Boyne opened in 1982, but its production level is well below the 100,000 tpy expected. Expansion plans to raise the plant's capacity to 206,000 tpy by 1984, and 412,000 tpy by 1985, may be postponed.[3]

Alcan Australia Ltd. operates the Kurri Kurri smelter in the Hunter Valley, New South Wales. In 1980 the capacity was raised to 90,000 tpy, and plans were made to increase capacity still further to 135,000 tpy by early 1983. In April 1982, construction at Kurri Kurri was slowed down and the startup plans for the new capacity were deferred.[14]

Of the two remaining new projects, only the Tomago Smelter in New South Wales is still under construction. The Lochinvar smelter, which had hoped to produce 236,000 tpy by 1984, was canceled when the Broken Hill Proprietary Co. Ltd.

could not find new partners after Alumax dropped out of the project.[14] The Tomago smelter is still under construction, although at a slower pace. The first 100,000 tpy potline is expected to be completed in late 1983, with another due in 1985.[15] In the long run, there are plans for a smelter at Bunbury, near Worsley, Western Australia. Feasibility studies are being undertaken, but the question of adequate electricity supplies to support one or two smelters has yet to be answered. Tables III-2-9 and III-2-10 list capacity figures for refineries and smelters, and Table III-2-11 lists companies involved in the bauxite, alumina, and aluminum industries.

Prospects for Future Production

The aluminum industry has had a history of rapid growth. In the near future, this trend will continue, although at a lower rate. Over the next 20 years it is expected that the demand for aluminum will grow at a higher rate than for other base metals. This increase in finished aluminum is expected to be 4 to 5% per year over the next 20 years, from 18 million tons in 1980 to 38 million tons in the year 2000.[6] Total capital requirements are projected to be $95 billion (1980 dollars) through the end of the century, $30 billion in Aus-

TABLE III-2-9   Changes in the Capacities of Australian Alumina Refineries

Facility	Ownership	Capacity
Pinjarra	Alcoa of Australia	Expansion from 2.2 to 2.5 mtpy by 1982
Wagerup	Alcoa of Australia	Initial 0.5 mtpy in 1981 but startup delayed, to be expanded to 2 mtpy
Worsley	Worsley Aluminum Pty. Ltd.	Capacity of 1 mtpy by 1984, expansion to 2 mtpy is possible
Gladstone	Queensland Alumina	Expansion from 2.4 to 2.75 mtpy by 1984

Source: Ref. 3 and updated.

TABLE III-2-10   Changes in the Capacities of Australian Aluminum Smelters

Location	Ownership	Capacity
Bell Bay	Comalco Ltd.	Expansion from 117,000 tpy to 129,000 tpy planned for 1983
Point Henry	Alcoa of Aust.	Expanded from 100,000 tpy to 165,000 tpy in 1981
Kurri Kurri	Alcan Australia	Expansion from 90,000 tpy to 135,000 tpy by 1983
Portland	Alcoa of Aust.	132,000 tpy delayed until mid-1985; later expansion to 528,000 tpy has been mooted
Gladstone	Boyne Smelters	Completion of 103,000 tpy in 1982, expansion to 206,000 tpy by 1984, later expansion to 412,000 tpy is possible
Gladstone	Alcan Queensland 55%; Nippon Light Metal 30%; Sumitomo 15%	Planned 100,000 tpy by 1984; canceled in 1982
Lochinvar	Hunter Valley Alu.	Planned 236,000 tpy by 1984; canceled in 1982
Tomago	Tomago Aluminum	110,000 tpy by 1983, 220,000 tpy by 1985

Source: Ref. 2 and updated.

tralia alone.[6] Real aluminum prices could rise by 15% over this period.[6]

As of the early 1980s the boom in the Australian aluminum industry had not been as great as was once hoped. This can be attributed to the decline in the demand for aluminum caused by the world recession of 1982. When demand picks up, so should the expansion plans of the new refineries and smelters.

With the expected growth of the Pacific Rim, especially in Korea, China, and Southeast Asia, Australia is in a strong position to become one of the world's leading aluminum producers. With Japan moving out of the aluminum industry, as seen in Figure III-2-5, decreasing Japanese imports of bauxite and alumina, and increasing imports of aluminum metal, there will be a void which increased Australian production can fill. This trend should continue as Japan closes down more of its smelters.

Several problems could hamper the growth of Australian aluminum production. The crucial question is the price and availability of electricity. Other problems involved in building and operating the new aluminum projects, such as transport facilities, will also have to be solved. In addition, there are environmental problems associated with mining bauxite and smelting aluminum, and there are those who believe that involvement with multi-national corporations is detrimental to the Australian economy. Large investments in a single industry may not be wise in the long run. These problems are discussed at length in Chapter III-5.

IRON ORE

Production, Resources, and Reserves

The USSR is by far the world's largest producer of iron ore. Brazil and Australia rank next, each producing less than half of the Soviet annual level. Australia has produced about 10% of the world's annual production in each year since 1973.[1] (Table III-2-12).

Australia has about 5% of the world's iron ore reserves, totaling more than 35,000 million metric tons (1981), of which 17,800 million metric tons are readily usable, low-phosphorus ore. Another 17,200 million metric tons are high-phosphorus ore. Over 95% of these reserves lie in the Hamersley Iron Provenance of Western Australia.[16] Total resources are large enough to support the iron ore industry for many decades to come.

Exports and Trade

About 15% of Australian annual iron ore production is consumed by the domestic steel industry; the rest is exported (Table III-2-13). Seven countries—Brazil, Australia, Canada, the USSR, Sweden,

TABLE III-2-11   Ownership in Australian Aluminum Companies

Company	Ownership
Alcan Australia	70% Alcan Aluminum Ltd.; 30% public
Alcan Queensland Pty. Ltd.	100% Alcan Aluminum Ltd.
Alcoa of Australia Ltd.	51% Aluminum Company of America; 20% Westminer Investments Pty. Ltd.; 13% Broken Hill South Ltd.; 12% North Broken Hill Ltd.; 4% other
Comalco Ltd.	45% Kaiser Aluminum and Chemical Corp.; 45% Conzinc Riotinto of Australia Ltd.; 10% public
Gove Joint Venture	70% Swiss Aluminum Australia Pty. Ltd.; (100% Swiss Aluminum Ltd.); 30% Gove Alumina Ltd. (51% CRS Ltd., 13% Pek-Wallsend Ltd., 36% Australian insurance companies and banks)
Hunter Valley Aluminum Pty. Ltd.	45% Alumax, Inc. (dropped out); 35% Dampier Mining Company Ltd. (100% the Broken Hill Proprietary Company Ltd.); 20% Alfarl Pty. Ltd. (20% Furukawa Electric, 27.5% Showa Aluminum, 20% Mitsiu and Company, 15% Mistisu Aluminum, 10% Toyota Motor)
Worsley Alumina Pty. Ltd.	40% Reynolds Australia Alumina Ltd. (100% Reynolds Metals Company); 30% Shell Company of Australia Ltd. (100% Royal Dutch Shell); 20% Dampier Mining Company Ltd.; 10% Kobe Alumina Associates Australia (40% Kobe Steel Ltd., 35% Nissho-Iwai Co., 25% C. Itoh and Co. Ltd.)

Source: Ref. 3.

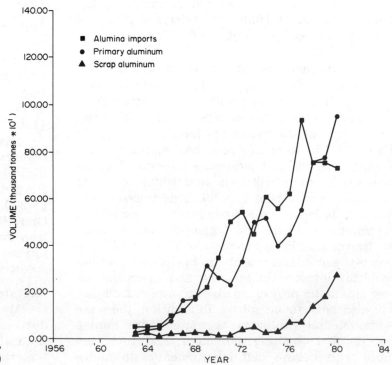

FIGURE III-2-5   Total Japanese imports of bauxite, alumina, and aluminum, 1963-1980. (From Ref. 48.)

TABLE III-2-12   World Iron Ore Production, 1976–1982 (Million Metric Tons)

	1976	1977	1978	1979	1980	1981	1982
World	884	838	843	891	888	861	788
USSR	239	240	244	246	245	242	240
Brazil	82	77	85	92	106	100	98
Australia	93	96	83	92	95	86	85
U.S.	78	55	84	86	71	74	36
Canada	57	55	44	60	49	51	36
China	65	60	70	65	75	70	71
India	44	43	39	40	41	41	42

Sources: Refs. 1 and 2.

TABLE III-2-13   World Iron Ore Exports, 1976–1982 (Million Metric Tons)

	1976	1977	1978	1979	1980	1981	1982
World	373	354	348	392	382	364	
Brazil	67	59	66	77	79	80	
Australia	81	79	75	78	80	72	73
USSR(e)[a]	44	41	41	39	38	38	
Canada	45	45	32	49	41	41	
India	42	44	43	39	25	23	

[a]e, estimated.

Sources: Refs. 1 and 2.

India, and Liberia—account for 80% of the world's iron ore export trade.[1]

The rapid growth of Australia's iron ore production since the mid-1960s paralleled the expansion of the Japanese steel industry. About 45% of Japan's iron ore requirements come from Australia, which constitutes about 70% of total exports[1] (Table III-2-14). Other markets for Australian iron ore include Europe (13% of exports), China (7%), South Korea (5%), and Taiwan (2%) in 1980.[17] Iron ore is exported as lump, fines, pellets, and run-of-mine. Of the 80.1 million metric tons of iron ore exported in 1980, 31.8 million were lump, 45.3 million were fines, and 3.0 million metric tons were in pellet form.[17]

Uses

Iron is used in greater amounts than any other metal, being the major ingredient in steel. In Japan and the developed countries of Western Europe, North America, and Australia, approximately 0.7 metric ton of steel is produced annually per head of population.[17] Steel is ubiquitous, and it is vital to the transportation, building, and consumer-goods industries.

The rate of future steel use depends on how healthy the world economy will be, as well as how rapidly the lesser-developed countries' economies grow. Substitution of steel by aluminum and plastics may continue. New high-technology materials being developed in Japan, such as fine ceramics,

TABLE III-2-14   Australian Iron Ore Exports, 1976–1982 (Million Metric Tons)

	1976	1977	1978	1979	1980	1981	1982
Total	81.4	78.9	75.3	78.3	80.1	72.0	72.6
Japan	62.6	62.3	52.2	55.3	58.8	55.0	
Europe	15.4	12.4	12.8	12.2	10.0	9.2	
China	1.4	1.4	6.2	5.7	5.3		
South Korea	1.2	1.6	1.9	3.3	3.8	4.1	
Taiwan	—	0.9	1.7	1.6	1.7		

Source: Ref. 1.

carbon fibers (which are lighter than aluminum, and stronger than steel), and amorphous metals, could also begin to replace steel.

History of the Industry

Attempts to establish an Australian iron industry date back to 1848, but it was not until 1915 that a successful large-scale iron works was initiated. It was located in Newcastle, New South Wales, and operated by the Broken Hill Proprietary Company Ltd. (BHP). Coking coal was mined near Newcastle, and the iron ore was shipped from Iron Knob in South Australia.[16]

Until 1960, all iron ore mined was used domestically. With the realization that there were literally mountains of high-grade iron ore in the Pilbara region of Western Australia, the Commonwealth Government relaxed its embargo on iron ore exports in December 1960, and the Western Australian Government lifted its restrictions on the granting of titles for the exploration and mining of iron ore in March 1961.[17] In the 10 years following 1965, annual iron ore production increased from 7 mtpy to 98 mtpy.[17]

As shown in Figure III-2-6, the growth in Australian iron ore production parallels the expansion of the Japanese steel industry. Several years worth noting are:

1972: Due to the recession in the steel industry, production of iron ore from Western Australia leveled off. Additional problems caused by a prolonged strike by Japanese seamen added to the slowdown of production.

1976: Production declined for the first time since 1962, when the industry was in its fledgling stage. This can be attributed to several factors. First the steel industry was still shaking off the effects of the 1975 recession; second and more important were the continuing industrial problems in the Pilbara area. The unreliability of supplies caused the Japanese to look to Brazil and India for a greater precentage of their iron ore requirements.

1978: The recession in the Japanese steel industry had a major impact on Australian iron ore production. The drop of 10% was the largest ever. The mining industry in the Pilbara region was so capital intensive that this unused capacity caused a serious reduction in the return on the capital investment. Australia realized that they must diversify their markets and seek new outlets.

Geology of the Major Deposits

Iron is the second most abundant metallic element in the earth's crust. Although iron is found in most rocks, only a few are suitable as ores (Table III-2-15). There are six iron provenances in Australia, of which four are in Western Australia (Figure III-2-7). The other two are the Middleback Ranges in South Australia and Savage River in Tasmania (Figure III-2-8).

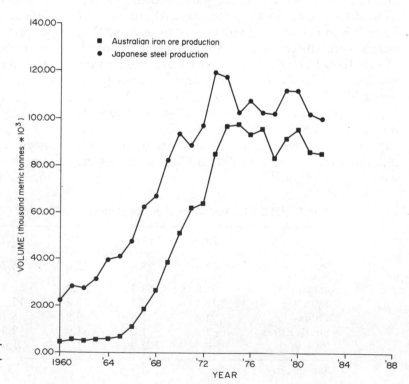

FIGURE III-2-6 Growth in Australian iron ore production versus Japanese steel production, 1960–1982. (From Ref. 48.)

TABLE III-2-15   Iron Minerals

Mineral	Chemical composition	Contained iron(%)
Magnetite	Fe_3O_4	72.4
Hematite	Fe_2O_3	69.9
Geothite	$Fe_2O_3H_2O$	62.0
Limonite[a]	Mixture of hydrated iron oxides	Variable
Siderite	$FeCO_3$	48.2

[a]Not generally recognized as a mineral.

Source: Ref. 16.

The Savage River provenance has the only magnetite mine in Australia. The deposits occur along a discontinuous mineralized zone up to 23 km in length. The central deposit consists of vertical, discontinuous lenses of massive and rhythmic-layered magnetite-pyrite ore.[9] It is mined by open-cut methods. Crude ore reserves are estimated to be 93 million metric tons, with an average grade between 36 and 39%.[13]

The Middleback Ranges have been strongly folded. The ore is located near or at the surface, where open-cut mining is used. Estimated in-situ

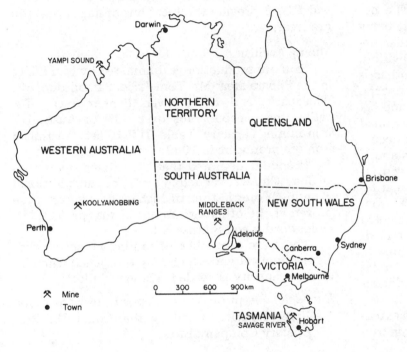

FIGURE III-2-7 Australian iron ore mines and treatment plants except the Pilbara region, 1981. (From Ref. 17.)

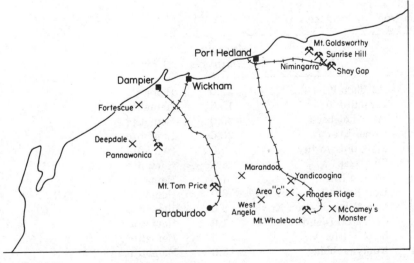

FIGURE III-2-8 Pilbara iron ore locations, 1981. (From Ref. 17.)

reserves are approximately 180 million metric tons of high-grade ore, and 170 million metric tons of low-grade ore.[9] The average grade of the ore, at a cutoff of 55% iron, is 63%.

Of the four iron provenances recognized in Western Australia, the Hamersley is the most important. The other three provinces are the South-Central, the North Pilbara, and the Kimberley.

The two island deposits in the Yambi Sound in the Kimberley province appear to be the product of direct sedimentation rather than the enrichment of preexisting sediments. The ore bed forms a distinct sedimentary unit in a surrounding succession of quartzite, shale, and conglomerate.[16] The ore has a very high grade of 65%, with hematite being the principal ore mineral.[16] There are similar deposits in other areas of the Kimberley Division. The largest occurs at Pompey's Pillar. Another is located in the Bandicoot Range near Kununarra. Neither prospect is being developed at this time.

Koolyanobbing is the only deposit being mined in the South-Central province. The deposit occurs in association with banded iron formations and is the product of secondary enrichment. Proven reserves are 60 million metric tons of hematite-limonite, which average slightly over 60% iron. There are numerous similar and smaller deposits in the province, but at this time none are being mined. The deposits of the North Pilbara are like those in the South-Central province: banded iron formations, interbedded with metamorphosed sedimentary and volcanic rocks.[16]

The Hamersley banded iron formation is extensive, spanning an 80,000-km² area. Three main iron

bands have been locally enriched to form zones of high-grade hematite. Total reserves of hematite have been estimated at 28,000 million metric tons. In addition, over 8000 million metric tons of low-grade limonite is available.[16] The hematite is mined in lump form, containing up to 68% iron.[16] In conjunction with these deposits, are found thousands of millions tons of hematite-geothite ore. At 50 to 65% iron these ores require beneficiation. The low-iron-content (about 55%) limonite ores can be heated to drive out the principal impurity, combined water. All the deposits in the Hamersley range are mined by open-cut methods. Figures III-2-7 and III-2-8 provide the locations of important iron ore areas.

Mining Facilities

Iron ore production is dominated by four mines in the Pilbara area: Mt. Tom Price, Paraburdoo, Mt. Whaleback, and Robe River (Pannawonica). Together they account for over 80% of Australia's iron mining capacity. Table III-2-16 lists Australian iron ore producers in 1980.

It should be noted that the mining operations at Shay Gap were ended in 1981, but an additional 32 million metric tons of high-grade reserves were proven at the other two mines in that project, Mt. Goldsworthy and Sunrise Hill.

There are a number of mining prospects that are economical and which may be opened when increased capacity is needed. These include:[17]

Deepdale: A pisolitic ore deposit; near the Robe River deposit; owned by the Broken Hill Proprietary Company Ltd.

TABLE III-2-16 Australian Iron Ore Producers

Project	Mines	Capacity (mtpy)	Ore type
Hamersley (W. Austr.)	Mt. Tom Prince	31.0	Hematite
	Paraburdoo	15.0	Hematite
Mt. Newman (W. Austr.)	Mt. Whaleback	40.0	Hematite
Cliffs Robe River (W. Austr.)	Robe River	20.0	Limonite
Goldsworthy (W. Austr.)	Mt. Goldsworthy	2.0	Hematite
	Shay Gap	4.5	Hematite
	Sunrise Hill	1.5	Hematite
Middleback Ranges (S. Austr.)	Iron Baron–Iron Prince	3.0	Hematite
	Iron Monarch–Iron Knob	3.0	Hematite
Yampi Sound (W. Austr.)	Cockatoo Island	1.0	Hematite
	Koolan Island	2.5	Hematite
Koolyanobbing (W. Austr.)	Koolyanobbing	3.0	Hematite, goethite
Savage River (Tas.)	Savage River	2.5	Magnetite
Wundowie (W. Austr.)	Koolyanobbing	0.1	Hematite, goethite

Source: Ref. 17.

Yandicoogina: Another pisolitic ore deposit, about 80 km northwest of Newman; owned by the Colonial Sugar Refinery Ltd.

Marandoo: A hematite-goethite deposit located 40 km northeast of Mt. Tom Price; owned equally by Conzinc Rio Tinto Australia Ltd. and Hancock & Wright.

MAC (Mining Area C): A hematite-goethite deposit about 100 km northwest of Newman; owned by group operating the Goldsworthy project.

West Angelas: A hematite-goethite deposit 100 km west of Newman; owned by Cliffs Robe River Iron Associates.

Processing Facilities

The steel industry in Australia is oriented to the needs and objectives of its domestic market. Its capacity is about 10 million metric tons. The major steel centers are located on the coast of New South Wales at Port Kembla and Newcastle. There are three other plants, at Whayalla, South Australia, and at Kwinana and Wondowie in Western Australia. The mill at Wundowie is of only minor importance, producing only small amounts of pig iron in its charcoal blast furnace. The other four plants are operated by the Broken Hill Proprietary Company Ltd (BHP).

The Australian steel industry has been in poor shape. During 1982, BHP cut back on both its production and work force at the Newcastle and Port Kembla mills. The health of the industry was reflected by statements by BHP as of late 1982 that threatened to shut down its mills within four years if it did not receive special aid in the form of tax breaks and a guaranteed share of the local market.[18] During the first quarter of fiscal 1984 (June–August 1983), BHP has shown signs of recovery as the company reported a 42% increase in profits over the previous year. BHP's steel division accounts for 80% of overall sales.

Iron ore preparation is done to a high degree, including the preparation of ore into sized lump (30 × 6 mm), fine product (minus 6 mm), extensive blending, and the operations of beneficiation and pellet plants. Table III-2-17 lists Australian processing plants.

Port Facilities

The building of infrastructure was the crucial element in the development of the Pilbara deposits. Approximately $2 were spent on infrastructure for every dollar spent on mine development. In addition to the rail and port facilities that were built, new towns and power stations had to be installed, all at premium costs due to the remote locale. Tables III-2-18 and III-2-19 provide information on rail distances and port facilities.

Ownership in iron ore ventures is primarily Australian, with a varying percentage of Japanese involvement. American ownership is found in several projects (Table III-2-20).

Prospects for Future Production

The future of Australian iron ore will depend heavily on the success of the steel industries of Asia, particularly of Japan. Since production at Australian iron mines has been well below capacity in the early 1980s, expansion of existing mines and the development of new ones does not seem likely. Of the 100-million metric ton annual capacity, Australian producers could expect to ship no more than 85 to 90 million metric tons in 1985, and may only approach 100 million metric tons by 1990.[13]

MINERAL SANDS

Production, Resources, and Reserves

Mineral sands products (rutile, ilmenite, zircon, and monazite) occur conjointly, although the composition varies from deposit to deposit. Australia is the leading producer of mineral sands, supplying the world with 65% of its rutile, 30% of its ilmenite, 75% of its zircon, and is a major producer of monazite[1] (Table III-2-21). Other producers of rutile include South Africa, Sierra Leone, and Sri Lanka. For ilmenite there are Malaysia, the United States, Sri Lanka, India, Canada, South Africa, and

TABLE III-2-17 Iron Ore Processing Plants

Plant type	Plant locations
Beneficiation	Savage River, Whyalla, Mt. Whaleback, Mt. Tom Price
Pellet	Port Latta, Whyalla, Dampier[a], Cape Lambert[b]
Sinter	Port Kembla, Newcastle, Kwinana
Blast furnace	Port Kembla, Newcastle, Kwinana, Whyalla, Wundunwie

[a]Placed on care and maintenance basis in January 1980.
[b]Operation suspended in April 1980.
Source: Ref. 17.

TABLE III-2-18 Mine-to-Port Distances

Mine	Distance to port (km)	Port
Mt. Tom Price	293	Dampier
Paraburdoo	393	Dampier
Mt. Whaleback	426	Port Hedland
Robe River	169	Port Walcott
Mt. Goldsworthy	112	Port Hedland
Sunrise Hill	180	Port Hedland
Iron Baron-Iron Prince	47	Whyalla
Iron Monarch-Iron Knob	52	Whyalla
Cockatoo Island	—	Cockatoo Island
Kooland Island	—	Kooland Island
Koolyanobbing	489	Kwinana
Savage River	85	Port Latta

Source: Ref. 17.

Norway. Note that ilmenite from Canada and South Africa is in the form of titaniferous slags, and from Norway as rock ilmenite. Zircon is also supplied by South Africa, the United States, India, and Sri Lanka. Australia has great resources of all four minerals, as shown in Table III-2-22.

Exports and Trade

Very little rutile and only a small amount of zircon is consumed domestically. About 20% of ilmenite production is processed locally for titanium dioxide pigment or beneficiated to synthetic rutile.[1]

The Australian mineral sands industry is substantially export-oriented, with the bulk of production shipped in concentrate form. The levels of export and destinations are shown in Table III-2-23.

Uses

Over 80% of ilmenite and 60% of rutile produced worldwide is used in the manufacture of titanium dioxide pigments.[8] These pigments are used in paints, papermaking, and plastics. The remaining rutile and ilmenite goes into the making of titanium metal. Rutile is the more valued of the two minerals, due to its higher titanium dioxide content, about 95 to 96% compared to 45 to 60% in ilmenite.

Titanium is desired for its high strength-to-weight ratio, and because of its resistance to corrosion. Alloys of the metal are utilized in the aircraft and aerospace industries, which uses about 75% of total titanium metal production. In addition, about 15% is used in chemical and desalination plants and 10% in heat exchanger tubing for power stations. The demand for titanium in these industries is expected to remain strong, with a growth rate of 6 to 8% per annum.[14] Zircon is used in refractories (furnace lining, molten metal, and glass vessels, etc.), foundry sands, in ceramic glaziers, and for the manufacture of abrasives.

History of the Industry

The mining of mineral sands was originally undertaken over 100 years ago. At that time east coast "Black Sands" were worked for the small amounts of gold, platinum, tin, and monazite they contained. It was not until 1895 that the presence of zircon was recognized, and 1925 before the rutile was identified.[8] However, during that period there was no commercial demand for rutile or zircon. With the development of electromagnetic and electrostatic separation techniques and the titanium boom of the 1950s, rutile and zircon production increased dramatically. Figure III-2-9 shows the

TABLE III-2-19 Australian Ports Handling Iron Ore

Port	Berth	Handling facilities	Vessel accommodation
Dampier	Parker Point	Single-sided jetty loader, 7200 metric tons per hour (tph)	120,000 dry weight tons (dwt)
	East Intercource Island	Single-sided jetty loader, 7500 tph	160,000 dwt
Port Walcott	Cape Lambert	Double-sided jetty single loader, 6000 tph	150,000 dwt / 100,000 dwt
Port Hedland	Finucone Island	Single jetty loader, 4500 tph	120,000 dwt
	Nelson Point	Double jetty, two loaders, 8000 tph each	160,000 dwt
Yampi Sound	Cockatoo Island	Single jetty loader, 1600 tph	65,000 dwt
	Kooland Island	Single jetty loader, 3000 tph	100,000 dwt
Port Latta	Savage River	Single jetty loader, 2700 tph	100,000 dwt
Whyalla	Ore Jetty	Traveling loader, 3000 tph	70,000 dwt

Source: Ref. 17.

TABLE III-2-20 Ownership in Iron Ore Projects

Project	Mine	Principal ownership
Hamersley	Mt. Tom Price, Paraburdoo	82.3% CRA Ltd.; 11.5% public; 6.2% Japanese trading companies and steel mills
Mt. Newman	Mt. Whaleback	30% Pilbara Iron Ltd. (68% CSR Ltd.); 30% Dampier Mining Co. Ltd. (BHP subsidiary); 25% Amax Iron Ore Corp.; 10% Mitsui-C. Itoh Iron Pty. Ltd.; 5% Seltrust Iron Ore Ltd.
Cliffs Robe River	Robe River	35% Robe River; 30% Cliffs Western Australian Mining Co. Pty. Ltd.; 30% Mitsui Iron Ore Development Pty. Ltd.; 5% Cape Lambert Iron Associates
Goldsworthy	Mt. Goldsworthy Sunrise Hill	46⅔% Consolidated Gold Fields (Aust.) Pty. Ltd.; 33⅓% Utah Development Company; 20% M.I.M. Holding Ltd.
Yampi Sound	Cockatoo Island, Kooland Island	100% BHP Ltd.
Koolyanobbing	Koolyanobbing	100% BHP Ltd.
Savage River	Savage River	50% Northwest Iron Co. Ltd. (U.S., Japanese, and Australian companies); 50% SRM Venture (Japanese companies)
Wundowie	Koolyanobbing	100% Agnew Clough Ltd.

Source: Ref. 17.

TABLE 2-21 World Mineral Sands Production, 1976–1982 (Thousand Metric Tons Concentrates)

	1976	1977	1978	1979	1980	1981	1982
				Rutile			
World	403	345	301	361	417	361	354
Australia	395	325	257	274	294	229	227
South Africa	—	—	18	42	48	50	50
Sierra Leone	—	—	—	10	47	51	50
Sri Lanka	—	—	11	15	13	13	14
				Ilmenite			
World	3385	4090	4828	4268	4745	4741	4118
Australia	959	1033	1255	1150	1254	1337	1215
Canada[b]	823	711	850	447	871	762	653
Norway	848	828	767	820	816	658	562
U.S.	591	580	535	580	499	462	236
USSR	NA	NA	NA	NA	417	426	426
				Zircon			
World	NA	425	454	549	577	547	437
Australia	420	398	392	447	459	424	339
South Africa	—	13	36	82	95	100	80

[a] $\frac{1}{n}$, below 10,000 metric tons.

[b] Titania slag containing 70 to 71% TiO_2.

Sources: Refs. 1 and 2.

TABLE III-2-22   Australian Resources of Mineral Sands[a]
(Thousand Metric Tons)

	Ilmenite	Rutile	Zircon	Monazite
Economic resources				
East coast	13,020	5864	5,700	36
West coast	32,695	3244	7,700	270
	45,715	9108	14,420	306
Subeconomic resources				
East coast	10,205	1047	NA	19
West coast	13,343	1180	NA	59
	23,548	2227	5,420	78

[a]Figures as of December 1979, except monazite 1975.
Source: Ref. 19.

TABLE III-2-23   Australian Mineral Sands Exports, 1976–1982 (Thousand Metric Tons)

	1976	1977	1978	1979	1980	1981	1982
Rutile							
Total	347	247	371	318	315	216	197
U.S.	155	101	195	121	130		
U.K.	60	40	48	81	55		
Japan	17	17	26	27	35		
Ilmenite							
Total	965	1095	1000	932	1084	923	818
U.K.	214	239	227	201	184		
France	152	127	156	82	54		
Japan	178	131	74	53	60		
U.S.	239	334	266	166	359		
Zircon							
Total	345	328	391	479	502	444	405
Japan	125	89	109	188	176		
U.S.	52	60	59	66	94		
Italy	43	32	43	65	81		
France	14	18	25	18	24		

Sources: Ref. 1; Australian Bureau of Mineral Resources.

impressive growth of world and Australian rutile production through the 1960s and the erratic behavior of the metal during the 1970s.

The two most obvious trends in this figure are first, the steady increases in rutile production between 1960 and 1970, and second, the decreasing percentage of world production contributed by Australia. The price of rutile increased sharply between 1973 and 1975, and again in 1978. This could be expected, since rutile is highly energy intensive to refine. To explain the reduction of the Australian share of the market from 98% in 1976 to 64% in 1982, there is the emergence of new producers such as Sierra Leone, South Africa, and Sri Lanka, which as a group contributed 32% in 1982, up from 9.5% four years earlier.

Geology of the Major Deposits

Mineral sands deposits occur on the east coast between Newcastle, New South Wales, and Gladstone, Queensland. They are also found along the southwest coast of Western Australia at Capel and Eneabba.[19] The sands are found in placer deposits adjacent to beaches, and inland along ancient shorelines.

East Coast deposits are high in rutile and zircon

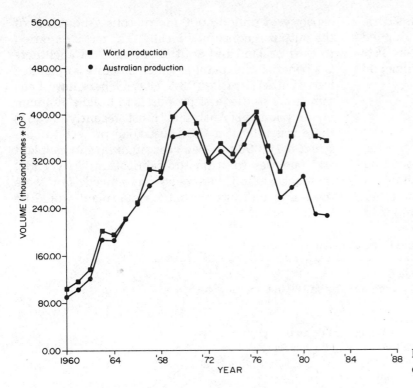

FIGURE III-2-9 Australian and world rutile production, 1960–1982. (From Ref. 48.)

and to a lesser extent ilmenite. Much of the east coast ilmenites contain too much chromium to be marketable in large quantities. Ilmenite is the major mineral in the west coast deposits, comprising between 70 and 90% of the heavy minerals.[8] Zircon and minor amounts of rutile are also recovered. Figure III-2-10 is a map of mineral sands deposits currently being mined.

Mining Facilities

In preparation of mining, the land is cleared of vegetation and a thin layer (15 to 30 cm) of top soil. The sands are mined by suction dredging or heavy earth-moving equipment. A concentrate of heavy minerals, which is obtained by wet gravity techniques, is transported to a central "dry plant," where electromagnetic and electrostatic processes

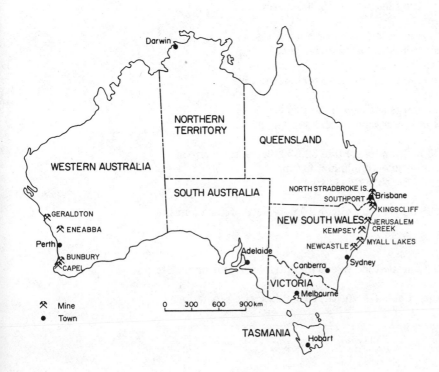

FIGURE III-2-10 Australian mineral sands deposits currently being mined, 1980. (From Ref. 19.)

are used to separate the individual mineral components. After the dry plant, the product is shipped out in bulk or packaged form. Table III-2-24 lists the principal companies involved in the mining and processing of mineral sands.

Processing Facilities

As is the case with many of its other minerals, Australia does very little in the way of processing mineral sands past the raw stage. Its two titanium dioxide plants, one at Burnie, Tasmania, the other at Bunbury, Western Australia, have a combined capacity of only 68,000 metric tons. About half of the output is consumed locally. The rest is exported to New Zealand and Southeast Asia. At Capel there is a beneficiation plant in which ilmenite is upgraded into synthetic rutile (93% TiO_2). There have been numerous studies and proposals to build a titanium metal plant in Australia. Most recently, (1981) there was talk of a U.S.-Australian project to construct a 10,000-ton per year titanium sponge plant and facilities for titanium fabrication.[20] By 1982, no construction had been announced, nor were there any indications that a metal plant would be forthcoming.

TABLE III-2-24   Mineral Sands Companies and Operations

Associated Minerals Consolidated Ltd.
  Capacity: ilmenite 390,000 metric tons, rutile 100,000 metric tons, zircon
      175,000 metric tons.
  Mines and separation plants
    North Stradbloke Is. (Queens.) mine and separation plant
    Southport (Queens.) separation and grinding plants
    Tingalpa (Queens.) grinding plant
    Jerusalem Creek (N.S.W.), mine
    Kempesley (N.S.W.) mines (2)
    Hexham (N.S.W.) separation and grinding plant
    Capel (W. Austr.) mine and separation plants
    Eneeabba (W. Austr.) mine and separation plants
Consolidated Rutile Ltd.
  Capacity: rutile 60,000 metric tons, zircon 50,000 metric tons
  Mines and separation plants
    North Stradbloke Is. (Queens.) mine
    Meeandah (Queens.) separation plant
Mineral Deposits Ltd.
  Capacity: rutile 40,000 metric tons, zircon 45,000 metric tons
  Mines and separation plants
    Crescent Head (N.S.W.) three mines and a separation plant
    Myall Lakes (N.S.W.) six mines and a separation plant
Rutile and Zircon Mines Ltd.
  Capacity: rutile 40,000 metric tons, zircon 45,000 metric tons
  Mines and separation plants
    Harrington (N.S.W.) two mines and separation plant
    Tomago (N.S.W.) three mines, separation and grinding plants
Allied Eneabba Ltd.
  Capacity: ilmenite 250,000 metric tons, rutile 50,000 metric tons, zircon
      130,000 metric tons, monazite 10,000 metric tons
  Mines and separation plants: Eneabba (W. Austr.)
Cable Sands Ltd.
  Capacity: ilmenite 165,000 metric tons, zircon 10,000 metric tons, monazite
      1,000 metric tons
  Mines and separation plants: Caple (W. Austr.)
Westralian Sands Ltd.
  Capacity: ilmenite 500,000 metric tons, zircon 40,000 metric tons, monazite
      2,500 metric tons
  Mines and separation plants: Caple (W. Austr.) two mines, two separation
      plants

Source: Ref. 19.

Prospects for Future Production

Australia's future in the mineral sands industry is in some doubt. Of course, Australia will continue to play a major role in production and exports, but the dominance it once had is being eroded away. As previously noted, there has been growing competition in the rutile export market. The increasing use of synthetic rutile, a product that Australia could produce more of, and titanium slag, has also cut into the Australian share of the market. What could really change the market would be the emergence of Brazil as a major producer of rutile. With a reserve base of over 85 million metric tons, compared to 9.1 million metric tons in Australia, and practically no production, Brazil has the potential to become the world's largest producer and exporter of rutile.[2]

Australia must also face the fact that most of its high-grade deposits have been mined out, causing mining companies to move to lower-grade deposits which require more expensive processing facilities and greater economies of scale. In addition, much of the rutile-rich east coast sands have been put off limits, due to environmental considerations. Mining is restricted in national park areas of northern New South Wales, denying access to an estimated 45% of the known high-grade reserves of the east coast.[13]

MANGANESE

Production, Resources, and Reserves

Australia ranks among the top five producers of manganese, with about 6% of annual world output. The USSR and South Africa are the two largest producers, with over 60% of total world production.[1]

Demonstrated economic resources of Australian manganese ore (averaging 40% Mn) are 490 million metric tons, as of 1980. These are located almost exclusively on Groote Eylandt, which lies just off the coast of the Northern Territory in the Gulf of Carpenteria. There are an additional 401 million metric tons of lower-grade marginal and sub-economic deposits scattered in Queensland and Western Australia.[21] Table III-2-25 gives production statistics for manganese.

Exports and Trade

Domestic consumption of manganese accounts for roughly 25 to 33% of production. Some 525,000 metric tons was consumed domestically in 1981 in the making of pig iron, manganese alloys, and manganese alloy sinter.[22] The low levels of ore produced at Peak Hill are used locally in brick making. Ore from Cloncurry is used in uranium processing.

Australia is the fourth-largest exporter of manganese ore, behind South Africa, Gabon, and the USSR. Over half of the Groote Eylandt production is exported as manganese ore; the balance is shipped to the Bell Bay treatment plant in Tasmania, and converted into ferroalloys. About 35% of the plant's ferromanganese has been exported in recent years.[1]

The largest consumer of Australian manganese ore is Japan, which imports about 50% of Australian production. Other important markets are the European Common Market (30%), South Korea (13%), and the United States (7%).[22]

Uses

Manganese is used almost exclusively in the manufacture of steel and cast iron. It acts as a desulfurant and deoxidant, as well as an alloying element to add strength, toughness, and hardness to special steels. With the use of manganese so closely tied to the production of steel, declines in manganese production can be traced to declines in worldwide production of steel. This was the case in 1978, and also 1981 and 1982.

History of the Industry

Australia began to produce manganese in significant quantities in 1965, with the opening of the Groote Eylandt mine. The initial capacity in

TABLE III-2-25 World Manganese Production, 1976–1982 (Million Metric Tons Actual Weight)

	Percent Mn	1976	1977	1978	1979	1980	1981	1982
World	43	24.6	22.8	22.3	24.5	23.5	23.6	20.1
Australia	37–53	2.2	1.4	1.3	1.7	2.0	1.4	1.1
USSR	35	8.6	8.6	9.1	9.5	10.0	9.4	9.4
South Africa	30–48	5.5	5.0	4.3	5.2	4.5	5.0	4.0
Gabon	50–53	2.2	1.9	1.7	1.8	1.6	1.5	1.3
Brazil	38–50	1.7	1.5	1.6	1.7	1.5	1.9	1.5
India	10–54	1.8	1.9	1.6	1.6	1.5	1.5	1.3

Sources: Refs. 1 and 2.

that year was 200,000 metric tons of lump ore.[22] By 1970, output had reached 750,000 metric tons annually. In 1972 a beneficiation plant, with a capacity of 1.2 mtpy, was commissioned so as to increase the range of manganese products available and to ensure consistent product grades. The plant was upgraded to a yearly capacity of 2 million metric tons in 1975.[22] Figure III-2-11 illustrates the development of the industry.

Geology of the Major Deposits

The manganese oxide deposits at Groote Eylandt are flat-lying tabular strata with a thickness that ranges from 0.5 to 15 m, averaging 3.5 m. It is overlain by lateritic overburden up to 24 m thick.[9] Figure III-2-12 is a map of Manganese mines and treatment plants.

Mining Facilities

As previously stated, the Groote Eylandt deposit is the source of most of Australia's manganese production and of all Australia's manganese exports. The deposit is mined using open-pit methods. The ore is then beneficiated to the following products:[21]

Premium: A lump and fine product with a manganese content of 51 to 53%. Because of the low silica content, the lump ore can be charged directly into a ferromanganese blast furnace. The fine ore, which is sintered before furnace use, is also marketable for chemicals and electrolytic manganese dioxide use.

Metallurgical: A lump and fine product of 48 to 50% manganese. The lump product can be used in both blast and electric furnaces, for ferromanganese production.

Siliceous: A lump and fine product with a manganese content of less than 48%. These are used in the production of silicomanganese.

In 1981, production was down 30% at Groote Eylandt due to the depressed steel industry. The drop in demand in 1981 and 1982 has adversely affected the plans to increase the capacity. The installation of tailings beneficiation plant in 1980 had raised capacity to 2.6 mtpy, but an additional increase, to 3 million metric tons, has been deferred until 1985 or 1986.[22] None of the other manganese mines or prospects are suitable for large-scale mining.

Processing Facilities

Australia's only manganese smelter is the ferromanganese plant at Bell Bay, Tasmania. Annual production capacity is 130,000 metric tons of alloy on a standard ferromanganese basis, 26,000 metric tons of 75% ferrosilicon, and 600 metric tons per day of manganese sinter.[22] The Bell Bay plant exports roughly 35% of its production to markets in the United States, Southeast Asia, New Zealand, and China.[22]

Trade in manganese is primarily in the form of ore. This is due to the limited smelter capacity of Bell Bay and also to the preference of most major

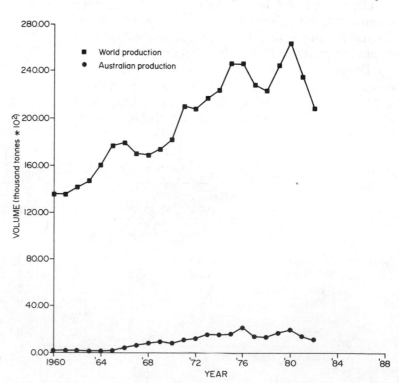

FIGURE III-2-11 Australian and world manganese production, 1960–1982. (From Ref. 48.)

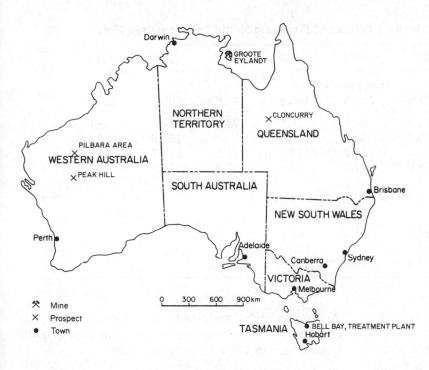

FIGURE III-2-12 Australian manganese prospects, mines, and treatment plants, 1981. (From Ref. 22.)

consumers to purchase manganese in its raw form. Japan and the EEC, Australia's two largest markets, have established ferromanganese industries and have high duties for manganese products past the concentrate stage.

Prospects for Future Production

The future of Australian manganese seems fairly stable. It will continue to reflect the health of the world's steel industry, especially Japan's. With the growth of the South Korean industry, there is a potential for additional increases for Australia, since the natural source of manganese for the Koreans would be Australia or India.

The growth of manganese use in other industries could also add strength to Australian production. The Broken Hills Proprietary Company Ltd. has built a pilot plant at their Newcastle research center to test the feasibility of producing manganese dioxide suitable for the battery industry. There is a chance that production from scattered manganese deposits could add to the potential capacity if better mining and beneficiation methods are developed or if the price of manganese rose dramatically.

NICKEL

Production, Resources, and Reserves

Australia ranks forth in world production of nickel, behind the USSR, Canada, and New Caledonia, with about 11% of total production.[1] As of 1980, the economic reserve base of Australian nickel was 1.72 million metric tons of contained metal. Total resources were 7.93 million metric tons.[21]

Exports and Trade

Domestic consumption of nickel is small, being only 5% of production. The rest is exported. All nickel mine products (except nickel-cobalt by-products) are processed to at least a matte (72% Ni, 6% Cu, 0.7% Co, and 20% sulfur), and more than 50% of exports are in the form of sintered nickel oxide or refined metal.[23] Australia ranks fifth in world production of refined nickel and third in exports (Table III-2-26). Tariffs and duties have an influence on the type of product sold. The European community is the second largest consumer of refined nickel and a large importer of nickel matte, and with Australia among the top three suppliers of unfinished nickel. On higher-processed forms of nickel, there are tariffs of up to 7.7%, compared to the duty-free status of low-processed forms.[21] Japan is the third-largest consumer of nickel. Australia supplies about 55% of her matte and 15% of her unalloyed unwrought nickel, but very little past that stage, due mostly to a 12% tariff on nickel past the unwrought stage.[21] The products that Australia exports to the United States (the largest consumer of nickel) are not processed beyond the unwrought stage. In this case, in addition to tariffs, the United States has Canada, the largest exporter of nickel, next door.[21]

Uses

The major use of nickel is in the making of stainless and alloy steels, for which nickel oxide (85 to 90% Ni) powders and briquettes have been developed. When the steel industry is depressed, a

TABLE III-2-26   World Nickel Production, 1976–1982 (Thousand Metric Tons Contained Metal)

	1976	1977	1978	1979	1980	1981	1982
			Mine production				
World	753.7	771.0	622.3	689.8	771.3	700.2	522.4
Australia	82.5	85.9	82.4	73.3	74.0	74.0	88.5
USSR	130.0	135.0	140.0	150.0	NA	157.8	154.2
Canada	225.5	221.3	111.8	123.9	194.9	159.7	81.6
New Caledonia	119.7	116.8	67.2	82.9	87.8	74.5	31.7
			Smelter-refinery production				
World	727.2	702.5	592.9	674.1			
Australia	30.8	34.4	33.8	40.1			
USSR	151.0	155.0	160.0	170.0			
Canada	94.8	93.9	79.2	105.9			

Sources: Refs. 1 and 2.

corresponding decline in nickel demand is usually seen.

Other nickel products are N-grade nickel powder and nickel briquettes. Both contain a minimum of 99.8% Ni.[23] The N-grade powder is used in high-purity nickel chemicals, nickel salts, welding rods, and in ceramic applications. The briquettes are used in making nickel-copper alloys.

History of the Industry

Until 1966, the only nickel operation in Australia was located in the Zeehan district of Tasmania. It was the 1966 discovery of high-grade mineralization in Western Australia that really started the

Australian nickel industry. Since then, annual production of nickel has grown to a high of 86,000 metric tons in 1977 (Figure III-2-13).

Geology of the Major Deposits

The two main types of nickel ores are laterites and sulfides, with laterites being much more abundant than sulfides, comprising 80% of the world's resources. Laterites require up to three times the amount of energy for refining than sulfides, making the sulfide the more valued ore. Australia has both types, but the most abundant are the sulfide deposits of the Kalgoorlie area of Western Australia. This area produces 7% of the world's nickel and has

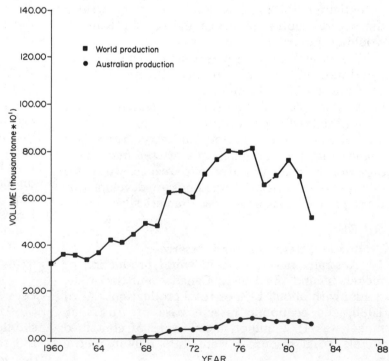

FIGURE III-2-13   Australian and world nickel production, 1960–1982. (From Ref. 48.)

14.5% of the known sulfide nickel ore of 0.8% or greater Ni content.[24] The Kambalda nickel deposits occur in ultramafic rocks in a greenstone belt.[25]

Figures III-2-14 and III-2-15 provide the locations of the important nickel mines, prospects, and treatment plants.

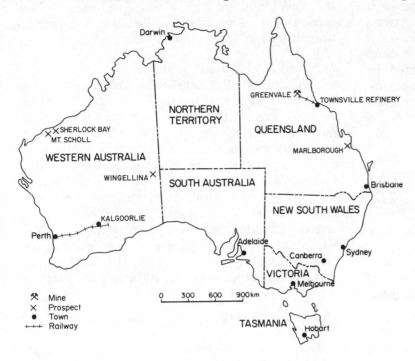

FIGURE III-2-14 Australian nickel prospects, mines, and treatment plants, 1981. (From Ref. 23.)

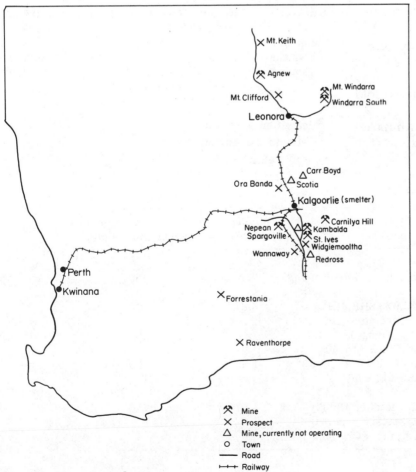

FIGURE III-2-15 Western Australian nickel locations, 1981. (From Ref. 23.)

Mining Facilities

The major producing mines in Western Australia are the Kambalda–St. Ives mines, and the Agnew mine. Important unmined deposits are located at Forrestania, Mt. Keith, Ora Banda, Wannaway, Widgemooltha, and Wingllina. Australia's other source of nickel is the Greensville mine in Queensland. This is a laterite deposit, being mined by open-cut methods. Table III-2-27 lists operations, ownership, and capacity of the various nickel mines.

Processing Facilities

Both capital and operating costs are higher for laterite processing than for the processing of sulfide deposits. The nickel sulfide ores can be concentrated by flotation and then treated by one of the following methods:[21]

Roasting and smelting, followed by electrolytic refining

Smelting followed by electrolytic refining

Hydrometallurgical process involving the use of ammonia (this process is used at the Kwinana refinery)

Since no effective physical means of concentrating nickel laterite/sillicate ores exist, processing must be done by one of the following methods:[21]

Smelting to produce an iron-nickel matte, which is then processed in a converter to eliminate iron, with the nickel cast into anodes and refined electrolytically

Smelting directly to ferronickel, which is sold directly to consumers

Leaching with ammonia or sulfuric acid, with nickel and cobalt precipitated from the solution (the method the Yabulu refinery uses)

Energy costs greatly determines the economic viability of the Queensland deposit. As a result, the Yabulu treatment plant has converted to coal firing for their smelter. The location, ownership, and capacities of Australian smelters and refineries is summarized in Table III-2-28.

Prospects for Future Production

The future of Australian nickel is tied to the demand for steel and the development of new uses for it. At present the only substitutes for nickel are generally more scarce and expensive. Australia certainly has the resources to increase nickel production, especially in the production of nickel sulfides. In 1982 the Agnew mine increased production by 5000 mtpy of concentrates, and the Kalgoorlie smelter increased its smelting capacity. There are several promising prospects, such as the deposit at Mt. Keith, but there have also been quite a few mine closures in the late 1970s. It is safe to state that any large increases in Australian production of

TABLE III-2-28   Australian Nickel Smelter and Refineries

Location	Ownership	Capacity
Smelter: Kalgoolie	Western Mining Corp. Holding Ltd. (MWC)	450,000[a]
Refineries Kwinana	WMC	30,000[b]
Yabulu	Metals Exploration Pty. Ltd. 50%; Freeport Qld Nickel, Inc. 50%	24,000[b]

[a]Metric tons of concentrates.
[b]Metric tons of contained metal.
Source: Ref. 23.

TABLE III-2-27   Nickel Mines in Australia

Location	Ownership	Grade (% Ni)	Annual capacity (thousand metric ton ore)
Western Australia			
Kambalda–St. Ives	Western Mining Corp. Holding Ltd. (WMC)	2.5	1650
Camilya Hill	WMC 55.9%; BHP 44.1%	3.87	60
Nepean	Metals Exploration Ltd.		100
Agnew	Seltrust Holdings Ltd. 60%; Mt. Isa Mines Ltd. 40%		500
Windarra (production suspended June 1978)	WMC 50%; the Shell Co. of Australia Ltd. 50%	1.63	1100
Queensland: Greenvale	Metal Exploration Qld Pty. Ltd. (MEQ) 50%; Freeport Qld Nickel, Inc. (FQN) 50%	1.57	2750

Sources: Refs. 23 and 13.

nickel would have to be the result of a sharp rise in the demand for stainless steel.

LEAD AND ZINC

Production, Resources, and Reserves

Australia is among the world leaders in the production of lead and zinc, with approximately 15% of the Western world's lead production and 10% of its zinc.[25] Only the USSR and the United States produce more lead, and only Canada, the USSR, and Peru produce more zinc (Tables III-2-29 and III-2-30).

The Australian Bureau of Mineral Resources estimated total Australian resources of contained lead at 31.2 million tons at the end of 1982. Of this, 14.6 million metric tons was considered to be

TABLE III-2-29   World Lead Production, 1976–1982 (Thousand Metric Tons)

	1976	1977	1978	1979	1980	1981	1982
Lead mine production[a]							
World	3866	3410	3456	3513	3520	3350	3450
Australia	381	414	380	402	382	392	450
U.S.	565	550	542	537	563	446	510
Canada	243	328	366	342	297	332	335
Refined lead[b]							
Western World	3785	4084	4116	4276	4004		
Australia	212	216	239	254	238	208	219
U.S.	1107	1169	1188	1226	1079		
West Germany	337	374	369	373	350		
U.K.	342	351	346	368	325		
Japan	342	351	346	368	325		
Canada	176	206	226	225	185		

[a]Lead in lead concentrates only.
[b]Including secondary recovery.
Sources: Refs. 1 and 2.

TABLE III-2-30   World Zinc Production, 1976–1982 (Thousand Metric Tons)

	1976	1977	1978	1979	1980	1981	1982
Mine production[a]							
World	6210	6049	5879	5998	5745	5844	6160
Australia	432	451	434	487	463	508	620
Canada	1145	1300	1245	1204	1059	1097	1165
U.S.	483	448	333	294	344	312	300
Peru	459	476	457	491	488	497	535
Refined Zinc[b]							
Western World	4116	4277	4293	4706	4467		
Australia	251	256	294	312	306	296	291
Japan	742	778	768	789	739		
U.S.	515	454	442	526	366		
Canada	472	495	495	580	592		
West Germany	305	355	307	356	365		
Belgium	244	274	234	253	248		

[a]Zinc in zinc concentrates only.
[b]Includes secondary zinc.
Sources: Refs. 1 and 2.

economic, demonstrated reserves.[21] Total zinc resources were assessed at 56.2 million metric tons, with 21.3 million metric tons being economic, demonstrated reserves.[25]

Exports and Trade

Domestic consumption of lead and zinc is low compared to total production. Only 28% of refined lead and 32% of refined zinc production are consumed locally.[25]

Exports of lead and zinc are in the form of ore, concentrates, bullion, and refined metal. Australia is the third-largest exporter of lead ore and concentrates, behind Canada and Peru, and the leader in lead bullion and metal exports.[1] In zinc exports, Australia trails Canada and Peru in ore and concentrates, and Canada and Belgium in refined zinc.[1] The major export markets of lead and zinc products are shown in Tables III-2-31 and III-2-32.

History of the Industry

Lead was first mined at Glen Osmond, South Australia, in 1841, but it was the discovery of silver, lead, and zinc at Broken Hill in 1883 that really started the Australian lead-zinc industry. The Mt. Isa area began sporadic production of lead and zinc in 1924, not making a profit until 1937. Today it is the single largest lead-zinc mine in Australia.[8] Since the 1950s, Australia has provided between 11 and 13.5% of the world's lead requirements (Tables III-2-33 and III-2-34).

The decrease in the world share of lead since 1950 is due to the increased production of the USSR and the U.S. Similarly, the decrease in zinc

TABLE III-2-31   Australian Lead Exports, 1976–1982 (Thousand Metric Tons Lead Content)

	1976	1977	1978	1979	1980	1981	1982
Ore and conc.	28.9	49.9	78.6	25.1	16.1		88.5
U.S	11.2	17.9	12.5	3.3	2.1		38.5
Belgium	3.6	4.5	22.1	2.2	4.0		
Japan	5.9	15.8	15.6	4.7	1.6		7.5
U.K.	—	6.7	2.9	13.3	5.1		28.8
Lead bullion	172.0	167.5	147.1	162.2	168.0		169.9
U.K.	147.5	140.5	117.1	142.3	132.5		156.4
Netherlands	17.6	12.8	13.0	7.8	11.2		
West Germany	2.0	2.8	7.5	4.5	3.6		3.5
Refined lead	164.1	132.4	158.3	158.3	154.0		194.8
Italy	22	19	15	15	17.5		9.0
U.K.	37	19	16	18	16.1		10.0
India	36	23	29	29	28.5		54.9
U.S.	4	19	16	10	7.3		10.9
China	5	3	9	22	19.0		8.0

Sources: Refs. 1, 21, and 25.

TABLE III-2-32   Australian Zinc Exports, 1976–1982 (Thousand Metric Tons Zinc Content)

	1976	1977	1978	1979	1980	1981	1982
Ore and conc.	235.3	175.0	243.0	183.8	259.3		569.0
Japan	114.1	112.8	103.7	109.3	156.6		324.8
Netherlands	70.8	40.6	45.7	6.8	30.3		
Belgium	31.2	5.7	2.6	11.6	15.3		23.6
U.K.	—	—	26.8	20.7	27.2		52.8
Refined zinc	157.7	151.8	215.4	197.9	219.1		223.4
Indonesia	22	24	35	37	52.6		58.9
U.S.	29	27	38	27	30.9		26.3
Thailand	16	18	22	24	28.6		18.9
New Zealand	19	19	12	20	16.0		16.5
Taiwan	15	16	36	26	23.7		21.9

Sources: Refs. 1, 21, and 25.

TABLE III-2-33 Australian Lead Production, 1950–1975 (Thousand Metric Tons)

Year	Tonnage	Percent of world production
1950	222.7	13.2
1960	313.1	13.2
1965	367.9	13.4
1970	456.7	13.2
1975	407.8	11.4

Source: Ref. 28.

TABLE III-2-34 Australian Zinc Production, 1950–1975 (Thousand Metric Tons)

Year	Tonnage	Percent of world production
1950	201.0	9.1
1960	322.5	9.6
1965	354.8	8.3
1970	478.2	8.3
1975	500.8	8.2

Source: Ref. 28.

is due to production increases in the USSR, Canada, and Peru.

Geology of the Major Deposits

The geology of the Broken Hill and Mt. Isa deposits were important in their mine development. The outcrop at Broken Hill extended over 4000 ft and consisted of cellular limonite and manganiferrous gossan. At a depth of 400 ft is the oxidized and secondary enriched zone, which contains lead, zinc, silver, copper, manganese, and rarer minerals.[26]

The ore at Mt. Isa is of two types: the silver-lead-zinc ore that is confined to definite bands in unaltered shales, and the ores that are contained in zones of disturbed material that may have been derived from the shale.[27] Figure III-2-16 indicates locations of lead and zinc deposits, mines, and treatment facilities.

Mining Facilities

In addition to the two main lead-zinc producers, Broken Hill (several mines) and Mt. Isa, there are many other smaller mines throughout Australia. Several of these mines are just coming on-stream (Elura, Que River, and Hilton), and a continued depressed lead and zinc market could adversely affect their continued operation.

There is potential for growth for the lead-zinc industry in Australia. Some of this increased production will come from the expansion of existing facilities, while new mines will also add to the expansion. Mt. Isa announced plans to expand its mine output by about 30,000 tons of lead concentrates during 1982–1983.[24] New prospects include Hilton (which has a trial mining program under way), Lady Loretta, and Dugald River, all in northwest Queensland. Hilton is expected to go into full production of 180,000 metric tons per year during the early 1990s.[13] Lady Loretta and Dugald River,

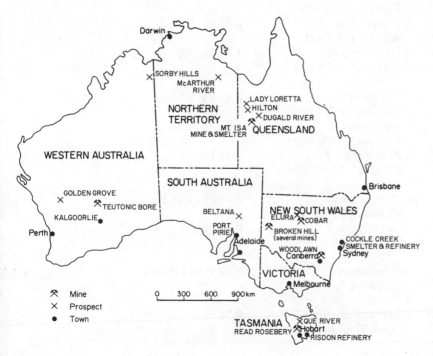

FIGURE III-2-16 Australian lead/zinc prospects, mines, and treatment plants, 1981. (From Ref. 25.)

as well as Golden River and Sorby Hills in Western Australia, will require more detailed exporation before mining programs can be started.

Perhaps the most promising lead-zinc deposit in Australia is McArthur River in the Northern Territory. With estimated resources of 190 million metric tons at 4.1% lead and 9.5% zinc, it is one of the world's largest undeveloped lead-zinc ore bodies. As of 1982 it was not feasible to develop under current economic conditions using known technologies. Tables III-2-35 and III-2-36 provide data about the various mining areas.

Processing Facilities

Most lead processing in Australia is done by the traditional method of smelting roasted concentrates in a blast furnace to produce lead bullion. The lead is then refined by the removal of copper, silver, and other impurities until it is 99.99% pure.[21]

The other smelting process used is the Imperial Smelting Process, which smelts sintered mixed lead-zinc concentrates.[24] In this process the molten lead has the zinc removed through a cooling launder to produce zinc of 98% purity. The lead bullion is then poured into "pigs" or blocks, in preparation for further treatment.[21]

Port Pirie is the locale of the world's largest primary lead refinery. It has a capacity of 230,000 metric tons of high-quality refined lead and lead-antimony alloys annually, of which about 80% is exported.[25]

Mt. Isa concentrates are smelted on-site to produce a silver-lead bullion, which is then railed to Townsville, on the Queensland coast, to be shipped to London. Concentrates not treated on-site, due to surplus mine production above the smelter capacity, are sold to Port Pirie or exported as concentrates.[29] There is an Imperial Smelting furnace at Cockle Creek, near Newcastle. Capacity is rated at 75,000 metric tons of zinc metal and 30,000 metric tons of lead-silver bullion.[25]

Electrolytic refining is the predominant method

TABLE III-2-35   Australian Lead-Zinc Mine Resources, 1980

Location	Principal resources (million metric tons)	Metal content (million metric tons)
New South Wales		
Broken Hill (NBHC)	7.0 mt averaging 8.7% lead, 9.7% zinc	0.61 Pb, 0.96 Zn
Broken Hill (ZC)	6.1 mt averaging 10.3% lead, 9.7% zinc	0.63 Pb, 0.59 Zn
North Broken Hill	6.3 mt averaging 13.0% lead, 10% zinc	0.82 Pb, 0.63 Zn
Woodlawn	9 mt averaging 3.5% lead, 9.7% zinc	0.32 Pb, 0.87 Zn
Elura	27 mt averaging 5.6% lead, 8.4% zinc	1.51 Pb, 2.27 Zn
Queensland		
Mt. Isa	54 mt averaging 6.4% lead, 6.5% zinc	3.46 Pb, 3.51 Zn
Hilton	37 mt averaging 7.7% lead, 9.6% zinc	2.85 Pb, 3.55 Zn
Lady Loretta	8.7 mt averaging 6.7% lead, 18.1% zinc	0.58 Pb, 1.57 Zn
Tasmania		
Read/Rosbery	8.19 mt averaging 5% lead, 16.7% zinc	0.41 Pb, 1.37 Zn
Que River	3 mt averaging 7% lead, 12.5% zinc	0.37 Pb, 0.66 Zn
Northern Territory:		
McAuthur River	190 mt averaging 4.1% lead, 9.5% zinc	7.80 Pb, 18.05 Zn
Western Australia:		
Teutonic Bore	2.5 mt at 9.5% zinc and 0.75 mt at 16% zinc	0.36 Zn

Source: Ref. 21.

TABLE III-2-36   Australian Lead-Zinc Mine Production

Location	Company	1981 production (metric tons metal)
Broken Hill	New Broken Hill Consolidated Ltd.	123,328 lead, 164,615 zinc (includes metal from Broken Hill ZC mine)
Broken Hill	Zinc Corporation Ltd.	55,000 lead, 40,000 zinc
North Broken Hill	North Broken Hill Ltd.	
Woodlawn	AM&S, Phelps Dodge Corp., St. Joe Mineral Corp.	17,594 lead, 59,578 zinc
Elura	EZ Industries Ltd.	1 million metric tons of ore per year beginning 1983
Mt. Isa	Mt. Isa Mines Ltd.	122,290 lead, 193,798 zinc
Hilton	Mt. Isa Mines Holdings	500 t/day trial mining project
Read/Rosebery	EZ Industries Ltd.	13,311 lead, 96,156 zinc conc. (includes metal from Que River mine)
Teutonic Bore	Seltrust Mining, Mt. Isa Mines Holdings	300,000 zinc ore

Sources: Refs. 21 and 13.

used in processing zinc concentrates. The concentrates are roasted to produce zinc oxide, which is leached in sulfuric acid to produce zinc sulfate, which is then electrolyzed.[21] The product is a special high-grade zinc (99.99% zinc minimum) or a high-grade zinc (99.9% Zn).[21] The other process used is the Imperial Smelting Process described earlier. This process produces a zinc (not less than 98% Zn) which is then further refined.[21]

The largest zinc refinery in Australia is the 220,000 metric ton per year electrolytic refinery at Risdon, Tasmania.[29] It is one of the largest facilities of its type in the world. In excess of 75% of its product is exported.[25] The other zinc refineries in Australia are the Cockle Creek smelter, which has a capacity of 75,000 metric tons of zinc per year, and the Port Pirie electrolytic refinery, which has a capacity of 40,000 metric tons of zinc metal annually.[25]

Prospects for Future Production

The demand for lead is closely linked to the automobile, since 60% of lead consumption is claimed by the auto industry, in the form of batteries and additives in gasoline.[29] The decline of car sales worldwide, and the use of calcium-lead "maintenance-free" batteries in the United States, has decreased the demand for lead considerably. The jump in the price of lead in 1979 created an excess supply. All of these factors have led to the depressed state of the lead industry.

Future increases in the demand for lead could be enhanced by the wide-scale use of electric cars, which are powered by lead-storage batteries. The USSR export level of lead has been declining in recent years, which might mean an increased share of the lead market for Australia.

With the major use of zinc being the galvanizing of steel, it is reasonable to expect a drop in zinc production when the steel industry is in a slump. There has also been a trend to substitute lighter materials for zinc applications in motor vehicle parts. In general, zinc demand is not expected to increase dramatically in its traditional uses.

Where growth is possible is in the demand for zinc oxide and zinc for dry-cell batteries. Zinc oxide is used as an activator in the rubber industry and as a trace element in fertilizers.[29] There are expectations that zinc dust, which is mainly used in zinc-rich primer paints, will increase in use.[29]

COPPER

Production, Resources, and Reserves

The copper industry of Australia is small in world terms. Current production represents only about 3% of world output. The major producers are the United States, the USSR, and Chile.[1] Australia has 5.66 million metric tons of demonstrated economic resources and total identified resources of 10.80 million metric tons of contained metal, according to the Australian Bureau of Mineral Resources, as of 1979.[21]

Exports and Trade

Domestic consumption of copper is approximately 60% of annual refined metal output. The copper is used in high-technology fabricated and semifabricated products.[30] Concentrate and blister copper exports were each less than 4% of world trade. Refined copper exports were less than 2% of world trade.[29] Tables III-2-37 and III-2-38 provide production and export statistics for Australian copper.

TABLE III-2-37 World Copper Production, 1976–1982 (Thousand Metric Tons Copper Content)

	1976	1977	1978	1979	1980	1981	1982
Mine production							
World	7856	7983	7873	7933	7656	8171	7780
Australia	219	222	222	235	232	223	250
U.S.	1457	1364	1358	1444	1171	1538	1106
USSR	1130	1100	1140	1150	1150	950	950
Chile	1005	1056	1036	1061	1072	1080	1190
Smelter production (primary and secondary)							
World	7966	8165	8093	8147	7939	8325	
Australia	170	171	167	173	182	165	
U.S	1439	1347	1343	1396	1053	1378	
USSR	1144	1100	1170	1170	1170	1045	
Japan	838	915	906	921	930	980	
Refinery production (primary and secondary)							
World	8790	9100	9201	9357	8971	9184	
Australia	188	185	175	174	183	191	
U.S.	1715	1677	1832	1980	1687	2038	
USSR	1420	1440	1460	1480	1480	1480	
Japan	864	934	959	984	1014	1050	

Sources: Refs. 1 and 2.

TABLE III-2-38 Australian Copper Exports, 1976–1982 (Thousand Metric Tons Copper Content)

	1976	1977	1978	1979	1980	1981	1982
Copper in concentrates							
Total	45.7	37.4	35.3	41.8	42.9		
Japan	43.3	32.4	32.1	37.6	33.4		
EEC[a]	1.8	4.4	2.6	3.1	3.7		
Blister copper							
Total	9.9	9.0	4.1	28.0	21.3		
Japan	8.6	7.0	—	3.9	0.9		
West Germany	0.3	2.0	4.1	0.3	0.3		
Belgium	—	—	—	20.9	8.7		
Refined copper							
Total	76.2	74.7	60.1	47.0	51.7		
EEC[a] (excl. U.K.)	43.5	43.8	34.0	23.4	31.6		
UK	17.3	20.1	19.4	14.2	8.5		
Japan	5.2	7.4	2.1	0.2	0.7		

[a]EEC, European Economic Community.

Source: Ref. 1.

Uses

Copper is used primarily in electrical applications (53%), as well as for construction (20%), machinery (13%), transportation (8%), and various other uses (8%). The demand for copper, and therefore the price, is affected by the growth of the construction, automobile, and capital-goods industries. Substitution for copper by aluminum in electrical applications, and by fiber optic cables in the telecommunication industry, may reduce the amount of copper required by those traditional users.

History of the Industry

Copper mining has historical importance in the mineral development of Australia. As early as 1845, copper mining was being established in South Australia, near Adelaide. Australia's largest copper mine, Mt. Isa, has been in operation since the 1940s, and still has 40 years of reserves left. Due to strong competition from the United States and South America, Australia has been limited to less than 3% of world production, going back to 1950. Only very early in her copper-mining history has Australia ever contributed substantially to world production. Table III-2-39 shows Australian copper production for 1950–1975. Figure III-2-17 shows the locations of important copper deposits.

Mining Facilities

The feasibility of mining a copper deposit is dependent on many factors. The size of the deposit, its grade, the by-product and co-product minerals, the location, and type of mining required to win the ore are all important in determining the profitability of an ore body. Australia has a variety of deposits, mined for different qualities and products. The most important are the Mt. Isa mines. Approximately 75% of Australian mine production, and 85% of primary refined copper, comes from this area. It is Australia's most important base-metal operation, yielding copper, lead, zinc, and silver. Tables III-2-40 and III-2-41 summarize the important copper mines of Australia.

Processing Facilities

Copper, due to the low grades mined and its association with other minerals, requires a great deal of smelting and refining to produce a finished product. Most copper mines in Australia have concentrators on-site that raise the grade to around 25% copper content. The concentrates are then shipped to a smelter, where blister copper (99% Cu) is produced. Australian smelters are located at Port Kembla, Mt. Morgan, Mt. Isa, and previously, at Tennant Creek. The copper is then refined, using an electrolytic process, and afterward fabricated

TABLE III-2-39   Australian Copper Production, 1950–1975 (Thousand Metric Tons)

Year	Tonnage	Percent of world production
1950	15.6	0.6
1960	72.2	1.6
1965	74.6	1.3
1970	157.8	2.4
1975	218.9	3.1

Source: Ref. 28.

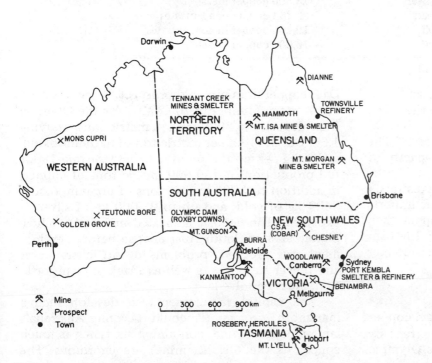

FIGURE III-2-17 Australian copper prospects, mines, and treatment plants, 1980. (From Ref. 30.)

TABLE III-2-40 Australian Copper Mine Reserves

Location	Demonstrated reserves (December 1979, million metric tons)
Queensland	
Mt. Isa	Primary — 121 mt at 3.2% Cu; secondary — 6.3 mt at 3.8% Cu
Gunpowder (Mammoth)	6.3 mt at 3.15% Cu
Mt. Chalmers	1 mt at 1.9% Cu
Mt. Morgan	Ore body mined out
Dianne	87,925 metric tons at 23.61% Cu
New South Wales	
Cobar	31 mt at 2% Cu
Woodlawn	6 mt at 1.7% Cu, 3.5 mt at 1.9% Cu
Tasmania	
Mt. Lyell	Measured 4.6 mt at 1.47% Cu; indicated 2.61 mt at 1.45% Cu
Rosebery	7.8 mt at 0.7% Cu
South Australia: Mt. Gunson	3.4 mt at 2.0% Cu
Northern Territory:	
Tennant Creek	Shut down operations in 1982
Western Australia: Teutonic Bore	2.5 mt at 3.5% Cu

Sources: Refs. 21 and 30.

TABLE III-2-41 Australian Copper Mine Production

Location	Type of mining by-products	1980–81 production (metric tons)
Mt. Isa	Underground; lead, zinc, silver	37,920 blister copper, 103,720 anode copper, 4,496 copper concentrates
Gunpowder	Underground	1,383 copper metal
Mt. Chalmers	Gold	605,115 ore
Cobar	Underground; lead, zinc	41,583 copper metal
Woodlawn	Open pit; lead, zinc	71,337 copper metal
Mt. Lyell	Gold, silver	19,500 copper metal
Rosebery	Zinc, lead	22,744 copper concentrates
Mt. Gunson	Open pit	10,500 copper in concentrates
Teutonic Bore	Open pit	10,000 copper metal

Sources: Refs. 21, 30 and 13.

into forms, such as cathodes, wire bar, rod, and wire. Australia has two copper refineries. The Townsville refinery has a capacity of 155,000 metric tons per year (tpy) of refined copper.[30] The other refinery, at Port Kembla, has a capacity of 60,000 tpy.[30]

Japan is the major consumer of Australian copper ores and concentrates, although it imports considerably less blister and refined copper than the European community. This is due to the high import tariffs that Japan has on finished copper products.

Prospects for Future Production

The long-range future of the Australian copper industry will be determined to a large extent by the development of the Roxby Downs/Olympic Dam copper-uranium-gold prospect. Estimates from the Western Mining Corp. (WMC) for the Olympic Dam deposit are 2000 million metric tons, assaying 1.6% Cu, 0.64 kg per metric ton of uranium oxide, and 0.6 g per metric ton of Au.[31] By the mid-1990s full production of 150,000 metric tons of copper, in addition to 3000 metric tons of uranium oxide, 3400 kg of gold, and up to 23,000 kg of silver, is expected. The deposit is located in an arid, inhospitable area of South Australia; so before development can proceed, problems of sufficient water and power supplies, as well as a lack of transportation, must be solved.

The capital requirements to develop such a mammoth project (it could become Australia's largest copper mine, containing six times as much copper as the Mt. Isa mine) are enormous. The developers, Western Mining Company and British Petroleum of Australia Ltd., are seeking $55 million toward a $220 million town at Olympic Dam, through an indenture bill being proposed in the South Australian state parliament. Opposition by the Labor Party to the uranium mining associated with this project could delay the bill. The total cost of the project is estimated at $1.4 billion (1981 dollars).[32] A $100 million dollar feasibility study is expected to be completed by 31 December 1984.

A more immediate prospect is the Golden Grove copper-silver deposit in Western Australia. It contains proven reserves of 15.5 million metric tons, at 3.6% Cu. The Goorumbla area near Parkes in central western New South Wales has high hopes to develop a porphyry deposit, totaling more than 300 million metric tons, with copper ranging between 1.82 and 0.62% Cu, plus 0.5 g of gold per metric ton, and 5.1 g of silver per metric ton.[13]

TIN

Production, Resources, and Reserves

Australia is the fifth largest producer of tin concentrates, accounting for 6% of 1981 world output.[33] The largest producers are Malaysia, Indonesia, and Thailand (Table III-2-42). The Australian Bureau of Mineral Resources reported that the total identified tin resources, as of December 1979, were 649,000 metric tons, of which 208,000 metric tons were classified as demonstrated, economic resources[34] (Table III-2-43).

Exports and Trade

Almost 80% of Australia's tin production is exported as concentrates, of which over 90% goes to the tin smelters of Malaysia. Most of the domestic smelter production is consumed locally, with only 25% being exported; the majority of that going to the United Kingdom (Table III-2-44).

Uses

Tin is used mainly in the production of tinplate (40 to 45%), as well as solder (25%) and alloys (15%). As an alloy, tin is mixed with a variety of other metals, including lead (electrical and electronic equipment), copper (ornamentation), zinc (hydraulic equipment), nickel (printed circuits), and aluminum (bearings).

History of the Industry

Australian tin mining began in the 1850s, and by the turn of the century annual production had reached 4354 metric tons of concentrates.[35] Output steadily increased until 1913 (7904 metric tons), then decreased until 1953 (1578 metric tons), then began a slow rise that has seen annual concentrate production of over 10,000 metric tons since 1976.[35] Table III-2-45 shown tonnages and world percent of tin production since 1950.

Geology of the Major Deposits

The ore body at Renison Bell (Tasmania) lies within a sequence of Lower Cambrian sediments and volcanics, whereas the Cleveland tin body (Luina, Tasmania) is found in a series of lenses, surrounded by volcanics. The deposits in Ardlethan (New South Wales) are located in tin-bearing granites.[9] Figure III-2-18 shows the locations of the major tin areas.

Mining Facilities

The majority of tin production, about 70%, comes from three mines: the Renison Bell and Cleveland mines of western Tasmania, and the Ardlethan mine in New South Wales. The balance comes from small aluvial operations in Queensland,

TABLE III-2-42   World Tin Production, 1976–1982 (Thousand Metric Tons)

	1976	1977	1978	1979	1980	1981	1982
			Concentrates				
World	231.0	230.7	241.1	245.9	246.5	253.0	238.0
Australia	10.6	10.6	11.8	12.6	11.6	12.0	12.0
Malaysia	63.4	58.7	62.7	63.0	61.4	60.0	55.0
Indonesia	23.4	25.9	27.4	29.4	32.5	35.0	32.0
Thailand	20.4	24.2	30.7	34.0	33.9	32.0	30.0
			Metal				
World		228.5	244.1	249.1	250.1	242.1	
Australia	5.6	5.6	5.1	5.4	4.8	4.2	3.1
Malaysia	78.0	66.3	72.0	73.1	71.3	68.5	
Indonesia	23.3	24.0	25.8	27.8	32.5	34.9	
Thailand	20.3	23.1	28.9	33.1	33.7	32.0	

Sources: Refs. 1 and 2.

TABLE III-2-43   Resources of Major Australian Tin Mines, 1980

Location	Indicated reserves (million metric tons)	Metal content (thousand metric tons)
Tasmania		
Renison Bell	Proven and probable reserves of 13.7 mt at 1.14% Sn	155.7
Luina	Measured reserves of 1.3 mt at 0.069% Sn	9.0
New South Wales: Ardlethan	Indicated reserves of 1.25 mt at 0.54% Sn	6.7

Source: Ref. 21.

TABLE III-2-44   Australian Tin Exports, 1976–1982 (Metric tons)

	1976	1977	1978	1979	1980	1981	1982
Concentrates							
Total	3360	5126	7023	6435	7423		
Malaysia	2226	4076	6886	6127	6912		
U.K.	1092	966	20	17	86		
Metal							
Total	2293	2267	2079	1618	1436		
U.K.	733	1261	1377	561	531		
Canada	594	—	—	—	—		
New Zealand	314	235	234	207	195		
U.S.	648	36	—	90	135		
Netherlands	3	631	468	469	242		

Source: Ref. 1.

TABLE III-2-45   Australian Tin Production, 1950–1975 (Thousand Metric Tons)

Year	Tonnage	Percent of world production
1950	1900	1.1
1960	2200	1.2
1965	3900	2.0
1970	8800	4.7
1975	9600	4.5

Source: Ref. 28.

New South Wales, and Tasmania. The Greenbushes area of Western Australia is also a minor producer of tin[13] (Table III-2-46).

In mid-November 1979, the federal government decided to phase out export controls on tin ore over the next two years. For three years after, bounty payments on refined tin would be required. This meant that Renison Ltd. could export their entire production after 1981. The Associated Tin Smelters, which by law received 75% of the 1978–1979 output in 1980 and 50% in 1981, would receive a bounty of $50 per metric ton for primary refined tin over the next three years.[29]

Processing Facilities

The Associated Tin Smelter Pty. Ltd. operates the nation's largest tin smelter, located at Alexandria near Sydney. It has a capacity of 6000 metric tons per year.[13] Other facilities include a small smelter at Greenbushes to smelt concentrates from Western Australia and the Northern Territory,[36] and also secondary tin plants in Sydney, Wollongong, and Melbourne.[36]

Prospects for Future Production

The future of Australian tin looks bright. It is expected that concentrate production will exceed the 15,000-metric ton mark in the mid- to late 1980s. In the forefront is the Renison Bell mine, which has recently increased ore capacity to 850,000 metric tons annually.[36] Hardrock reserves will continue to be the main source of tin production over the next 10 years. Improved recovery

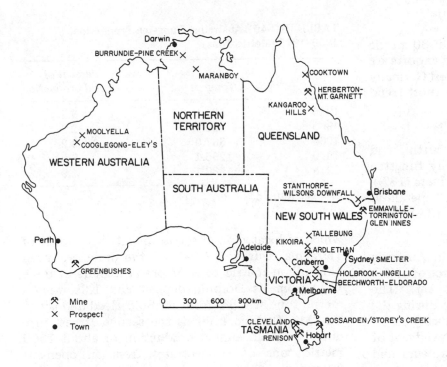

FIGURE III-2-18 Australian tin prospects, mines, and treatment plants, 1981. (From Ref. 34.)

TABLE III-2-46 Australian Tin Production from Selected Mines

Location	Ownership	Capacity (metric tons)
Renison Bell	Renison Ltd.	850,000
Luina	Aberfoyle Ltd.	400,000
Ardlethan	Aberfoyle Ltd.	600,000

Source: Ref. 13.

through preconcentration techniques, flotation, and fuming will also help to increase output. The high cost of labor, energy, and construction, as well as the exchange rate of the Australian dollar, all favor the expansion of existing facilities over the development of new mining areas. Any new developments are likely to be open-pit mines, at grades of 0.5% or better.[37]

In general, Australia expects to be able to main-tain increased tin production levels past the turn of the century, utilizing her present mine facilities. This, of course, is dependent on the demand for tin over the next 20 years.

TUNGSTEN

Production, Resources, and Reserves

Australia is part of the second tier of tungsten producers (Australia, Bolivia, U.S., South Korea, and Thailand), with about 7% of world production. China (23%) and the USSR, together account for over 40% of world output[38] (Table III-2-47). The Australian Bureau of Mineral Resources assessed total resources of tungsten at 462,000 metric tons of tungsten oxide (WO_3) as of December 1979. Of these, 254,000 metric tons were in the demonstrated economic category. King Island, Tasmania, has indicated reserves of 56,000 metric tons of tungsten oxide, and Mt Carbine, Queensland, has indicated reserves of 25,000 metric tons.[21]

TABLE III-2-47 World Tungsten Production, 1976–1982[a] (Metric Tons Tungsten Content)

	1976	1977	1978	1979	1980	1981	1982
World	39,825	40,914	44,619	43,895	54,112	48,785	42,720
Australia	1,935	2,358	2,680	3,168	3,331	3,322	2,313
China(e)[a]	9,000	9,000	9,980	9,980	15,011	13,514	11,338
USSR(e)	8,000	8,200	8,500	8,620	8,707	8,843	8,843
U.S.	2,644	2,725	3,128	3,013	2,737	3,544	1,451

[a]e, estimated.

Sources: Refs. 1 and 2.

Exports and Trade

Domestic consumption is only about 80 metric tons per year of concentrates. Tungsten exports are destined mainly for Western Europe, West Germany in particular.[38] Table III-2-48 shows export trend in recent years.

Uses

The largest use of tungsten is in cutting and wear-resistant materials (65%), primarily tungsten carbide. Other end uses are in mill products (18%), specialty steels (5%), chemicals (4%), superalloys (2%), and hard-facing rod and materials (2%).[39]

History of the Industry

Tungsten mining was established in Australia by 1905, with Queensland being the source of 81.5% of the 953 metric tons produced in that year.[40] The demand for tungsten has varied greatly during the years after World War I, caused by the heavy use of the metal for military purposes. With the advent of tungsten carbide bits, used in the petroleum and tool and dye industries, tungsten demand has stabilized. Table III-2-49 provides statistics on Australian tungsten production for the years 1950-1975.

Geology of the Major Deposits

The ore bodies on King Island occur in skarn rocks which have been formed by selective metasomatic replacement of limestone beds with the scheelite occurring as fine disseminated grains. The other important deposit is at Mt. Carbine, where wolframite is the most valued mineral.[9] Figure III-2-19 has the locations of the important tungsten areas.

Mining Facilities

The two major producers of tungsten are King Island Scheelite Pty. Ltd. and Queensland Wolfram Pty. Ltd. Together they account for some 90% of production.[13] The larger of the two, King Island Scheelite, operates a hard rock mine on King Island,

TABLE III-2-49 Australian Tungsten Production, 1950-1975 (Metric Tons)

Year	Tonnage	Percent of world production
1950	587.5	3.2
1960	895.7	2.9
1965	867.0	3.2
1970	1265.1	3.9
1975	1880.1	4.9

Source: Ref. 28.

which is located in the Bass Strait, just north of Tasmania. During 1980-1981 the mine produced 1.3 million metric tons of ore at a grade of 0.9% WO_3 from its Dolphin deposit and 1.35 million metric tons at 0.8% from its Bold Head deposit.[13] The Mt. Carbine mine is the second largest producer, with an annual production of about 1500 metric tons of concentrates from an open-cut operation.

Minor producers are located in Tasmania, New South Wales, Queensland, and the Northern Territory. There are high hopes for the Mt. Mulgine prospect in Western Australia, in which reserves of 37 million metric tons plus of 0.19% WO_3 have been estimated. Part of the deposit also contains economically recoverable quantities of molybdenum.[13]

Prospects for Future Production

The future of Australian tungsten will depend on several factors. Continued development of metallurgical processes to further reduce the grade that can be mined, and the development of the Mt. Mulgine deposit, should bolster Australian reserves so that a high level of production can continue. The most important factor is the level of future production from China. With almost half the world's reserves, the Chinese certainly have the capability of increasing their exports.

TABLE III-2-48 Australian Tungsten Exports, 1976-1982 (Metric Tons Tungsten Content)

	1976	1977	1978	1979	1980	1981	1982
Total	1722	2105	2770	2865			
West Germany	544	988	1598	1778			
U.K.	312	311	162	40			
U.S.	40	66	115	137			
Japan	164	315	93	247			
Sweden	61	140	308	299			
USSR	436	259	353	139			

Source: Ref. 1.

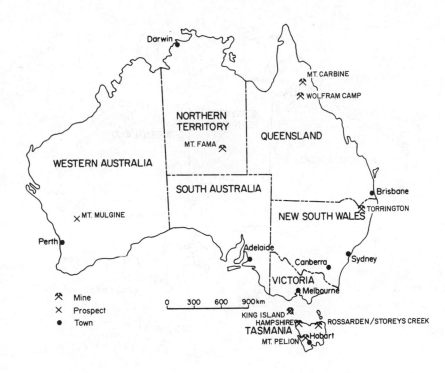

FIGURE III-2-19 Australian tungsten prospects, mines, and treatment plants, 1981. (From Ref. 39.)

DIAMONDS

Production, Resources, and Reserves

The USSR, South Africa, and Zaire are the dominant producers of diamonds. Australia is a newcomer in the diamond field, and although it mines only a small fraction of that of the major producers, has twice the resource base of any other nation.[2] (Table III-2-50).

History of the Industry

The entrance of Australia into the diamond market was initiated by the discovery of a large number of kimberlite pipes in the Kimberley region of Western Australia in 1977. By 1979, two main prospects (Ellendale and Argyle) had been identified. The Argyle area, with its four deposits, is being developed first (Figure III-2-20). The main deposit, Argyle Kimberlite 1 (AK-1), is a kimberlite pipe that covers a surface area of 45 ha and has in excess of 100 million metric tons of kimberlite ore. The other three deposits are composed of aluvial residue associated with the pipe.[41]

The Argyle deposits are being developed by the Ashton Joint Venture (AJV), which in turn is owned by Conzinc Riotinto Australia (CRA 56.7%), Ashton Mining Group (38.7%), and Northern Mining Corporation (5%). Initial output, which was scheduled to begin by the end of 1982, is planned to be 2 million carats per year. Mining of the AK-1 pipe is expected to begin by 1985 or 86, with production to increase to 20 million carats per year.[42]

The marketing of the diamonds is a controversial issue. The Fraser government, fervent in its stand against South Africa's racial policies, had expressed reservations in marketing Australia's diamonds through the De Beers' Central Selling Organization (CSO).[43] Eventually, in the fall of 1982, an agreement was reached between the AJV partners, except the Northern Mining Corp., and De Beers. As it stands, the CSO will market all of the gemstones (about 10% of the total output) and 75% of the near-gem and industrial-quality diamonds during the first five years of full production from AK-1.[44] The agreement also has provisions for the establishment of a pilot cutting and polishing facility in Western Australia.

TABLE III-2-50 World Industrial Diamond Production, 1980-1982 (Million Carats)

	1980	1981	1982	Reserve base
World	30.9	29.0	28.0	870
Australia	—	0.2	0.3	400
USSR	8.6	8.5	8.2	50
Zaire	9.9	7.2	7.5	200
South Africa	5.1	6.1	5.5	50
Botswana	4.3	4.2	4.0	125

Source: Ref. 2.

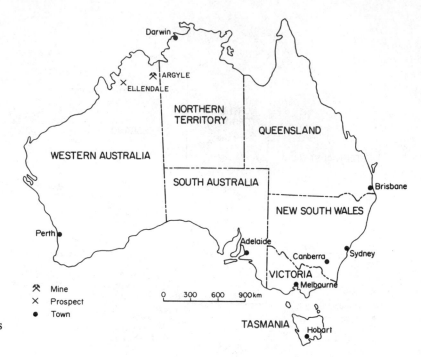

FIGURE III-2-20 Australian diamond prospects and mines, 1982. (From Ref. 44.)

Prospects for Future Production

The depressed state of the diamond market during the early 1980s may alter the development schedule of the Argyle mine. De Beers has had to stockpile excess diamond production and buy surplus stones in order to support the market. By the end of 1981, the stockpile had reached $1.35 billion, up from $393 million in 1979.[45] There have been closures and output reductions at several De Beers mines. Should the Argyle mine flood the maket with new stones, the already weak gem price could be adversely affected.

GOLD

Production, Resources, and Reserves

Australia, once a leader in production, presently plays only a minor role in the gold industry of the world. In the decade to 1860, Australia produced 40% of the world's gold, but today, with less than 2% of world production, is dwarfed by the output of South Africa and the USSR (Table III-2-51).

Due to the unstable nature of the price of gold, reserve figures vary depending on the price of bullion. The Bureau of Mineral Resources estimate of December 1979 listed total identified resources at 381.6 metric tons of metal, of which 277.9 metric tons were in the demonstrated category.

Exports and Trade

The main market for Australian gold is the United Kingdom, with exports also going to Hong Kong, Singapore, Japan, and New Zealand.

TABLE III-2-51 World Gold Production, 1979–1982 (Thousand Ounces)

	1979	1980	1981	1982
World	39,000	39,000	40,780	41,000
Australia	655	600	618	607
USSR(e)[a]	8,100	8,200	10,400	10,500
South Africa	22,600	21,700	21,120	21,200
U.S.	970	960	1,380	1,400

[a]e, estimated.

Sources: Refs. 2 and 46.

Uses

Gold is used as a basis for monetary systems, in foreign trade, as an ornamental metal, and in the electronics industry.

History of the Industry

Although its importance has declined in the past few decades, gold played a significant role in the development of Australia. In their search for gold, prospectors traveled to every colony, through Queensland, then west along the coast all the way to Perth. When gold was first discovered at Bathurst, New South Wales, in 1823 by James McBrien, the deposit was thought to be too small to be of any real importance. It was not until February 1851 that the Australian Gold Rush began, as gold was found in quantity within several miles of the original find. The excitement spread quickly into Victoria as the towns of Ballarat and Bendigo became the sites of major rushes. The quest for gold soon covered the entire continent, with the last of the great finds at Coolgardie and Kalgoorlie in Western Australia early in the 1890s.[47]

Within 10 years after the 1851 rush, Australia's population tripled, to more than 1 million people. This spurred the development of inland towns, and communication and trade between the colonies and the rest of the world. The search had additional benefits, in that other valuable minerals deposits, such as copper, silver, lead, and zinc, were found while prospecting for gold. Western Australia remains the most prolific gold-producing state, followed by the Northern Territory and Queensland.

Prospects for Future Production

There was renewed interest in gold by Australian mining firms in 1983. Several promising small gold deposits have been discovered in Western Australia, Victoria, and Queensland. Gold has the advantage of having a high unit value so that small increases in mining can have a relatively large increase in earnings. It is hoped by many mining interests that increased gold mining will make up for declines in base-metal mining. This could certainly be the case, but any "gold fever" must be kept in proper perspective. Gold mining in Australia accounted for only 1.7% of mineral exports by value in 1982. Therefore, it would require a massive gold find to affect significantly the entire mineral picture.

REFERENCES

1. Australian Department of Trade and Resources. *Australian Resources in a World Context*. Canberra: Australian Government Publishing Service, 1982.
2. U.S. Department of the Interior, Bureau of Mines. *Mineral Commodity Summaries 1983*. Washington, D.C.: U.S. Government Printing Office, 1982.
3. Australian Department of Trade and Resources. *Australia's Mineral Resources: Bauxite/Alumina/Aluminum*. Canberra: Australian Government Publishing Service, 1980.
4. Australian Information Services. *Fact Sheet on Australia: Aluminum Industry*. Canberra: Department of Administrative Services, February 1980.
5. *Aluminum*. Perth: Geological Survey of Western Australia, 1982.
6. Fitzgerald, M. D., and G. Pollio. "Aluminum: The Next Twenty Years." *Journal of Metals*, December 1982.
7. "Use of Aluminum in Auto Wheels Expected to Triple by 1983." *Modern Metals*, 16 August 1982.
8. Barnett, D. W. *Minerals and Energy in Australia*. Melbourne: Cassell Australia Ltd., 1979.
9. Knight, C. L., ed. *Economic Geology of Australia and Papua New Guinea, Volume 1: Metals*. Melbourne, Victoria: Australian Institute of Mining and Metallurgy, 1975.
10. "Aussie Aluminum: Chilled by the Recession's Pall." *Chemical Week*, 15 September 1982.
11. "Queensland Has Integrated Aluminum Industry." *World Mining*, November 1982.
12. Regan, R. J. "A Leaner Aluminum Industry Searches for the Comeback Trail." *Iron Age*, 4 October 1982.
13. Stewart, G. N. "Australia." *Mining Annual Review*, June 1982.
14. "Light Metals." *Mining Annual Review*, June 1982.
15. McCawley, F. X., and P. A. Stephenson. "Aluminum." In *Minerals Yearbook 1981*. Washington, D.C.: U.S. Department of the Interior, 1983.
16. *Iron Ore*. Perth: Geological Survey of Western Australia, 1983.
17. Australian Department of Trade and Resources. *Australia's Mineral Resources: Iron Ore*. Canberra: Australian Government Publishing Service, 1981.
18. Walsh, M. "BHP's Threat: We'll Close the Steel Industry." *The Bulletin*, 28 December 1982.
19. Australian Department of Trade and Resources. *Australia's Mineral Resources: Mineral Sands*. Canberra: Australian Government Publishing Service, 1980.
20. Hough, R. A., and L. E. Lynd. "Titanium." In *Minerals Yearbook 1981*. Washington, D.C.: U.S. Department of the Interior, 1983.
21. Australian Trade Development Council. *Minerals Processing, A Comparative Study*. Canberra: Australian Government Publishing Service, 1980.
22. Australian Department of Trade and Resources. *Australia's Mineral Resources: Manganese*. Canberra: Australian Government Publishing Service, 1981.
23. Australian Department of Trade and Resources. *Australia's Mineral Resources: Nickel*. Canberra: Australian Government Publishing Service, 1981.
24. "Exploration Roundup." *Engineering and Mining Journal*, March 1982.
25. Australian Department of Trade and Resources. *Australia's Mineral Resources: Lead and Zinc*. Canberra: Australian Government Publishing Service, 1983.
26. Woodward, O. H. *A Review of the Broken Hill Lead–Silver–Zinc Industry*. Sidney: West Publishing Corporation, 1965.
27. Ridge, J. *Annotated Bibliographies of Mineral Deposits in Africa, Asia (exclusive of the USSR) and Australia*. Oxford: Pergamon Press Ltd., 1976.
28. Schmidt, Helmut, and Manfred Kruszona. *Regional Distribution of Mining Production and Reserves of Mineral Commodities in the World*. Hanover, West Germany: Federal Institute for Geosciences and Natural Resources, January 1982.
29. Alexander, J., and R. Hattersley. *Australian Mining, Minerals, and Oil*. Sydney: The David Ell Press Pty. Ltd., 1980.
30. Australian Department of Trade and Resources. *Australia's Mineral Resources: Copper*. Canberra: Australian Government Publishing Service, 1983.
31. "Olympic Dam/Roxby Downs Mineralization Valued at $140 Billion." *Engineering and Mining Journal*, 21 January 1983.
32. "Roxby Will Cost $1400 Million to Develop." *Engineers Australia*, 21 January 1983.

33. The International Tin Council. "Tin." *Mining Annual Review*, June 1982.

34. Australian Department of Trade and Resources. *Australia's Mineral Resources: Tin.* Canberra: Australian Government Publishing Service, 1981.

35. Rich, P. J. H. "The Future of Tin as a Tonnage Commodity." *Institute of Mining and Metallurgy*, January 1980.

36. Wyche, C. "The Mineral Industry of Australia." In *Minerals Yearbook 1980*. Washington, D.C.: U.S. Department of the Interior, 1982.

37. Chadwick, J. R. "Tin, What of the Future?" *World Mining*, November 1981.

38. Ho, C. E. "Tungsten." *Mining Annual Review*, June 1982.

39. Australian Department of Trade and Resources. *Australia's Mineral Resources: Tungsten.* Canberra: Australian Government Publishing Service, 1981.

40. Bain, G. W. "Resources of Tungsten." *Institute of Mining and Metallurgy*, April 1980.

41. Australian Department of Trade and Resources. *Australia's Mineral Resources: Gemstones.* Canberra: Australian Government Publishing Service, 1982.

42. "AJV Agreements—Northern Pursues Its Own Markets." *World Mining*, December 1982.

43. "Aussie Diamonds May Alter the Market." *The Wall Street Journal*, 22 February 1982.

44. "Argyle Mine, Australia Will Be One of the World's Major Diamond Producers." *Mining Magazine*, October 1982.

45. "DeBeers May Be Cutting Diamond Output in Renewed Bid to Support Ailing Market." *The Wall Street Journal*, 8 April 1982.

46. Lyons, L. A. "Worldwide Survey—Australia." *World Mining*, August 1982.

47. Blainey, G. *The Rush That Never Ended.* Melbourne: Melbourne University Press, 1963.

48. Australian Department of Trade and Resources. *ABS Overseas Trade Statistics 1971-72 through 1981-82.* Canberra: Australian Government Publishing Service, 1982.

Chapter III-3

Fuel Minerals

Australia is blessed with abundant energy resources. Coal is the nation's largest energy commodity, as 84% of Australia's remaining recoverable demonstrated economic nonrenewable energy resources is in the form of brown and black coal (Figure III-3-1). While oil and gas account for only a small portion of Australia's resources, they are the country's principal source of energy. Over 50% of Australian energy demand is supplied by oil (Table III-3-1). The industrial and transport sectors are the primary energy consumers, as shown in Table III-3-2.

BLACK COAL

Production, Resources, and Reserves

Coal is Australia's principal mineral export on both a dollar and tonnage basis. Black coal [coals having a gross specific energy value of more than 20.00 MJ/kg on a dry ash-free (d.a.f.) basis[3]] comprises between 70 and 75% of Australian coal mining, and all of its coal exports. Australian coal output is small in comparison to the major coal-mining nations: the United States, the USSR, and China (Table III-3-3).

The bulk of Australian coal resources are found in New South Wales and Queensland (Figure III-3-2). Although the nation's coal resources are very large, demonstrated economic reserves represent only a small portion. This reflects a reluctance to invest in exploration to identify resources beyond those necessary to satisfy short-term requirements. Table

TABLE III-3-1 Australian Energy Mix, 1980–1981

Commodity	Percent
Oil	53
Natural gas	13
Electricity	14
Black coal	48%
Brown coal	24%
Natural gas	9%
Hydro	15%
Other	5%
Other	20

Source: Ref. 2.

TABLE III-3-2 Energy Use by Consumer, 1980–1981

Consumer	Percent
Industrial	38
Commercial	6
Transport	38
Household	12
Other	6

Source: Ref. 2.

III-3-4 presents a comprehensive view of Australian black coal resources.

Export and Trade

Australia, together with the United States and Poland, have historically been the world's largest coal exporters (Table III-3-5). Japan remained the largest market for Australian coals in 1981–1982, accounting for 71% of coking coal exports, 59% of steam coal exports, and 68% of total coal exports.[5] Tables III-3-6 and III-3-7 detail coal exports by destination, and Figure III-3-3 shows trade routes for Australian coal exports.

The trade in steam coal has grown rapidly since the mid-1970s, due to the sharp increase in oil prices and the long-term unreliability of oil supplies. As seen from Table III-3-7, Australian steam coal exports more than doubled between 1977–1978 and 1980–1981. Continued export increases in the

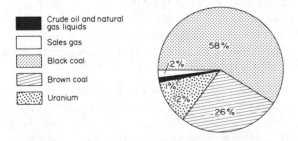

Crude oil and natural gas liquids
Sales gas
Black coal
Brown coal
Uranium

58%
2%
1%
12%
26%

Percentage breakdown of Australia's remaining recoverable demonstrated economic non-renewable energy resources (expressed in terms of energy content).

Percentages do not add due to rounding

FIGURE III-3-1 Australian energy resources, 1980. (From Ref. 1.)

TABLE III-3-3   World Coal Production, 1976–1982
(Million Metric Tons Anthracite and Bituminous Coal)

	1976	1977	1978	1979	1980	1981	1982
Total	2394.0	2521.9	2616.2	2729.5	2840.0	2789.5	2865.7
Australia[a]	84.2	87.3	89.3	92.8	93.6	110.9	110.2
U S	598.5	606.9	609.0	703.7	711.1	700.9	697.5
China[a]	440.0	550.0	618.0	663.2	620.0	620.0	618.0
USSR[a]	494.4	499.8	501.5	495.0	553.0	544.0	558.0
Poland	179.3	186.1	192.6	201.0	193.1	163.0	189.3
U.K.	122.3	120.7	121.7	120.6	128.2	124.3	121.4
India	101.9	101.3	101.8	103.3	114.0	124.9	130.0
South Africa	74.6	85.4	90.4	103.4	115.1	130.4	137.3
West Germany	96.3	91.3	90.1	93.3	87.1	88.4	96.3

[a]Raw coal basis.

Sources: Refs. 4, 5, and 33.

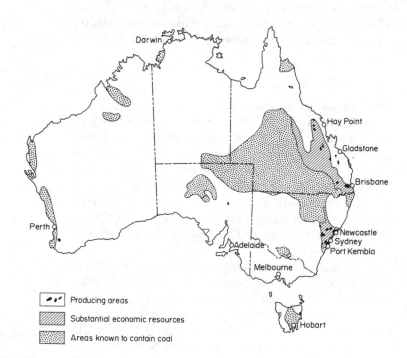

Producing areas

Substantial economic resources

Areas known to contain coal

FIGURE III-3-2   Australian black coal resources, 1981. (From Ref. 5.)

future will depend on the continued growth of Western Pacific economies, as well as being more price competitive in European markets. The Australian Department of Trade and Resources estimated in 1980 that the demand for Australian steaming coal would probably be between 30 and 40 million metric tons by 1985, and 50 to 70 million metric tons by 1990.[6] Although these figures may bear out, the world recession during the early 1980s, and the leveling off of oil prices, has ebbed steaming coal demand.

Australian steaming coals vary in quality, but typical characteristics are (air-dried basis):[6]

Calorific value between 6000 and 7200 kcal/kg
Sulfur content from 0.4 to 0.8%

Hardgrove grindability ranging from 40 to 70
Ash levels between 10 and 19%
Nitrogen from 1.3 to 2.0% d.a.f. basis
Volatiles generally between 25 and 36%
High ash fusion temperatures: deformation above 1380°C, flow above 1590°C

Recent trade in metallurgical coals has shown little growth, reflecting the state of the Japanese steel industry. This dependence on Japanese trade has decreased somewhat since the late 1960s. Since 1969, when 99% of Australia's black coal exports were to Japan, the level of Japanese exports has dropped to about 75% of total metallurgical coal exports.[7] Australian coking coals available for export cover the full range of rank. All are character-

TABLE III-3-4   Australian Black Coal Resources (Million Metric Tons)

	New South Wales[a]	Queensland[b]	Australia
Measured and indicated reserves			
In-situ (raw coal)			
Bituminous	22,243	23,780	46,553
Subbituminous	500	5,173	8,002
Total	22,743	28,953	54,555
Recoverable (raw coal)			
Bituminous	11,672	13,841	25,759
Subbituminous	450	3,970	4,932
Total	12,122	17,811	30,691
Open-cut mining	3,071	6,544	10,054
Underground mining	9,051	11,267	20,637
Coking	2,740	7,959	10,699
Noncoking	9,382	9,852	19,992
Inferred resources			
In-situ (raw coal)			
Bituminous	480,036	Very large	Very large
Subbituminous	10,000	Very large	Very large
Total	490,036	Very large	Very large

[a]From assessment by Joint Coal Board 1978 and 1979.
[b]From information published in Queensland Government *Mining Journal*, July 1982.
Source: Ref. 5.

TABLE III-3-5   World Coal Trade, 1976–1982 (Million Metric Tons)

	1976	1977	1978	1979	1980	1981	1982
Total	190.7	201.5	197.9	229.0			
Australia	34.1	36.2	38.7	40.9	42.8	51.0	
U.S.	54.5	49.3	37.0	59.9	83.2	102.0	
Poland	38.9	39.3	40.1	41.3	30.9	15.0	

Sources: Refs. 4 and 5.

TABLE III-3-6   Australian Coking Coal Exports, 1977–1978 to 1981–1982 (Thousand Metric Tons)

	77–78	78–79	79–80	80–81	81–82
Total	33,634	33,257	36,144	36,854	37,399
Japan	25,754	24,354	27,260	27,305	26,451
U.K.	432	750	781	566	863
Other Europe	5,362	4,459	3,970	4,070	3,970
South Korea	1,204	1,933	2,273	3,110	3,447
Taiwan	780	1,126	874	872	1,151
Elsewhere	102	635	986	931	1,517

Source: Ref. 5.

TABLE III-3-7 Australian Steaming Coal Exports, 1977-1978 to 1981-1982 (Thousand Metric Tons)

	77–78	78–79	79–80	80–81	81–82
Total	4,277	5,021	7,017	10,585	9,753
Japan	535	820	1,746	5,501	5,753
U.K.	942	985	1,125	1,265	1,385
Other Europe	2,071	2,640	3,184	2,673	1,358
South Korea	—	—	11	—	247
Taiwan	51	236	899	1,062	469
Elsewhere	678	340	52	84	541

Source: Ref. 5.

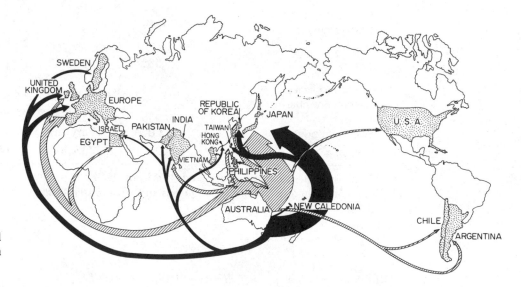

FIGURE III-3-3 Australian coal exports, 1981-1982. (From Ref. 5.)

istically low in sulfur, ranging between 0.4 and 0.7%. Below are the general characteristics of the main producing areas:[7]

NSW—Southern coal fields: Low- to medium-volatile coking coals with rank generally decreasing westward; ash specifications generally 8.5 to 10.3%

NSW—Western District: High-volatile low-swell coals largely used for briquetting; high-volatile, low-rank coking coals awaiting development

NSW—Northern coal fields: High-volatile low-rank coking coals generally 7.5 to 9.5% ash and high-volatile low-swell briquetting coals of 9.5% ash

Qld—Bowen Basin: Low- to medium-volatile coking coals; ash specifications in the range 7 to 10.5%

Future demand for coking coals will depend on the growth and technological changes in the world steel industry. Australia is in a strong position by virtue of low-cost resources, efficient mining, and proximity to the steel industries of Japan, the Republic of Korea, and Taiwan.[8]

Over 40% of the black coal mined in Australia is consumed domestically (Table III-3-8). New South Wales is the largest consumer, with about 65% of total Australian consumption. Queensland is second with almost 20% (Figure III-3-4). Although the percentage varies from state to state (i.e., New South Wales, 62%; Queensland, 75%; Western Australia, 93%), the majority of coal consumed is used to generate electricity.[5] Table III-3-9 breaks down consumption by major users.

Uses

The principal uses for black coal are in steelmaking, steam generation, and gaseous and liquid fuel conversion. Coal is also used in the cement industry, in the manufacture of brick, tile, pottery, and glass, and in paper, milk, and meat processing.[9]

Metallurgical-grade coals are used to make coke. The coke functions as source of fuel and reducing agent in the blast furnace. A proximate analysis is essential to determine the quality of the coal and the amount of preparation required.

Coking and caking tests are used to determine the swelling index and the petrographic analysis.

TABLE III-3-8   Australian Coal Supply and Disposal, 1977–1978 to 1981–1982 (Thousand Metric Tons)

	77–78	78–79	79–80	80–81	81–82
Net production	71,014	73,373	73,594	87,387	91,077
Exports	37,911	38,278	43,161	47,439	47,152
Consumption	32,555	33,431	35,624	37,712	36,788
Year-end stocks	15,502	17,166	11,974	14,216	21,343

Source: Ref. 5.

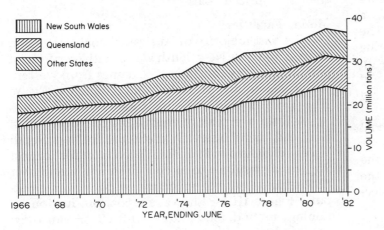

FIGURE III-3-4 Consumption of black coal in Australia by states, 1966–1982. (From Ref. 5.)

TABLE III-3-9   Consumption of Black Coal in Australia by Major Consumers, 1977–1978 to 1981–1982 (Thousand Metric Tons)

	77–78	78–79	79–80	80–81	81–82
Electricity generation	20,754	21,032	23,673	24,964	24,483
Iron and steel	8,284	8,738	8,257	9,054	8,417
Metallurgical coke	325	313	332	352	311
Railways and gas works	97	79	39	30	17
Cement industry	839	956	911	904	1,109
Other	2,256	2,313	2,412	2,408	2,451

Source: Ref. 5.

Other factors for coking coals are the ultimate analysis and the coal's grindability.

The other major use for black coal is as a fuel for steam generation. Australian power stations are designed to burn raw coal, and normally the coal is cleaned only when a coking coal fraction can be extracted. In contrast, the majority of exported steam coal has to be cleaned to some extent. Steam coals are analyzed for size distribution, proximate analysis, calorific value, sulfur and chlorine content, caking properties, grindability, and ash composition and fusion properties.[9]

The coal market with the greatest growth potential is for gasification and liquefaction. With Australian dependence on foreign sources for petroleum on the rise, there has been a great deal of interest in producing liquid fuels from domestic coal. At present, studies are in progress to understand better the maceral composition of Australian coals in regard to hydrogenation.[10]

History of the Industry

Coal was first discovered at Newcastle in 1791, only five years after the founding of the colony of New South Wales. By 1801, coal was being mined and the first exports (to India) were made.[3] The coal lay close to the surface and was easily won. In being close to port, the coal could be mined and transported quickly and cheaply to the wharves at Newcastle. The readily available black coal provided for industrial development and the establishment of an iron and steel industry sufficient for the country's needs. Coal soon became the leading tonnage mineral commodity export, as the Pacific rim began to industrialize.[11] Queensland coal was first discovered in the Bowen Basin in 1845, and mining has been carried out sporadically since the early part of this century.[12,13]

Two developments in the first quarter of the twentieth century slowed the development of Australian coal trade. One was the completion of the Panama Canal in 1914, which allowed eastern U.S. coals to compete in the Pacific Rim market more favorably. Second, oil began to replace coal as a fuel for ships, factories, and power stations. The competition from oil was particularly strong during the 1950s and 1960s. In 1953–1954 coal made up 66% of Australia's primary energy balance; 10 years later the value had dropped below 50%, and throughout the 1970s it was close to 40%.[3]

During the past 20 years the production of coal has increased rapidly. The lion's share has been for exports, but the rise also reflects the growth in the domestic demand for electricity. Since 1960, the use of coal as a percentage of all fuels used for electricity generation rose from 36% to 67%.[5] Figure III-3-5 shows the growth of the Australian coal industry.

Geology of Major Deposits

The Bowen, Surat, and Galilee basins of Queensland and the Sydney basin in New South Wales are the major black-coal-bearing areas with large economical resources. Coal is also found in the other states to a lesser extent. Victoria is the exception in that it contains only brown coal. Figure III-3-6 is a map showing the coal-bearing areas. It also has the locations of active and proposed mines, railways, and ports for coal transport.

Australian coal deposits vary in their depth and thickness, as well as the number of seams and the thickness of the partings between them. The ratio of coal thickness to overburden dictates whether a deposit will be economical to develop. In addition, the quality and grade of the coal will also help determine the feasibility of mining a coal deposit.[13]

Australian black coals do not have the metallurgical qualities that northern hemisphere coals, and in particular eastern U.S. coals have. This means that U.S. coals bring a higher price on the international market. Japan, a large consumer of coal, will often purchase Australian coals to blend with the higher-priced coals.

Mining Facilities

The vast majority of Australian coal is mined in Queensland and New South Wales, with over 95% of production. The remaining 5% comes from Western Australia, South Australia, and Tasmania. Figure III-3-7 shows production by state.

A little over half of Australia's coal production is mined using open-cut methods. In New South Wales underground mining accounts for almost 75% of that state's total coal production and over 88% of the nation's underground coal mining output.[5] Figure III-3-8 shows raw coal production by mining method, and Table III-3-10 provides raw coal production figures by state and mining method.

There are over 130 black coal mines in Australia, with over half being underground operations in New South Wales. Table III-3-11 lists the distribution of coal mines in production for the period between June 1981 and June 1982. The largest of these mines belong to the open-cut class. Table III-3-12 provides data on the 20 largest coal mines in Australia for 1981–1982.

Metallurgical coal production is confined to the Bowen Basin of Queensland and the Sydney Basin of New South Wales.[7] The Central Queensland Coal Association (CQCA) dominates Queensland's coking-coal production. Between CQCA's four open-pit mines, located on the western side of the

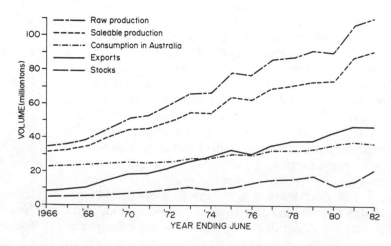

FIGURE III-3-5 Growth in Australian coal production, 1966–1982. (From Ref. 7.)

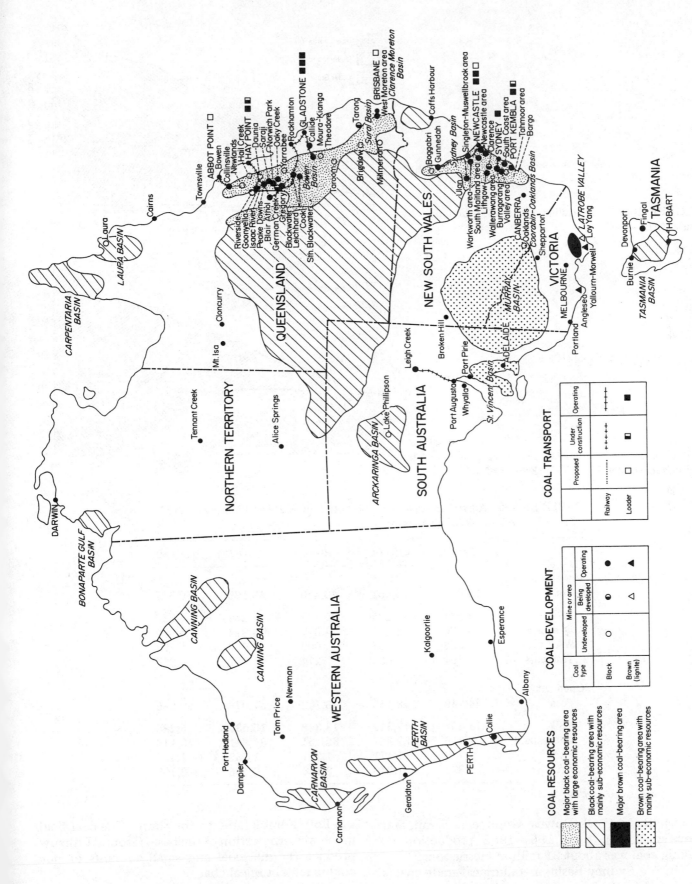

FIGURE III-3-6 Australian coal-bearing areas, mines, and transport facilities, 1981. (From Ref. 7.)

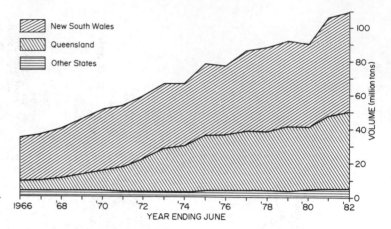

FIGURE III-3-7 Australian raw coal production by state, 1966–1982.

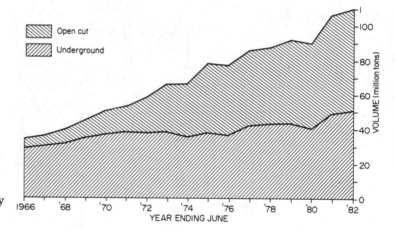

FIGURE III-3-8 Australian raw coal production by mining method, 1966–1982. (From Ref. 5.)

TABLE III-3-10   Australian Raw Coal Production by State and Mining Method, 1977–1978 to 1981–1982 (Thousand Metric Tons)

	77–78	78–79	79–80	80–81	81–82
Underground					
Total	43,041	43,451	40,456	49,126	50,617
NSW	37,875	38,205	35,362	43,257	44,530
Queensland	4,420	4,426	4,277	4,914	4,990
WA	552	567	603	653	707
Tasmania	194	253	214	302	390
Open cut					
Total	44,749	48,450	49,742	57,073	59,580
NSW	11,397	12,312	13,348	15,035	15,288
Queensland	29,718	32,827	32,251	37,841	40,148
SA	1,780	1,471	1,717	1,732	1,436
WA	1,872	1,840	2,426	2,465	2,708

Source: Ref. 5.

Bowen Basin (Peak Downs, Goonyella, Saraji, and Norwich Park), their 1981–1982 production of coking coal was about 21 million metric tons.[5]

In the Sydney Basin, metallurgical-grade coal is produced in the Newcastle area in the north and in the Port Kembla area to the south. The coal fields in the western portion of the basin (around Lithgow) produce thermal coal and small amounts of non-coking metallurgical coal.[7]

Many of Australia's coal mines are owned by or

TABLE III-3-11   Australian Coal Mines in Production, 1981–1982

| | June 1981 | | | June 1982 | | |
State	Underground mines	Open cuts	All mines	Underground mines	Open cuts	All mines
Australia	98	37	135	94	47	141
NSW	74	14	88	72	19	91
Queensland	21	20	41	19	24	43
WA	1	2	3	1	3	4
SA	—	1	1	—	1	1
Tasmania	2	—	2	2	—	2

Source: Ref. 5.

TABLE III-3-12   Twenty Largest Coal Mines in Australia, 1981–1982

Mine	Company	State	Raw production (thousand metric tons)
Peak Downs	Central Qld Coal Ass.	Qld	5950.8
Norwich Park	Central Qld Coal Ass.	Qld	5365.9
Blackwater	Utah Development Co.	Qld	5210.2
Goonyella	Central Qld Coal Ass.	Qld	5195.2
Saraji	Central Qld Coal Ass.	Qld	4368.9
Ravensworth No. 2	Electricity Comm. of NSW	NSW	4368.9
Gregory	BHP Minerals Ltd.	Qld	3151.8
Moura	Thiess Dampier Mitsui Coal	Qld	2878.0
Callide	Thiess Bros. Pty. Ltd.	Qld	2316.0
Hunter Valley No. 1	Coal & Allied Ind. Ltd.	NSW	2066.4
South Bulli[a]	The Bellambi Coal Co. Ltd.	NSW	1995.2
Muja	Griffin Coal Mining Co.	WA	1831.0
Coal Cliff[a]	Kembla Coal & Coke Pty.	NSW	1824.1
Buchanan Lemington	Buchanan Borehole Colieries	NSW	1756.7
Clarence[a]	Clarence Colieries Pty.	NSW	1738.2
Howick	Clutha Development Pty.	NSW	1706.3
West Cliff[a]	Kembla Coal & Coke Pty.	NSW	1529.5
Leigh Creek	Electricity Trust of SA	SA	1436.0
S. Blackwater	Thiess Bros. Pty. Ltd.	Qld	1435.6
Wyee State[a]	Elcom Collieries Pty. Ltd.	NSW	1257.8

[a]Indicates an underground mine; all others are open-cut mines.
Source: Ref. 5.

captive to the consumer. The Electricity Commission of New South Wales owns the state's largest coal mine, as well as several other large coal mines. The mines in South Australia and Western Australia produce thermal coal exclusively for their state energy commissions. Much of the steam coal mined in Queensland is sold for electricity generation at state-owned power stations and for local industry (i.e., alumina refining at Gladstone). Coking-coal production for domestic uses is also extensively controlled by the consumer, as 80% of the coking coal used by Australian industries comes from mines owned by or otherwise captive to the final user.[7]

Processing Facilities
Processing of Australian coal generally entails washing to lower the coal's ash content. Broken Hill Proprietary operates various collieries in which the coal is used to make coke for their steel plants in New South Wales and South Australia. At present no coke is produced for the export market, although there have been proposals for export coke plants at Gladstone and Bowen.[14]

Studies on the feasibility of producing liquid fuels from coal have been undertaken. An Australian–Federal Republic of Germany feasibility study, the Imhausen Report, called for the commissioning of conversion plants in Victoria, New South Wales, and Queensland. The New South Wales and Queensland plants would use black coal, while the Victorian facility would utilize brown coal.[14] Japan has been working on a 50-metric ton per day brown coal liquefaction plant at Morwell, Victoria.[11] A large Australian group in association with Fluor Engineering plans to set up a Sasol-type conversion plant west of Brisbane.[14] Government support for these programs has been modest, as the intense capital nature of such a venture has created many skeptics concerned about the economic feasibility of coal liquefaction.

Transport Infrastructure

The production of coal involves the transport of massive amounts of material (more than 110 million metric tons of coal in 1981–1982) to port or to the domestic consumer. Although coal that is mined at a captive collary (for, say, a power station) can be moved by conveyor, it is the railroads that move the bulk of Australian coal. Roads are used only in a supplementary role, usually through short-haul deliveries in areas not serviced by rail. Coastal shipping accounts for only 2 million metric tons a year, with half of that being coking coals to the steelworks in South Australia.[3]

The development of the rail system in New South Wales was quite different from that of Queensland, even though both are operated by their respective state governments. In New South Wales, industry grew around the coal fields, which were close to the coast. This concentration of population and industry meant that new coal fields were close to established rail systems and ports. The negative side is that since the railways have many users, when exports of coal began to rise the rail system was stressed to keep pace with the increased demand. To ease these problems the New South Wales government has committed itself to improving and extending the rail system to meet present and future growth. Upgrading and track extensions, electrification of tracks, construction of new coal-loading facilities, upgrading of rolling stock, and the introduction of new coal waggons with a 75-metric ton capacity are all in the works.[3]

Queensland coal transport faced an entirely different problem. Because of the small population of the Bower Basin, a new rail system, suitable for a high level of traffic, had to be built. This required massive amounts of capital for major upgrading of the existing railways, and more commonly for the construction of new rail links largely dedicated to coal traffic. The financing of the new lines was an additional problem. In New South Wales, the improvements on existing lines benefited more than just the mining sector, due to the multipurpose nature of its rail system. Therefore, the costs were spread to all sectors. In Queensland, most of the new lines would serve the coal mines exclusively, even though the ownership and maintenance belonged to the state government. To account for this, the Queensland government collects a railway security deposit, and generally charges a higher freight rate for mine products.[3] As new mines are planned and developed, new rail lines are financed and built in the same manner. At this time Queensland railway development plans include a new line to connect the Boundry Hill mine with the Moura line near Annandale, and one to connect the Curragh mine with the Central Railway near Blackwater. The Riverside mine is being linked with the Goonyella railway, and the Blair Athol mine is being connected to the same rail system near Mallawa.[15] The typical Queensland coal train is comprised of up to six diesel locomotives and 150 cars of up to 56 metric tons capacity each.[15]

To develop and sustain a world-scale coal exporting industry, large ports and rapid coal-loading facilities are needed. In the past, common-user port facilities have kept pace with new mining projects by expansion and upgrading existing ports and building new ones. However, when the export market swelled, requiring vessels with capacities of over 200,000 dwt, adequate port facilities lagged behind. Of the two exporting states, Queensland is in a better position to meet the growing demand on its port facilities. In New South Wales, shipping suffers from shallow ports and inadequate loading equipment, and although improvements are planned, serious problems may continue to plague future export plans. This issue, as well as other infrastructure problems, are dealt with in detail in Chapter III-5. Table III-3-13 provides information on the principal coal-loading facilities at Australian ports.

In an effort to relieve port congestion, a number of projects are under way. The principal large-scale programs in progress are:[3,6,7]

Adjacent to Hay Point Qld, a complete new multiuser coal loading facility (Dalrymple Coal Terminal), capable of handling vessels of at least 150,000 dwt, is being developed by the Queensland government. The 8-million metric ton/year facility will handle coal from new Northern

TABLE III-3-13   Principal Coal-Loading Facilities at Australian Ports, 1980

Location	Nominal vessel size (dwt)	Annual capacity (million metric tons)	Maximum load rate (metric tons)	1979–1980 output (million metric tons)
Queensland				
Hay Point[a]				
No. 1 Berth	120,000	20	4,000	14.75
No. 2 Berth			6,000	
Gladstone				
Auckland Point	55,000	5	1,600	3.43
Barney Point[a]	60,000	8	2,000	3.38
Clinton[b]	60,000	10	4,000	0.25
Brisbane	35,000	1.0	1,200	0.03
Bowen	16,000	0.5	700	0.01
New South Wales				
Newcastle				
Basin	70,000	18	2,500	11.92
Channel	70,000[c]		5,000	
Sydney				
Balmain	55,000	4.5	1,250	2.74
Balls Head[a]	35,000	1.0	2,000	0.68
Port Kembla: Inner Harbour	58,000[d]	7.5	2,500	7.03

[a]Privately owned, not available on a common-user basis.
[b]Started operations in May 1980.
[c]Vessel size limited by depth at channel mouth.
[d]In 1979–1980 more than 50 vessels in the range of 60,000 to 100,000 dwt were handled, generally by part-loading.
Sources: Refs. 3, 6, and 7.

Bowen Basin metallurgical and steam coal mines. Completion is scheduled for late 1984.

In Newcastle NSW, installation of a second coal berth at Channel Loader was expected to be completed in 1982. A harbor-deepening program which has been in progress since 1977 was expected to be completed by December 1982 and will enable vessels of 110,000 dwt to be fully loaded at the channel loader, with further upgrading, including the provision of a second loading berth at the channel loader. The 1982 effective combined capacity was 25 million metric tons.

In Port Kembla, NSW, construction of the first stage of a new loading facility capable of handling vessels of 110,000 dwt was expected to be completed by August 1982. It is intended to have an annual capacity of 15 million metric tons by 1985.

Deepening of the approaches to the Clinton loader in Gladstone Harbour, Qld, which will enable vessels of 120,000 dwt to be fully loaded and raise capacity to about 12 million metric tons per year. Completion was expected in 1984.

Construction by MIM Holdings of a new port at Abbot Point, Qld, capable of handling vessels of 120,000 dwt. Completion was expected in 1984 to serve mines being developed by that company. The terminal will provide facilities to export 6.5 million metric tons per year and later expansion plans could bring this figure to 24 million metric tons per year.

Construction of a third facility in the port of Newcastle on Kooragang Island, capable of handling vessels in excess of 110,000 dwt and with an ultimate capacity of 50 million metric tons per year. The first stage (15 million metric tons per year) is currently expected to be completed in late 1985.

Construction of an interim facility at the mouth of the Brisbane River, Qld, to an annual capacity of 5 million metric tons to be completed by mid-1983 and initially capable of handling Panamax vessels.

None of these projects are immune to delays or setbacks. For example, the coal loader planned for Kooragang Island has encountered serious problems in respect to the terms of lease proposed by the state government. As proposed, the state would be entitled to a base charge of 25 cents per metric ton, linked to the Consumer Price Index, with a minimum charge of A$2.5 million for the loader site on the government-owned land on Kooragang Island.[16] The trouble is that when the project is complete and running at its annual capacity of 50 million metric tons per year in 1992, the levy (after indexing) could be as high as 50 cents per ton, meaning a minimum annual payment of A$25 million.[16] In addition, the state government appears to be proposing a lease of only 20 years for the loader. This would leave little incentive to upgrade loading facilities after the initial installation is completed. Some of the investors in the project, which includes private investors, coal exporters from the Hunter Valley, and Japanese coal consumers, feel that their investments would be better served elsewhere, as the proposed changes would make their coal less attractive on the international market.[16] The price of coal is sensitive to the world economy, so that recessions in the world economy, such as in 1982–1983, will lower coal demand and the feasibility of port expansions. Cutbacks in contracted shipments of coal creates pressures to delay port and loader upgradings. This in turn lowers the return on investment on a port project, which could cause the project's cancellation.

Prospects for Future Production

The future for Australian black coal looks good as long as potential infrastructure problems can be avoided, and current expectations of coal use continue to hold true. According to various experts, coal exports could be as high as 170 million metric tons per year by the end of the century.[15] More conservative estimates would place exports closer to 100 million metric tons per year.

The production levels in New South Wales and Queensland over the next 20 years will be limited more by external factors (infrastructure, demand) than by exhaustion of reserves. Either of the two states could produce enough coal for exports of 100 million metric tons per year. The trouble is that to produce at that level with combined state production, three things must occur. One is that demand for coal, probably steam coal, must continue to grow. Second is that infrastructure must keep pace with mine output. Third, the Australian coal industry must remain competitive in its pricing.

As of 1982–1983, the demand for coal remains depressed. As a result, the Commonwealth government authorized an 18% cut in the price of Australian coal exported to Japanese steel mills. How the market for coal will act over the next five years is difficult enough to project, much less over 20 years. For the Australian coal industry to grow, trade must increase to Japan, as well as to other Asian markets such as South Korea and Taiwan.

The development of infrastructure continues as new rail and port facilities are built. Whether this growth will be adequate will depend on the world, particularly the Western Pacific region, economy. The concessions by the state and commonwealth governments will also affect infrastructure growth.

As far as remaining competitive in coal trade, it will take an effort by all the players. Australia has enormous reserves of easy-to-mine coal, but if the government is not willing to let mining be profitable, and if labor relations remain poor at mines, rails, and ports, and if companies cannot better coordinate their marketing efforts, Australia will lose parts of her markets to competitors such as the United States, Canada, and South Africa.

BROWN COAL

Production, Resources, and Reserves

Lignite or "brown coal" (coals having a gross specific energy value of less than 20.00 MJ/kg d.a.f. basis) is found in abundance in Australia. Victoria contains 98% of the nation's demonstrated economic brown coal resources and is the only state that produces it (Table III-3-14).

Brown coal production has remained close to 30 million metric tons per year since 1976, although production in 1981–1982 was over 37 million metric tons (Table III-3-15). Of this, more than 95% came from the State Electricity Commission of Victoria's mines at Yallourn and Morwell.[18] The remaining production came from privately operated mines at Anglesea and Bacchus Marsh. Anglesea coal is used by Alcoa of Australia to provide electricity for the company's aluminum smelter at Point Henry. The coal from Bacchus Marsh is used to supply steam and electricity at the Australian Paper Manufactures mills at Faurfield and Broadford.[18]

Due to the low heat value of brown coal, its high moisture content (55 to 75%), and problems with spontaneous combustion, exports have not taken place; nor are they likely to occur in the near future.

Uses

At present brown coal is used almost exclusively for thermal purposes. Victoria obtains 85% of her electrical power from a series of state-owned brown-coal-fueled power stations in the Latrobe Valley.

TABLE III-3-14   Australian Brown Coal Resources, 1981 (Million Metric Tons)

Location	Measured and indicated		Inferred		Total identified
	Eco	Sub-eco	Eco	Sub-eco	
Victoria					
Latrobe Valley	35,129	72,093	—	—	107,222
Other	4,210	430	4,300	7,500	16,400
	39,339	72,523	4,300	7,500	123,662
South Australia		1,909		780	2,689
Australia	39,400	74,400	4,300	8,200	126,300
Recoverable	36,200	68,600	3,900	7,600	116,300

Source: Ref. 6.

TABLE III-3-15   Australian Brown Coal Production, 1976–1977 to 1981–1982 (Million Metric Tons)

	76–77	77–78	78–79	79–80	80–81	81–82
Production	31.0	30.5	32.1	32.9	32.1	37.5

Sources: Refs. 3 and 5.

The newest, a 2000-MW facility, is currently under construction at Loy Yang.[19]

As mentioned previously, the Japanese have been working to develop a coal liquefaction plant using Latrobe Valley coals. The Japanese have also been experimenting with brown coal in the production of solvent-refined coal (SRC). The SRC is a hydrogen-enriched product that could be used in the production of metallurgical-grade coal for steel making. In addition, SRC could possibly be used in making anodes for aluminum smelting, as a fuel oil substitute, and as a clean fuel for thermal power stations.[20]

History of the Industry

Australian brown coal was first discovered in Victoria in 1857, and over the next 20 years coal was reported in 32 localities. During the 1890s several of the deposits were worked, but few were economically successful. Renewed interest occurred during World War I, due to a shortage of New South Wales black coals. Production began on a large scale in 1916, and cumulative output to June 1976 amounted to 557 million metric tons.[21]

Geology of Major Deposits

Brown coal occurs in thick seams over vast areas. In the Latrobe Valley, seams vary in thickness from 90 to 230 m, and underlie an area 10 to 30 km wide extending over a length of 70 km.[21] The coal is characterized by high moisture contents, low calorific values, and generally low ash. Table III-3-16 provides information for some major coal fields, and Table III-3-17 has data on coal quality.

Mining Facilities

Morwell and Yallourn are the largest mines producing brown coal. Both are surface operations that use huge bucket-wheel excavators that can mine 30,000 metric tons of coal a day. The nature of lignite and lignite deposits, being low in value and thick-seamed, requires that a deposit be able to be mined using surface methods. The Yallourn mine alone produces about 13 million metric tons of coal annually. The coal from these mines is then transported by conveyor to a nearby power station, where the coal is burned. Figure III-3-9 shows brown coal mining areas.

Future of the Industry

The development of Loy Yang, also owned by the State Electricity Commission of Victoria, will eventually become the state's largest coal operation. The project, which is still under construction, will include two 2000-MW power stations. The first of these facilities should be in operation between 1984 and 1987. When fully operational, the complex could lead to an annual output from the Loy Yang mine alone of 32 million metric tons in the 1990s.[22]

Additional use of brown coal would occur if liquefaction programs planned prove to be eco-

TABLE III-3-16   Australian Brown Coal Field Resources, 1981
(Million Metric Tons)

Coal field	Resources		Overburden ratio	Seam thickness (m)
	Recoverable	Identified		
Anglesea	160	400	1:1	30
Bacchus	20	100+	1.5:1	25
Gelliondale	200	1,000	3:1	50
Gormandale	600	2,400	1:1	200
Loy Yang	4,700	12,500	4:1	250
Morwell	3,300	6,700	4:1	100
Yalloun	2,800	9,500	3.5:1	60

Source: Ref. 21.

TABLE III-3-17   Latrobe Valley Coal Quality

	Yalloun	Morwell	Loy Yang	Gormandale
Moisture (%)	66.1	60.3	62.8	56.6
Ash (%) (dry basis)	1.1	3.4	1.4	2.6
Volatile (%) (dry basis)	51.2	47.9	51.5	51.8
Fixed carbon (%) (dry basis)	47.7	48.7	47.1	45.6
Carbon (%) (dry basis)	67.3	68.1	68.4	66.1
Hydrogen (%) (dry basis)	4.8	4.8	4.9	4.9
Sulfur (%) (dry basis)	0.3	0.4	0.4	0.9
Chlorine (%) (dry basis)	0.1	0.1	0.1	0.1
Calorific value (J/kg)				
Gross dry	26.2	26.7	26.5	25.8
Gross wet	8.9	10.6	9.9	11.2
Gross net	6.9	8.7	7.9	9.3

Source: Ref. 21.

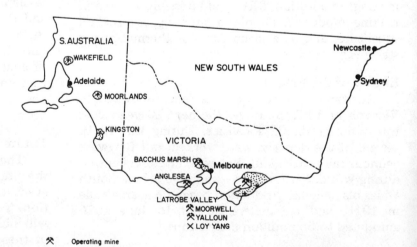

FIGURE III-3-9   Brown-coal-mining areas of Victoria and South Australia, 1981. (From Australian Department of National Development and Energy.)

nomically feasible. Australia, which at present imports 30 to 35% of her oil requirements, would benefit enormously from such a program.

OIL AND GAS

Production, Resources, and Reserves

Australia is a net importer of oil, with about 30 to 35% of her requirements coming from abroad (the Middle East and Indonesia). Consumption is 590,000 barrels per day, and only 400,000 barrels per day is derived from indigenous crudes.[22] It is expected that the amount of oil supplied from domestic sources will decline over the next 10 to 15 years due to declining reserves in the Bass Strait. Australia is self-sufficient in natural gas, with production between 9 and 10 billion cubic meters annually.[4]

The Australian Bureau of Mineral Resources estimated in June 1979 that demonstrated recoverable crude oil and condensate resources were 2620 million barrels.[23] Natural gas demonstrated economically recoverable resources were 21.7 trillion cubic feet as of September 1980.[25]

Uses

The transportation sector constitutes the largest percentage of Australian oil use, between 55 and 60%. The balance of consumption is divided between industry (35%) and domestic and commercial use (6%).[17]

History of the Industry

The first Australian commercial oil field to begin production was the Moonie field in the Surat Basin, Queensland, in the year 1964. It was followed by the Barrow Island field in Western Australia in 1967, and the Cooper Basin, South Australia, in 1969. 1969 was also the year that Australia's largest producing area, the Gippsland Basin in the Bass Strait, Victoria, was discovered. Since the Bass Strait discoveries, no major oil fields have been found.[22] The first natural gas field to be developed was at Roma, Queensland, in 1969, when a 438-km pipeline was completed to Brisbane. It was soon followed by production from the Cooper Basin fields, the Dongara-Mondarra-Gingin fields in Western Australia, and from the Bass Strait.[17] A major gas discovery occurred at Scott Reef, off the northwest coast of Western Australia in 1971. Pipelines to Perth have been built, but the developers are more interested in exporting liquefied natural gas (LNG) to Japan. This would require massive amounts of capital to build LNG plants, and with the current oil glut, demand for LNG has dropped. Figure III-3-10 shows oil- and gas-producing areas, as well as pipelines and refineries.

Production

The Gippsland Shelf accounts for over 90% of Australian oil production with 134 million of the 143 million barrels produced in 1981. The remainder came from the Barrow Island field, the Cooper

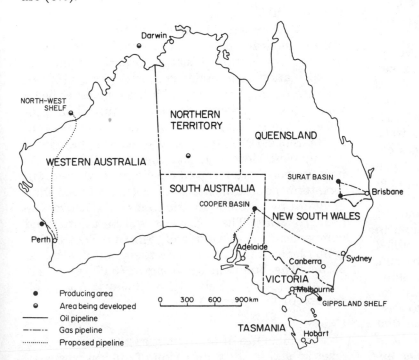

FIGURE III-3-10 Australian oil- and gas-producing areas, 1982. (From Australian Department of National Development and Energy.)

Basin, and from the Moonie and Alton fields in Queensland.[14] The Gippsland Shelf, together with the Cooper Basin, account for about 85% of total gas production.

Australia has 11 oil refineries, which produce the bulk of the country's needs of gasoline, diesel fuel, fuel oil, and other products, including feedstocks for the petrochemical industry. About 12% of Australia's refinery product requirements are imported.

One of the problems faced in developing the various oil and gas fields in Australia is the great distances between the fields and the market where the oil is needed. Australia is fortunate that the Bass Strait fields are relatively close to Melbourne. However, moving gas from the Cooper Basin requires a 800-km pipeline to Adelaide and a 1300-km pipeline to Sydney. New pipelines have been proposed to transport gas from the Mereenie and Palm Valley gas fields in the southwest Northern Territory, and to move oil and gas condensates from the Cooper Basin to Redcliff and on to Adelaide.

Prospects for Future Production

Exploration and development programs continue to grow, partly as a response to government incentives, which include taxation and financial concessions to exploration companies, and also the government's import parity oil-pricing policy. The focus of onshore explority drilling has been on Queensland, South Australia, and Western Australia. Offshore exploration is concentrated in the Gippsland Basin and off the Western Australian and Northern Territory coast.

The North-West Shelf is still in the development stage. Plans call for the gas to be produced from the North Rankin, Goodwyn, and Angel fields, then shipped through a submarine pipeline to an onshore gas treatment plant. Starting in 1984, the State Energy Commission of Western Australia will purchase 10.3 million cubic meters of gas per day. It is hoped by 1986 that LNG exports to Japan will begin, building up to 6 million metric tons per year.[23]

It is hard to estimate Australian oil and gas potential, due to the low level of exploration. In the United States over 80,000 wells were drilled in 1982. Only 309 wells were drilled in Australia in the same year.[1] With this in mind, such statements as "Australia is running out of oil" or "there is little oil in Australia compared to gas" lose their impact. Although it may be true that Australia has more gas than oil, there have been recent discoveries of oil where only gas was expected. The important concern for Australia is whether the new discoveries will be in areas where they can be economically exploited.

URANIUM

Production, Resources, and Reserves

Australia's position in the international uranium market is complicated by several factors unique to the nuclear industry. Concern with respect to proliferation of nuclear weapons, and environmental problems, as well as insecure markets and price fluctuation, have hindered the development of the Australian uranium industry.

Australia contains about 16% of the Western world's reasonably assured uranium resources, recoverable at up to $80 per kilogram of uranium. The 30 June 1980 estimate of 299,000 metric tons of uranium ranks second after the United States, which has over 500,000 metric tons.[4] Total Australian uranium resources (reasonably assured resources and estimated additional resources recoverable up to $130/kg U) are estimated at 547,000 metric tons uranium.[24]

Australian production of uranium oxide is small when compared to the United States, Canada, and South Africa (Table III-3-18). The 1980 production of 1840 metric tons of U_3O_8, represented only 3% of Western world output.[4] In 1981 production nearly doubled as the Ranger mine started up. This trend should continue as the Alligator Rivers uranium province is further developed.

Exports and Trade

At the present time there are no firm plans to utilize nuclear power in Australia; therefore, uranium production is slated toward the international market. Spot sales comprise about 10% of the international market, with the remainder being sold under long-term contracts. Australian producers have to date sold only under long-term contracts. Since 1977, the primary markets have been Japan, West Germany, and the United States. As new projects have received developmental approval, long-term commitments have been made for shipments of over 45,000 metric tons of U_3O_8 over the next 15 years. The recipients will be utilities and companies in West Germany, Finland, France, Belgium, Japan, South Korea, Sweden, and the United States. The bulk will be to Japan (37%) and West Germany (38%).[25] Table III-3-19 shows Australian exports of U_3O_8 since 1977.

Uses

The main use of uranium is as a fuel for nuclear reactors and in the production of nuclear weapons. Depleted uranium is used as ballast, radiation shield-

TABLE III-3-18   World Uranium Production, 1977-1982 (Metric Tons U_3O_8)

	1977	1978	1979	1980	1981	1982
Australia	420	607	832	1,841	3,373	5,251
U.S.	13,561	16,745	16,992	19,410		
Canada	6,828	8,022	8,039	8,313	7,746	
South Africa	3,962	4,671	5,639	7,247	7,234	
Namibia	2,758	3,160	4,481	4,717	4,681	
Niger	1,897	2,429	4,263	4,575	4,500	
France	2,473	2,574	2,783	3,125	3,000	
Gabon	1,660	1,205	1,298	1,218	1,360	

Sources: Ref. 4; Australian Bureau of Statistics.

TABLE III-3-19   Australian Uranium Exports, 1977-1983 (Metric Tons U_3O_8)

	1977	1978	1979	1980	1981	1982
Exports	1545	1114	1317	1120	1625	5277

Sources: Ref. 4; Australian Bureau of Statistics.

ing, as a catalyst and glass colorant, in alloys and electrical components, and as a fuel for fast-breeder reactors.[17] Twenty-two countries have installed nuclear power stations with a total of 258 operational reactors and a capacity of 151,000 MW. Of these reactors, the United States has 75, Britain 32, Japan 23, and West Germany 11. Altogether they represent from 10 to 12% of the total installed electricity generating capacity of these countries. In addition, there are 227 reactors under construction and 41 on firm order.[27]

With the public outcry against nuclear power, many of these new reactor projects have been canceled or had their startup dates postponed. In addition, the move away from nuclear power has been accelerated by the stabilization of oil prices during the early 1980s.[26]

History of the Industry

Uranium has been known to exist in Australia since the 1890s, and in the 1930s was mined at Radium Hill, South Australia. Small amounts of radium were recovered for medical purposes, and a few hundred kilograms of uranium was obtained as a by-product.[28] Serious exploration for uranium, stimulated by government tax and guaranteed prices, began in 1947. The discovery of ore at Rum Jungle (south of Darwin) in 1949 was followed by a prospecting boom in the early 1950s. During this period five independent producers were operating in the Mt. Isa-Cloncurry region of northwest Queensland, and the Katherine-Darwin area of the Northern Territory. Between 1954 and 1964, Australia exported roughly 7900 metric tons of U_3O_8.[28]

As the demand for uranium declined during the early 1960s, mining operations were either closed or mothballed. This situation remained until 1967, when forecasts of the demand for uranium by the commercial power industry spurred an upsurge of exploration. As a result, in 1970, the Ranger, Koongarra, and Nabarlek deposits in the Alligator Rivers region, Northern Territory, and the Beverly deposit in South Australia were found. These were followed by the Jabiluka deposit, NT, in 1971, and the Yeelirrie deposit, WA, in 1972.[28]

With the December 1972 election of a Labor government, development of new ore bodies was halted until policies concerning foreign ownership, Aboriginal land rights, environmental safety, and marketing could be formulated.[28] On 25 August 1977, the government announced that development of uranium resources could continue under strictly controlled conditions. Using the findings of the Ranger reports, as well as comments and suggestions made during public debate, the government's decision reflected four fundamental considerations:[24]

1. The need to reduce the risk of nuclear proliferation
2. The need to supply essential sources of energy to an energy-deficient world
3. The need to protect the environment in which mining development will take place
4. The need to ensure that proper provisions are made for the welfare and interests of the Aboriginal people in the Alligator Rivers region, and of all other people living in the region and working on the development projects

Additional encouraging news for Australian

uranium came in 1977, with the announcement of the discovery of a huge low-grade uranium deposit at Roxby Downs, South Australia. The deposit also contains considerable amounts of gold and copper.

Geology of Major Deposits

Uranium is found in association with acid igneous rocks and in sedimentary rocks. Table III-3-20 is a classification of uranium minerals. Table III-3-21 is a classification of uranium deposits.

Most of the important Australian uranium deposits are of the vein type (including those of the Alligator Rivers province). Australian uranium deposits contain, by world standards, relatively high-grade ores. The deposits, being large and close to the surface, can be mined using open-cut methods. Both these factors are important since, in the case of the Alligator Rivers deposits, the uranium is found in isolated locales where the cost of labor is high. Figure III-3-11 shows the locations of Australian uranium mines and prospects.

Mining Facilities

Between 1974 and 1980, the only Australian uranium mine in operation was Mary Kathleen. Producing up to 800 metric tons of U_3O_8 per year, the deposit was mined out in 1982.[14] More than

making up the loss in production have been the Nabarlek and Ranger mines, both located in the Alligator Rivers province of the Northern Territory. The Nabarlek deposit contains an estimated 10,200 metric tons of U_3O_8. Mining operations began in 1979 and mill production started in 1980. By 1980 the ore had been mined and stockpiled. The mill is producing yellow cake at the rate of 1400 metric tons per year, and the mine has total sales contracts with overseas utilities for over 6000 metric tons of U_3O_8.[14] The Ranger mine has been producing at or above its design capacity of 3000 metric tons of U_3O_8 per year since its startup in October of 1981.[30] With uranium oxide resources of over 100,000 metric tons, and long-term contracts for over 42,000 metric tons, a doubling of plant capacity is envisaged by the end of the decade.[14]

Several other projects and prospects are in varying stages of planning and government approval. The Yeelirrie project, which was scheduled to begin production in 1986, was side-tracked when Esso Exploration of Australia pulled out in May 1982. The Jabiluka and Koongarra projects are waiting for government clearance on environmental and Aboriginal land issues. The Jabiluka deposit, which has resources of over 175,900 metric tons of uranium oxide, has been given governmental approval to develop. Table III-3-22 lists Australian uranium deposits in their various stages of development.

Processing Facilities

In an effort to advance the upgrading and processing of Australian minerals prior export, the Commonwealth government initiated a feasibility study on the establishment of a domestic commercial uranium enrichment industry. The study was undertaken by the Uranium Enrichment Group of Australia. The group consisted of the fol-

TABLE III-3-20 Classification of Uranium Minerals

Primary uranium minerals
  Uraninite (pitchblend)
  Davidite
Secondary uranium minerals
  Carnotite
  Torbernite and metatorbernite
  Autunite
  Uranium ochres and gummite
  Thorogummite

Source: Ref. 29.

TABLE III-3-21 Classification of Uranium Deposits

Type of deposit		Minerals
Associated with acid igneous rocks	Veins, replacement bodies, pegmatites	Primary minerals (uraninite, pitchblend, etc.)
In sedimentary rocks	Conglomerates	Primary minerals
	Sandstones	Primary and secondary minerals (characteristically carnotite)
	Calcrete	Carnotite
	Clayey lacustrine sediments	Secondary minerals (carnotite)
	Bituminous shales, lignites, and phosphorites	(Enriched with uranium)

Source: Ref. 29.

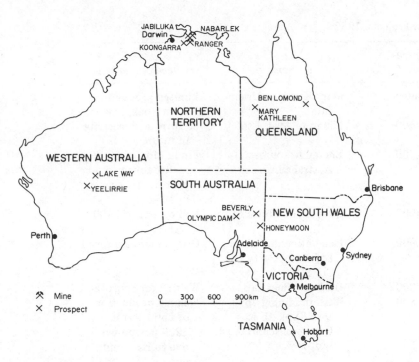

FIGURE III-3-11   Australian uranium mines and prospects, 1982. (From Ref. 24.)

lowing companies: BHP Ltd., CSR Ltd., Peko-Wallsend Operations Ltd., and Western Mining Corp. Ltd.

The group recommended that sites near Adelaide and Brisbane be further evaluated as they satisfy the requirements of a plant based on Urenco-Center centrifuge technology.[32] Urenco Center (a consortium comprising West Germany, the Netherlands, and U.K. interests) has estimated that such a plant would cost over $1 billion. A detailed engineering and feasibility study will still have to be undertaken before the government can make any firm plans.

Future of the Industry

The future of the Australian uranium industry will depend on how the issues of nuclear proliferation, aboriginal land rights, environmental safety, and acceptable levels of foreign participation are handled on a continuing basis. While in office, the Liberal-National Country Party coalition eased many of the restrictions imposed by the Labor government of 1972-1975. The 100% Australian ownership requirement for new uranium discoveries was reduced to 75%, and in special cases a 50% ownership minimum would be considered. With the election of a Labor government in March 1983, uranium policies are likely to change again, although Prime Minister Hawke's government will have the reality of high unemployment to contend with when they make any decisions to close or hold back uranium projects. The best example is that of the Roxby Downs prospect. The South

Australian government had traditionally taken an antinuclear stand, but with unemployment in that state being the highest in the federation, and the prospect of Roxby Downs being one of the largest mines in the country, the Adelaide government is now pushing for the deposit's development, as well as to build Australia's first uranium enrichment plant in their state. In addition, uranium-producing companies must contend with shipping problems caused by the Maritime/Wharfside union's and the Australian Trade Union Council's antiuranium policy.

Regardless of the government in office, Australia has maintained strict requirements on uranium exports and nuclear safeguards. In an effort to ensure that Australian uranium is used only for peaceful purposes, the government retains the right to choose to which countries uranium exports will be permitted. In trade involving non-nuclear-weapon countries, sales are made only to parties of the Treaty on the Non-proliferation of Nuclear Weapons (NPT). Nations with nuclear-weapon capability can receive and use Australian uranium only under the International Atomic Energy Agency (IAEA) safeguards. As a backstop to the IAEA safeguards, all recipient nations must accept bilateral agreements with Australia. These agreements stipulate that Australian uranium cannot be used for any explosive or military purposes, that appropriate physical security measures are taken, and that prior Australian consent is obtained before Australian uranium may be transferred to others, enriched beyond 20% ^{235}U, or reprocessed.[24]

TABLE III-3-22 Australian Uranium Projects, 1982

Deposit	Resources (metric tons U_3O_8)	Ownership	Stage of development
Mary Kathleen (Queens.)	Depleted	Mary Kathleen Uranium Ltd.	Limited production from stockpiled ore
Ben Lomond (Queens.)	3,400	Minatore Aust. Ltd.	Sulfuric acid leaching in future
Ranger (N. Terr.)	105,730	Energy Resources of Australia Ltd.	Production of 3000 metric tons per year (t/yr) began in 1981
Nabarlek (N. Terr.)	10,200	Queensland Mines Ltd.	Mill output of 1400 t/yr
Jabiluka (N. Terr.)	175,900	Pancontinental Mining Ltd. 65%; Getty Oil Co. 35%	Government approved
Koongarra (N. Terr.)	11,300	Denison Mines Ltd.	Waiting for approval
Yeelirrie (W. Austr.)	39,800	Western Mining 75%; Esso Expl. 15%; Urangesellschaft Aust. Ltd. 10%	Planned production of 2500 t/yr in 1986 postponed due to Esso pull out
Lake Way (W. Austr.)	3,400	Delhi Inter. Oil Corp. $53\frac{1}{2}$%; VAM Ltd. $46\frac{1}{2}$%	Planned open-pit mine
Beverly (S. Austr.)	11,500	Western Nuclear Aust. Ltd. 50%; Transoil $16\frac{2}{3}$%; Petromin $16\frac{2}{3}$%; Oilmin $16\frac{2}{3}$%;	Production set for 1984 using insitu leaching process
Honeymoon (S. Austr.)	2,900	Mt. Isa Mines 49%; ARR Ltd. $25\frac{1}{2}$%; Teton Exploration Drilling $25\frac{1}{2}$%	Insitu leaching on a small scale in 1983 to reach 450 t/yr by 1986
Olympic Dam (S. Austr.)	1,200,000[a]	Western Mining 51%; BP Aust. Ltd. 49%	With copper and gold, uncertain startup date

[a]Subject to change due to mineral relationships.
Sources: Refs. 24, 27, 14, and 31.

In the event of a breach of these bilateral agreements, Australia has reserved the right to suspend or cease uranium shipments to the defaulting importer, and to require the return of any Australian uranium in that importer's possession. Countries that have signed such agreements include Finland, the Philippines, South Korea, the United States, the United Kingdom, France, Canada, Sweden, and the European Atomic Energy Community. Negotiations with Japan and Switzerland were still in progress as of 1981.[25]

The greatest threat to the Australian uranium industry is the world trend away from nuclear power. As fewer nuclear power plants are planned and built, the competition from other uranium-exporting nations, such as Canada and South Africa, increases. This could impede the development of new mining areas as declining uranium prices cuts into their economic feasibility. The luxury of stringent export agreements may no longer be possible. This could cause the government to rethink its policies, or it may be that they decide to suspend exports altogether.

REFERENCES

1. Australian Department of National Development and Energy. *Australian Energy Statistics: 1982*. Canberra: Australian Government Publishing Service, 1982.
2. Carrick, John. "Outlook for Energy Source Development Promising." *Mining Review*, December 1982.
3. Australian Department of Trade and Resources. *Australian Coal*. Canberra: Australian Government Publishing Service, 1981.

4. Australian Department of Trade and Resources. *Australian Resources in a World Context*. Canberra: Australian Government Publishing Service, 1982.

5. Joint Coal Board. *Black Coal in Australia 1981–82*. Sydney: Joint Coal Board, 1982.

6. Australian Department of Trade and Resources. *Australia's Mineral Resources: Steaming Coal*. Canberra: Australian Government Publishing Service, 1981.

7. Australian Department of Trade and Resources. *Australia's Mineral Resources: Metallurgical Coal*. Canberra: Australian Government Publishing Service, 1981.

8. Matsuoka, H. "Requirements for Coals in Japanese Coking Blends." In *Australian Black Coal*. Edited by A. C. Cook. Illawarra, Australia: Australasian Institute of Mining and Metallurgy, 1975, pp 252-263.

9. "Markets Available to Black Coal." In *Australian Black Coal*. Edited by A. C. Cook. Illawarra, Australia: Australasian Institute of Mining and Metallurgy, 1975, pp. 102-106.

10. Durie, R. A. and I. W. Smith. "Production of Gaseous and Liquid Fuels from Coal." In *Australian Black Coal*. Edited by A. C. Cook. Australia: Australasian Institute of Mining and Metallurgy, Illawarra, Australia: 1975, pp. 161-172.

11. Broinowski, R. P. *Australian Energy Resources; Their Marketability in Japan and the Pacific Region*. Paper presented at the Conference of Australian Energy Resource Development, Canberra, 5-7 December 1982.

12. Milligan, E. N. "The Geology of the Bowen and Galilee Basin Coalfields." In *Australian Black Coal*. Edited by A. C. Cook. Illawarra, Australia: Australasian Institute of Mining and Metallurgy, 1975, pp. 10-18.

13. Blayden, I. D. "Geology of Australian Coalfields." In *Australian Black Coal*. Edited by A. C. Cook. Illawarra, Australia: Australasian Institute of Mining and Metallurgy, 1975, pp. 19-30.

14. Stewart, G. N. "Australia." *Mining Annual Review*, June 1982.

15. "Dispute over New South Wales' Third Coal Loader." *Mining Journal*, 22 January 1982.

16. "Coal: Focus on Queensland." *Mining Journal*, 6 August 1982.

17. Alexander, J., and R. Hattersly. *Australian Mining, Minerals, and Oil*. Sydney: The David Ell Press Pty. Ltd., 1980.

18. "Loy Yang: A Major Brown Coal Project." *Mining Magazine*, October 1982.

19. Scambary, R. "Brown Is Bountiful for Local Use." *Australia Now*, 1982.

20. Bowen, K. G. "Brown Coal: Victoria's Prime Energy Resource." *World Coal*, July 1978.

21. Ford, J. "New Energy Sources to Restore the Balance of Power." *Australia Now*, 1982.

22. Scambary, R. "Oil and Gas: High Hopes in Ever-Deeper Waters." *Australia Now*, 1982.

23. *Petroleum in Western Australia*. Perth: Government of Western Australia, Department of Mines, Petroleum Branch, January 1982.

24. Australian Department of Trade and Resources. *Australia's Mineral Resources: Uranium*. Canberra. Australian Government Publishing Service, 1981.

25. Australian Department of Trade and Resources. *Energy Australia*. Canberra: Australian Government Publishing Service, 1981.

26. Govett, G. J. S. and M. H. Govett. "Australian Minerals." *Resources Policy*, December 1982.

27. Miline, J. "Uranium to Fuel the World's Reactors." *Australia Now*, 1982.

28. Hampson, D. C. "Australia's Uranium." *Resources Policy*, June 1980.

29. *Prospecting for Uranium*. Perth: Geological Survey of Western Australia, 1978.

30. "Ranger Mine Producing above Design Capacity." *World Mining*, November 1982.

31. "Estimates of Olympic Dam Reserves Finally Released." *World Mining*, July 1982.

32. "Uranium Enrichment: Report Outlines Details." *Mining Review*, October 1982.

33. British Petroleum Company. *BP Statistical Review of World Energy: 1982*. London: Dix Motive Press Ltd., 1982.

Chapter III-4

International Trade

EXPORT DEPENDENCE

Australia is important to the world's mineral trade not only because it is a major mineral producer, but also because it is a major mineral exporter. Australia accounts for about half of the world's alumina trade. It is the largest exporter of rutile, zircon, and monazite, as well as refined lead and lead bullion. Australia is also a key exporter of coal, iron ore, nickel, manganese, zinc, ilmenite concentrates, and synthetic rutile. In addition, Australia exports copper, tin, and tungsten.[1] Uranium exports, which until recently had been restricted by the government, are on the rise, and large exports of liquefied natural gas are planned for in the near future.

Many of the mining projects developed during the 1960s and 1970s have been, by economic necessity (size of deposit, location), large operations. Reflecting this and the general high level of mineral exports, Australia has become export dependent. The danger of this type of export dependence is that a small reduction in trade can severely affect the mining industry. As exports decline, mines that depend on economies of scale could be forced to close. In industries such as iron ore and bauxite mining, this could mean that a large segment of production could be closed at one time. These problems are multiplied when mineral trade is highly dependent on a single nation, as is the case with Australia and Japan.

Export Destinations

Although Australia exports to nations all over the world, the majority of her trade is with Japan. This is due to the proximity to Japan and to the growth of Japanese metal production (especially steel) over the past 20 years. For the most part Australia has benefited from this arrangement, but she has also been vulnerable to recessions in the Japanese economy. Table III-4-1 illustrates the distribution of exports by major countries and import groups.

Level of Foreign Investment

Australia has historically relied on foreign capital in developing her mineral resources. It was the case during the nineteenth-century gold bonanza,

TABLE III-4-1 Distribution of Australian Mineral Exports, 1981–1982

Japan	43.7%
U.S.	10.6%
ASEAN[a]	6.3%
EEC	15.3%
South Korea	3.4%
New Zealand	3.7%
Other	17.0%

[a]Association of South East Asian Nations: Thailand, Singapore, Malaysia, Indonesia, and the Philippines.
Source: Ref. 2.

and it is still true today. During the decade 1968–1978, total capital expenditures for the mining sector (mining and quarrying, excluding smelting, refining, and secondary processing) were A\$11,155 million (in 1978 dollars).[3] During the period only 15% of new capital for exploration and development was raised by Australian Stock Exchange–listed companies, 33% came from overseas investments.[3] Due to the high level of foreign ownership and control, many Australians have complained that foreign interests were being placed above those of the host nation.[4] This issue has become politicized, and as a result attracting new capital became more difficult. If plans to build new processing facilities (e.g. aluminum smelters) are to be realized, massive amounts of foreign capital will be required.

Export Levels

A large percentage of Australian mineral production is exported. In some cases (rutile, zircon, and alumina) nearly 100% goes to the international market. Table III-4-2 shows the percentage of Australian mineral production that is exported. These figures emphasize the dependence of Australia on overseas markets and the relative inability of the domestic market to absorb reductions in exports. Table III-4-3 details the value of Australian mineral exports over the past five years.

The level of mineral processing in Australia is minimal when compared to that of other developed countries. The processing that is done is typically for an intermediate product. This reflects the small

TABLE III-4-2   Percentage of Australian Mineral Production That Is Exported, 1981 and 1982

	1981	1982
Alumina	96	90
Aluminum	15	39
Iron ore	84	86
Iron, ingot steel, ferroalloys	6	3
Black coal	45	39
Rutile concentrates	101	91
Zircon concentrates	102	90
Ilmenite concentrates	78	71
Tungsten	87	101
Tin	79	61
Refined lead	65	
Refined zinc	72	

Sources: Refs. 5 and 6.

domestic market for processed minerals in Australia, as well as the desire of most developed nations to import minerals in raw form. Table III-4-4 shows the degree of processing of selected minerals as a percentage of mine output, and Figure III-4-1 compares the income earned from various mineral exports over the past 10 years.

JAPANESE TRADE

The importance of Japanese trade to the Australian mineral industry cannot be overstated. Japan is not only the largest importer of Australian minerals, but has also become one of the principal sources of venture capital for Australian mineral projects. This is not to say that the producer–consumer relationship between the two nations has been frictionless. The competitive nature of long-term mineral contract negotiations has often left bitter feelings between the two countries. Regardless of negotiation after-affects, both nations realize the advantage of trading with each other.

The growth of Australia as a mineral power, especially in coal,[7] iron ore, and alumina, was triggered by the emergence of Japan as a major industrial power. Raw-material-hungry Japan provided the market needed to develop the large-scale mines in remote areas of Western Australia, Queensland, and the Northern Territory. Figure III-4-2 shows

TABLE III-4-3   Value of Australian Mineral and Metal Exports, 1977–1978 to 1981–1982[a] (Million Australian Dollars F.O.B.)

	77–78	78–79	79–80	80–81	81–82
Coking coal	1365	1384	1481	1612	1875
Steam coal	117	140	197	358	415
Uranium and thorium ores and conc.	87	81	98	124	203
Iron ores and conc.	921	968	1076	1739	1252
Copper ores and conc.	28	44	89	76	55
Lead ores and conc.	42	32	57	39	42
Tin ores and conc.	45	57	77	81	115
Manganese ores and conc.	b	b	b	b	8
Tungsten ores and conc.	40	51	53	58	40
Rutile ores and conc.	62	64	89	63	63
Ilmenite ores and conc. (excl. beneficiated)	16	17	20	20	22
Zirconium ores and conc.	31	30	34	38	38
Copper matte	2	3	12	3	5
Iron and ferro alloys	49	84	113	65	13
Copper unworked	86	103	142	172	80
Aluminum and aluminum alloys unworked	69	82	68	56	166
Lead unworked	189	252	538	298	256
Zinc and zinc alloys unworked	96	111	116	132	167
Tin and tin alloys unworked	28	21	29	19	13
Diamonds rough	NE	1	b	b	b
Diamonds cut (excl. industrial)	NE	NE	14	12	10

[a]NE, not exported.
[b]Less than 1.

Source: Ref. 19.

TABLE III-4-4   Degree of Processing of Select Minerals, 1979

Ores and Conc.	Mine output (metric tons)	Processed (metric tons)	Processed as percent mine output
Nickel conc.	356,563	365,563	100
Lead ores and conc.	713,648	663,693	93
Bauxite	27,583,429	21,239,240	77
Copper ores and conc.	852,429	613,928	72
Zinc ores and conc.	928,898	594,495	64
Tin conc.	23,203	10,209	44
Manganese ore	1,697,671	322,557	19
Ilmenite conc.	1,181,010	188,962	16
Iron ore	91,717,011	11,923,211	13
Zircon conc.	444,979	11,000	2
Tungsten conc.	5,623	56	1
Rutile conc.	274,533	2,200	1
	125,779,242	35,926,114	29

Source: Australian Bureau of Mineral Resources.

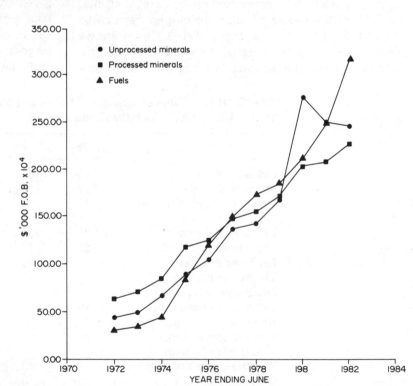

FIGURE III-4-1   Value of Australian mineral exports, 1971–1982. (From Australian Bureau of Statistics.)

the parallel between the growth in Australian mineral production and Japanese imports.

Level and Form of Exports

From Table III-4-5 it is evident just how dependent Australia is on Japanese trade. Coal and iron ore trade, which represents almost 50% of Australia's mineral export income, is almost entirely dependent on exports to Japan. Bauxite, copper, zinc, and zirconium ores and concentrates are also mined in large part for the Japanese market.

Australian mineral exports to Japan are typically in the form of ores and concentrates. As seen from the various trade tables in Chapter III-2 and from Table III-4-6, although Japan is the leading importer of Australian minerals in the ore or concentrate stage, Japan imports a disproportionately small share of refined metals. This is due to Japan's policy of importing minerals in their rawest form reasonable so as to maximize earnings inherent in smelting and refining. Japan is able to enforce this policy by having higher duty rates than found in the United

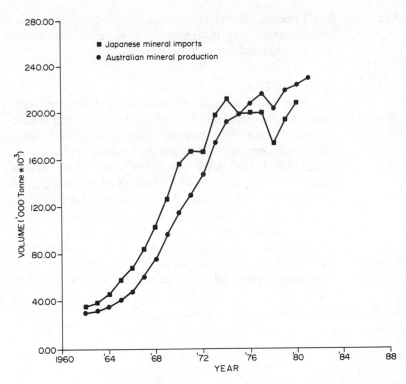

FIGURE III-4-2 Growth in Australian mineral production versus total Japanese mineral imports, 1962–1982. (From U.S. Bureau of Mines.)

TABLE III-4-5 Japanese Percentage of Australian Mineral Exports, 1976-1977 to 1981-1982[a]

	77-78	78-79	79-80	80-81	81-82
Coking coal	77	73	75	74	71
Steam coal	13	16	25	52	59
Uranium and thorium ores and conc.	44	6	15	—	—
Iron ores and conc.	80	72	76	73	74
Copper ores and conc.	87	78	72	89	80
Lead ores and conc.	15	10	4	18	4
Zinc ores and conc.	53	54[b]	65	62	54
Tin ores and conc.	—		—	—	—
Manganese ores and conc.	—	—	—	—	71
Tungsten ores and conc.	9	5	7	7	6
Rutile ores and conc.	6	9	10	16	15
Ilmenite ores and conc. (excl. beneficiated)	8	16	7	5	5
Zirconium ores and conc.	23	36	37	34	40
Copper matte	—	—	—	—	19
Iron and ferroalloys	6	22	8	11	18
Copper unworked	16	6	4	16	28
Aluminum and aluminum alloys unworked	72	59	42	—	NA
Lead unworked	2[b]	3	1	2	5
Zinc and zinc alloys unworked	—	—	—	—	—
Tin and tin alloys unworked	—	3	1	—	—
Diamonds rough	NE	—	—	—	—
Diamonds cut (excl. industrial)	NE	NE	[b]	[b]	—

[a]—, nil; NE, not exported; NA, not available.

[b]Less than 1.

Source: Ref. 19.

TABLE III-4-6 Selected Japanese Imports of Australian Minerals, 1981–1982

	Million dollars	Percent of total exports
Iron ore and concentrates	855.5	72
Iron and steel ingots	Nil	Nil
Zinc ores and concentrates	62.7	52
Zinc and zinc alloys, unwrought	Nil	Nil
Copper ores and concentrates	43.9	80
Copper, refined, unwrought	5.5	10

Source: Ref. 1.

States and the European Economic Community (EEC) (Table III-4-7).

This may change in the future as Japan faces a growing comparative cost disadvantage and increas-ing concern about the protection of the environ-ment. Japan may then show a greater willingness to import minerals in a more processed form.

Terms of Trade

The bulk of the mineral trade between Aus-tralia and Japan has historically been accomplished through long-term supply contracts rather than direct ownership. There are several reasons for this. Until 1971 the Japanese government had severe restrictions on overseas investment. Also due to the rapid expansion of Japanese metal production dur-ing the 1960s, metal-producing companies were not able to spare the investment funds for resource development. In addition, Australian mining com-panies were seeking equity contributions from companies (typically American or European) that

TABLE III-4-7 Tariff Levels for the EEC, Japan, and the United States

Commodity	U.S.	EEC[a]	Japan
Copper			
Ores and conc.	0	0	0
Blister and anode	1.3%[b]	0	8.1%[c]
Refined	1.0%	0	4.5/lb
Aluminum			
Bauxite	0	0	0
Alumina	0	7.6%	6.5%[c]
Aluminum	3.0%[b]	6.6%	9.0%[c]
Iron and steel			
Ores and conc.	0	0	0
Pig iron	0	3.9%	3.9%
Structural steel	0.9%[b]	5.7–6.7%	5.7%
Lead			
Ores and conc.	0.75/lb	0	0
Bullion	3.5%[b]	0	1.6/lb
Other unwrought	3.5%[b]	3.5%[b]	6.9%[c]
Manganese			
Ores and conc.	0	0	0
High-carbon ferromang.	1.6%[b]	4.0%	11.69%[c]
Nickel			
Ores and conc.	0	0	0
Unwrought	0	0	24.48/lb
Tin			
Ore	0	0	0
Metal unwrought	0	0	0
Tungsten			
Ores and conc.	17/lb W	0	0
Ammonium paratungstate	12.1%[b]	NA	5.7%
Zinc			
Ores and conc.	0.53/lb Zn	0	0
Unwrought other than alloy	1.8%[b]	3.5%	1.6/lb Zn

[a]NA, not available.
[b]Ad valorum.
[c]C.i.f. value.

Source: Ref. 8.

had the mining and marketing experience they did not possess themselves.

Case in point is the development of the iron ore deposits of Western Australia. Following the lifting of the Australian iron ore export embargo in 1960, American and European investors poured in capital to develop the deposits. Since the Americans and Europeans were not generally seeking new sources of raw materials, the finance for these ventures was obtained with the assurance provided by the long-term contracts between the Australian mines and Japanese steel producers.[10]

The long-term contract usually extended over 5 to 15 years, with the tonnage delivered for a specific yearly quote (+/− 10% at the purchaser's option), at a fixed dollar U.S. price subject to renegotiations within a predetermined range (+/− 7½% in the case of iron ore), once every three to five years.[11] For some minerals, such as copper, the price is based on the London Metal Exchange. Others, such as coal, have their prices renegotiated on nearly a yearly basis. The advantage of these long-term contracts are that they provide a guaranteed market over a relatively long period. Fixed prices ensure the profitability of resource developments. The Japanese on their end are guaranteed long-term access to large supplies of raw materials at a price that allows them to maintain a competitive position in the metals industry.

From an Australian viewpoint, the effectiveness of these contracts was eroded during the 1970s due to changes in exchange rates and higher rates of inflation.[11] Problems also arose due to the penchant of the Japanese to demand large decreases in contracted prices during recessionary periods. For example, the decline in Japanese steel production in 1982 and 1983 has caused sharp reductions in contracted tonnages. Mt. Goldsworthy's 7.75-million metric ton per year contracts, which expired in April 1983, were extended for two years at a reduced annual tonnage of 4.25 million metric tons per year.[12] Australia's two largest iron ore exporters, Hamersley Iron and Mt. Newman, accepted a 12.5% cut in the price of exports to Japan. Under the new agreement the price of fines delivered until 31 March 1984 will fall from 34.76 cents per Fe unit to 30.45. Lump prices were cut by 12.9%.[13]

The Japanese complaint is that the high level of industrial disputes have left suppliers short on contracted tonnages and have disrupted shipping arrangements.[11] The Japanese also claim that Australian mining companies use periods of mineral shortages to unfairly raise prices.

In negotiating contracts, Australian producers, especially during the 1960s, were played off against one another by Japanese buying cartels. In iron ore, the "Committee of Ten," a cartel comprised of Japan's 10 largest steel companies, allowed the Japanese to negotiate as a single unit, while the Australian mining companies competed with each other on contract terms.[10] The state governments of Australia derive their revenue from royalties, based on quantities produced, not overall profitability. Therefore, they are less likely to put up a fight against lower prices than the Commonwealth government, which levies taxes on profits.[10]

In addition, the Australians feared that the Japanese would turn to India or Brazil for long-term contracts. The effect of the Japanese monopsony on the area iron ore market was that the prices obtained by the Australians was on average 20% lower than those paid for imports from other sources in Asia, Africa, and America.[10] Moreover, despite the insistence of the Australians for c.i.f. prices, the contracts were stipulated in f.o.b. terms, which allowed the Japanese to reap the benefits of future declines in freight costs.[10]

The trade in coal has also been influenced by the Japanese monopsony in the western pacific. Australia, the United States, Canada, and South Africa all export coal to Japan, the world's largest importer. During recessionary periods Japan has reduced orders and cut prices agreed to under long-term contracts, although Australia has suffered less than "swing suppliers" such as the United States.[13]

Japanese contracts for bauxite have followed the same pattern as those for iron ore. Between 1970 and 1976 the average real price (in yen) paid by Japanese importers to Australian producers has declined by about 70% due to inflation and the devaluation of the Australian dollar on which the contracts were based.[9]

Following the lifting of Japanese restrictions on overseas investments in 1971, Australia has seen increasing levels of equity participation by Japanese firms. The reasoning was that the growing instability of long-term contracts had caused the major lending institutions (which had financed the large iron ore, coal, and bauxite projects) to seek stronger guarantees of future market outlets.[9] For its part the Japanese government was anxious to reduce balance-of-trade payments surpluses.

Japanese direct investment typically takes the form of minority participation by a group of metal companies or trading firms. The industries with the greatest amount of Japanese participation have been: iron ore, coal, aluminum, and most recently uranium. Japanese investment in iron ore projects include 6.2% in Hamersley, 10% in Mt. Newman, and 30% in Cliffs Robe River.[14] Coal projects with

Japanese equity are mainly large open-cut mines in Queensland. Japanese investment in the Australian aluminum industry is primarily in new and proposed smelters (Table III-2-11). Foreign investment in the Australian uranium industry had been eliminated under the Labor goverment of 1972–1975. The return of the Liberal-NCP government reduced these restrictions so that today foreign participation can be up to 25%, and more than that in special cases. The Ranger uranium project is partially owned (11%) by Japanese utilities.[15] The new Labor government has not yet placed any new restrictions on uranium projects, although the stance it eventually takes is most likely to be less restrictive than those under the previous Labor government.

Future trade with Japan is assured, although the amount and form are uncertain. The direction that Japanese industry takes over the next 5 to 10 years will in many ways dictate the fate of the Australian mineral industry. When Japan grew into an industrial power during the 1960s, Australia tagged along and in the process, due as much to the Japanese as to her own bountiful resources, became a world leader in raw mineral production. Now as Japan evaluates which direction its mineral-processing industry will take, Australia must make plans that will lessen any adverse affects.

WESTERN PACIFIC TRADE

Trade Partners

Australian trade prospects in the future will be greatly influenced by the emergence of new industrial nations in the western Pacific. The Western Pacific Region (WPR) includes Japan, Australia, South Korea, Indonesia, Taiwan, the Philippines, Thailand, Malaysia, New Zealand, Hong Kong, Singapore, Papua New Guinea, and China. It is the home of 32% of the world's population.[16]

During the 1970s, economic growth in the WPR was the most dynamic in the world. The world as a whole had an average annual real growth rate of 4% between 1970 and 1979. During the same period South Korea averaged 10% real growth, and Hong Kong, Singapore, and Taiwan averaged over 9%.[16] Table III-4-8 lists the individual growth rates.

Trade Levels

Australian mineral trade in the WPR has been dominated by Japan. Over the past 10 years this trend has begun to change. South Korea, Taiwan, Hong Kong and Singapore are increasing their share of Australian mineral imports. For example, in 1970–1971 Japanese coal imports represented 86% of Australian exports. By 1981–1982 that figure had dropped to 68%. During the same period Ko-

TABLE III-4-8   Average Annual Growth Rates of Western Pacific Region Countries, 1970-1979

	Average annual real economic growth
Developed economies	
Australia	3.3
New Zealand	2.1
Japan	5.1
Developing or industrializing economies	
Hong Kong	9.1
Singapore	9.0
South Korea	10.0
Taiwan	9.9
Indonesia	7.4
Malaysia	8.0
Philippines	6.6
Thailand	7.5
Papua New Guinea	4.3 (1970-1977)
China	5.8

Source: Ref. 16.

rean imports rose 7.5% and Taiwanese imports rose 3.3%. Exports of iron ore shows a similar pattern.

The WPR contains the entire gamut of economies, from an industrial giant such as Japan, to agrarian cultures such as Papua New Guinea. Australia benefits from this in two ways. By providing the raw materials to industrial economies, as has been the case with Japan, Australia has been able to develop her mineral resources. This trade should continue, as industrial newcomers such as South Korea, Taiwan, Hong Kong, and Singapore are natural resource deficient and will need increasing amounts of raw and semiprocessed materials. Just as important, by being in a corner of the world that is growing rapidly, Australia can increase her processed mineral production and have a ready market.

The amounts and types of commodities traded between Australia and the various WPR nations depends on the economies of those countries. Whereas South Korea imports primarily ores, metal scrap, and coal, a developing nation such as the Philippines imports more processed minerals and finished metals. Tables III-4-9 through III-4-11 provide detailed export data to WPR nations.

Trade Agreements

Associated with developing closer trade between Australia and the WPR have been a series of political and social issues. For years there has been talk of the Pacific Rim concept, in which the advanced industrial states (Japan and the United States) would provide capital, technology, and planning;

TABLE III-4-9 Australian Trade with Korea, Taiwan, Hong Kong, and Singapore, 1979–1980 (Thousand Dollars)

Commodity	South Korea	Taiwan	Hong Kong	Singapore
Iron ore and conc.	11,900	26,106	—	—
Iron and steel	7,272	5,133	31,797	24,005
Coal and coke	91,534	59,306	—	—
Metallic ores and conc.	71,534	—	—	1,111
Copper and copper alloys	9,548	—	3,098	7,865
Lead and lead alloys	—	7,799	—	5,748
Zinc and zinc alloys	—	15,375	10,130	4,044
Aluminum and aluminum alloys	—	—	4,774	7,226

Source: Ref. 16.

TABLE III-4-10 Australian Trade with Indonesia, Philippines, Thailand, and Malaysia, 1979–1980 (Thousand Dollars)

Commodity	Indonesia	Phil.	Malaysia	Thailand
Iron and steel	25,337	23,293	9,120	4,330
Iron ore and conc.	—	11,990	—	—
Coal and coke	—	1,962	—	—
Copper and copper alloys	2,480	2,134	12,564	1,021
Lead and lead alloys	5,024	5,856	6,701	13,295
Zinc and zinc alloys	23,392	3,880	6,405	16,437
Aluminum and aluminum alloys	2,508	12,866	—	3,263

Source: Ref. 16.

TABLE III-4-11 Australian Trade with New Zealand, Papua New Guinea, and China, 1979–1980 (Thousand Dollars)

Commodity	New Zealand	PNG	China
Iron and steel	62,080	13,523	106,276
Iron ore and conc.			51,146
Copper and copper alloys	23,279	—	—
Lead and lead alloys	6,971	—	13,905
Zinc and zinc alloys	11,522	—	—
Aluminum and aluminum alloys	5,498	—	24,039
Nonferrous metals		2,025	

Source: Ref. 16.

Australia, New Zealand, and Canada would act as sources of foodstuffs, raw materials, and energy; and the developing cheap labor states of ASEAN and East Asia would provide manufactured goods.[17]

Although the concept may seem sound, none of the nations involved are willing to forgo their own self-interests to implement it. Cries of protectionism and limited market access by all parties seriuolsy limit the chances of implementation.

Australia would seem to benefit from such an arrangement and has committed itself to increased economic interdependence with Japan, ASEAN, and the fast-developing nations of East Asia, through tariff and quota arrangements.[17] At the same time Australia has expressed its concern that the increased economic growth in many WPR nations has not been followed by improved standards of living. Considerable evidence suggests that in most developing nations of the WPR, unemployment, rural poverty, maldistribution of income, and social inequality persist or are getting worse. Australia's concern over these problems, above purely humanitarian ones, is that the widening gap between the rich and the poor could lead to growing instability in these countries and therefore in Australian trade.

On a smaller scale Australia has looked to remove trade barriers by forming a South Pacific Common Market with New Zealand. Such an agreement would open the respective markets to each others' goods by abolishing all trade restrictions between the two countries.[18] But even such a small measure (compared to the huge concept of a Pacific Rim market) has come under fire, as some Australians feel that they give up too much by opening up Australia's 15-million-person market while only gaining access to New Zealand's 3.2-million-person market.

Future Trade Prospects for the Region

The future of Australia as a raw and processed mineral producer will depend on the growth of the WPR. With the prospect of the Japanese moving away from the steel and base-metal industries, it becomes clear that Australia will need to increase her trade in the WPR if it hopes to maintain even current levels of mineral production. There is also the prospect of stepped-up competition from the Canadians. As Canada looks to diversify her export destinations, it has begun to look very seriously at the attractive market to the west. The greatest effect this will have is on Australia's coal industry. It means that Australia will have to improve its competitive position to meet the Canadian challenge. Other players that will also influence the market are South Africa and the United States in coal, and Brazil and India in the iron ore export business.

WESTERN EUROPEAN TRADE

Trade Levels

Western Europe has traditionally been one of Australia's most stable markets. As a trade partner, the European Economic Community (EEC) is second only to Japan in the level of mineral exports. Over the past 10 years, exports to the EEC have never dropped below 15% of total mineral exports, and in three years were over 20%.[19]

Another important aspect is that there is a higher level of processed mineral exports to the EEC than anywhere else. This will be an important consideration if Australia plans to increase her level of domestic processing. Figure III-4-3 is a time series of EEC mineral imports from Australia. Figure III-4-4 compares processed mineral imports for the EEC, Japan, and the United States. Figure III-4-5 shows where the majority of EEC exports go.

Mineral markets in which Australia has strong European connections include lead (both raw and processed), zinc (ore and concentrates), mineral sands, tungsten (ore and concentrates), and copper (matte and worked). Fuel is another area in which Australia does a lot of business with the EEC, as steam coal and uranium exports have been high. Table III-4-12 details the levels of mineral exports to the EEC.

The data in Table III-4-12 hint at Australia's long-standing mineral relationship with Europe. As

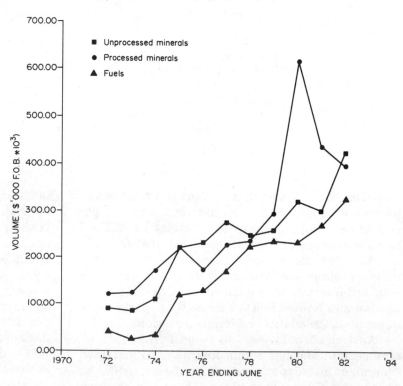

FIGURE III-4-3 Form of Australian mineral exports to the EEC, 1971-1982. (From Australian Bureau of Statistics.)

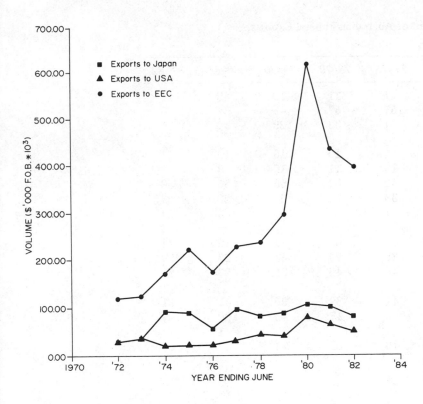

FIGURE III-4-4 Comparison of processed mineral exports to the EEC, Japan, and the United States, 1971–1982. (From Australian Bureau of Statistics.)

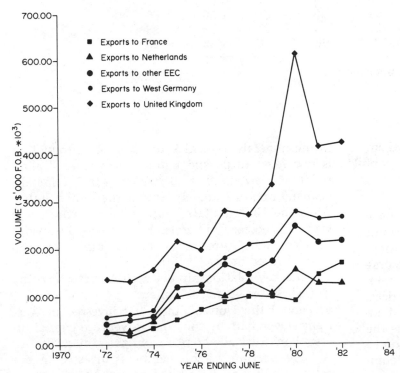

FIGURE III-4-5 European destinations of Australian mineral exports, 1971–1982. (From Australian Bureau of Statistics.)

interest in Australian mining grew during the mid-1800s, European capital (typically British) poured in. Silver, lead, and zinc production from mines at Broken Hill and later, Mt. Isa was sent to local smelters, then shipped to European refineries. The

Japanese-spawned iron ore and coal boom in the 1960s had little effect on the European market, as most of the new mining operations exported largely to the Japanese. What did renew European interest in Australian minerals was the energy crisis of the

TABLE III-4-12 EEC Percent of Australian Mineral Exports,
1977-1978 to 1981-1982[a]

	77–78	78–79	79–80	80–81	81–82
Coking coal	11	11	10	9	11
Steam coal	57	57	46	34	26
Uranium and thorium ores and conc.	31	32	29	32	50
Iron ores and conc.	11	10	10	10	11
Copper ores and conc.	10	11	10	3	8
Lead ores and conc.	42	53	66	55	50
Zinc ores and conc.	33	20[b]	22	28	23
Tin ores and conc.	7		3	3	4
Manganese ores and conc.	—	—	—	—	16
Tungsten ores and conc.	64	63	31	72	71
Rutile ores and conc.	35	29	41	27	29
Ilmenite ores and conc. (excl. beneficiated)	45	40	32	27	25
Zirconium ores and conc.	41	32	35	41	31
Copper matte	66[b]	17[b]	75	100	47[b]
Iron and ferro alloys			1	1	
Copper unworked	79[b]	82[b]	71	62	64
Aluminum and aluminum alloys unworked			—	—	NA
Lead unworked	70	68	74	75[b]	67
Zinc and zinc alloys unworked	4	6	3		1
Tin and tin alloys unworked	88	80	77	56	30
Diamonds rough	NE	53	19	53	28
Diamonds cut (excl. industrial)	NE	NE	34	33	20

[a]—, nil; NA, not available; NE, not exported.

[b]Less than 1%.

Source: Ref. 19.

early 1970s. Then, in addition to base-metal imports, many European countries increased imports of Australian coal and uranium.

Future Trade Prospects

Australia will remain important to European mineral trade in the future, though this does not mean that there will not be increased competition. In exporting to Europe, Australia must overcome the distance disadvantage it has compared to mineral producers in southern Africa and the Americas. Australia would like to at least maintain her share of base-metal exports, and hopefully increase the level of its iron ore, coal, and uranium exports to Europe. This may be difficult. Australia must compete with lower-priced South African coal and additional iron ore exports from Brazil.

U.S. TRADE

Trade Level

The United States is Australia's third largest mineral export market. Since the United States is also a large mineral producer, she is most interested in the minerals for which her domestic production is low (e.g., rutile and bauxite). From Australia's point of view, neither of her two "big" minerals, coal and iron ore, are exported to the United States. Also, the United States imports very little in the way of processed minerals. Table III-4-13 provides detailed trade figures for Australian mineral exports to the United States.

The key relationship between Australia and the United States, as far as minerals go, is the amount of capital the United States has invested in Australia. Over half of the foreign venture capital invested in Australia comes from the United States. A large portion of this capital is invested in the bauxite industry, as all the major American aluminum companies have part-ownership in Australian aluminum companies (Table III-2-11). The bauxite industry has been singled out by many Australians as not paying their fair share of taxes as a result of manipulating "transfer prices" by American as well as other major aluminum companies. This dilemma is covered in Chapter III-5.

TABLE III-4-13   U.S. Percent of Australian Mineral Exports,
1977-1978 to 1981-1982[a]

	77-78	78-79	79-80	80-81	81-82
Coking coal	b	b	b	—	—
Steam coal	11	8	b	b	b
Uranium and thorium ores and conc.	44	36	30	29	20
Iron ores and conc.	b	b	—	b	b
Copper ores and conc.	3	7	7	8	6
Lead ores and conc.	31	28	31	23	45
Zinc ores and conc.	—	3	—	—	—
Tin ores and conc.	—	—	3	—	4
Manganese ores and conc.	2	—	—	—	—
Tungsten ores and conc.	3	5	3	4	2
Rutile ores and conc.	47	47	30	36	38
Ilmenite ores and conc. (excl. beneficiated)	32	19	21	31	36
Zirconium ores and conc.	19	13	15	14	14
Copper matte	—	—	—	—	—
Iron and ferroalloys	13	6	7	14	17
Copper unworked	4	6	12	9	—
Aluminum and aluminum alloys unworked	b	—	—	—	NA
Lead unworked	6	4	4	2	3
Zinc and zinc alloys unworked	20	18	16	14	12
Tin and tin alloys unworked	—	—	7	22	35
Diamonds rough	NE	—	26	17	35
Diamonds cut (excl. industrial)	NE	2	5	7	7

[a]—, nil; NA, not available; NE, not exported.

[b]Less than 1%.

Source: Ref. 19.

FUTURE TRADE PROSPECTS FOR MINERAL EXPORTS

The prospect for Australian mineral exports, like those of any mineral exporter, are greatly influenced by world economic conditions. Economic recessions, which affect the basic industries (farming and mining), eventually end. The lessons to be learned from them are what measures can be taken so as to lessen the effects of the next recession.

What Australia must gleen from its current economic problems is the necessity of export-destination diversification. Japan, through its industrial boom, helped Australia become the world's fifth-largest nonfuel mineral producing country and the second-largest exporter of coal. Today Japan is in the midst of an industrial shift, as more of an emphasis for their future growth will be in "high-tech" products. If the void that this shift creates in base-metal production is filled by other countries in the Western Pacific region (including Australia), Australia would probably benefit from the change. By being able to export to a number of consumers, Australia would improve her bargaining position. She would no longer face a single powerful combatant at the negotiating table, but many eager up-and-comers in competition to build their new industries.

This same shift could also mean disaster to Australia if the Japanese decide to move their basic metal production capacity out of the "neighborhood." A case in point is the Japanese aluminum industry. Due to the high energy costs of aluminum smelting, Japan began a policy to relocate its smelting capacity by phasing out much of its own facilities and investing in foreign projects. Unfortunately for Australia, a large percentage of the new processing facilities are located in North and South America (Table III-4-14). The result is that much of the Japanese aluminum smelting that Australia once serviced is now located in countries that either have their own bauxite supplies or are closer to other sources of raw materials.

It is easy to postulate a similar scenerio occur-

TABLE III-4-14 Japanese Share of Overseas Aluminum Production

Project	Equity (%)	Investment (million dollars)	Production (thousand metric tons)
NZAS (New Zealand)	50	89	75
Venalum (Venezuela)	20	49	160
Alpac (Canada)	50	84	45
Asahan (Indonesia)	75	1713	225
Amazon (Brazil)	49	772	160
Gladstone (Australia)	50	287	102
South Carolina (U.S.)	25	82	45

Source: Ref. 20.

ring with Japanese steel production. This would cripple Australia's output in her two largest mineral exports: iron ore and coking coal. Whereas Australia exports a large share of her bauxite and alumina to the United States and the EEC, her iron ore and coking coal exports are highly dependent on the Japanese market.

To ease the potential loss of the Japanese market, Australia must enlarge its customer base. This will include increasing her share of the U.S. and EEC markets. The drawback is that Australia's greatest competitive advantage in supplying the Japanese market will be the greatest obstacle to overcome in supplying the U.S. and EEC markets. This is shipping distance. To be more competitive, Australia will have to reduce costs. This could come from a variety of directions and is the topic of the next chapter.

REFERENCES

1. Bambrick, Susan. "Australian Mineral Resources Trade and Policies." Paper presented to the Energy and Mineral Resources seminar, The University of Texas—Austin, Austin, Tex., 29 November 1982.
2. Brown, G. R., N. E. Guernsey, and T. M. Li., eds. *Mineral Resources of the Pacific Rim.* New York: American Institute of Mining, Metallurgical, and Petroleum Engineers, Inc., 1982.
3. Govett, G. J. S., and M. S. Govett. "The Role of the Australian Minerals Industry in World Mineral Supplies." *Resources Policy*, June 1980.
4. Harris, Stuart. "Resource Policy in Australia." *Resources Policy*, June 1980.
5. Govett, G. J. S., and M. S. Govett. "Australian Minerals." *Resources Policy*, December 1982.
6. "Commodity Review 1982." *Australian Mineral Industry Review.* Canberra: Australian Bureau of Mineral Resources, 1983.
7. Joint Coal Board. *Black Coal in Australia 1981–82.* Sydney: Joint Coal Board, 1982.
8. Nahai, L., and C. Wyche. *Future Trends and Prospects for the Australian Mineral Processing Sector.* Washington D.C.: U.S. Department of the Interior, 1982.
9. Rodrik, D. "Managing Resource Dependency: The United States and Japan in the Markets for Copper, Iron Ore, and Bauxite." *World Development*, July 1982.
10. Krause, L. B., and H. Patrick, eds. *Mineral Resources in the Pacific Area.* San Francisco: Federal Reserve Bank of San Francisco, 1978.
11. "Japanese Market Move Challenge Australia's Mines." *Engineering and Mining Journal*, February 1983.
12. "In the Pacific: Australia." *Engineering and Mining Journal*, July 1983.
13. Bayless, A. "Japan Wants Coal Producers to Cut Prices and Deliveries As Steel Market Declines." *The Wall Street Journal*, 21 January 1983.
14. Australian Department of Trade and Resources. *Australia's Mineral Resources: Iron Ore.* Canberra: Australian Government Publishing Service, 1981.
15. "Ranger Mine Producing above Design Capacity." *World Mining*, November 1982.
16. Australian Department of Trade and Resources. *Survey of Major Western Pacific Economies.* Canberra: Australian Government Publishing Service, 1982.
17. Bell. R. "Problems in Australian Foreign Policy, July–December 1979." *The Australian Journal of Politics and History* 26, No. 1 (1980).
18. Galante, S. "Australia, New Zealand Near Accord to Cut Trade Barriers, Merge Markets." *The Wall Street Journal*, 20 October 1982.
19. Australian Department of Trade. "ABS Overseas Trade Statistic 1970–71 through 1981–82." Canberra, 1983.
20. Australia/Japan Joint Study Group on Raw Materials Processing. *Australian and Japanese Aluminum Smelting Industries: Future Development and Relationship.* Canberra: Australian Government Publishing Service, 1980.

Chapter III-5

Policy Issues

COMPETITIVENESS

Australia's Current Position

The world mineral industry has been in a severe recession since early 1981. Commodity prices in real terms are depressed, some at their lowest levels in 30 years. Even those that have not been greatly affected (coal, iron ore, and nickel) have suffered severe price pressures in 1983. Steel production, a crucial indicator of the shape of the mineral economy, is about half of what it was in 1974, with the U.S. steel industry operating at about 40% of capacity.

As a result the competitive position of the Australian mineral industry has deteriorated. The promise of a "resource boom" in the 1970s led to excessive demands by labor, unrealistic taxes and payments to the government, and the imposition of overly restrictive regulations. Even if Australia were the most cost-efficient and competitive mineral producer in the world, the effect of low mineral prices would mean a drop in revenue, delays in new ventures, and a reduction in exploration. Without a competitive edge, in addition there are also losses of markets to other mineral producers, such as Brazil, Canada, and South Africa.

The Australian economy is in trouble. The inflation rate, which is running at an annual rate of over 12%, is approximately double that of other major developed countries. This is particularly troubling since if Australian costs are rising faster than those of other producing nations, their competitive position will clearly be affected. At the same time unemployment is near 11% and is expected soon to exceed 12%. In March 1983, the currency was devalued by 10%, which should encourage mineral exports in the short run but could lead to long-term difficulties. Despite massive inflows of capital, the mineral industry is in a depressed state, and this in turn has caused a slump in profits and layoffs in the manufacturing sector. To make matters worse, the country suffered through its worst drought in 200 years, which has crippled farm and pastoral output.

The 1980-1981 Coopers & Lybrand survey of the Australian Mining Industry, which included 100% of bauxite, copper, lead, zinc, and nickel production, 91% of iron ore production, and 80% of black coal production, documented the decline in Australian mining profitability. Industry net profits on a constant basis declined 33% between 1979-1980 and 1980-1981. In the following year there was an additional drop of 70%.[1] Over the same two-year period, dividends fell 41%. In 1981-1982 the effective after-tax return on average funds employed was only 3.7% compared to 13.3% in 1979-1980.[1]

At the time costs were up 13.1% for 1981-1982 on a constant group basis compared to the preceding year. This reflected in the increase in labor costs up 16%, government service costs up 16%, purchases from suppliers up 12%, and interest expenses increased by 32%.[1] State and federal income and resource-based tax imposts as a proportion of preimpost profits increased in 1981-1982 to 72%, compared to 56% in 1980-1981 and 54% in 1979-1980.[1] Table III-5-1 is a simplified income statement for the Australian mineral industry for 1981-1982. Figure III-5-1 shows the distribution of the revenue dollar for the same period.

On the positive side, Australian mineral production has held up reasonably well in the year to June 1982. Buoyed mostly by coal and iron ore revenue, mineral exports were worth a record A$7541 million, accounting for some 38% of the country's total export receipts.[2] This trend is likely to change markedly in 1983 as declines in export demands have caused significant production and price cutbacks (i.e., price cuts of $12 per metric in ton coal and $12 to 13 in iron ore prices). A number of new projects have been canceled or delayed as a result.

Capital expenditures have continued to increase. In 1981-1982 they exceeded $3000 million, more than double that of the preceding year on a constant group basis.[1] This trend can be misinterpreted. Although capital expenditures are on the rise, much of the current spending was planned and dedicated for long-term projects in 1980 or earlier, when the outlook for the industry was far more encouraging than at the present.

Exploration expenditures, often a clue to how

TABLE III-5-1   Australian Mineral Industry Income Statement, 1981-1982 (Million Dollars)

	79-80	80-81	81-82
Mining sales	4554	5047	5779
Smelting and refining sales	2460	2583	2667
Other revenue	147	228	186
Total revenue	7161	7858	8632
Supplies	2526	2960	3393
Labor costs	1477	1843	2249
Government services	307	372	507
Depreciation and amortization	641	847	1120
Interest	262	327	468
Exchange losses (gains)	19	(2)	19
Sundry taxes	30	44	32
Costs before resource based taxes and income taxes	5262	6393	7788
Mineral royalties, tenements, and mining license fees	234	202	220
Coal export duty	79	78	88
Income tax	711	545	296
Direct taxes	1024	825	604
Extraordinary losses (gains)	(8)	(22)	92
Tax on extraordinaries	4	(3)	(49)
Net profit	879	665	197

Source: Ref. 1.

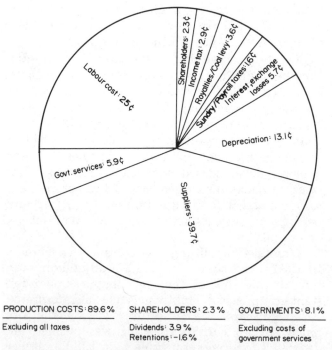

PRODUCTION COSTS: 89.6%	SHAREHOLDERS: 2.3%	GOVERNMENTS: 8.1%
Excluding all taxes	Dividends: 3.9%	Excluding costs of
	Retentions: -1.6%	government services

FIGURE III-5-1 Distribution of revenue dollar, 1981-1982. (From Ref. 1.)

well the industry on the whole is doing, reached A$463 in 1981-1982. However, since spending is closely allied to corporate profitability, the industry forecast is for a 13.7% decline in exploration in 1982-1983.[2]

Stability

One of the most pressing issues in determining the competitiveness of the Australian mineral industry is stability. Due to the long lead times involved with developing a mineral property, changes in government regulations and tax laws, and radical increases in labor costs can often turn a profitable mining operation into a money loser or shut one down for legal reasons.

Frequent changes not only affect the profitability of existing operations but also deter exploration. After all, a company has to question the feasibility of spending millions to explore and delineate a mining area if when they are done, the area is closed to mining due to newly imposed restrictions. The classic example is uranium mining. Before the election of a Labor government in 1972, abrupt changes were made to the government uranium policy. All new projects were shelved, and development was delayed for five years, costing mineral producers untold millions of dollars.

A similar argument can be made for mineral sands, although in this case it was the Liberal-NCP's ban on the production of Frasier Island sands on environmental reasons. The Northern Territory Land Rights Act removed vast amounts of land from mining, much of which had already been explored and targeted for development.

Tremendous instability is caused by altering tax laws and other types of governmental payment provisions. The state governments are particularly guilty of this, as royalty rates and infrastructure charges often change many times during the life of a mine.

As Australia emerges from the "resource boom" mentality and struggles through a world economic recession, the key to the recovery of her mineral industry will be renewed competitiveness. This will mean a revaluation of governmental and labor policies and a goal of increased stability. In the same vein, the general public must understand that just because their nation is blessed with vast mineral resources, there is no guarantee of prosperity. Australians must learn that mineral development cannot proceed if (1) demand is not forthcoming, and (2) if mineral projects are uncompetitive. The current situation was best described by the former president of the Australian Mining Industry Council, Hugh Morgan, during the 1983 National Economic Summit Conference:

> The universal recognition that today's national economic conditions, in part arise from international market conditions carries with it the inherent recognition that the nation is an irrevocable participant in that marketplace. That marketplace is a ruthlessly competitive component of the race of life for any nation, a race from which there is no escape, which deals harshly with nonperformers and which dispenses its discipline irrespective of political philosophies of nations or administrations.

MINERAL FINANCE AND TAXATION

The financing and taxation methods used for Australian mineral projects have markedly influenced the development and profitability of the industry. In turn, the growth of the mineral industry has greatly affected other domestic industries and the national economy as a whole. This section deals with these aspects of mineral finance, the effects of rapid mineral export growth, the cost and availability of capital, and the impact of taxation on the mineral industry.

Mineral Growth

The promise of a resource boom in the 1970s led to the question of whether increased mineral exports would automatically improve the Australian standard of living. Some claimed that increased exports would adversely affect the nation's rural and manufacturing sectors. The main proponent of this theory was R. G. Gregory of the Australian National University.[3] His contention was that an investment surge in a single industry would make imports comparatively less expensive. This would be due to an exchange-rate appreciation caused by inflows of foreign capital and later by rising export receipts. The result would be a reduction in rural exports and increased competition by imports in the domestic manufacturing industry. In turn, the resulting adjustment away from labor-intensive manufacturing to capital-intensive mining would raise the level of unemployment and that the increase in the money supply would lead to inflation. The Gregory Thesis has been used to argue against growth in mining, and the Australian Labor Party (ALP) has used it in opposition to unbridled mining growth.

Whether the theory would hold true (partially or completely) is a matter of debate. The Australian Department of Trade and Resources' study "Resource Investment in Prospect and the Balance of Payments: Some Implications for the Economy," released in December 1980, envisaged a 10.3% appreciation of the exchange rate by 1989–1990, due to resource investment alone.[4]

Neville Norman argued against the theory, claiming that it did not allow for:[4]

1. Positive income effects that benefit a wide range of domestic industries
2. Amelioration of the extent of currency appreciation because of a number of items in the balance of payments to offset capital inflow and increased exports
3. Positive intersectional leakages

Norman concluded that the positive effects of mineral growth outweighed any negative outcome. The Australian Bureau of Industrial Economics 1982 report suggested that leakages to imports are unlikely to reduce the impact of mineral exploration on the balance of payments. It went on to state that the eventual balance will be influenced by other factors, such as repatriation of profits and/or repayments of loans, further capital inflows, and induced demand for imports resulting from higher real incomes for Australian consumers. The final outcome will also depend on the policy stance chosen by the government in their response to any increase in foreign investment.[4]

The unfortunate aspect of these studies is that, due to the current mineral slump, Australia has not been able to benefit from the resource boom. The Australian dollar–U.S. dollar exchange, shown in Figure III-5-2, which rose until mid-1980, dropped sharply thereafter. Not shown in the figure is an additional 10% devaluation that the APL government enacted in March 1983 in an effort to boost mineral exports. As of October 1983 the Australian dollar was worth only 0.9 U.S. dollar.

Finance

An often misunderstood aspect of the mining industry is the availability and cost of capital. Even a high-grade deposit is worthless if the required venture capital cannot be obtained or the deposit may be only marginally economical if the interest expenses are high. Australia is fortunate to be blessed with many valuable mineral deposits, but due to her small capital market, many mining projects have had to look overseas for funding. This meant that Australian mining companies have had to pay a premium to finance their operations.

Two major considerations relevant to investment decisions are the willingness to invest and the ability to invest. The first is determined by the risk

and quality of the project. The second reflects estimated profits, internal funds, and access to external finance.[5] Funding of Australian mineral projects is highly geared; that is, they have a high debt–equity ratio. This can be attributed to a number of factors:[5]

1. The capital intensity and large capital requirements of these industries
2. The taxation treatment of interest compared with that of dividends
3. The declining contribution of internally generated business savings, such as depreciation and undistributed income as a source of funds for gross capital accumulation in the corporate sector

The Coopers & Lybrand survey showed that total assets increased 19% in 1981–1982 on a constant group basis over the previous year,[1] the major portion of the increase being on coal- and energy-based projects. As a consequence of the higher capital expenditure and negative retained earnings, borrowing rose sharply in 1981–1982, with a corresponding increase in the industry debt–equity ratio. This is crucial, as it limits the extent to which the industry may borrow new funds for expansion of operations and employment, and in years of low sales revenue exposes companies to high interest burdens and liquidity problems.[1] Table III-5-2 is the Coopers & Lybrand Mineral Industry Aggregate Balance Sheet, and Table III-5-3 shows the sources and application of funds for 1979–1980 to 1981–1982.

The source of funding for a mineral project will depend not only on the availability of domestic capital, but also on the comparative terms of domestic versus overseas loans. For example, in late 1978 and early 1979 it was possible to borrow more favorably from the United States than in Australia, even though comparable interest rates were higher in the United States. This was due to the "covered interest rate differential," which takes into account both the difference between comparative nominal interest rates and the cost of purchasing forward exchange rate cover against the other currency. In the U.S.-Australia example, U.S. dollars could be purchased forward at a discount that outweighed the difference between nominal rates.[5] This risk involving exchange rates has been extremely relevant to Australia in 1983 in light of the major devaluation of the Australian dollar.

Sources of domestic funds have included major trading banks, merchant banks, life insurance companies, and savings banks. In addition, two other organizations have funds available for mining investments. The Australian Industry Development Cor-

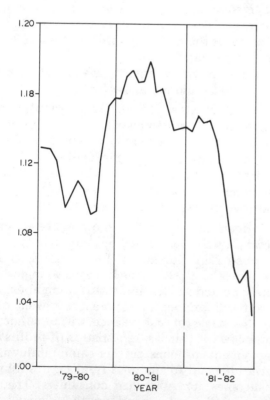

FIGURE III-5-2 United States dollar–Australian dollar exchange rate. (From Ref. 1.)

TABLE III-5-2   Mining Industry Aggregate Balance Sheet,
1979–1980 to 1981–1982 (Million Dollars)

	79–80	80–81	81–82
Shareholders funds	4,960	6,906	8,927
Borrowing	3,388	4,811	6,781
Provision for income tax	1,309	1,304	1,349
Other provisions	274	408	491
Trade creditors and accruals	788	1,063	1,299
Other liabilities	149	256	238
Total equity and liabilities	10,868	14,748	19,085
Fixed and deferred assets	7,595	10,748	14,098
Loans and advances	315	477	479
Current assets	2,764	3,252	3,969
Other assets	194	271	539
Total assets	10,868	14,748	19,085
Asset turnover ratio	1.10	0.91	0.77
Funds turnover ratio	0.96	0.76	0.62
Debt–equity ratio	0.68	0.70	0.76

Source: Ref. 1.

TABLE III-5-3   Sources and Applications of Funds (Million Dollars)

	79–80	80–81	81–82
Net profit	879	665	197
Non cash items			
Income tax expense	715	542	247
Depreciation and amortization	641	847	1120
Provisions	40	130	94
From equity to share holders	180	420	680
New borrowing	718	1670	3267
Net movement between mining and other business sectors	214	(143)	262
Total sources	3296	4131	5867
Expenditures on:			
Exploration	173	337	463
Mining assets	885	1621	2070
Smelting and refining assets	206	437	982
Borrowing repaid	734	499	1048
Income tax paid	723	595	275
Dividends paid	444	404	333
Net movement in other assets/liabilities	131	238	696
Total applications	3296	4131	5867

Source: Ref. 1.

poration is a statutory corporation established to promote both Australian ownership in the industry and the development of Australian resources. The other is the Australian Resource Development Bank Ltd., which is not a government instrumentality, but rather, owned by trading banks, and is accorded bank status.[6]

To aid further in resource development, state governments have been allowed to raise funds, in addition to "normal" borrowing for infrastructure, particularly electrical energy supplies, projects. Under a Loan Council program overseas borrowing would be assessed case by case under guidelines which include that the necessary funds are unavailable on the domestic market and that specific Loan Council approval is given.[1]

Total borrowing by the Australian mineral industry stood at $6781 million in 1981–1982, a 35% increase over the previous year. Interest expenses rose 32%, to $468 million. Overseas sources accounted for 52.3% of total borrowing in 1981–1982.[1] This illustrates the importance of foreign capital to Australian mining. Table III-5-4 breaks down borrowing by length of investment and source of capital.

An important source of capital has been through consortiums. The form of a consortium (new company, unincorporated joint venture, etc.) is to some extent influenced by taxation provisions and regulations under the Companies Act. Benefits from such an arrangement include:[5]

1. *Cost of finance:* New projects may necessitate fund raising in excess of, or disproportionate to, a single firm's assets. Thus by spreading the burden of project finance, consortia arrangements can lessen average financial costs.
2. *Access to markets:* Consortium participation by major consumers can assist in gaining market access by "underwriting" a proportion of output, which may then be used as collateral for loans.
3. *Cost of equipment and technology:* When a firm enters a new area, even though it may have finance and markets, it may be advantageous to enter into a consortium to obtain technical skills, design, and equipment.

Consortia arrangements are particularly advantageous when foreign participation is sought. The limitations of this avenue's use in Australia is studied later in this chapter.

The cost of venture capital can be affected by changes of both a broad (i.e., inflation, exchange rates) and a specific (depreciation schedules, investment allowances) nature. From a macro point of view, inflation contributes to the cost of capital, but the real danger is when your own inflation rate is higher than that of your competitors. This is a problem that Australian mining companies are facing due to relative inflation shifts in 1983 that has left Australia with an inflation rate that is higher than those found in most developed countries.

Exchange rates can have a varied impact on the cost and availability of capital. When the exchange rate is declining, as has been the rate between U.S. and Australian currency, foreign investors lose confidence in Australian investments since the value of the Australian currency will be less when paid back. From the Australian side, currency devaluation makes her exports more attractive, but at the same time they make imports (i.e., necessary foreign equipment and technical expertise) more expensive. Methods available to lessen the risk include forward exchange covers, arrangements such as the "spreading of foreign currency" in a loan package, or the negotiation of long-term sales contracts in the currency borrowed.[5]

Australian capital costs are higher than those of other developed countries for a number of reasons. Lower labor productivity caused by poor industrial relations make for longer lead times. A higher cost structure as a result of tariffs, and the extra burden imposed by high infrastructure costs, also increase the cost of capital. Infrastructure is considered the variable component of determining capital costs, as it is dependent on the project site and the extent to which the private-sector finance is required.[5]

There are many fiscal measures that if adopted or altered could affect the rate of return on capital. In comparative terms the depreciation schedules available to Australian firms are not as attractive as those found in other developed countries. The Australian Industries Development Association (AIDA) found in its 1980 International Survey on

TABLE III-5-4   Australian Borrowing Patterns, 1979–1980 to 1981–1982 (Million Dollars)

	79–80	80–81	81–82
Repayable within:			
1 year	406	882	1149
2–3 years	612	643	774
4–5 years	593	677	945
After 5 years	672	1254	2158
Term not specified	1105	1355	1755
From overseas sources	1876	2225	3547
From Australian sources	1512	2586	3234
	3388	4811	6781

Source: Ref. 1.

Depreciation Rates, Tax Rates, and Investment Incentives that the principal feature distinguishing Australia from other countries is the lack of accelerated depreciation provisions.[5] To improve this, the 1980–1981 budget called for a 20% increase in depreciation rates. On the positive side, Australia's 20% investment allowance particularly counteracts the depreciation discrepancy. Since mineral projects are usually capital intensive, companies have a strong desire to accelerate cash flows in the early stages of a project. The 20% investment allowance is used to shift cash flows forward.

In a hypothetical mineral project, the Commonwealth/State Joint Study Group on Raw Minerals Processing analyzed the effects of altering a number of variables on the project's internal rate of return. The base case involved a 15-year project working under Australia's 1980 tax scheme (10% yearly depreciation, 20% investment allowance, 46% corporate income tax, loan premium payments in the last year of the project). The sensitivity analysis showed that the selling price of the commodity was the most sensitive variable, much more so than tax rates or depreciation schedules (Figure III-5-3).

In a world mineral industry where shortages

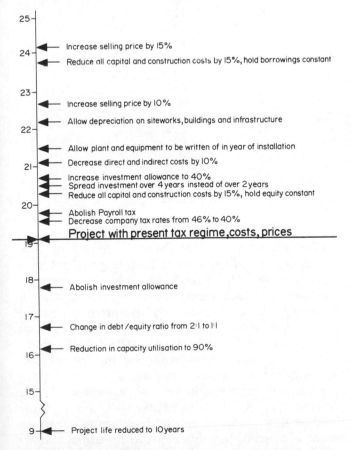

FIGURE III-5-3 Effects of variables on a project's feasibility. (From Ref. 5.)

and surpluses can substantially alter commodity prices, this example should point out the need for tax laws that compensate for price fluctuations. A sliding-scale tax rate is one possible answer. Here the tax rate depends on the profitability of the industry, so in years where mineral prices are high, the tax rate increases, thus soothing the public's fear that the mining companies are robbing the nation's mineral wealth. Also if mineral prices decline, companies would receive a tax break, thus reducing risks and thereby creating a more stable investment climate. Another advantage is that a sliding-scale tax would eliminate "high-grading," since more lucrative deposits would be taxed at a higher rate.

Taxation

The Australian tax system is complicated as the government "take" includes a variety of payments. This section focuses on the federal income tax and its components, while the "Federal versus State Charges" section examines the remaining payments (royalties, freight charges, etc.).

Due to the general decline in mineral industry profits, the level of income tax fell to $247 million in 1981–1982, representing a 51% decline from the year before. At the same time there was a 56% decline in pretax profits.[1] Table III-5-5 shows the level of government payments for 1979–1980 to 1981–1982.

The Australian corporate tax rate is currently 46%. With the exception of the years 1970–1973, this is the highest rate in the past 30 years (Table III-5-6).

Although the Australian tax rate may seem low compared with other developed nations (e.g., United States 48%, Japan 56%), Australian companies do not enjoy many of the tax benefits found elsewhere. For example, the AIDA study cited that of the countries surveyed, only Australia and West Germany did not have accelerated depreciation provisions. In addition, Australia's depreciation rates, which are determined by the Taxation Commissioner, and are typically 5 to 15%, are below the average rates found in other countries. Also, research and development expenditures can not be expensed as they occur, but have to be capitalized.[5]

Present taxation provisions allow for the depreciation of mining transport infrastructure over 10 or 20 years and limited write-offs for nonplant items of an initial nature for ports, but not towns. Mine site plant and equipment, including buildings, are depreciated over a 10-year period. All depreciation is on a straight-line basis.[7]

The Australian processing sector faces even more restrictions, as depreciation is allowable only

TABLE III-5-5 Mining Industry Payments to the Government,
1979-1980 to 1981-1982 (Million Dollars)

	79-80	80-81	81-82
Mineral royalties, tenements, and mining license fees	234	202	220
Coal export duty	79	78	88
Income tax	715	542	247
Land taxes and rates	22	31	41
Payroll tax	61	77	98
Other taxes	8	13	(9)
Employees: group tax	339	403	547
Lenders and shareholders for withholding tax	18	24	14
	1476	1397	1246

Source: Ref. 1.

TABLE III-5-6 Australian Corporate Tax Rate, 1950-1983

	Type of company and income				
	Public		Private		
Year	First $10,000	Balance	First $10,000	Balance
---	---	---	---	---
1950	25	35	25	30
1951	25	45	25	35
1952	35	45	25	35
53-55	30	35	20	30
1956	35	40	25	35
57-59	32.5	37.5	22.5	32.5
60-63	35	40	25	35
64-67	37.5	42.5	27.5	37.5
68-69	40	45	30	40
1970	42.5	47.5	32.5	42.5
71-72	47.5	47.5	37.5	42.5
1973	47.5	47.5	45	45
1974	45	45	45	45
75-76	42.5	42.5	42.5	42.5
77-83	46	46	46	46

Source: Ref. 5.

for plant and equipment, thereby excluding buildings, infrastructure, and site works. This includes all basic processing procedures, such as cleaning, grinding, and gravity or electrostatic concentrating. Processing plants for pelletizing, sintering, calcining, producing alumina, and so on, are specifically excluded. This has clear ramifications if Australia is to increase the level of domestic processing.[7] Table III-5-7 compares depreciation rates for selected countries.

One measure that has eased the tax burden is the investment allowance (IA). The IA improves the cash flows of projects by providing a deduction from taxable income in addition to normal depreciation allowances.[5] The current IA rate of 20% has the effect of tripling the first-year depreciation allowance on a 10% straight-line item. The program has been criticized, most strongly by the Australian Treasury, on the grounds of resource allocation (i.e., using more capital intensive techniques than the market will justify). A 1979 OECD study showed that industries with the highest capital-intensity ratios included iron and steel, nonferrous metals, and nonmetallic mineral processing.[5] Since Australia is in international competition in these industries, a study of market forces must take into account the tax benefits enjoyed by other nations in determining resource allocation.

TABLE III-5-7   Comparative Depreciation Rates, 1980[a]

Country	Buildings (factories)	Plant and equipment	Vehicles
Australia	Nil	6-18%SL	18-24%SL
Brazil	4%SL	4-20%SL	20%SL
Canada	30%RB for mining and processing buildings	30%RB	30%RB
Columbia	5%SL	40-60%SL	20%SL
West Germany	2.5-6.66%SL	12-20%SL	25%SL
Indonesia	2.5-6.66%SL	8.33-12.5%SL	10-20%SL
Japan	1.5-4%SL	7-17%SL	12%SL
South Africa	2%SL	10-25%SL	20%RB
U.S.	1.7-3.3%RB	20%RB	50%RB

[a]SL, straight line; RB, reducing balance.

Source: Ref. 7.

Australia is the only major industrial nation without some form of accounting for tax purposes of the effects of inflation on stock values.[7] Australia does not provide investment inducements offered by other countries, especially those of developing nations. These "tax packages," such as tax holidays, regional processing allowances, and low-interest finance, help attract both foreign and domestic investment. Table III-5-8 illustrates various tax incentives for selected countries. Where Australia does have an advantage in attracting investment capital is in her stable political and economical climate and the availability of a skilled work force.

FEDERAL VERSUS STATE CHARGES

State and Federal Conflicts

Government payments by the mineral industry are complicated by the separate powers of the state and federal governments. The ownership of mineral rights is held by the federal government, whereas the right to exploit is vested in the state governments. With each level of government trying to maximize its resource revenue, mining companies have been hard-pressed to maintain a profit during the current recession. Figure III-5-4 illustrates the increase in tax percentages versus the decrease in net profits.

This dilemma is caused by charges and payments such as royalties and service charges not directly affected by operating profits. As seen in Table III-5-5, between 1979-1980 and 1981-1982, income tax decreased from $715 million to $247 million. This was caused by general decline in the industry's profitability. During the same period there were increases in royalties, license fees, and government service charges.

The high level of charges stems from the way in which the federation government was set up. Unlike the United States, where the states have specific powers and Washington, D.C., maintains all residual powers, Australia has an evolutionary type of feder-

TABLE III-5-8   Comparative Tax Incentives

	Australia	Brazil	Canada	South Africa
Tax rate	46%	35%	36%[a]	42%
Investment allowance	Yes		Yes	Yes
Relative infrastructure	Mining only		Yes	Yes
Accelerated depreciation	No	Yes	Yes	Yes
Tax holidays	No	Yes	No	
Processing incentives	No	Yes	Yes[b]	
Regional tax concessions	No	Yes	Yes	Yes
Withholding tax	15%	25%	15%	15%

[a]Federal tax rate only.
[b]Province tax only.

Source: Ref. 7.

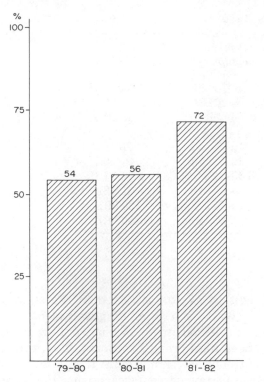

FIGURE III-5-4   Tax as a percent of pretax operating profit, 1979–1982. (From Ref. 1.)

ation in which the states confer on the central government (Canberra) certain powers and reserve the residual powers for themselves. While the Canberra government controls exports, the states control mineral exploration, development, and production. Because of this separation of power, the two levels of government are constantly fighting over who receives the largest portion of the "mineral pie." The states complain that the federal government is parsimonious with funds they require for infrastructure development. Since the charges on the state level are deductible against federal income taxes, the states use these fees to siphon profits before the Canberra government can tax them.[8]

In addition, since the states do not have equal mineral endowments, there are even more incentives to reap as much as possible on the state level. Mineral-rich states such as Queensland and Western Australia do not want to see their wealth transferred to other states. This is further complicated in that federal tax collections are computed on a state basis according to where the tax return is filed, bearing no relation to where the income was earned. Therefore, Victoria and New South Wales, who suffer the most from "head office syndrome," pay taxes on income collected outside their state and receive no benefits from the local charges collected.[8] Unfortunately for the mining industry, most of the state charges are based on production,

not profits, so during times when the mineral prices are low, the companies pay the same amount regardless of the profitability of the operation.

On the federal level, in addition to income taxes, there is the coal export tax and other export duties. Until July 1982 the coal export duty was generally applicable at a rate of $1 per metric ton on all export coal except for certain hard coking coal, which attracted a $3.50 per metric ton duty.[9] Since then the $1 duty has been removed, although the coking coal duty remains in place. Being a volume charge, the duty remains the same regardless of the price received for the coal. An excise duty of $0.20 per metric ton of coal production is used to finance the coal-mining industry long service leave; another $0.05 per metric ton levy is used to finance coal research.[9]

State Charges

State taxation of the mineral industry falls into three forms: royalties, front-end payments, and charges for state services. To compare Australian royalty rates to those of other nations is not especially meaningful since they vary not only among the states but also within each state according to the particular mineral.[6] In Australia there are seven different royalty systems, with royalty rates in each dependent on the type of mineral produced. On top of this, rates are often altered as the mineral climate and political perception of the industry changes. They also reflect the philosophy of the political party in charge of the state government. The highest royalty is charged by N.S.W. for the lead and zinc produced from Broken Hill, where the royalty can rise to 50% of the company's profits (before the mid-1960s the maximum royalty was 100% of profits). The lowest royalty is probably by the Northern Territory on Gove Peninsula bauxite, where locally processed bauxite incurs a royalty of 30 cents per metric ton. Common royalties for metallic minerals are between 2 and 10%.[10]

To illustrate the turbulent condition of state royalty rates, it is helpful to use coal as an example. Coal was chosen for two reasons: (1) because it is Australia's most important mineral export, and (2) because of the two diverse systems used by the coal-exporting states, N.S.W. and Queensland.

In N.S.W. royalty rates (1982) relate to the volume of production (unit royalty), payable at the rate of $1.70 per salable metric ton. No distinction is made between royalties for underground and open-cut mines, despite the significantly different economies. Super royalties are charged at between $1.05 and $1.91 per salable metric ton on coal from new leases.[9]

The Queensland government imposes an ad valorem royalty on export coal at 5% of the f.o.r. value of coal from open-cut mines and 4% of the value for underground coal. Domestically consumed coal has a royalty of only $0.05 per metric ton.[9] Table III-5-9 provides selected royalty rates.

Front-end payments are required for the granting of new leases. The form of these payments varies from state to state. In N.S.W., front-end payments are used as direct contributions for infrastructure financing. The basis for determining mining lease front-end payments depends on the mining circumstances and whether the lease is an extension of an existing lease or an entirely new lease. The N.S.W. government is developing a new policy where the proponents of new mining projects would be required to provide the capital necessary for the development of the industrial and social infrastructure.[9]

In Queensland the government charges a railway security deposit which provides for the finance of new railway facilities and rolling stock for state railway systems. On top of this, the government then charges mining companies excessive rates to use the lines.[9]

State governments have developed additional methods to increase their share of mineral rent. The Queensland, Western Australian, and South Australian governments charge mineral producers a higher rail freight than other users. The 1982 Queensland rail profit from recent rail agreements was approximately equivalent to a 15% ad valorem levy. The N.S.W. State Rail Authority made a profit of $1.36 per metric ton of export coal hauled in 1979–1980.[9]

The typical rail charge in Australia is high by international standards, about four to five times greater on a metric ton per kilometer basis than those applicable in Canada, the United States, and South Africa. Recent concessions to U.S. coal buyers from rail proprietors in exchange for tonnage guarantees will further erode Australia's competitive position. Other taxes and payments charged by state governments include payroll taxes, land taxes, stamp duties, road levies, and license fees.[9]

As stated previously, the reason why there are so many methods to extract rent from the mining industry is the conflict between the federal and state governments. The states look for means to raise their income before the federal government takes its share. When the Canberra government threatens to regard royalties as nondeductible for income tax purposes, the state governments respond in turn by increasing service charges.[8] When the federal government reduces taxes, as with investment allowances, the states are quick to absorb any new rent created by increasing royalty payments.

Resource Rent Tax

This conflict between the different tiers of government has been the major stumbling blocks in efforts of tax reform. One of the best examples is the current talk of imposing a profit-based resource rent tax at the federal level. A resource rent tax is not a new idea. Early Australian discussion of the resource rent tax assumed that it would be levied by the Commonwealth on behalf of both the state and federal governments, in place of income taxes and state payments.[6] Although the Liberal-NCP government gave up on the idea after opposition from the industry leaders, who feared that the states would not relinquish their rights to levy royalties, the resource rent tax has resurfaced now that the ALP has come into power.

The principle behind the resource rent tax is that it would tax only monopolistic profits. The economic rent above normal profits (normal profits include a margin to service the firm's capital and provide a real return to the owners) can be taxed away without affecting production decisions. Normal profits are usually calculated as the difference between operating revenues and all costs incurred in earning income, including exploration and operating costs and the cost of capital.[9] This type of tax could be applied on a project or a company basis.

The main proponents for a resource rent tax are Labor's Minister of the Treasury, Paul Keating, and their Minister for Resources and Energy, Peter Walsh. For the same reasons mentioned earlier, the chances of an all-encompassing resource rent tax that would replace all taxes, royalties, and so on, are slim. During a May 1983 interview,[13] Senator Walsh proposed a tax that would apply to oil and coal in addition to the company income tax. It would be applied only after net earnings exceed

TABLE III-5-9   Selected Royalty Rates, 1980

New South Wales	
Iron ore	$0.35/metric ton
Alumina	$0.70/metric ton
Bauxite	$0.35/metric ton
Western Australia	
Ilmenite concentrates	2% f.o.r. (or f.o.b. if exported)
Iron ore (direct shipping ore)	7½% f.o.b. minimum $0.60/ metric ton
Alumina	$0.50/metric ton
Bauxite	$0.50/metric ton

Sources: Refs. 11 and 12.

(1) infrastructure and exploration costs, and (2) a threshold rate of return. Senator Walsh did not envisage applying the tax until the 1984–1985 fiscal year. On this level the resource rent tax would replace oil and coal levies. Opponents of a resource rent tax say that it would be unfair unless it replaces all existing taxes and provides for losses to be fully offset against income. Also, if the tax were implemented on a project basis, it would not account for the risk of failed exploration programs.

Taxation's Effect on the Mining Industry

The mineral industry has complained bitterly about how the continuing increase in federal and state charges has threatened the ability of companies to survive. The then Shadow Minister of Minerals and Energy, Paul Keating, said that the proliferation of taxes on the mineral industry would finally undermine the industry if not checked.[15]

The three main complaints about the present tax system are (1) its complexity, (2) its instability, and (3) its suboptimal nature. The nature of the Federation, with its separate national and state taxation powers, the diverse systems of payments on a state-to-state basis, makes for an extremely complex tax system. For these reasons the Australian tax system has been characterized by instability, with each level of government out to maximize its rent at the expense of the others. In trying to decide on mineral project investments, mining companies have had to add taxation risks to existing market risks, thereby raising acceptable rates of return. Examples of ad hoc amendments to the taxation system include:[9]

1. The life-of-mine deduction for Division 10 allowable capital expenditures has been changed three times since May 1, 1981.
2. The basis for calculating the coal export duty has been amended six times since it was introduced.
3. Royalties payable to the N.S.W. government increased by 70% in 1981.
4. Freehold coal rights in N.S.W. were removed under the Coal Acquisition Act (1981).
5. In N.S.W. coal freights were increased by 25% and port handling charges by 47% on January 1, 1982; rail freights have increased by 80% since 1979.
6. In Queensland average coal rail freights have increased by about 37% over the past two years.
7. The level of front-end payments for new projects typically increases as each new project commences development.
8. The Queensland government arbitrarily imposed a road levy on the state's coal companies in an attempt to extract $140 million from the industry.

The use of production-based taxes distorts the coal industry by disregarding the nature and location of the ore body worked. Although ad valorem taxes are somewhat more equitable than unit taxes, they can also cause economic distortions. Both methods cause underutilization of lower-grade ores in existing mines, the premature closure of established mines, and a "bonanza" mentality among mining and exploration companies as lower-grade deposits are disregarded.

Although profit-based royalties would ease the problem of underutilization, opponents of the system claim that it would discriminate against more efficient producers. This has been a major complaint of Queensland coal producers, who fear that a profit-based tax would cause a shift in the proportion of revenue paid by their state and would in effect subsidize less-cost-efficient operations.

To develop an equitable tax system for Australia will require a coordination of taxation between the various tiers of government. The Australian Coal Association has suggested a joint federal/state/industry coordination committee to review taxation policies and make recommendations.[9] This type of group would help add stability and predictability to the industry.

In such a framework of cooperation, a system that would work to eliminate the problems of unit versus ad valorem royalty schemes, these would be a two-tier method of royalty payments. One proposed by the Queensland Opposition Leader, Mr. Wright, would combine a profit-based royalty with an ad valorem system.[14]

If such a system could be adopted by all the states, the federal government could consider using a resource rent tax to replace the income tax and other federal charges. This kind of combination, a single system of royalties and a federal resource rent tax, would appeal to all sides. The federal government would be allowed to implement the resource rent tax it wants to. The states would be assured a percentage of mining revenue before it reaches the federal government. The mining industry would benefit by new stability in taxes and would be assured a minimum rate of return, and it could exploit lower-grade deposits. The nation would benefit as Australia would become a more competitive exporter in the world mineral trade.

FOREIGN INVESTMENT

The Need for Capital

Because of her small population and domestic capital market, Australia has historically relied on foreign capital to develop her mineral industry. This foreign investment is seen as one of the major problems in managing resources development for the benefit of national residents. At issue is how to finance mineral development with the aid of foreign capital, while ensuring an acceptable profit level for Australia.

The nineteenth-century Australian gold rushes were financed primarily with British capital. Today, American companies have invested heavily in Australian bauxite, coal, and oil. During the mineral boom of the 1960s and early 1970s foreign ownership and control of mineral projects nearly doubled. Until 1974–1975, the Australian Bureau of Statistics published annual figures showing the level of foreign participation in the Australian mineral industry (Tables III-5-10 through III-5-12).

Whereas foreign ownership can be accurately measured, foreign control figures are inherently arbitrary and liable to be overestimated.[16] The Australian Bureau of Statistics (ABS) defines an operation as being entirely under foreign control if the foreign direct ownership exceeds 50%, or if direct ownership by a foreign entity exceeds 25% and there is no Australian parent company holding a greater equity interest in the venture.[10] This classification has come under fire by the Australian Min-

TABLE III-5-10   Foreign Ownership in the Australian Minerals Industry

Year	Metallic minerals (%)	Fuel minerals (%)	All minerals (%)
1963	39.8	11.5	27.3
1964	40.6	14.2	30.0
1965	43.6	20.2	34.0
1966	46.3	23.1	37.6
1967	50.0	25.6	40.9
1968	51.0	32.3	44.0
71–72	53.4	51.1	48.7
72–73	52.3	54.0	49.6
73–74	50.8	55.8	49.8
74–75	49.8	59.6	51.8

Source: Ref. 10.

TABLE III-5-11   Foreign Control in the Australian Minerals Industry

Year	Metallic minerals (%)	Fuel minerals (%)	All minerals (%)
1963	53.6	15.5	36.8
1964	52.8	18.4	39.1
1965	57.7	25.5	44.6
1966	62.1	28.7	49.8
1967	64.9	32.5	52.9
1968	68.8	39.7	58.1
71–72	52.8	68.2	55.0
72–73	54.2	71.4	57.7
74–75	56.4	74.5	60.2

Source: Ref. 10.

TABLE III-5-12   Direction of Foreign Ownership

Year	U.K. (%)	U.S. (%)	Total foreign (%)
1963	16.3	9.2	27.3
1964	19.5	8.9	30.0
1965	20.8	9.3	34.0
1966	18.7	15.2	37.6
1967	19.2	18.2	40.9
1968	17.2	23.9	44.0
71–72	16.2	25.5	48.7
72–73	15.7	26.4	49.6
73–74	13.2	27.7	49.8
74–75	11.6	28.7	51.8

Source: Ref. 10.

ing Industry Council (AMIC), as the ABS plans to resume the collection of ownership and control statistics. The AMIC complains that the concept of "foreign control" is not a black-and-white issue but a matter of degree of foreign influence.[17]

Foreign ownership may take the form of a portfolio investment in local companies or by direct foreign investment. Portfolio investments by geographically diverse companies increase the funds available for resource development and are unlikely to lead to actual control.[16] Direct foreign investment involves the establishment in Australia of a wholly owned subsidiary or joint venture operation owned or controlled by overseas parents. Often it is in the form of a minority, but controlling interest in an established Australian company, or a joint project with Australian interests.[16] It is this direct foreign investment that has caused the uproar in Australian political circles.

Effects of Foreign Ownership

Foreign investment not only increases the capital available for resource development, but can also increase the host nation's access to up-to-date technology and managerial skills, and open new market opportunities.[18] The level and form of foreign investment is one of the major factors in determining the rate of domestic resource development and the degree to which that resource benefits Australians.[8]

The real or perceived disadvantage of foreign ownership can be either economical or political.[10] Australians, like the general populations of most resource-rich countries, have developed a distrust of large companies, especially of multinational companies in the mineral and energy industry. For example, when a major oil company reports large profits for a given period, countless Americans from all walks of life cry out that the conglomerate giant is cheating the public. The same is true in Australia, and even more so when the company in question has foreign ownership. What the public, and for that matter many politicians, do not realize is that before a company (foreign or domestic) can make a profit it must pay taxes. With the tax structure for the mineral industry the way it is in Australia, a company must pay a considerable amount of taxes, royalties, charges, and levies before a profit can be made.

Although the problem above is political in nature, tax avoidance is an economic one in which the Australian public has a legitimate grievance. Tax avoidance can be accomplished in the mineral industry through the transfer price mechanism. When a multinational is vertically integrated in a number of countries, it can control where the bulk of its taxes are paid. A case in point is the aluminum industry. The output from each stage of processing, bauxite to alumina to aluminum, has a value termed the "transfer price," which represents the price at which it can be sold on the open market.[10] When a company is involved in several or all stages of processing, it can establish its own transfer price. When a product leaves the country it is this transfer price on which taxes are set. Using the aluminum example, an American company, or Canadian, French, or Swiss, may own the bauxite mine. If the company feels that it will get a better tax treatment in the country in which its refinery and/or smelter is located, it can set an abnormally low transfer price, thereby reducing its Australian tax liability.

Changes in Australian Policy

As a result of the threat, real or imagined, of too much foreign control, the ALP government of 1972–1975 enacted specific guidelines for Australian participation in the minerals and energy sector. As it stands today, projects involving investment in excess of $5 million will be given government approval only where there is at least 50% Australian equity and Australian member on the controlling board of directors. This applies to mining, apart from uranium, and to primary industry projects. If the government deems a project to be in the national interest, and if the government judges that the unavailability of sufficient Australian equity capital on reasonable terms and conditions would unduly delay the development, the project can proceed without 50% domestic equity.[4] For uranium projects (mining, enrichment, or any activity connected with the nuclear fuel industry) 75% Australian equity and control is required. These requirements do not apply in the mineral exploration phase, although foreign companies must notify the Foreign Investment Review Board, the government agency which must approve a project that involves foreign capital if the project goes past the exploration stage.

The goal of a program like this is to reduce the tax losses caused by the transfer price dilemma. If 50% of the project is owned and controlled by Australian interests, it can be expected that they will want to make as high a profit as possible before the mineral is exported. Through this measure the government is ensured its "rightful tax."

During the period when the ALP controlled the federal government, these regulations were strictly enforced. Companies with a high degree of Australian control but with a majority of overseas ownership, such as CRA, felt discriminated against. An argument was mounted that the guidelines offered little or no incentive to foreign-owned companies to offer equity to Australian share holders or to increase the degree of Australian control over its operations.[4]

When the Liberal-NCP coalition returned to office the rules of foreign ownership, although not abolished, were relaxed. In May 1978 the government announced its "naturalization" policy, under which a "naturalized" or "naturalizing" company can enter new projects, either on its own or in partnership with Australian companies, naturalized or naturalizing; or in any combination jointly with foreign companies, provided that the 50% equity and control guidelines are observed.[4]

Naturalized status may be granted if a company is at least 51% Australian owned, if its articles of association provide that a majority of the members of the board are Australian citizens, and if general understandings have been reached about the exercise of voting powers in respect to the company's business in Australia. A company with a minimum of 25% Australian equity and a public commitment to increase equity to 51% may be granted naturalizing status if the company agrees to the other requirements of a naturalized company.[4] For purposes of calculating total "Australian" equity, a naturalizing company is given anticipatory credit for 51% Australian ownership and is regarded as Australian-controlled. A naturalized company is credited with its actual level of Australian ownership and is regarded as Australian-controlled.[4]

Future Policy

The federal government's policy toward foreign investment is aimed at maximizing the economical benefits of mineral exploitation for the Australian people. If in the effort to do this, the politicians deter mineral development by driving away needed sources of capital, they are defeating their own purpose. The best way to ensure the maximum Australian benefit is through taxation, not ownership regulations. The problem of transfer losses can be handled through export controls, a power that the federal government already has. If the transfer price is considered too low, the federal government can intervene and withhold an export license until price adjustments are made. Since the transfer price maneuvers are used for tax purposes, the companies involved would not be burdened with higher costs, only a shifted tax payment.

If Australia is to expand its mineral production, and more important, if it plans to shift into processing, there will have to be massive amounts of foreign capital. This is an economic reality that Australia cannot escape. Considering the technology and markets that this capital entices, the idea of foreign ownership is even more palatable.

INDUSTRIAL RELATIONS

Labor relations in Australia have been a problem for the mining industry and in many ways have contributed to the decline of Australia's competitive position in the mineral export business. Wage inflation is deeply rooted, as Australian labor have been pioneers in the search for a progressively shorter work week without offsetting reductions in labor costs. Australians' propensity to strike has now reached the proportions of a national pastime.[19] Labor costs account for 25.2% of the revenue dollar, and despite the fact that mining earnings were down over the past two years, the average labor cost per employee increased by 14.7% in 1981–1982 alone. Total employment in the mineral industry was down 2% between 1980–1981 and 1981–1982.[1] Of the total number of employees, 30% were in N.S.W., 27% were in Western Australia, 27% in Queensland, and 8% in Tasmania. Table III-5-13 details employ-

ment and cost statistics, Tables III-5-14 and III-5-15 compare wage rates with those of developed and developing countries.

Philosophy Behind Labor

To comprehend the psyche governing the labor union actions requires an understanding of Australia's roots and history. In addition, it is necessary to realize the geography of many of the deposits being mined today.

Being part of the British Empire, Australians brought with them many of the customs and political beliefs of the mother country. Whereas the American colonies gained their independence by

TABLE III-5-14 Comparison of Wage Rates in Basic Metal Industries Between Australia and Developed Countries (Equivalent Australian Dollars per Hour)

	1976	1977	1978[a]
Australia	4.77	5.36	5.63
U.S.	5.53	6.67	7.16
Canada	5.36	6.04	5.82
Japan	3.89	5.21	6.72
Belgium	4.21	5.46	6.38
France	2.66	3.22	3.78
West Germany	3.56	4.58	5.37
Italy	2.42	3.14	NA
U.K.	2.41	2.80	3.49

[a]NA, not available.

Source: Ref. 7.

TABLE III-5-13 Australian Mineral Employment and Labor Costs, 1979-1980 to 1981-1982 (Million Dollars)

	79-80	80-81	81-82
Employment by activity			
Exploration	2,428	2,545	3,579
Mining	52,609	61,613	59,735
Smelting and refining	16,573	18,263	17,887
	71,610	82,330	81,201
Labor cost			
Gross wages and salaries	1,255	1,567	1,896
Payroll tax	61	77	98
Other labor costs	161	201	255
	1,477	1,845	2,249
Recipient of labor costs			
Employees			
Net wages	916	1,137	1,349
Other benefits	161	201	255
Government			
Group tax	339	430	547
Payroll tax	61	77	98
	1,477	1,845	2,249

Source: Ref. 1.

TABLE III-5-15   Comparison of Wage Rates in Basic Metal Industries Between Australia and Developing Countries[a] (Equivalent Australian Dollars per Hour)

	1976	1977	1978
Australia	4.77	5.36	5.63
Argentina	0.33	0.29	NA
Bolivia[b]	0.41	0.52	NA
Brazil[b]	1.18	NA	NA
Chile[b]	0.32	0.65	0.64

[a]NA, not available.
[b]Based on a 48-hour week.
Source: Ref. 7.

revolt, the Australians, like the Canadians, were gradually given self-rule through decree. As a result, Australians have maintained many of the social and political characteristics of the United Kingdom. This includes a strong labor movement and powerful unions.

The Australian population is small, and many of the major deposits are in remote and not entirely pleasant locations. Recruiting a skilled labor force has required attractive pay and/or benefits. This is especially true for Western Australia's iron ore, where to entice the needed labor force, Pilbara mining companies provided low-cost housing and other measures of social infrastructure, such as shopping centers and leisure facilities.[20]

With the proclamation of the "resource boom" in the 1970s, many Australians let it be known that they expected their share of the spoils. This led to wage demands for dividing the prosperity before the boom was given a chance to happen. The softening of the economy and the weak commodity prices in the early 1980s, compounded by demands by both labor and the government, reduced the resource boom to a bust.

Past and Present Problems

Australia has had a long history of labor problems in its mineral industry: from the violent confrontations over dangerous conditions at Bendigo gold mines in the last century, to the 15-month general strike by N.S.W. coal miners in 1929 over wage cuts, to strikes called in the Pilbara over the number of ice cream flavors. Lost labor-hours at the mine and mill site, together with frequent loading and transport strikes, has given Australia the image of an unreliable trading partner.

Australian docks and wharves are controlled by powerful and militant unions. This has caused numerous problems in materials and equipment transport along coastal routes and delays at the dock for mineral exports. Australia almost annually chalks up the worst record in the world for shipping disputes. It is common that 30 or so ships will be waiting off Newcastle or Port Kembla to load coal, with loading delays of 35 days at a time. Work stoppages can occur for almost any reason, from the dismissal of a popular worker, to a means of pressuring a company to use more Australian shipping.[21] The overall effect is that many countries fear that contracted tonnages of Australian minerals will be held up and delayed due to dockside labor disputes.

The coal miners of Queensland and N.S.W. suffer the greatest amount of lost time due to industrial disputes, as mining projects in Western Australia lose only a fraction of the time than that of its eastern counterparts. Figure III-5-5 shows the level of lost time for Australian coal mines. In Western Australian iron mines much of the production lost to strikes is over "town amenities." Between 1970 and 1974, of the strikes in which 5000+ hours of production was lost, 5 to 20% were disputes over town amenities.[20]

Recent events involving labor relations have revolved around the method used to determine wage levels.[22] The Coopers and Lybrand survey showed that between 1980–1981 and 1981–1982, the average labor cost per employee increased by 14.7%[1] (Table III-5-16).

The principal wage increase was a flow on from the metal industry award in December 1981, which resulted in a $25 per week increase and a 38-hour week, followed by an additional $13 per week increase in June 1982. Coal miners' wages and traveling allowances rose by $63 per week, without a change in working hours.[1]

In December 1982, Prime Minister Fraser imposed a 12-month wage freeze on all federal employees and was granted a six-month freeze by the private sector, although he did not get union support. By January 1983, the president of the Australian Council of Trade Unions (ACTU), Cliff Dolan, warned that unions would push for pay demands despite the freeze.[23] With the election of a ALP government in March 1983, the relations between the government and the ACTU should improve, seeing that the new Prime Minister, Bob Hawke, is the former president of the council. Hawke's policies call for consensus and cooperation with the unions rather than the confrontation typical under the Fraser government. He advocates centralized wage fixing, tougher legislation governing trade union practices, and control over nonwage incomes such as salaries, rent, and dividends.[24]

At the National Economic Conference, a high-level forum involving the heads of governments,

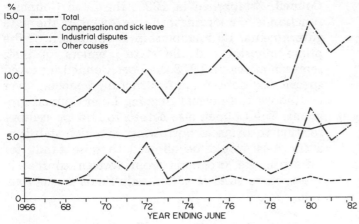

FIGURE III-5-5 Worker-shifts lost as a percent of worker-shifts possible in Australian coal mines, 1965–1982. (From Ref. 23.)

TABLE III-5-16 Australian Labor Costs per Employee, 1980–1981 to 1981–1982 (Dollars per Week)

	80–81	81–82
Employees' net wages	285	318
Other employee benefits	50	60
Group P.A.Y.E. tax	107	130
Payroll tax	19	21
	461	529

Source: Ref. 1.

labor, and industry held in April 1983, a fragile consensus was reached. It would call for a return to centralized wage fixing overseen by the Australian Council and Arbitration Commission.[25] Within the ranks of business, employers operating solely or largely within the domestic market are inclined to favor the centralized system because it affords a simple universal nondiscriminatory way of adjusting wages. Employers in the export sector claim that their competitive position would be damaged by such a system.[25] It is expected that the government will support a modest (around 3%) increase in wages before the end of 1983.

The basic form of the centralized wage system has yet to be determined, as differences remain about the basis of adjustment (Consumer Price Index, movements versus productivity), frequency, form of increases (flat plateau or percentage), discounting factors, and control of other improvements in wages and conditions of employment.[26]

The labor issue will not be solved easily since many of the problems are deeply rooted in the Australian way of life. Instead of any sweeping changes in the way wages are determined or adjusted, what is truly needed is restraint and maturity from all parties involved. Cooperation and communication between labor and management is essential. Both groups have a vested interest in increasing productivity and remaining competitive. Management should seek and utilize more input from labor and develop a better rapport with the unions. Labor needs to adopt a more responsible attitude during times of low commodity prices. These measures could go far in improving the competitive position of the Australian minerals industry; and if the industry is going well, a high level of employment and decent wages can be maintained.

INFRASTRUCTURE

The Importance of Infrastructure Costs

In a nation such as Australia, with its small concentrated population and vast area, the cost and finance of infrastructure costs can make or break a mineral project. During the mining boom of the 1960s, mines were established in remote, inhospitable locations which required the installation of entire communities as well as the physical infrastructure needed at typical mine sites. The support costs of maintaining a town in a remote area were often the deciding factor in starting development or continuing operations.

In the 1960s, for every $1 spent on mine development for iron ore, coal, and nickel deposits, $1.8 had to be spent on infrastructure.[6] Table III-5-17 shows the results of a survey of 11 mining projects developed during the period 1960–1969.

Providing for Costs

Who pays for these costs? Should the infrastructure be funded from the public purse or from private sources? Under the Australian Constitution, the provision of infrastructure within state borders is primarily a state government responsibility. It is usual practice for state governments to construct and operate infrastructure services such as railways, ports, and electricity generation and transmission.[18] In many resource projects the states have provided infrastructure with or without

TABLE III-5-17 Infrastructure Costs for Eleven Mining
Projects, 1960–1969

	Cost (million dollars)	Percent of cost
Ports	125	15.7
Railways and transport	200	25.0
Towns	93	11.6
Power and water	73	9.2
Off mine roads	13	1.6
Local authorities	9	1.1
Airstrips	2	0.3
Subtotal	285	64.5
Mine development	515	35.5
Total	800	100.0

Source: Ref. 6.

direct recoupment of costs. In remote areas, and
during times of closely controlled government
spending, mining companies themselves are frequently expected to furnish the required infrastructure. Where the infrastructure is to be shared
with the states, agreements made typically require
significant up-front payments by the companies
involved.[8]

It is often argued that if infrastructure has no
purpose other than to service a particular mine that
has a limited life, the cost of such infrastructure
should rest on the mining company. Based on this
logic, an isolated project that cannot meet its own
infrastructure costs should not be developed. This
dilemma of whether the government should contribute to these costs comes down to the question:
Will an adequate return on investment be realized?
Will the benefits accrued in financing a new mineral
project (i.e., taxes, employment) outweigh the cost
of a government subsidy for the infrastructure? In
many projects the benefits are substantial, but due
to the risk involved in mining ventures and due to
the limited amount of funds available, many projects
have had to find other sources of funding.

Iron ore mining in Western Australia is the
classic example. Without federal funds (or coordination), each of the area mining interests had to secure
foreign capital. This private ownership of infrastructure led to duplication of rail and port facilities,
as in the Pilbara there are three deep-water ports
and five separate loading berths (Figure III-2-8).
The most unfortunate aspect is that the deepest
port can accommodate vessels only up to 160,000
dwt. This reduces the economies of scale that using
"super" carriers, which can be loaded in excess of
200,000 dwt, would bring.

The federal government is involved in state
infrastructure projects through the Australian Loan
Council. Established in 1927, the Loan Council's
function is to coordinate the borrowing by the state
governments. Its membership is comprised of the
prime minister and the state premiers, or their
representatives. In 1978 the Loan Council approved
special additions for infrastructure financing over
and above the normal ongoing Loan Council programs. The Council also agreed to new procedures
relating to overseas borrowing, in which statutory
authorities would be allowed to raise funds for
infrastructure programs from foreign sources in
cases where sufficient domestic funds are unavailable.[18]

Physical Infrastructure

Physical infrastructure includes such items as
transport mechanisms (rail lines, roads, ports, air-
strips, and pipelines), power generation (power
stations, electrical grids), and water supplies. All
are essential to a mining area and can often be the
major expenditure in developing a project.

Overland transport (rail, roads, and pipelines)
provide the means of moving the ore from the
mine site to a processing plant or port. Railways
are the most economical means of moving massive
amounts of material over long distances. Land
transport of over 100 miles is common in Australia;
therefore, transportation charges can seriously affect the profitability of a mineral project. Table
III-5-18 lists rail distances to move ores and concentrates to principal destinations. In areas where
rail lines do not exist, roads are usually the only
means of transport. Figure III-5-6 is a map showing
railroads and port locations.

Australia has numerous port facilities servicing
the mineral industry. The largest of these handle
coal and iron ore exports. In shipments of these
two high-volume commodities, the quality of a
port can be measured in the size of vessel accommodation and the speed of its handling facilities. In
today's shipping industry, the larger the carrier
used, the lower the delivery price to the consumer,
therefore the greater the exports. Although Australia has a number of ports (Tables III-2-19 and
III-3-13), they are limited in size, especially for
coal. South Africa, with whom Australia is in
competition for both coal and iron ore exports, has
several of the deepest export harbors in the world.
Richards Bay, which handles coal exports, will
accommodate vessels up to 250,000 dwt and handle
44 million metric tons of coal per year when its
expansion program is completed in 1985–1986.[27]
South Africa's large metallic ore port at Saldanha
Bay can receive 250,000-dwt carriers and has provisions for ships up to 350,000 dwt.[27]

Shipping is carried out with both liner ships

TABLE III-5-18   Rail Transport Distances for Selected Mines

Commodity and location	Destination	Distance (km)
Copper		
Mt. Isa, Queens.	Townsville, Queens.	960
Cobar, N.S.W.	Port Kempla, N.S.W.	600
Tennant Creek, N. Terr.	Mt. Morgan, Queens.	2500
Iron ore		
Mt. Newman, W. Austr.	Port Hedland, W. Austr.	427
Mt. Tom Price, W. Austr.	Dampier, W. Austr.	293
Lead		
Mt. Isa Mine, Queens.	Townsville, Queens.	960
Broken Hill, N.S.W.	Port Pirie, S. Austr.	500
Broken Hill, N.S.W.	Cockel Creek, N.S.W.	1200
Nickel		
Kalgoorlie, W. Austr.	Kwinana, W. Austr.	640
Greenvale, Queens.	Yabula, Queens.	200
Coal		
Moura, Queens.	Rockhampton, Queens.	250
Singleton, N.S.W.	Newcastle, N.S.W.	100

Source: Ref. 27.

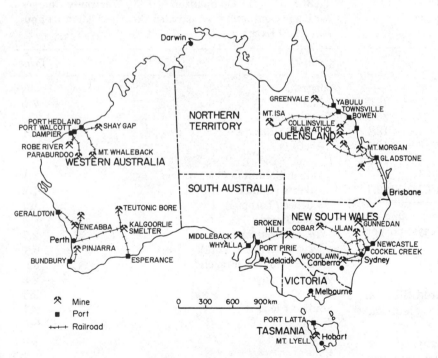

Mine
Port
Railroad

FIGURE III-5-6   Australian mineral rail and port facilities. (From Australian Department of National Development and Energy.)

and bulk carriers. A number of factors affect freight rates. One is the type and size of the vessel; others include port delays, vessel turnaround, availability of backloading, and the cost and availability of bunkers.[7] Tables III-5-19 and III-5-20 provide comparisons of shipping rates for both liner ships and bulk carriers.

Australia is in a good position in terms of developing adequate electric supplies for her mining and processing industries. There is an abundant fuel supply in practically every state. In Queens-land, N.S.W., and Victoria there is coal in large blocks. Western Australia, with the completion of the gas pipeline from the North-West shelf to Perth, should be able to produce reasonably priced electricity. In addition, hydroelectric power is the main source of electricity in Tasmania and could be an important source of power for N.S.W. and Victoria. The total electrical power supply capacity in Australia is estimated at 25,000 MW in 1980. It is expected to increase to 34,000 MW by 1985 and to 41,000 MW by the year 2000. Of this increase,

TABLE III-5-19   Indicative Liner Freight Rates, June 1980 (Australian Dollars per Metric Ton)

Exporting country	Commodity	Importing country		
		Japan (Yokohama)	U.S. (New Orleans)	Europe (Antwerp)
Australia	Lead ingots	90.96	95.56	97.20
(Townsville)	Zinc ingots	58.83	95.56	73.45
	Copper ingots	78.43	84.77	88.25
Canada	Lead ingots	86.71	—	103.58
(Vancouver)	Zinc ingots	86.71	—	103.58
	Copper ingots	84.88	—	84.75
Chile	Lead ingots	78.08	69.10	116.34
(Antofagasta)	Zinc ingots	78.08	69.10	123.44
	Copper ingots	71.02	76.68	131.41
		(Alcapulco)		(Tampico)
Mexico	Lead ingots	68.21	—	62.77
	Zinc ingots	68.21	—	62.77
	Copper ingots	62.58	—	93.97

Source: Ref. 7.

TABLE III-5-20   Comparative Freight Costs in a Handy-Sized Bulk Carrier[a] (Australian Dollars per Metric Ton)

	Distance (miles)	1/1/79	1/1/80
Bell Bay/Hamburg	11,350	24.5	39.7
Rio de Janeiro/Hamburg	5,535	12.8	20.8
Bell Bay/Yokohama	4,920	11.3	18.4
Rio de Janeiro/Yokohama	11,517	24.8	40.3
Groote Eland/Yokohama	3,225	8.1	13.1
Vancouver/Yokohama	4,262	10.3	16.6
Port Elizabeth/Yokohama	8,320	18.3	29.7

[a]Assume: cargo weight, 23,000 metric tons; round-trip voyage with return in ballast, seven port days; no port charges included.

Source: Ref. 7.

15,000 MW will be generated by coal-fired thermal power stations.[7] Table III-5-21 compares electrical rates of several countries.

The real question for Australia in terms of electricity supplies is whether there will be enough generating capacity. During the late 1970s, when the outlook for the Australian aluminum industry looked particularly bright, there were plans calling for the construction of a number of new power plants. Most of these would be necessary to encourage the construction of aluminum smelters at Gladstone, Queensland, Portland, Victoria, Tomago and Lochinvar, N.S.W., and refineries at Worsley and Wagerup, W.A. Unfortunately, at the same time there was public pressure for heavy energy users to be subjected to penalty tariffs rather than price

TABLE III-5-21   Comparison of 1980 Electricity Charges for Large Consumers[a] (Australian Cents per Kilowatt-hour, May 1980 Exchange Rate)

Country	Price
Australia	
N.S.W.	2.0
Queensland	2.8
South Australia	2.1
Victoria	1.9
Canada	
British Columbia (Vancouver)	1.1
Alberta (Edmonton)	1.3
Quebec (Toronto)	1.3
United States	
Texas (Houston)	2.7
New York (New York)	6.1
California (San Francisco)	3.6
West Germany	2.5
U.K.	5.5
South Africa	2.7
Japan	6.6
Brazil	1.3
Chile	3.5–6.8
Mexico	1.5
Philippines	2.2–4.3

[a]Based on 50 MW at 90% load factor.

Source: Ref. 7.

discounts. Victoria's State Electricity Commission was publicly criticized for its proposed lower charges for the proposed aluminum smelter at Portland.[6] After many hearings the project was eventually canceled.

In many ways the issue of relative energy costs

is similar to whether the government should subsidize transportation. It becomes a matter of the cost of providing low-cost electricity versus the return that the smelter would provide when in operation. Bulk consumers of electricity feel that they should pay less, since the cost of providing one block of electricity is less than providing it for a number of smaller users. Some public groups claim that everyone should pay the "real resource cost" of the energy. But as with transportation, there are additional benefits in pricing electricity low to encourage new facilities: namely, new jobs and export revenue.

Water is often an overlooked requirement for mining and processing projects. In remote arid locations such as the Pilbara, the cost of water storage and pipelines is quite high. Water is required for dust suppression on haul roads, and massive amounts of water are used in mineral washing and beneficiating. For example, 200 liters of water is required on average to produce 1 metric ton of washed coal.[28] Securing an adequate supply of water is therefore a high priority and often a high-cost item in developing a new project.

Social Infrastructure

Social infrastructure is in many ways as important as physical infrastructure. This is especially true where the mine site is hundreds, even thousands of kilometers removed from "civilization." Facilities taken for granted in urban areas, such as housing, sewage treatment, schools, and recreation complexes, have to be built from the ground up in frontier areas.

Since 1967, the mining industry has built 24 new towns with a total population of more than 38,000, ranging in size from a few hundred to about 4500.[29] The source of funds for these facilities varies. For projects located near existing municipalities, the federal, state, and local governments provide most of these infrastructure costs. As the project is located farther away from populated areas, the level of government funding decreases. Comalco has spent $47 million by 1983 on the township facilities at its Weipa operation. These facilities include kindergarten, primary, and secondary schools, library, hospital, police station, cinema, tennis courts, supermarket complex, swimming pool, and sporting complex.[29]

During periods of mineral prosperity these costs can be managed by the companies. In times of low commodity prices companies may require additional help from the government. Again the merit of funding these costs must be balanced against the return that an operating project will provide.

ABORIGINAL AND ENVIRONMENTAL PROBLEMS

Aboriginal land rights and environmental protection are related issues in that each will affect the availability of land for mineral development. Although both can add costs and delays to mining projects, each issue has a degree of legitimacy.

Aboriginal Land Rights

The impetus for aboriginal land rights legislation is based on the government's recognition of the basic rights of aboriginals to retain their racial identity and traditional life-style where they desire. The 1977 Aboriginal Land Rights (Northern Territory) Act conferred reserve land on the aboriginal population and established procedures whereby land claims could be made on unalienated land.[4]

Aboriginals and Torres Straite Islanders, who represent approximately 1.2% of the population, now hold title to about 533,000 km^2 in the Northern Territory and South Australia, of which 470,000 is held under unalienable freehold title. When the 214,000 km^2 of aboriginal reserves in Western Australia and Queensland is added, the total "aboriginal land" in Australia accounts for some 9.7% of the total land area of the nation. This includes some 30% of the Northern Territory and 10% of South Australia.[30]

The main impact of the Act has been in the Northern Territory uranium province, where exploration and development have been subject to delays and uncertainties. The Australian Mining Industry Council (AMIC) has opposed the Northern Territory Aboriginal Land Right Act because of four basic deficiencies:[31]

1. An absence of time limits on land claims, which is concerned about how much land (i.e., Northern Territory) will come under aboriginal control.
2. Inadequate land claims procedures, which questions if the land is being granted without proper regard to the national interest or the rights of other users.
3. Inappropriate right of veto over mineral activities, which disagrees with the right of the aboriginals to veto mineral activity on their land.
4. Inappropriate compensation rights, which is said to have led to unrealistic compensation claims for minerals by the aboriginals.

The government's view is that there should be some type of limit on claims, but they see no reason to alter the veto powers or the compensation rights of the aboriginals. The major concern of the Commonwealth is that there is a fair distribution of any royalty payments among the aboriginals.

Environmental Protection

Environmental protection is an issue that first captured the general public's attention during the 1970s. Worldwide concern over the disruption of the environment by industries such as mining spread to Australia and led to the 1974 Environmental Protection (Impact of Proposals) Act. It was confined to projects in which the Commonwealth had some interest or where it was required to give approval (e.g., export approval).

The industry on which environmental protection has had the greatest affect is mineral sand mining. Dredging operations at Fraser Island (1976) and Myall Lakes (1983) were shut down for environmental reasons, thereby closing a large portion of east coast mineral sands production and consequently much of Australia's rutile production. Other noteworthy cases of environmental issues affecting mining have been bauxite mining near Perth and uranium mining in the Alligator Rivers Region in the Northern Territory. Both have received strong criticism for environmental reasons. Most recent has been the concern for valuable topsoil, which it is feared may be lost due to open-cut mining in the Hunter Valley, N.S.W., and in central Queensland.[32]

Environmental controls have led to improved reclamation procedures, better protection of water and soil resources, and a greater concern for wildlife. In the past there have been problems in coordination between state and federal regulations; however, changes have been made so that it is now possible for developers to meet simultaneously both state and federal requirements for environmental impact statements on specific projects.[8]

Both the aboriginal land rights and the environmental protection issues will continue to exist and affect the Australian minerals industry. Whether their affect has been positive or harmful to the nation as a whole has been a matter for debate and will continue to be so for years to come. What is important, regardless of one's point of view, is that there must be stability and certainty in government actions. This is necessary so that mining interests do not waste valuable resources in developing an area only to have it shut down for environmental or aboriginal reasons.

PROCESSING

An important objective of Australian mineral development policy is to encourage further processing of raw minerals domestically as long as it is eco-

nomically feasible with sound industrial development. Mineral processing is a field where Australia, from a resource point of view, is particularly well suited to compete. The nation not only has abundant raw materials (bauxite, iron ore, manganese, copper, lead, zinc, etc.), but also has vast resources of energy minerals to generate electricity. Compared to most developing nations, with whom Australia will have to compete for international investment, Australia has a relatively low rate of inflation as well as a developed technological base and skilled work force.[33] Presently, Australia is more involved in raw mineral production than in refined metals. Those that it does process, bauxite, copper, lead, zinc, and nickel, are typically only to an intermediate form.

Expected Growth for Processed Minerals

Future world demand for processed minerals will depend largely on the growth of the world economy. Predictions made in the 1970s for future mineral production growth often seem unreasonable when looked at today (1983). In 1979, the U.S. Bureau of Mines published *Mineral Trends and Forecasts*, in which they estimated the world demand for selected commodities in 1985 and 2000 (Table III-5-22). From these figures it was determined what the required Australian production capacity would be (1) if Australia were to maintain her 1979 share of processed mineral production, and (2) if Australia were to double her share. Since these studies were made the world has suffered a severe economic recession, and it does not seem likely that the U.S. Bureau of Mines predictions will be realized, at least for the 1985 estimates. To compensate, a third scenario, "half of the maintained share," can be added. By doing this we can (1) still use the "doubled share" figures as an ultra-optimistic level of Australian production; (2) use the "maintained share" figures as either an optimistic level considering present economic conditions or as a pessimistic value of a doubled share if the world economy does not improve; and (3) we can use the "half of maintained share" as the maintained share for a low-growth scenario. Table III-5-23 shows the predictions of the U.S. Bureau of Mines for Australian mineral processing with this third scenario included.

Benefits of Increased Processing

The benefits of further processing can be measured using several criteria:[33]

Value of output, which is an indirect measure of increased opportunities for employment and capital

Employment, direct and indirect

Increased value of exports

TABLE III-5-22 Projected World Demand for Selected Commodities in 1985 and 2000 (Thousand Metric Tons)

Commodity	1976	1985	2000	Percent increase 1976–2000
Aluminum	16,000	31,000	58,000	263
Copper	7,250	10,250	17,500	143
Iron ore	518,000	700,000	1,000,000	93
Lead	3,345	4,300	6,500	94
Manganese	9,954	13,000	20,000	101
Nickel	595	900	1,500	152
Tin	241	250	300	24
Titanium pigment	1,562	2,400	5,000	220
Zinc	5,795	7,000	9,500	64
Steel	678,000	900,000	1,444,000	129

Source: Ref. 33.

TABLE III-5-23 Estimated Capacity of Selected Raw Mineral Plants in Australia (Thousand Metric Tons)

Commodity	Production 1979	Australian production capacity required by 1990 to: Half share	Maintain share	Double share
Aluminum	270	321	642	1,284
Copper	138	136.5	273	546
Lead	216	156.5	313	626
Nickel	46	42.5	85	170
Steel	8,125	5,500	11,000	22,000
Titanium pigment	60	47	94	188
Tin	4.9	3.2	6.3	12.6
Zinc	305	165.5	331	622

Sources: Ref. 33; author.

An evaluation of a number of raw material processing industries in Australia indicates that on average, the increase in value of output that results from processing is approximately 300% of the value of the prime raw mineral. In some instances the increase can be considerably greater (e.g., bauxite to aluminum is around 1400%).[33] Using the "maintained share" (MS) scenario, the exportable mineral surplus in 1990 would be approximately double (at 1979 values) the value of exports of the commodities in 1977 (at 1979 values).[33]

The direct employment benefit is relatively low, as about 12,000 new jobs would be created under the MS scenario. This is due to the capital-intensive nature of the business. Studies of the multiplier effect of a large resource-based project estimated that the direct, indirect, and induced employment multiplier effect is around 4. This would increase the total of new jobs to 48,000, roughly a 1% increase in present employment levels.[33]

Constraints to Increased Processing

The major limitation to future processing in Australia will be the demand in minerals and metals. As the 1981–1983 recession made clear, without demand the prospect for new facilities is next to nil. This was certainly the case for many of the new aluminum smelters that were canceled. If demand does grow, and if Australia is able to capture a share, there are still challenges that have to be faced.

The new facilities required for the MS scenario would call for investments of over $5 billion.[33] This does not take into account inflation, which could greatly affect this value. Australia with its limited capital base would have to attract a major portion from overseas sources. This would test the foreign exchange rate and could create pressure to revaluate the Australian dollar. As a result of the overseas capital influx, the level of direct foreign ownership would probably increase. This would put pressure on the federal government to relax its restrictions on minimum Australian ownership requirements.

Australia would be hard-pressed to keep up with the necessary electricity requirements. As the lead time in building a processing plant is roughly half that of a major power station, Australia may be limited by not being able to meet the electricity demand. Building new power stations would also strain the capital market. To maintain its share of processing, Australia will need at least 1000 MW of new electrical capacity, with a capital cost of about $600 million.[33] Table III-5-24 lists the requirements for the MS case in investment expenditures and electricity generating capacity.

It has been questioned if Australians will be able to provide enough skilled trades persons. If not, it could lead to inflationary wage pressures. There is also the problem that the new facilities will have on the environment. There will either have to be a general relaxation of pollution controls, which is not likely to occur, or there will have to be substantial expenditures on pollution abatement.

Although the benefits of increased processing are great, so are the risks. Building these new facilities will not guarantee a market or the required return on investment, especially if the product is not competitive with other sources. Australia's first big move into further processing may have left it wary, as many of the alumina refinery and aluminum smelter projects had to be canceled. It is argued that the failure of these projects was a result of the recession and that when the economy recovers these projects can continue. This is probably true, but it also points out how vulnerable Australia is to the

ebb and flow of world mineral demand. It also illustrates why its mineral industries must strive to become more competitive, because without government subsidies, only the most efficient producers will survive mineral surpluses.

REFERENCES

1. Australian Mining Industry Council. *Minerals Industry Survey 1982.* Canberra, 1982.
2. "Australian Mining Analysed." *Mining Journal,* 14 January 1983.
3. Gregory, R. G. "Some Implications of the Growth of the Mineral Sector." *The Australian Journal of Agricultural Economics,* August, 1976.
4. Bambrick, Susan. "Australian Energy Policy." Paper presented at the Conference on New Developments in Foreign Trade and Investments, San Francisco, 12 August 1982.
5. Commonwealth/State Joint Study Group on Raw Materials Processing. *Discussion Paper I: The Impact of Fiscal Measures on the Cost and Availability of Capital on Raw Materials Processing in Australia.* Canberra: Australian Government Publishing Service, 1981.
6. Bambrick, Susan. *Australian Minerals and Energy Policy.* Canberra: Australian National University Press, 1979.
7. Australian Trade Development Council. *Minerals Processing, A Comparative Study.* Canberra: Australian Government Publishing Service, 1980.
8. Bambrick, Susan. "Energy Resource Development: Economic Growth and Socio-political Priorities." Paper presented at the Conference on Australian Energy Resource Development, State College, Pa., 5–7 December 1982.
9. Australian Coal Association. "The Present State and Prospect of the Australian Export Coal Industry with Particular Reference to Its Contribution to the Economy: Taxation Policy." Submission to the Senate Standing Committee on Trade and Commerce, Canberra, November 1982.
10. Barnett, D. W. *Minerals and Energy in Australia.* Melbourne: Cassell Australia Ltd., 1979.
11. Western Australian Department of Resources Development. *Minerals and Mineral Development 1981.* Perth: Government of Western Australia, 1982.
12. New South Wales Department of Mineral Resources. *New South Wales Mineral Industry 1981 Review.* Sydney: Government of New South Wales, 1982.
13. "How Labor Plans to Tax the Miners." *The Bulletin,* 3 May 1983.
14. Lancer, John. "New Concessions on Taxation System." *Mining Review,* May 1983.
15. "Miners Focus on Stability." *Mining Review,* June 1982.
16. Australian Department of Trade and Resources. *Australia's Mineral Resources: Development and Policies.* Canberra: Australian Government Publishing Service, 1981.
17. Lawriwsky, Michael. "Some Issues in Foreign Rela-

TABLE III-5-24   Investment Expenditures and Electricity Requirements by 1990 for Raw Mineral Processing

	Investment (million dollars)	Electricity (MW)[a]
Aluminum	1000	630
Copper	215	11
Lead	90	3
Nickel	560	NA
Steel[b]	3050	266
Titanium pigment[c]	130	17
Tin	4	Neg.
Zinc[d]	103	38
	5152	965

[a]NA, not available.
[b]Electric arc furnace.
[c]Chloride route.
[d]Electrolytic.

Source: Ref. 33.

tions and Control of Australia's Mineral Resources." *The Australian Quarterly*, Winter 1982.

18. "Foreign Investment Statistics: Need for Objectivity in New Definitions." *Mining Review*, August 1982.

19. "Australia's Key Issue." *London Financial Times*, 4 February 1983.

20. Thompson, Herb. "Normalisation: Industrial Relations and Community Control in the Pilbara." *The Australian Quarterly*, Spring 1981.

21. Carr, Bob. "When Will the Waterfront Mess Be Cleaned Up?" *The Bulletin*, 26 January 1982.

22. Pritchard, Chris. "Australian Labor Scene, Now Tranquil, Shows Signs of Growing More Turbulent." *The Wall Street Journal*, 5 January 1983.

23. Joint Coal Board. *Black Coal in Australia 1981-82.* Sydney: Joint Coal Board, 1982.

24. "Australia Looks to Labor." *Mining Journal*, 11 March 1983.

25. Walsh, M. "How Hawke and the ACTU Conned Big Business." *The Bulletin*, 26 April 1983.

26. "Restraint Needed." *Mining Review*, July 1983.

27. Nahai, L., and C. Wyche. *Future Trends and Prospects for the Australian Mineral Processing Sector.* Washington D.C.: U.S. Department of the Interior, 1982.

28. South African Department of Foreign Affairs and Information. *Official Yearbook of the Republic of South Africa: 1982*, 8th ed. Johannesburg, South Africa: Chris van Rensburg Publications, 1982.

29. Lewis, John. "Water Resources and Coal Mining." *Mining Review*, July 1983.

30. Vardabasso, Monica. "Mining Towns: An Extra Dimension to Regional Development." *Mining Review*, June 1983.

31. "Aboriginal Land Rights: The Search for a National Concensus." *Mining Review*, March 1982.

32. Lewis, John. "Miners vs. the Farmers and Graziers." *Mining Review*, October 1982.

33. Commonwealth/State Joint Study Group on Raw Materials. *Processing Discussion Paper 2: The National Benefits Arising from Raw Materials Processing.* Canberra: Australian Government Publishing Service, 1981.

Chapter III-6

Conclusions

Where does Australia stand, and what future does she have in the mineral industry? Australia is often called "the lucky country," and considering its vast mineral wealth, it certainly is. Despite the world economy or political pressures or industry competitiveness, Australia still has huge resources of an assortment of minerals. Its bauxite, iron ore, coal, and mineral sand deposits generally occur in large blocks near the surface and relatively close to the coast. Australia has undeveloped base mineral deposits that would be economical to mine in many parts of the world. Australia has huge deposits of diamonds and its gold resources are again looking attractive. Besides its abundant coal reserves, Australia has almost 20% of the world's uranium resources and may soon have natural gas to export. The magnitude of its mineral resources is enormous relative to its domestic needs. Australia is one of the most richly endowed countries in the world.

Australia has done well in developing its minerals. This is reflected by its position in world mineral production. However, the key to its recent growth as a mineral producer has been the emergence of Japan as a major industrial power. Because of the Japanese hunger for raw materials, Australia has become one of the world's largest iron ore and coal exporters but has also become highly dependent on a single market. Unfortunately, the Japanese, in response to higher energy costs and cheaper labor in neighboring countries, are moving out of basic mineral processing. This has left Australia with three options. The first is to hope that most of the market lost due to Japanese industrial changes can be made up by greater exports to other Western Pacific nations. The second would be to try to compete more effectively in the U.S. and EEC markets. Third, Australia can increase its own processing capacity and export a larger percentage of processed metals. The road Australia takes will be different for each mineral, and each option has its own advantages and risks. In most cases it will take a combination of all three.

For all mineral commodities Australia hopes that there will be an increasing demand by the developing nations in the WPR. There has been

some growth, but will it cover the losses of markets in Japan? Australia has historically been a nation with a "wait and see" attitude. Although this may have kept the country out of some questionable projects, it has also meant losses of markets to more aggressive nations, such as South Africa. If Australia decides to wait and see how much the WPR grows, it may be disappointed to find that Japan has moved much of its processing capacity into South America or Canada.

If Australia tries to expand its share of the U.S. and European markets, it may find that the costs inherent in its mineral industry structure will price it out of the market in some instances. Unfortunately for Australia, its labor and government costs are geared to exporting to Japan, where Australia has lower shipping distances than its competitors. With American and European markets, Australia is the nation with the longer shipping distances, but other costs remain high.

To become more involved in processing, Australia faces some of the same problems. However, in addition there will be massive capital requirements. Being unable to meet these domestically, Australia will have to look to overseas sources. Moving into processing, increases the value received for minerals but also increases the risks, and mineral recessions could be even more disastrous to the Australian economy.

The question is: What should Australia do? Instead of looking for real or imaginary scapegoats, Australia must improve its competitive position. To market minerals successfully internationally, a nation must compete with countries that may have competitive advantages over Australia. Although it is unrealistic to expect Australians to work for the poverty-level wages found in South America and southern Africa, relations between labor and management must improve. Often it is not the wage level that affects trade the most, but the lost time caused by strikes. Australia also suffers high costs because of the relation between the state and federal governments. While both compete for the maximum resource rent, the eventual loser is the mining industry, which has to bear these costs. Better cooperation between each tier of government and industry could eliminate many unnecessary taxes, which would be more than made up for by increased production and increased exports. The same holds true for a number of regulations imposed on production and trade.

For minerals such as bauxite, Australia is set to increase its processing capacity, and due to its rich deposits and cheap energy sources can expect that when the world economy improves, so should Australia's export position. With energy minerals such as coal and uranium, Australia can expand exports as much as demand will allow. The limiting factor will be internal costs. Since Australia will have to compete with other developed economies that also want to increase exports, it must be able to maintain reasonable labor and government costs. When the steel slump ends, Australia's future success will depend on greater diversity of markets. Australia has already reached an equilibrium for most of its base-metal exports, although new mar-

kets for processed base metals could be expanded in the Western Pacific.

The two commodities that may hold the key to Australia's mineral future are diamonds and gold. Australia has already become the fourth-largest producer of diamonds, and new interest in gold is likely to improve the country's production vastly. Since both of these commodities have a high unit value and need little in the way of processing, they could greatly increase Australia's mineral revenue while not depending on high production levels.

The two words that sum up Australia's mineral future are competitiveness and exports. If Australia can improve its competitive position, thus opening new and increasing old market shares, the nation can expect increased exports. If there is one element that has been true for all of Australia's important commodities, be they mineral, agricultural, or pastoral, it is that without exports the national economy will stagnate.

PART IV

LATIN AMERICA

Chapter IV-1

Introduction

Latin America is among the most important mineral-producing regions in the world. Only the USSR, Canada, southern Africa, and Australia produce comparable volumes of such a wide range of mineral commodities. Latin American nations are among the world's largest producers of metals. Sizable shares of the world's reserves and resources of a number of strategic minerals are located in Latin America, and it is the principal source of U.S. imports of 10 mineral commodities.[1] The important mineral-producing countries in Latin America rank among the four or five largest sources of U.S. imports of over 30 different mineral commodities.[1]

Mexico was the leading supplier of bismuth, fluorspar, natural graphite, and strontium to the United States in the period 1978–1981.[1] Brazil was the largest beryllium, columbium, and quartz crystal supplier to the United States in the same period. Chile supplies the largest share of U.S. copper and rhenium imports, and Bolivia has been the most important supplier of antimony.[1] The United States is totally reliant on imports of columbium, natural graphite, and strontium.[1]

Latin America produces and exports agricultural and mineral commodities to many Western nations. Brazil is particularly important to Western security interests. Brazil is Latin America's largest and most populous country, South America's largest military power, and the world's eighth-largest market economy.[2] Brazil is in a position to control the South Atlantic sea lanes that bring oil from the Persian Gulf, and strategic minerals from southern Africa, to the United States and Western Europe.[2]

Recently discovered oil and natural gas reserves have increased U.S. interest in developing a "special relationship" with Mexico.[3] The United States is interested in the political and economic stability of Mexico.[3] Mexico's increased role as a spokesman for Third World countries, and the necessity of Mexican input into a peace proposal for war-torn Central America, have also increased Mexico's importance to the United States.[3]

The present world oil glut has underlined the interdependence of Mexico and the United States. Depressed world oil prices and demand have increased pressure on Mexico to modify its November 1980 decision not to export more than half of its oil to any one country.[3] Mexico's August 1981 commitment to sell oil to fill the U.S. strategic petroleum reserve is viewed as a response to the world oil glut. A prolonged oil glut would force Mexico to sell its oil to anyone willing to buy it.[3]

The present world recession and low mineral commodity prices have seriously affected the U.S. domestic mining industry. The 1981 productivity index for total U.S. extractive industries was almost 20% below the 1977 level.[4] These factors contribute to increased interest in Latin America's mineral resource potential.

REFERENCES

1. U.S. Department of the Interior, Bureau of Mines. *Mineral Industries of Latin America*. Mineral Perspectives Series. December 1981.
2. Fishlow, Albert. "The United States and Brazil. The Case of the Missing Relationship." *Foreign Affairs* 60 (Spring 1982).
3. Purcell, Susan Kaufman. "Mexico–U.S. Relations: Big Initiatives Can Cause Big Problems." *Foreign Affairs* 60 (Winter 1981–1982).
4. U.S. Department of the Interior, Bureau of Mines. *Mineral Commodity Summaries 1983*. Washington, D.C.: U.S. Government Printing Office, 1983.

Chapter IV-2

Brazil

Brazil is important to the world mineral economy as Latin America's major producer of a number of mineral commodities. Brazil's 120 million people and 8.5 million km² of territory represent half of South America's inhabitants and total area.[1] Brazil is the world's largest producer of iron ore, columbium, ferrocolumbium, and electronics-grade quartz crystal.[1]

Brazil as a consumer imports large quantities of mineral products, including oil and coal, fertilizer minerals, copper, lead, and zinc. Brazil has achieved notable success in the domestic processing of its mineral output, and in developing domestic mineral resources to reduce dependence on foreign supply sources.[2]

Despite the world recession, Brazil is proceeding with ambitious energy and mineral development plans and is completing projects that would be canceled or delayed in other countries.[3] Investments in energy projects are responsible for about one-third of Brazil's total foreign debt, the world's largest, at over $90 billion.[3] As these projects begin generating revenues, Brazil's foreign debt and balance-of-payments situations may improve.

Brazil's debt problems stem from a combination of factors. Among these are the second oil price shock of 1979–1980, the historically high interest rates, world recession, increased protectionism in Western nations, and precipitous drops in world mineral and agricultural commodity prices.[3] Many of these factors are interrelated.

Brazil is the world's eighth-largest industrial economy.[3] State-owned companies accounted for 70% of all economic activity in the period 1977–1981.[3] Brazil's economy has grown steadily for over 30 years, with an average growth rate in its gross national product (GNP) of 7% per year from 1951 to 1980.[3] Inflation is a recurring problem for Brazil's economy, as is chronic underemployment in Brazil's largest metropolitan centers, Rio de Janeiro and São Paulo.

Brazil's exports are dominated by agricultural commodities, such as sugar, orange juice, soybeans, and coffee, and by mineral commodities, such as iron ore, manganese, steel, aluminum, and ferroalloys.

MINERAL RESERVES AND RESOURCES

The mineral potential of Brazil's large land area has yet to be fully evaluated. Numerous geologic similarities link Brazil to the major mineral-producing areas in Zambia and the Cape Province in southern Africa.[5] Geologists consider that Brazil contains numerous undiscovered mineral deposits.

Brazil has experienced a measure of success in expanding the country's mineral reserves. New coal, copper, nickel, lead-zinc, tin, and phosphate deposits have been discovered in recent years.[2] Brazil's mineral reserves, as of the end of 1980, are presented in Table IV-2-1. More recent information is included where available. The primary source is the *Anuário Mineral Brasilerio* (Brazilian Minerals Yearbook) *1981*.

TABLE IV-2-1 Brazil: Mineral Reserves (Thousand Metric Tons)

Mineral	Reserves
Asbestos	5,000
Beryllium	15,000
Bauxite	4,700,000
Columbium ore	267,642
Chromite ore	10,000
Gold	30
Iron ore	10,831,863
Lithium	8,800
Manganese ore	273,000
Nickel ore	340,000
Phosphates	1,540,233
Quartz crystal	8,264
Tin	99,588
Tungsten	1,836
Zinc	18,744
Coal	23,000,000
Natural gas[a]	60,000,000
Petroleum[b]	1,495,000

[a]Unit of measure is million cubic meters.
[b]Unit of measure is thousand barrels.

Source: Anuário Mineral Brasileiro 1981; Ref. 1.

GEOLOGY

Brazil is geologically similar to the important mineral-producing areas in Zambia and the Cape Province of South Africa.[5] The Carajas mineral province contains deposits of a wide variety of metallic minerals.

The Serra dos Carajas is a mountain range consisting of a group of ridges rising out of the Amazon jungle. Sparse underbrush (*canga*) is the surface manifestation of underlying iron ore deposits.[6] This ore is blue to dark gray in color, and in mainly $3/8$-inch to 60-mesh-size material.[6] Hematite is the major iron ore mineral in Brazil.

Manganese deposits in the Carajas region are typically layered, with a surface layer of detrital ore, a manganiferous laterite layer, and an underlying carbonate protore.[6] The minerals in the Carajas manganese deposit (Igarape Azul) include cryptomelane, brinessite, lithiophilite, gibbsite, goethite, and kaolinite.[6]

Carajas copper deposits are of the stratiform type. The deposits show lithologic control, and copper sulfides are associated with schists containing magnetite, amphibole, garnet, olivine, and biotite.[6] Copper oxide minerals are absent from the deposits.[6] The deposits are of lower Proterozoic age, and at least one (Inco's Chapada prospect) contains associated gold mineralization.[5]

Brazil has large nickel laterite deposits, and one sulfide deposit has been identified. The Carajas nickel deposits form long hills, composed of mafic and ultramafic rocks.[6] The deposits' lateritic profiles have four zones: a laterite layer on top, clay, saprolite, and serpentinite.[6] The sulfide deposit contains copper and nickel.[5]

Lead-zinc deposits in Brazil are of the stratiform type. Some deposits possess copper mineralization.[5] Deposits occur in carbonate rocks. Major ore minerals include sphalerite and galena, with some pyrite and pyrrhotite.[5] The Upper Proterozoic deposits may also contain barite.[5] Large areas of Brazil exhibit sulfide mineralization, and copper-lead-zinc deposits are likely to occur in several areas.[5]

Gold deposits are alluvial and are usually associated with metasedimentary volcanogenic sequences.[6] The deposits exhibit lithologic, tectonic, and structural control.[6] Gold is concentrated in siltstones, at the gray-red siltstone interface, in manganiferous breccias, and in zones of hinge folding and faulting.[6] Proterozoic quartz-pebble conglomerates, of the Witwatersrand type, have been identified.[5]

Brazilian tin deposits are mainly alluvial, derived from phaneritic, granitic, plutonic bodies of the Protozoic age.[5] Brazil also has low-grade porphyry tin potential. Tin and tungsten occur in granitoid intrusive suites near Carajas.[6] The tungsten mineral wolframite occurs principally in alkaline granites.[6]

Most of the world's known columbium reserves are in the Araxa carbonatite complex in Brazil.[7] Zones of deep weathering and strong residual enrichment contain the highest-grade ore.[7] The principal ore mineral is pandaite, a bariopyrochlore. Rare earths are present in the deposit in the minerals monazite and gorceixite, and ilmenite and barite are also present.[7] The carbonatite complex forms a ring structure that was arched into a dome by an alkaline intrusion in the Upper Cretaceous.[7] Large phosphate deposits occur in fine-grained clastic rocks. Phosphates are also extracted from carbonatite deposits.[5]

Brazil's geologic similarity with southern African mineral provinces suggests that many as yet undiscovered mineral deposits probably exist in Brazil. As the world economy recovers, exploration and development work should improve geologists' understanding of Brazil's geology and mineralogy.

LOCATIONS OF MINES AND PROCESSING FACILITIES

Mining and mineral-processing facilities are spread over most of Brazil. The Amazon basin is now being opened up to extensive mineral development. About half of Brazil's mineral output is from the southeast part of the country, and about one-third of all mineral output is from the northeast part of the country. Brazil's major mines and mineral processing centers are shown in Figure IV-2-1.

Mining activity began in coastal areas and is now spreading to the large, undeveloped interior of the country. Among the development areas in Brazil's interior are the Rondonia tin district near the Bolivian border, the bauxite mines near the Amazon River, and the Grande Carajas project in the Amazon basin.[1] The most important of these is the Carajas project, which will open up a sizable area to development and which will produce a wide range of mineral commodities.

Mineral-processing facilities are located in coastal areas and in and around the large cities in southeastern Brazil, where the demand for mineral commodities is large. New mineral-processing centers are under construction in northeastern Brazil

FIGURE IV-2-1 Brazil: Mines, processing facilities, and infrastructure. (From Ref. 1.)

and the Amazon region. Mineral-processing plants are often sited near ore-producing areas.

Major steel and iron pellet plants are located near the Iron Quadrangle and the rail lines and ports that service the area.[1] Proximity to transport and port infrastructure is important in determining the location of mineral-processing centers. Alumina and aluminum projects are concentrated near the bauxite producing and exporting area on the Amazon River.[1]

Existing mineral-processing facilities and plants currently under construction in Brazil are listed in Table IV-2-2. Mineral-processing centers are becoming more important to Brazil's economy as mineral exports shift from ores and concentrates to refined metals and semimanufacturers.

PHYSICAL INFRASTRUCTURE

Energy

Brazil is about 60% self-sufficient in energy, the balance being supplied by oil, natural gas, and coal imports.[4] Brazil has a wide variety of energy resources, including coal, natural gas, petroleum, hydropower, oil shale, alcohol, bagasse, wood, and uranium. Brazil's 1980 energy balance was as follows: oil (41%), hydropower (28%), wood (17%), bagasse and coal (5% each), alcohol (2%), and gas and charcoal (5% combined).[1] Oil should decline in importance in the future as domestic coal, hydroelectric, and uranium resources are developed.

Brazil has developed about 12% of its 200,000-MW hydroelectric potential. In 1980, 86% of

electric generating capacity was based on hydropower, with the balance from fossil-fuel-fired plants. Hydroelectricity will supply close to 40% of Brazil's energy requirements by the mid-1980s.[4]

The two large hydropower projects at Itaipu and Tucurui will give Brazil over 20,000 MW of new hydroelectric generating capacity by the late 1980s. Itaipu is a joint Brazilian-Paraguayan project on the Parana River generating electricity in 1983, with 12,600 MW of capacity to be achieved later in the decade.[8] The Tucurui project on the Tocantins River will supply 8000 MW to the Amazon region, with a large part going to power the Grande Carajas mineral project.[8]

Brazil's nuclear energy development plans call for the construction of three power stations (3000 MW total) at Angra do Reis,[9] a site near Rio de Janeiro. Nuclebras, the state nuclear energy company, is also planning two reactors near São Paulo and a total of 15 nuclear reactors by the end of the century. Shipments of yellow cake from the Osamu Utsami mine to the first reactor at Angra do Reis began in 1981.[9] Pechiney Ugine Kuhlmann (PUK) is constructing a 500-million metric ton per year (mtpd) (later 2000 mtpd) uranium enrichment plant at Resende, utilizing West German technology, with startup set for 1985.[9] The two reactors near São Paulo have been shelved indefinitely.

Coal resource development has been given priority status by the Brazilian government. Current output of 5 million metric tons per year is projected to increase to 16 or 17 million metric tons

TABLE IV-2-2 Brazil: Major Mineral-Processing Facilities

Mineral-processing facility	Capacity (metric tons/year)[a]
Aluminum	
Pocos de Caldas	90,000
Mairingue	90,000
São Luís	100,000p
Barcarena	80,000
Candeias, Bahía	58,000p
Recife	50,000
Santa Cruz, R.J.	80,000
Copper	
Camacari	150,000p
Rio Grande do Sul	60,000p
Ferroalloys	
Simões Filho	150,000
Araxa	18,000
Pojuca	90,000
Iron and steel	
Tubarão (pellets)	72,000,000
Belo Horizonte (steel)	3,500,000
Volta Redonda (steel)	2,500,000
Piacaguera (steel)	2,500,000
Lead	
Bahía São Paulo	35,000
Manganese	
Porto Santana	1,500,000
Phosphate	
Uberaba	900,000
Tin	
Volta Redonda	7,000
São Paulo ⎫	
Manaus ⎭	9,000
Zinc	
Tres Marias	50,000
Juiz de Fora	70,000

[a]p, preliminary.

Sources: Refs. 1, 9, and 15.

per year by 1985.[9] These goals are generally considered to be somewhat ambitious; Brazil's coal is generally of low quality (low Btu and high ash), and therefore the country will probably continue to import coking coal to meet the needs of the expanding steel industry.[8]

Brazil's oil production has been increasing slowly in recent years. Imports in 1979 were 370 million barrels (86%) against domestic output of just 62 million barrels (14%).[4] Large oil import bills have promoted Brazil's development of a wide range of domestic energy resources. Natural gas production doubled from 1976 to 1980, reaching 50 billion cubic feet. Gas plays a minor role in Brazil's mineral supply position, accounting for less than 1/2 of 1% of the country's total primary energy consumption.[4]

Brazil's energy-import situation promoted the Alcohol Fuels Program, begun in 1976.[8] Production in 1982 was 4.2 billion liters, a 13% increase over 1981.[8] Some automobiles run on hydrated alcohol. Anhydrous alcohol is mixed with gasoline in a ratio of 5 liters of gasoline to 1 liter of alcohol, for combustion in cars and trucks. The alcohol production goal for 1985 is 10.7 billion liters.[4]

Brazil is developing a synthetic fuel program based on extracting oil from shale. The pilot plant at Irati operated smoothly in 1980, producing almost 20 million cubic meters of oil and 2000 tons of sulfur.[4] Engineering studies have been completed for a large-scale shale oil plant at São Mateus in Paraná state.[4] Petrobras' Research and Development Center performed the engineering and design work for the project.

Brazil still obtains over one-fourth of its energy needs from wood, bagasse, and charcoal. This share should remain fairly constant. Trends in Brazil's energy supply position should see oil's importance decline as hydroelectric capacity is increased in the 1980s. Coal's share of Brazil's balance is forecast to double between 1980 and 1985.[1] Nuclear power should make some contribution to Brazil's energy supply by 1985.

Water

Brazil has abundant water resources. These resources are used to supply water to residential and industrial users, to irrigate agricultural lands, to produce hydroelectricity, and as the primary means of transportation in the Amazon region.[2] Brazil's system of navigable inland waterways is extensive.

Brazil recognizes the need to study and plan for the best use of the country's water resources. The Companhia de Pesquisa de Recursos Minerales (CPRM) has established 2000 hydrometric stations throughout the country to monitor water flows and establish a data base from which a national water resource plan can be developed.[2]

The development of mining, agriculture, and forestry in the Amazon basin should increase the importance of Brazil's water resources. The Tocantins and Xingu Rivers both flow near the Carajas mineral development area.[1] The Araxa niobium mine uses mainly recycled water, with makeup water from a nearby freshwater lake.[7] Northeastern Brazil has been experiencing a severe drought since 1919.

Transportation

Brazil's mineral production is transported by a number of different haulage systems. The shipping of minerals by way of Brazil's inland waterway and rail systems is the mining industry norm. Some mine and mill output is shipped by slurry pipeline, conveyor belt, and truck. Mineral transport infrastructure and ports are shown in Figure IV-2-1.

Brazil's efficient and reliable system of railroads serving the Iron Quadrangle is among the factors responsible for the growth of the iron and steel industry.[1] The Vitoria-Minas Railroad (EFVM) operates a 550-km double-track rail line from producing areas to three sea terminals on the coast.[1] The Central Railroad System operates one rail line from the Iron Quadrangle to a port on the Atlantic coast in Sepetiba Bay.[1] Ferrocolumbium (ferroniobium) and niobium oxide produced at the Araxa mine complex are shipped by rail to Rio de Janeiro.[7] A short rail line connects the Serra do Navio manganese deposit and the port at Porto Santana.[4]

The Grande Carajas mineral project will be linked to the new port near São Luís by a 900-km rail line currently under construction. Iron, manganese, and bauxite ores will be shipped on this rail line. Future development of the copper and nickel deposits, located 60 km and 20 km, respectively, from the main rail terminal will utilize existing rail and port facilities.[6] A 150-km branch line will be constructed from the bauxite deposits near Paragominas to the main Carajas-Ponta da Madeira rail line.[6] A total of 11 km of bridges, including a 2.4-km span across the Tocantins River, will be erected during the construction of the main rail line.[6]

Brazil has an extensive road system, covering about 1.5 million kilometers, of which only about 5% is paved.[1] Major highways include the Trans-Amazon road, which stretches west from Recife for over 3000 miles across the Amazon basin.[1] It is the major east-west highway in Brazil. Another major highway is the Belém-Brasilia road.[1] These two major highways connect the north, center, and south of the country to Maraba, about 200 km northeast of the Grande Carajas project.[6]

Roads are used to transport a portion of Brazil's mineral output. Truck shipment is reserved primarily for mineral products that will be consumed domestically, and for semimanufactures and finished goods. The Trans-Amazon road has experienced many problems since construction began in 1973, earning the project the nickname *trans-amargura* or "road of bitterness."

Samarco Mineração (Samarco) operates a 400-km slurry pipeline to transport iron ore from its mine in the Iron Quadrangle to the port at Ponta Ubu.[1] The pipeline was completed in 1977. Oil, natural gas, and petroleum products pipelines connect import terminals and areas of internal consumption.

Brazil's mineral production is shipped through a number of ports along the Atlantic coast and along the Amazon River. Bauxite exports are shipped from Porto Trombetas, and manganese ore exports are shipped from Porto Santana.[1] The port at Vila do Conde (near Belém) should be completed in 1985, with a 60,000-dead weight ton (dwt) capacity.[6] When completed, Vila do Conde will join Porto Santana and Porto Trombetas as the major mineral ports in the Amazon region.

The principal port for exports of Carajas mineral production will be the 300,000-dwt port at Ponta da Madeira.[6] A secondary mineral transport system would entail rail shipment from Carajas to the Tocantins River, and water transport to Vila do Conde.[6] The port of Itaqui near São Luís can handle 60,000 dwt vessels and could be used to ship some of the Carajas region's mineral output.[6] The mineral export ports on the Atlantic coast are Atalaia, Paul, Tubarão, and Ponta Ubu.[1] Other ports on the Atlantic coast that handle mineral imports and exports include Porto Alegre, Santos, Rio de Janeiro, Salvador (Bahía), and Recife.[1]

Brazil's long coastline and extensive system of navigable inland waterways facilitate the transport of bulk mineral commodities. The rail lines that serve mineral-producing regions are both efficient and reliable.

MINING LABOR

The mining industry labor force in 1980 was 83,200.[10] Another 60,000 workers are employed by the state oil company Petrobras.[1] The Brazilian steel industry employs about 150,000 workers.[1] Over one-third of all mining industry workers are in the state of Minas Gerais.[10]

The mineral industries employing the greatest number of workers are iron (12,400), coal (12,000), tin (4200), and manganese (2500).[10] Gold prospectors (*garimpeiros*) are not included in these statistics.

Most mining activity in Brazil utilizes capital-intensive methods.[1] The demand for mining industry laborers is relatively low compared to Brazil's total labor force. Supply and demand for workers are basically in balance in the mining industries.

Mining industry literature has reported no major strike activity or work stoppages in the past several years. Mining industry labor relations are relatively good compared to the mining industries of other Latin American countries.

DOMESTIC MINING INDUSTRY

Overview

Minerals and mineral fuels production in 1980 was valued at U.S.$5.4 billion.[10] In constant cruzeiros, the value of Brazil's mineral and fuel production grew over 40% in 1980.[10] Mining's share of GNP was about 2.2% in that year, and mineral exports accounted for about 10% of total exports.[10,1] Mining labor represents less than 1% of Brazil's total work force.

Brazil's 1980 mineral output, on a value basis, was divided among fuels (41.6%), metallic minerals (31.6%), nonmetallics (25.8%), and gems and diamonds (1%).[10] The 1980 value of Brazil's mineral production, by commodity, is presented in Table IV-2-3.

Although Brazil's mineral and fuels production does not represent a large part of the country's GNP, mineral exports figure prominently in export totals. Brazil's economy is dominated by the agricultural and manufacturing industries.[1]

Foreign investment capital for mining industry projects comes mainly from the United States, Japan, and Western Europe.[1,4] As new mining and mineral processing projects come on-stream, minerals will play a more important role in Brazil's overall export and balance-of-payments positions. Foreign investment in the mineral and fuels industries was concentrated in basic chemicals and metallurgy projects.[4] Mineral extraction, iron and steel, and petroleum products industries also attracted foreign investment capital.[4] About one-fourth of all foreign investment in Brazil in 1979 was in minerals, fuels, and related industries.

Trends Among Major Mineral Producers

Brazil's output of most mineral commodities has grown steadily in recent years. Despite precipitous drops in many commodity prices, the real value of the output of the mineral and fuel indus-

TABLE IV-2-3   Value of Brazil's 1980 Mineral Production by Commodity (Percent)

Oil	35.0
Iron	17.6
Gold	4.5
Natural gas	3.9
Natural phosphates	3.3
Coal	2.7
Columbium	2.1
Aluminum	1.9
Tin	1.8
Manganese	1.8
Other	25.4

Source: Anuário Mineral Brasileiro 1980.

tries increased by 13% in 1980.[4] Brazil became a uranium producer with the opening of the Osamu Utsumi mine in 1981.[9] Output in 1982 was about 350 tons of U_3O_8.

The most important trend in Brazil's mineral trade is a large increase in the level of processing of mineral exports.[4] From 1977 to 1979, the portion of mineral output exported as semimanufactures (crude steel, ferroalloys) increased from 28% of the total to 45% of the total.[4] The annual rate of increase for those two years was 30%. Higher levels of processing enable Brazil to capture the value added by the processing, increasing foreign exchange earnings and employment.

In 1980, Brazil achieved significant growth in the output of steel (10.3%), ferroalloys (14%), coal (14%), and phosphates (60%).[4] Primary aluminum metal production has increased from 56,000 tons in 1970 to 260,000 tons in 1980.[4] Brazil's tin metal output is expected to continue to grow rapidly.[11] Mine production of zinc ore, manganese ore, and bauxite has grown steadily in the last five or so years.[4] Oil and natural gas output grew 74% and 99%, respectively, in terms of real value of production in 1980.[10] Gold production increased by 262% in 1980, largely due to increased mining activity by *garimpeiros*, gold prospectors.[10]

Brazil's iron and steel and ferroalloys industries have exhibited rapid growth in recent years. Brazil produces over 15 different ferroalloys, including ferrocolumbium, ferrochrome, ferromanganese, ferrotitanium, ferrotungsten, and ferrovanadium.[4] Brazil produces about five-sixths of the world's ferrocolumbium.[12] Ferrocolumbium production grew 36% in 1979 and 26% in 1980.[4] From 1971 to 1980, total ferroalloy production grew from just over 125,000 metric tons to over 550,000 metric tons. In the same decade, ferroalloy exports increased by a factor of about 7.5, to over 175,000 metric tons.[4]

Brazil's mineral and fuels industries have posted, and probably will continue to post, impressive annual growth rates. Base-metal production, especially copper, lead, and zinc, will grow to reduce Brazil's reliance on imports of concentrates of these metals.[4] Ten of Brazil's 21 ferroalloy producers are expanding their capacities, specifically in ferrocolumbium, ferrosilicon, ferrosilicomanganese, and silicon metal.[4] Brazil's mineral output will continue to grow through the 1980s.

Industry Structure

The Brazilian government dominates the country's mining and fuels industries and economy in general. State-owned companies accounted for about 70% of all economic activity in Brazil from 1977 to 1981.[3] The two dominant state mineral and fuels companies are Companhia Vale do Rio Doce (CVRD), the world's largest iron ore producer and exporter, and Petroleo Brasileiro S.A. (Petrobras), the state oil company.[4] Companies associated with CVRD produce other mineral commodities, including bauxite, phosphates, and manganese.[4] Both of these companies are mixed companies, with private Brazilian interests holding 20% of CVRD and 25% of Petrobras. Petrobras ranked second among Brazil's most profitable companies in 1982, and CVRD ranked eighth.[13]

A total of about 670 mostly small and medium-sized mining companies are active in Brazil.[1] About three-fifths of total mineral output is produced by the largest 50 companies. Of these 50 companies, three are state firms, 22 are private Brazilian, and 25 are foreign owned. Brazil's mineral output is divided among state companies (one-fifth of the total), private Brazilian firms (two-fifths), and foreign firms (two-fifths).[1] Foreign firms are important in the extraction of iron ore, manganese, columbium, gold, tungsten, nickel, beryllium, and tin.[1]

About three-fifths of Brazil's steel output is by Siderúrgica Brasileira S.A. (Siderbras), a group of mixed steel companies.[4] A total of about 43 companies are engaged in steelmaking in Brazil.[1] Brazil's 21 ferroalloy producers are organized into the Associação Brasileiro dos Productores de Ferro-Ligas (Abrate).[4]

Ferrous Metals. Brazil's iron ore and pellets industry is dominated by CVRD, the state iron ore company. The second-largest iron ore producing company is Mineração Brasileiras Reunidas (MBR), a private Brazilian company. The shares of 1980 iron ore output produced by the leading eight firms are presented in Table IV-2-4.

Brazil's iron ore industry is highly concentrated, with a four-firm concentration ratio of over 85% and an eight-firm ratio of 99.6%. When Minas da Serra Geral S.A.'s Capanema project reaches its 11.5-million ton per year capacity, the iron ore industry will be even more concentrated.

Brazil's steel and ferroalloy industries are less concentrated. Forty-three firms produce steel in Brazil. Five firms produce iron ore pellets: CVRD; Nibrasco, Itabrasco, Hispanobras (three ferroalloy producers), and Samarco.[4] Production has grown steadily, increasing from 67,000 metric tons in 1978 to over 100,000 metric tons in 1980.[4] Brazil's two ferrocolumbium producers have also increased output in recent years. Companhia

TABLE IV-2-4   Shares of Iron Ore and Pellet Shipments, 1980

	Percent of total shipments
1. Companhia Vale do Rio Doce (CVRD)	52.9
2. Mineração Brasileiras Reunidas (MBR)	15.3
3. Nibrasco, Itabrasco, and Hispanobras[a]	9.3
4. Ferteco Mineração S.A.	8.2
5. SA Mineração da Trinidade (Samitri)	5.2
6. Samarco Mineração S.A.	4.4
7. Companhia Siderurgica Nacional (CSN)	3.1
8. Wm. H. Mueller S.A.	1.2
Other	0.4
Top 4 firms	85.7
Top 8 firms	99.6

[a]Ferroalloy producers.

Source: U.S. Bureau of Mines, Mineral Industries of Brazil; reprint of 1980 U.S. Bureau of Mines, Minerals Yearbook.

Brasileira de Metalurgiae Mineração (CBMM) produces about five-sixths of total output, while Mineração Catalão de Goias SA produces the remainder. 1980 production of 17,530 metric tons was over 70% above 1978 output.[4]

Brazil's titanium industry is dominated by a handful of firms. One firm, Nuclebras de Monazita, produces ilmenite and monazite from sands, and is Brazil's sole producer of ilmenite and rutile concentrates.[4] Titanio do Brasil is the only TiO_2 producer, and output is being increased to 50,000 tons per year.[4] Three firms produce ferrotitanium in Brazil; their production increased about 350% in the period from 1967 to 1979. The firms are Electrometalur S.A., Industriae Comercio, and Termoligas Metalúrgicas S.A.[4]

Two firms dominate the manganese industry in Brazil. Industries e Comercio de Minerios S.A. (Icomi) produces about three-fifths of Brazil's total output of manganese ore.[1] The company is owned by Brazilian interests (51%) and Bethlehem Steel Corp. (49%) of the United States.[1] The other large manganese producer is the CVRD subsidiary Urucum Mineração S.A.

Brazil's nickel industry is dominated by one firm. The major nickel producer is Morro do Niguel S.A. Owned by private Brazilian interests and the Hochschild Group, the firm produces about 70% of Brazil's nickel output.[1]

Other Metals. Two firms, Mineração Boquira S.A., and Plumbum S.A., produce all of Brazil's lead output, and Companhia Mineira de Metais (CMM) produces most of Brazil's zinc.[1] Other refined zinc producers are Companhia Industrial e

Mercantil Inga S.A. and Companhia Paraibuna de Metais.[4] The two largest copper firms in Brazil are Caraiba Metais and Fluma (in which Noranda Mines has a minority equity interest).[1]

Four firms dominate the Brazilian tin industry. Paranapanema S.A. is the largest producer, followed by Mineração Brumadinho S.A., Mineração Brasilense S.A., and Mineração e Prospeccões Minerais S.A.[4] Among the firms producing tungsten ores and concentrates are Mineração Tomaz Silustino, Tungstenio do Brasil Minerais e Metais, and Mineração Geral.[9] All of these firms are engaged in expansion programs.

Most of Brazil's gold output is produced by prospectors (garimpeiros).[4] Two mines produced about 4.5 tons of gold in 1980.[4] The two operating companies are Morro Velho Mineração S.A. and Mineração Tejucana S.A.[4] The latter company also produced about 65,000 carats of diamonds in 1979.[4]

Fuels. Brazil's oil industry is in the hands of Petróleo Brasileiro S.A., a mixed company owned by the government (75%) and private Brazilian interests (25%).[1] Crude oil and natural gas production and petroleum refining are under the control of Petrobras and its subsidiaries.

Two firms produce about three-fourths of Brazil's coal output. Carbonífera Próspera S.A. (a subsidiary of Cia. Siderúrgica Nacional) produces almost half of Brazil's coal, and Companhia Riograndense de Mineração (CRM) produces close to one-fourth of total output.[1] CRM is owned by the government of the state of Rio Grande do Sul.[1] A total of about 16 firms are active in coal mining in Brazil.[4]

Bauxite/Alumina/Aluminum. Brazil's bauxite production has increased from under 1 million metric tons per year in 1976 to 6.7 million metric tons in 1980.[12] The major producer is Mineração Rio do Norte S.A. (MRN), a consortium of CVRD and other domestic and foreign companies.[4]

The alumina-aluminum industry is in the hands of a few companies with Brazilian, Japanese, and U.S. ownership. The largest companies are Companhia Brasileira de Aluminio S.A. (CBA), Companhia Mineira de Aluminio S.A. (Alcominas), Aluminio Brasileiro S.A. (Albras), Alúmina do Norte do Brasil (Alunorte), Valesul, and Mineração Vera Cruz S.A. (MVC).[4,1]

Two U.S. firms already involved in Brazil's aluminum industry, Alcoa and the Hanna Mining Corp., plan to increase alumina and aluminum production by building an integrated complex at São Luís.[4] The Canadian firm Alcan operates a smelter at Bahia.[9] Japanese involvement in the Brazilian aluminum industry is in the Albras-Alunorte project, being developed in a consortium with CVRD.[9]

Vertical/Horizontal Integration. With most mining industries dominated by a few state, mixed, private Brazilian, and/or foreign firms, levels of horizontal integration are high. The only exceptions are the steel and ferroalloy industries, which have a few dozen firms, and the gold industry, where the bulk of output is from small-scale prospecting operations.

All of the state and mixed companies that produce iron and manganese ore and steel are in a sense integrated. The government-controlled steel companies that form Siderbras are 80% owned by the government.[1] Carbonífera Próspera S.A., the largest coal producer in Brazil, is a subsidiary of one of the state steel firms.[1] CVRD is integrated into both iron and manganese ore production.[4] CVRD is also integrated into the production of natural phosphates.[4] Ferbasa (private) is the country's largest chromite ore producer and sole ferrochrome producer.[4]

The country's dominant columbium (pyrochlore) producer, CBMM, no longer exports concentrates, shifting production to ferrocolumbium and technical grade (minimum 98% Nb_2O_5) columbium oxide.[4] CBMM is owned by Moly Corp. (U.S. 33%), Moreira Salles (51%), and Pato Consolidated Gold Dredging Ltd. (16%).[1] The integrated complex at Araxa produces 86% of the world's columbium.[12] 1980 production was 30,700 metric tons of concentrates (18,380 metric tons of contained metal).[10]

Brazil's major zinc producer, Companhia Mineira de Metais S.A., produces zinc ore, concentrates, and refined zinc at the integrated mine-mill-smelter complex at Vazante, M.G.[4] Caraiba Metais is constructing a smelter to complement existing mine and concentrator facilities in Bahía.[9]

The tin-producing company Mineração Brumadinho S.A. is integrated into tin mining, concentration, and smelting.[9] The lead company Plumbum is integrated into mining, refining, and smelting at operations in Bahía.[9]

Two large integrated alumina-aluminum complexes are in the construction or planning stages. The Albras-Alunorte (CVRD-Japanese) project at Barcarena near Belém will eventually expand to 320,000 metric tons per year of aluminum (75,000 metric tons initially) and 800,000 metric tons per year of alumina (600,000 metric tons initially).[9]

Minero Rio Norte is planning an integrated alumina-aluminum complex to produce initially

100,000 metric tons of aluminum and 500,000 metric tons of alumina annually, with expansion to 300,000 metric tons of aluminum metal and 1.5 million metric tons of alumina by the end of the decade.[9]

Government Role in the Mining Industry

The Brazilian government controls the mining and fuels industries in a number of ways. In addition to owning many producing and processing firms, the country's tax policy and mining law and policy provide the mineral industries with the "rules of the game." State governments are also active in mining. The government sponsors mining and metallurgical research and technology development.

The Departmento Nacional do Produção Mineral (DNPM) oversees all mineral and fuels activity in Brazil and keeps mining industry statistics. DNPM grants prospecting licenses (*alvará de pesquisa*), exploration licenses (*decreto de pesquisa*), and exploitation licenses (*decreto de lavara*).[1] DNPM also implements mining policy and enforces mining law.[1] The DNPM is a department within the Ministry of Energy and Mines, the policymaking body for Brazil's mining industry.[1]

Mining Taxation. The mineral industries in Brazil are subject to a number of taxes. Mining companies must pay the profits tax (*imposto de renda*).[12] Other taxes include the import tax (*imposto de importação*) and the industrial products tax [*imposto sobre produtos industrializados* (IPI).][12]

The only tax on minerals [*imposto único sobre minerais* (IUM)] is a tax levied on mineral production.[12] The portion of the IUM corresponding to selected mineral commodities is presented in Table IV-2-5. Revenue generated by the IUM in 1980 was 10.2 billion cruzeiros ($193.2 million).[12] The tax amounts to about 3.6% of the total value of mineral production. Total mining taxation accounts for only 1% of Brazil's total government revenues.[1]

Mining taxes in Brazil are stable and are clearly set out in the mining code. Mining taxation policy is considered liberal.[1] Tax concessions authorized for mining companies have reduced the already small tax burden. The mining tax burden is modest and will remain so in an effort to attract private domestic and foreign investment capital.

Tax Subsidies and Government Incentives. The Brazilian government has authorized tax concessions for virtually all of the taxes to which mining

TABLE IV-2-5   Brazil: Selected Commodities Participation in the IUM

Mineral	Rank	Share of IUM (%)
Iron ore	1	21.5
Coal	2	9.6
Tin	3	6.4
Asbestos	6	3.9
Manganese	8	2.8
Bauxite	11	1.7
Gold	13	1.3
Columbium	14	1.3
Tungsten	16	1.1
Lead	17	0.9
Chrome	18	0.9

Source: *Anuário Mineral Brasileiro 1981.*

companies are subject. Prospecting expenses can be written off against the profits tax.[12] Reductions of up to 80% of the IPI and the import tax on equipment, and from 10 to 90% of the IUM, are among the fiscal incentives authorized by the government.[12] Reductions in the IUM saved the iron, aluminum, columbium, and manganese exporters 2950 cruzeiros ($56 million 1980 U.S. dollars).[12]

The government provides energy and mineral transport infrastructure. Construction of the rail line to the Grande Carajas project, and of the enormous Itaipu and Tucurui hydroelectric facilities, has been undertaken by the government. The state electric energy company Eletrobras is in charge of the Itaipu project.[8]

Mining Law and Policy. Brazil's mining code dates to 1967 and is set out in Decree Law No. 227.[1] Decree No. 62934 of July 1968 implemented the mining law. Brazil's clear and concise mining law provides a firm foundation for further mineral development.[1] The DMPM implements mineral policy developed by the Ministry of Energy and Mines.

The government's Carajas mineral development project is a major policy move. (CVRD will develop iron and manganese ore deposits, and copper, nickel, bauxite, tin, and gold deposits will be left to private firms.)[14]

In the energy sector, the two hydroplants at Tucurui and Itaipu stand out, as do the ambitious oil, coal, alcohol, and synfuels projects that constitute the major elements of Brazil's energy policy.

Alcohol production is a government priority. The bulk of the 4.2 billion liters produced in 1981 was consumed by automobiles.[8] Coal and alcohol fuels policies are geared to limiting oil imports. Oil policy calls for holding imports steady while substituting alcohol, coal, shale oil, and other synfuels for oil.[4] Increases in Brazil's energy consumption will be met by these energy sources rather than by increased oil imports.

Brazil's mineral policy is oriented toward increasing output to meet domestic demand, and to export where possible. In an effort to reduce imports, coal, phosphates, copper, lead, and zinc have been designated priority minerals for exploration and development.[15] Other priority minerals include tin, tungsten, gold, nickel, cobalt, fluorspar, and asbestos.

Government Role in Research and Development. The Brazilian government supports mining exploration, research, and technology development through a number of agencies. The Companhia de Pesquisa de Recursos Minerais (CPRM) is an autonomous state firm that does geologic mapping, exploration, and geophysical and geochemical work for mining companies.[2] Petrobras has its own research and development center, where work on synthetic shale oil production and processing as well as oil industry research is centered.[4] CVRD has its own research and technology development branch.

Domestic Supply–Demand Dynamics

Brazil is self-sufficient in almost all ferrous metals but continues to be import dependent for many other mineral commodities. Brazil imports the fuels petroleum and coal and a number of basic metals, including aluminum, copper, lead, zinc, and nickel.[4] Brazil exports a number of mineral commodities, chiefly iron ore, manganese ore, ferrocolumbium and other ferroalloys, bauxite, and tin.[4]

The size of the Brazilian economy ensures a large domestic market for Brazil's mineral industry output. Exports of several mineral commodities will increase in the coming years, and Brazil's dependence on imports of aluminum, copper, lead, zinc, and nickel should decline as projects come on-line in the 1980s.[1]

Productive Capacity. Brazil's crude oil productive capacity is about 70 million barrels per year, about 60% of which is offshore.[4] Brazil is capable of producing about 5 million metric tons of coal per year.[12] Natural gas capacity is about 50 billion cubic feet per year.[4] Brazil produces over 4 billion liters of alcohol per year, and installed uranium capacity is 500 metric tons of yellow cake per year.[8,9]

Brazil's beneficiated iron-ore capacity is about 140 million metric tons, about 70% of Latin America's total.[12] Manganese capacity is 3 million metric tons of ore (2 million metric tons beneficiated).[12] Brazil's crude steel capacity is about 15 to 16 million metric tons per year.

Brazil's ferroalloy capacity is about 600,000 metric tons per year. Ferroalloy production in Brazil in 1980 is presented in Table IV-2-6. Ferrotitanium capacity is about 400 metric tons per year.[4]

Brazil's chromite ore capacity is close to 1 million metric tons per year.[4] Nickel output is about 2500 metric tons per year.[4] Metallurgical coal output is around 1.3 million metric tons per year.[13] Columbium concentrate capacity is about 30,000 metric tons per year.[12] Brazil is a sizable bauxite-alumina-aluminum producer. Bauxite capacity is close to 7 million tons; alumina capacity is on the order of 500,000 metric tons per year, and primary and secondary smelter capacity is about 310,000 metric tons per year.[12]

Brazil can produce about 25,000 metric tons per year of lead from mines, with smelter capacity 85,000 metric tons per year.[4] Zinc capacity is about 100,000 metric tons per year mine output and about 90,000 metric tons per year of smelter capacity.[4] Secondary copper refining capacity is 60,000 metric tons per year.[4] Tin mine capacity is about 13,000 metric tons per year of contained metal, and metal capacity is about 16,000 metric tons per year.[12,9] Tungsten capacity is about 1200 metric tons per year of contained metal. Brazil produces large volumes of asbestos (2.4 million metric tons of ore per year) and natural phosphate.[4] Phosphate capacity is about 1 million metric tons of contained P_2O_5 per year.[4] Brazil also produces close to 5 million pounds of natural quartz crystals for electronics applications, about half of total world production.[1]

TABLE IV-2-6 Brazil: Ferroalloy Production, 1980 (Metric Tons)

Ferroalloy	1980 production
Ferrochromium	93,443
Ferrocolumbium	17,530
Ferromanganese	140,496
Ferronickel	11,280
Ferrosilicon	109,410
Ferrosilicomanganese	134,243
Ferrotitanium	698
Ferrotungsten	217
Ferrovanadium	807
Other	43,625
	551,749

Source: Anuário Mineral Brasileiro 1981.

Domestic Supply–Demand. In 1980, oil imports were 930,000 barrels per day, compared to domestic consumption of 1,100,000 barrels per day (85%).[4] 1980 oil imports of $9.8 billion was the largest single import item in Brazil's foreign trade.[4] Six of Brazil's 10 leading mineral import sources in 1979 were OPEC countries.[4]

Brazilian iron and steel producers consumed about 18% of 1980 iron ore production, about 17.4 million metric tons.[4] Annual per capita steel consumption is about 115 kg. Steel production and domestic consumption are roughly equal at about 15 million metric tons per year.[15] Domestic consumption of manganese, about 1 million metric tons per year, is about half of beneficiated ore output.[12]

Coal imports, principally for the steel industry, are about 5 million metric tons per year, about the same as Brazilian production.[12] Brazil is self-sufficient in chromite ore and in ferrochrome.[12] Brazil's domestic ferroalloy consumption was 327,000 tons in 1980.[4]

Copper imports come primarily from Chile, Peru, and Zaire.[4] Brazil imported about 38% of its 1980 zinc requirements, mainly from Mexico, Peru, and Canada.[12,4] Lead imports from the United States, South Africa, Greenland, and Ireland provide about half of Brazil's total requirements.[9]

Aluminum production has grown considerably in recent years, but Brazil must still import about 1 million tons of aluminum per year.[12] Brazil also imports nickel to meet domestic consumption of about 10,000 tons per year.[15] Brazil produces no silver, and imported about 7 million troy ounces to meet domestic requirements in 1979.[4]

Brazil imports about 40% of its fertilizer mineral needs, a considerable improvement over the 85% import dependence of 1974.[4] Brazil imports all potassium fertilizer needs. Domestic consumption was about 1.3 million tons of K_2O in 1980.[4] Phosphate rock production in Brazil has grown considerably, reducing Brazil's import dependence.[12]

Balance of Production Available for Export. A large proportion of Brazil's mineral production is available for export. The bulk of Brazil's iron ore, beryllium, quartz crystal, and tungsten output is exported.[1] Brazil's major metal exports for 1978 to 1980 are presented in Table IV-2-7.[12] Ferroalloy exports for the same period are presented in Table IV-2-8. Steel exports in 1979 were about 1.5 million metric tons.[4]

Manganese ore exports of about 1 million metric tons are about half of domestic production.[12] Ferroalloy exports are about one-third of domestic production.[4] About one-third of chromite and bauxite ore production is exported.[1] Iron ore exports are about 80% of total domestic production.

As Brazil's steel industry grows in the 1980s, the surplus manganese ore available for export will fall steadily, to only 275,000 tons in 1989.[4] As the Carajas project and other new projects and expansions come on-stream in the 1980s, Brazil's iron ore exports should grow from about 80 million metric tons per year to well over 100 million metric tons per year by the end of the decade.

POLITICAL ENVIRONMENT

Domestic

Brazil's military took control of the government in a bloodless coup d'etat on March 31, 1964.[16] The military has been the dominant political actor since then, outlawing political parties until the elections in November 1982.[16] The authoritarian military regime had total control over the government from the 1964 coup up to the 1982 elections. The military government was anticom-

TABLE IV-2-7   Brazil: Mineral Exports, 1978–1980 (Metric Tons)

Commodity	1978	1979	1980
Bauxite	4,005	516,152	2,679,429
Beryllium	739	452	500
Chromite	81,858	45,214	—
Iron ore	66,371,317	75,588,306	78,958,057
Lithium (petalite)	—	1,074	2,000
Manganese ore	894,458	1,187,309	1,037,437
Niobium (columbite/tantalite)	131	326	455
Columbium (pyrochlore)	1,983	1,660	2,430
Tungsten (all forms)	1,724	1,655	2,806
Quartz crystal	2,405	4,743	5,753
Tin (all forms)	4,341	4,758	20,171

Source: Anuário Mineral Brasileiro 1981, pp. 39–41.

TABLE IV-2-8   Brazil: Ferroalloy Exports, 1978-1980 (Metric Tons)

Ferroalloy	1978	1979	1980
Ferrochrome	42,392	48,205	45,921
Ferrocolumbium	10,851	12,786	14,566
Ferromanganese	40,115	40,256	37,832
Ferromolybdenum	842	425	—
Ferrosilicon	12,308	19,641	29,664
Ferrotitanium	—	5	7
Ferrotungsten	251	284	65
Ferrovanadium	164	279	170
Other	43,442	50,232	48,196
	150,365	172,113	176,421

Source: Anuário Mineral Brasileiro 1981.

munist, and Brazilian security concerns paralleled those of the United States.[16] The Washington government responded with financial aid, modern military equipment, and advanced training.[16]

The authoritarian government repeatedly exercised its right to "change the rules" under which firms operated in Brazil, and to dismiss Congress in order to change the Constitution.[16] Brazil's government has allowed a considerable "political opening" in recent years. This policy is known as the *abertura* policy, and it culminated in the 1982 city, state, and federal elections.[16]

The military government allowed political leaders, banned from Brazil in 1964, to run in the 1982 elections.[16] Two of these leaders won state governorships: Iris Rezende (Goias) and Wilson Barbosa Martins (Mato Grosso do Sul).[8] The right of habeas corpus has been reinstated, after being removed in December 1968.[16] The judiciary branch has become more important in balancing the power of the executive branch. The government has given the public the right to freedom of information, and formal censorship of the press has ceased.[16] The public is now able to criticize more openly, and the government is more tolerant of protest activity.[16] Analysis of the 1982 election results leaves one with an awareness of the discontent with the growing centralization of power within the federal government over the last two decades.[8] Opposition parties captured close to half of the state governorships, including all of the states where Brazil's major cities and politically active upper- and middle-class population are located.[16]

Tax reform and some form of economic and political decentralization akin to Reagan's New Federalism are priorities in the new government.[8] The federal government would like to relocate education, health, and sanitation services at the state level, with corresponding funds also being transferred to the state level.[8] Economic and political decentralization is politically attractive to the federal government (PDS) in these times of austerity and economic crisis. The federal government is able to transfer part of the responsibility for dealing with the economic crisis to the state level as decentralization progresses.[8]

Opposition parties captured state governorships in 10 of Brazil's 22 states, responsible for about three-fourths of Brazil's GNP, 60% of the territory, and 58% of the country's population.[8] Opposition parties won all of the important industrial states of south-central Brazil: São Paulo, Rio de Janeiro, Minas Gerais, and Paraná.[8] These states also contain the most productive agricultural activity, the most active segment of the Brazilian business community, and the most sophisticated portion of Brazil's large working class.[8] Nine of the 10 governorships won by opposition parties went to the Brazilian Democratic Movement Party (PMDB). The PMDB is made up of moderate elements (former Brazilian Democratic Movement) and more liberal elements from the Popular Party (PP).[8] The illegal Brazilian Communist Party (PCB) supported many of the PMDB candidates in the 1982 elections.[8] The other opposition party to win a state governorship was the Democratic Labor Party (PDT), whose candidate Leonel Brizola won in Rio de Janeiro.[8]

Brizola despised the regime elected in 1964 and was one of the key instigators of the 1964 military coup.[8] He is now politically aligned with the European Social Democrats.[8] The Socialist Workers Party (PT), led by Metal Workers' Union leader Luíz Inacio da Silva (Lula), will probably close ranks with Brizola's PDT.[8] The other major party is the moderate Brazilian Worker's Party (PTB), with whom the government would like to establish dialogue and a good relationship.[8]

The government's PDS party came out the big

winner in 1982 elections, with control over the electoral college and power to choose the country's next president in 1985.[8] The other big winners were the moderates within the PMDB.[8] The governors of Paraná and São Paulo, the most important states in all of Brazil, are both PMDB moderates. The PMBD governors of Goias and Mato Grosso do Sul are radical, whereas the PMDB leaders of Para and Acre are more moderate. The far-left-wing candicates in general did poorly, garnering only about 10% of the total vote. Brazil's return to democratic rule and moves toward a freer economy with a smaller role for the federal government are promising signs of a progressive economic and political opening.

Brazil's "changing the rules" before the 1982 elections to require that voters vote a straight party ticket at all levels of government portends to a backlash against recently acquired freedoms.[11] This move ensured government control over the legislature and selection of the president through the 1980s.[8]

Major areas of policy debate will include the role of the federal and state governments in the economy, tax policy, political and economic decentralization, and policies to deal with the economic crisis. The government may allow opposition politicians to take charge of important ministries to open up debate on economic and political matters.[8] One political problem in Brazil that directly relates to the mining industry is the plight of primitive Amazon Indians. Mining development brings the technology and civilization of the 1980s and a number of diseases against which the Indians lack immunity into remote areas.[6] The government established Funai to deal with Indian problems and relocation. In general, mining interests are given priority when conflicts arise.[6]

International

Former Deputy Assistant Secretary of State for Inter-American Affairs, Albert Fishlow, expressed some thoughts on Reagan administration policy toward Brazil in the Spring 1982 *Foreign Affairs*.[16] Fishlow feels that U.S. policy toward Brazil, based on a shared perception of a security threat from the USSR, "misrepresents present reality."[16] The U.S. government is preoccupied with domestic priorities, ignoring the impact of its policies on the international economy and relations with Brazil.[16] Fishlow sees U.S. policy as "completely unresponsive to the significant changes now under way in Brazil."[16]

Brazil is important to U.S. strategic interests. Brazil is South America's largest military power

and is located near South Atlantic sea lanes that transport petroleum from the Persian Gulf and strategic minerals from southern Africa.[16] The United States has eliminated tariffs on some products imported from Brazil.[16] Largely symbolic visits have not improved Brazilian-U.S. relations. The U.S. government needs to develop a better understanding of the political environment in Brazil and the economic and political transition occurring there.

Major stumbling blocks in U.S.-Brazilian relations have been nuclear nonproliferation, subsidies to export industries, human rights, and trade-related problems. The United States has allowed shipments of nuclear material from Urenco (a Dutch-German-British consortium), despite previous opposition.[16]

Brazil refused to participate in American grain embargoes against the USSR and sent a commercial mission to Moscow in 1981.[16] Brazil has broken diplomatic relations with Cuba, but a high-level private commercial mission to Havana indicates a desire to resume trade.[16] Brazil has political and commercial ties with Angola and other black African nations, and is not concerned with the security of South Atlantic sea lanes.[16]

Brazil and Paraguay are cooperating in the enormous Itaipu hydroelectric project on the two countries' border, the Paraná River. Brazilian technology and capital are being used in the project, with Brazil and Paraguay to share the projects' 12,600-MW planned output.[8] Brazil is a rapidly growing industrial power and one of Latin America's most stable nations. The success of the 1982 elections, and the smooth return to democratic rule after almost two decades of authoritarian rule, signal moves to a freer, more open political and economic environment in Brazil. More segments of the Brazilian population will have a voice in the formulation of Brazil's policy in the future.[8]

REFERENCES

1. U.S. Department of the Interior, Bureau of Mines. *Mineral Industries of Latin America*. Mineral Perspective Series. December 1981.
2. República Federativa do Brasil, Ministério das Minas e Energia. *Contribução da CPRM ao Sector Mineral*, May 1982.
3. "Brazil: Where Growth Is a Necessity." *The Wall Street Journal*, 27 January 1983.
4. U.S. Department of the Interior, Bureau of Mines. *Minerals Yearbook 1980*. Washington, D.C.: U.S. Government Printing Office, 1982.
5. Schiller, E. A. "Mineral Exploration and Mining In Brasil—1977." *Mining Magazine*, September 1977.

6. "Brazil Battles the Jungle to Mine Carajas Minerals." *World Mining*, January 1983.

7. "Araxa Niobium Mine." *Mining Magazine*, February 1982.

8. "Brazil Is Making Massive Minerals Investments." *The Wall Street Journal*, 28 January 1983.

9. Spooner, John, ed. *Mining Journal Annual Review 1982*. London: Mining Journal, June 1982.

10. República Federativa do Brasil, Ministério das Minas e Energia. *Anuário Mineral Brasileiro*, 1981.

11. "Decisive Year for Brazil." *Mining Journal* 298 (1 January 1982).

12. U.S. Department of the Interior, Bureau of Mines. *Mineral Commodity Summaries 1983*. Washington, D.C.: U.S. Government Printing Office, 1983.

13. "Brazilian Leaders Discuss the Economy." *The Wall Street Journal*, 31 January 1983.

14. "CVRD Obtains Mining Concession for Carajas Manganese Reserves." *Engineering and Mining Journal* 182 (November 1982).

15. Spooner, John, ed. *Mining Journal Annual Review 1980*. London: Mining Journal, June 1980.

16. Fishlow, Albert. "The United States and Brazil: The Case of the Missing Relationship." *Foreign Affairs* 60 (Spring 1982).

Mexico

Interest in Mexico's potential as a supplier of strategic minerals to the United States has grown in step with Mexico's new-found oil reserves. These oil reserves are estimated by the Mexican government at 60 billion barrels proven, 38 billion barrels probable, and 250 billion barrels potential.[1] Recently, these estimates have been revised downward. Mexico is currently the world's fourth-largest petroleum producer, with a 1980 production of 2.6 million barrels per day. Mexico is among the world's five largest producers of at least 13 minerals.[1]

The Mexican mining industry has a long history, dating to pre-colonial days. Metals mined by the native Indians provided the impetus for the rapid conquest of Mexico by Hernán Cortés from 1519 to 1521. Spanish mining began in 1525. From the time of the conquest to Mexican independence from Spain in 1821, production from Mexican silver mines equaled the total previous world production. The state received all rights to the subsoil in the constitution of 1917. President Lázaro Cardenas nationalized the Mexican oil industry in 1938. The Mexican government reserves the right to exploit all oil and gas, uranium, sulfur, potassium, and phosphate rock.

Mexico's total exports increased 74% in 1980, to $15.3 billion, due largely to enormous increases in oil export revenue. Although oil exports have recently declined, the period from 1976 to 1980 saw an increase in crude oil's value share of exports from 13.3% to 61.7%.[2] Mexican mining and metallurgical industry's value share of exports fell from 13.3% to 8.5% over the same period.[2] However, the total value of Mexican mining and metallurgical exports has tripled.

Mexico is the world's largest silver, fluorspar, and strontium (celestite) producer.[2] It ranks second in world arsenic and natural graphite production, and third in antimony and bismuth production.[2] Mexico also produces significant amounts of cadmium, selenium, lead, zinc, mercury, and sulfur.[2]

Mexico's consumption of mineral commodities has been increasing in recent years. As the Mexican steel industry expands, iron ore and coke production increase, and consumption of fluorspar and manganese also increase.[3] Steel production grew just over 10% in 1981.[3] As domestic mineral demand increases, the balance of Mexico's mineral production available for export decreases. Most mineral and fuels exports are down, with copper and crude oil the only major exceptions.[2]

Mexico should continue to be one of the world's important mining countries. Mineral production will first be used to satisfy domestic needs, with the remainder of production available for export. Foreign and private Mexican investment capital will be essential to the further development of the mining industry in Mexico.

Fluctuations in the world oil price, and resultant foreign debt and trade problems, also affect the rate of expansion of Mexico's mining industry. The state of the world economy and mineral commodity prices are important in the expansion of Mexico's mining industry.

MINERAL RESERVES AND RESOURCES

Mexico is endowed with a wide range of mineral deposits and a significant share of world reserves of a number of important minerals. Table IV-3-1 presents Mexico's minerals and fuels reserve base. Mexico's fluorspar reserves are particularly important to the United States given the essential nature of fluorspar to steelmaking, the proximity of the deposits to the United States, and the 85% United States dependence on imported fluorspar.[4]

Mexico's diverse mineral endowment assures self-sufficiency in most ferrous and base metals.[2] New projects have reduced imports of fertilizer minerals. Oil and gas reserves are large enough to provide for domestic needs and support an aggressive export policy (market conditions permitting). Other important nonmetals occurring in Mexico include barite and strontium minerals (celestite).

Reserves of silver, lead, and zinc have declined, due primarily to precipitous drops in silver and other mineral commodity prices.[3] A much-higher-grade ore is required to be classified as reserves when prices fall. Silver and lead-zinc usually occur in the same deposit. Some zinc mines with minor silver values have been forced to curtail operations.[3]

Prices are one of the chief determinants of the

TABLE IV-3-1 Mexico: Mineral Reserve Base
(Thousand Metric Tons)

Mineral	Reserve base[a]
Antimony	265
Arsenic	Large
Barite	10,000
Bismuth (mt)	5,500
Cadmium (mt)	20,000
Coal	2,400,000
Copper	18,000
Fluorspar	62,000
Gold (1000 troy oz)	12,000
Graphite (natural)	3,600
Iron ore	510,000
Lead	5,000
Manganese ore	29,400
Molybdenum (mt)	70,000
Mercury (76-lb flasks)	250,000
Natural gas (million m^3)	1,743,000
Petroleum (1000 42-gal bbl)	60,000,000
Phosphates	44,000
Selenium (mt)	5,400
Silver (1000 troy oz)	1,060,000
Strontium (celestite)	NA
Sulfur	19,000
Zinc	3,000
Uranium	150

[a]NA, not available.

Source: Ref. 7.

volume of mineral reserves. Other factors affecting the classification of mineral deposits as reserves include size, location, and depth of burial of the deposit; ore grade; by-product and co-product relationships; mining method; proximity to infrastructure, processing facilities, and areas of end use; and mineral supply–demand dynamics.

GEOLOGY

Mexico's major copper deposits are similar to porphyry-type deposits in the southwestern United States.[5] Both of Mexico's two largest copper mines contain pyrite, chalcopyrite, sphalerite, galena, and bornite.[5] Supergene enrichment is important to the economics of the La Caridad deposit.[5] The relatively enriched zone contains chalcopyrite, chalcocite, covellite, pyrite, and digenite.[5]

The main ore body at La Caridad is about 1 km by 3 km.[5] The enriched zone is about 1 mile in diameter and is up to 250 m thick in the center (90 m average thickness).[5] The upper oxidized zone has had most of the copper leached from it. The oxide cap averages only 50 m thick.[5] Molybdenite occurs mainly as aggregates of fine-grained crystals in the eastern part of the deposit only.[5]

A large number of mines in northern and central Mexico exploit polymetallic sulfide deposits. These massive deposits typically contain lead, zinc, silver, and copper in varying proportions. Some deposits also contain recoverable bismuth and cadmium values.[3] Some deposits support operations that mine mainly for silver; others produce as many as five or six minerals in significant volumes. Some small operations are primarily gold and silver producers.

The Real de Ángeles mine exploits a deposit that yields silver, lead, zinc, and cadmium.[6] In general, mines in the states of Chihuahua and Zacatecas produce a number of minerals in large volumes. Mine output is typically balanced with significant output of copper, silver, lead, and zinc.[1] Mines in the state of Guanajuato produce mainly gold and silver. The large open-pit copper mines in the state of Sonora produce copper with small volumes of by-product molybdenum, gold, and silver.[1] La Caridad and Cananea are in the extension of the porphyry copper belt that runs through Arizona and New Mexico.[5] The mineralization at La Caridad and Cananea is the result of supergene chalcocite enrichment.[5]

The mineralization and alteration in the deposit is associated with quartz monzonite porphyries.[5] The texture and composition of the porphyries is typical of those associated with porphyry copper deposits.[5] The copper deposits of Mexico typically have hypogene sulfide mineralization.

Manganese deposits in Mexico contain carbonate ores.[2] The ore must be calcined and nodulized for use in the manufacture of ferroalloys and steel.[2] The best deposits are near Molango in Hidalgo state.[1] Coal and coke production are centered in northern Mexico in the state of Coahuila.[1] Major iron ore deposits are in the states of Guadalajara and Morelia near the Pacific coast, and in northern Mexico in Chihuahua.[1]

Mexico's fluorspar deposits occur around San Luis Potosí along the contacts between massive limestone beds and Tertiary rhyolite flows.[1] The fluorspar ore bodies are typically tabular and several hundred meters long and thick, and several tens of meters wide.[1] Because the ore has a high fluorspar content, it can be shipped as metallurgical-grade fluorspar after crushing and screening.[1]

Mexico's major strontium (celestite) producer mines a very high grade (92% $SrSO_4$) deposit in the state of Coahuila.[1] The deposit is relatively flat lying over a distance of 10 km and is about 5 m thick.[1] This one mine produces about one-third of the world's strontium minerals.[7]

Mexico has made progress in reducing imports

of fertilizer minerals, mainly through the development of mining of phosphatic beach sands in Baja California. The deposits contain about 24% P_2O_5 in oolites that make up the bulk of the medium-grained sandstone in the Humboldt bed, the best deposit identified in the area.[8] The best beds are about 1.8 m thick.[8]

LOCATIONS OF MINES AND PROCESSING FACILITIES

The locations of Mexico's major mineral-producing areas, mines, and mineral-processing facilities are shown on Figure IV-3-1. Major mines are located where the best mineral deposits occur, and mineral processing is concentrated near mines and in the large metallurgical centers, Monterrey and San Luis Potasí.[7] Other mineral-processing plants and steel mills and ferroalloy plants are spread throughout the country, with preference given to locating near mines, ports, and areas of domestic mineral consumption. Steel, aluminum metal, and ferroalloys production is principally in coastal cities, near ports that import raw materials and export finished goods.[7]

As mining is tied to specific locations where minerals occur, there is no choice about the location of mines. The location of concentration, processing, and manufacturing facilities depends on the considerations mentioned above and other factors, such as the availability and price of both labor and energy and proximity to raw materials sources and transport infrastructure.

PHYSICAL INFRASTRUCTURE

Energy

Mexico's large oil reserves have led to an energy balance that is dominated by oil and gas. Oil and gas together provide about 88% of Mexico's primary energy requirements.[7] The remainder is supplied by coal (9%) and hydropower (3%).[7] Mexico's installed electricity generating capacity of 15,130 MW is based on thermal-fired plants (57%), hydropower (42%), and geothermal plants (1%).[2]

Geothermal capacity of 150 MW in 1980 is set to increase to 620 MW by the end of the decade.[2] Coal is projected to provide about 12% of Mexico's future energy needs, and nuclear power should begin to contribute to Mexico's overall energy supply in the medium term.[2] Electricity output has been growing about 10% per year.[2]

Oil and gas will continue to dominate Mexico's energy balance. A wide variety of other sources will become more important to Mexico's overall energy supply. Among these are coal, nuclear, hydropower, and geothermal. New coal-fired power plants are planned and under construction, as are several nuclear reactors.[2]

Only about one-tenth of Mexico has been systematically surveyed and explored for uranium potential.[9] Proven reserves are 150,000 metric tons, and exploration for additional reserves continues in seven states.[9] The National Nuclear Energy Institute is funding projects to increase Mexico's uranium production.[9]

FIGURE IV-3-1   Mexico:
Mines, processing facilities, and
infrastructure.

Water

The Ministry of Patrimony's Commission on Energy Sources is undertaking a study of Mexico's water resources.[2] The study will determine the country's hydroelectric potential and investigate the use of Mexico's water resources.[2] Mexico's long coastlines facilitate the transport of bulk commodities by barge and ship.

Water is generally available in Mexico, although large areas in the northern part of the country are very dry. The country's water resources are poorly distributed and are often of inferior quality. Mexico's large population and agriculture and industry are all large consumers of water.

Mexicana de Cobre's La Caridad copper mine and mill near Nacozari, Sonora, use water from the Angostura Dam.[5] Three pumping stations can pump up to 500 liters per second through a 24-inch concrete and steel pipeline the 24 km from the dam to storage ponds near the mine site.[5] Drought has limited output at the La Caridad mine.[10]

Transportation

Mexico's minerals and mineral fuels are transported by a number of different methods. Most of Mexico's mineral output is shipped by rail. Trucks are used to transport ores and concentrates in more remote areas. Mexico has an extensive road system. Pipelines transport crude oil, gas, and petroleum products from producing areas to refineries, areas of domestic consumption, and export terminals. Mexico has a number of ports for the importation and exportation of mineral commodities.

Rail. Mexico's largest railroad, Ferrocarriles Nacionales de México (FNM), operates about 70% of Mexico's 20,000-km railway system.[2] Mexico's rail system is antiquated and inadequate, and rolling stock is insufficient to meet the needs of the mining and metallurgical industries.[2] Steel plants have been hurt by delays in the shipment of coke and iron ore pellets, and Pemex has been affected by delays in barite deliveries to drilling operations.[2]

The World Bank has granted a $150 million loan to the country to rehabilitate its rail system.[5] Most of Mexico's rail system is composed of single-track lines.[2] Bulk mineral freight such as iron ore, coke, pellets, and manganese represent over one-fourth of total rail traffic.[2] The land bridge across the Isthmus of Tehuantepec is being reestablished. A 300-km rail line based on a railroad built by the British in about 1890 will link the Gulf of Mexico and the Pacific Ocean.[7] The Las Truchas steel plant at Lázaro Cardenas on the Pacific coast is linked by rail to Mexico City.[2]

Rail lines link most of the country, and mineral-producing regions are linked by rail to ports, industrial and metallurgical centers, and other end-use areas. Some mineral output is shipped by rail to the United States. Rehabilitation of Mexico's rail system is a prerequisite to the further development of the country's mining and steel industries. Mexico's rail system constitutes a serious bottleneck in the country's mineral transport system.

Roads. Roads are used to transport mineral production in remote parts of Mexico. Most new mining projects require considerable investment in the construction of roads from the mine to rail or major highway transport systems. Mexico's road system totals over 200,000 km, and about 30% is paved.[7]

The development of the La Caridad copper mine by Mexicana de Cobre required the construction of 93 km of paved roads.[5] Investments in the transport infrastructure necessary to link a new mine with existing transport facilities can represent a large share of total mine development expenses.

Pipelines. Mexico's system of crude oil, natural gas, and petroleum products pipelines total over 13,000 km.[2] Mexico's oil and gas pipeline system is being expanded to about 60,000 km by 1990.[2] The most important project completed is the 1344-km natural gas pipeline from the Reforma fields in Chiapas to the industrial center at Monterrey and on to McAllen, Texas.[7] Expanding natural gas pipeline service should increase the industrial and commercial use of gas. Crude oil pipelines cross the Isthmus of Tehuantepec, linking the oil tanker terminals at Salina Cruz and Coatzacoalcos.[7]

A slurry pipeline is being built from the Hercules iron ore mine to the pellet plant at Muzquiz.[11] The 385-km pipeline is currently under construction.[11]

Ports. Other major oil export terminals are at Tuxpan, near the cities of Tampico and Veracruz on the Gulf coast, and near Acapulco on the Pacific coast.[7] These four tanker terminals and the two facilities at Salina Cruz and Coatzacoalcos handle almost all of Mexico's oil exports. The port with the largest volume of crude oil exports is Coatzacoalcos.[2]

The port at Coatzacoalcos can also load up to 9000 metric tons per hour of bulk and molten sulfur into ocean-going vessels.[1] Output of copper and molybdenum from the La Caridad mine is exported at the Pacific port of Guaymas.[5] Besides Guaymas and Salina Cruz, other important Pacific ports include Lázaro Cardenas, Acapulco, and Mazatlán.[7]

Veracruz and Tampico have long been Mexico's major ports on the Gulf of Mexico. Two new ports are being built on the Mexican Gulf coast. A 100,000-dead weight ton (dwt) port is being built at Altamira, north of Tampico, to handle iron ore imports.[2] Another port is being built at Laguna de Ostión.[2]

The Pacific ports at Lázaro Cardenas and Salina Cruz are being expanded. Phosphate production in Baja California is shipped from a 40,000-dwt facility at Punta Belcher to the port at Lázaro Cardenas.[8] Mexico's transport infrastructure is shown in Figure IV-3-1.

MINING LABOR

Mexico's mining and mineral fuels industries face shortages of skilled laborers, technicians, and engineers.[1] The mining industry employs about 180,000 workers, about 3% of Mexico's labor force.[7] Pemex employs an additional 130,000 workers in the petroleum and petrochemical industries.[7]

The Council of Mineral Resources employs about 750 technical people, including geologists, mining engineers, and economists.[1] CRM is training technical specialists in mining-related fields to help reduce the shortage of skilled labor in mining. The La Caridad mine has an in-house training program for mine workers.[5]

Most labor demands have been met without resorting to strike activity in recent years. Labor contracts signed in 1981 contained an average 30% increase in basic wages.[3] Mexico's only copper refinery was closed by a work stoppage that lasted about two months in 1981.[3] The Santa Bárbara and Santa Eulalia units were on strike for 10 weeks, and Frisco's San Francisco unit was shut down by strikes for a month.[3] Strike activity reduced silver, lead, and zinc production in 1981.[3] Mexico's labor relations climate is generally good. Strike activity has been relatively minor in Mexico's mining industry in recent years in comparison with other Latin American countries.

DOMESTIC MINING INDUSTRY

Overview

Mining and Energy in Mexico's Economy. Mexico's gross national product (GNP) has grown about 7% per year until the recent recession, reaching U.S. $167 billion in 1980.[2] Petroleum accounted for 4.5% of total GNP, about $7.5 billion.[2] Metallic and nonmetallic minerals provided about 1.1% of Mexico's 1980 GNP, earning about $1.8 billion.[2] The 1980 inflation rate was officially 30%.[2]

Mexico's exports are dominated by mineral fuels and mining and metallurgical products. Mexico's total 1980 exports were $15.3 billion.[2] Hydrocarbon exports by Pemex were valued at $10.3 billion that year, about two-thirds of total exports.[2] Mining and metallurgical exports were valued at $1.3 billion, about 8.5% of total exports.[2] Mineral and fuel exports together accounted for 76% of Mexico's total 1980 exports.[2] Mexico's Cámara Minera estimated 1981 mining investment at $850 million.[2]

Trends Among Major Mineral Producers

Economic growth resulting from Mexico's increased oil revenues has increased domestic production and consumption of mineral commodities. A number of new projects and expansions should increase Mexico's output of minerals in the near future. The minerals posting the largest increases in output in 1981 were manganese (30%), copper (31.4%), and silver (9%).[11] Oil production increased about 20% in 1981, to about 2.5 to 2.6 million barrels per day.[3] Coal production was up over 0.5 million metric tons in 1981.[11]

The strong growth of Mexico's steel industry in recent years has created additional domestic demand for steel industry minerals. This increased demand has induced additional iron ore, manganese, coke, and fluorspar production.[3] High demand by Pemex has spurred higher barite output.[3] Despite falling prices, silver output increased in 1981 as mines concentrated on rich veins and ore bodies.[3] Lead and zinc output fell in 1981 due to decreased activity at silver-lead-zinc mines.[3] Delays in concentrate shipments hurt Mexico's lead and zinc smelters and refineries.[3] Copper output increased substantially due to increased output at La Caridad and Cananea, but refined copper output was reduced by strike activity at Cobre de México's refinery.[3] Bismuth and cadmium output both fell in 1981.[3] Frisco's Cumobabi mine began production as Mexico's first molybdenum mine in 1981, producing about 295 metric tons of molybdenum and some by-product copper.[11]

Mexico is capable of increasing mineral output with existing installed capacity as the industrialized countries emerge from the recession and demand for mineral commodities increases. Mexico's manganese and fluorspar producers are operating well below capacity and will increase output as market conditions dictate.[3]

Copper production will continue to increase as output from the La Caridad and Cananea mines expand through the 1980s.[1] The Real de Ángeles mine reached capacity in August 1982 at 10,000 tons of ore per day.[6] This mine will greatly increase silver output (by 7 million ounces per year); lead (31,000 tons per year), zinc (26,000 tons per year), and cadmium (415 tons per year).[6] This one project will increase output of these minerals by 12 to 30% over 1981 levels.[11] This open-pit mine is the largest silver mine in the world.[6] Completion of Mexicana de Cobre's 180,000-metric ton per year flash smelter at the La Caridad mine will substantially increase Mexico's refined copper production.[11]

The most important trend in Mexico's mineral and fuel exports is the rise in importance of oil exports and the decline in the mining and metallurgical share of exports.[2] Crude oil and mining-metallurgical products each provided 13.3% of Mexico's 1976 exports.[2] By 1980, the crude oil share had risen to 61.7% of the total while the mining-metallurgical share had fallen to just 8.5%.[2] Another important trend is the increase in the level of processing of Mexico's mineral exports.

Industry Structure

Mexico's long mining history has resulted in an industry composed of a combination of fully state-owned companies, companies in which the government is either a minority or a majority partner, and fully privately owned companies.[7] The Comisión de Fomento Minero (CFM) holds an interest in all mineral companies which are partially or totally state owned.[2] The CFM also promotes and finances small and medium-sized mining companies.

The mineral industry in Mexico is highly oligopolistic. Approximately 90% of all mining activity is conducted by the five largest mining companies.[2] The remaining 10% of mining activity is divided among the remaining 900 or so mining companies. There are over 15,000 separate mining ventures in Mexico.[2]

Mexico's 1979 mineral output was divided among large foreign and private domestic companies (55%), mining companies with government equity (30%), and medium-sized and small mining companies (15%).[7]

Actual foreign investment in mining companies is substantial. The United States, Canada, and Japan are among the largest sources of foreign mineral investment capital.[7] Foreign mining companies own shares of a number of large mining projects in Mexico. Canada has the largest share of foreign investment in Mexico's mining industries.[12]

Mexico's mineral industries exhibit varying degrees of concentration and vertical and horizontal integration. The bulk of Mexico's mineral output is processed, smelted, and refined in Mexico before entering domestic and export markets.

Concentration Ratios. The degree of concentration of Mexico's mineral industries varies widely. At one extreme is the oil and gas industry, where Pemex is a fully vertically and horizontally integrated firm that produces oil and gas, operates refineries, makes petrochemicals, and markets petroleum products. At the other extreme, many small and medium-sized silver and fluorspar miners are active, selling ores to nearby processing centers.[2]

Petróleos Mexicanos (Pemex) has had a state-granted monopoly on oil and gas production and refining since the nationalization of the oil industry in 1938.[7] The most productive of Pemex's oil fields are in the Campeche Sound, where offshore production that began in 1980 reached over 1 million barrels per day during the first year of production.[2] Mexico's oil refinery capacity is 1.3 million barrels per day, but is set to increase to 1.75 million barrels per day.[2] Pemex operates Mexico's seven major and two minor refineries and all of the country's major petrochemical complexes.[7]

Several of Mexico's mineral industries are highly concentrated, with one or a few firms owning the majority of all producing mines. The Cia. Minera Autlán S.A. de CV produces almost all of Mexico's manganese ore and manganese ferroalloys.[2] Autlán produces 90% of Mexico's manganese, with the remainder being produced by two small firms, Cia. Minera Buena Vista S.A. (6%) and Cia. Minera San Martín y Anexas (4%).[2] Autlán also produces 90% of Mexico's ferroalloys.[7]

Mexico has an integrated iron ore, pellet, and steel industry dominated by a small group of firms. A group of four large steel companies together form the Consorcio Minero Benito Juárez, the owner of the largest share of Mexico's iron ore and pellet production.[2] Three of the firms, Hojalata y Lámina S.A. (HYLSA), Altos Hornos de México (AHMSA), and Fundidora de Monterrey S.A. (FMSA), individually own other large iron ore mines and pellet plants in Mexico.[2] The fourth member of the consortium is Tubos de Acerode México S.A. (TAMSA), producer of 5% of Mexico's steel output.

Siderúrgica Lázaro Cardenas–Las Truchas S.A. (SICARTSA) also owns major iron ore and pellet-producing facilities.[2] The four largest steel producers account for about 80% of Mexico's total steel output.[2] The leading steel producers and their shares of total output are as follows: AHMSA (32%), HYLSA (22%), FMSA (14%), SICARTSA (11%), and semiintegrated steel plants (16%).[2] Some of Mexico's large steel producers are joined together to form Siderúrgica Mexicana (Sidermex).

Two firms dominate Mexico's growing copper

industry. The Cananea mine produces about 24% of Mexico's copper output.[7] The mine is owned by private Mexican interests (53%), Atlantic Richfield (34%), and the Mexican government (13%).[7] Mexicana de Cobre (joint government–private Mexican) produces about half of Mexico's copper output.[7] Industrial Minera México S.A. (IMMSA) controls about 15% of the Mexican copper industry.[1] Minera Frisco S.A. de CV's Cumobabi mine is Mexico's only molybdenum mine.[3]

Mexico's lead-zinc industry is also highly concentrated. Penoles's metallurgical complex at Torreon produces 60% of Mexico's lead and 35% of the country's zinc.[7] México Desarollo Industrial Minera S.A. (Medimsa) produces 40% of Mexico's lead and 25% of the country's zinc.[7] These two firms produce all of Mexico's refined lead and two-thirds of the country's zinc metal.[7]

Many firms mine silver in Mexico. Some mine primarily for silver, while others obtain silver in the processing of copper, lead, and zinc ores. Industrias Penoles S.A. de CV is Mexico's largest silver producer, with about two-thirds of 1981 production.[11] Penoles is part-owned by AMAX, Inc.[7] IMMSA is the second largest silver producer, with about 15% of total 1981 silver output.[11]

Cia. Minera La Valenciana is Mexico's largest producer of strontium minerals (celestite).[1] The United States consumes about half of the world strontium output, and imports from Mexico provide 99% of U.S. imports.[4] The San Agustín mine is the world's largest source of strontium minerals.

Mexico's barite industry is dominated by four firms. The largest is Barita de Sonora S.A. de CV (state), with about 52% of mine output (165,000 metric tons per year).[2] Baramín de México S.A. and Barita de Apatzingán are the second- and third-largest producers, at about 60,000 metric tons per year, about 38% of Mexico's output when combined.[2,11] The fourth barite producer is Cia. Minera Guadalcazar S.A., whose 20,000-metric ton per year capacity represents 6.3% of Mexico's 1981 output.[2] These four firms together produce over 96% of Mexico's barite output.

Two firms dominate Mexico's sulfur industry. The state-owned Azufrera Panamericana S.A. (APSA) produces over half of Mexico's sulfur.[2] Mexico's second-largest producer is Compañía Explotadora del Istmo S.A. (CEDI), owned by Texasgulf, Inc. (34%) and the government (66%).[2] Both firms use the Frasch process at operations on the Isthmus of Tehuantepec.[1]

Vertical and Horizontal Integration. State companies are highly integrated. Cia. Minera Autlán is Mexico's largest ferroalloy and manganese pro-

ducer. The company is also integrated internationally, with a ferroalloy plant in Mobile, Alabama.[2] Autlán produces both battery- and metallurgical-grade manganese ore, ferromanganese, and silico-manganese.[2]

Mexico's largest silver producer, Penoles, has a mine output of 12 million troy ounces per year, and refinery capacity of about 35 million ounces per year.[11] Penoles is also horizontally integrated into lead, zinc, fluorspar, and sodium sulfate production and silver-lead-zinc smelting and refining.[7]

Cia. Minera de Cananea has an integrated copper mine and smelter complex at its Cananea mine.[7] The La Caridad mine will soon have a 180,000-metric ton per year flash smelter and a molybdenum recovery plant to complement the existing mine and concentrator facilities at the mine site.[5]

Government Role in the Mining Industry

The Mexican government controls the mining industry by setting tax and subsidy guidelines, owning producing companies, formulating mining law and policy, and providing support services. The government also regulates domestic and external trade through foreign exchange and currency policy. The government also creates the institutional framework in which mining companies operate.

Mining Taxation. Mining firms operating in Mexico are subject to the standard 42% corporate income tax and a number of other taxes that apply only to the mining industry.[2] Mining companies pay flat concession and exploration taxes of $1.33 per hectare (nonmetals) or $2.65 per hectare (metals), and an additional $0.44 per hectare per year.[2] The production tax that mining companies pay varies among minerals. The production tax on gold, silver, and sulfur is 9%.[2] Iron, magnesium, and coal are subject to a 4% production tax, and other minerals are taxed at 7%.[2] Mining taxation is relatively stable, and comparable to mining taxation in Canada and the United States.[12]

Government Subsidies and Incentives. Article 16 of the 1977 mining tax and development law provides for at least three types of subsidies.[2] First, upon payment of the production tax, mining companies receive a subsidy of 2% of the value of mineral production.[2] For producers with annual sales of less than $1 million, this subsidy is increased to 3%. Mining companies are also eligible for a subsidy equal to 75% of the import tax on mining machinery and equipment.[2] Mining companies can also receive a 75% subsidy on import taxes on spare parts.

The Mexican government has granted a 30% discount on natural gas, liquid hydrocarbon fuels, and electricity to processing plants that export at least 25% of their output for at least three years.[2] Mexico's Consejo de Recursos Minerales (CRM) is an agency established in 1955 to undertake mineral exploration efforts and provide geologic studies and technical support to the country's small and medium-sized mining companies.[2]

Government Ownership of Producing Companies. The Comisión de Fomento Minero, founded in 1974, holds the state interest in large mining projects.[5] Mining companies with government equity accounted for 30% of the value of Mexico's 1979 mineral output.[2] The state owns all or part of a number of mining and mineral processing facilities.

Mining Law and Policy. Mexico's mining law was approved in late 1975.[5] The law reserves the exploration and exploitation of iron ore, sulfur, potassium, phosphate rock, and uranium for the federal government.[5] Oil and gas are also reserved exclusively for state exploitation. This law also gave the government the right to take over unworked mining concessions and develop them or turn their development over to private companies.[5] The law gives the government a legal mandate for the establishment of state companies for large-scale mineral exploration and development.[5]

Mexico adopted the policy of "Mexicanization" of the mining industry in 1961. "Mexicanization" refers to the reduction of foreign ownership of mineral-producing companies.[5] Specifically, mining operations in Mexico must be majority-owned by Mexican citizens, and must be at least 66% owned by Mexicans if located on state-owned land.[5] Mining operations initiated after 1961 are all Mexicanized, and mining operations that were already in production in 1961 have until 1986 to comply with Mexicanization regulations.[5]

The goal of government policy is to reduce the state's ownership role in the mining and mineral-processing industries.[5] Mexico's mining law and policy are relatively stable. The present mining law has not been changed since its adoption. The government will still take an equity role in large mineral development projects to ensure their timely completion.

Government Role in Research and Development. The Mexican government sponsors mining and metallurgical research through several state agencies. The Consejo de Recursos Minerales (CRM) is active in mineral exploration and supports private exploration efforts with geologic studies and financial aid.[2] The CRM also publishes Mexico's statistics on mineral production and trade.[2] The CRM has designated bauxite, iron ore, coal, and nickel as

priority minerals for exploration in an effort to reduce imports of these minerals.

The Comisión de Fomento Minero provides technical assistance to small and medium-sized mining companies. Pemex has its own independent research and technology development staff and facilities. Large steelmakers also do extensive research and technology development. HYLSA developed the HYL III continuous direct reduction process after over a decade of research.[2] The HYL III process utilizes a continuous direct reduction technology, an improvement over previous batch technologies, yielding lower capital and operating costs than do other direct reduction processes.[2]

Domestic Supply–Demand Dynamics

Mexico produces and consumes a wide range of mineral commodities in large volumes. Mexico's productive capacity and domestic consumption determine the portion of its mineral fuels output available for export. Mexico's large population and relatively high level of industrialization are responsible for the large demand for mineral commodities and energy.

Productive Capacity. Mexico has installed capacity to produce a number of important minerals. The present recession has lowered the demand for most mineral commodities; hence a sizable share of Mexico's mineral productive capacity is unutilized or underutilized.[3] Fluorspar and manganese production in recent years have been well below total installed capacity. Mexico's mineral and mineral fuels productive capacity is presented in Table IV-3-2. Where specific figures for productive capacity were unavailable, estimates are given based on output in recent years.

Mexico produces other minerals not presented in Table IV-3-2. Barite productive capacity is about

TABLE IV-3-2 Mexico: Mineral and Fuels Productive Capacity (Metric Tons)

Mineral	Productive capacity
Coal	7,000,000
Copper	230,000
Fluorspar	1,000,000
Iron ore (pellets)	8,400,000
Lead	175,000
Manganese	500,000
Natural gas (million ft^3)	1,300,000
Petroleum (bbl)	1,000,000,000
Silver (troy oz)	52,000,000
Strontium minerals	40,000
Sulfur	3,200,000
Zinc	265,000

Sources: Ref. 2; *Mining Annual Review 1982.*

275,000 metric tons per year.[3] Mexico's capacity to produce coke is about 2.5 million metric tons per year.[3] Cadmium capacity is just under 2000 metric tons per year.[3] Mexico's molybdenum output should become significant as molybdenum recovery plants are completed at major copper mines, and as the country's only molybdenum mine reaches full capacity by the mid-1980s.[3]

Domestic Supply–Demand. Mexico's domestic demand for mineral commodities, especially steel industry minerals, has been increasing in recent years.[2] Growth in domestic demand for iron ore and manganese has brought about increased production.[3] Mexico consumes about half of domestic production of several minerals, including arsenic, barite, cadmium, lead, manganese, selenium, and silver.[5] Mexico consumes small amounts of its antimony, fluorspar, mercury, and strontium minerals output.[5] Domestic consumption of sulfur and zinc is between 40 and 50% of total production.[5] Mexico is basically self-sufficient in the minerals it produces but must import some minerals which are not produced in the country.[2]

Mexico's largest mineral imports include coal, coke, chromite, nickel, tin, titanium, and aluminum metal.[2] Mexico's reliance on imported fertilizer minerals has been greatly reduced by increased domestic production of phosphatic and potassic fertilizers.[2] Mineral imports are about 5% of total imports.[2]

Balance of Production Available for Export. Mexico's productive capacity for a number of minerals is sufficient to meet domestic demand and allow sizable exports of mineral commodities when international market conditions permit them. Mexico's strontium minerals, mercury, and fluorspar industries primarily produce for export markets.[5] The majority of Mexico's antimony, sulfur, and zinc output is exported.[5] Mexico's leading mineral exports on a value basis are presented in Table IV-3-3. Silver is the nation's most important mineral export, accounting for one-third of the value of the country's total mineral exports.[5]

As Mexico's steel and other industries expand in the 1980s, less of its mineral output will be available for export. Manganese and fluorspar are two possible exceptions, as productive capacity for these two minerals greatly exceeds present production.[3] The expansion of Mexico's copper industry throughout the decade should increase exports. Increasing capacity should also increase the amount of silver available for export in the near future. Mexico's 1979 mineral exports are presented in Table IV-3-4.

The United States is the leading importer of

TABLE IV-3-3   Mexico: Value of Mineral Exports, 1979

Mineral	Percent (rounded)
Silver	33.4
Lead	12.4
Zinc	12.6
Sulphur	12.6
Fluorspar	7.6
Copper	5.4
Iron ore	1.9
Manganese	1.9
Other	11.8
	100.0

Source: Ref. 7.

TABLE IV-3-4   Mexico: Mineral Exports, 1979 (Short Tons)

Mineral	Volume
Antimony	2,030
Arsenic	3,605
Barite	148,042
Bismuth	266
Cadmium	696
Celestite	43,959
Copper	22,466
Fluorspar	779,481
Graphite	10,027
Iron ore	19,668
Lead	84,271
Manganese	242,992
Mercury	202
Selenium	807
Silver	30,285,876[a]
Sulfur	1,324,400
Zinc	150,680

[a]Unit of measure is the troy ounce.
Source: Ref. 2.

Mexico's mineral commodities, representing two-thirds of Mexican nonfuel minerals exports.[2] The United States imports virtually all of Mexico's exports of barite, celestite, graphite, gypsum, iron, and arsenic.[2] U.S. imports of sulfur, silver, antimony, and fluorspar are all between 70 and 85% of total Mexican mineral exports.[2] The U.S. imports between 40 and 50% of Mexican exports of lead, cadmium, and zinc.[2]

The major export destinations for Mexico's mineral exports are presented in Table IV-3-5. Besides the United States, the other main destinations for Mexico's mineral exports are other Latin American countries, Western Europe, and Japan.[2] Mexico's manganese exports are spread among a number

TABLE IV-3-5 Mexico: Mineral Exports, 1979,
by Selected Country of Destination

Mineral	Country of Destination
Antimony	U.S., 77%; Brazil, 14%; Japan, 6%
Arsenic	U.S., 89%; Brazil, 9%; Japan, 2%
Barite	U.S., 100%
Bismuth	U.S., 0%; Benelux, 100%
Cadmium	U.S., 46%; Brazil, 40%; France, 4%; Benelux, 4%
Celestite	U.S., 100%
Copper	U.S., 0%; West Germany, 26%; South Korea, 24%; Spain, 18%; Japan, 17%; Bulgaria, 15%
Fluorspar	U.S., 70%; Canada, 17%; Benelux, 6%
Graphite	U.S., 97%
Gypsum	U.S., 98%
Iron ore	U.S., 97%
Manganese	U.S., 21%; France, 23%; Japan, 22%; Argentina, 11%; Canada, 10%; Brazil, 5%
Mercury	U.S., 3%; Brazil, 82%; Argentina, 13%
Selenium	U.S., 9%; England, 66%; Brazil, 13%; Benelux, 12%
Silver	U.S., 80%; England, 9%; Switzerland, 5%
Sulfur	U.S., 83%; England, 7%; Brazil, 4%
Zinc	U.S., 40%; Benelux, 30%; Brazil, 16%

Source: Ref. 2.

of countries, because the country produces large volumes of special grades of manganese, such as battery- and chemical-grade ores.[2]

With most of Mexico's mineral exports going to the United States and other western countries, the country is important as a potentially secure source of minerals during periods of supply disruption or international conflict. Overland and Gulf of Mexico supply routes are short and would be relatively easy to defend in times of war or other emergencies.

The level of processing of Mexico's mineral exports has been increasing in recent years. Manganese exports are all in the form of concentrates.[2] Iron exports are 95% finished products.[2] All cadmium exports are refined metal, and 88% of lead exports are refined metal.[2] Zinc exports are divided among concentrates (46%) and refined metal (54%).[2] Mexico's metallurgical industry is very advanced for a relatively undeveloped country.

POLITICAL ENVIRONMENT
Domestic

Mexico's domestic political arena has been dominated by the Partido Revolucionario Institucional (PRI) since the Mexican Revolution in 1910. The PRI selected each of the presidential candidates, all of whom easily won election. One of the elements of PRI's political philosophy since the Revolution

has been the practice of not allowing politicians to run for reelection. The party secretly decides on the successor to the presidency.[13] All of Mexico's living former presidents form an advisory committee within the party to provide counsel and assure the continuity of Mexico's domestic and foreign policies.[13]

The PRI incorporates all major organized interests into a pluralistic system, although power is not shared equally among the groups.[14] The PRI has included a wide range of political philosophies in its power base since the Revolution. In addition to moderate groups, leftist ideologues and their lower-class supporters have been integrated into the PRI power elite.[14] The PRI has followed a policy of providing symbolic and concrete benefits to leftists in return for their promise to operate within the political system.[14] Opposition political parties are few, and their activities are restricted.[13]

The military has not been an important political actor in Mexico since the Revolution. Instability in Central America has created concern in Mexico that the military may increase its power in response to hostilities in the region.[14] Mexico's purchase of F-5 jet fighters from the United States is viewed by some observers as evidence that an increased role for the military is already a reality.[14] The military buildup is a response to the fighting in El Salvador, Honduras, Guatemala, and Nicaragua and the increased flow of refugees into Mexico from neighboring countries.[14]

Other major political actors in Mexico include business and labor groups, the Catholic Church, and other special-interest groups. Mexico's relatively open political environment allows for considerable dialogue between these other political actors and the federal government and its representatives. Although Mexico's political system is not democratic, it is pluralistic, with most interested political groups able to obtain access to decision makers.

Mexico's domestic policies have historically been oriented toward protecting domestic industry and reserving the country's energy and mineral resources for exploitation by the state.[14] Trade barriers have protected domestic industries, and Mexico's energy wealth has been in the hands of the state since the nationalization of the oil industry in 1938.[14] The Mexican government is generally more conservative in its domestic policies than it is in foreign policy matters.[14]

International

Several factors contribute to Mexico's importance in the North-South dialogue. Among these are Mexico's revolutionary heritage, importance to the world energy and mineral economy, and relatively high level of industrial development.[14] Mexico's newly discovered oil reserves give it an even more important role as a spokesman for the Third World. Mexico's foreign policy is based on the principles of nonintervention and self-determination.[14] Mexican political leaders often support revolutions publicly, in defiance of U.S. foreign policy initiatives in the region.[14]

Relations between the United States and Mexico improved during the Reagan administration.[14] President Reagan and former Mexican President López Portillo have more similar personalities than did Presidents Carter and López Portillo.[14] The tone of relations between the Reagan administration and the government of Mexico's President Miguel de la Madrid Hurtado, who took office in December 1982, was fairly cordial. New initiatives aimed at creating a special relationship between the United States and Mexico are "on hold" until the new Mexican government consolidates its power.[14]

Mexico and the United States both view Central America as within their respective spheres of influence. The two nations differ in their views as to the causes of political instability in the region. Mexico views the instability and unrest in Central America as the result of profound social and economic changes inevitable in the region.[14] The United States sees leftist activity in Central America as the result of communist attempts to control the region, and believes that the guerrillas in El Salvador and other Central American nations lack the popular support of the majority of the population.[14]

Mexico and the United States also differ on the possible forms of government for Central American nations. Mexico (and a number of other nations) supports a center-left coalition government incorporating elements of the radical left for El Salvador.[14] The Reagan administration rejected this solution, believing that the radical elements would eliminate moderate elements and convert El Salvador into another Marxist-oriented socialist regime in alliance with Cuba and the USSR.[14]

In the realm of United States–Latin American relations, Mexico usually sides with the other Latin American nations, clearly feeling that in the past the relationship has been characterized by "imperialist, hegemonic and interventionist" actions on the part of the United States and its allies in the region.[14] Mexican policy toward Latin America probably will continue to fall to the left of U.S. policy.

Mexico and Venezuela have entered into an arrangement with oil import–dependent Latin American nations to provide low-interest loans for up to one-third of the value of their oil imports.[14] Mexico supports center-left groups in El Salvador, in an effort to end political violence and provide a moderating and unifying force in the country.[14] Mexico will continue to be a spokesman for Latin America in international dialogues in the North-South debate and in the United Nations.[14]

Successful revolutions in El Salvador and Guatemala could bring the Mexican political system even more strain.[14] Should these revolutions prove successful, radical leaders in Mexico would probably begin pressing demands for a better income distribution.[14] Mexico's large numbers of unemployed and underemployed would be the targets of increased political rhetoric from the left. Right-wing political groups would view successful revolutions in Central American nations as a threat to the status quo and political stability in the region.[14] The importance of the military in the Mexican political arena would be likely to increase in such a situation. The most likely result would be a polarization of left-wing and right-wing political factions such as occurred during the Cuban Revolution.[14] Continued disagreements between Mexico and the United States with respect to tensions in Latin America should not preclude significant agreements or cooperation between the two nations.[14]

Mexico would like to spread its oil exports among a number of nations, with a policy goal of selling no more than half of total exports to any one nation.[14] The present world oil glut and weakness in the world oil price may force Mexico to modify its commitment to this policy.[14] President López Portillo took a large political risk when he authorized the construction of the natural gas pipeline to the United States.[14] Disagreements over the price of the gas created anti-U.S. rhetoric in Mexico.[14]

The illegal flow of Mexicans into the United States is another important topic of debate between the two countries, as are Mexican trade barriers and other protectionist policies, which have drawn criticism from firms in the United States.[14] Mexico and the United States have disagreed in recent years on the 1954 vote condemning communism in Guatemala, the inter-American peacekeeping force sent to the Dominican Republic in 1965, and on sanctions against Cuba approved by the Organization of American States.[14]

New initiatives in Mexican-U.S. relations should be forthcoming. Susan Kaufman Purcell states in the Winter 1981–1982 *Foreign Affairs*[14] that new initiatives should emanate from Mexican leaders. This would indicate their acceptability to Mexico and would also indicate that Mexico recognizes a willingness among U.S. policymakers to compromise and hear new proposals that spur increased dialogue on the problems facing the two countries in the region.[14] Domestic and international concerns in both countries must be balanced in order for them to develop a closer bilateral relationship.

REFERENCES

1. White, Lane. "Mining in Mexico." *Engineering and Mining Journal* 180 (November 1980).
2. U.S. Department of the Interior, Bureau of Mines. *Minerals Yearbook 1980*. Washington, D.C.: U.S. Government Printing Office, 1982.
3. Spooner, John, ed. *Mining Journal Annual Review 1982*. London: Mining Journal, June 1982.
4. U.S. Department of the Interior, Bureau of Mines. *Mineral Commodity Summaries 1983*. Washington, D.C.: U.S. Government Printing Office, 1983.
5. Sisselman, Robert. "La Caridad Copper." *Engineering and Mining Journal* 179 (October 1979).
6. U.S. Department of the Interior, Bureau of Mines. *Minerals and Materials*, August–September 1982.
7. U.S. Department of the Interior, Bureau of Mines. *Mineral Industries of Latin America*. Mineral Perspectives Series. December 1981.
8. Harris, Robert. "ROFOMEX Moves Mexico toward Phosphate Self-Sufficiency." *Engineering and Mining Journal* 181 (July 1981).
9. "Mexico: Reasons for Optimism." *Mining Journal* 290 (2 June 1978).
10. "Worldwide Survey: Latin America." *World Mining*, August 1981.
11. "Worldwide Survey: Latin America." *World Mining*, August 1982.
12. "Eyes on Mexico." *Mining Journal* 296 (15 May 1981).
13. Loprete, Carlos A., and Dorothy McMahon. *Iberoamerica*, 2nd ed. New York: Charles Scribner's Sons, 1974.
14. Purcell, Susan Kaufman. "Mexico-U.S. Relations: Big Initiatives Can Cause Big Problems." *Foreign Affairs* 60 (Winter 1981-1982).

Chapter IV-4

Chile

Chile is important to the world mineral economy largely because it is the world's largest copper producer and exporter.[1] The copper belt in Chile constitutes the largest, highest-grade copper occurrence in the world. The copper belt contains about 30% of the world's identified copper resources.[2] Chile is also the world's second-largest molybdenum producer, with 24% of the world's reserves.[3] Molybdenum is produced as a by-product of copper. Chile has one-third of the world's iodine reserves and ranks second in world production.[4] Iron ore and by-product vanadium production are of some importance in the world market.[4]

Chile produces many other minerals in quantities that figure prominently in world markets. Among these are metallurgical lithium, rhenium, and selenium.[4] Chile has over half of the world's lithium reserves.[1] Gold, silver, titanium (rutile), and tungsten occur in some of the Chilean porphyry copper deposits.

Gold and silver are currently produced as by-products of copper, lead, and zinc mining and refining.[4] Chilean mineral fuel production is limited to small amounts of coal, natural gas, and oil. Chile imports sizable volumes of both coal and petroleum.[1]

MINERAL RESERVES AND RESOURCES

Statistics on Chile's mineral reserves and production are presented in Tables IV-4-1 and IV-4-2. Reserve figures for vanadium and selenium are not available. Vanadium is present in Chile's iron ore (magnetite) deposits. Reserve estimates for the gold, silver, titanium, and tungsten contained in the Chilean copper are also unavailable.

Chile has a long mining history. The Chilean economy has historically been dominated by a series of mineral products. Gold dominated Chilean mining from the 1500's to about 1830.[5] Silver was the principal mineral product from then until the end of the nineteenth century. Nitrate production was very important to the Chilean economy from about 1880 to about 1925.

Since then, copper and by-product molybdenum production has dominated Chilean mining. Iron ore

TABLE IV-4-1 Chile: Mineral Reserves (Million Metric Tons)

Mineral	Reserves
Lithium	4.3
Copper	95
Rhenium	779
Coal	5500
Gas[b]	129
Iodine	1000
Iron ore	910
Molybdenum	2.45
Oil[c]	570
Gold	NA
Silver	NA
Vanadium	NA
Titanium	NA

[a]NA, not available.
[b]Unit of measure is billion cubic meters.
[c]Unit of measure is million barrels.

Sources: Ref. 1; Comisión Chilena del Cobre, *Boletín 4; Estadisticas del Cobre*, May 1982; Comisión Chilena del Cobre, *Reseno Sobre el Litio*, taken from Industrial Minerals.

has become Chile's second most important mineral product, and its by-product vanadium earns additional foreign exchange for Chile. Chile nationalized the copper industry in 1966, and most other mining activity is in the hands of a group of state-owned companies.[1] Some mineral deposits are being developed as joint ventures between private international resource companies and Chilean state companies. Foreign investment capital has been returning to Chile since the adoption of the foreign investment law in 1974, and new projects approved in 1981 have a combined value of $25 billion.[4]

GEOLOGY

This section relies heavily on the article by R. H. Sillitoe, "Regional Aspects of the Andean Porphyry Copper Belt in Chile and Argentina," which appeared in the February 1981 issue of *Applied Earth Science*.[2] The copper belt in Chile, and adjacent Argentina and southern Peru, is the largest and

TABLE IV-4-2   Chile: Mineral Production, 1977-1981[a] (Thousand Metric Tons)

Mineral	1977	1978	1979	1980	1981
Copper					
Mine output	1,056	1,036	1,060	1,068	1,081
Smelter output	888	927	947	953	954
Refined	676	749	780	811	776
Gold (1000 troy oz)	116	102	111	220	389
Iron and steel					
Ore and concentrate	4,670	4,336	4,634	5,440	5,242
Pig iron	432	539	611	644	582
Ferroalloys (mt)	8,604	8,224	11,432	11,555	10,550
Crude steel	548	598	657	712	641
Semimanufactures	395	446	503	572	504
Lead: Mine output (mt)	116	431	252	461	223
Manganese: Ore and concentrate	18	23	25	28	26
Mercury (76-lb flasks)	20	—	—	—	—
Molybdenum: mine output	11	13	14	14	15
Selenium (kg)	8,297	8,165	28,290	17,100	15,000
Silver (1000 troy oz)	8,461	8,210	8,740	9,958	10,927
Vanadium: Mine output (mt)	860	690	450	272	127
Zinc: Mine output (mt)	3,918	1,814	1,847	1,134	1,516
Barite	65	182	227	226	229
Iodine: elemental (mt)	1,856	1,922	2,410	2,601	2,678
Nitrates					
Sodium	482	423	467	469	471
Potassium	81	107	154	160	153
Coal	1,342	1,148	957	1,024	904
Coke	220	220	195	190	195
Natural gas (million ft^3)	133,857	123,588	138,094	135,000	130,000
Petroleum (1000 bbl)					
Crude	7,119	6,281	7,561	12,159	15,104
Products	32,581	34,671	36,809	35,204	31,759

[a]Unless otherwise specified, all figures are for metal content.
Source: Ref. 1.

highest-quality copper deposit in the world. Up to 30% of the world's identified copper resources are contained in the major porphyry deposits in this part of the world[2] (see Figure IV-4-1).

The Andes copper deposits typically contain valuable by-product and co-product minerals. Molybdenum is a co-product of copper production at all of Codelco's operating divisions.[6] Recoverable gold and silver values are present in some Chilean copper deposits. The El Indio mine is primarily a gold and silver mine, with by-product copper. Other minerals recovered in the refining of copper include rhenium, selenium, and tellurium.[3]

Iron ore mining featured more prominently in Chile's overall mining value and export earnings.[5] Vanadium is produced as a by-product of magnetic ores.[3] Most of Chile's porphyry copper deposits are associated with isolated stocks or in the roof zones of comagmatic batholiths.[2] Tourmaline breccia pipes are another source of Chilean copper. The

largest mine exploiting breccia pipe deposits is Exxon's 40,000-metric ton per year Disputada mine.[2]

The primary source for the minerals contained in Chilean porphyry copper deposits is material subducted at the plate margin in the Peru-Chile trench. The material is primarily subcrustal and is subducted where the east Pacific plate collides with the Americas plate.[2] Most of the deposits consist of linear, longitudinal subbelts, which gradually migrated eastward between the upper Carboniferous and the Mid-Upper Miocene.[2] About one-fourth of the deposits show no obvious structural control, while 30 to 40% have had their emplacement influenced by north-south zones of structural weakness.[2]

At Chuquicamata, the primary ore mineral in the sulfide part of the deposit is chalcocite. Other ore minerals include enargite, chalcopyrite, covellite, molybdenite, and sphalerite.[6] Mineralization

FIGURE IV-4-1 Chile: Mines, processing facilities, and infrastructure. (From Ref. 1.)

at El Teniente (primarily bornite and chalcopyrite), is centered in the Lower Tertiary Farellones Formation and in the underlying, intensely folded Coya-Machali Formation, which dates to the Upper Cretaceous.[6]

The El Teniente ore body experienced both hypogene and supergene alteration and was intruded by quartz-dacite-diorite intrusives.[6] Hypogene mineralization produced a central bornite zone surrounded by chalcopyrite.[6] The El Salvador and Andina mines produce copper from disseminated dacite porphyry and tourmaline breccia deposits. Mineralization is in the form of freckles and veins and fingers of copper and molybdenum mineralization.[6] The major ore minerals are chalcocite, bornite, chalcopyrite, and molybdenite, with lesser amounts of sphalerite, galena, cuprite, and native copper.[6]

Faulting was an important factor limiting the alteration and mineralization at both Chuquicamata and El Salvador.[6] At Chuquicamata, a major fault limits the ore body to the west, while at El Salvador long periods of both normal and inverse faulting gave porphyritic intrusives easy avenues for intru-

sion.[6] Secondary sulfides at many deposits are the result of the leaching of primary sulfide deposits by acid meteoric water.[6] Mineralized channels and sheets containing secondary sulfide ore minerals are found near Chuquicamata (Exótica) and El Salvador.[6]

The most fundamental factor controlling the formation of Chilean porphyry copper deposits is the eastward subduction of oceanic lithosphere at the plate margin in the Peru-Chile trench.[2] Faults and zones of weakness trending north-south aided in the ascent and localization of stocks and plutons containing copper minerals.[2]

LOCATIONS OF MINES AND PROCESSING FACILITIES

Major mines and mineral processing plants, as well as most mineral deposits are in the northern half of Chile. The Chilean copper belt (Figure IV-4-2) and the Salar de Atacama are the sites of most mining activity in Chile. Mineral fuels, including oil, natural gas, coal, and lignite, are found in far

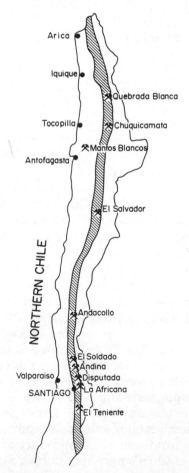

FIGURE IV-4-2 Chile: Copper occurrences and mines. (From Alexander Sutulov. *Chilean Mining.* Santiago: University of Concepcion Press, 1978.)

southern Chile.[1] CODELCO's largest mine, Chuqui-camata, is located at an elevation of 2700 m and is about 1500 km north of Santiago.[6] Chuqui is the northernmost of the CODELCO mines. The mine's location in the Atacama desert places it in the driest place on earth.[6] The concentrator at Chuqui has a 96,000-metric ton per day capacity, with smelter-refinery capacity at 500,000/375,000 metric tons per year, and a molybdenum flotation plant with the capacity to handle up to 3200 metric tons per day of bulk concentrates.[6]

Proceeding in a southerly direction, the next CODELCO mine is El Salvador. The mine is also in the Atacama desert at an elevation of 2400 m, about 1000 km north of Santiago.[6] The division's smelter and refinery are 50 km southeast of the mine at Potrerillos (2900 m above sea level).[6] Barquito and Llanta are two nearby towns that house some of El Salvador's support functions.[6] With the exception of Mantos Blancos, which is located west of the main copper belt near Anto-fagasta, the remaining copper mines are all near Santiago.

CODELCO's Andina division, and the two Exxon properties at El Soldado and SMP's Lo Aquirre and La Africana mines, are all near Santiago, slightly northeast of the capital. Andina is located about 50 km from Santiago, and facilities are at elevations from 3000 to 3600 m above sea level.[6] Production of over 50,000 metric tons per year of copper contained in concentrates is shipped to Japan for smelting.[6] When Exxon bought the bulk of Cia. Minera Disputada de Las Condes in 1977, the company acquired two producing mines and concentrators and one smelter. The Los Bronces deposit is high in the Andes at 3500 to 4150 m above sea level.[6] Production is about 40,000 metric tons per year and may be increased later in the decade.[4]

The remaining CODELCO division, El Teniente, is located about 100 km southeast of Santiago. The nerve center of El Teniente is the city of Rancagua (population 85,000) in Chile's central valley at an elevation of 500 m.[5] El Teniente's elevation is between 1900 and 2300 m above sea level. The distance between the mine and Rancagua is about 50 km. Production at El Teniente is approaching 300,000 metric tons per year of contained copper, making it CODELCO's second-most-prolific producer.[6] Two concentrators and one smelter handle El Teniente's ore production.

Chile's natural nitrates and metallic lithium are produced from brines in the Atacama desert.[1] All of Chile's major copper mines produce co-product molybdenum, and gold and silver are primarily ob-

tained as by-products in the refining of copper.[4] Iron and steel production are centered north of Santiago in Coquimbo province.[1] Virtually all of Chile's coal mining and oil and gas production areas are in far southern Chile.[1]

PHYSICAL INFRASTRUCTURE

Energy

Chile's energy infrastructure is limited. Small volumes of coal, natural gas, and oil are produced, but Chile continues to rely on oil imports for about 60% of its primary energy requirements.[5] The two-fifths of primary energy demand provided from domestic sources is made up of primarily oil (48%), coal (32.5%), and hydroelectric power (18%).[5]

Chile's energy resources are dominated by coal, which accounts for about 90% of the total.[5] The remaining 10% of Chile's energy production is made up of oil and natural gas (4%) and hydroelectric power (6%).[5] Chile's hydroelectric potential is over 10,000 MW.[5] A small (30-MW) geothermal plant is being built to provide electric power to the Chuqui copper-molybdenum mine.[3]

Chile's energy demand has grown 4 to 5% per year in recent years.[3] Energy demand is projected to increase at 6% per year throughout the 1980s.[3] Chile's energy balance is dominated by oil (77%), with coal (15.6%) and hydropower (7.4%) supplying the balance.[5] Installed electric generating capacity is about 3400 MW, over half of which is from hydro sources.[1] New projects include the 300-MW Antuco and the 500-MW Colbun-Machicura plants.[3] The development of Chile's hydroelectric and coal resources will be necessary to meet Chile's growing energy demand.

Chile's coal production comes primarily from the ENACAR mines at Punta Arenas in far southern Chile.[1] The total capacity at ENACAR's Lota, Schwager, and Aruaco mines is about 1.1 million metric tons per year.[1] Chile must import coal, primarily from the United States and Poland, to satisfy domestic requirements.[1] The bulk of Chilean oil production stems from fields in the Straits of Magellan.[1] Capacity is growing steadily and production topped 15 million barrels in 1981, more than twice the production level prevailing from 1977 to 1979.[4] Three new offshore platforms in the Straits of Magellan are the source of much of the increased oil production.[4]

Gross natural gas production has fallen in recent years, but the volume of marketed natural gas has remained fairly level at 125 to 140 billion cubic meters per year since 1977.[4] Almost 40%

of Chile's gross natural gas production is reinjected.[3] The bulk of Chile's natural gas is located in the Straits of Magellan area.[4]

Chile may have exploitable uranium deposits, but exploration activity has been limited. The state agency for nuclear energy, the Comisión Chilena de Energía Nuclear (CCEN), granted a uranium exploration license to Essex Mineral Company of the United States in 1981.[4] The recovery of uranium at Chuquicamata is under study.[4]

Chile has a network of crude oil, natural gas, and petroleum products pipelines. Oil production and imports are piped to major consumption centers. Crude oil pipelines total 755 km and petroleum product pipelines total 785 km.[1] Natural gas pipelines cover a total of 320 km.[1] Chile's dependence on oil imports is decreasing as domestic production increases and as alternative energy sources are developed. Chilean oil now provides close to half of the country's oil consumption.[3] Chile is developing coal, hydroelectric, and oil resources to lessen the country's dependence on mineral fuel imports. Chile's oil import bill is approaching $1 billion per year ($890 million in 1979).[3]

Water

Chile has sufficient water resources to meet the needs of domestic users, but the distribution of water resources is uneven. Large parts of far northern Chile are very dry, desertlike, and inhospitable. The Salar de Atacama, where lithium, iodine, and natural nitrates are mined, is one of the driest places in the world. Water availability should not constitute a barrier to further mineral development in Chile, but the cost of providing water to remote deposits could render them uneconomic.

Transportation

Chile has a well-developed transportation infrastructure. The Chilean railway system, one of Latin America's best, extends for over 3100 km.[1] Chile's road system covers over 75,000 km, with about 9000 km of this paved.[1] The Chilean road and rail systems run in similar patterns. Chile's mineral production is transported by a wide range of interconnected transport systems. Most of the country's mining output is shipped by rail, truck, conveyor belt, and slurry pipeline.[1] Exxon's Los Bronces copper mine utilizes an aerial tramway to transport ore 6 km to feed the San Francisco mill.[6]

Chile's transport infrastructure is most extensively developed in the northern two-thirds of the country. The long coastline and abundance of ports give Chile a well-integrated transport system.[1] The major ports for mineral exports are Arica, Antofagasta, and Valparaíso. Mineral exports from Bolivia use the same rail and port system, leaving South America at Arica and Antofagasta.[1] The mail rail lines runs from Iquique on the coast in far northern Chile, through the capital, Santiago, to near Puerto Montt, about two-thirds of the way down the coast.[1] A number of short connecting rail lines run east and west. There are two rail connections to Bolivia and two to Argentina.[1]

CODELCO's largest mine, Chuquicamata, uses mainly trucks for in-pit haulage.[6] The second largest producer, El Teniente, uses a rail system.[6] The other two CODELCO mines, Salvador and Andina, utilize a combination of rail, load-haul-dump vehicles, and pipelines and conveyors to transport mining output.[6]

CODELCO's Salvador mine suffers from a combination of expensive infrastructure and poor mineral transportation systems.[6] The mine is located in the Atacama desert, where labor and physical infrastructure are not readily available.[6] Electricity for the facilities is generated at the coastal shipping town of Barquito.[6]

In contrast, some of Chile's mines are well situated with respect to transport infrastructure and available laborers. The Chuquicamata mine is located near rail line connections. Soc. Minera Pudahuel Ltda. (SMP)'s Lo Aguirre mine is situated on a deposit near Santiago, with its large labor force, and Valparaíso, with the country's best port facilities. The deposit is barely 30 km from a Santiago bus terminal, and the Santiago-Velparaiso highway runs past the plant.[6] The climate is mild, the ore grade high, and the prospects for low operating costs are good.[6] Production at Lo Aguirre in 1981 was about 14,000 metric tons.[4]

The country's long coastline and advanced rail system facilitate the flow of mineral output. There are some variations among individual mines. A reduction of Chile's dependence on imported oil and coal would improve CODELCO's chances of increasing copper and molybdenum output. The bills for fuel imports cost Chile a large portion of the foreign exchange earned by copper and other mineral exports, thus limiting the further development of large-scale mining activity in Chile.

MINING LABOR

Chile introduced a new mining labor code in June 1979 (Decree-Laws No. 2756 and No. 2758), giving the workers the right to bargain collectively and to strike.[1] Other provisions of the law include an open shop, multiplant unions, inflation indexing of wages, a 60-day strike limit, and the right of employers to stage lockouts.[1]

CODELCO's El Teniente and Chuquicamata mines experienced labor unrest in 1979.[7] No work stoppages resulted from the unrest in 1979. The strike staged by eight mine workers' unions, affecting over 10,000 workers at El Teniente from late April to mid-July 1981, cut CODELCO's total copper production for the year.[3] Record first-quarter (1981) production mitigated the strike's effect on copper supplies.[3] The Chilean government has given the country's mine workers a new, more liberal labor law. Despite the reforms, the strike that occurred at the El Teniente mine in 1981 was the largest and most serious labor strike in Chile since the military took over in a 1973 coup.[3]

DOMESTIC MINING INDUSTRY

Overview

Mining Role in the Economy. The mining industry is important to Chile's GNP, exports, foreign exchange earnings, and ability to attract foreign investment capital. Mining contributed 7% of Chile's 1981 GNP, which was about $32 billion.[4] Mineral exports constituted 58% of Chile's total export value in 1981.[4] About 5% of all new private foreign investment in Chile is in the mining industry.[4] The mining, mineral-processing, and mineral fuels industries employ about 76,000 workers, about 2% of Chile's labor force.[1] Of the total, 38% are employed by CODELCO, 16% by ENAP and ENACAR, and 15% and 32% by the small and medium-sized mining companies, respectively.[1]

The mining industry will continue to be important to the Chilean economy for foreign exchange earnings and income.[3] Copper with other mineral production is a most important element in Chile's economic future. The inflow of private foreign investment capital should help fuel the expansion of Chile's mineral industries. An in-depth article on all of CIPEC's major producers in the *Engineering and Mining Journal* concluded that Chile was the country "where major new copper output can materialize faster than anyplace else."[6] Virtually all of this new output will be from privately owned mines.

Trends Among Mineral Producers

Chile's copper production has remained fairly constant since 1977 at between 1.03 and 1.08 million metric tons per year copper content.[4] Copper exports increased 10% from 1978 to 1980.[4] By-product molybdenum production increased 40% from 1977 to 1981.[4] Increases in gold production, mainly from St. Joe's El Indio mine, have resulted in a 335% increase in gold production from 1977 to 1981.[4] Silver production also posted sizable increases over the five-year period ending in 1981. Specific mineral production statistics are presented in Table IV-4-2.

Iron ore, pig iron, steel and ferroalloy production have all posted moderate increases over the same time period.[4] Selenium production increased 240% between 1977 and 1979, but fell 47% the following two years. Vanadium production in 1981 fell to 15% of the 1977 level.[4]

Production trends in the energy minerals industries are mixed. Over the last five years, coal production has fallen by one-third, and natural gas production has fallen by one-fourth.[4] In sharp contrast, Chilean oil production more than doubled over the same time period.[4]

Industry Structure

The Chilean mineral industry is dominated by six large state companies. The largest of these is the Corporación Nacional del Cobre de Chile (CODELCO). Another large state company that controls the mining industry is the Corporación de Fomento de la Producción (CORFO), the state industrial holding company.[3] The other state mineral companies include the Empresa Nacional de Minería (ENAMI), the parent company of medium-sized and small mines and smelters, and Cia. de Acero del Pacifico, S.A. (CAP), which operates the country's iron ore mines and only major steel mill. The remaining three state mining companies are the Empresa Nacional del Carbón S.A. (ENACAR), the state oil and gas company, and the Sociedad Química y Minera de Chile (SOQUIMICH). SOQUIMICH produces iodine, lithium, and nitrates, and refines nitrates.[3]

The small and medium-sized mines produce about 17.3% of Chile's copper production.[4] The four largest copper producers, CODELCO and the Disputada de Las Condes, Mantos Blancos, and Soc. Minera Pudahuel Ltda. y Cia. C.P.A. mines, produce 90.5% of Chile's copper production.[4] CODELCO's production was 82.7% of the country's total in 1981. Copper production by division within CODELCO, and by small and medium-sized mines for 1977-1981, is presented in Table IV-4-3. CODELCO produces all of the country's molybdenum output.[4] CODELCO's mines are probably the world's lowest-cost producers.[8] Chile has made some progress in increasing the domestic refining of its copper production. Between 1977 and 1981, the portion of the country's copper production refined domestically increased from 64% to 71.8%.[4] The copper industry is both vertically and horizontally integrated.

The energy industry in Chile is dominated by

TABLE IV-4-3   Chile: Copper Production, 1977–1981 (Thousand Metric Tons Copper Content)

	1977	1978	1979	1980	1981
Large mines (CODELCO)					
Chuquicamata	477.8	500.7	507.2	520.9	472.4
El Teniente	275.7	250.6	278.2	266.0	291.9
El Salvador	80.7	77.5	78.1	74.8	76.5
Andina	58.5	47.7	46.7	52.8	52.8
	892.7	876.5	910.2	914.5	893.6
Small and medium-sized mines					
Empresa Nacional Minería (ENAMI)	80.6	77.0	82.4	102.8	99.7
Mantos Blancos	28.5	38.2	36.0	29.7	31.7
Disputada de Las Condes	36.1	28.8	20.9	28.6	39.2
Other	18.3	15.0	11.5	2.1	16.6
	163.5	159.0	150.8	163.2	187.2
Grand total	1056.2	1035.5	1061.0	1077.7	1080.8

Source: Comisión Chilena de Cobre, Annual Reports.

the state coal company ENACAR and the state oil and gas company ENAP.[1] The recently formed Comisión Chilena de Energía Nuclear (CCEN) has control over the exploration, development, and exploitation of uranium and thorium minerals.[4] Over 70% of Chile's gold production emanates from the small and medium-sized mining companies, with an additional 26% recovered by CODELCO as a by-product of copper.[4,3] The El Indio mine, owned by St. Joe International Corp. (80%) and a Chilean company, produced 54% of Chile's 1981 gold production.[4] CAP, the state iron and steel company, is composed of seven subsidiaries. The firm is integrated into many phases of the steel industry, including iron ore mining and pelletization, production of ferromanganese and steel, and the marketing of iron ore, pellets, and steel products.[3]

Virtually all of Chile's mineral producing and processing facilities are owned by the Chilean government. Two private foreign firms own copper producing and processing facilities. Exxon Minerals Corp. owns about 3.6% of Chile's copper productive capacity, and the Hochschild Group owns about 2.9% of the capacity.[1,4] Private Chilean investors own 5% of CAP's iron ore and steel producing facilities.[1]

Government Role in the Mining Industry

The Chilean government is the dominant actor in the country's mineral economy.[1] Control over mining takes the form of owning mineral-producing companies and mineral-processing facilities, developing mining laws and policy, and coordinating the mining industry in general and mineral marketing specifically. The Chilean government therefore affects mining in many ways. Government-sponsored research is centered in the Centro de Investigaciones Minero-Metalúrigcos (CIMM). Virtually all of Chile's active mines and processing plants are owned by the state. Corfo, the state industrial holding company, oversees and coordinates Chilean mining activity.

Mining Taxation. The mining sector generates sizable revenues for the Chilean government. CODELCO tax and profit contributions for 1980 totaled about $1 billion.[10] The Chilean government and U.S. firms operating in Chile have had disputes over the taxation of income and profits obtained from mining activity. These disagreements were brought about by a series of policies adopted by the Chilean and U.S. governments in the mid-1950s. The United States unilaterally fixed copper prices at about half the market rate during World War II and again during the Korean War.[5] In response to this market distortion, Chile fixed the official peso–dollar exchange rate to 19.37 pesos to the dollar. Black market rates were on the order of 100 to 200 pesos per dollar.[5] Very low rates of return on investment were also imposed.

The result of these policies was a real tax rate increase from 6% to about 70%.[5] In return for a decrease in tax rates from 75% to 50%, the foreign companies agreed to double production with respect to the 1949–1953 average level.[5] With depressed copper demand and unutilized capacity, the production increase required little additional investment.[5] The agreement between the government and the mining companies came to an end when Congress imposed an additional 13% tax on company profits to upgrade the armed forces.[5]

New foreign investment that has taken place in

Chile since the adoption of the new Foreign Investment Decree-Law No. 1748 is subject to similar tax provisions as other mining activities in Chile. Investors have the option of paying a fixed 49.5% profits tax for 10 years or paying the same 48.8% profits tax to which Chilean companies are subject.[1] New tax laws favor private foreign and domestic investment in the mining of all minerals that have not been reserved for exploitation by official state monopolies.[1]

Government Subsidies. The Chilean government subsidizes the mining industry in many ways. Technical and financial support are given to the mining companies owned by the state. Large investments in processing facility expansion and infrastructure improvement are financed by the government through Corfo or the state mineral companies. Chile's mineral output is marketed by several state agencies, including two CAP subsidiaries, Pacific Ores and Trading and Acero Comercial S.A. and Molymet, a molybdenum marketing company.[4]

Government Ownership of Producing Companies. As stated before, the Chilean mining industry is almost completely state owned. Corfo oversees mining activity. Six state companies dominate mineral production and processing. ENAMI purchases and processes most of the output of small and medium-sized mining companies.[1] CODELCO is the largest of these state mineral companies and is the world's single largest copper-producing company.[9] The state has chosen to develop some as yet undeveloped copper deposits as joint ventures with private firms.

Mining Law and Policy. Chilean mining law and policy have been reoriented toward attracting foreign investment capital and reducing the role of the state government in mining activity and the economy in general. This reorientation toward expanding mining with private foreign investment capital is the result of Chile's realization that desired investments in Chilean mining are too large to be met by domestic investment capital.[5]

Chile adopted Decree-Law No. 1748 concerning foreign investment in 1977. Among the provisions of this law are clear definitions of the contractual obligations of the state and the mining companies, provisions for repatriation of profits, a guarantee of nondiscriminatory access to foreign mineral markets, and a clause giving Chile the opportunity to purchase strategic minerals at prevailing market prices.[1]

Chile approved a new mining law in December 1981.[4] The law was designed to promote investment in new mining ventures and to streamline existing mining legislation.[4] The ultimate goal of this policy is a doubling of copper production by the end of the 1980s.[4] The new law establishes exploration and exploitation concession regulations. Exploration concessions are for two years with one two-year extension.[4] Exploitation concessions have no fixed time limit and are extended as long as all license fees are paid. Offshore areas, strategic areas, oil-and-gas-and-lithium-producing areas are not subject to these concession provisions.[4] The new mining code grants private citizens the right to own mineral concessions. Concession owners are protected against expropriation without compensation by government guarantees. All expropriations would entitle the concession owner to full compensation as determined by judicial review.[4] This provision should reduce fears of expropriation among foreign investors, as a stable, guaranteed arrangement for compensation is in place.[4]

Domestic Supply–Demand Dynamics

Analysis of the availability of Chile's mineral production for export to overseas markets must begin with an examination of Chile's productive capacity for major minerals. From there, a knowledge of the dynamics of mineral supply and demand allows the estimation of the portion of production that is available for export. The detection and analysis of trends is important here, as is a good knowledge of the country's dependence on imports of energy and mineral resource commodities.

Chile has a large productive capacity for a wide range of minerals, many of which are produced in quantities that are important on a world scale. Of these, copper, molybdenum, iron ore, and several others are important to both world supply and the Chilean balance of trade. However, many of Chile's mineral and energy industries produce at levels well below their full capacity. Sizable declines in overall ore grades at all major copper mines signal a challenge to the Chilean copper industry to maintain present productive capacity. Productive capacities for some of Chile's major minerals are shown in Table IV-4-4. Specific capacities for minerals produced as by-products and co-products are generally unavailable, and have been estimated based on the last several years' production levels and trends.

Chile's mineral production is primarily for export. Only minor volumes of Chilean mineral production are destined for domestic consumption. Some of Chile's mineral production is processed domestically, but most is ultimately exported. Chile is about 45% self-sufficient in mineral fuels; hence

TABLE IV-4-4   Chile: Mineral Productive Capacity, 1981
(Metric Tons)

Mineral	1981
Copper	
Mine output	1,100,000
Smelter	960,000
Refinery	800,000
Molybdenum: mine output	15,000
Iron	
Ore	11,000,000
Pellets	3,500,000
Steel	750,000
Ferroalloys	12,000
Manganese: ore and concentrate	25,000
Selenium (kg)	28,000
Vanadium	800
Gold (troy oz)	400,000
Silver (troy oz)	10,900,000
Barite	230,000
Iodine	2,700
Nitrates	
Sodium	470,000
Potassium	150,000
Coal	1,000,000
Coke	200,000
Natural gas (million ft^3)	130,000,000
Petroleum (bbl)	
Crude	15,000,000
Products	35,000,000

Sources: Refs. 3 and 4.

none of the Chilean production is exported.[3] Chile imports coal and oil to meet domestic demand, although increased domestic oil production has checked the growth of Chile's oil import bill.[3] Domestic iron ore consumption by the CAP steel plant at Huachipato in 1981 was about 1 million tons.[4] This represented about 11.6% of 1981 production. Silver is another mineral produced in Chile and consumed domestically in sizable volumes. Some manganese and gold are also consumed domestically, as is about one-fourth of iodine and natural nitrate production.[1]

All of Chile's molybdenum production is exported as concentrates and molybdic oxide, and 95% of copper production is exported.[1] These high export levels are relatively stable. Table IV-4-5 shows Chile's copper production by level of processing. In 1980, Chile's copper exports were mainly refined copper (73.6% of total exports), blister copper (14.5%), and ores and concentrates (11.9%).[4] The domestic consumption of Chile's mineral production should remain relatively stable at low levels.

Copper production is relatively stable at just over 1 million metric tons, and molybdenum production is in the range 13,000 to 15,000 metric tons per year. Almost all new copper capacity in the 1980s will be added by the small and medium-sized mines. Declining ore grades at all CODELCO mines, except El Salvador, should reduce molybdenum output by about 10%.[11]

Completion of the lithium plant under construction in the Salar de Atacama by Foote Minerals and the Sociedad Chilena del Litio Ltda. will add 1000 tons per year of metallic lithium to the world market. This is about 20% of all freely traded lithium.[12] Output began in 1984.

Coal production is forecast to decline, but the drop will be offset by reduced domestic consumption. Iron ore production levels are about 80% of capacity, with almost 90% of the 8.5 million metric tons of ore produced in 1981 available for export.[4] By-product vanadium production has fallen dramatically since 1977.[4] High prices in recent years have fostered increased gold and silver production, although little is available for export.

The South American share of CIPEC's "big four" copper production has increased substantially since 1973.[6] Peru and Chile were responsible for 44% of 1973 production, with the remaining 56% coming from Zambia and Zaire.[13] In 1978, the two continents' respective importance had reversed itself, with Chile and Peru producing 57% of the total while the two African nations added 43%. Producer stocks can be important factors in

TABLE IV-4-5   Chile: Copper Production, 1977-1981, by Level of Processing
(Thousand Metric Tons)

Type	1977	1978	1979	1980	1981
Refined	676.0	749.1	781.8	810.7	775.6
Blister	212.4	178.3	167.8	142.4	178.3
Other[a]	167.8	108.1	111.4	114.6	126.9
	1056.2	1035.5	1061.0	1067.7	1080.8

[a]Includes concentrates, slags, and minerals.

Source: Comisión Chilena de Cobre, Annual Reports.

international commodity markets. The ore inventory at El Teniente in 1978 totaled about 38 million metric tons, about two years' production.[6] The stockpile at Chuqui typically contains about 400,000 metric tons of crushed ore.[6] Large stocks in the hands of producers can also mitigate the effects of work stoppages at CODELCO's major mines.

Most of Chile's mineral production will continue to be exported. As the backbone of the Chilean economy, copper will continue to dominate mineral exports. Continuing as important sources of foreign exchange earnings will be molybdenum and other copper by-products, natural nitrates, and iron ore.[4] Revenue generated from molybdenum exports added over $260 million to CODELCO's 1981 revenues, pulling the company out of the red with a $97 million profit that year. Revenue from other copper by-products, such as rhenium, sellenium, and tellurium, as well as the traditional molybdenum, gold, and silver, will affect the economics of most Chilean mining ventures in the 1980s.

POLITICAL ENVIRONMENT

Chile has been ruled by General Augusto Pinochet since Marxist President Salvador Allende was overthrown by the military in September 1973.[14] President Allende had a leftist and minority coalition government. Pinochet has kept control of the army as president, eliminating potential rivals in the process. Pinochet's term in office is scheduled to continue until 1989. Chile follows the traditions of parliamentary rule. Military takeovers have been relatively rare, with only two occurring in the twentieth century.[14]

Prior to the overthrow of Allende, Chile had failed to develop large, modern, mass-based, Western-oriented political parties.[14] Little attention was given to economic and political realities, and there was no consensus on policies to manage the economy.[14] The Pinochet administration adopted a coordinated economic policy advocating free trade.

The failing Chilean economy has increased political dissent and pressure on Pinochet to improve economic conditions. In 1982, Pinochet fired his economic advisors, the "Chicago Boys," in what many political observers viewed as an effort to find a scapegoat for Chile's economic problems. The advisors' free-market-oriented economic policies, based on the theories of Milton Friedman, had not improved the Chilean economy. Pinochet's extravagant life-style and new public-financed pres-

idential palace have drawn criticism in a period of austerity and government calls for sacrifice.

In another effort to find a scapegoat for the country's economic problems, Pinochet decided that the two largest financial conglomerates in Chile, and their large foreign debts, were the cause of the economic problems. He suspended debt payments by the two firms in January 1983, resulting in a drain of $800 million in foreign reserves in the following two months and the creation of a black market for dollars at up to 80% above the official exchange rate. Pinochet has been accused of avoiding Chile's economic crisis by vacationing outside Santiago and touring southern Chile for two weeks while the foreign bankers who hold Chile's $17 billion foreign debt were forced to wait.

Opposition political parties are not a significant political force in Chile. The Socialist Party is split by internal dissent. The Communist Party is trying to get its exiled members back into Chile, and the Christian Democrats are discredited by the military government.[14] The leftist supporters of former President Allende are mostly in jail or in exile.[14] Politicians have increased their calls for a return to democracy, and labor and political groups have increased their criticism and protests directed toward the Pinochet government.

Pinochet has also drawn criticism for his handling of two recent political expulsions. The president of the Wheat Farmers Association was expelled in January 1983 for complaining about agricultural policy. Pinochet provoked a serious confrontation with the Catholic Church by sending 50 armed police to arrest two priests, accused of being political agitators, and transport them to the airport. Pinochet has been more lenient toward political opponents recently, although he is viewed as unpredictable in Chile. In the past he used the secret police to arrest and exile all political opponents.

Important political issues in Chile include the labor code, income distribution, and timetables for Chile's return to civilian rule.[14] Important economic issues include the management of Chile's $17 billion foreign debt, 1982's 20% inflation rate, and the combination of low mineral commodity prices and high real interest rates.[12] Economic recovery in Chile will be a prerequisite for Pinochet's completion of his term in office.

Large numbers of Chile's usually apolitical middle-class and upper-class citizens now oppose Pinochet and his nepotism in filling high government posts and awarding government contracts. Labor and political leaders have opposed Pinochet all along. Without a significant economic recovery soon, many Chilean political observers feel that Pinochet's days in office are numbered.

REFERENCES

1. U.S. Department of the Interior, Bureau of Mines. *Mineral Industries of Latin America*, Mineral Perspectives Series. December 1981.
2. Sillitoe, R. H. "Regional Aspects of the Andean Porphyry Copper Belt in Chile and Argentina." *Applied Earth Science* 90 (February 1981).
3. U.S. Department of the Interior, Bureau of Mines. *Minerals Yearbook 1980*. Washington, D.C.: U.S. Government Printing Office, 1982.
4. U.S. Department of the Interior, Bureau of Mines. *Minerals Yearbook 1981*. Washington, D.C.: U.S. Government Printing Office, 1983.
5. Sutulov, Alexander. *Chilean Mining*. Santiago: University of Concepción Press, 1978.
6. Dayton, Stan. "Chile." *Engineering and Mining Journal* 179 (November 1979).
7. Spooner, John, ed. *Mining Journal Annual Review 1980*. London: Mining Journal, June 1980.
8. "Chilean Copper: Plans for the Future." *Mining Magazine*, October 1980.
9. "Codelco Costs Now Seen as Topping Current Cu-Mo Prices." *Engineering and Mining Journal* 182 (July 1982).
10. "Codelco-Chile Meeting the Challenge." *Mining Journal* 248 (18 June 1982).
11. "Chile Will Maintain Policy of Stepped Up Copper Production during 1982." *Engineering and Mining Journal* 182 (June 1982).
12. Spooner, John, ed. *Mining Journal Annual Review 1982*. London: Mining Journal, June 1982.
13. "Chilean Copper: Major Expansion." *Mining Journal* 292 (29 August 1980).
14. First International Bancshares, Economics Division. "Chile." April 1980.

Chapter IV-5

Peru

Peru is one of the major mining countries in the world, with a long mining history that dates to pre-Columbian times. Peru is the world's fourth-leading silver and zinc producer, ranks fifth in lead output, and ranks sixth in world copper production.[1] Exports of these four mineral commodities alone represented over two-fifths of Peru's total exports in 1981.

Peru produces and exports a wide range of mineral commodities. In addition to those minerals mentioned above, other minerals produced in large volumes include bismuth, cadmium, indium, selenium, tellurium, gold, tungsten, antimony, molybdenum, and iron ore.[2]

Peru also produces sizable volumes of crude oil and natural gas. Exports of crude oil earn important foreign exchange for the country.[2] The nation's energy balance is dominated by oil and natural gas.[1] Peru has a large hydroelectric potential.[1]

MINERAL RESERVES AND RESOURCES

Peru publishes no general mining statistics; therefore, estimates of the country's mineral reserves and resources are somewhat suspect. Peru's mineral deposits typically contain several minerals, such as copper-lead-zinc, lead-zinc-silver, and so on. One deposit (Toromocho) contains 10 different metals. This complicates the estimation of Peru's mineral reserves. Table IV-5-1 shows the U.S. Bureau of Mines estimates for Peru's reserve base for a number of minerals.

Peru's diverse mineral endowment is the basis for a major mining industry. Important by-product and co-product relationships affect the economics of mining ventures. In recent years, high silver prices have kept mining operations in operation during periods of extremely low commodity prices.

GEOLOGY

Peruvian copper deposits are of the porphyry type.[1] Most are disseminated deposits. Deposits typically are composed of a cap or top layer of oxide ores over a richer deposit of sulfide ores.[3] Ore grades range from 1.5 to 2% copper, with a cutoff of 0.45 to 0.5% copper.[3] Average grades are 0.9% to 1.1% copper, slightly lower than deposits in neighboring Chile.[3]

The large deposits also contain associated molybdenum values averaging 0.023% MoS_2.[3] The Cobriza underground mine exploits a replacement-type deposit that grades 1.8% Cu with 0.7 oz per ton of silver.[3] The Cerro Verde mine treats oxidized ore (1.1% Cu) and has a large low-grade sulfide reserve grading 0.6% Cu.[3]

The Cuajone deposit is composed of three layers that are roughly flat lying and are laterally continuous for an area of about 1200 m by 900 m.[3] There is a primary sulfide ore body overlain by an enriched sulfide zone and an oxide layer. Chalcocite is the main copper mineral in the entire deposit, and molybdenite appears in small quantities in an irregular distribution.[3]

The Toquepala copper deposit occurs in and around a large breccia pipe. The major ore minerals are chalcocite and hypogene chalcopyrite. Mineralization is highest in the diorite portion of the deposit and in most of the breccia.[3]

The ore body at the Cobriza mine is an area of mineralization localized in a series of anticlines in the central Andes as a replacement of a large calcareous formation. The vein is composed of hornblende, diopside, augite, quartz, pyrrhotite, and pyrite.[3] The two ore bodies at Cerro Verde have a cap layer of oxide ore that is being mined and a very large sulfide ore body that underlies it. The major oxide mineral is brochantite.[3]

LOCATIONS OF MINES AND PROCESSING FACILITIES

The locations of Peru's major mines and mineral processing plants are shown in Figure IV-5-1. Peru's major copper and polymetallic deposits are in the southern half of the country. Most oil production is in the eastern Amazon basin region, and in northwestern Peru near Talara.[1] A wide range of metals are produced from the large mineralized zone in Peru's central Andes.

The country's primary metal-processing center is Centromín's La Oroya metallurgical complex.[3]

TABLE IV-5-1   Peru's Mineral Reserve Base[a] (Thousand Metric Tons)

	Reserve base	Share of world reserves (%)
Mineral		
Bismuth	6,300	5
Copper	28,000	5
Gold (kg)	2,000,000	3
Iron ore	1,600,000	1
Lead	3,600	2
Molybdenum	242	3
Selenium	11,800	6
Silver (kg)	18,000,000	10
Tellurium (mt)	3,000	5
Tungsten	85,000	3
Zinc	10,700	4
Fuels		
Coal	NA	NA
Natural gas (million m^3)	36,000	1
Petroleum (1000 bbl)	700,000	1

[a]NA, not available.

Source: Ref. 1.

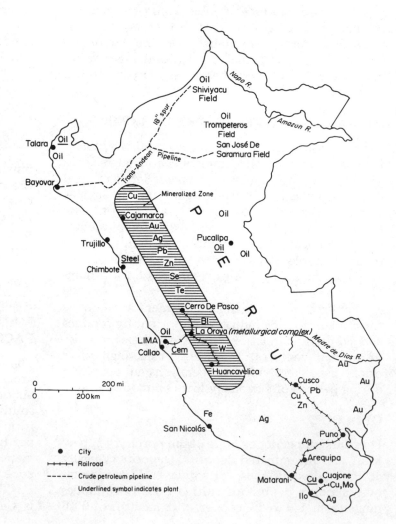

FIGURE IV-5-1 Peru: Mines, processing facilities, and infrastructure. (From Ref. 1.)

Oil refineries are located near the producing fields or on the Pacific coast.[1] Southern Peru Copper Corp. (SPCC) has a large smelter at Ilo, on the coast near their large open-pit mines at Cuajone and Toquepala.[1] The country's only steel mill at Chimbote is closed due to depressed demand and large stocks of finished products.[4]

SPCC's Cuajone and Toquepala mines produce most of the country's copper and molybdenum contained in concentrates. Other major mines are Mineroperú's Cerro Verde mine and Centromín's Cobriza and Morococho mines.[1] Most of Peru's refined metal production comes from the La Oroya complex. In addition to copper, La Oroya refines lead, zinc, silver, gold, bismuth, cadmium, indium, selenium, and tellurium metals.[1]

Most of Peru's gold is produced from placer deposits in the southeastern Amazon basin region.[1] Centromín produces most of Peru's silver, lead, and zinc, with the remainder from a number of small and medium-sized mining companies.[2] Peru's major tungsten mines are at Pasto Bueno, San Cristóbal, and a mine owned by Minera Regina S.A.[2]

PHYSICAL INFRASTRUCTURE

Energy

Peru's energy balance is dominated by oil and natural gas, which together represent 60% of total energy demand.[1] Large mining and mineral processing sites have their own energy sources, as a part of the project's overall infrastructure development plan.[1]

Peru has an extensive and largely undeveloped hydroelectrical potential. Over 70% of Peru's 2860 MW of installed generating capacity is based on hydro sources.[1] Several new hydroelectric projects totaling 1200 MW and a 500-MW coal-fired plant are in the planning and construction phases.[1]

The major electricity generating facility for SPCC's copper facilities is located at the firm's Ilo smelter complex. Waste heat from the furnaces and boilers powers four steam turbines, with a total installed capacity of 176 MW.[3] The electricity travels a 230-km closed loop of 138-kV transmission lines.[3] The plant supplies power to nearby towns and to the Cuajone mine (42 MW), the Toquepala mine (33 MW), and the Ilo processing center (15 MW).[3] Electric power is also sent to the pump station at Lake Suches and to other nearby facilities. Power for the Cobriza mine is transported 55 km by a 30-MW transmission line from the Mantaro hydroelectric plant.[5] Two substations at Campo Armino and Pampa de Coris distribute power to the mine's different facilities.

Water

The scarcity of water does constitute a barrier to Peru's further mineral development. Parts of the country are very dry for part of the year and very wet during other parts of the year. Drought affects copper production, and heavy rains hamper mineral transportation on mountain roads. Heavy rains also cause mine flooding and severe land and mud slides. Northern Peru suffered a severe drought in 1980. Production at the Minas Aguila S.A. mine in the Ancash Department was held to just half of the mine's 60,000-metric ton per year capacity.[2] In an effort to avoid future problems, the firm extended its water supply to a lake higher in the Andes.[2]

Water scarcity has limited the production from existing facilities, and it may affect future mineral development. The availability of water will in part determine the size of the planned coal mine at the Alto Chicama deposit.[2] Proposed output levels range from 1 to 3 million metric tons per year, and water availability will be an important determinant of the project's final capacity.

SPCC's Cuajone and Toquepala mines receive their water supplies from Lake Suches, about 25 km from the Cuajone mine.[3] A human-made reservoir near the Cuajone mine stores water for use at the mine and concentrator. The water is pumped and in some places flows by gravity. Lake Suches is about 2500 ft above the open pit at Cuajone.[3] Water requirements at the concentrator are about 300 gal per short ton milled.[3]

Mineroperú's Cerro Verde mine obtains its water from the Río Chili.[3] Three pump stations move the water to the mine complex. Water for the Cobriza mine is supplied by gravity feed from the nearby Huaribamba River. The Tintaya copper project will utilize water from the Salado River and a group of eight water wells located southwest of the mine complex.[6]

The Ilo smelter and refinery complex uses a desalination plant to obtain its fresh water from seawater.[3] Boilers fired by fuel oil power the desalination plant and also power turbines to generate electricity for the smelter, concentrators, and townships.

In the development of the Toquepala mine, about half of SPCC's expenditures were for infrastructure, including the provision of a new fresh water supply.[7] The bulk of the remaining expense was for long railroad tunnel boring.

The main water problem in Peru stems from the seasonally heavy rains and long periods of drought. Water is often available in the wrong places and is of inferior quality. Water is in short supply in the dry months, while roads become dangerously impassable during wet periods. Water supplies are sufficient, but seasonal variation in rainfall and river flow volumes create problems for large industrial water consumers.

Transportation

Most of the output of Peru's mines and processing centers is transported by rail and truck. Three short, unconnected rail lines link mineral-producing regions to shipping and processing centers on the Pacific coast. The Peruvian rail system covers close to 2000 km.[1] Peru's transport infrastructure is shown in Figure IV-5-1.

Peru's road and highway system totals 53,000 km, about 10% of which is paved.[1] Travel by road in Peru necessitates huge changes in elevation. Roads climb over mountain ranges and drop into intermountain valleys, and fall off to the Amazon basin to the east and to the Pacific Ocean to the west.

A total of five ports handle Peru's mineral exports. The port at Bayovar in northern Peru is the export terminal for oil produced in the Amazon basin oil fields.[1] Most of Peru's minerals exports leave the country through the ports at Ilo, Matarani, San Nicolás, and Callao.[1] These ports are all in the southern half of Peru.

The rail line serving SPCC's Cuajone and Toquepala mines runs a total of 145 miles.[8] The existing infrastructure serving Toquepala improved the economics of the Cuajone project. A series of five railroad tunnels totaling 17 miles links the Cuajone mine and the rail line from the Toquepala mine to the port and smelter at Ilo.[3]

The longest of the tunnels was driven over 9 miles through a mountain. When this tunnel was holed through in November 1975, alignment at the junction was within 2 inches.[8] The steep sides of the valleys in this part of the Andes led to the building of very short spans between the tunnels. The tunnels also transport mill tailings to the Toquepala tailings disposal area.[8] A modern rotary car dumper unloads concentrates from Cuajone and Toquepala at the Ilo smelter.[8] Blister copper from the Ilo smelter is shipped by rail the 5 km to the Mineroperú refinery or to the docks for export.[8]

Centromín's expanded Cobriza mine and concentrator have poor infrastructure to handle their increased output. The transportation of concentrates constitutes a bottleneck in the transportation of Cobriza's mineral output.[9] Small trucks carrying concentrates must travel over 200 km on rough mountain roads that twice climb above 13,500 ft.[9]

Centromín has begun construction of a 50-km truck bypass that will keep the road at high elevations until the final descent into Huancayo.[9] This bypass will cut the round-trip travel time by about three hours.[9] The state rail company is also building a rail spur that will transfer the truck/rail transfer station outside the city of Huancayo.[9] The new road will permit the use of larger trucks.

Concentrates produced at Cobriza and trucked to Huancayo are transported by rail to the La Oroya metallurgical complex for smelting. La Oroya smelts most of the country's small and medium-sized mines' output. Most of the lead, zinc, silver, copper, and other metals contained in concentrates are shipped to the La Oroya complex by truck.

The mineral production of northern Peru, mainly from small and medium-sized mines, is transported by truck over mountain roads to Cerro de Pasco and by rail on to the La Oroya complex. The Cerro Verde mine is linked by a 12-km road to the Pan-American Highway and the port at Matarani, where concentrates are exported.[10]

The Tintaya copper project of the state company Tintaya S.A. has completed some of its mineral transport infrastructure.[10] A total of 84 km of highways had been completed as of August 1982, and the link between Tintaya and Arequipa was nearly complete.[10] Concentrates will be transported by rail to Matarani for export, and production was begun around September 1984.[10]

Crude oil production from the Shiviyacu, Trompteros, and San José de Saramura fields is shipped to the export terminal at Bayovar through the Trans-Andean pipeline.[1] An 18-inch-diameter spur connects the Shiviyacu field to the main pipeline.[1] The Bayovar terminal can handle ocean-going oil tankers.

MINING LABOR

Peru's mining labor force is about 80,000, roughly 2% of the country's total labor force.[1] The mining workers are highly unionized and have the ability to affect international trade flows. Strikes at major mines and processing centers have lowered Peruvian mineral production in recent years.

The year 1979 was characterized by labor unrest and a long teachers' strike.[11] Copper output was down 8% in 1980 as the result of strikes at the country's two largest mines.[2] The mine and concentrator at Cuajone were closed by strikes for a month, and the Toquepala mine was hit by an 18-day strike. The La Oroya metallurgical complex also suffered from labor problems in 1980.[2]

Total production losses from strikes in 1981 totaled 5.7 million worker-hours, over twice the large 1980 loss.[12] The Cobriza mine experienced problems in attracting and keeping skilled workers for the concentrator expansion project.[5] The mine's isolated location was the main cause of this labor shortage. Labor disputes lowered Peru's 1981 steel output.[2]

The Peruvian government adopted measures in 1982 to aid the laborers in the small and medium-sized mining sector. Provisions include a six-month freeze on layoffs and wages, with cost-of-living adjustments the only exception.[13] These wage controls and the freeze on firings ran through January 1983. In another decree, the miners' right to strike and to have collective bargaining was suspended.[14]

Continued labor unrest is likely in Peru. The unionized mining labor force knows that it has considerable power to affect Peru's mineral production, exports, and foreign trade balance. Recent reforms to the mining law include a profit-sharing clause that requires that mining labor receive up to 10% of new income before taxes.[15] Gold producers must distribute 7% of their net income before taxes to their workers and employees.[15]

Reforms of mining labor contracts, with provisions for labor participation in management, profit sharing, seniority, and job tenure, are in order. With cooperation between government, management, and unionized labor, the potential for future labor unrest in Peru may decline. Significant reforms have yet to be fully realized.

DOMESTIC MINING INDUSTRY

Overview

Minerals and mineral fuels play an important role in the Peruvian economy. Minerals exports earned $1.86 billion in foreign exchange in 1980,

and oil exports brought in another $810 million.[2] Together oil and mineral exports accounted for 77% of Peru's total 1980 exports.[2] Foreign investment in nonferrous mining in 1981 was just under $200 million, about 20% of total direct foreign investment.

The mining industry accounts for about 10% of Peru's gross national product (GNP).[2] Petroleum adds another 5% of total GNP.[2] Peru's GNP has grown at a moderate rate (3 to 4%) in recent years.[2] The share of Peru's GNP provided by mining has fallen slightly in recent years, as services play a larger role in the economy.

Trends Among Major Mineral Producers

Production of copper, Peru's most important foreign exchange earner, peaked in 1979 at 400,000 metric tons.[3] Production fell 8% in 1980 and another 10.6% in 1981.[2] The primary reasons for the production declines were labor problems, drought in northern Peru, and falling metal content in the country's concentrate output from SPCC's two open-pit mines.[16]

Molybdenum recovery has improved, and production has increased substantially in the last several years. Improved recovery of selenium and tellurium in the refining of copper has increased output of these mineral commodities.[2] Production of bismith, tin, lead-zinc, and gold and silver have all increased in the past few years.[2] Peru's mineral production for the period 1976–1981 is shown in Table IV-5-2.

The most important trend in Peru's mineral

TABLE IV-5-2   Peru: Mineral Production, 1976-1981 (Metric Tons)

Mineral	1976	1977	1978	1979	1980	1981[a]
Antimony	168	819	745	762	1,050	721
Arsenic	1,042	1,367	1,257	3,222	3,205	NA
Bismuth	597	644	611	527	431	782
Cadmium	429	380	350	424	—	—
Copper	228,400	327,070	366,400	400,385	367,560	328,850
Gold[b]	121	104	103	142	149	157
Indium (kg)	3,721	3,734	3,302	3,484	3,675	NA
Iron ore	4,722,000	4,064,000	3,199,000	3,622,000	3,722,000	4,000,000
Lead	174,700	175,700	182,700	174,000	177,000	187,000
Molybdenum	446	456	729	1,086	1,616	2,700
Selenium (kg)	8,743	15,936	12,927	18,320	27,535	25,070
Silver[b]	38,661	38,812	37,045	43,415	40,500	46,810
Tellurium (kg)	12,306	18,370	15,382	21,202	20,104	20,400
Tin	299	366	800	929	755	1,380
Tungsten	845	526	582	755	700	NA
Zinc	456,069	405,384	457,700	432,000	403,291	496,700
Petroleum[c]	27,936	33,276	55,079	69,952	71,369	70,400

[a]NA, not available.
[b]Unit of measure is thousand troy ounces.
[c]Unit of measure is thousand barrels.
Sources: Refs. 2 and 16.

production and trade position is the resumption of oil exports in 1979. Oil surpassed copper as Peru's largest foreign exchange earner in 1980, when oil exports were valued at $810 million.[2] Peru's most important mineral exports in terms of value are shown in Table IV-5-3.

Industry Structure

Peru's large-scale copper mining and smelting capacity is owned by a group of foreign companies. The majority of the remainder of Peru's mineral industries are state owned.[1] Most of the country's remaining processing plants are also owned by the government. Some small and medium-sized mines are privately owned.[1]

The country's largest copper producer, the Southern Peru Copper Corp. (SPCC), is owned by a group of five international corporations. Asarco, Inc., owns the largest share of SPCC, with 52.3%.[1] The other firms' shares are as follows: Phelps Dodge Overseas Capital Corp. (16.3%), the Marmon Group, Inc. (20.7%), and Newmont Mining Corp. (10.7%).[1] Billiton BV owns an 11.5% interest in SPCC's Cuajone mine.[1]

The two large state companies engaged in copper mining and processing are Empresa Minera del Centro del Perú (Centromín) and Empresa Minero del Perú (Mineroperú).[2] Centromin processes most of the mine output of the 250 or so small and medium-sized mining companies.[2]

Centromín produces a wide range of refined metals at the La Oroya metallurgical complex. La Oroya produces a dozen or so different refined metals, and tin and zinc concentrates for export.[2] Refined metal production at the La Oroya complex from 1976 to 1980 is shown in Table IV-5-4.

The iron and steel industries are both in the hands of state companies. Empresa Minera del Hierro del Peru (Hierroperú) owns all the country's iron ore producing facilities and the shipping facilities at the port of San Nicolás.[1] The state steel company Empresa Siderúrgica del Perú (Siderperú) closed the country's only steel works at Chimbote for three months in 1982, due to excess stocks equal to eight months' production, and depressed world markets.[4]

The better part of Peru's silver, lead, and zinc is produced by Centromín.[1] The small and medium-sized mining companies also produce large volumes of these minerals. Two medium-sized mining firms, Málaga Santoalla and Minera Regina S.A., produce most of Peru's tungsten.[2] Centromín is Peru's second largest gold producer after the private firm Minas Ocona S.A.[2]

Peru's petroleum industry is dominated by the Peruvian subsidiaries of Occidental Petroleum Corp. and Belco Petroleum Corp., who together account for about two-thirds of total Peruvian production.[2] The state oil company Petroperú was responsible for most of the remainder of domestic output. Uranium is reserved for state exploitation under the authority of the Instituto Peruano de Energía Nuclear.[1]

An analysis of the concentration ratios in six mineral industries is presented in Table IV-5-5, together with the share of total industry output produced by the first, fourth, and eighth largest producers.

Judging from these figures, the Peruvian oil, tungsten, copper, and zinc industries exhibit high degrees of concentration. The silver industry is somewhat concentrated, and gold production is spread among many small and medium-sized companies.

The state companies active in Peru's minerals industries exhibit high levels of both vertical and horizontal integration. Large private firms (SPCC) and the small and medium-sized mining companies are much less vertically integrated. The state controls virtually all of Peru's mineral-processing capacity. Centromín, Mineroperú, and Hierroperú are the mining companies with the highest level of both vertical and horizontal integration.

Government Role in the Mining Industry

Mining Taxation. Mining enterprises in Peru are subject to a variety of taxes. Mining taxes were reduced in mid-1982.[13] Mining companies are exempt from export taxes (previously 17.5%), payroll taxes, goods and services taxes, and import duties on spare parts and equipment.[16] A temporary 7% tax on sales and a 5% increase in income tax rates have been enacted.[16]

The income tax scheme to which Peruvian mining companies are subject is progressive and is

TABLE IV-5-3   Peru: Value of Mineral Exports
(1981 Dollars)

Mineral	Export value	Export share (%)
Oil	710,000,000	22
Copper	526,400,000	16
Silver	315,500,000	10
Zinc	285,700,000	9
Lead	216,400,000	7
Gold	72,000,000	2
Other	338,000,000	11
Total minerals	2,464,000,000	77
Total exports	3,220,000,000	100

Sources: Ref. 16; *Minerals Yearbook 1981.*

TABLE IV-5-4   Peru: Refined Metal Production
La Oroya Metallurgical Complex, 1976–1980 (Metric Tons)

	1976	1977	1978	1979	1980
Metals					
Copper	45,786	55,022	51,897	54,291	54,104
Lead	74,066	79,243	74,255	85,112	81,976
Zinc	64,382	66,949	68,852	68,196	63,829
Silver[a]	19,228	21,572	20,897	25,448	23,971
Gold[a]	35	36	32	43	42.5
Bismuth	456	516	611	523	498
Cadmium	174	182	170	190	172
Indium (kg)	3,271	3,734	3,302	3,845	3,610
Selenium (kg)	8,743	15,936	12,941	18,320	22,908
Tellurium (kg)	12,306	18,370	15,418	21,233	20,920
Subproduct					
Antimony	315	569	—	629	427
Arsenic trioxide	797	1,405	1,322	1,415	6,038

[a]Unit of measure is thousand troy ounces.
Source: Centromín Annual Report, 1980.

TABLE IV-5-5   Peru: Concentration Ratios in
Peru's Mineral Industries[a] (Percent)

Mineral	One firm	Four firms	Eight firms
Copper	70.6	88.8	92.7
Gold	7.3	17.8	26.7
Silver	25.3	43.2	57.8
Tungsten	42.3	92.8	100.0
Zinc	50.5	70.3	85.2
Crude oil	54.2	100.0	100.0

[a]The table shows the share of Peru's total mineral production produced by the largest firms in the country.
Source: Ref. 2.

based on return on investment as well as income.[15] At lower income levels, a simple progressive scale with tax rates from 20 to 40% is used.[15] Firms with income above $300,000 are subject to an income tax ranging from 29 to 55%.[15] The tax rate depends on the firm's income (gross income less operating costs) as a percent of net investment in fixed assets.[15] Other taxes include a 1% tax on pretax income for funding the Government Mining Scientific and Technological Institute.[15] The Institute funds and promotes mining research and technology development.

Government Subsidies and Incentives. The Peruvian government has adopted a number of generous incentives and tax reductions to mining companies, especially those engaged in gold mining activity. Gold mines are eligible for reductions of up to 100% on a number of taxes, depending on a mine's location.[15] Gold mining

firms are exempt from all taxes, and gold miners are exempt from personal income taxes.[15] These provisions extend through 1993.[15]

The Peruvian government has also adopted generous provisions allowing for the reinvestment of mining income. All reinvested mining income is exempt from income taxes. In some parts of the country, mining companies can reinvest up to 100% of their nonmining income in the mining industry tax free.[15] Other provisions apply in different parts of Peru and for different minerals, with limits of 30 to 50% on the amount of income that can be reinvested tax free.[15]

Gold miners can write off all investments in roads, transport, and communications infrastructure against income tax liabilities.[15] The firms can deduct their expenditures on infrastructure from income taxes for the first five years of production.[15] This is a large incentive to the development of new gold mines.

Minero Perú Comercial (Minpeco) held a monopoly on the marketing of Peru's mineral output until mid-1981.[16] Up to that time, all mining companies had to market their minerals through Minpeco, paying a 2% sales commission.[2] Minpeco will continue to market the output of state firms and the private firms that want to use Minpeco's services.[16] The sales commission has been reduced to 0.8%, bringing about a favorable response from mining firms.

Government Ownership of Producing Companies. With the exception of the large-scale copper and oil industries, most mineral production in

Peru is in state hands. The government's operating companies include Centromín, Mineroperú, Hierroperú, and Petroperú. The other state companies that serve the mining industry are the state rail company Enafer and the state electrical energy company Electroperú.[1]

Two state agencies give financial assistance to the mining industry. The Banco Minero del Perú assists small and medium-sized mining companies in financial and technical matters.[1] The state financial development corporation (Cofide) provides financial assistance to Peru's mining industry. Cofide owns some mineral producing companies and provides loans to private mining companies, in conjunction with Mineroperú and Centromín.[1]

Mining Law and Policy. Peru's present mining law was adopted in 1981, in an effort to stimulate mining activity and foreign investment in the minerals industries.[16] Peru defined a coal policy in 1981 with the creation of the Empress Promotora del Carbón S.A. (Procarbón).[2] Procarbón will evaluate Peru's coal potential and promote the substitution of coal for petroleum in industrial applications. The agency will not actually own or operate coal mines. Procarbón is also charged with developing an integrated coal policy for Peru.[16]

Recent modifications to Peru's mining and hydrocarbon law are aimed at attracting joint venture capital from both foreign and private domestic investors. The new oil law, No. 23231, allows tax credits of from 40 to 50% for oil firms investing over $20 million in oil activities.[2] The tax credit is applicable the year the investment is made and for the next three years.[2] Other policy moves include the tax cuts and subsidies mentioned earlier and some attractive guarantees on tax stability. All new plants and expansions (20 to 30%) receive the same guarantee for a five-year period.

Government Role in Research and Development. The government agency responsible for geological, mining, and metallurgical research is the Institute Geológico Minero Metalúrgico (Ingemmet).[1] Ingemmet is funded by a tax on mining companies. The state funds the bulk of the research on mining and metallurgical topics that is undertaken in Peru.

Domestic Supply–Demand Dynamics

Analysis of Peru's domestic mineral supply and demand must begin with an examination of the country's mineral productive capacity, domestic consumption, and the balance of mineral output available for export. Virtually all of Peru's mineral output is for the export market, with limited industrialization and hence low domestic demand for strategic minerals. Information about Peru's mineral production and the portion of that output that was exported in 1979 and 1981 is presented in Table IV-5-6.

Peru exports large shares of its mineral production to the industrialized Western countries. The United States, Japan, and Western Europe have long been the major destinations for Peruvian mineral exports. Table IV-5-7 presents information regarding the destinations of Peru's mineral exports. Peru exports over half of its total exports of every mineral commodity, with the exception of tin and arsenic, to the industrialized Western countries.[2]

TABLE IV-5-6 Production and Exports of Selected Commodities, 1979 and 1981 (Metric Tons)

Mineral	Production	Exports	Exports/ production
Copper (81)	328,850	319,400	0.97
Gold (81)	4,880 kg	4,769 kg	0.98
Lead (81)	186,735	154,900	0.83
Silver (81)	1,455,850 kg	909,500 kg	0.63
Zinc (81)	496,707	477,500	0.96
Bismuth (79)	530,000	365,000	0.69
Iron ore (79)[a]	5,444,000	5,619,000	1.03
Molybdenum (79)	1,240	1,240	1.00
Selenium (79)	18,000	11,880	0.66
Tellurium (79)	21	12	0.57
Tungsten (79)	560	560	1.00

[a]Stock reductions in 1979 were responsible for exports exceeding production.

Sources: Refs. 1, 2, and 16.

TABLE IV-5-7   Peru: Mineral Exports, 1979, by Selected Country of Destination[a] (Metric Tons)

Mineral	To U.S.		To Japan	To W. Europe
Antimony				
Ore and concentrates	—		—	563 (76%)
Metal	102	(100%)	—	—
Arsenic	450	(30%)	—	—
Bismuth: metal	220	(59%)	—	75 (20%)
Cadmium: metal	141	(67%)	—	38 (18%)
Copper				
Ore and concentrates	2,781	(2%)	19,798 (17%)	75,320 (65%)
Metal	54,474	(17%)	63,557 (18%)	52,297 (15%)
Indium: metal	3	(75%)	1 (25%)	—
Lead				
Ore and concentrates	45,236	(32%)	33,512 (24%)	9,538 (7%)
Metal	29,187	(34%)	—	22,359 (26%)
Molybdenum: ore and concentrates	372	(17%)	—	1,276 (60%)
Silver: metal[b]	14,211	(60%)	3,569 (15%)	—
Tellurium: metal	1	(8%)	—	—
Tin: ore and concentrates	263	(11%)	—	6 (50%)
Tungsten: ore and concentrates	597	(53%)	348 (31%)	91 (8%)
Zinc				
Ore and concentrates	47,304	(7%)	183,498 (28%)	192,469 (29%)
Metal	12,321	(22%)	—	—

[a]This does not represent Peru's total mineral exports for 1979; only the most important minerals and export destinations are presented.
[b]Unit of measure is thousand troy ounces.

Source: Ref. 2.

Peru has no domestic demand for most of the minerals it produces. Small amounts of lead, and about one-third of silver, selenium, tellurium, and bismuth production, are consumed domestically.[1] The domestic demand for Peru's mineral production is not expected to increase dramatically in the short to midterm; hence Peru's mineral exports will be closely tied to productive capacity, the actual level of mineral production, and demand in consuming countries.

Future increases in Peru's mineral production capacity will hinge on the Peruvian government's ability to create an attractive climate for foreign investment. Price trends among major mineral commodities will also determine the rate at which Peru's large identified, but as yet undeveloped mineral potential is brought into production. Tax reforms have increased foreign interest in Peruvian mining, but announcements of new projects and investments have not been forthcoming.

POLITICAL ENVIRONMENT

Domestic

The election of Acción Populaire party candidate and former president Fernando Belaúnde Terry in May 1980 marked the return of Peru to civilian control after 13 years of military rule.[17] Belaúnde Terry was president from 1964 to October 1968, when he was overthrown in a military coup d'etat led by General Juan Velasco Alvarado.[17] Velasco Alvarado shared the left-wing ideology of his subordinates, who were mainly from the underprivileged classes.[17] A bloodless coup in 1975 gave power to a military regime led by General Morales.[17]

The election of the Peruvian Constituent Assembly in 1979 and the 1980 presidential elections have revived the open political process in Peru.[11] Besides President Belaúnde Terry's slightly right-of-center Acción Populaire (AP) party, which holds a substantial majority, other major political parties are Partido Populaire Cristiano (PPC), a centrist party; Alianza Popular Revolucionaria Americana (APRA), left of center; and a number of small, fragmented extreme-left-wing parties.[11]

The leader of the PPC is 1980 presidential candidate Luis Bedoya.[11] The APRA party is without a clear leader after the 1979 death of the party's founder Victor Raúl Haya de la Torre.[11] President Belaúnde Terry has an unstable power base and must try to balance the desires of the extreme left wing and the right wing to solidify it.[17] The APRA party is divided by internal power struggles, and the extreme-left parties are not serious contenders for power in Peru.[11]

President Belaúnde Terry's AP party is generally pro-business, and his election must be seen in a positive light by mining companies.[17] Laws and regulations that were arbitrarily changed by the military government should be much more stable under the new civilian government.[3] The new government will need the support of the military to maintain its power.[11] Industry representatives hope that the new government will promote mineral development and foreign investment in mining projects.

International

Peru has a border dispute with Ecuador. The area in dispute has some energy and mineral resource implications, as it is near oil-producing regions in Peru.[1] The tensions between the two countries erupted into a brief war in January 1981. Relations with other Latin American countries are relatively good.

The election of President Belaúnde Terry should improve relations between Peru and the United States. The military government that he replaced was defiant of North American political leaders and embraced Cuba's leader Fidel Castro.[18] The new government is oriented toward Western economic interests and recognizes the importance of mining to the country's economy.[17]

The Belaúnde Terry administration will have to improve the international business and financial climate in order to foster the expansion of the mining industries. The military had established a reputation for violating the Mining Law and drafting confiscatory and often contradictory laws and regulations affecting the mining industry.[3] The new administration should be a force for stability in Peru's mining taxation and regulation system.

REFERENCES

1. U.S. Department of the Interior, Bureau of Mines. *Mineral Industries of Latin America*. Mineral Perspectives Series. December 1981.
2. U.S. Department of the Interior, Bureau of Mines. *Minerals Yearbook 1980*. Washington, D.C.: U.S. Government Printing Office, 1982.
3. Sisselman, Robert. "Peru." *Engineering and Mining Journal* 179 (November 1979).
4. U.S. Department of the Interior, Bureau of Mines. *Minerals and Materials*, October–November 1982.
5. *Engineering and Mining Journal* 182 (August 1982).
6. Todd, Joan C. "Peru's Tintaya Copper Project to Go on Stream in September 1984." *Engineering and Mining Journal* 177 (July 1977).
7. "Production Starts at Minero Peru's Cerro Verde Copper Mine." *Engineering and Mining Journal* 177 (July 1977).
8. "Cuajone." *Mining Magazine*, August 1976.
9. Wippman, Larry. "Centromin Invests $261 Million to Expand Cobriza Production." *Engineering and Mining Journal* 182 (August 1982).
10. Ridley, Regina Starr. "Peru's Cobriza Brings New Capacity on Stream." *World Mining*, September 1982.
11. Spooner, John, ed. *Mining Journal Annual Review 1980*. London: Mining Journal, June 1980.
12. "Peru: Taking the Long View." *Mining Journal* 299 (30 July 1982).
13. "Peru Adopts Measures for Small/Medium Mines." *Engineering and Mining Journal* 182 (September 1982).
14. U.S. Department of the Interior, Bureau of Mines. *Minerals and Materials*, August–September 1982.
15. Banco Minero del Perú. *Metallic Gold Mining Law Regulation*. Lima: Banco Minero del Perú, 1979.
16. Spooner, John, ed. *Mining Journal Annual Review 1982*. London: Mining Journal, June 1982.
17. "Peruvian Promise." *Mining Journal* 296 (13 February 1981).
18. Loprete, Carlos A., and Dorothy McMahon. *Iberoamerica*, 2nd ed. New York: Charles Scribner's Sons, 1974.

Chapter IV-6

Venezuela

Venezuela is important in the world energy and mineral economy for its oil and gas production, iron and steel industry, and its aluminum metal production. The petroleum sector has earned about 95% of Venezuela's foreign exchange earnings in recent years.[1] In 1980, Venezuela provided about 6.5% of U.S. imports of crude oil and petroleum products.[2] Aluminum metal has recently surpassed iron ore as the nation's second-largest export commodity.

The largely state owned minerals and fuels industries are integrated into the production of ferroalloys, iron and steel, and refined petroleum products. Within the next several years, domestic bauxite and alumina capacity will reduce Venezuela's current dependence on alumina imports from Jamaica and Suriname. Venezuela should also become a significant coal producer in the 1980s.[2] Gold and diamonds are also currently being obtained from alluvial deposits.

Mining activity is concentrated in northern Venezuela, where fuel, electric power, infrastructure, and the labor force are readily available at low prices. The potential for further mineral discoveries is high in the virtually unexplored and sparsely populated southern part of Venezuela. Inexpensive hydroelectricity has contributed to the development of the Venezuelan aluminum industry. The geographical distribution of mineral producing regions, processing facilities, and related infrastructure are presented in Figure IV-6-1.

Minerals and fuels are transported by ship, rail, and truck in Venezuela. An integrated network of oil and natural gas pipelines connects producing regions to ports and refineries. The country's two rail lines were built specifically to service the rich iron ore belt in eastern Venezuela.

MINERAL RESERVES AND RESOURCES

Venezuela's proven oil reserves stood at 20 billion barrels as of December 1981.[3] Natural gas reserves are placed at about 1.2 trillion cubic meters.[1] Venezuela also has large reserves of iron ore and bauxite, and sizable reserves of coal, gold, and diamond placer deposits. Venezuelan reserves of minerals and mineral fuels are presented in Tables IV-6-1 and IV-6-2. Other minerals found in Vene-

FIGURE IV-6-1 Venezuela: Mines, processing facilities, and infrastructure. (From Ref. 1.)

427

TABLE IV-6-1 Venezuela: Mineral Reserves[a]

Natural gas	1.2 trillion m^3
Oil	20 billion bbl
Iron ore	2.116 billion mt
Coal	700–900 million mt
Bauxite	3 billion mt
Gold	NA
Diamonds	NA
Nickel	55 million mt
Magnesite	6.6 million mt
Titanium	5 million mt
Phosphates	40 million mt
Copper/lead/zinc	4.5–6.5 million mt

[a]NA not available.

Source: La Industria Minera en Venezuela: Situación y Perspectivas para el Desarrollo Nacional. Ministerio de Energía y Minas, direccion de Minas y de Geología, December 1980.

TABLE IV-6-2 Venezuela: Proven Iron Ore Reserves, (1 January 1980) (Million Metric Tons)

Deposit	Ore grade (%)	Reserves
San Isidro	63	392
Cerro Bolivar	63	223
Los Barrancos	63.4	232
Cerro Redondo	55	165
Cerro Altamira	55	135
Cerro Arimagua	55	136
Cerro El Trueno	63.5	110
Cerro Las Pailas	63.5	80
Cerro San Joaquin	62	65
Cerro El Pao	61.4	50
Punta de Cerro	63	50
Piacoa	45	181
Maria Luisa	47	258
Other	50	55
		2132

Source: Dr. Gustavo T. Ascanio, *El Mineral de Hierro en Venezuela,* Foletos Técnicos del Ministerio de Energía y Minas 2.

zuela include magnesite, nickel, titanium (ilmenite), and copper, lead, and zinc.

GEOLOGY

An examination of the geology of Venezuela must begin with a general knowledge of its physiography. Venezuela can be divided into regions of high and low relief. The areas of low relief include the Maracaibo depression, the delta of the Orinoco, and the central and southern plains areas. The areas of high relief include the Andes and coastal mountain ranges, the Sierras of Perija and Imataca, and the savannah and mesa region of southeastern Venezuela and the Escudo of Guayana.

This division is useful because different commodities are found in the two areas. Virtually all oil production originates in the areas of low relief, where sediments have accumulated for many years. Most production comes from the Maracaibo basin. The area is characterized by widespread subsidence and a very thick accumulation of sediments.

The areas of high relief are of greatest interest. All iron ore activity, coal mining, and future bauxite mining take place in the areas of high relief. Coal is mined in the Sierra of Perija and the Andes regions in the states of Zulia and Tachira. Coal has been found in the mountains in western Venezuela, near the border with Colombia. Most of the coal dates to the Tertiary period and is classified as subbituminous and lignite. The coal has a high volatile content. Other deposits have been located in the north-central part of the country in the states of Aragua and Guarico.

Bauxite occurs as a residual ore, caused by the weathering of intrusive rocks in favorable conditions. Most bauxite and aluminiferous laterite deposits are found in the highlands of the Escudo of Guayana. Principal deposits occur in the states of Bolivar and the Federal Territory of the Delta Amacuro. The best ores have developed in areas where there has been a prolonged, continuous exposure to the forces of weathering. The deposits are typically high in both iron and titanium content.[4]

Gold and diamonds are found as alluvial deposits in the rivers that divide the Escudo of Guayana. Gold and diamonds frequently occur in the same deposits. The best diamond deposits can be traced to quartz conglomerates in the region. The richest gold deposits are in ancient rocks that formed at high temperatures. The mineralization is of the hydrothermal type. Gold is usually found in association with plutonic rocks of the Caripacho-Pastora series.[4]

The most important mineral deposits in the areas of high relief are the rich iron ore deposits of the Venezuelan iron belt. The belt runs for about 500 km in an east-west direction along the south bank of the Orinoco River. The belt varies in width from 80 to 100 km. The iron ore belt is called the Imataca Complex and is composed mainly of folded fault blocks composed of gray, white, and pink quartz-feldspatic gneisses. Typically, the rocks contain silica-rich bands that are light in color and iron-rich bands that are dark in color.

The fault blocks, or *fajas* as they are known locally, exhibit different chemical compositions, textures, and grain sizes. Deposits to the north of the El Pao fault are typically derived from silica-poor ores by metamorphism and are of large grain size. Deposits in the fault blocks between the El Pao and the Guri-Río Caparo fault are of a medium grain size and were derived from preexisting ore bodies by tropical weathering processes. Deposits located south of the Guri–Río Caparo fault are typically fine-grained.[4]

The iron ore district is covered mainly by savannah, where iron ore deposits exhibit the following geomorphology. Deposits occur as rounded hills up to 100 m in elevation and as mesas up to 400 m high. Most deposits are of a sedimentary origin and have experienced intense metamorphism. The dominant oxides are hematite and magnetite, and the major ferromagnesian silicate present is orthopyroxene.[4]

In conclusion, mining and petroleum activity take place in different geological and geographical areas in Venezuela. Oil and gas activity is centered in the lowlands, and virtually all other mining activity is in the highland regions. The heaviest concentration of this activity occurs in the Precambrian Escudo of Guayana area. For more exact locations, see Figure IV-6-1.

PHYSICAL INFRASTRUCTURE

Infrastructure is well developed in the mineral-producing regions of Venezuela. Oil- and gas producing regions are connected by both crude oil and natural gas pipelines to refineries, ports, and areas of internal demand. Two large rivers, the Caroní and Orinoco, are used to transport iron ore and aluminum. Two rail lines, totaling 107 km of track, connect the iron ore belt to the river/ocean transport system.

One rail line runs from the Cerro Bolívar and Altamira mines to Puerto Ordaz on the Caroní river. From there, ore is loaded onto ships for further transport. The other rail line connects the Cerro El Pao mine to the C.V.G. Siderúrgica del Orinoco S.A. (SIDOR) steel complex at Matanzas, near Ciudad Guayana. The SIDOR plant utilizes iron ore to produce crude steel and manufactured products and has a capacity of 4.8 million metric tons per year.

The same infrastructure serves the alumina and aluminum facilities at Ciudad Guayana and will be used to transport bauxite when the Los Pijiguaos deposit is exploited. Gold and diamond production is transported by river and truck from producing areas near the Caroní river and the Guayanese border. From Puerto Ordaz and Ciudad Guayana,

mineral products are transported down the Orinoco and out the river's Bora Grande to Europe and to U.S. ports such as Mobile and Philadelphia.

The existence of a good transport network, coupled with readily available labor, water, and hydroelectric energy, have resulted in the concentration of mining activity near the Caribbean coast and in the area served by the Orinoco and Caroní Rivers in eastern Venezuela. The interior of the country is still relatively undeveloped. Because the interior of the country is sparsely populated and has not been extensively studied for mineral deposits, exploitation of mineral deposits in the area would require substantial investments in infrastructure.

Venezuela has abundant energy available in the form of liquid and gaseous hydrocarbon fuels, coal, and hydroelectricity. Electric generating capacity is about 8000 MW, of which 35% is from hydro sources. The industrialization of the steel and aluminum center at Ciudad Guayana is largely due to cheap hydroelectricity from the Guri dam. Current capacity is about 2000 MW and will be expanded to 9000 MW by the late 1980s.[1] Coal will be used in Zulia state to fire new thermal electric generating capacity and to fuel a steel plant planned south of Maracaibo.

Even with falling oil production, the energy outlook for Venezuela is bright due to large reserves of heavy oil and sizable coal reserves, which will augment cheap hydroelectricity and declining production of lighter crudes. Unfortunately, Venezuela has experienced a lack of stability in its electric power supply.

MINING LABOR

Venezuela's population was roughly 14 million in 1982. The employed labor force was only about 4 million, of which only about 1.1% was engaged in the mining and mineral fuels sector. The petroleum sector employs 30,000, and the mining sector employs about 16,000 workers, of which 4500 are involved in iron mining.[1,4] Labor supply and demand are in a state of imbalance in the undeveloped interior of the country.

Labor relations are relatively good, with no major labor unrest or strike activity reported in the last few years. Labor relations are particularly good in the iron and steel industry. Venezuela still suffers from a shortage of trained technicians in the mining sector.

The iron and steel industry is very active in providing support for its labor force. Six schools for some 2700 students are currently operated by the iron and steel industry. The industry also operates six hospitals and provided medical services to 150,000 patients in 1980.[4] The industry has also constructed about 8350 housing units in the last 15 years.

These services and programs are all provided by the iron and steel industry and are probably typical of the services and programs provided by the other state-owned minerals companies.

DOMESTIC MINING INDUSTRY

Overview

Venezuela has a largely petroleum-dependent economy. Other products important to the Venezuelan economy are iron and steel, aluminum metal, and coffee and cacao. The mining industry accounts for about 10% of the country's GDP, with virtually all of this from the petroleum sector. In addition, oil provides over 95% of all foreign exchange earnings and about 65% of all government revenues.[2] These oil revenues have fostered the formation of a solid industrial base and have provided funds for badly needed social programs.[1]

The general economic stagnation that has prevailed in Venezuela since 1979 has given rise to downward trends in exports of major mineral commodities. Oil exports in 1981 were 1.757 million barrels per day at an average price of $29.72 per barrel, owing largely to the worldwide oil glut.[2] Iron ore exports have fallen from a high of over 26 million metric tons in 1974 to 11.8 million tons in 1981.[4] This level of exports, coupled with 1980 exports of 11.75 million metric tons, represent the lowest level of iron ore exports since the early 1960s.[4] Severe damage at the country's largest aluminum metal plant held 1982 exports well below 1981 levels. Export levels for other mineral products, including gold and diamonds, have also been adversely affected by the domestic economic stagnation and the general world recession.

Industry Structure

Virtually all mining and mineral fuels activity in Venezuela is in the hands of state-owned companies. Petróleos de Venezuela S.A. (Petrovén) is the largest company in the Third World and the fifteenth largest company in the world, excluding the United States.[2] Petrovén operates through four subsidiary companies: Lagovén, with 41% of production; Maravén (28%); Corpovén (15%); and Menevén (16%). Petrovén invested about $2 billion in exploration, production, and refining in 1980.[2]

The aluminum industry is organized into several mixed companies. Three of these companies are partially owned by private foreign firms. Aluminio del Caroní S.A. (ALCASA) and Industria Venezolano de Aluminio C.A. (VENALUM) are the two aluminum-producing companies. Reynolds International, Inc., owns 27.9% of ALCASA, and a consortium of six Japanese firms owns 20% of VENALUM.[1]

Iron ore and steel are completely controlled by two government-owned operating companies. The iron ore company is C.V.G. Ferrominera del Orinoco C.A., and the steel concern is C.V.G. Siderúrgica del Orinoco C.A. (SIDOR). Ferrosilicon is produced by the state company Fesilvén, and ferromanganese is produced by Hornos Eléctricos de Venezuela S.A. The latter firm is owned by Mexican (50%), Venezuelan (25%), and Japanese (25%) companies.[2]

Gold production is divided among state-owned Minervén (26%), the Miner's Syndicate (50%), independents (18%), and concessionaires (6%). Virtually all diamond mining is by independent miners.

Mining activity is dominated by the state, with the notable exception of the aluminum and ferroalloy industries, where some private foreign capital has been invested. Gold and diamond production is primarily by independent miners. Revenues generated by the petroleum sector have provided capital for the activities of most of the other state-owned mining companies.

Vertical/Horizontal Integration

Venezuela currently has a vertically integrated iron and steel industry and petroleum–petroleum products industry. Petrovén operates producing fields in five states: Anzoátegui, Falcón, Guárico, Monagas, and Zulia. Petrovén also owns major refineries at Amuay Bay and Cardón, both in Falcón state. The state has also integrated forward into petrochemicals, with operations managed by Petroquímica de Venezuela S.A. (Pequivén). Additionally, Petrovén operates several smaller refineries, including El Palito and Puerto La Cruz.

Venezuela is on the verge of developing a fully integrated bauxite–alumina–aluminum industry with two projects now under way. ALCASA and VENALUM already operate two large aluminum refineries at Ciudad Guayana. At present, they process alumina that has been shipped from Jamaica and Suriname. This will change when Interalúmina opens its smelter at Ciudad Guayana. Startup was scheduled for 1983. Interalúmina is owned by the Venezuelan government (88.7%) and Alusuisse (11.3%).

The development of the Los Pijiguaos bauxite deposit is proceeding but has fallen behind schedule. The project is being developed by the iron company C.V.G. and by the operating company Bauxita Venezolana C.A. (Bauxivén). The state mineral financing agency FIV holds a 50% interest in Bauxivén.[2]

Feasibility studies were conducted by Alusuisse and are currently under study together with alternative plans for transportation and infrastructure. Funding delays pushed the schedule date for the first ore shipments to sometime in 1984.[3] With the development of bauxite mining and alumina smelting, Venezuela should have a fully integrated aluminum industry within the next two or three years.

Vertical integration by other minerals-producing companies is minimal. The state processes most of the gold and diamonds produced, but this is on a relatively small scale. The ferroalloy industry is somewhat vertically integrated, with all manganese ore imported by Hevensa. Fesilvén operates a silica-rich quartzite mine in eastern Venezuela. The company then trucks the ore to Puerto Ordaz, where it produces ferrosilicon and silicon metal.

The level of vertical integration in the mining and mineral fuels sector is very high. Virtually all activity in a given industry is undertaken by one or more large state-owned firms. Some foreign equity capital is present, but the government still controls practically all activity in the mining and mineral fuels sectors.

Government Role in the Mining Industry

The Venezuelan government exercises control over the mining and mineral fuels sectors using its mining laws and relevant policies as well as through its ownership of mineral-producing companies. In addition, most expenditures for physical and social infrastructure are undertaken by the state through its operating companies. The Venezuelan government has almost total control over the mining and mineral fuels sectors.

Mining Taxation. The current mining taxation scheme dates to 1979. Mining ventures in Venezuela are subject to a variety of taxes at different stages of exploration and exploitation. An exploration tax of 1 bolívar (B) per hectare is payable during exploration. In addition, a surface tax is also payable during exploration. Rates are 1 B per hectare the first year, increasing to 2 B during the third year. Once exploitation has begun, the tax reduces to 1 B per hectare per year. These rates apply to alluvial concessions.[7]

Mantle- or lode-type concessions are subject to the same taxes, but the tax rates are slightly different. The surface tax rate for these concessions is 2 B per hectare the first year, 2.5 B the second year, 3 B the third year, 3.5 B the fourth year, and 4 B the fifth year. Once exploitation begins, the tax rate falls back to 2 B. per hectare per year. There is no exploration tax for lode- or mantle-type deposits.[7] In situations where exploitation has not begun at the end of the fifth year, the surface tax rate becomes 1.5 times the rate prevailing in the fifth year.[7]

Another tax that becomes payable when production begins is the exploitation tax. Refined metal is taxed at 3% of its commercial value in Caracas. Ores are taxed at 6% of their value at the mine. This value at the mine includes all costs up to the point where the ore is delivered to a refining or treating plant, regardless of where the plant is located. In addition, income from mining is taxed under the Venezuelan income tax law.

On the whole, Venezuelan tax rates appear to be stable. However, Venezuela began taxing technical services performed in other countries, but used to generate income in Venezuela in 1980.[2] This could cause problems of double taxation for U.S. companies, placing them at a disadvantage compared with European competitors offering engineering and technology contracts.

Tax Concessions and Subsidies. Venezuela offers generous tax concessions to firms that perform activities that will foster mineral development. Tax rates may be reduced every three years for firms that increase proven reserves above the level delineated in the feasibility study. Tax concessions can also be extended to firms that train and employ Venezuelan nationals and construct highways, roads, schools, and medical clinics. Any activity that improves the physical, cultural, social, or economic conditions within Venezuela may receive unspecified tax concessions.[7]

The government also grants subsidies to its operating companies when economic conditions warrant such action. An example of this type of action occurred with ALCASA in 1980. After two consecutive years of operating at sizable losses ($7.9 million in 1978 and $16.5 million in 1979), ALCASA registered a $2.5 million profit in 1980. This profit was the direct result of a government subsidy of $22 million.

Without this subsidy, ALCASA's losses in 1980 would have been close to $20 million. The other large aluminum company, VENALUM, also received an export subsidy of $23 million in 1980.[2]

Other forms of government subsidy include government training programs, research and development programs, technical support, and injections of capital for firms experiencing economic hardship. An example of the last of these subsidies is the $58 million investment in ALCASA by the government financing agency FIV.

Government Ownership of Producing Companies. The extremely high level of government ownership of minerals-producing companies has been discussed previously. With the exception of a handful of companies in which foreign firms hold minority equity interests, all mining activity is undertaken by state-owned companies.

Venezuela's small coal-mining industry is run by three companies that are all state-owned. The firms are C.A. Minas de Carbón de Lobatera, Corp. de Desarrollo de la Región Nor-Oriental (Corporiente), and Carbones del Zulia C.A. (Carbozulia). Coal production has historically been insignificant. The development of the Guasare deposit by Carbozulia may change this situation. The deposit will be mined from an open-pit facility at a projected rate of 4 million metric tons per year. Mining of the 700- to 900-million metric ton block began in 1985.[8]

The ownership of the companies in the bauxite-alumina-aluminum industry has been discussed previously, as has that of the companies in the iron and steel industry. The aluminum sector has allowed some minority foreign capital, but the companies are still controlled by the state. The aluminum and iron ore industries are Venezuela's second- and third-most-important export commodities.

Mining Law and Policy. Mining activity in Venezuela is governed by the Mining Law of 28 December 1944.[1] This law has been amended by decrees and resolutions several times. Venezuela's policy regarding foreign investment is in strict accord with Andean Pact investment codes. Foreign investment capital is allowed under the present Mining Law.

Companies owned by foreigners are expected to convert to mixed companies, with majority ownership by Venezuelan nationals, within a period of 15 years. The government would prefer an even shorter conversion period.[1]

In the petroleum sector, Venezuela is a member of OPEC and supports the cartel's policies publicly. Foreign investment is in the hands of the Superintendencia de Inversiones Extranjeras (SIEX). SIEX favors the entrance of foreign capital and makes a fairly liberal interpretation of Andean Pact Decision 24, regarding foreign investment capital.

Mining concessions that transfer technology to Venezuela and promote the processing of Venezuela's minerals within the country are given priority by the government. Projects that include the construction of major physical and social infrastructure are considered in a favorable light.

Exploration concessions are granted for 36 months for lode-type deposits and for 24 months for alluvial deposits.[7] Applications for mining concessions are subject to the provisions stipulated in Article 2, Decree No. 2039, dated 15 February 1977.[7] Exploitation concessions are granted for 48 months for lode-type deposits and for 30 months for alluvial deposits.

The terms of the Venezuelan tax and concession arrangement depend on such factors as the level of utilization of domestic materials and labor, the level of domestic processing and refining, and the contribution to the development of the Venezuelan economy.

Prospecting licenses are granted by the Directorate General of Mines and Geology. Licenses may not be transferred, are for an areal extent not to exceed 5000 hectares, and are for one year, with one extension. These time and extension limits may be modified by the Ministry of Energy and Mines in special cases. This license permits only general geological survey and search methods to be used. Typical exploration and mining equipment is not allowed, and test pits are not allowed to exceed 4 m². Quarterly reports of prospecting activity must be submitted to the Ministry of Energy and Mines.[7]

Government Role in Research and Development. Most research and development activity in the Venezuelan minerals and mineral fuels sectors is undertaken by the government through one of its subsidiaries. Petrovén has a research branch called the Instituto Tecnológico Venezolana del Petróleo (Intevep), which conducts energy research and provides technical support for the petroleum industry.[2] Research is geared primarily toward developing new technologies to facilitate the extraction and refining of Venezuela's heavy crudes (8 to 12° API). Many refineries have already been modified to process the increasing volumes of heavy crudes.

Mining sector research is conducted by departments within the large state-owned producing companies. C.V.G. Ferrominera del Orinoco does the research for the iron and steel industry. The aluminum industry also conducts its own research. A department of the Ministry of Energy and Mines also conducts investigations in the areas of planning and mineral economics. This is located within the department of Mines and Geology and is known as the Dirección de Planificación y Economía Minera.[4]

Despite the activities of these groups, there is still a serious lack of technology and trained personnel at all skill levels in Venezuela. This situation continues to place Venezuela's minerals sector in an incipient stage of development. Venezuela's growth must be based on mining as well as the oil industry, in order to establish a firm industrial base for future development. Venezuela's oil wealth will not last forever; thoroughly planned and well-balanced industrialization will be necessary for the survival of Venezuela.

Domestic Supply–Demand Dynamics

Trends in production and productive capacity and in domestic mineral consumption determine the level of Venezuela's mineral exports. Analysis of these trends points to two characteristics of Venezuela's mineral economy. Venezuelan mineral and fuels production, and in some cases productive capacity, is falling. In addition, more of Venezuela's mineral production is being consumed and processed domestically.

Nowhere are these trends more apparent than in the iron and steel industry. In the decade from 1972 to 1981, domestic sales of iron ore increased from 0.05% to over 20% of total sales.[4] During the same decade, total production fell to just 60% of its all-time high in 1974. Productive capacity is projected to climb back to the 23 million metric tons per year level shortly from its current level of 16 million metric tons per year.[2]

Most of the increase in domestic sales occurred in 1980, with the bulk of this iron ore going to the Matanzas steel plant. The plant, with its 4.8-million metric ton per year capacity, is the world's largest integrated steelworks utilizing solely direct reduction technology.[2] Trends in production, domestic sales, and exports of iron ore are presented in Table IV-6-3.

TABLE IV-6-3 Venezuela: Iron Ore Production and Sales Domestic and Export, 1972–1981 (Thousand Metric Tons)

Year	Production	Exports	Domestic sales
1972	18,499	17,005	8
1973	23,110	21,660	127
1974	26,424	26,277	237
1975	24,772	19,405	230
1976	18,685	15,672	282
1977	13,684	11,936	250
1978	13,515	12,828	514
1979	15,260	12,976	716
1980	16,102	11,752	2,952
1981	15,900	11,800	3,080

Source: Ref. 5.

In the oil sector, production has been falling very slowly. Productive capacity is presently about 2.5 million barrels per day and is set to increase to about 2.8 million barrels per day in the next few years. Production fell slightly to 2.11 million barrels per day in 1981.[2] Since total production has been falling, the share of production available for export has also fallen slightly. Venezuela's production levels are in accord with OPEC's production rationing scheme.

Venezuela's coal production is currently insignificant, about 60,000 metric tons per year. This is set to change drastically when the open-pit project is Zulia state is completed. Carbozulia plans to mine up to 4 million metric tons per year of steam and metallurgical coal, beginning in 1985.[8] This project is Venezuela's first major coal operation, and Fluor Mining and Metals of the United States will perform the basic engineering services for the project.

Venezuela does not currently have any bauxite production, but this situation will change when production begins at the Los Pijiguaos deposit, owned by Bauxivén. Scheduled to begin operations in 1985, production is expected to be about 3 million metric tons per year. Interalúmina's alumina smelter at Ciudad Guayana will give Venezuela 1 million metric tons per year of alumina capacity from the time it opened in 1983. Venezuela's aluminum metal refining capacity is about 400,000 metric tons per year. Venezuela consumes about 40% of total production domestically. Damage to the country's largest refinery has reduced productive capacity by about 25%.[3] The Venezuelan aluminum industry will become more important to the world mineral economy as vertical and horizontal integration proceed in the mid-1980s. Aluminum production has shown an average annual increase of over 100% in the years 1978–1980.[4]

More of Venezuela's iron ore production is being consumed domestically. This trend, together with increasing aluminum metal production, has combined to move aluminum into second place behind oil as a foreign exchange earner. The aluminum industry is likely to become increasingly important to the Venezuelan economy, and coal may also be produced for export in the future.

Most of Venezuela's mineral production is available for export. Between 60 and 70% of oil and petroleum products production is for export.[1] Venezuela exports about 80% of its iron ore production and about 60% of its aluminum metal production.[1] With the exception of iron ore, where domestic consumption has been rising rapidly, these export levels are relatively stable.

POLITICAL ENVIRONMENT

Domestic

Venezuela has a conservative democratic government. The current president, Dr. Luis Herrera Campins, was inaugurated in March 1979. His Christian Democrat party must form alliances with left-wing parties in order to pass legislation in Congress.[6] Opposition parties include the Social Democrats and labor and business groups. Unrest within the military led to speculation about a possible coup attempt in 1980. That unrest has, for the most part, subsided in the last year or two.

Herrera Campins' predecessor, Carlos Andrés Pérez, was one of Venezuela's most corrupt presidents. Provisions of the country's public credit law enabled state agencies to borrow unlimited amounts without approval provided that the loans were for one year or less.[9] The country's public debt currently stands at about $35 billion.

Unrest in the country is based primarily on the inequitable distribution of income. Labor, business, religious, and intellectual groups and the leftist political parties oppose the president's policies. The president is trying to cut government spending and increase revenues.

Actions announced to date include a 10% reduction in the federal budget, an import ban on clothing and shoes, and a "Buy Venezuelan" decree. Herrera Campins has also increased artificially low gasoline prices and reduced or eliminated subsidies on basic commodities. Most of these actions have drawn criticism from the opposition.

International

Venezuela and the United States have long been both friends and trading partners, but policy toward Cuba and Central America has created a rift in U.S.-Venezuelan relations. Venezuela feels that the proper way to counter Cuban influence in the region is by massive economic aid. The U.S. feels that military aid is the way to stop the spread of communism in the region. Although Venezuela supports U.S. activity in El Salvador, they support the Marxist Sandinista government in Nicaragua against U.S. wishes.

Venezuelan relations with other Latin American nations are difficult to characterize. Venezuela has good relations with Mexico, and the two countries have signed a joint agreement to finance part of the oil needs of Central American and Caribbean nations.[2] Cooperation between Suriname and Venezuela may lead to the integration of the two countries' aluminum industries.[6]

Venezuela is involved in border disputes with both Colombia and Guyana. The Cubans are trying to exacerbate this problem with Guyana by supplying Guyana with MiG fighters and troops.[10] Discussions between the two countries will probably continue.

The border dispute with Colombia is in the process of being resolved. The presence of oil reserves in the area of dispute has compounded the situation. The two countries have basically agreed on a draft agreement, but domestic squabbling has delayed the signing of a treaty. The areas of discussion include the ownership of the Los Monjes islands and the rights of Colombia in the Gulf of Venezuela.[2]

The government of President Herrera Campins has had success in reducing the country's inflation rate, but the country still suffers from high unemployment. The president's austerity programs may keep the balance of trade in surplus but will not alleviate the unemployment problem. To climb out of its current economic stagnation, Venezuela must put to rest its reputation as an unreliable borrower.

REFERENCES

1. U.S. Department of the Interior, Bureau of Mines. *Mineral Industries of Latin America*, Mineral Perspectives Series. December, 1981.
2. U.S. Department of the Interior, Bureau of Mines. *Minerals Yearbook 1980*. Washington, D.C.: U.S. Government Printing Office, 1982.
3. Spooner, John, ed. *Mining Journal Annual Review 1982*. London: Mining Journal, June 1982.
4. Zozaya F., Dionisio. *La Industria Minera en Venezuala: Situación y Perspectivas para el Desarrollo Nacional*. División de Información Técnica Geológica-Minera. Ministerio de Energía y Minas, 1978.
5. República de Venezuela, Ministerio de Energía y Minas. *Hierro y Otros Datos Estadísticos Mineros*. Caracas, December 1980.
6. Spooner, John, ed. *Mining Journal Annual Review 1980*. London: Mining Journal, June 1980.
7. U.S. Department of State. *Mineral Trade Notes*, 76, No. 3 (March 1979).
8. "Venezuela's $200 Million Carbozulia Coal Project." *Mining Magazine*, May 1982.
9. *Newsweek* 89 (6 July 1981).
10. *Business Week* 53 (6 July 1981).

Chapter IV-7

Colombia

Colombia stands ready to join the ranks of Latin America's major mineral producers. Several major projects that have come into production, or are about to do so, will raise Colombia to a significant position in the world mineral economy. Long known as the world's major source of gem-quality emeralds and as a minor gold producer, Colombia will soon be a significant world producer of ferronickel, high-grade steam coal, asbestos, phosphates, and possibly copper and bauxite.[1]

MINERAL RESERVES AND RESOURCES

Colombia does not have a clear picture of its mineral resource endowment because mineral exploration has been limited by a lack of capital, personnel, infrastructure, and equipment. Large areas in Colombia have not been adequately explored, and the country's mineral inventory does not accurately reflect its potential.

Colombia has the largest and highest-quality coal reserves in Latin America. These reserves totaled 16.5 billion metric tons as of January 1981.[2] Iron ore reserves are estimated at 235 million metric tons.[1] Colombia also has about 90% of the world's gem-quality emerald reserves. Colombia's mineral and mineral fuels reserves are presented in Table IV-7-1. The actual reserves of minerals and mineral fuels in Colombia may be much larger than Table IV-7-1 indicates. Only after a major exploration effort will its mineral endowment be known with any degree of certainty.

GEOLOGY

Colombia's topography is dominated by the Andes mountains, which run along the western side of the country, and by the Amazon basin region, which occupies large parts of eastern Colombia. Coal is found in about 35 basinal structures in Colombia. The coal is extremely high in quality, with less than 1% sulfur and less than 1% ash content.[2] Most deposits are tabular and are at relatively shallow depths, lending themselves to open-pit exploitation. Only seven of the 35 coal-bearing basins have

TABLE IV-7-1 Colombia: Mineral and Fuels Reserves (Metric Tons Ore)

Mineral	Ore reserves[a]
Asbestos	10,000,000
Bauxite	375,000,000
Chromium	180,000
Copper/molybdenum	450,000,000
Gold	6,000,000
Iron	235,000,000
Nickel	40,000,000
Phosphates	38,000,000
Platinum	1,000,000
Silver	NA
Coal	16,500,000,000
Natural gas (million ft^3)	4,700,000
Petroleum (bbl)	960,000,000
Uranium (U_3O_8)	40,000

[a]NA not available.

Sources: Refs. 1, 3, 4, 5, 6, and 8.

been fully explored, and large deposits of coal undoubtedly remain to be discovered.

Colombia's nickel reserves occur in a large laterite deposit in western Colombia.[2] The nickel is located over a peridotite intrusive stock that cuts Miocene sedimentary beds.[3] The Cerro Matoso deposit forms a prominent landform that rises above the surrounding terrain. Nickel and iron ores overlay one another at Cerro Matoso.[4] Iron ores at Paz del Río are oolitic and date to the Oligocene.[3]

Virtually all of Colombia's gold, silver, and platinum are obtained from alluvial deposits. All of these metals are obtained from the same deposits, principally in the Antioquia and Choco departments.[5] Some silver is obtained from small lode deposits, and the remainder is recovered as a by-product of gold operations. Copper deposits in the Colombian Andes are of the porphyry type. Copper ores average 0.7% copper, with associated molybdenum values. Colombian copper deposits are similar to porphyry deposits in the Chilean and Peruvian Andes. The geology of porphyry copper deposits is discussed in more detail in the sections on Chile and Peru.

LOCATIONS OF MINES AND PROCESSING FACILITIES

Mining activity in Colombia is concentrated in the western part of the country. Most deposits that are currently being exploited are in the mountains or near the coast. The country's limited infrastructure has limited mineral exploration and development. The locations of major mines and processing centers are shown in Figure IV-7-1. Coal mines have historically been developed near population centers, utilizing underground mining methods.[3] Foreign firms have concentrated exploration activity near the coast. The El Cerrejón deposit, located on the Guajira peninsula, contains at least 2 billion metric tons of steam coal.[3] The El Cerrejón North open-pit mine is the first large-scale coal project in Colombia. Development is proceeding apace, with production having begun in 1983.[6]

Nickel mining and smelting began in June 1982 at the Cerro Matoso project, located about 400 km northwest of Bogota.[3,7] Cerro Matoso produces ferronickel, primarily for export. Iron ore mining,

and Colombia's only major integrated steelworks, are located near Paz del Río, northeast of Bogota.[3]

Gold, silver, and platinum mining is undertaken by dredging alluvial sands and gravels in the Nechi, San Juan, and Atrato Rivers. These rivers are in the Antioquia and Choco departments. The major emerald mines are in the Boyaca department at Muzo and Coscuez.[3]

Colombia's petroleum and natural gas reserves occur in basins located near the Caribbean coast and offshore. Other oil and gas fields are in isolated river basins, the most important of which are the Magdalena and Putumayo river basins.

PHYSICAL INFRASTRUCTURE

Energy

Colombia has a promising energy future with its huge coal reserves and tremendous hydroelectric potential. Colombia's hydroelectric potential is in areas where the bulk of the population resides. Present electric generating capacity is about 4000 MW, and about 70% of this is from hydro sources.[1]

FIGURE IV-7-1 Colombia: Mines and processing facilities. (From Ref. 1.)

Colombia has exploited only about ½ of 1% of its total coal resources.[6] Only about 3% of Colombia's 93,000-MW hydroelectric potential has been developed.[2]

Colombia's oil production is only three-fifths of the 1971 level, but production has been relatively stable at 45 to 50 million barrels per year since 1977.[2] Gas production is about 50% above the 1971 level.[2] Colombian oil imports were about 18.6 million barrels in 1981.[2] Colombia predicts that oil production will remain at present levels through 1990.[2] Natural gas production will continue to increase as demand conditions warrant.[2] Colombia flared about 14% of total 1981 gas production, a decrease from previous years.[2] Colombia is reinjecting some of the gas it previously used to flare.

The availability and low price of electricity, coal, and natural gas give mineral development projects in Colombia an advantage when compared to projects in most other countries. The magnitude of Colombia's coal and hydroelectric resources constitute a tremendous incentive to further mineral development and industrialization.

Water

Colombia has abundant water resources. Even in periods of low rainfall, Colombia has sufficient water resources to meet its needs for both water and hydroelectricity. The availability of water should not constitute a barrier to further mineral development and industrialization. Water is available near most major population and industrial centers.

Transportation

Colombia's limited transport infrastructure is a contributing factor to the relatively low level of mineral development in the country. Most mineral commodities are transported long distances by truck.[1] The nation's rail system is limited, and connections between truck transport and rail and river transport have not been developed and constitute a bottleneck in the mineral transport system.[1] Colombia's mineral transport system is ill equipped to handle bulk commodities such as coal and iron and nickel ores. Colombia's energy and mineral transport infrastructure is shown in Figure IV-7-1.

A rail line connects the iron and steel center at Paz del Río and the capital, Bogotá. The country's major rail lines run in a north-south direction: from Santa Marta on the Caribbean coast, up the Magdalena river valley to the capital; and from Popayan in the southwest part of the country through the cities of Cali and Medellín, to connect with the Santa Marta–Bogotá rail line.[1] A short rail line connects Cali and the Pacific port at Buenaventura.[1]

Oil- and gas-producing regions, both onshore and offshore in the Caribbean, are connected to refineries and consumption areas by crude oil and natural gas pipelines.[1] The major oil refineries are located at Barrancabermeja and Cartagena. The Barrancabermeja complex has a capacity of 140,000 barrels per day, and the capacity at Cartagena is 47,000 barrels per day.[2]

DOMESTIC MINING INDUSTRY
Overview

Mining activity makes a small contribution to Colombia's gross national product; in 1980 and 1981 the contribution was only 1%.[2] This figure includes all hydrocarbon fuels, metallic, and nonmetallic minerals. Colombia's national development plan, the Plan de Integración Nacional, sees the mining and energy industries as the foundation for further economic development. Government investments in mining and energy increased 240% between 1978 and 1980.[4]

As ferronickel, steam coal, and asbestos projects reach full productive capacity in the mid-1980s, mining's share of gross national product should increase. Colombia's low level of industrialization should allow the bulk of the output of these projects to be exported. Government revenues from the mining industry will increase, and Colombia's balance of trade situation should improve.

Industry Structure

Since the adoption of Decree No. 1275 in 1970, all minerals have been under state ownership.[1] This mining law specifically excluded those concessions granted prior to 1970 and still being exploited. Colombia recognizes that foreign capital is desirable to expedite the development of the country's large undeveloped mineral potential.

The state will enter into joint venture projects with foreign firms, provided that the terms of the contract conform to Colombia's rather liberal interpretation of Andean Pact investment codes, specified in Decisions 24 and 103.[1] Colombia will allow foreign investment, provided that local labor, management, and materials are utilized. Other requirements include technology transfer, infrastructure development, and increased export potential.

The provisions of the contract between Exxon Minerals' subsidiary Intercor and the state coal company Carbocol regarding the development of the El Cerrejón coal deposit illustrate the type of joint venture agreement Colombia desires. The two firms will share all development costs and will each be allowed to market half of total mine output for the first 23 years of production.[4] After this period, 75% of the deposit will still be in place.[8] Carbocol will gain ownership of the mine and related infrastructure and the technology bound up in the project at no cost.[8]

A similar agreement exists between the state nickel company, Econiquel, and Billiton International Metals BV of the Netherlands and the Hanna Mining Company of the United States.[1] Billiton Overseas has agreed to buy all the ferronickel produced by Cerro Matoso for the first 12 years of production.[7] After the first 25 years of production, ownership will revert entirely to Econiquel.[3]

Colombia is therefore willing to make some short-term sacrifices to attract the investment capital needed to develop its mineral resources. Colombia is trying to attract investment capital to further develop the country's coal resources and to begin the development of its copper-molybdenum resources.

Colombia has created separate companies to develop and exploit each of the minerals occurring in the country. These firms are authorized to enter into joint ventures with both private domestic and foreign concerns to facilitate the development of mineral resources.[4] One state company, the Empresa Colombiana de Minas (Ecominas), has control over several minerals.[1]

The concentration ratios in Colombia's minerals and mineral fuels industries are very high. All of Colombia's mining activity is undertaken by a handful of state and mixed companies. The state owns all or part of every mineral project in the country. The oil industry is the only exception; it is much less concentrated.

Vertical integration in the minerals and mineral fuels industries is limited. The state owns an integrated iron and steel complex and owns and operates three refineries that refine Colombian and imported oil. The completion of the Cerro Matoso nickel smelter will give Colombia South America's first integrated ferronickel complex.[1] The remainder of the mineral industry is not vertically integrated. Chromite ore is produced and exported without beneficiation.[4] Activity in the precious metals industries is limited and utilizes primitive methods. Industrial minerals have not been developed in Colombia.

The low level of industrialization in Colombia accounts for low levels of domestic consumption for most mineral commodities. As Colombia's role in the world mineral economy increases, the level of vertical integration in the minerals industries may increase. Rapid changes are unlikely in the short to medium term.

Government Role in the Mining Industry

The Colombian government influences the mining industry in a number of ways. The state owns part or all of every mining venture in Colombia. Mining laws and policies affect the institutional framework in which mining firms operate, and the mining taxation system is also important to the firms. As in most Latin American countries, Colombia grants the ownership of minerals to the state.[1]

Mining Taxation. Information regarding mining taxation in Colombia is difficult to obtain. Statistics regarding taxes and revenues are generally not published in Colombia. The tax provisions of the agreements for the development of the Cerro Matoso and El Cerrejón projects are somewhat clearer. Exxon Minerals' subsidiary Intercor has agreed to pay a 15% royalty on its half of production at El Cerrejón.[4] The royalty will be computed on the mine-head production at El Cerrejón, and will be payable either in cash or in kind.[4]

The concession agreement signed by the government and Billiton and Hanna Mining for the Cerro Matoso nickel deposit provides for a royalty of 8%.[4] This royalty is also payable in kind and is projected to provide about $12 million annually in government revenues.[4] Royalty payments for emerald mining provide Ecominas with a steady cash flow for investment in other mining projects. Ecominas receives $6.6 million annually from emerald mining concessionaires and collects an additional 23% royalty on the value of the gems.[9]

The 5% tax on coal production will fund the Fondo Nacional del Carbón. This government entity was organized to improve coal transport infrastructure and to provide assistance to small and medium-sized coal miners.[1] Tax provisions for other minerals are not available.

Government Subsidies. The government does not levy the 5% coal tax on coal produced in Colombia and used domestically to produce electricity. To stimulate capital investment and private participation in mining projects, Colombia has created some tax incentives. Tax deductions for investments in coal, phosphates, copper, and bauxite were promulgated in Law 20 and Resolution 30 of 1979.[2] In addition, Resolution 56 of 1979 gives some flexibility to the disposition of income derived from mineral exports.[2]

Mining Law and Policy. Colombia's Mining Law No. 1275 of 1970 gave ownership of all minerals and mineral fuels to the state, except for those concessions previously authorized.[1] Decrees No.

2181 and No. 1620 modified the mining law in 1972, changing the regulations governing mineral exploration and exploitation.[1] The mining law is relatively stable. Colombia has adopted several new mining laws to facilitate coal exploration and development. Exploitation of the El Cerrejón deposit is regulated by Law No. 61 of 1979.[2] This law provides for the 5% tax on coal to further coal exploration activity.

Resolution No. 29 of 1976 was intended to expedite imports of equipment, machinery, and capital for mining activity.[2] Strict foreign exchange controls hindered the importation of spare parts and machinery prior to the act. Emerald mining activity is under the authority of the Ministry of Energy and Mines, as specified by Decree No. 293 of 1964.[2] Decree No. 1411 of 1978 gave certain emerald mining functions to Ecominas.[2] Gold mines in the Caldas and Risaralda departments passed to government hands with the adoption and implementation of Decree No. 2064 in late 1980.[2] The reserves in the areas were designated special reserves and are under the authority of the Ministry of Energy and Mines, and Ecominas.

Colombia adopted a series of Laws and Resolutions between 1969 and 1976 to promote the exploration and development of oil and gas resources. Law No. 20 of 1969 and Decree No. 2310 of 1974, regarding association agreements for oil and gas activity, have increased activity in the industry.[2] Resolution No. 50 of 1976 established a price equal to the international price for all oil discovered in association with the state oil company, Ecopetrol.[2] An indication of the success of these measures is the increase in oil exploration activity in the country. The number of exploratory wells drilled increased from 29 to 1979, to 100 in 1981.[2] Of the 100 wells drilled in 1981, 61 were exploratory wells and 39 were stratigraphic test wells.[2]

Government Role in Research and Development. The Colombian government undertakes energy and mineral research under the auspices of the Ministry of Energy and Mines, the Instituto Nacional de Geología y Minas (Ingeominas), and the Instituto de Investigaciones Technológicas (IIT).[1] Colombia has been involved in joint research efforts with Belgium, Spain, West Germany, Italy, Mexico, Brazil, and Japan.[1]

Domestic Supply–Demand Dynamics

Colombia's relatively low level of industrialization leaves a large portion of the country's mineral production available for export. Large-scale mining projects that are presently being developed, or are about to do so, will substantially increase Colombia's mineral export potential. Colombia should be exporting large volumes of high-grade steam coal, ferronickel, and possibly asbestos by the mid-1980s. Copper and molybdenum from porphyry deposits, and bauxite and alumina, may be produced and exported by 1990.

Colombia is completely dependent on imports of several basic minerals and highly import dependent in some others. Colombia imports all the tin, tungsten, and zinc required to meet domestic demand.[4] Colombia must import virtually all its modest domestic copper and lead requirements. Colombia also imports most of the primary steel, semimanufactures, pig iron, and ferroalloys needed for domestic consumption.[4] Colombia is also very highly dependent on imports of fertilizer minerals and is 40 to 50% dependent on imported sulfur.[4]

Colombia does not presently export large volumes of mineral commodities. However, virtually all of the world's gem-quality emeralds are produced and exported in Colombia. Large-scale exports of steam coal and ferronickel will enter the market in the next few years. The country also exports minor volumes of platinum.[1]

Colombia stands on the brink of entering the world mineral economy as a significant participant. The development of steam coal and ferronickel export potential will affect the world supply of these commodities. Domestic consumption of these commodities will remain low. The bulk of the planned production of 15 million metric tons per year of steam coal and 40 to 50 million pounds of nickel contained in ferronickel ingots will therefore be available for export.[8]

The Ministry of Energy and Mines has conducted studies that indicate that after 1990, El Cerrejón coal will have a lower cost ($4.1 per million Btu) delivered to Europe than either South African ($4.5 per million Btu) or Australian ($4.7 per million Btu) steam coal.[2] Statistics regarding Colombia's exports and imports of mineral commodities are fragmentary at best.

POLITICAL ENVIRONMENT

Domestic

Turbulence in Colombia's domestic political environment has given the country a negative image in the eyes of foreign investors. The M-19 guerrilla group took control of the Dominican Republic embassy in February 1980, holding a sizable group of diplomats hostage.[5] Some guerrilla activity is directed against the government, and some is between rival rebel groups.[6]

Colombia's political arena is dominated by two

parties, the Liberals and the Conservatives.[10] The parties have alternately held the presidency for most of the past 50 years, and alternated presidents from 1958 to 1970.[10] The Liberals are more powerful and held the presidency from 1974 to 1982. The Liberals favor business, liberal trade policy, the promotion of industry, and foreign investment controls.[10] Both the Liberals and the Conservatives are split by internal dissent. The Conservatives are allied with agriculture interests and favor foreign investment.[10] Other political parties include the Communists and the Anapo party.[10]

The current president is the Conservative Belisario Betancur Cuartas. His predecessor was the Liberal Dr. Cesar Turbay Ayala, whose term ended in May 1982. President Betancur is seen by the business community as more likely to create a favorable investment climate.[11] Recently announced monetary and credit controls are designed to lower interest rates and to increase the money supply to fuel increased public spending.[11] Betancur has pledged to reduce inflation and create jobs and housing, especially for the poor.[11]

Other major political actors include the military, who have not recently intervened in politics, and the church, a conservative force for stability.[10] The moderate labor groups are the Colombian Workers Union and the Colombian Workers Confederation. The Consicol labor group is communist-influenced.[10]

The National Coffee Growers Federation (NFCC) supports both major political parties.[10] Two business groups, the National Industrialists (ANDI) and the National Merchants Federation (Fenalco), support a larger private role in the economy.[10] Left-wing extremist groups include the M-19 guerrilla group and the Armed Forces of the Colombian Revolution (FARC).[10] Narcotics traders, including marijuana growers and cocaine processors, contribute to the atmosphere of political cynicism and distrust.

Colombia is politically stable, relative to the rest of Latin America, and prospects for war or a coup d'etat are presently considered low.[10] Guerrilla violence will probably persist in rural areas. Labor unrest, a disenchanted middle class, and university graduates who sympathize with urban and rural guerrillas are all problems.[10] Transport problems, foreign loans, high interest rates, unemployment, and inflation all plague Colombia.

Colombia presents a positive climate for foreign investment. The country is nationalistic and pro-Western, risks of nationalization are presently low,

and contracts are generally enforceable.[10] Considerable tax subsidies are granted in mining and manufacturing, and the state plays a small role in the economy.

International

Colombia is involved in two territorial jurisdiction disputes, one with Nicaragua and the other with Venezuela.[2] Nicaragua rejected previous agreements with Colombia when it extended its maritime limit to 200 miles.[2] Nicaragua has claimed ownership of three cays—groups of islands and reefs composed of sand and coral. Nicaragua specifically claimed ownership of the minerals in the area. Nicaragua has stated that it will pursue its claims by legal means.[2]

The border dispute with Venezuela centers on the maritime boundary within the Gulf of Venezuela.[2] This border dispute also concerns mineral resources. A proposed agreement would give ownership of the Colombian Los Monjes Islands to Venezuela in return for limited Colombian rights in the Gulf.[2] The two countries may share hydrocarbons which are believed to lie under the Gulf. The two countries have made progress on the issues, but internal disagreements have so far delayed the signing of the draft proposals.

With the exception of these border disputes, Colombia has relatively good foreign relations. Colombia has viewed the international political situation from a somewhat isolationist perspective. Politics are typically dominated by national, not international, issues. Colombia's relative political stability compared to the Latin American region as a whole makes it an attractive place for both foreign and private domestic investment.

REFERENCES

1. U.S. Department of the Interior, Bureau of Mines. *Mineral Industries of Latin America.* Mineral Perspectives Series. December, 1981.
2. Rodado Noriega, Carlos. *Transición Energética y Minera.* Bogotá: Ministerio de Minas y Energía, June 1982.
3. Schiller, E. A. "Mineral Exploration and Mining in Colombia." *Mining Magazine,* January 1980.
4. U.S. Department of the Interior, Bureau of Mines. *Mineral's Yearbook 1980.* Washington, D.C.: U.S. Government Printing Office, 1982.
5. Spooner, John, ed. *Mining Journal Annual Review 1980.* London: Mining Journal, June 1980.
6. Spooner, John, ed. *Mining Journal Annual Review 1982.* London: Mining Journal, June 1982.
7. U.S. Department of State. *Mineral Trade Notes* 76, No. 5 (May 1979).

8. Scheiderman, Stephen J. "Colombia Joins Latin American Mining's Big League." *World Mining*, September 1981.

9. "Colombia Unveils Plans to Step Up Mineral Development." *Engineering and Mining Journal* 180 (November, 1980).

10. "Nickel Mine Dedicated in Northern Colombia." *Mining Engineering* 34 (August, 1982).

11. De Shields, Anne P., Council of the Americas. "Colombia."

Chapter IV-8

Bolivia

Bolivia is one of the major mining countries in the Latin American region. Bolivia is one of the world's largest producers of antimony, tin, and tungsten, and also produces significant amounts of silver, lead, and zinc. Crude oil and natural gas are produced in modest quantities, with natural gas exports to Argentina ranking second to tin in importance to the Bolivian economy.[1]

Bolivia nationalized its mining industry in 1952, forming the state-owned Corporacion Minera de Bolivia (Comibol). The state-owned Empresa Nacional de Fundiciones (ENAF) operates virtually all of the refineries and smelters in Bolivia.[1] Bolivia has the capacity to process all its tin concentrate production into refined tin metal. The oil industry is dominated by the state oil company, Yacimientos Petroliferas Fiscales Bolivianos (YPFB).[1] Oil production is declining in Bolivia, while natural gas production continues to increase.

MINERAL RESERVES AND RESOURCES

Estimates of Bolivia's mineral reserves and resources are difficult to obtain. The Bolivians themselves are unsure of the extent of their mineral endowment. In addition to its harsh, mountainous topography, Bolivia has an area of over 1 million square kilometers and a relatively low population density. Bolivia's unexplored mineral potential is therefore probably large.[1] Statistics on Bolivia's mineral reserves are sketchy and fragmentary at best. Bolivia's mineral reserve base, as estimated by the U.S. Bureau of Mines in December 1981, is presented in Table IV-8-1.[1]

Bolivian mining firms have not adequately explored and delineated known mineral deposits. The firms prefer to mine what they know exists (even when total reserves amount to only two or three years' production), with an emphasis on earning profits immediately. The companies hope to expand their reserves but will not do so until they literally run out of the ore they presently mine.

The U.S. Bureau of Mines estimates Bolivia's natural gas reserves at about 162 billion cubic meters and oil reserves at about 170 million barrels.[1] Bolivia is roughly self-sufficient in energy, but declin-

TABLE IV-8-1   Bolivia: Mineral Reserve Base (Metric Tons)

	Reserve base
Mineral	
Antimony	310,000
Bismuth	6,900
Gold[a]	19,000
Lead	500,000
Silver[a]	61,000
Tin	1,150,000
Tungsten	39,000
Zinc	2,000,000
Fuels	
Natural gas[b]	162,000
Petroleum[c]	170,000

[a]Unit of measure is thousand troy ounces.
[b]Unit of measure is million cubic meters.
[c]Unit of measure is thousand barrels.
Source: Ref. 1.

ing oil production may necessitate imports by the end of the 1980s.[1] Bolivia has significant undeveloped hydroelectric potential.

GEOLOGY

Bolivia's most significant geologic feature is the Andes mountain range. Most mining activity takes place high in the Andes or on the high plains (altiplano) within the range. A large belt of tin and tungsten deposits runs through Bolivia's central Andes.[1] The mineralized area also contains the bulk of the country's antimony, bismuth, lead-zinc, gold, and silver deposits. Bolivia's mineral deposits are shown in Figure IV-8-1.[1]

Another major feature of the Bolivian topograpy is the Amazon basin area of eastern Bolivia. This region is the source of most of Bolivia's oil and gas production. The Mutún iron ore deposit is also in the Amazon basin region. Oil and gas activity is centered around Santa Cruz and extends to the southeast, toward Argentina and Paraguay.

Many of Bolivia's mines extract ores that contain more than one mineral. Tin and tungsten often occur together. Some deposits also contain associated gold, silver, and lead-zinc, in addition to the

FIGURE IV-8-1 Bolivia: Mines, processing facilities, and infrastructure. (From Ref. 1.)

tin content.[2] The country's major zinc producer is the Matilde mine on the shores of Lake Titicaca.

Major mines, areas of hydrocarbon production, and processing facilities are shown in Figure IV-8-1. The infrastructure linking these producing and processing centers to areas of domestic consumption and overseas markets is also shown in Figure IV-8-1.

The major tin mineral in Bolivia is cassiterite, and the most important tungsten mineral is wolframite. Most gold occurs in alluvial deposits, especially in the Tipuani River area.[2] A large part of the tin lost in the concentration process is in the form of tin slimes.

The country's largest metallurgical complex is located at Vinto, near Oruro, in the heart of the tin and tungsten belt. ENAF smelts virtually all of Bolivia's tin, antimony, silver, and lead-zinc output at this large complex. ENAF has a state-granted monopoly for smelting.[1] ENAF also has a tin volatilization plant that was built with Soviet materials and financing.[2]

This plant, and a new lead-silver smelter being constructed by a Belgian-German consortium, are located in Potosí. The new smelter should have an annual capacity of 24,000 metric tons (mt) lead, 33,000 mt zinc, 2000 mt antimony, 800 mt tin, and 200 mt silver.[2] Capacities at Vinto are about 30,000 mt tin and 5000 mt antimony, with additional lead-zinc and silver recovery circuits.

PHYSICAL INFRASTRUCTURE

Energy and Water

At present, Bolivia is basically self-sufficient in petroleum products and natural gas. Gas production has been increasing slowly in recent years. Production of crude oil and condensate has fallen steadily since its peak in 1973.[2] Over 75% of Bolivia's total energy consumption is in the form of oil. Electric generating capacity of 460 MW provides about 8% of Bolivia's energy needs.[1] Of the total, 60% is from hydro sources and 40% is fossil fuel fired.[1] Firewood makes up the remainder of Bolivia's energy balance.[1]

YPFB has had a monopoly on refining and domestic sales of petroleum since the 1950s. Domestic energy prices were raised in 1979 and in 1981, increasing one of mining's largest operating costs.

Infrastructure is in place for oil exports to Argentina, and through Chile to international markets, although no exports have been made since 1979.[2] Natural gas exports to Argentina amount to over 200 million cubic feet (MCF) per day, and when the pipeline linking Bolivia's gas fields and Brazil is completed (1985–1986), the export total may reach 600 mcf per day.[2]

Coal is not used as an energy source in Bolivia. Geological survey and exploration efforts aimed at locating uranium deposits have been unsuccessful. Bolivia has sizable untapped hydroelectric potential. Some substitution of indigenous natural gas for imported oil is likely, especially in residential and commercial applications.

The availability of water in Bolivia does not appear to constitute a significant barrier to further mineral development. Large parts of the altiplano are very dry, but the country's overall water supply is adequate. As in many parts of the world, water is available in Bolivia, although it is often in the wrong places and of a questionable quality.

Transportation

Bolivia's mineral transport infrastructure is severely limited by the prevalent mountainous topography. The infrastructure serving mineral deposits in remote areas has been financed by state and private mining companies.[1]

As Bolivia is a land-locked country, all of its mineral exports must pass through neighboring countries to ports and areas of mineral consumption. Most mineral exports travel by rail to the Chilean ports of Arica and Antofagasta. The mineral production of the La Paz area is transported by rail and truck to Arica, Chile; and Matarani, Peru. Bolivia's mineral transport infrastructure is shown in Figure IV-8-1.

Mineral production from Vinto-Oruro and Potosí is shipped by rail to Antofagasta, Chile. This British-owned rail line also serves the mammoth open-pit copper mine at Chuquicamata, Chile. Another rail line links Santa Cruz with Corumba, Brazil, passing the Mutún iron ore deposits. Some mineral production is shipped by rail and truck to Argentina. Natural gas exports to Argentina travel through a 24-inch pipeline that crosses the border at Yacuiba-Pocitos.[1]

Within Bolivia, mineral production is transported by rail and truck from producing areas to the Vinto processing complex. Bolivia has an extensive railway system, linking mineral producing and processing areas to ports and areas of end use in neighboring countries. Less than 10% of Bolivia's 37,000 km of highways is paved.[1]

Oil and gas are shipped by pipeline to both domestic and export markets. Bolivia has a total of 13 crude oil and natural gas pipelines with a combined length of about 3400 km.[1]

MINING LABOR

The mining industries in Bolivia employ about 79,000 workers, equivalent to about 5% of the labor force.[3] The mining labor force has been declining, falling by about 5% in 1981.[3] Of the total, about 30% each are employed respectively, by Comibol, the small miners, and mining cooperatives, with the remaining 10% employed by the medium-sized miners.[1] YPFB employs an additional 4500 workers in oil and gas exploration and development.

Miners and railway workers staged strikes during the political unrest of late 1979 and 1980.[2] Most mining activity was halted for two to three weeks following the Army coup d'etat in 1980.[4] No significant strike activity affecting the mining industry occurred in 1981, and the government gave all workers an additional holiday to ease labor tensions.[3] Although the costs of fuel and food have increased in recent years, salaries and wages have not. Workers have thus experienced a decrease in real purchasing power.

The medium-sized mines provide scholarship funds for about 25 students in mining and metallurgical engineering and in related technical fields.[3] Comibol trains many of its employees internally. Because badly needed capital investments have not been made, the skill level of Bolivian miners is low and laborers are readily available at modest wage rates.

DOMESTIC MINING INDUSTRY

Overview

Mineral fuels contribute about 7% of Bolivia's GNP.[2] Exports of minerals and mineral fuels brought in about 85% of Bolivia's total foreign exchange earnings in 1980.[2] Tin is responsible for about two-thirds of the total value of Bolivia's mineral output. Mining activity employs about 5% of Bolivia's labor force.

Industry Structure

Bolivia's largest mineral producing company is the state mining company Comibol. Comibol operates through a dozen operating companies and a few mixed companies. Comibol totally dominates the Bolivian mining industry and is the country's largest tin, tungsten, silver, and lead-zinc producer.[1]

Next in importance to the Bolivian mineral

economy is a group of about 25 medium-sized mining companies organized into the NAMM. The NAMM produces about 25% of Bolivia's total mineral output and about 80% of all antimony output.[2] The remaining mineral production is by small mining companies, mining cooperatives, and a few concessionaires. The contribution of these miners to Bolivia's total mineral production is on the order of 5 to 15% for different minerals.[2]

The shares of Bolivia's total production of various minerals by Comibol and the small and medium-sized mining sectors in 1981 are presented in Table IV-8-2.[5] The material presented in this table suggests a relatively high level of concentration in the mineral industries. For example, two companies produced 69% of Bolivia's 1980 antimony production.[2] As the state controls almost all mining and mineral processing activity, levels of both vertical and horizontal integration are very high.

The bulk of Bolivia's mineral production is processed and smelted by ENAF. The trend is toward higher levels of domestic mineral processing to enable Bolivia to earn more foreign exchange. Unfortunately, ENAF purchases tin concentrates at artificially low prices which are well below those quoted on international metal exchanges.[5]

YPFB's oil and gas production has become more important to the Bolivian economy as gas exports to Argentina have increased. The oil company brought in about 24% of Bolivia's foreign exchange earnings in 1980 and accounted for 2.7% of the country's GNP.[2] Completion of the proposed natural gas pipeline to Brazil will bring in additional revenues and increase the importance of Bolivia's energy sector to the country's economy.

Government Role in the Mining Industry

The Bolivian government plays an extensive role in the mining and mineral fuels industries. Control over the mining sector includes tax policy, the ownership of mineral-producing companies, and mining laws. The government draws between one-third and one-fourth of its revenues from its taxation of the mining industries.[2] Mining activity is taxed in many ways and at all levels of processing.

Mining Taxation. Mining in Bolivia is subject to many types of taxes. Mining companies must pay a land tax of 6 pesos per hectare per year. Mining companies must also pay a research tax to the fiscal treasury. The medium-sized miners alone paid $2365 million dollars research taxes in 1981.[3] Until March 1980 tin exports were subject to a special export levy of 7.5%.[1]

Mining companies' income tax rate was increased from 38% to 53% in March 1980.[2] This tax is levied on the difference between the sales price and the presumed cost specified by the government. For tin, this presumed cost ranges from $3.93 per pound for the medium-sized mining companies to $4.48 per pound for Comibol mines.[2] The notion of a fixed presumed cost per unit of production does not allow for a very realistic taxation scheme.

The tax rates are not stable, and this must be considered detrimental from a planning perspective. A Harvard University study has recommended a move away from export and income taxes to some form of profits tax.[1] All elements of the mining industry express dissatisfaction with the present mining taxation scheme. The World Bank has recommended the use of periodic and systematic adjustments of the presumed cost allowed by the government.[3]

Another tax has been proposed to siphon off revenue from those holding mining concessions for purely speculative reasons. This tax is called the Mining Patent Tax and has not yet been implemented by the government.

The 1980 *Mining Annual Review* stated that the level of royalties and export taxes prevailing at the time were "threatening to kill the goose that lays Bolivia's tin if not golden egg."[5] The special

TABLE IV-8-2   Bolivia: Mineral Production 1981, by Sector (Metric Tons)

Mineral	Comibol	Medium-sized mines	Small mines	Other	Total
Antimony	—	12,155	2,473	673	15,301
Bismuth	9	—	—	2	11
Copper	2,623	—	14	—	2,637
Lead	9,598	4,359	2,442	358	16,757
Silver	171	24	4	6	205
Tin	20,828	6,281	2,621	50	29,780
Tungsten	1,424	1,765	235	25	3,449
Zinc	27,218	17,731	139	1,941	47,029

Source: Ref. 3.

export levy, a type of windfall profits tax imposed when tin prices are high, has been removed. Simultaneously, the income tax was increased by 15%. The government-specified presumed cost figure for Comibol's tin is $0.50 per pound higher than for the NAMM, and somewhat higher than that allowed for the small miners and mining coops.

Experts at the World Bank and Harvard University agree that reforms to the Bolivian mining taxation scheme are in order.[1,3] The problems that should be addressed when a new mining taxation system is devised include differences in and adjustments to the presumed cost figure, moves toward a profit tax, tax stability, and the low prices paid for tin concentrates purchased by ENAF.

Government Subsidies. Government subsidies to the mining industry in Bolivia are most commonly in the form of tax concessions. With production volumes and prices of many mineral commodities down, the high dependence on mining for fiscal revenues does not permit many tax concession subsidies.

Tin producers who increase output through new mine investments may receive an income tax credit of up to 40% of the amount invested.[1] This investment tax credit is also available to antimony, copper, and tungsten producers who increase output through new investment.

Two supreme decrees issued in 1979 offer unspecified incentives to mine mechanization and modernization projects.[2] These unspecified incentives have attracted little interest from mining firms. Government involvement in infrastructure projects is limited to the expenditures made by the state minerals, fuels, and smelting companies.

The government's role as the marketer of Bolivian mineral production is a sore point for the industry. ENAF purchases ores and concentrates at prices well below the prices prevailing in major metal markets. ENAF markets the country's mineral exports (sometimes at a loss), and smelts at a lower recovery rate and at a higher cost than competing smelters located in other parts of the world.

Evidence of significant mining education and training programs is unavailable. Given the ready surplus of miners in Bolivia and the low skill level needed for most mining industry jobs, training programs are not a top prioirity in the Bolivian mining industry.

Government Ownership of Producing Companies. The government is the dominant owner in the minerals and mineral fuels industries in Bolivia. Comibol is the country's largest mining compnay. YPFB is the only oil and gas firm, and ENAF has a state-granted monopoly in the processing and smelting of minerals. Comibol produces roughly two-thirds of Bolivia's mineral production.

Mining Law and Policy. Bolivian mining law is based on the Spanish colonial law that gave ownership of all minerals to the state. Individuals can undertake mining activity through state-authorized mining concessions. Bolivia's mining industry was nationalized in 1952, and the state mining company Comibol was formed.[1]

Bolivia has no concrete mining law or policy. Various mining codes and supreme decrees govern mining activity. Some mineral deposits have been designated fiscal reserves and have been withdrawn from development. The recent trend is to the re-opening of the fiscal reserves in an effort to encourage their development.

Investment in mining is a part of the government's five-year development plan that runs through 1985. The estimated investment needs for the period are over $600 million, with more than half to be invested in minerals processing.[2]

Comibol has been actively seeking foreign investors for joint venture projects since 1979.[6] The terms specified include a maximum duration of 20 years and an eventual divestiture of all equipment and facilities at the end of the contract period. Investors have adopted a wait-and see-attitude, and interest has been modest.[2]

Recent mining legislation has been ineffective and in some cases even ignored. The lifting of the ban on mining concessions in the Precambrian shield area near Santa Cruz in 1981 was a positive step.[3] Other legislation includes the creation of security areas around state-owned industrial plants and a decree aimed at discouraging clandestine mineral sales.[3]

A stable mining law and a clear, purposeful mining policy will be necessary to enable Bolivia to deal with the severe problems facing its budget. Economizing on many areas of expenditures should help increase a favorable climate for foreign investment. Efforts to provide incentives for mining activity and to provide guarantees against nationalization culminated in a new pro-mining law, approved in December 1981.[2]

Government Role in Research and Development. The Bolivian government provides support and technical services to the mining industry through the Servicio Geológico de Bolivia (Geobol) and the Instituto de Investigaciones Minero-Metalúrgicas (IMM).[1] These two entities are geared primarily toward helping the small and medium-sized mining companies. A mining industry accounting project will use three companies to form a data base from which a uniform accounting system for mining companies will be devised.[2]

Domestic Supply–Demand Dynamics

Productive Capacity. Bolivia's productive capacity for a number of minerals is difficult to determine. Falling ore grades and antiquated and inefficient mining equipment and procedures complicate the process of determining Bolivia's productive capacity for some minerals.

Bolivia's tin capacity at the mine head is about 30,000 metric tons (mt) per year, and ENAF's tin smelting capacity is also about 30,000 mt per year.[1] Lead and zinc capacities are about 20,000 mt per year and 60,000 mt per year, respectively.[1] Tungsten capacity is just over 3000 mt per year.[1]

Bolivia's antimony metal-refining capacity is about 5000 mt per year and mine capacity is between 15,000 and 16,000 mt per year.[2] Comibol's silver capacity is about 200,000 troy ounces per year. Bolivia also has a bismuth smelting capacity of 1000 mt per year.[2]

Producer Stocks. Bolivian tin producers have enormous stocks of tin contained in old mine, mill, and concentrator tailings. These tailings total about 300,000 mt at an average grade of 0.44% tin.[5] The importance of these stocks becomes evident when one considers that the average ore grade at the Catavi mine has fallen below 0.38% tin.[5]

Producer stocks of other minerals are much smaller. Most processing and smelting facilities have small stocks of ores and concentrates on site to ensure a steady supply to the plant. Tailings at these plants also contain moderate amounts of minerals that may or may not be recoverable.

Domestic Supply and Demand. With the exception of gold and mineral fuels, the majority of Bolivia's mineral production is for export. Bolivia does not export gold or crude oil on a regular basis, and exports only about one-third of all marketable natural gas production.[1] Small amounts of manganese ore are exported, with about 70% of production consumed domestically.[1] Bolivia's hydrocarbon consumption fell by 10% in 1980.[2]

Almost all of Bolivia's production of other minerals is exported. The low level of industrialization in Bolivia does not lead to high levels of domestic consumption of strategic minerals such as antimony and tungsten. Bolivia exports at least 80% of antimony, bismuth, lead, silver, tin, tungsten, and zinc.[2] Bolivia's mineral exports, as a percentage of total production for 1980, are presented in Table IV-8-3.

The world economic recession has adversely affected both the domestic and international demand for Bolivia's mineral production. Downward trends in mineral prices in 1980 and 1981 have adversely affected Bolivian mineral producers. A study undertaken by the NAMM, presented in their 1981 Annual Report, indicates price effects on Bolivia's mineral production. Some of the findings of that study are presented in Table IV-8-4.[3]

As the figures in Table IV-8-4 indicate, despite significant increases in tin and silver production, these two minerals earned $130 million less in 1981, due to adverse price fluctuations. On a value basis, these two minerals are the most important mineral commodities in Bolivia. The gross value of Bolivia's

TABLE IV-8-3 Bolivia: Mineral Production and Exports, 1980 (Metric Tons)[a]

Mineral	Production	Exports	Exports/production (%)
Antimony	15,465	12,623	81.6
Bismuth	11	NA	78.0e
Gas, natural[b]	78,644	NA	36.0e
Gold[c]	52,075	—	—
Lead	17,747	15,936	89.8
Manganese ore	1,350	425	31.5
Silver[c]	6,099,000	5,684,000	93.2
Tin	27,272	22,530	82.6
Tungsten	3,359	3,435	102.3
Zinc	50,260	46,237	92.0

[a]e, estimated; NA, not available.
[b]Unit of measure is million cubic meters.
[c]Unit of measure is troy ounces.

Source: Ref. 5.

TABLE IV-8-4   Bolivia: Value of Mineral Production, 1980–1981 (Thousand Dollars)

Mineral	1980	1981	Percent change 1980–1981
Antimony	32,041	28,072	-12
Copper	4,008	4,593	+15
Tin	457,849	419,526	-8
Silver	124,127	71,561	-42
Lead	15,645	12,191	-22
Tungsten	46,727	30,353	-35
Zinc	39,779	42,509	+7
Other	2,091	1,843	-12
	$722,267	$610,648	-15

Sources: Ministry of Mines; National Association of Medium Miners.

mineral production fell by 15% in 1981 ($19.6 million).[3] The continuing and deepening world recession and continuing weakness in world metal prices in 1982 were responsible in part for an unfavorable climate for mining activity in Bolivia.

POLITICAL ENVIRONMENT

Domestic

One of the main factors contributing to Bolivia's lack of mineral development is the country's unstable political environment. The lack of political stability has led to a number of associated problems. Among these are the short-term perspective of both the government and Comibol, monetary and fiscal inefficiencies, domestic economic crisis, and a dangerously high dependence on exports of minerals and mineral fuels for foreign exchange earnings.

A digression into Bolivia's recent political history is useful at this point in order to understand better the crisis situation in the country's mining industry. President Luis García Meza seized power in a violent coup d'etat in July 1980. He survived frequent coup attempts by elements of the military until September 1981, when the current president, Celso Torrelio Villa, was sworn in.[2] From 1978 to 1980, Bolivia had seven different Ministers of Mines and Metallurgy.[2]

This climate of political instability has led to an extremely short-term perspective by the government, the Ministry of Mines and Metallurgy, and Comibol. Bolivia's political instability and high level of debt servicing have made the United States, private investors, and international banking institutions reluctant to inject investment into its economy.

The domestic economic crisis is largely the result of fluctuations in commodity prices, inefficient foreign exchange controls, and ineffective monetary policy.[3] Elements of the crisis include inflation, a lack of loanable funds, a collapse of the dollar supply, and the foreign exchange problems mentioned above. The prices of inputs to the mining industry rose sharply in 1981, led by a 64% increase in gasoline and petroleum products prices, a 60% rise in transport costs, and a 35% increase in electric rates.[3] With no available sources of loan funds, mining firms have experienced severe cash flow problems. The cash flow problems were compounded by a restrictive monetary policy that reined in the money supply growth rate from 40% in 1980 to just 3.7% in 1981.[3] The Central Bank also imposed rigid foreign exchange controls and then stopped sales of foreign exchange altogether.[3]

A parallel currency market for recycled cocaine dollars arose as a result of these Central Bank policies, but was incapable of meeting the demand for dollars for consumer, intermediate, and capital goods imports.[3] The exchange rate on the parallel market was about 40 bolivars to the dollar at the end of 1981.[3]

The crisis in monetary policy, combined with high inflation and fixed wages for miners, has caused increased hardship among mining company employees. The NAMM reports that the economic crisis has caused declines in miners' standards of living and in their productivity.[3]

Most of these problems are of a short-term, tractable nature. More fundamental, long-term problems face Bolivia's mining industry. Commodity price fluctuations will continue to have serious adverse effects on the Bolivian economy as long as the country is dependent on a handful of mineral exports for the bulk of its foreign exchange.

The recent history of economic and political instability in Bolivia will not be quickly forgotten by international lenders. The short-term perspective of the many short-lived administrations has created a shortage of investment in the mining industry for the last two decades.

Consequently, Bolivian mining equipment is outdated, inefficient, and even unsafe. Reserve-production ratios for many mineral commodities are as low as 3 to 1 because badly needed exploration work has not been undertaken. The antiquated Bolivian mining industry will require massive injections of foreign investment capital to become competitive once again in international metal markets.

Bolivia's large cocaine trade is somewhat of a deterrent to the further development of Bolivia's mineral industry. Many people in the producing areas near Santa Cruz work all night in the clandestine drug labs and are unavailable for any other type of work. *The Wall Street Journal* placed 1982 cocaine exports at about $1.5 billion, about twice the country's tin exports.[7] With mineral prices low, the incentive to engage in this activity is large.

International

Bolivia is a member of the International Tin Council (ITC) and the International Association of Primary Tungsten Producers.[3] A Bolivian delegation went to China in 1981 to discuss the tungsten market and problems facing antimony producers.[3]

The NAMM is widely recognized as the spokesman for the Bolivian mining industry. Bolivia maintains a hard-line stance in the ITC, advocating the position that buffer stock floor and ceiling prices be tied to market prices as opposed to production costs.[2] Bolivia's position in the ITC has been damaged by the expansion of tin production in Indonesia and Thailand. Bolivia is now the fourth- or fifth-largest tin producer, after holding second place behind Malaysia for most of the twentieth century.[5]

Bolivia is the strongest defender of high tin prices.[5] The United States has been cast as the principal threat to tin producers. Bolivia lobbied hard in the U.S. Congress to prevent passage of a bill authorizing the sale of 35,000 long tons of tin by the General Services Administration (G.S.A.). These G.S.A. sales have proceeded at an annual rate of 10,000 long tons since July 1980, with an additional 5000 long tons being transferred to the ITC buffer stock.[2] Bolivia feels that these sales threaten its economy and the economies of all other tin producers.

Bolivia has shown a desire to obtain foreign technical and financial assistance. The Soviets have constructed a tin volatilization plant in Bolivia and have signed a contract for another.[2] Belgian and West German interests are constructing a lead-silver smelter in Potosi.[2] An agreement has been reached with Brazil for the construction of a natural gas pipeline linking Santa Cruz and São Paulo. Bolivia will have to change the attitudes of international financial institutions if investment funds and technical assistance are to flow into the country. Bolivia is also a member of the Andean Pact and follows the group's foreign investment guidelines.[1] Bolivia does not have the capital and human resources to develop its mining industry. Therefore, the creation of a climate that is favorable to foreign investment must be considered a priority.

REFERENCES

1. U.S. Department of the Interior, Bureau of Mines. *Mineral Industries of Latin America.* Mineral Perspectives Series. December, 1981.
2. Spooner, John, ed. *Mining Journal Annual Review 1982.* London: Mining Journal, June 1982.
3. National Association of Medium Miners. *Annual Report 1981.* La Paz: Inderpa Ltd., 1981.
4. "Worldwide Survey: Latin America." *World Mining*, August 1981.
5. U.S. Department of the Interior, Bureau of Mines. *Minerals Yearbook 1980.* Washington, D.C.: U.S. Government Printing Office, 1982.
6. U.S. Department of State. *Mineral Trade Notes* 76, No. 4 (April, 1979).
7. Martin, Everett G. "A Little Town in Bolivia Is Thriving as a Financial Center." *The Wall Street Journal*, 17 February 1983.

Chapter IV-9

Bauxite Producers

Three small western hemisphere countries are important to the world mineral economy for their alumina and bauxite production. The three countries, Jamaica, Suriname, and Guyana, accounted for 18.8% of estimated free world bauxite production in 1982.[1] When Brazil's 1982 bauxite output is added to the output of these three countries, the total is 27% of all freely traded bauxite.[1]

Bauxite and alumina exports provide 10 to 30% of the three countries' gross national products and between 45 and 80% of each country's total foreign exchange earnings.[2] The three countries are the world's third-, fourth-, and fifth-largest bauxite producers.[2] Jamaica is the world's fourth-largest alumina producer.[2] Jamaica and Suriname provide over half of U.S. bauxite imports and close to one-fourth of the U.S. alumina imports.[1]

With about 15% of the world's bauxite reserves, Guyana, Jamaica, and Suriname will continue to play an important role in the world bauxite market.[2] Of the three, Jamaica has the largest bauxite productive capacity and reserves.[2]

MINERAL RESERVES AND RESOURCES

Jamaica has the largest share of the three countries' bauxite reserves, with about 2 billion metric tons (mt), 57% of their total reserves.[2] Guyana has 1 billion mt of bauxite reserves, about 29% of their reserves.[2] Suriname's 500,000,000 mt of bauxite reserves are 14% of the three countries' total reserves.[2]

Guyana's bauxite reserves are important to the world mineral economy because they are of special grades.[3] Guyana produces some 80% of the world's calcined bauxite and calcined refractory-grade bauxite.[5]

GEOLOGY

The bauxite deposits of Guyana, Suriname, and Jamaica are of the low-level peneplain type that are found in tropical areas in South America and Australia.[6] The deposits typically have a boehmitic composition and are generally less than 30 ft thick.[6] The deposits in Guyana are separated from the underlying parent rock by a layer of kaolin (clay).

This type of deposit is usually associated with detrital bauxite horizons and is produced by the actions of fluvial or marine processes.[6] Jamaica's bauxite is of medium quality but of a simple molecular structure that requires less heat and pressure to process.[7]

LOCATIONS OF MINES AND PROCESSING FACILITIES

Most of Jamaica's bauxite-producing areas are near the island's coast. Some mines and alumina refineries are closer to the center of the island.[2] Jamaica's mines are at Lyford and Water Valley in St. Ann Parish.[2] Major alumina refineries are at Clarendon Nain, Kirkvine, and Ewarton.[2] Jamaica's bauxite mines and processing facilities and transport infrastructure are shown in Figure IV-9-1.

Guyana's most important bauxite mine is at East Montgomery, near Linden.[5] Two smaller mines are located at Kwakwani and Ituni.[8] The country's only major alumina refinery at Linden has a rated capacity of 240,000 mt per year.[5] A major bauxite calcining plant is located near the coast at Everton.[2] The locations of major mining operations and processing facilities in Guyana are shown on Figure IV-9-2.

All of the bauxite deposits, mines, the only alumina refinery, and the only aluminum smelter are located in the northern coastal area in Suriname. The Suriname Aluminum Co. (SURALCO) operates the country's only alumina refinery and aluminum smelter; both at Paranam.[2] Mines and processing facilities in Suriname are shown on Figure IV-9-3.

Suralco and NV Billiton Maatschappij Suriname, a Royal Dutch/Shell subsidiary, produce almost all of Suriname's bauxite from mines at Moengo and Lelydorp (Suralco) and Kankantrie and Para (Billiton).[2] Gold production is centered in the northern part of the country near Kwakoegron and near the Litani river in southern Suriname.[2]

PHYSICAL INFRASTRUCTURE
Energy

All of the oil consumed in Jamaica, Guyana, and Suriname is imported.[2] Guyana imports all of its oil requirements in the form of refined petro-

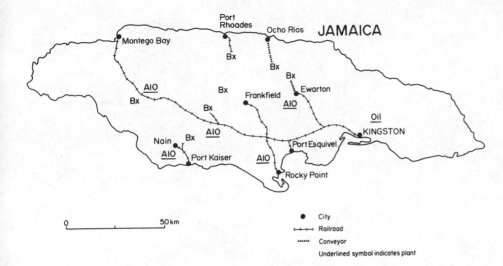

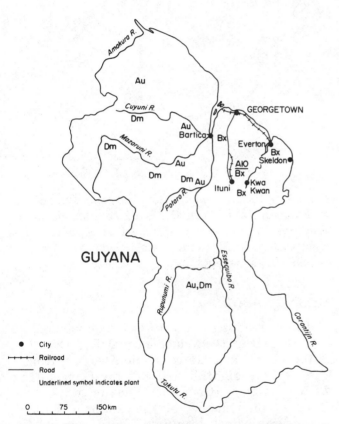

FIGURE IV-9-1 Jamaica: Mines, processing facilities, and infrastructure. (From Ref. 2.)

FIGURE IV-9-2 Guyana: Mines, processing facilities, and infrastructure. (From Ref. 2.)

leum products from the oil refineries on Trinidad and Tobago and the Netherlands Antilles, Jamaica has a large (10 million barrels per year) refinery at Kingston, owned by Esso West Indies Ltd.[2] Suriname has no oil-refining capacity.[2]

Guyana's installed generating capacity of about 175 MW is all based on thermal units.[2] Guyana has substantial hydroelectric potential, estimated at 10,000 to 14,000 MW.[2] These three bauxite pro-

ducers all import oil to power their bauxite and alumina industries. Present weaknesses in world oil prices must be viewed as a very positive event for these three highly oil-import-dependent bauxite-producing companies.

Suriname imports oil primarily for the bauxite industry (65%) and for transportation (10%).[2] Suriname has no production of mineral fuels.[2] The petroleum needs of the bauxite industry in Jamaica are handled by the producing companies and do not enter into the country's oil accounts.[2]

Suriname has one major hydroelectric plant, the 180-MW Afobaka plant, which supplies over 95% of the country's total electric generating capacity.[2] Jamaica has no hydroelectric capacity and limited hydroelectric potential.[2] Jamaica's 1400 MW of installed generating capacity is all based on oil-fired sources.[2]

Water

The availability of water should not constitute a barrier to further mineral development in any of these three countries. Guyana has the largest hydroelectric resources of the three, but Suriname has the only installed hydroelectric generating capacity.[2]

Transportation

Bauxite and alumina production is transported by rail, truck, and conveyor belt from producing areas to ports and alumina refineries.[2] Water transport is also important in both Suriname and Guyana.[2] Mineral transport in Jamaica is over short distances, due to the island's size and proximity to its major market in the United States.[7]

In Jamaica, short rail lines and conveyor belts transport mine output to port facilities.[2] The ports through which Jamaican mineral output leaves the country are Port Rhoades, Ocho Rios, Port Esquivel, Rocky Point, and Port Kaiser.[2] All of the major

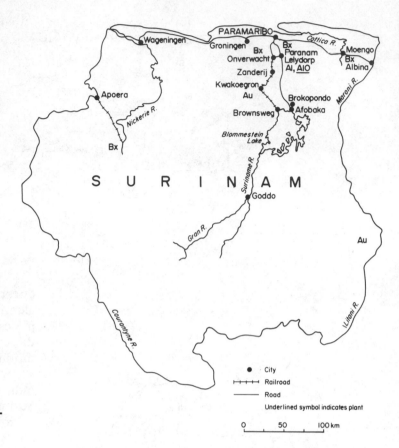

FIGURE IV-9-3 Surinam: Mines, processing facilities, and infrastructure. (From Ref. 2.)

dock facilities for bauxite and alumina exports are company owned.[2] The main rail line runs from Montego Bay in northwest Jamaica to the capital, Kingston, in the southeast part of the country.[2] Jamaica's highway system totals 11,300 km of which about two-thirds is paved.[2]

Guyana's bauxite and alumina output is transported by a combination of rail and water transport.[2] Guyana has two short rail lines. One line connects the bauxite-producing center at Ituni and the processing center at Linden, and the other links the capital Georgetown and the city of Everton, where bauxite is calcined.[2] Bauxite produced at Kwakwani is barged down the Demerara River. Other transport infrastructure in Guyana includes 5700 km of roads (10% paved) and an equal distance of navigable inland waterways.[2]

Suriname's mineral production is mainly transported in barges and ships.[2] Navigable inland waterways total 4500 km.[2] Suriname's alumina and aluminum output, as well as the country's direct shipping ore output, is transported in small vessels to Trinidad, where it is loaded into large vessels for international shipment.[2] Some alumina is shipped to Venezuela, as the two countries are attempting to integrate their aluminum industries.[3]

Two short rail lines, totaling 150 km, run from Bronsweg to Onverwacht near the processing center at Paranam, and from the bauxite deposits in the Bakhuys Mountains to the port at Apoera.[3] The road system in Suriname totals about 2500 km, of which 20% is paved.[2]

MINING LABOR

The mining industry labor force is highly unionized in all these bauxite-producing countries.[3] Over 85% of the bauxite labor force is unionized in Guyana.[8] The workers are affiliated with the Guyana Mine Workers Union (GMWU).[8] Bauxite workers in Jamaica are members of the National Workers' Union.[8] Suriname's highly unionized bauxite labor force demands appreciably higher wages than do mine workers in other bauxite-producing countries.[3]

The mining industries in Jamaica and Guyana each employs about 8000 workers, 2 to 3% of each country's total labor force.[2] In Suriname, the mining industry employs about 6000 workers, about 6% of the country's labor force.[2] In Guyana an estimated additional 8000 workers are indirectly employed in the transportation of mineral output.[2]

Two strikes in Guyana in 1979 lowered the country's total bauxite production.[8] A three-week strike in February was observed by about 80% of the mining labor force, and a five-week strike in late July and August 1979 led Guymine, the Guyana Mining Enterprise Ltd., to declare *force majeure* on alumina and bauxite shipments for part of this time.[8]

Wage demands by the bauxite workers in Suriname held 1981 production to about 20% below the 1980 level.[3] The unionized bauxite workers in Suriname have considerable impact on the industry. The bauxite industry labor force in Guyana has a great deal of economic power and is therefore a significant political actor in the industry. Jamaica's bauxite workers union has seen some of its power undermined by very high (30%) unemployment on the island.[8]

DOMESTIC MINING INDUSTRY

Overview

The bauxite-alumina-aluminum industry is very important to the economies of these three countries. Production of these mineral commodities accounts for 10 to 30% of each country's gross national product and government revenues.[2] The mining industry employs between 2 and 6% of each country's total labor force.[2] Each country receives a large part of its total foreign exchange earnings from bauxite and alumina exports.[2]

The major trend that these three countries share is a downward trend in bauxite production and exports. Guyana's bauxite and alumina output have fallen each year since 1977.[3] Production in 1981 was about 30% below the target set by the state bauxite company.[3] Declines in bauxite output have adversely affected the country's economy and foreign exchange position.

The bauxite-alumina industry in Jamaica is in a depressed state, despite the sale of 1.6 million long tons of calcined bauxite to the United States for the strategic stockpile.[8] Bauxite production in 1982 was estimated at 8 million metric tons, about one-third below capacity.[1] Labor problems and the world recession, which has reduced demand for most mineral commodities, are the chief causes of the declines in output in Jamaica's bauxite-alumina industry.

Mining output in Suriname has fallen to about two-thirds of 1978 levels. The most important trend in the mining industry is a shift in alumina exports from primarily western Europe to the United States between 1980 and 1981.[3] Exports of alumina to the United States increased over 300% in 1981.[2]

The primary owner of Suriname's bauxite-alumina-aluminum industry is the Aluminum Company of America (Alcoa). The shift in exports to the United States is indicative of the depressed state of the U.S. mineral-processing industries. Alcoa processes its bauxite output in Jamaica to avoid the high labor costs at U.S. mineral-processing plants.

Industry Structure

The governments of these three countries have chosen different industry structures for the development of their bauxite resources. The state is the sole owner of bauxite-producing facilities in Guyana.[2] Jamaica has chosen to form mixed companies, with multinational aluminum companies as the minority partners, to develop its bauxite-mining industry.[2] International firms own the majority of the country's alumina refineries.[2] In Suriname, two large multinational companies own all bauxite and alumina capacity.[2]

A group of five large North American aluminum companies own about half of Jamaica's bauxite mine capacity and about 95% of the island's alumina refinery capacity.[2] The government owns 51% of all bauxite-mining activity and a minor share of each of the country's alumina refineries.[2]

All activity in the bauxite-alumina industry in Guyana is controlled by the Guyana Mining Enterprise Ltd. (Guymine).[8] Guymine is also in charge of all other mining activity in the country. Guymine is part of the Bauxite Industry Development Co. Ltd. (Bidco), a holding company that coordinates and provides services to the bauxite industry.[8]

In sharp contrast to the state monopoly in neighboring Guyana, the bauxite industry is dominated by two large multinational corporations (MNCs) in Suriname. Suriname Aluminum Co. (Suralco) is a wholly owned subsidiary of the Aluminum Co. of America (Alcoa). Suralco owns bauxite mines and the country's aluminum smelter and alumina refinery.[2] A subsidiary of the Royal/Dutch Shell group operates one mine in Suriname. The state has formed a third company to develop the bauxite resources of the Bakhuys Mountains region.[5]

The bauxite-alumina industries in these three countries exhibit high degrees of both vertical and horizontal integration, with a total of just seven firms in the industry in all three countries taken together.[2] Suralco has an integrated bauxite-alumina-aluminum operation in Suriname.[2] The same firms mine bauxite and produce alumina in Jamaica.[2] The bauxite-alumina industry in Guyana is completely integrated, both vertically and horizontally, with all activity in the hands of one state company.[2] These three countries have very concentrated aluminum industries that are highly integrated. These three bauxite-producing countries exhibit industries with total, partial, and no state ownership. Regardless of the level of state ownership, the governments of these three countries control the bauxite industry in a number of other ways.

Government Role in the Mining Industry

Mining Taxation. Jamaica has long had difficulty in taxing the integrated bauxite-alumina-aluminum industry because it is hard to determine at which stage of processing the value of the contained minerals should be taxed. Jamaica's bauxite levy is a flat 7% tax on the price of aluminum ingots in the North American market.[7] The 1979 reduction of the levy was greeted with increased output by the bauxite producers.

Taxation of the bauxite industry is an important source of government revenues in Suriname. The two large producers pay over $2 billion per year in taxes to the fiscal treasury.[8] Details of the bauxite levy in Suriname have been unavailable. The levy presently in effect was signed in 1979.[8] Information regarding the mineral taxation scheme in Guyana is unavailable.

Government Subsidies and Incentives. The Guyanese government provides support services to the state-owned mining company. Bidco is charged with the development of the country's bauxite resources. Bidco evaluates mineral deposits and does some mineral exploration work.[9] The government of Suriname is providing the infrastructure required to develop bauxite deposits in the Bakhuys Mountains.[9] A rail line and two major bridges have been completed, and a large hydroelectric power plant is planned for the area.[9] The bulk of the output of the plant will be used by the bauxite industry and by a proposed 150,000-metric ton per year alumina refinery.[9]

Government Ownership of Producing Companies. The Guyanese government owns the entire bauxite industry. The government of Jamaica is the majority owner of all bauxite-producing facilities and a minority owner of the country's processing facilities.

Mining Law and Policy. Jamaica's principal mining legislation is Mining Law No. 41 of 1947, as supplemented by the Mining Regulation of 1947 and amendments.[2] Petroleum resources are regulated by the Petroleum Law of 1953 and amendments made in Law 10 of 1955, Law 47 of 1956, Law 18 of 1957, and Law 59 of 1960, and supplementary regulations.[2]

Jamaica actively encourages foreign investment that is in association with Jamaican partners.[2] The government requires that bauxite companies reclaim mined-out areas for agricultural use.[2] Since the mid-1970s, the government has been purchasing land from mining companies and re-leasing it to them for long periods.[2]

The Guyanese government allows foreign investment in manufacturing, tourism, fishing, forestry, and nonbauxite mining.[9] Foreign investment may be, but need not be, in association with Guyanese firms.[9] Guyana has no laws or regulations regarding technology transfer, royalty payments, or the repatriation of profits.[9] The Overseas Private Insurance Corp. (OPIC) insures private investments in Guyana against loss due to expropriation or war.[9] Guyana's private investment code was issued in 1979.[2] This code is a set of policies and does not constitute legal regulations.

Suriname created a Bauxite Bureau to develop an effective mineral policy.[9] New mining laws have been proposed but have not yet been adopted. Recent trends in the extractive industries indicate a preference for joint venture projects between the government and private investors.[2] A government agency has been established to promote industrial development and exports.[9] Foreign energy firms are exploring offshore concessions, and several other areas off Suriname's Caribbean coast will be opened up for oil exploration.[9]

Domestic Supply–Demand Dynamics

Jamaica exports all of its alumina output and over half of its bauxite output.[2] Bauxite that is not refined in Jamaica is exported directly. Productive capacity at Jamaica's bauxite mines is about 12 million metric tons per years, but reduced world bauxite demand has forced some facilities to operate at below 25% of capacity.[10] Jamaica imports all of its oil requirements as crude oil and refines the oil on the island.[2] Jamaica also imports most of domestic requirements of a number of basic minerals and mineral commodities.[9] Jamaica will continue to import oil to fuel its export-oriented bauxite and alumina industry.

Guyana's output and exports of bauxite and alumina have fallen in recent years. Although total productive capacity is over 5 million metric tons per year, 1981 output of bauxite and alumina was under 2 million metric tons.[3] Labor, technical, and equipment problems plague Guyana's bauxite-alumina industry.[9] Low domestic demand leaves the bulk of Guyana's bauxite production available for export—either as ore or alumina. All of Guyana's alumina output is for export.[3] Guyana is dependent on imports for almost all of its domestic mineral requirements. In addition to importing all the country's oil needs, Guyana imports iron and steel products, nickel, copper-lead-zinc, and fertilizer minerals.[9] Imports of these commodities will increase as population and demand grow. The principal destinations for Guyana's bauxite exports are the United States and Canada, each with about one-third of the total.[9] Venezuela and West Germany each receive about 5% of Guyana's total bauxite exports.[9]

Suriname's bauxite productive capacity is about 7.5 million metric tons (mt) per year, but output in recent years has been between half and two-thirds of total installed capacity.[5] The country's alumina productive capacity is about 2.2 million mt per year.[3] Suriname has installed capacity to smelt about 60,000 mt per year of aluminum metal.[2]

The domestic consumption of Suriname's alumina and aluminum output is small. About 10% of the country's alumina output is smelted domestically, with the remainder for export.[3] All of the aluminum metal produced in the country is exported. The bauxite-alumina-aluminum industry in Suriname will always be oriented toward the export market. Suriname relies on imports for all its oil needs and also imports virtually all its basic metals requirements.[9] The development of the country's hydroelectric resources could reduce reliance on imported oil.

These three countries together supply about one-fifth of the free world's bauxite. With idle capacity, an improvement in the world demand for mineral commodities can be met by sizable production increases without large investments. The health of the economies of these three countries depends on an improved demand for bauxite and alumina in the major industrialized countries. Labor union bargaining power must be considered when analyzing the alumina and bauxite industries in Jamaica, Guyana, and Suriname. Strikes are likely to affect total mineral output in these bauxite-producing countries.

OTHER MINERALS

Guyana and Suriname have small gold reserves and minor production.[3] Guyana also has some minor alluvial diamond deposits.[3] Other minerals which occur in Guyana include iron ore, manganese, copper, and molybdenum.[2] Guyana's Omai gold deposit is believed to contain over 1 million troy ounces of gold.[4]

The only other mineral activity in these countries involving foreign firms is some gold exploration activity in Guyana. Norman Mines and Duncan Gold Reserves Inc. (both Canadian) are exploring for gold deposits and evaluating other identified prospects.[8] Guyana's minor gold and diamond production is obtained by individual prospectors.[2] Gold and diamonds are obtained from alluvial deposits in the Cuyuni, Mazaruni, Potaro, and Rupuruni river areas.[2]

Gold and diamond production in Guyana and Suriname is by small mining companies that utilize primitive, labor-intensive mining methods.[3] There is no unionization among gold and diamond miners in these two countries. Guyana exports about two-thirds of its annual diamond production of about 10,000 carats.[2] Gold production is increasing, although little is exported.[3] Roughly 40% of the country's diamond output is gem quality.[3]

POLITICAL ENVIRONMENT

Domestic

Jamaica's Prime Minister, Edward Seaga, faces economic troubles. Jamaica suffers from unemployment, a shortage of capital and foreign investment, balance-of-payments deficits, and shortages of foreign exchange. Jamaica's dependence on imported oil is one of the priorities of Seaga's Jamaica Labour Party administration.[8] Seaga is viewed as pro-business and investment and is an advocate of a free market.[8]

Guyana's President Forbes Burnham faces a number of severe economic problems. Guyana suffers from negative economic growth, falling per capita GNP, high unemployment and inflation, low international reserves, and low labor productivity.[5] The president has imposed public austerity measures, including public sector layoffs, import restrictions, and increased taxes.[5]

Suriname's domestic political climate is highly unstable. President Henck Arron was overthrown in a February 1980 military coup d'etat instigated by the Army.[3] President Henk Chin a Sen was removed by the military in February 1982.[5] He was attempting to return the country to parliamentary rule when he was overthrown.[5] The National Military Council then took direct control of the government, but their leader, Lt. Col. Deysi Bouterse, was removed in a military coup later in the year.[5] The political situation in Suriname has been characterized by instability since the former Dutch colony gained independence in 1975. The military will continue to figure prominently in the course of political events in Suriname.

Bauxite workers in these three countries have

a considerable amount of economic and bargaining power and play a role in the country's political environment. Suriname is the most unstable of these three countries. The military plays an important political role in both Guyana and Suriname. Jamaica has the most democratic government of the three countries.

International

The United States is encouraged by the cancellation of post-1983 alumina shipments from Jamaica to the USSR by the Seaga administration.[8] Jamaica is now generally pro-Western, and President Reagan's Caribbean Basin Initiative pointed to Jamaica as the new model for the region.[8] Jamaica's close relationship with the United States is crucial to the island's future, and G.S.A. bauxite purchases eased the decline in the country's bauxite industry in 1982.[8]

Suriname and Venezuela have made efforts to integrate their aluminum industries.[9] The only foreign firms active in Suriname are Billiton, Alcoa, and Gulf Oil Co. Suriname has disputes with both Guyana and French Guiana over the countries' international boundaries.[2] The present boundary with Guyana is the Courantyne River, and the boundary with French Guiana is the Litani River.[2]

Guyana's international boundaries with Suriname and Venezuela are in dispute.[2] The border dispute with Venezuela is important from a mineral resources perspective, as geological surveying efforts indicate uranium potential in the disputed area.[9] Cuba is trying to aggravate the border dispute by supplying Guyana with MiG fighters and troops.[11]

REFERENCES

1. U.S. Department of the Interior, Bureau of Mines. *Mineral Commodity Summaries 1983*. Washington, D.C.: U.S. Government Printing Office, 1983.
2. U.S. Department of the Interior, Bureau of Mines. *Mineral Industries of Latin America*. Mineral Perspectives Series. December, 1981.
3. U.S. Department of the Interior, Bureau of Mines. *Minerals Yearbook 1981*. Washington, D.C.: U.S. Government Printing Office, 1983.
4. "Worldwide Survey: Latin America." *World Mining*, August 1981.
5. Spooner, John, ed. *Mining Journal Annual Review 1982*. London: Mining Journal, June 1982.
6. Evans, Anthony M. *An Introduction to Ore Geology*. New York: Elsevier North-Holland, Inc., 1980.
7. "Importing Oil to Export Bauxite." *The Courier* 72 (March-April 1982).
8. Spooner, John, ed. *Mining Journal Annual Review 1980*. London: Mining Journal, June 1980.
9. U.S. Department of the Interior, Bureau of Mines. *Minerals Yearbook 1980*. Washington, D.C.: U.S. Government Printing Office, 1982.
10. U.S. Department of the Interior, Bureau of Mines. *Minerals and Materials*, October-November 1982.
11. *Business Week* 69 (6 July 1981).

Chapter IV-10

Ecuador

Ecuador is of some importance to the world mineral economy as an OPEC member, and as Latin America's fourth-largest oil producer.[1] Mining activity is very limited in Ecuador. Ecuador's known mineral resources include gold, silver, titanium, cadmium, lead, zinc, copper, and molybdenum.[2] The main barriers to mineral development in Ecuador are a lack of adequate geological exploration and a mining law that is too conservative to attract investment capital.[3] Both foreign and private Ecuadoran investors feel that current incentives and guarantees need to be liberalized to make mining in Ecuador profitable.

MINERAL RESERVES AND RESOURCES

Energy

Ecuador's oil reserves are estimated at about 1.2 billion barrels.[2] Possible natural gas reserves in the southern Gulf of Guayaquil amount to 670 million cubic feet.[3] Coal and uranium may be present in Ecuador, and small-scale exploration continues in the southern Andes Mountains.[4]

Nonfuel Minerals

Ecuador's identified mineral resources include copper, molybdenum, lead, and zinc. Other minerals occurring in copper and lead-zinc ores include gold, silver, cadmium, tungsten, and possibly bismuth and tin.[3]

Proven ore reserves at the Chaucha copper deposit are 72 million tons.[3] Ore grades average 0.7% copper and 0.04% molybdenum.[5] Other mineral values in the area are 0.15 ounce per ton silver, 0.2 gram per ton gold, and 0.004% tungsten.[5] Ecuador's mineral reserve statistics are difficult to obtain and do not reflect the country's actual mineral endowment, because mineral prospecting and exploration have been very limited.

GEOLOGY

More is known about the Chaucha copper-molybdenum deposit than about any other mineral deposit in Ecuador. The deposit is a typical porphyry-type deposit and is characterized by zones of hydrothermal alteration.[5] The mineralization is both in a network of veins and disseminated.[5] The minerals found in the deposit include pyrite, chalcopyrite, bornite, covellite, and molybdenite.[5] The ore in this deposit has recoverable copper, molybdenum, gold, silver, and tungsten values.[5]

Other areas of copper-lead-zinc mineralization, trending northeast-southwest, are found in the Balzapampa and Chaso Juan deposits.[5] The mineralization is located in intrusive granodiorites that have been structurally controlled by intense faulting.[5] Veins and contact zones typically contain copper sulfates and the minerals pyrite, molybdenite, and possibly chalcopyrite.[5]

The British have discovered 1 to $1\frac{1}{2}$-m-thick phosphate rock layers in sedimentary beds of Upper Cretaceous age.[6] The formation is in eastern Ecuador and is known locally as the Napo Formation.[6] Ecuador's diverse geology is only now being explored in detail, but sulfur and titanium are known to occur in the country. Most other known mineralization is copper-nickel, copper-lead-zinc, or copper-molybdenum.[5] Anomalies typical to these types of mineralization have been found in northern Ecuador near the Colombian border.[5]

PHYSICAL INFRASTRUCTURE

Energy

Energy availability does not constitute a barrier to mineral development in Ecuador. Hydroelectric potential accounts for 78% of the country's energy resources and hydrocarbons account for 19%.[4] In 1980, oil provided over half of Ecuador's energy, while hydroelectricity contributed only 5%.[4]

Ecuador plans to exploit its hydroelectric potential with several new projects due to come on-line in the mid-1980s. Current electric generating capacity is 660 MW, but only about 25% is from hydro sources.[1] Among the new projects are Daule-Peripa (550 MW), Naiza (1980 MW), and Agoyán (150 MW).[4] Expected investment in these projects amounts to U.S.$4 billion.[1] The Ecuadoran Atomic Energy Commission is studying the possibility of building a nuclear power plant on the Pacific coast, in the Manabí province.[1]

The National Planning Board indicated that oil

production fell from 77 million barrels in 1981, to 59 million barrels by 1984, a 23% decline.[2] Ecuador may be the first OPEC member to become a net importer of oil, as early as 1985.[2]

Water

Water availability should not constitute a significant barrier to further mineral development in Ecuador. Present mining industry water requirements are small, due to the limited mining activity in the country.

Transportation

Ecuador's highways total 22,000 km, of which less than 4000 km are paved.[1] The state rail network is 1121 km in length, and the main line runs from San Lorenzo to Quito and Guayaquil.[1] Minerals are transported to processing centers, primarily by rail and truck.[1]

Crude oil and petroleum products pipelines total about 2000 km and run from the oilfields near Lago Agrio to the shipping terminal at Esmeraldas and from just south of Quito to Guayaquil on the Pacific coast.[1] The major export facility is located off Esmeraldas.[1] The country's other major ports are at Guayaquil (owned by the state oil company), Manta, and Puerto Bolívar.[1] Most imported goods enter the country through the port at Guayaquil. Ecuador's mineral and fuels transport infrastructure, ports, and mineral and hydrocarbon-producing areas are shown in Figure IV-10-1.

MINING LABOR

Ecuador's labor force is about 2 million, and less than 1% are employed in the petroleum and mining industries.[1,4] Unskilled labor is readily available.[1] About 15% of the labor force is unionized, with over half the total in agriculture.[1] The minimum wage in 1979 was $160 per month.[5] Information regarding specific mining labor training programs is unavailable. No reports of major labor disturbances or strike activity affecting the minerals industries are evident in the mining industry literature. The 1980 U.S. Bureau of Mines *Minerals Yearbook* reported that labor and other problems had driven some mining companies close to bankruptcy.[4]

DOMESTIC MINING INDUSTRY
Overview

Ecuador's very limited mining industry does not make significant contributions to the country's economy. The petroleum and refining industries account for about 17% of Ecuador's gross domestic product.[4] Current mining activity is minor, with only a handful of mines operating. The petroleum and mining industries together grew 1% in 1981, reversing a trend of negative growth of 5% in 1980.[3]

Industry Structure

The most important energy company in Ecuador is the state oil company, Corporación Estatal Petrolera Equatoriana (CEPE). The CEPE-Texaco

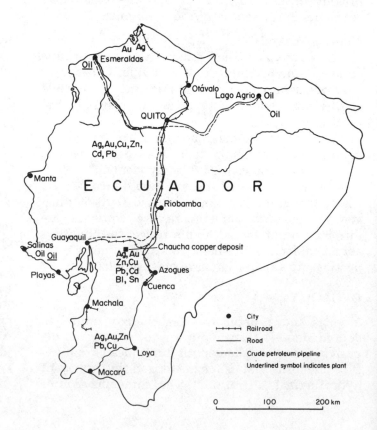

FIGURE IV-10-1 Ecuador: Mines, processing facilities, and infrastructure. (From Ref. 1.)

Inc. consortium produces virtually all of Ecuador's oil. CEPE owns 67.5% of the consortium and Texaco owns the remainder.[1] CEPE also owns the Esmeraldas refinery, where capacity is 55,000 barrels per day, and is planned to increase to 90,000 barrels per day in 1987.[3]

The other two refineries are located on the Santa Elena peninsula. The 32,000-barrel per day Anglo facility is owned by Clyde Latino América S.A. and CEPE.[3] The Petróleos Gulf de Ecuador refinery, owned by private Ecuadoran interests, has an 8000-barrel per day capacity.[3] CEPE is planning to construct a new 75,000-barrel per day refinery at Atahualpa.[4]

CEPE's refinery expansion plans will increase Ecuador's refinery capacity from the present 96,000 barrels per day to 210,000 barrels per day by 1988.[3] The design and construction of the new refinery and the expansion projects are in the hands of UOP, Inc., of the United States.[3]

A 400,000-ton direct reduction steel mill, utilizing local natural gas, is being planned near Puerto Bolívar by Compañía Ecuadoriana de Siderúrgica S.A., a quasi-governmental steel company.[3] Ecuador imports large volumes of primary steel and semimanufactures.[3]

The most important mine in operation in Ecuador is the Cia. Minera de Toachi S.A. La Plata mine near Santo Domingo.[1] This mine is owned by Cia. de Minas Buenaventura, a Peruvian firm with 60% of the mine, and Outokompu Oy of Finland with the remaining 40% of the mine.[1] The mine produces primarily copper and zinc, but also produces gold and silver and may produce cadmium in the future.[1]

The Ecuadoran oil industry does exhibit some vertical integration, with CEPE controlling 60% of Ecuador's crude oil production and 57% of the country's refinery capacity.[1] Ecuador's minor minerals industry is not vertically integrated.

Typical indicators of industry structure, such as concentration ratios, are almost meaningless when applied to Ecuador's mining and energy industries. The Texaco-CEPE consortium produces over 95% of Ecuador's crude oil.[1] The Cia. Minera de Toachi S.A. produces about half of the country's gold and silver.[1] Another mining company, Cia. Industrial Minería Asociada S.A., recently went bankrupt.[6] Other mining activity in Ecuador does not enter into the world economy and merits no further discussion.

Government Role in the Mineral Industries

The Ecuadoran government exercises control over the energy and mineral industries through its mining law, tax policy, and ownership of producing companies. Information regarding the taxes, to which mining companies are subject, is difficult to obtain. The principal state mineral agency is the Dirección General de Geología y Minas (DGGM).[1]

Mining Law. Ecuador's current mining law is the Ley de Fomento Minero (Mining Promotion Law), adopted by the military government in 1974.[1] This law offers few incentives for mineral development and has attracted very little interest among investors.[4]

The DGGM and the Ministry of Natural Resources have been discussing a new, more liberal mining law for about five years, but no new legislation has as yet been approved.[4] Debate over the new mining law has been vigorous in Congress, with many members trying to make the legislation too protective and favorable to Ecuador.[4] The resulting law may not be a significant improvement in the eyes of foreign investors.

The new law would establish a state mining development corporation, the Corporación de Fomento Minero (COFEMIN), to oversee mineral prospecting, exploration, and exploitation.[4] The role of the DGGM would be reduced somewhat.[4]

A new law covering hydrocarbons was approved by the military government in 1978 but has yet to be implemented.[4] Ecuadoran President Hurtado favors liberalizing contract provisions for association agreements with foreign oil companies.[7] Development of the natural gas reserves in the Gulf of Guayaquil has been held up by contractual differences between the government and the foreign oil companies involved there.[7] Congress has not passed legislation to stimulate oil exploration and development activity.[7]

Ecuador offers significant incentives to industrial firms locating outside Quito and Guayaquil.[1] The measure is designed to spread industrialization throughout the country.

Domestic Supply–Demand Dynamics

The development fostered by oil export revenues has led to large increases in domestic oil consumption. The National Planning Board projects a 16% annual growth rate for domestic oil consumption.[2] At this rate, Ecuador's domestic oil consumption will double every five years.

When these factors are combined with flat or falling oil production, the balance of oil production available for export will probably fall. However, Ecuador is expanding refinery capacity and may be able to export up to 50,000 barrels per day of surplus No. 6 fuel oil.[3]

The five-year national plan of 1981 gives very low priority to mining development.[1] Little development should be expected in Ecuador's nonfuel minerals industry. Ecuador continues to import most of its metals and nonmetals, iron and steel products, and fertilizer minerals.[4]

POLITICAL ENVIRONMENT

Domestic

The Republic of Ecuador has a democratic government. The current President, Osvaldo Hurtado, assumed power in a smooth transition in 1981 after the death of President Jaime Roldos Aguilera.[7] Hurtado is pursuing Roldos' goals of economic development in a democratic government.[7]

The Concentration of Popular Forces (CPF) is Ecuador's largest and dominant political party. The CPF is a populist party. The second largest political party is the Radical Left Party. Other political parties include the Conservative Party of Ecuador, the Democratic Left, the Communist Party of Ecuador, and the Revolutionary Socialist Party.

The center-right coalition of the Radical Liberal and Conservative parties received 28% of the vote in the April 1979 election, while Roldos' Populist Party received the majority of the votes cast (62%). Civilian government replaced military rule when Roldos took power in August 1979.

Current domestic political problems include inflation, capital flight, falling investment levels, and falling foreign exchange reserves.[7] Another serious problem is declining oil production and exports, which provide about 60% of Ecuador's foreign exchange earnings and account for 17% of the country's GNP.[4]

International

Ecuador and Peru have a boundary dispute. The area of dispute is near oil-producing regions in Peru.[1] Relations with other Latin American nations are generally good. Tensions between Ecuador and Peru erupted into a brief war in January 1981.

Great Britain has provided technical assistance to Ecuador in developing a metallogenic map of Ecuador.[4] The World Bank and a West German bank are providing funds for expanding the La Plata copper-zinc mine.[4]

The United States receives about half of Ecuador's 45 million barrels of annual oil exports.[4] The remainder is exported to other Latin American nations, principally Panama and Colombia.[4] The Ecuadoran government has estimated that at least 5000 barrels of gasoline per day are illegally exported to Peru and/or Colombia.[4]

REFERENCES

1. U.S. Department of the Interior, Bureau of Mines. *Mineral Industries of Latin America*. Mineral Perspectives Series. December 1981.
2. Spooner, John, ed. *Mining Journal Annual Review 1980*. London: Mining Journal, June 1980.
3. U.S. Department of the Interior, Bureau of Mines. *Minerals Yearbook 1981*. Washington, D.C.: U.S. Government Printing Office, 1983.
4. U.S. Department of the Interior, Bureau of Mines. *Minerals Yearbook 1980*. Washington, D.C.: U.S. Government Printing Office, 1982.
5. Ministerio de Recursos Naturales y Energéticos, Dirección General de Geología y Minas. *Informativo Geológico-Minero*, December 1981.
6. Spooner, John, ed. *Mining Journal Annual Review 1982*. London: Mining Journal, June 1982.
7. U.S. Central Intelligence Agency, National Foreign Assessment Center. *The World Factbook—1981*. Washington, D.C.: U.S. Government Printing Office, 1982.

Chapter IV-11

Argentina

Argentina is a minor contributor to the world mineral economy. Argentina produces about ½ million barrels per day of crude oil and is Latin America's largest producer of boron minerals.[1] Argentina has not been widely explored for minerals, largely due to shortages of capital, skilled personnel, and equipment. Recent liberalization of Argentina's mining law has attracted considerable foreign and domestic interest. An inflow of investment capital for exploration, machinery, and infrastructure will be a prerequisite to growth in Argentina's mineral industry.

MINERAL RESERVES AND RESOURCES

Argentina does not have an accurate inventory of its energy and mineral resources. Large portions of the country have not been surveyed or explored for mineral deposits. Statistics on Argentina's mineral reserves are presented in Table IV-11-1. Where exact figures are unavailable, estimates are given.

Argentina has over 1200 million metric tons of copper-molybdenum ore reserves in two large

TABLE IV-11-1 Argentina: Mineral Reserves
(Thousand Metric Tons Ore)

	Reserves
Mineral	
Copper	1,200,000
Iron	1,000,000
Lead	6,600
Manganese	4,650
Silver	11,250
Zinc	6,600
Fuel	
Coal	480,000
Lignite	7,500,000
Natural gas[a]	640,000
Petroleum[b]	2,900,000
Uranium[c]	28,000

[a]Unit of measure is million cubic feet.
[b]Unit of measure is thousand barrels.
[c]Unit of measure is metric tons U_3O_8.
Sources: Refs. 1, 2, 3, and 4.

deposits.[2] Since large-scale copper mining occurs in Chile within 400 m of the Argentine border, Argentina's copper resources are probably much larger than its proven reserves. The two large deposits are at El Pachón and Bajo de la Alumbrera.[2] Molybdenum is also present in these deposits, although reserve estimates for this mineral are unavailable.

Iron ore reserves at Sierra Grande and Nueve de Octubre total close to 1000 million metric tons.[3] The lead-zinc-silver reserves of Cia. Minera Aguilar S.A. are 6.6 million metric tons, with an average grade of 6.2% lead, 7.2% zinc, and 3.7 ounces per ton silver.[4]

Manganese reserves at Farallón Negro, with an average grade of 12.2 to 15.9% manganese, are 1650 million metric tons.[2] Gold and silver values are very high, ranging from 3 to 9 grams per ton gold, and from 72 to 161 grams per ton silver.[2] Further lower-grade resources at Alto de la Blenda add another 3 million metric tons of 8.7% Mn ore.[2] Argentina also has boron minerals and fluorspar reserves. Bauxite has not been discovered in Argentina.

GEOLOGY

The world's highest-grade porphyry copper belt is in the central portion of the Andes in Chile, Argentina, and southern Peru.[5] Up to 30% of the world's identified copper resources may lie in the major porphyry deposits and occurrences in this area.[5]

The copper deposits in the Andes porphyry copper belt contain valuable by-product and co-product minerals. Molybdenum is produced as a co-product at major open-pit operations in Chile and southern Peru, and many deposits also contain recoverable gold and silver values.[6] Although Argentina's porphyry copper potential was discovered later, the deposits are the result of the same geologic conditions and contain a similar suite of minerals as in Peru and Chile. Tellurium and selenium values are also present in some deposits.[5]

Two types of manganese mineralization are found in the deposits of the Catamarca province.[3] The Farallón Negro and Alto de la Blenda deposits exhibit very long, high-grade veins of mineralization.[3] The Bajo de la Alumbrera, El Durazno, and San Lucas deposits are in areas of hydrothermal alteration, with disseminated mineralization.[3]

Only the Farallón Negro and Bajo de la Alumbrera deposits have been explored in sufficient detail to make them appear economically feasible.[5] High gold and silver values make the Farallón Negro deposit particularly promising.[3] The Bajo de la Alumbrera deposit contains copper, molybdenum, gold, and silver mineralization.[3]

Iron ore deposits in the Sierra Grande province contain primarily magnetite.[3] The Mina Ángela de Cerro Castillo produces copper, lead, zinc, and silver in concentrates from a massive disseminated sulfide and vein-type deposit.[4]

Most copper-bearing deposits are associated with copper-bearing stocks and plutons.[5] The source of the components of the deposits is probably subcrustal material subducted at the plate margin between the east Pacific plate and the Americas plate at the Peru-Chile Trench.[5] Many large deposits exhibit concentric zoning of hypogene mineralization.[5] Supergene enrichment occurred at some of the deposits.[5] Tourmaline breccia pipes with copper mineralization also occur near, and within, porphyry copper deposits.[5] The breccia pipe deposits also contain gold and some titanium.[5]

Most of the deposits consist of linear, longitudinal subbelts which gradually migrated eastward.[5] About one-fourth of the deposits show no obvious structural control, while 30 to 40% of the deposits have evidently had their emplacement influenced by north-south zones of structural weakness.[5]

PHYSICAL INFRASTRUCTURE

Energy

Argentina has large resources of energy minerals. At present, the country is 95% self-sufficient in oil.[2] Large gas reserves and uranium deposits complement Argentina's energy potential. Coal and lignite deposits are also present in Argentina, and the country has hydroelectric potential that remains to be developed.

Argentina's energy resource potential is divided among oil (13.5%), hydroelectricity (46%), coal (8.5%), gas (20.6%), and charcoal (2.3%).[3] Despite the large potential, hydroelectricity provides only 8.4% of Argentina's total energy supply.[1] Argentina's 12,300 MW of electric generating capacity is derived from thermal (51.2%), hydro (42.2%), and nuclear (6.6%) sources.[3]

Argentina has an extensive network of oil, gas, and petroleum products pipelines. Crude oil pipelines total 4100 km, petroleum products pipelines total 2200 km, and gas pipelines amount to over 8100 km.[7] The capacities of most major gas pipelines have been increased and links extended to major areas of gas consumption in Mendoza, San Juan, San Luis, and Córdoba provinces and the Santa Fe and Buenos Aires areas.[3] Feasibility studies are proceeding on a proposed pipeline link for the shipment of 10 million cubic meters per day of gas from Brazil.[3] Pipelines also connect Chilean supply regions to the Argentina market.[1]

Major oil and gas pipelines are shown on Figure IV-11-1. Figure IV-11-1 also shows the energy mineral producing and processing areas. The country's major hydroelectric facilities at the Yacireta-Apipe Dam and the Salto Grande Dam are shown in Figure IV-11-1.

Water

Argentina has abundant water resources. Navigable rivers have a total length of 11,000 km.[7] Major rivers that provide water, hydroelectricity,

FIGURE IV-11-1 Argentina: Mines and mineral transport infrastructure. (From Ref. 1.)

and transportation include the Pilcomayo, Paraná, Uruguay, Plata, Chubut, Negro, and Colorado.[1] Water availability and price should not constitute a barrier to mineral development in Argentina. Local water availability problems may occur in the exploitation of isolated mineral deposits in inaccessible areas.

Transportation

Rail. Argentina's rail system has a total of about 40,000 km of operating rail lines.[1] Most long-distance transport of heavy freight and commodities is by rail. Coal from the Río Turbio mines is transported by rail to the port at Río Gallegos.[3] The port at Río Gallegos is a bottleneck in Argentina's coal transport infrastructure. The maximum capacity for ships at Río Gellegos is 6000 metric tons.[2] Virtually all of Argentina's coal reserves are located in the Río Turbio area.

Rail connections with Chile and Bolivia are good. One line connects Buenos Aires and the Pacific port at Valpariso, Chile.[1] The major rail lines used to transport mineral production in Argentina are shown in Figure IV-11-1.

Roads. The Argentine road and highway system comprises about 207,000 km.[1] Close to half of the roads are paved or graveled. Major roads used to transport minerals are shown in Figure IV-11-1.

Pipelines. A slurry pipeline connects the Sierra Grande mine and the 2-million metric ton per year pellet plant to the port of Puerto Buitrago.[1] The development of these projects has greatly improved the Argentine iron and steel industry. St. Joe Minerals Corp.'s El Pachón copper-molybdenum deposit may be serviced by a 90-km slurry pipeline linking rail and ship facilities. Concentrates are exported to Germany for smelting.[4]

Ports. The major port for bauxite imports (largely from Guyana) and alumina shipments is Puerto Madryn.[1] Most other mineral and mineral fuel exports and imports pass through ports in Santa Fe province.[1] The Río Plata, and the port at Buenos Aires, also handle some mineral activity. Iron ore is imported to meet the demand of domestic steelmakers, and the volumes imported will increase port activity. Ports are also shown on Figure IV-11-1.

MINING LABOR

Approximately 29,000 workers are employed in mining activity, and another 21,000 are employed in oil production.[1] Coal miners account for the largest share of mine employees, totaling 4100.[1] Iron and manganese miners total 2600 and lead miners total 2000.[1] Employment in the coal industry has fallen by about 30% in recent years. Argentina's poor general economic health has caused relatively high unemployment and less overall economic activity.

The labor supply in Argentina is adequate to meet future demands, with many miners unemployed as a result of the world economic recession. The availability of labor should not constitute a barrier to mineral development. About 35% of the employed labor force is in the Buenos Aires area.[1]

Information regarding public and private education and training programs for the mining industry is difficult to obtain. Most mining activity is under the direct or indirect control of the Army, through the Dirección General de Fabricaciones Militares (DGFM), and industry training programs are not publicized.[1]

DOMESTIC MINING INDUSTRY

Overview

Argentina's mineral industry makes minor contributions to the domestic and world economies. The nonfuel mineral industries account for less than 1% of Argentina's gross national product.[3] The bulk of the country's mining activity is concentrated in construction materials. The energy and mining industries together employ about 50,000 workers.[1]

Trends Among Major Mineral Producers

Several important trends are apparent, when examining Argentina's mineral imports, exports, and production. The most striking of these trends is in aluminum. Argentina went from importing over 85,000 tons of aluminum in 1974 to exporting 67,000 tons in 1980.[3] The 1974 figure represented 90% of apparent domestic consumption. A planned aluminum smelter would double the present capacity of 140,000 metric tons.[3] All of this increased production would be for export.

Argentina exported minor gold volumes in 1979 and sold 9500 tons of zinc concentrates to the USSR.[2] The country's total mineral exports were up 100% in 1980.[3] Other mineral exports are mainly lead, zinc, tin, and silver concentrates.[3] Exports of nonmetallic minerals fell by 70% in 1980.[3] Imports of natural gas from Chile and Bolivia have increased to cover domestic production declines.[1] Coal production fell by 46% in 1980, due to low demand and high imports.[3]

Industry Structure

Argentina's mineral industry is dominated by state companies, although some commodities are produced by private firms.[1] State control is executed through the Army's DGFM, under the direction of the Mining Secretary of State.[1]

All energy minerals are produced by state-owned companies. The uranium-producing company is Empresa Nuclear Mendoza.[1] Coal-mining activity is undertaken by Yacimentos Carboníferas Fiscales (YCF), which is 100% state owned.[1] Oil and gas exploration and exploitation are in the hands of the state companies Yacimientos Petrolíferas Fiscales (YPF) and Gas del Estado.[1]

Other oil activity is undertaken by 35 private domestic and international firms operating under exploration and development concessions.[3] The largest of these private producers are Amoco Argentina Oil Co. (30% of private output), Petrolera Pérez Compania S.A. (27%), Argentina Cities Services (18%), and Bridas SAPIC (10%).[3]

Certain commodities are produced by private companies. Cement, gypsum, and all boron minerals are produced by private companies.[1] The country's 30,000-metric ton per year lead smelter is owned by private Argentine interests.[1] The Argentine steel industry is dominated by the three private firms Acindar S.A., Dalmine-Siderca SAIC, and Somisa.[3] However, iron ore and pellet production is under Army administration.[1] Sidersur, another steel producer, is a consortium of small Argentine steel mills. Sidersur has signed an agreement with Hojalata y Lámina (HYL) of Mexico for a 500,000-metric ton per year HYL III direct reduction steel plant which started operation in lage 1983–1984.[3] The plant is being built in a coastal city.

Virtually all of the country's lead, zinc, and silver mine output is produced by St. Joe Minerals Corporation's subsidiary Cia. Minera Aguitor S.A. (CMA).[3] One of the country's zinc smelters at Comodoro Rivadavia is owned by Cia. Metalúrgica Austral Argentina S.A. and has a capacity of 47,000 metric tons per year.[1] The other large smelters are in the northern part of Argentina.

Aluminio Argentino SAIC operates the country's only large aluminum smelter at Puerto Madryn, where a capacity of 140,000 metric tons per year is based on local hydroelectric resources.[1,3] Although there is some small- and medium-scale mining activity in Argentina, the bulk of mineral output emanates from a few companies and a few mines.

Recent demands by the Army's DGFM for a 30% carried interest in the Farallón Negro, Nevadas de Famatina, and Bajo de la Alumbrera deposits have left few parties interested in these prospects. The military wants a hand in most major mineral development projects.

Argentina's mineral industry exhibits some vertical and horizontal integration. With virtually all mineral fuel activity, and sizable portions of total mining activity under state control, levels of horizontal integration are very high. Argentine efforts at vertical integration have been on a smaller scale, with the high levels of state ownership and control. Argentina's mineral industries exhibit high concentration ratios. Concentration ratios are also high in industries dominated by private firms. This is especially true in the lead-zinc-silver industry, where CMA produces all the domestic lead-zinc output and 62% of the country's total silver output.[1]

Government Role in the Mining Industries

The mining industry is largely controlled by the military. This control takes many forms, all under the administration of the Army's DGFM, which implements policies put forth by the Mining Secretary of State.[1] The government's role in mining includes the ownership of producing companies, control over the taxation, concession, and contract regulations, and a general policy control over virtually all mining activity. Information regarding specific state-owned companies is presented in the industry structure section, and tax and concession policies are discussed in the sections on mining taxation and mining law.

Mining Taxation. Mining activity in Argentina is organized along the lines of the Spanish system of mining concessions in exchange for royalties.[3] Firms engaged in mining activity in Argentina are also subject to a wide variety of taxes. Corporate income, capital gains, and dividends are taxable in Argentina.[2] Imports of machinery and spare parts are taxed, and net worth is also taxable.[3] The government also imposes a value-added tax at all levels of processing.[2] Recent modifications to the Argentine Mining Code have granted very liberal tax concessions and tax holidays.

The government has not specified the duration of tax concessions and holidays. However, some tax reductions can be for up to 51 years, so the present mining taxation scheme is relatively stable and generous. Changes in the government should not erase the country's recognition of the need to provide incentives for mineral development.

Government Subsidies and Incentives. Reforms to Argentina's Mining Code, which took effect in November 1980, grant specific tax holidays to private domestic and international firms that develop the country's mineral resources.[3] Provincial laws, approved in 1981, grant full 100% tax holidays on a number of taxes.[8]

Argentina's concern that tax benefits accrue to overseas investors, as opposed to their governments, has prompted Argentina-U.S. discussions concerning a double taxation agreement.[9] The two countries are close to finalizing an agreement that would entail some type of tax credit. Specific tax benefits include reductions ranging from 10 to 100% on capital, profits, and net worth taxes for up to 51 years.[3] Another benefit is an exemption from import duties on equipment and spare parts for the duration of a given project.[3] Mines will be allowed scaled reductions in the value-added tax for up to 15 years. The reductions will begin at 100% and scale down to 10% in the last year.[2]

The Mining Promotion Law allows sizable income tax deductions for most preproduction expenses.[2] In addition, income tax on dividends and the capital gains tax are eliminated on investments of at least five years' duration.[2] Corporate taxes may also be reduced or deferred to provide an additional incentive in special cases.[2] Argentina is also offering research and exploration loans, preferentially to small and medium-sized Argentine mining companies, through the Mining Promotion Fund. The government will allow up to 80% of the loan to be written off when no commercially exploitable deposits are found.[3]

The Argentine government may provide services such as electricity to companies that invest heavily in infrastructure.[9] Argentina lacks the necessary support systems for mining and is trying to remove the lack of infrastructure from the list of barriers to mineral development.

Government Ownership of Producing Companies. The mineral fuels and mining companies that are owned by the state are discussed in the industry structure section. The Argentine government owns virtually all mineral fuel production areas and many minerals-producing companies.

Mining Law and Policy. In an effort to attract domestic and foreign investment capital to the mining industry, Argentina adopted the Mining Promotion Law (No. 22095) in October 1979.[4] The law extends and grants new tax, import duty, loan, and research and exploration incentives for the mining industry. These have been discussed above. This law also modifies the regulations governing contracts, concession agreements, prospecting claims, and international agreements.[1] Another new law (No. 20551) allows additional tax and expense deductions. The value of some mineral reserves may be capitalized and depreciated under this law.[2]

The Risk Contract Law (No. 21788) of 1978

has resulted in the opening of large offshore areas to exploration by the private sector.[3] Argentina is also promoting private prospecting and exploration for uranium under the authority of the state Nuclear Energy Agency (CNEA).[3] Argentina is trying to develop petrochemical complexes to utilize domestic oil and gas resources.[3]

The extent and duration of concessions and the classification of minerals have been modified by recent reforms to the Mining Code. The Department for Mining Promotion, which is within the Mining Secretariat, has offered several promising mineral deposits for international tenders.[3] Foreign investors have expressed more interest in exploration agreements for unexplored areas than in these previously identified deposits.[9]

Government Role in Research and Development. The Argentine government will soon underwrite up to 80% of the risk of new mineral exploration and research projects. Loans through the Mining Promotion Fund for research are administered through the Department of Mining Promotion.[3]

Domestic Supply–Demand Dynamics

The Argentine economy continued its downward path in 1981 and in 1982 as a result of the world economic recession, the exchange rate policy, and the South Atlantic crisis. Argentine industry suffered hardships caused by the exchange rate, which overvalued the peso in early 1981.[4]

Argentina produces enough nonmetallic minerals to meet domestic demand.[3] The country continues to import large volumes of iron ore, manganese, bauxite, chromium, and coal.[3] Current consumption of manganese is about 75,000 metric tons per year.[3] Argentina imports copper, alumina, steel, nickel, molybdenum, sulfur, and fertilizer minerals.[3]

Oil is imported principally from Saudi Arabia and Iraq.[3] Gas imports from Bolivia and Chile are increasing, and the price of the gas has been steadily increasing.[3] The current import volume is 84 billion cubic feet (mcf) per year, at $4.06 per mcf.[3] The United States supplied about two-thirds of Argentina's 1980 bituminous coal imports.[4]

Argentina's exports of aluminum metal increased by 40% in 1980, to 67,000 metric tons.[3] With present consumption in the 70,000- to 80,000-ton per year range, and the productive capacity now at 140,000 tons per year and set to increase to 280,000 tons per year, Argentina has sizable domestic production available for export.[3] Argentina also exports some boron minerals to Brazil.[3] The only other mineral exports are lead-zinc, tin, and silver concentrates.[3]

Argentine coal production, which fell 46% between 1979 and 1980, rebounded with a 28% increase in 1981.[3,4] Increased imports from Poland and the United States have contributed to the low demand for coal.[3] The state coal company's only purchasers, Agua y Energía, SEGBA (electricity generating companies), Ferrocarriles Argentino (the state rail company), and Somisa (a major steel producer), have reduced their purchases to 1966 levels.[2] The slack demand in 1979 resulted in the stockpiling of over 100,000 metric tons of coal on the beaches at Santa Cruz.[2]

Iron-ore production fell during 1981, due largely to operational problems at Hierro Patagónico S.A.'s pellet plant.[4] Metallurgical problems at the Somisa steel plant have created a need to import iron ore from CAP of Chile for blending.[4] Ore stocks are over 3 million metric tons.[2]

Copper, molybdenum, and manganese imports should decrease in the 1980s as several domestic deposits are brought into production. The El Pachón deposit could produce 100,000 metric tons per year of electrolytic copper, 1700 metric tons per year of molybdenum concentrate, and 200 metric tons per year of anodic bars with gold, silver, selenium, and tellurium contents.[3] The ore has an average grade of .6% copper and .16% molybdenum, with some gold.[2] Manganese production of 25,000 metric tons per year of 38% Mn concentrate, with sizable gold and silver co-product production, is being planned for the Alto de la Blenda deposit.[3] This would meet about one-third of total domestic demand.

The prospect of Argentina becoming a major mineral exporter in the short term is highly unlikely. As a political and economic climate favorable to large-scale investment is created, Argentina's mineral industries may become more important to the world mineral economy, sometime in the future.

POLITICAL ENVIRONMENT

Domestic

Argentina has been under military rule since 1976.[10] Military governments in Argentina have a poor record of stability, with over 40 coup d'etat attempts between 1930 and 1976.[10] Argentina's economic policy is devised by Economic Minister Alfredo Martínez de Hoz.[10] He favors a policy of gradual economic liberalization, reduced inflation, low unemployment to neutralize labor unions, and sensible foreign exchange policy.[10] Subsidies protecting domestic industry are being phased out.[10]

The Perón administration was overthrown in March 1976 amid widespread labor disruptions, rural and urban terrorist activity, and economic upheaval.[10] The military government has reduced leftist guerrilla violence, imposed civil order, and improved a previously fragmented and often counterproductive economic strategy.[10]

A September 1980 uprising by 3rd Army General Luciano Menéndez was not supported by the armed forces.[10] A general strike called in April 1980 received little support.[10] The public experienced a state of civil peace in 1979, although Argentina has experienced considerable unrest since then. Discussion of a return to civilian rule increased.[10] Moderates dominate the military.[10]

Other political actors in Argentina include the Peronists (supporters of former President Juan D. Perón and his wife), labor unions (75% of the labor force is unionized), the Catholic Church, and business groups.[10] The Peronists are the largest political group in Argentina, with a political philosophy based on economic nationalism and blue-collar welfare.[10]

The midterm prospects for political stability are relatively good in Argentina.[10] Given Argentina's history of political instability under military rule, long-term investments and projects are more questionable.[10]

International

Companies from a wide range of countries have expressed a desire to become actively involved in mining and mineral-related industries in Argentina.[9] Cooperative mineral programs with other Latin American countries are commonplace in Argentina.

Firms in Canada, the United States, South Africa, Japan, and Europe have all expressed interest in Argentine mining.[9] Two South African firms, Anglo-American and General Mining, are interested in opportunities in Argentina.[9] The Germans and Japanese provide funding, nuclear systems, steel industry equipment, and machinery to Argentina.[3] The Japanese government is looking into two copper prospects, with associated gold and molybdenum values, in the Catamarca province.[9]

Major corporations have visited Argentina to assess mineral prospects, including Exxon, Amoco, Noranda Mines, Duval, and Río Algón.[9] Many other North American, South African, and European firms have requested more information on mining opportunities and mining law liberalization.

Argentina and Brazil are cooperating in efforts to establish a bidding consortium for offering West German Kraftwork Union heavy water nuclear plants to the rest of Latin America. Mexico and Chile have expressed interest in the consortium.[3] Argentina is also helping Bolivia and Uruguay to develop nuclear research programs. Argentina will provide a 10-MW research reactor and trained nuclear engineers and technicians to Bolivia and Uruguay.[3]

Argentina and Brazil are studying the feasibility of constructing a 2200-km 10-million cubic meter per day natural gas pipeline between the two countries.[3] Other pipelines being planned may connect Argentina to markets in Paraguay and Uruguay.[3] Argentina and Paraguay are cooperating in the development of the hydroelectric resources of the Paraná river at the Yacireta-Apipe Dam, where output is projected at 5400 MW by 1990.[1]

The implications of the Falklands/Malvinas Islands war on foreign involvement in Argentine mining are difficult to assess. Whether or not the recent crisis will delay or prevent foreign investment in Argentine mining remains to be seen. Initial anti-U.S. sentiments have not persisted or become more widespread.

REFERENCES

1. U.S. Department of the Interior, Bureau of Mines. *Mineral Industries of Latin America*. Mineral Perspectives Series. December, 1981.

2. Spooner, John, ed. *Mining Journal Annual Review 1980*. London: Mining Journal, June 1980.

3. U.S. Department of the Interior, Bureau of Mines. *Minerals Yearbook 1980*. Washington, D.C.: U.S. Government Printing Office, 1982.

4. Spooner, John, ed. *Mining Journal Annual Review 1982*. London: Mining Journal, June 1982.

5. Sillitoe, R. H. "Regional Aspects of the Andean Porphyry Copper Belt in Chile and Argentina." *Applied Earth Science* 90 (February 1981).

6. U.S. Department of the Interior, Bureau of Mines. *Minerals Yearbook, 1981*. Washington, D.C.: U.S. Government Printing Office, 1983.

7. U.S. Central Intelligence Agency, National Foreign Assessment Center. *The World Factbook—1981*. Washington, D.C.: U.S. Government Printing Office, 1982.

8. "Worldwide Survey: Latin America." *World Mining*, August 1981.

9. Welna, David. "Argentina Meeting with Success in Securing Foreign Funds for Mineral Development." *Minerals Week*, 9 January, 1981.

10. First International Bancshares, Economics Division. "Argentina." February 1980.

Chapter IV-12

Conclusions

Several factors are responsible for the varying levels of mineral development among Latin American countries. The most highly developed mining industries are in countries where the development was the result of several of these factors. In some countries, a combination of state ownership of mining companies and extensive foreign borrowing have fostered well-developed mining industries. Protectionist trade barriers have also been used to foster the development of the mining industries in some countries. Other Latin American countries have well-developed mining industries because mines and mineral processing plants were built by multinational corporations (MNCs) which have since been expropriated or nationalized.

Other factors that are responsible for the varying levels of mineral development among Latin American countries include political and economic stability, the resource endowment of the country, and the quality of existing infrastructure. Politically stable countries, such as Brazil and Mexico, tend to have more developed mining industries. A stable economy and well-developed infrastructure are both prerequisites to mineral development. As infrastructure expenses can be much greater than capital costs for the mine itself, a dependable energy and transport infrastructure in-place is a tremendous incentive to investments in mineral development.

Latin American countries have chosen to develop their mining industries with varying degrees of state ownership. Some countries have fully nationalized mining industries, with the state owning virtually all mines and mineral-processing plants. Countries using this mineral development strategy include Bolivia, Brazil, and Chile. In other countries, the state enters into joint venture agreements with private domestic and foreign companies in order to develop their mineral resources. Mexico, Colombia, and Venezuela have chosen this mineral development strategy. The historical involvement of MNCs is one of the major factors controlling the level of mineral development in Latin America.

The historical imbalance in the bargaining power of Latin American mineral-producing countries, and the MNCs active in their mining indus-

tries, have resulted in the development of several types of joint venture arrangements, with varying degrees of state involvement. The inequitable bargaining position of the host governments with respect to the MNCs have created a number of problems. The MNCs investments were allowed into the countries in the belief that they were additional, not substitute, capital.[1] Most Latin American countries have insufficient capital formation to develop fully their mineral resource potential.

The MNCs domination of the mining industries in Latin American countries has created several problems. When mineral commodity prices were high, profits went disproportionately to the MNCs.[1] The income generated by mining activities was not reinvested in the host countries but rather, was concentrated in the industrialized Western countries.[1] The MNCs also utilized artificially low transfer prices to move revenues outside the host countries, thus escaping their tax regimes.

Another problem created by the large role of MNCs in Latin America's mineral development was the high level of dependence on Western industrialized countries as the markets for their mineral exports.[1] U.S. government policies freezing strategic mineral commodity prices during World War II and the Korean War cost mineral-producing countries several hundred million dollars in lost foreign exchange earnings.[1] MNCs complained about high tolls charged for smelting and refining in the host countries, where processing charges were well above the international average.[1] Sales of minerals from the U.S. strategic minerals stockpile, especially tin and tungsten, have drawn criticism from mineral-producing countries.[1]

The experiences of three South American mineral-producing countries since the nationalization of their mining industries in the last half of this century form a good case study in the role of MNCs in mineral development. Bolivia nationalized its mining industry in 1952, and its case is very different from those of its neighbors, Chile and Peru. The Revolutionary Nationalist Movement Government took this action two decades before Chile and Peru made similar moves, long before the concepts of the "Third World" or a "New

International Economic Order" were topics of international discussion.[1]

Several other important considerations set the Bolivian experience apart from those of its mineral-producing neighbors. Large-scale mining in Bolivia was originated by Bolivians, such as Simon I. Patino, who began tin mining around 1910.[1] The mineral industries in Chile and Peru were originally developed by the multinational mining companies. The most important of these mining companies were Anaconda, Kennecott, American Smelting and Refining Corporation (ASARCO), Phelps Dodge International, Cerro Pasco Corporation, and Newmont Mining.[1]

The nationalization of the Bolivian mining industry was more widespread than the nationalizations of the Chilean and Peruvian mining industries. In Bolivia, the nationalization extended into all stages of mining, mineral processing, and marketing.[1] Bolivia's nationalization extended to all minerals, while some mining activity was left in private hands in Chile and in Peru. Peru's large-scale copper mining industry was not nationalized. Chile allows foreign energy and minerals companies to form joint ventures with the government for the exploitation of some copper deposits, although most large-scale mining activity is undertaken by the state mining companies. Bolivia had no previous experience in state intervention or supervision of mining enterprises.

The nationalization of the Bolivian mining industries occurred before extensive mine expansions and modernizations had taken place, as was the case in Chile.[1] By delaying the nationalization of the copper industry until the 1970s, Chile inherited a more efficient and modern mining industry than did Bolivia.[1] Chile also had long experience working with the multinational mining companies and a labor force with an increasing number of university-trained mining engineers, scientists, and technicians.[1] The Bolivian nationalization underlined the importance of domestic political considerations in actions of this nature, and indicated that a socialist government was not a prerequisite for this type of action.[1]

Bolivia's mining industry nationalization was followed by a period when production, productivity, and profits declined as the government learned to manage and supervise large-scale mining activity.[1] The major causes of these productivity and profitability declines were overexploitation of the deposits by the "tin barons," declining ore grade, declining tin prices, and the exodus of about 85% of the country's mining engineers after the mining industry nationalization.[1]

Bolivia constructed tin smelting plants at Vinto in the 1970s, giving the country sufficient capacity to process all tin mine output domestically.[1] This gave Bolivia control over where it sold its tin exports. Bolivia's dependence on the U.S. market for its tin concentrates was replaced by the freedom to sell refined tin metal to a more diverse group of countries, including the USSR and other socialist countries (27% of 1979 tin exports) and the expanding Latin American market (10% of 1979 exports).[1] The domestic smelting and refining capacity effectively eliminated the multinational mining companies from any role whatsoever in the Bolivian tin industry.[1] The international tin market is considerably different from the world copper market. Because the tin market is much less oligopsonistic, Bolivia can market its tin metal production without the use of multinational intermediaries.[1]

The nationalization of large-scale mining activity in Chile was the result of an inequitable distribution of the income generated by mining enterprises between the Chilean government and the multinational mining companies (Anaconda and Kennecott). Chile based its decision to nationalize the mining industry on United Nations General Assembly Resolution 1803 (XVII), adopted on December 14, 1962.[1] This resolution recognizes the "sovereign right of every state to dispose of its wealth and natural resources." The expropriated companies received no compensation from the Chilean government, which had deducted previous "excess profits" from the book value of the companies' assets.[1] An agreement was signed by Anaconda, Kennecott, and the Chilean government in 1974, authorizing compensation totaling $377 million for the companies.[1]

While Chile nationalized its large-scale copper mining industry, consolidating it into the Corporación Nacional del Cobre (CODELCO), Peru left its large-scale copper mining industry under the control of ASARCO.[1] However, the act of nationalizing the mining industry did not result in rapid changes in the relative bargaining power of the host governments and the MNCs.[1]

Although Chile, Bolivia, and Peru have chosen varying levels of state involvement for their mining industries, all three have achieved success in increasing mineral production, the level of domestic mineral processing, the diversification of mineral export destinations, and the share of mining industry profits remaining in the host country. Peruvian exports of refined copper increased by a factor of 4.4, to over 160,000 metric tons per year, between 1975 and 1978.[1] The share of the value of copper production retained in Chile grew from 68% prior to nationalization to about 90% in the decade since the nationalization of the large-scale copper industry.[1] Bolivia has experienced the greatest success of the three in diversifying the destinations of its mineral exports.

Despite Chile's previous mining industry nationalizations, it has one of the most favorable climates for mining industry foreign investments. Present 10-year tax stability guarantees and provisions for compensation in the event of expropriation, subject to international arbitration, contribute to Chile's favorable mining investment climate. Chile's Foreign Investment Decree-Law No. 1748 grants generous tax benefits and holidays, and clearly defines the role the Chileans would prefer to see the multinational mining companies play in their further mineral development. To date, the MNCs appear to be more interested in grass-roots mineral exploration contracts than in the handful of identified copper deposits being offered by the Chilean government.

State involvement in the Peruvian mining industry has been concentrated in the smelting and refining of the country's mineral production and in developing some new mineral deposits. Enami, the state mineral processing company, purchases the concentrates of the small and medium-sized mining companies, where large shares of Peru's production of several minerals emanates (see Chapter IV-5). The concentrates produced by Peru's large-scale copper industry are processed at state-owned processing plants.

Bolivia's low level of mineral development relative to Chile and Peru is the result of several factors. The low skill level of Bolivia's labor force has hindered mineral development. Bolivia's position as a land-locked country, with poor infrastructure, remote from the markets for its mineral output, places it at a comparative disadvantage compared to other major tin producers. The increase in the importance of Southeast Asian tin producers has also adversely affected Bolivia's tin industry and position in the International Tin Council. Sales from the U.S. strategic minerals stockpile have hurt the Bolivian tin and tungsten industries. The mining industry plays a smaller role in the Bolivian economy than do the Chilean and Peruvian industries. Bolivia's gross national product per capita of $570 is well below Peru's $930 and Chile's $2150 1980 gross national product. However, the value of 1979 Bolivian mineral exports per capita, at $137, is well above Peru's $86 per capita. This difference reflects the greater importance of the service economy in Peru.

The experiences of Brazil and Mexico with varying degrees of state involvement in their mining industries provide a good case study in the effects of state involvement on mineral development. Brazil and Mexico are Latin America's most populous, most highly industrialized countries. The mining industries make minor contributions to the two countries' gross national products (2.2% in Brazil and 1.3% in Mexico in 1980).[1] Mexico's gross national product per capita of $2090 was just $40 more than Brazil's 1980 gross national product per capita.[2]

The Brazilian government plays a much larger role in the economy in general, and specifically in the mining industry, than does the Mexican government. The Brazilian mining industries are dominated by a few large state mining companies. The most important of these is the Companhia Vale do Rio Doce (CVRD), the world's single largest iron-ore producer and exporter. Most other minerals and energy resources are under state control. Only Mexico's oil industry exhibits such a high degree of state control.

While Brazil has chosen to use a combination of foreign borrowing and joint ventures between MNCs and state and private Brazilian companies, Mexico has utilized a combination of protectionist policies and joint ventures with multinational energy and mining companies to further mineral development. Multinational minerals companies are active in the Mexican aluminum, copper, fluorspar, lead, silver, and zinc industries.[1] In both Mexico and Brazil, foreign firms can own no more than 49% of any mining enterprise.[1] Only the Mexican oil and sulfur industries are completely government owned. The level of state ownership is higher in most Brazilian mineral industries than it is in Mexico.

Mexico has utilized protectionist policies, including trade barriers and restrictions, and foreign exchange controls, in conjunction with heavy foreign investment, to further develop its mineral resource potential. Mexico and Brazil have the world's largest foreign debts; both are over $80 billion. Foreign investment in Brazil's mineral industries has come from a variety of countries, including the United States, Canada, Japan, South Africa, West Germany, and others.[1] The largest share of foreign investment in Mexico's mineral industries is by U.S. and Canadian companies.[1]

The Brazilian government enters directly into joint ventures with MNCs. The Mexican government prefers to act as an intermediary between private Mexican and international mining companies when securing the financing of large-scale projects. The state has retained an interest in Mexico's two largest copper mines. MNCs are active in Mexico's aluminum, copper, fluorspar, lead, silver, and zinc industries.

World recession has had serious effects on mineral and energy resources development in Brazil and Mexico. Foreign exchange earnings have been reduced by low prices in most agriculture, energy, and mineral commodity markets. Mexico and Brazil share the world's largest foreign debts, and historically high interest rates have greatly increased their annual debt service payments.

An important difference between the two countries' economies lies in their energy minerals trade positions. While Mexico's 1980 crude oil exports were valued at $9.5 billion, Brazil's net oil import bill was $10 billion.[2] This fundamental difference between the two countries' economies affects the rate at which they can develop their mineral resource potential. Income from Brazil's mineral exports must be used to pay for imported oil, whereas the profits of Mexican mining firms can be reinvested in the industry. Mineral development incentives authorized by the Mexican government offer subsidized oil and electric power to companies that increase productive capacity and exports.

Latin America offers international mining companies a wide range of investment opportunities.

Chile and Peru presently have the most favorable foreign investment climates in the region, due to recent modifications to mining and foreign investment laws. Venezuela and Colombia also have favorable foreign investment climates, and both countries should play a more important role in the world mineral economy in the future.

Foreign investment opportunities are limited in Bolivia by the country's expensive infrastructure and remoteness from international metals markets. Investment opportunities in Ecuador and Argentina are uncertain at present, as neither country's mineral potential has been evaluated systematically.

The resilience of the Brazilian and Mexican economies in the face of the present world recession will determine the rate at which their mining industries expand. Renewed economic growth in Mexico and Brazil is dependent on the recovery of the world economy. As the world economy emerges from the present recession, improvements in mineral markets should begin to lift Brazil and Mexico from their current economic stagnation.

REFERENCES

1. Knakal, Jan. "Transnationals and Mining Development in Bolivia, Chile, and Peru." *CEPAL Review*, August 1981.
2. U.S. Department of the Interior, Bureau of Mines. *Minerals Yearbook 1981*. Washington, D.C.: U.S. Government Printing Office, 1983.

Chapter V-1

Chinese Mineral Development: Overview and Constraints

China is often described as a country with great potential for development of energy and mineral resources. By its own accounts, China has large reserves of a number of very important materials. However, the Chinese must contend with conditions that constrain resource development, including the structure of the Chinese economy and growing internal demand that threatens to outstrip economic output. These problems make it questionable whether China can emerge as a large exporter of strategic materials in the future.

China has traditionally been one of the world's largest exporters of such commodities as antimony, tin, and tungsten. Recently, China has become important as a supplier of titanium and bauxite to the United States. Based on its own estimates, China has the world's largest reserves of tungsten, tin, antimony, zinc, titanium, tantalum, and rare earth metals, the second-largest reserves of lead, nickel, mercury, molybdenum, and niobium, while reserves of copper and aluminum are fourth and fifth in the world, respectively.[1] Additionally, China has the third-largest reserves of coal in the world. Although reserve estimates on oil and gas vary widely, it appears that China also has potential in these areas.

There are several problems that must be considered when examining Chinese reserve figures. One is that the country has not been explored completely in terms of mineral potential. Also, China does not release much data about the economy in general or about minerals specifically. This condition appears to be improving, partially as a result of China's membership in international financial organizations. Finally, the standard problems of reserve estimation must be considered. In short, data about China are still generally fragmentary, aggregated, and secondary in origin.

Copper[2] has noted some of the advantages of an adequate resource base to the Chinese: a stable supply of inputs for industrial development, protection from resource disruption, potential for generating investment capital, and possible leverage to influence the policy of other countries. However, he has also noted that China's reserves do not appear large when evaluated in relation to China's population base on a per capita basis.

China, traditionally a minor participant in world trade, has expressed an intent to expand the export of mineral commodities. Results will be mixed. China could become a major supplier of energy and strategic minerals. This would provide an additional source of supply for resource-dependent countries. Alternatively, Chinese exports could prove to be disruptive to world markets. China has massive reserves of certain minerals and could assume a more dominant role to the detriment of the market share of other producers. Additionally, withdrawal from certain import markets is possible as the Chinese improve their mining and processing capability.

If China is to become a significant, reliable, and secure supplier of a variety of minerals, certain policy and investment commitments must be made. Important policy decisions range from the degree of outside assistance to be sought to the relative priority of resource development in the Chinese economic planning programs.

Although information presented by Chinese sources has improved in recent years, data reported are rather limited. Production levels are primarily estimated; imports/exports are based on reports of trade partners. Subsequent analysis reflects the limitations of the data and the problems of accounting for the vagaries of Chinese policy. The analysis is descriptive rather than quantitive.

ROLE OF RESOURCES

Strategic and Economic Relationship of Resources

An important factor is the Chinese perception of the strategic role of resources. An article in the *Beijing Review*[3] describing the existence of a "resource war," characterized the USSR as trying to exploit the vulnerability of the United States and Western Europe to resource supply disruptions. A resource war as Soviet policy would be consistent with China's viewpoints on Soviet expansionism. However, China appears willing to export strategic materials. The idea that strategically important materials should not be exported is regarded as old-fashioned.[4] In some cases, notably tungsten, it appears that China withholds exports on a selective basis for economic rather than strategic reasons.[5] Minerals are exported not only for currency but as repayment for assistance in development. Petroleum exports are particularly important as a source of foreign currency. China has emphasized trading resources for goods and services whenever possible to reduce cash outflow. It has been suggested that the United States could follow a similar policy, trading weapons for Chinese military modernization for strategic minerals to promote United States security.[6]

This promotion of resources exports is a relatively recent view. The decision to export oil in the 1970s was characterized by controversy. An article in the *Red Flag* described oil exports as a concession of sovereignty and a selling out of China's natural resources to others.[7] A strong campaign by the Chinese leadership to promote the benefits of outside trade was necessary to secure support for oil exports. Essentially, there are competing viewpoints regarding resource development in China. On one hand, there are those who would conserve Chinese resources for its future needs, at the expense of experiencing a somewhat slower economic growth. On the other hand, the recently dominant faction prefers export of resources with the objective of speeding up the development process.[8]

Resources in Economic Development

It is difficult to analyze resource development in isolation from the overall Chinese economic development scheme. The centralized planning characteristic of a socialist economy means that the priorities established will determine which economic sectors receive larger shares of limited investment funds. A second reason why one must examine the comprehensive economic development scheme is that natural resource development projects are feasible only with the existence of a large-scale infrastructure. China is noticeably deficient in both physical and social infrastructure. By examining economic development plans, it is possible to determine which problems China will be addressing and what the time frames for resolution are.

Resources have been important components of development from the beginning of the People's Republic of China. A strong mining base was established by the implementation of the initial Five-Year Plan with Soviet assistance. The Soviet model was based on building up heavy industry, with emphasis on iron and steel. China has modified this approach to take into account the special needs of producing food for a large population without becoming overly reliant on imports. Therefore, agriculture as well as industry must be developed.

One of the weakest sectors is that of nonferrous metals. During the 1960s, there was little investment in this sector. Petroleum was the preferred resource sector because it generated foreign exchange.[9] There was a 20% drop in output of 10 major nonferrous metals from 1967 to 1977, largely as a result of the Cultural Revolution.[10]

The late 1970s witnessed a rather substantial shift in economic policies. A plan prepared in 1978 proposed numerous heavy industrial projects, emphasizing energy and raw material projects. The plan included 48 major energy projects, 19 major mineral projects, and 30 transport projects.[11] This was a departure from previous plans only in its scope. Soon after the plan was announced, there was a realization that China would be overextended in attempting this ambitious effort. A period of readjustment followed. A number of the projects were canceled or delayed. The most substantial policy revision, as a result, was added emphasis on light industry. There were several reasons for this. Light industry is less capital intensive and more labor intensive. This was more in line with the strengths and weaknesses of China. Additionally, light industrial output served to increase consumer welfare. China is not immune to rising consumer expectations. A recent announcement from China indicated that the retrenchment has run its course. As a result, some previously canceled projects are expected to be revived. Areas of emphasis in the future are expected to be energy, other natural resources, transport, and communications.[12] Energy has in fact never been deemphasized, as about 21% of investment since 1979 has been devoted to energy projects.[13]

China has signaled an interest in and a commitment to economic and resource development. Problems must still be overcome. A list of constraints is included in Table V-1-1. These are indicative of the types of problems that Chinese planners must address. In the following discussion, certain key potential constraints are noted.

CONSTRAINTS TO DEVELOPMENT

Political Factors

Political factors have had a very important influence on Chinese development, most often as a constraint, but more recently as a stimulus. The policy of self-sufficiency, primarily for reasons of

TABLE V-1-1 Some Constraints to Development in China

Infrastructure
  Insufficient capacity
  Geographically concentrated
Undeveloped nature of the society
  Outdated technology
  Not enough managerial and technical personnel
Economic organization
  Bureaucratic inertia
  Planning and pricing problems
  Conflicts between national and provincial interests
Other
  Insufficient capital and foreign exchange
  Population pressures
  Raw material shortages

Sources: Refs. 4, 10, 13, 19, and 27; Vaclav Smil, "Energy Development in China: The Need for a Coherent Policy," *Energy Policy*, 9 (June 1981); pp. 113-126.

security, occurred at the expense of rapid development.[14] The policy was in effect under the leadership of Mao Zedong. Following his death, a more pragmatic leadership emerged, and economic rather than ideologic development became the predominant societal goal.

Analysis of political risk is an important determinant of the willingness of international corporate and financial institutions to invest in a country. Additionally, at the domestic level, there is a circular relationship between political stability and economic growth. Political stability provides a basis for economic growth, which contributes to improved living standards, thereby promoting stability. Politics in China has traditionally been a negative influence, due to ideological movements that disrupted the economy with periodical policy shifts contributing to uncertainty and failures to take advantage of international trade.

An example of a politically inspired economic movement that in fact hampered economic growth was the Great Leap Forward. This occurred from 1958 to 1960. The theory was to use political zeal to stimulate the workers, together with some decentralization of industry. The objective was to promote rapid economic development. A symbol of this time was the "backyard furnace." These small, inefficient operations produced pig iron for local consumption. The results of the Great Leap Forward were large initial increases in output, followed by much smaller increases in industrial output and a precipitous drop in agriculture. The overall effect was a weaker rather than a stronger economy.[15]

The Cultural Revolution, which occurred from 1966 to 1968, provided another example of the negative effects of politics on the Chinese economy. The objective of this movement was "to rekindle revolutionary fervor through mass actions outside the formal party organizations."[15] The movement soon got out of control, resulting in disruptions of work and transportation. People with managerial and/or technical expertise were often replaced with persons with political credentials. A most serious result was the closing of the universities for a number of years.[15]

More recent disruptions, specifically in the coal-mining sector, have been associated with the overthrow of the "Gang of Four" and with the anti-Confucian movement.[16] Supporters of the Gang of Four opposed the modernization policies of the leadership replacing Mao Zedong. Since that movement was thwarted, there has been a consistent dedication to the goals of economic growth. One result is a willingness to depart from the socialist line by experimenting with market signals on a limited basis. This is labeled "market socialism." Additionally, there has been a dramatic change from economic isolationism to a willingness to engage in international trade, primarily to gain access to state-of-the-art technology. Mixed feelings concerning the need to turn outward continued, as summarized by an article on Chinese energy development: "China Debates Foreign Involvement in Energy Development—Energy Planners Want Self-Reliance but Need Foreign Technical Assistance.[17]

Chinese pragmatism defines the rationale for international trade—to gain the advantages of the more advanced economies. "There is no large constituency in China for the view that turning outward is good in itself."[18] This may be a stage of transition until China can build up its own capabilities. Meanwhile, failure of the economy to respond to an outside technology "fix" or a consistent balance of trade deficits could result in a move toward self-reliance once again.[19]

Chinese Railways

China's railway system is a bottleneck, affecting industrial and power output. The Chinese railway system is among the world's largest. However, service is geographically concentrated in the east and northeast, preventing large-scale development of resources in the west. Much of the system is outdated. The biggest single problem is that the railways are overused relative to capacity. Within the limits of system capabilities, it appears that the railroads are efficiently run.[15]

One of the biggest components of demand for transport is moving raw materials, particularly energy materials. Approximately 50% of all bulk freight in China is shipped by rail.[20] Resource-related freight is a large component of total rail volume as follows: coal, 38%; ore, 17%; petroleum and products, 6%. Recent improvements in pipelines have shifted oil transport from the rail to pipeline mode, easing the demand somewhat in the rail sector. Currently, 80% of the oil is shipped by pipeline.[20] Coal is particularly important as a determinant of transport demand. High-quality coal is concentrated in the north and must be transported. One approach is the utilization of smaller decentralized mines. Problems with such an approach include inefficiency of operations and the lower-quality coal that must be used.

Current policy calls for development of electrical generating plants near sources of coal or hydropower. This transforms the problem from one of long-distance coal transport to long-distance power transmission. Potential energy sources and power-consuming centers tend to be distant. Regarding transport of other raw materials, the rail system has been described as "largely unsuitable for heavy bulk ore transport," and the observation was made that "ore deposits can be at the end of something like a thousand miles of difficult-to-maintain, single-line railway track."[21] China has been involved in a program to upgrade and extend the rail system. This includes electrification and double tracking, as well as import of some equipment.

Power Generation

Electrical power generation is a weak link in the Chinese economy. Current capacity is 68,000 MW and will need to increase to 220,000 MW by the year 2000.[22] The industrial sector is the primary consumer of power, using 62% of the available supply.[23] Problems with the industry include relatively small scale plants, low-capacity transmission lines, and lack of interconnected systems.[23] Much of the equipment is outdated by at least 20 years. The importance of this for industrial production is that insufficient capacity results in curtailments of power, as well as unscheduled outages. Industrial capacity often cannot be fully utilized. Power shortages are particularly important constraints in such energy-intensive operations as aluminum smelting. In one instance, it was discovered that a steel rolling mill when run at full capacity could consume the entire electrical capacity of the province where it was located.[21]

Electrical demand in China is more directed toward industrial consumption than would be anticipated for a developing country, due to historical emphasis on heavy industry.[24] Metals pro-

cessing is quite energy intensive. Nonferrous metals account for 11% of the total consumption, and ferrous consumes 10.4%.[25] Due to energy conservation, closure of inefficient plants, and changes in the composition of heavy industrial output, "there have been increases in the relationship of value of output in metallurgy per unit of energy consumed."[24]

Additional electrical capacity is a necessity to enable China to use current industrial capacity efficiently, to allow for industrial growth, and to promote consumer welfare. China is making a concerted effort to increase generating capacity through import of equipment from a variety of countries. Additionally, although location and the lengthy lead times may be a problem, China has an impressive hydropower potential that has barely been exploited.

Finance

China must address a relative lack of investment funds and foreign exchange. China's commitment to economic modernization necessitates large-scale investment that cannot be met completely by internally generated funds. Although China has invested a large proportion relative to GNP, this is not enough, given the poverty of the country. In 1978, investment in China was comparable to that of France, despite China's much larger needs due to population, size of the country, and stage of development.[15] Access to large amounts of investment capital is necessary for resource development. "Compilation of a suitable financial package for a large-scale mining venture typically involves the entire world."[26]

Foreign exchange is another potential problem. China has not traditionally been a significant participant in world trade. Minerals have not been large generators of outside funds, but energy exports have been quite important. Woodard[27] has noted the very close relationship between funds generated by oil exports, and funds expended on large-scale imports of foreign technology.

Commercial and international financial institutions, as well as numerous countries, have responded quite favorably to China's need for outside finances. Credits, commercial loans, and low-cost loans have been offered. China, known for its conservative economic policies, has taken advantage of only a fraction of what was offered. Policies have been devised that minimize payback in cash, and maximize payback in kind. China has kept a low debt/service ratio in relation to exports. This involves a trade-off. More expansionist economic policies could speed up the development process. On the other hand, unlike many developing countries, China is not near bankruptcy and, in fact, enjoys an excellent credit rating.

Skilled Workers

The modernization movement changing many traditional societies finally became a worldwide phenomenon only when China embarked on a modernization effort.[28] Problems with late entry into this process have been noted. One is a lack of skilled labor. With its large labor pool, China is suffering from unemployment, while at the same time there are "severe shortages of skilled workers, engineers, technicians, scientists, and managerial personnel."[15] The advanced technology that China is importing to improve economic conditions will be of little use without competent personnel to direct and to operate such technology.

There are several factors contributing to the shortages. Closing of universities as a result of the Cultural Revolution left a gap in the supply of educated workers. Prior to the initiation of modernization, "intellectuals" were considered elites not in accordance with the objectives of egalitarianism. Additionally, education was more likely to emphasize political and practical learning, at the expense of theory. A practical result of the inadequacy of education and training is that sophisticated imported equipment, such as mining machinery, is not being properly operated or maintained.

China has been addressing this problem by upgrading its schools, opening new universities, and sending students to other countries for their education. China has received financial aid from the World Bank to increase technical education programs, and has increased budgetary allocations to education.

REFERENCES

1. "Progress in Nonferrous Metals." *Beijing Review* 25 (15 February 1982), pp. 7-8.
2. Copper, John F. *China's Global Role.* Stanford, Calif.: Hoover Institution Press, 1980.
3. "The Battle for Resources." *Beijing Review* 24 (4 May 1981), pp. 14-15.
4. "On China's Economic Relations with Foreign Countries." *Beijing Review* 25 (31 May 1982), pp. 13-16.
5. Tan, C. Suan. "China's Tungsten Pricing Strategy: Why the PRC Has Reduced Its Tungsten Supplies." *The China Business Review* 6 (March-April 1979), pp. 28-33.
6. Stuart, Douglas T., and William T. Tow. "China's Military Modernization: The Western Arms Connection." *China Quarterly*, June 1982, pp. 253-270.
7. Japanese External Trade Organization. *China: A Business Guide.* Tokyo: Japanese External Trade Organization and Press International Ltd., 1979.
8. Meyerhoff, A. A. "China's Petroleum Industry: Geology, Reserves, Technology and Politics." In *China Trade: Prospects and Perspectives.* Edited by David C. Buxbaum, Cassondra E. Joseph, and Paul D. Reynolds. New York: Praeger Publishers, 1982, pp. 94-112.
9. Szuprowicz, Bohdan O., and Maria R. Szuprowicz. *Doing Business with the People's Republic of China.* New York: John Wiley & Sons, Inc., 1978.
10. U.S. Department of Commerce, International Trade Association. *China's Economy and Foreign Trade 1979-81,* by Nai-Ruenn Chen, May 1982.
11. Szuprowicz, Bohdan O. *How to Avoid Strategic Material Shortages: Dealing with Cartels, Embargoes and Supply Disruptions.* New York: John Wiley & Sons, Inc., 1981.
12. "China Says It Won't Cancel Any More Projects in the Next Few Years as Retrenchment Is Over." *The Wall Street Journal,* 27 July 1982.
13. "China Trying to Maintain Slow Growth." *The Asian Wall Street Journal,* 2 September 1982, p. 3.
14. Willrich, Mason. *Energy and World Politics.* New York: The Free Press, 1975.
15. Bunge, Fredericka M., and Rinn-Sup Shinn, eds. *China: A Country Study.* Washington, D.C.: U.S. Government Printing Office, 1981.
16. U.S. Central Intelligence Agency, National Foreign Assessment Center. *Chinese Coal Industry: Prospects over the Next Decade,* February 1979.
17. "China Debates Foreign Involvement in Energy Development: Energy Planners Want Self-Reliance but Need Foreign Technical Assistance." *World Business Weekly,* 13 April 1981, pp. 15-16.
18. Perkins, Dwight H. "The Chinese Economy in the 1980's." In *China among the Nations of the Pacific.* Edited by Harrison Brown. Boulder, Colo.: Westview Press, Inc., 1982, pp. 1-14.
19. Feintech, Lynn. *China's Modernization Strategy and the United States.* Washington, D.C.: Overseas Development Council, August 1981.
20. Peterson, Albert S. "China: Transportation Developments, 1971-80." In *China under the Four Modernizations,* Part 1. Joint Economic Committee, U.S. Congress, 13 August 1982, pp 138-170.
21. "China Poised to Become the World's New Supplier of Platinum Group Metals." *Metals and Materials,* December 1980, pp. 27-30.
22. "Electricity for the People's Republic of China." *Electric Light and Power* 60 (February 1982), pp. 11-12.
23. Clarke, William. "China's Electric Power Industry." In *Energy in the Developing World—The Real Energy Crisis.* Edited by Vaclav Smil and William E. Knowland. New York: Oxford University Press, Inc., 1980, pp. 145-166.
24. Field, Robert Michael, and Judith A. Flynn. "China: An Energy Constrained Model of Industrial Performance through 1985." In *China under the Four Modernizations,* Part 1. Joint Economic Committee, U.S. Congress, 13 August 1982, pp. 334-364.
25. Cox, Madeline. *China Economic Overview—A British Columbia Perspective.* Province of British Columbia,

Ministry of Industry and Small Business Development, no date.

26. Radetzki, Marian. "Regional Development Benefits of Minerals Projects." *Resources Policy* 8 (September 1982), pp. 193–200.

27. Woodard, Kim. "China and Offshore Energy." *Problems of Communism* 30 (November 1981), pp. 32–45.

28. Wilson, Thomas W. *Science, Technology and Development: The Politics of Modernization.* New York: Foreign Policy Association, 1979.

Chapter V-2

Resources and Trade

Evaluating the Chinese resource position is difficult. The lack of information, and the selective release of what appears to be optimistic assessments, have caused some observers to observe that the Chinese deliberately overestimate their resource potential. For example, it is possible that the Chinese may have overstated offshore oil potential to attract Japanese investment to China rather than Siberia.[1] Copper[2] indicates that while the Chinese may have underestimated their resources previously, it is likely that this has been replaced by a tendency toward overestimation.

A large resource potential in China should be compared with the population base that must be served. An estimate is that China has approximately 4% of the world's mineral reserves.[3] However, the population of China is about one-fourth of the total world population. Evaluated on a per capita basis, China's resource situation is not that promising.[2,4]

EXPLORATION AND IMPORTANCE OF INFORMATION

The importance of an accurate data base describing natural resources has been emphasized as a prerequisite for optimal decision making and policy formation. Odingo has stated that "the availability of resources for development may not be as serious as the lack of reliable planning data for that development."[5] China has done a great deal of work in mineral exploration. For example, based on their accounts, the Chinese have discovered enough minerals to adequately support industrial growth to 1990, with few exceptions. A national survey of coal will be essentially complete by 1990.[6] Considerable exploration for oil and gas remains to be done, however. Almost half of China shows promising formations for oil occurrence; about 10% of this area has been explored.[7] China is seeking foreign assistance and more sophisticated technology. The U.S. Geological Survey and their Chinese counterpart have signed an agreement on geological cooperation between the two countries.

There are several reasons why a more complete data base would be advantageous to the Chinese:

As a basis for setting economic development priorities: Some of the previous problems experienced by the Chinese apparently were due to misunderstandings about the existing situation, at least partially a result of an inadequate data base.

As a basis for attracting outside investment and participation: The failure to make information available, except on a very limited scale, encourages speculation that the Chinese are manipulating information. A detailed, disaggregated data base would be advantageous as basic information for feasibility studies.

As a basis for negotiations: The Chinese are increasingly committed to long-term contractual trade agreements. It is quite important that such agreements are based on accurate data on mineral quantity and quality.

TRADE

China's traditional policy of self-reliance has prevented any large impact by the Chinese on world trade. Until recent emergence as an energy exporter, China was minimally involved in world energy trade. For nonenergy minerals, and metal trade, Chinese exports have not matched import needs. Over a period in the 1970s the Chinese experienced a $2 billion deficit in such trade.[8]

Japan provides a large market for Chinese mineral exports. In 1980, the Japanese imported more than half of China's total exports for a number of mineral products, including elemental antimony, manganese ore and concentrate, fluorspar, coal, and crude oil.[10] China was an important supplier to the United States of antimony, beryllium, barite, tungsten, titanium, and germanium in 1982.[11] Clearly, the most significant trend to emerge in Chinese mineral trade is the cessation of close trade links with the USSR. China has become a supplier primarily to Japan, the United States, and Europe. In 1957, the dollar value of metals and minerals exports to the USSR was $160 million; by 1965 that had declined to less than $17 million. Meanwhile, over the same time period, the value of exports to Japan increased from less than $7 million to $44 million. Additionally, there was a mineral trade surplus in 1957.[12]

NONFUEL MINERALS

Aluminum

The Chinese have ample sources of raw materials with which to produce aluminum. However, they are significant importers of aluminum metal. At the same time, the Chinese are emerging as an exporter of calcined bauxite (Figure V-2-1). Constraints to increased aluminum production are insufficient processing capacity and electrical power shortages.

Reserve estimates for bauxite vary considerably (Table V-2-1). This is indicative of the problem of information regarding Chinese resources. Earlier information generally is descriptive, while more recent information is quantitative. However, there is considerable variation among sources. China produced 1.8% of the world's bauxite in 1980.[3]

Chinese bauxites are mainly diaspores.[10] They contain gallium and germanium.[13] Chinese aluminum materials require large quantities of chemicals in processing. Soviet technology developed for Chinese bauxites is quite energy intensive. Recent Canadian technology requires from 12,000 to 14,500 kwh to process a ton of aluminum. In comparison, Chinese technology requires 15,000 to

20,000 kwh.[14] China is considering the import of Japanese technology for aluminum processing. Due to the costs of energy imports, the Japanese have

TABLE V-2-1 Aluminum Resource Estimates

Source	Estimate	Date
Wang[a]	Sizable reserves, off-grade, aluminum	1975
Bureau of Mines[b]	150 million metric dry tons, bauxite reserves	1979
Chinese sources[c]	1 billion tons, bauxite reserves	1979
Federal Institute Hannover[d]	180 million tons, bauxite reserves	1980
Business International[e]	9 billion tons, bauxite reserves	1980

[a]U.S. Department of Interior, Bureau of Mines, *The People's Republic of China: A New Industrial Power with a Strong Mineral Base*, by K. P. Wang. 1975.
[b]U.S. Department of Interior, Bureau of Mines, *Mineral Facts and Problems, 1980 Edition*.
[c]K. P. Wang, "China," *Mining Annual Review*, 1980; pp. 435-456.
[d]Helmut Schmidt and Manfred Kruszona. *Regional Distribution of Mining Production and Reserves of Mineral Commodities in the World*. Hannover, West Germany: Federal Institute for Geosciences and Natural Resources, January 1982.
[e]Business International Asia/Pacific Ltd., *Business Strategies for the People's Republic of China*. Hong Kong: Business International Asia/Pacific Ltd., 1980.

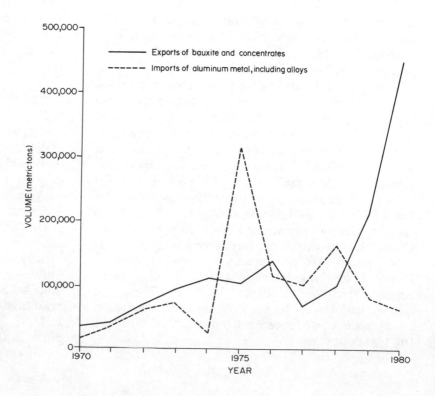

FIGURE V-2-1 China: Apparent aluminum imports and bauxite exports, 1970-1980. (From U.S. Bureau of Mines, *Minerals Yearbooks*.)

developed energy-efficient processing methods.[15] China currently has 12 major aluminum plants with a known capacity of over 315,000 tons per year. This does not include the capacities of two additional plants, which are unknown. China produced an estimated 360,000 metric tons of aluminum metal in 1981.[10] Figure V-2-2 shows the trend of Chinese aluminum metal production.

Perhaps the most interesting trend to emerge is China as a significant exporter of calcined bauxite. For example, the United States received 122,000 metric tons in 1981, providing almost half of the U.S. imports.[16] Improvements were made to these products, following complaints about uniformity of calcination and aluminum grade in early shipments.[16] The U.S. General Services Administration has purchased Chinese calcined bauxite for the U.S. stockpile.

Antimony

Antimony is one of the traditional metals exported by China. China has large reserves that could be developed to meet export demand.[17] However, demand has not been strong enough to justify large-scale expansion. China's reserves of antimony are estimated to be 2,400,000 short tons of antimony content.[18] These are the largest reserves in the world. The principal mining zone is the Xikuangshan, Henan, area.[10] In this area stibnite is found in sandstone in veins, seams, pockets, and lenses.[19] Additional deposits occur in Guangdong province. China produced 15.5% of the total world production of antimony in 1980.[5] Antimony production exhibits a downward production trend (Figure V-2-3). The main destinations of Chinese elemental antimony in 1980 included Japan, France, and West Germany.[10] Antimony is one commodity of which China exports significant quantities of metals in addition to ores (Figure V-2-4). Generally, China is not an exporter of processed materials.

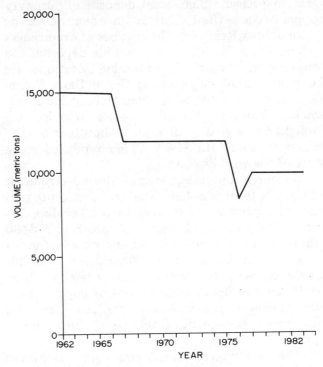

FIGURE V-2-3 China: Estimated antimony production, mine output, metal content, 1962–1981. (From U.S. Bureau of Mines, *Minerals Yearbooks*.)

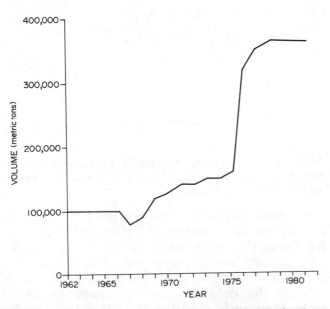

FIGURE V-2-2 China: Estimated aluminum production, refined metal, 1962–1981. (From U.S. Bureau of Mines, *Minerals Yearbooks*.)

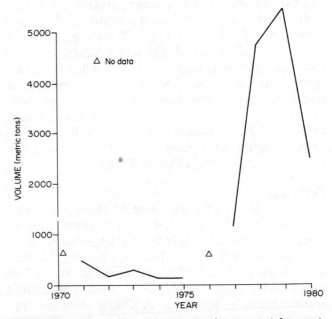

FIGURE V-2-4 China: Apparent antimony metal exports, 1970–1980. (From U.S. Bureau of Mines, *Minerals Yearbooks*.)

Copper

Copper is another metal that the Chinese currently import. Production limitations appear to be due to inadequate mining and processing capacity rather than reserves, as China has significant copper reserves. Based on Chinese accounts, one-third of the reserves occur on the middle and lower Chanjiang River.[2] Significant deposits of porphyry copper occur in the Jiangxi Province and the Xizang Autonomous Region.[20] Other types of occurrences include skarn, nickel-copper, sulfide deposits, and iron-copper. Major copper-mining locations are Tongling, Anhui; Baiyanchang, Gansu; Daye, Hubei; Dexing, Jiangxi; Zhongtiaoshan, Shanxi; and Dongchuan, Yunnan. China's copper reserves are thought to be greater than 50 million tons of contained copper.[10] In 1980, China produced some 2.1% of the world's copper.[5]

Resource diplomacy seems to be a determinant of China's foreign policy. The need for secure supplies of copper seems to have been a deciding factor in the Chinese decision to construct a $400 million railroad connecting mining areas in Zambia to a port in Tanzania.[21,22] Associated with this assistance was an arrangement whereby China could acquire up to 50,000 tons of copper annually. However, political problems have prevented this trade agreement from being fully implemented.[14]

The awareness of China's large copper resources is relatively recent, and efforts to improve domestic mining and processing capacity will replace efforts to secure outside resources. Current constraints included small mining operations, outdated technology, and insufficient numbers of trained personnel. The capacity of China's eight major smelting-refining operations was 290,000 tons per year. In addition, there were five known smaller operations of unspecified capacity.[10] Production trends for copper are shown in Figure V-2-5. Major sources of Chinese metal imports are Zambia, Chile, and Peru. Additionally, the Philippines has been a source of copper concentrates. China imports primarily metal (Figure V-2-6).

Fluorspar

China is a major exporter of fluorspar. China has reserves of some 1.9 million short tons of fluorine.[18] Provinces producing large amounts of fluorspar include Fujian, Hunan, Liaoning, Shanxi, and Zhejiang.[23] In 1980, China produced 8.5% of the world's fluorspar.[5] Due to relatively limited consumption of fluorspar in China, there are substantial amounts left for export.[17] Japan was by far the major importer of Chinese fluorspar in

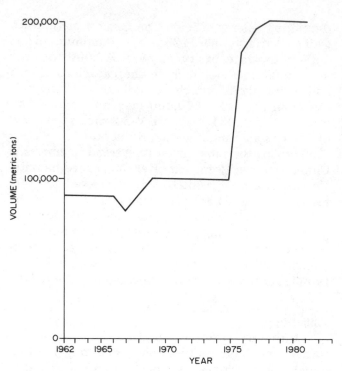

FIGURE V-2-5   China: Estimated copper production, mine output, 1962–1981. (From U.S. Bureau of Mines, *Minerals Yearbooks*.)

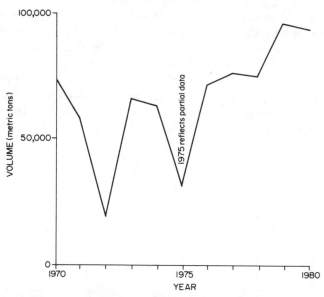

FIGURE V-2-6   China: Apparent copper metal imports, including alloys, all forms, 1970–1980. (From U.S. Bureau of Mines, *Minerals Yearbooks*.)

1980, with West Germany and the United States also significant consumers.[10] China exhibits increasing production of fluorspar (Figure V-2-7). Fluorspar exports are reported with related materials and the undifferentiated information makes it difficult to determine exact trends. In 1980, China apparently exported 359,078 metric tons of fluorspar, feldspar, leucite, and nepheline.[10] This

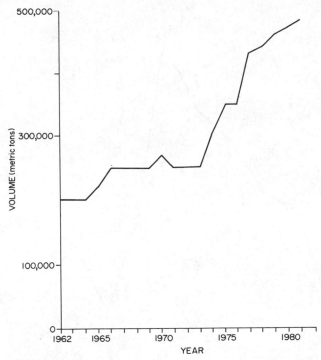

FIGURE V-2-7 China: Estimated fluorspar production, 1962-1981. (From U.S. Bureau of Mines, *Minerals Yearbooks*.)

was over two and a half times the undifferentiated feldspar/fluorspar exports in 1970.

Rare Earths

Rare earths are an example of a resource where the Chinese have a comparative advantage due to the absolute size of the reserve base. China is thought to have larger reserves than the rest of the world combined. China has an estimated 36 million short tons of reserves[11] (rare earth oxide contents) of the world total of 46 million tons. Reserves are concentrated in Nei Mongol. The most significant mine is the Bayan Obo. The Chinese are prepared to develop this resource. There is a Rare Earth Research Institute. Additionally, the Chinese have a research commitment to utilize fully their complex mineral deposits.

Increased export of rare earths is a good prospect as new uses are discovered that take advantage of the unique reactive characteristics.[24,25] Samarium-cobalt magnets, used in electronics, are a recent use of rare earth materials. Future markets include "ceramic cutting tools, oxygen sensors, chemical catalysts, . . . fluorescent phosphors . . . and amorphous intermetallic products."[25]

Tin

Tin is one of the traditional export materials of China. Major areas of tin production are Gejui, Yunnan, and Hechi, Guangxi.[10] Tin occurs in cassiterite. A large proportion of the tin ore is in massive rocks rather than in placers.[20] Regarding tin reserves, the U.S. Bureau of Mines estimates that Chinese tin resources include 1,500,000 metric tons of reserves and 1,000,000 metric tons of other resources. The Chinese resource potential is second only to Indonesia.[18] In 1980, the Chinese produced 7% of the world's total tin.[3] Smelting capacity is 14,600 tons per year.[10] Tin production and exports both have been decreasing since the mid-1970s (Figures V-2-8 and V-2-9). Major Chinese uses include tinplate, solder, brass, and bronze. The possibility that additional tin would be consumed by the steel industry[13] may have been affected by readjustment. Chinese sources attribute slowing down of exports to increasing internal demand.[20] The United States purchased $22,262,960 worth of unwrought unalloyed tin from the Chinese in 1981.[26]

Titanium and Vanadium

Development of titanium and vanadium evidently has a high priority. The Chinese plan to become significant world producers.[22] A primary source of titanium is the titaniferous magnetite deposit at Panzhihua. Titanium is recovered from the slag from iron production. Additionally, ilmenite is mined off the coast of the provinces of Guangdong and Guangxi. Estimated reserves of titanium are 80 million tons.[27] The use of the Panzhihua deposit is indicative of Chinese commitment to improve capabilities to utilize fully complex ores that occur in China. The titaniferous magnetite also yields vanadium. Other complex deposits include the iron ore and rare earths of Baotou, and the nickel sulfide of Jinchuan, Gansu.

The United States was the main consumer of Chinese titanium metal in 1980; Japan was the

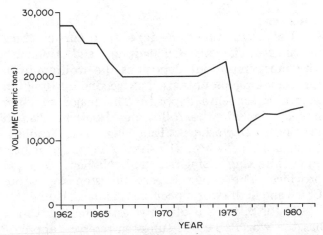

FIGURE V-2-8 China: Estimated tin production, 1962-1981. (From U.S. Bureau of Mines, *Minerals Yearbooks*.)

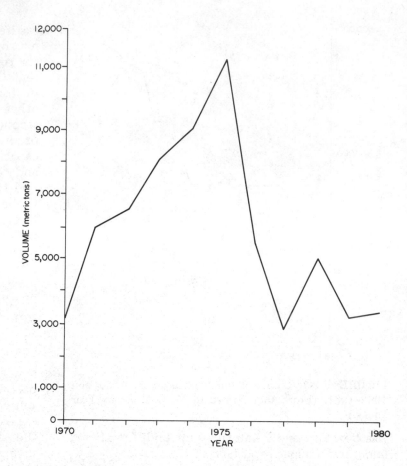

FIGURE V-2-9 China: Apparent tin exports, 1970-1980. (From U.S. Bureau of Mines, *Minerals Yearbooks.*)

main importer of titanium oxide.[10] The dollar value of U.S.-Chinese titanium trade in 1981 was as follows: unwrought titanium sponge, $10,007,270; unwrought titanium, $1,742,236, and titanium waste and scrap, $812,066.[26] The relatively rapid growth of exports is exhibited in Figure V-2-10. The major trend is the move to export titanium metal, as well as titanium oxide. The Chinese produced about 10,000 metric tons of vanadium in 1981 and in 1982.[11]

Tungsten

The main tungsten deposits occur in three provinces: Jiangxi, Guangdong, and Hunan.[10] The most important deposit is the wolframite in the Nan Ling mountains.[23] A secondary tungsten source is scheelite deposits. The major tungsten mines[28] include the following: Huashan, Jiangxi province; Dachishan, Tangjiang, and Yaoling, Guangdong province; Dajishan, Zhejiang province; and Dongxing, Lingma, and Nashan, Guangxi province. The single largest tungsten deposit in China and in the world occurs at Xikuang, Hunan.[28]

The U.S. Bureau of Mines estimates that China has the world's largest tungsten reserves, approxi-

mately 53% of the world's total. This amounts to some 3 billion pounds of tungsten reserves.[18] In 1980, China produced about 28% of the world's total tungsten output[3] (Figure V-2-11). The case of tungsten demonstrates how the unique characteristics of a mineral influence the price and quantity of the commodity. China's share of tungsten reserves worldwide gives it the opportunity to assume leadership in this market. The massive reserves of China have discouraged other would-be producers, who fear China's ability to undercut them in price.[29] In the early 1960s, the relationship between the USSR and China defined the tungsten market.[30] The Soviets were importing Chinese tungsten and exporting their own lower-grade tungsten at a low price. China continued to trade with the USSR and to ship tungsten to the USSR after withdrawal of technical assistance, probably to meet commitments. However, by 1963 China entered the world market, and two years later, the major European customers and the Japanese were dealing directly with the Chinese for their tungsten supplies. This is indicative of the initial emergence of China from close trade ties with the USSR to more of a world trader. Another

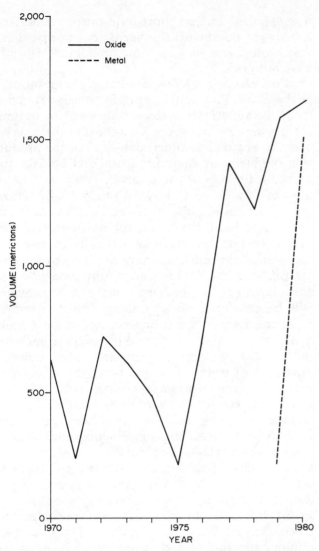

FIGURE V-2-10 China: Apparent titanium exports, 1970–1980. (From U.S. Bureau of Mines, *Minerals Yearbooks.*)

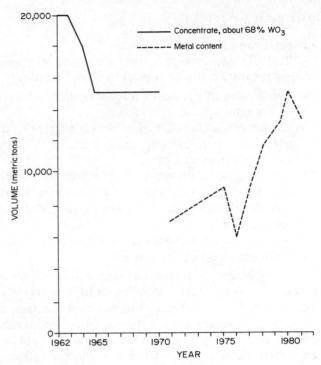

FIGURE V-2-11 China: Estimated tungsten mine production, 1962–1981. (From U.S. Bureau of Mines, *Minerals Yearbooks.*)

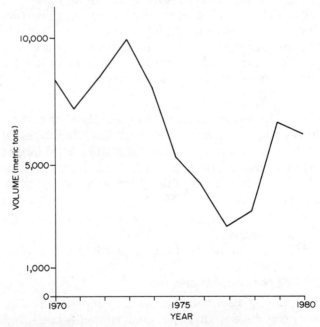

FIGURE V-2-12 China: Apparent tungsten exports, 1970–1980. (From U.S. Bureau of Mines. *Minerals Yearbooks.*)

factor affecting the tungsten market in the 1960s was the U.S. stockpile sales of tungsten.[29]

The question arises as to why the Chinese, with their resource base, do not export more tungsten. This question was examined by C. Suan Tan.[30] One determinant seemed to be the Chinese wish to keep supply down to influence price. Additionally, export decisions are influenced by the anticipated revenue that will be needed to finance import needs. Tungsten has traditionally provided only a small share of this income. Recent trends show increased exports of tungsten (Figure V-2-12). This could reflect an increased need for foreign currency, or an increased role for tungsten as a revenue generator, or both. However, the increase is not

large and less tungsten is being exported than in the early 1970s. Tungsten has shown a general trend of increasing production since 1976. However, production dropped in 1981 (Figure V-2-11).

ENERGY MINERALS

Energy Problems

The Chinese face problems supplying adequate energy resources due to several factors, including:

Declining rates of production for the major sources of energy

Wasted energy as a result of relatively high rates of use due to outdated, small-scale, and other inefficient operations

The importance of energy both as a domestic input and as an export commodity to generate foreign exchange and to meet terms of trade agreements

Expanding uses for energy minerals related to the modernization of the economy

The Chinese response, as reflected in the most recent Five-Year Plan, includes both conservation and production measures. Increased production is necessarily a longer-term process. Quite impressive gains have already been made as a result of energy conservation efforts. This involved closing of some inefficient operations, as well as changing industrial priorities to emphasize less energy intensive light industry. Subsequent improvements in the energy situation as a result of conservation will be more difficult to achieve.[31] Given the falling production and not too promising near-term outlook, the key energy issue will be the trade-offs involved in allocating energy minerals as raw materials for Chinese consumption and for export.

Coal

China has an estimated 600 billion tons of reserves and 1.5 trillion tons of total resources of coal.[23] China therefore has the world's third-largest reserves and is the third-largest producer of coal. Smil[32] characterized the Chinese coal-mining industry as follows as of 1980:

6.3% of operations in major mines completely mechanized

4.6% of total production originates from surface mines

17% of coal output cleaned

Coal mining's contribution to the Chinese economy has been stable. In 1952, coal mining represented 2.4% of the total value of industrial output; in 1979 it was 2.6%.[33]

In recent years, China has made improvements to productive capacity. However, the available investment capital has been limited by the amount devoted to oil. China's historical policy has been to emphasize improvement of existing mines and the opening of rural mines.[34] Current policy revolves around opening large new mines. Previous policy was deficient in that short-term output considerations were dominant; the necessary improvements that would ensure long-term productive capacity were not made.

Coal reserves are overwhelmingly concentrated in the north. Production is not as concentrated due to deliberate efforts to decentralize mining to limit the pressure on transport. One-third of the reserves occur in Shanxi Province;[8] one-sixth of the production originates in Shanxi.[10] Based on CIA data, the north, with 70% of the reserves, produces 28% of the coal. Southern China, with only 1% of the reserves, produces 10% of the country's coal.[35] Besides being more abundant, the northern coals are of higher quality. Previous attempts to invest in southern coal production have not been overly successful, and the Chinese now fully recognize the advantages of the northern coals. Shanxi province will be developed in accordance with its reserve position. Transport can be expected to be a problem. The mountainous terrain limits track improvements that would accommodate unit trains.[36] Shortages of water in Shanxi make it unlikely that coal slurry pipelines can be constructed.[37] The following are the major Chinese coal mines: Datong, Shanxi; Fengfeng and Kailuan, Hebei; Benxi and Fushun, Liaoning; Hegang, Heilongjiang; Weibei, Shanxi; the northern slope of the Tianshan mountains, Xinjing; Pingdingshan, Henan; and Huainan and Huaibei, Anhui.[10] Underground mining is predominant, and the long-wall technique is most prevalent. Not all of the long-wall faces are completely mechanized.[28] China has shown interest in surface mining, and the United States would be expected to benefit in terms of equipment sales opportunities. Concerns about effects of strip mining in China are expressed due to possible loss of productive agricultural land.[2]

China has often pursued a policy of smaller-scale, locally oriented plants, in addition to the regular, more conventional industrial units. The small coal mines are such an operation. There are over 20,000 such mines, and they provide about 44% of China's coal production.[38] These numbers fluctuate quite widely. Large numbers were opened as part of the Great Leap Forward and the Cultural Revolution. The numbers ranged from 110,000 such mines in 1958 to 15,000 in 1959.[35] The recent trend has been to increase production from existing small mines rather than to open new ones. The disadvantages of the small mines are obvious. They are inefficient operations and produce lower-quality coal. Less sophisticated extraction methods could result in much waste. However, small mines do provide certain advantages:

The output of smaller mines can be used for meeting local needs, leaving the output of the larger more efficient operations for export or for use in other provinces. This is the case in the Shanxi province.[35]

Local mining for local demand reduces transportation demand.

Rural mines provide seasonal employment opportunities for agricultural workers.[21]

The local mines lessen the overall demand on central planning and managerial expertise.[34]

The ability to very rapidly increase the number of small mines, and thereby increase output, means that these mines provide an important buffer in meeting demands.[31]

Chinese coal production is labor intensive and therefore quite susceptible to political disruption.[21] Political instability in China has always negatively affected coal production (Figure V-2-13). An impact of increased mechanization would be to reduce this tendency. Coal mining is plagued by labor problems, including difficulties in recruiting workers and high absenteeism. Part of the problem could be attributed to working conditions that are quite unsafe by Western standards.[37]

Coal exports have increased quite substantially,

from 344,353 metric tons in 1971 to 2,795,000 tons in 1980. Japan has been the main recipient of Chinese coals (Figure V-2-14).

Petroleum

The problems of Chinese oil development are very similar to other Chinese resource issues. Some of the characteristics of this resource are as follows

Production is geographically concentrated.

Previous investment and development overemphasized increasing output to meet production targets. As a result, insufficient attention was devoted to exploration. Reserves were not maintained at a level that could support previous rates of rapid growth in production.

Exploitation of some oil resources is made difficult due to the distance from markets, geographical constraints, and lack of infrastructure in place.

The Chinese, encouraged perhaps by outside speculation, have tended to overstate their capability to make rapid improvements in the oil industry. It is unclear whether this represents an erroneous understanding of the situation, or an example of strategic behavior on the part of the Chinese to deceive outsiders as to the true

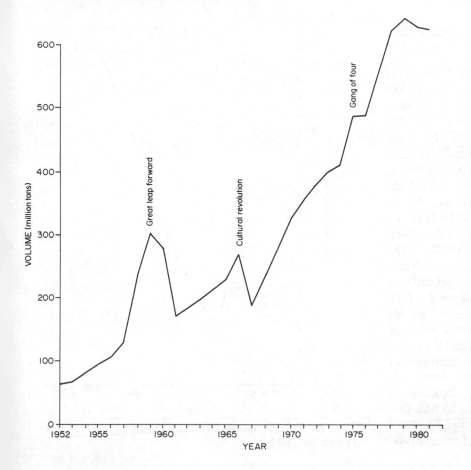

FIGURE V-2-13   China: Coal production, 1952–1981. (From Ref. 40.)

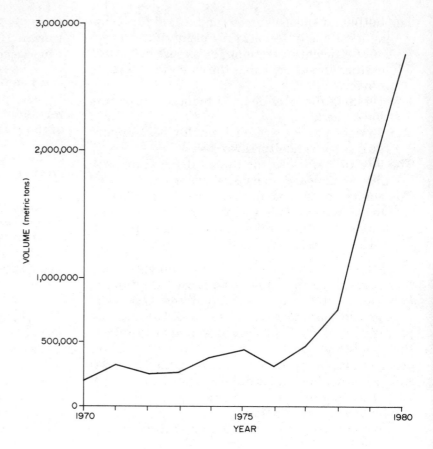

FIGURE V-2-14 China: Apparent coal exports, 1970–1980. (From U.S. Bureau of Mines. *Minerals Yearbooks.*)

potential, and the problems in developing their oil.

In 1981, China was the sixth-largest producer in the world, producing some 3.6% of the world's total output of oil.[39] Growth in Chinese oil production has been impressive. In 1949 China produced 0.121 million tons of crude oil; by 1981 China produced 101.220 million tons[40] (Figure V-2-15). China was forced to develop oil capability following the loss of imports from the USSR in the 1960s. Another indication of the growth is the share of gross value of oil in relation to total industrial output. Oil represented 0.5% of the value in 1952 and 5.4% of the output in 1979.[33] The problem is that this growth has stopped. From 1970 to 1975, the growth rate of oil production was 19.4%; however, between 1980 and 1981, there was a 5.7% decline in production.[41] This is a serious problem due to growing internal demand, the need to export oil, and the possibility of uses of crude oil other than for energy requirements. China had begun the transition from energy use based primarily on coal to increasing dependence on oil. However, the Chinese are trying to halt this process.[32] The Chinese hoped to completely phase out all oil-burning operations by 1985, for example.

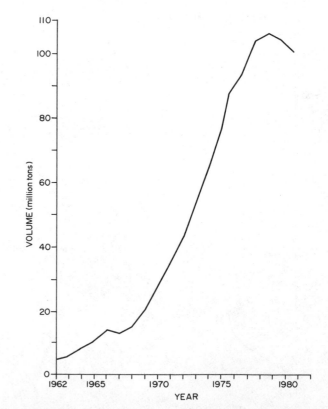

FIGURE V-2-15 China: Crude oil production, 1962–1981. (From Ref. 40.)

Oil burning was reduced 50% between 1978 and 1980.[31] Given the importance of oil as an export earner, an observer has described burning oil in China as equivalent to burning foreign exchange.[14]

Another problem is that oil is a source of materials for fertilizers[32] and the export-oriented textile industry.[32,42] The following are the major oil-producing areas in China and their approximate yearly production figures: Daqing, 365 million barrels; Shengli, 131 million barrels; Jizhong, 117 million barrels; Sichuan, 73 million barrels; Liaohe, 37 million barrels; and Karamay, 29 million barrels.[10] China's problem in oil production is that the largest fields show declining rates of production. Oil exports became an important factor following the 1973 oil embargo. Exports grew rapidly through 1979. However, exports declined in 1980, reflecting the Chinese oil production problems (Figure V-2-16).

Security of supply is an important consideration in oil production. China's oil fields, concentrated as they are in the northeast, would be easily accessible to large numbers of Soviet troops in the event of a war between these two countries. Additionally, offshore operations would be subject to disruption by the Soviet navy.[41] Another problem is that there is considerable disagreement over the extent of territorial claims of China and neighboring countries in the offshore waters, including the South China Sea and the Yellow Sea.[43] The possibility for conflict over these waters cannot be discounted. These factors create additional uncertainty for potential outside investors.

The Chinese have developed small operations in coal, steel, fertilizers, and machine building. However, oil production requires too much capital and technical expertise to allow such small-scale operations. As a result of not being as labor intensive as coal, oil production is less susceptible to political disruptions. With the exception of recent declines, which are related to economic rather than political influences, the only previous decline in oil production was associated with the Cultural Revolution.

POTENTIAL FOR INCREASED MINERAL EXPORTS

Given a strong reserve position for a number of minerals, the Chinese should be able to increase exports if the necessary investment and policy commitments are made. Tin, antimony, and tungsten exports obviously could be increased. However, the

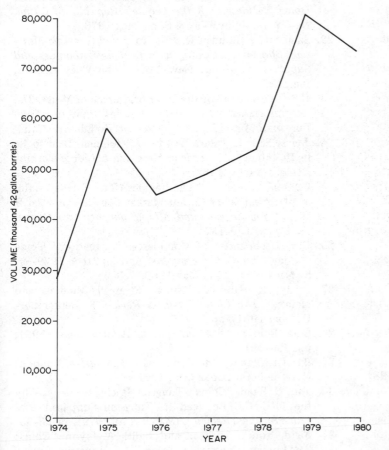

FIGURE V-2-16 China: Apparent oil exports, 1974–1980. (From U.S. Bureau of Mines. *Minerals Yearbooks.*)

first two are limited by competition, while tungsten exports are apparently limited by Chinese policy. China has an advantageous reserve position in coal and adjacent markets in the rapidly growing Pacific Basin countries. However, mining and infrastructure problems, and North American and Australian competition, could prevent the Chinese from realizing this potential. The future of oil exports is uncertain due to slower-than-anticipated offshore development, and growing internal demand. The Chinese appear committed to titanium and rare earth development, and the market for these exports seems promising due to aerospace and electronics applications. Some of China's future commodity exports may be unexpected. China rapidly became an exporter of germanium in response to price increases in 1981. Additionally, China supplied significant amounts of chromite to the world market in 1980. The previous assessment was that China was deficient in chromium reserves.[27] Quite simply, at this stage it is impossible to predict completely Chinese production potential and supply trends. This should improve markedly if the Chinese continue the trend of more complete data reporting.

REFERENCES

1. Barnett, A. Doak. *China's Economy in Global Perspective*. Washington, D.C.: The Brookings Institution, 1981.
2. Copper, John F. *China's Global Role*. Stanford, Calif.: Hoover Institution Press, 1980.
3. Schmidt, Helmut, and Manfred Kruszona. *Regional Distribution of Mining Production and Reserves of Mineral Commodities in the World*. Hannover, West Germany: Federal Institute for Geosciences and Natural Resources, January 1982.
4. Hsu Chieh. "China's Mineral Resources." *China Reconstructs* 30 (April 1981), pp. 62-64.
5. Odingo, R. S. "New Perspectives on Natural Resources Development in Developing Countries." *GeoJournal* 5 (1981), pp. 521-530.
6. *Xinhua*, 9 December 1982. Foreign Broadcast Information Service (FBIS). *China Daily Report*, 9 December 1982, p. K26.
7. Wang, K. P. "China." *Mining Annual Review*, 1980, pp. 435-456.
8. Japanese External Trade Organization. *China: A Business Guide*. Tokyo: Japanese External Trade Organization and Press International Ltd., 1979.
9. "Sleeping Giant, but Not in Mining." *The Northern Miner*, 16 September 1982, p. 2.
10. U.S. Department of the Interior, Bureau of Mines. *The Mineral Industry of China*. Preprint from the 1981 Bureau of Mines *Minerals Yearbook*. By Edmond Chin.
11. U.S. Department of the Interior, Bureau of Mines. *Mineral Commodity Summaries 1983*. Washington, D.C.: U.S. Government Printing Office, 1983.
12. U.S. Central Intelligence Agency. *Research Aid: China's Minerals and Metals Position in the World Market*. ER 76-10150. March 1976.
13. Wang, K. P. "Mineral Industries of China." U.S. Department of Interior, Bureau of Mines. Reprinted from *Mining Annual Review 1979*.
14. Cox, Madeline. *China Economic Overview: A British Columbia Perspective*. Province of British Columbia, Ministry of Industry and Small Business Development, no date.
15. "Chinese Envisage Big Project for Production of Aluminum." *The Japanese Economic Journal*, 25 January 1983, p. 7.
16. "China, A Major U.S. Supplier of Calcined Refractory Bauxite." *Engineering and Mining Journal* 183 (October 1982, 158.
17. U.S. Department of the Interior, Bureau of Mines. *The People's Republic of China: A New Industrial Power with a Strong Mineral Base*, by K. P. Wang. 1975.
18. U.S. Department of the Interior, Bureau of Mines. *Mineral Facts and Problems, 1980 Edition*.
19. Jensen, Mead L., and Alan M. Bateman, *Economic Mineral Deposits*, 3rd ed. New York: John Wiley & Sons, Inc., 1981.
20. Zhong Hongren. "Mineral Exploration and Development in China." Paper presented at a workshop on Mineral Policies to Achieve Development Objectives, East-West Center, Honolulu, 9-13 June 1980.
21. Szuprowicz, Bohdan O., and Maria R. Szuprowicz. *Doing Business with the People's Republic of China*. New York: John Wiley & Sons, Inc., 1978.
22. Szuprowicz, Bohdan O. *How to Avoid Strategic Materials Shortages: Dealing with Cartels, Embargoes and Supply Disruptions*. New York: John Wiley & Sons, Inc., 1981.
23. U.S. Department of the Interior, Bureau of Mines. *The Mineral Industry of China*. Preprint from the 1980 Bureau of Mines' *Mineral Yearbook*. By Edmond Chin.
24. Jones, Sam L. "Rare Earths' Development Said to Be in Its Infancy." *American Metal Market Metalworking News*, 8 August 1982, p. 29.
25. Cannon, Joseph G. "Lanthanides (Rare Earths): Shipment Lower in 1982, but Market Growth Potential Is High." *Engineering and Mining Journal* 189 (March 1983), pp. 136-137.
26. U.S. Department of Commerce, International Trade Agency. *China's Economy and Foreign Trade 1979-81*, by Nai-Ruenn Chen, May 1982.
27. Brady, E. Sabrina. "China's Strategic Minerals and Metals." *The China Business Review* 8 (September-October 1981), pp. 55-69.
28. Chin, Edmond, "China." *Mining Annual Review* 1981, pp. 385-391.
29. Barbier, Claude. *The Economics of Tungsten*. London: Metal Bulletin Books Ltd., 1971.
30. Tan, C. Suan. "China's Tungsten Pricing Strategy: Why the PRC Has Reduced Its Tungsten Supplies." *The China Business Review* 6 (March-April 1979), pp. 28-33.
31. Field, Robert Michael, and Judith A. Flynn. "China: An Energy Constrained Model of Industrial Perfor-

mance through 1985." In *China under the Four Modernizations*, Part 1. Joint Economic Committee, U.S. Congress, 13 August 1982, pp. 334-364.

32. Smil, Vaclav. "Energy Development in China: The Need for a Coherent Policy." *Energy Policy* 9 (June 1981), pp. 113-126.

33. Field, Robert Michael. "Growth and Structural Changes in Chinese Industry: 1952-79." In *China under the Four Modernizations*, Part 1. Joint Economic Committee, U.S. Congress, 13 August 1982, pp. 303-333.

34. U.S. Central Intelligence Agency, National Foreign Assessment Center. *Chinese Coal Industry: Prospects over the Next Decade*, February 1979.

35. Taylor, Robert P. *Rural Energy Development in China.* Washington, D.C.: Resources for the Future, 1981.

36. Jones, Dori. "Breaking the Bottlenecks." *The China Business Review* 7 (May-June 1980), pp. 47-48.

37. Weil, Martin. "China's Troubled Coal Sector." *The China Business Review* 9 (March-April 1982), pp. 23-34.

38. "Developments in China Signal Serious Efforts for Coal." *World Coal* 8 (November-December 1982), pp. 92-93.

39. American Petroleum Institute. *Basic Petroleum Data Book, Petroleum Industry Statistics, Volume III, No. 1.* Washington, D.C.: API, January 1983.

40. Kraus, Willy. *Economic Development and Social Change in the People's Republic of China.* Translated by E. M. Holz. New York: Springer-Verlag New York, Inc., 1982.

41. Woodard, Kim. "China and Offshore Energy." *Problems of Communism* 30 (November-December 1981), pp. 32-45.

42. Yuan, Paul C. "China's Energy Image Slightly Distorted." *Offshore* 41 (September 1981), pp. 97-99.

43. Earney, Fillmore C. F. "China's Offshore Petroleum Frontiers: Confrontation? Conflict? Cooperation?" *Resources Policy* 7 (June 1981), pp. 118-128.

Chapter V-3

Energy and Mineral Resources Development: Chinese Efforts and Foreign Assistance

China has departed from a policy of self-sufficiency, producing primarily for internal consumption, and has begun seeking outside assistance to boost production of selected energy and mineral resources for export and for domestic consumption. China proposes to develop mineral resources that are of high value, or would result in import substitution, or have export potential.[1] From the Western perspective, Chinese resource development projects could provide a market for extractive technology, as well as providing additional raw materials.

The formidable constraints to resource development in China have been noted. As Figure V-3-1 indicates, the sheer size of the country is one such constraint. Resource development is characterized by long lead times, involving lengthy feasibility studies and large-scale physical improvements. At this point in its development, China apparently does not have the capability to be a large exporter. Present actions will determine whether China will have such a capability in the future. One area worthy of investigation is that of Chinese plans and policies for resource and infrastructure development. The most recent Chinese economic development plan is analyzed here, and specific resource development policies are described.

A second area of interest is China's increasing reliance on assistance from foreign entities. Selected examples of outside assistance are described. As a prerequisite to attracting foreign involvement, the Chinese are making significant changes in their legal system. Additionally, the Chinese are promot-

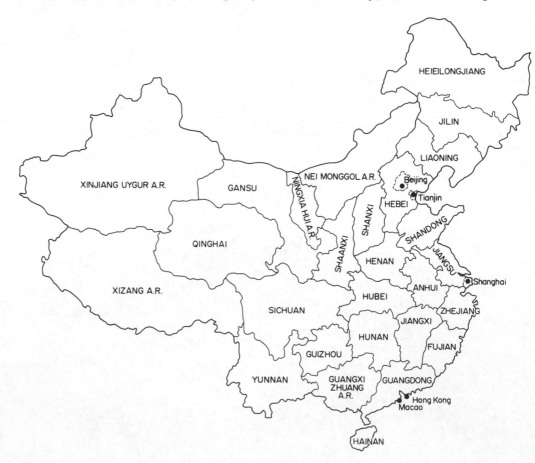

FIGURE V-3-1 The People's Republic of China.

ing use of those institutional arrangements most in line with their strengths. The Baoshan steel plant is examined as a planning failure, indicative of the problems of importing whole-plant technology in the metals processing section. Japan is the country most involved in recent Chinese resource development efforts, and the Sino-Japanese resource agreements are summarized. Coal development is crucial to overall economic development due to the linkages to power and transport, and Chinese policies and foreign assistance are described.

The role of outside engineering and planning expertise, in addition to technology import, is discussed in the hard minerals sector, specifically copper. Offshore oil has attracted the most interest from foreign investors. Additionally, it provides the best single case of Chinese attempts to capture maximum benefits from outside involvement. Finally, an account of the role of international agencies is provided. Although the total amount of financial aid is not large in relation to China's massive needs, it is strategically concentrated, and accompanied by technical assistance, perhaps more important than the funding.

WHY THE CHINESE WISH TO EXPORT RESOURCES

Resource exports have been viewed unfavorably recently as a source of revenue for developing countries. There are a number of reasons for this. Resource markets are sensitive to economic conditions and therefore subject to significant fluctuations in prices. A stable economic development policy does not follow from an uncertain revenue expectation. Even if minerals are processed, the value added that can be captured by exporting resources is generally not as great as that which can be realized through the export of manufactured goods. While the energy market has been favorable for exporters, following an upward trend until recently, the metals and minerals market has often been plagued by both relatively low prices and fluctuating market conditions.

Despite these observed problems, the Chinese have expressed interest in increasing exports of minerals, including both energy minerals and nonferrous metals.[2-4] There are a number of possible explanations for this policy:

1. Due to previous failures to develop its mineral base fully, China probably enjoys comparative advantages in terms of grades of remaining mineral deposits. This is, of course, conjectural, due to lack of detailed information. Efficient resource development will be feasible as China continues to gain access to technology. The disadvantage of the current level of development is that necessary infrastructure is not in place.

2. Other goods that China wishes to export face problems. For example, textiles are subject to increasing protectionism abroad, as well as more competition among labor-intensive producers. Other Chinese products suffer in international markets due to poor quality and marketing.

3. Given the Chinese need for currency, all possible sources, including raw materials, will be exploited.

4. Natural resource development, through the mechanisms of compensation trade or joint ventures, can attract outside sources of funding, expertise, and technology that the Chinese need. In addition, mineral resources are widely distributed in China (Figure V-3-2), allowing for regional economic development through this exploitation.

THE 1981–1985 PLAN

Perhaps the best way to understand the current Chinese economic development strategy regarding resources and related infrastructure is to analyze the recent economic plan, covering the period from 1981 through 1985. The following analysis is based on excerpts of the plan presented in *Renmin Ribao* (*People's Daily*).[5] In contrast with the reluctance or refusal of China to release basic statistics, this plan provides a detailed look at the Chinese approach to modernization. The size, stage of development, and population of China causes large-scale needs across the board. The plan places some parameters on what the leadership feels can realistically be achieved within the existing constraints. The plan provides budgetary and investment commitments, as well as targets for increases in capacity. The budget is perhaps the best overall indication of the priority assigned to the various problem areas.

The previous attempt to make improvements too rapidly, with a preponderance of capital-intensive programming, is an indication of the difficulties the Chinese face in attempting to improve their economic situation. The result was a retreat (readjustment) from the plan soon after it was announced. Although the Chinese apparently have learned a great deal about their capabilities as a result of previous mistakes, it is difficult to prejudge the likely success of the current planning effort.

Proposed Investment Targets

A particularly pressing problem is that there is simply not sufficient capital to address all the needs. Accordingly, the plan emphasizes the impor-

FIGURE V-3-2   Geographic distribution of mineral deposits in China.

Legend:
- ■ Coal
- ✳ Petroleum
- ✿ Natural gas
- ▲ Iron
- ▨ Manganese
- ▣ Chromium
- ○ Copper
- ⬗ Bauxite
- ◐ Lead zinc
- ◑ Gold
- □ Wolfram
- ◆ Antimony
- ⊖ Tin
- ⊘ Molybdenum
- △ Nickel
- ◈ Magnesium
- ⊖ Mercury
- ● Phosphorus
- ◪ Asbestos
- ▲ Mica
- ◈ Fluorspar

tance of realizing improved return on investment funds. Chinese development has been hampered in the past by many poorly conceived projects, including those that duplicate existing facilities and locally developed projects inconsistent with national objectives. The plan stresses the need for more effective utilization of investment funds from project inception to project management phase, to conclusion, and final inspection.

Emphasis on adequate feasibility studies and on planning authority review could have mixed results. On the one hand, decision making ideally would be more rational. However, lead times, already long for infrastructure and resource development projects, can be expected to increase.

China's budgetary commitment for investment in new capacity is 230 billion yuan over the planning period. Table V-3-1 documents investment targets for resource and infrastructure improvements. The allocation for modernization of existing equipment ("updating equipment and technical transformation") is some 130 billion yuan. The share of investment that is anticipated to be secured from foreign loans is 26.2 billion yuan (1.9809 yuan = 1 U.S. dollar, May 1983).[6]

TABLE V-3-1   Sixth Five-Year Plan, People's Republic of China, 1981–1985 (Capital Construction Funds, Energy/Minerals/Infrastructure)

Sector	Budget (billion yuan)	Percent total investment
Fuel and power		
Coal	17.93	7.8
Petroleum	15.47	6.7
Power	20.73	9.0
Energy saving	4.50	1.9
	58.63	25.4
Railways	17.29	7.5
Metallurgical	17.51	7.6
Geological exploration	1.49	0.6

Source: Ref. 5.

Proposed Improvements in the Minerals Sector

The Chinese apparently have large expectations for the minerals sector. However, only limited insight into how they will achieve their objectives to increase exports is provided by analyzing the plan. The target for growth in output of the 10 nonferrous metals over the planning period is 12.7%. The metals included in this category are copper, aluminum, lead, zinc, nickel, tin, antimony, mercury, magnesium, and titanium. To achieve this, new capacity in the metallurgical industry is expected to be funded by 17.51 billion yuan, about 7.6% of the total expenditures in the investment category. One of the objectives of the plan is to promote the ability to utilize complex ores occurring at Jinchuan, Panzhihua, and Baotou. The commitment to export mineral commodities is also reflected in the plan—specifically, it is noted that "it is necessary to organize the export of nonferrous metals, rare metals, nonmetallic mineral products ... which are plentiful in China or have great production potential."[5]

Energy Supply Improvements

Energy supply is identified in the plan as one of the weak links in the economy. The commitment to this program is impressive. Some 25% of the investment for new construction over the planning period will be devoted to energy-related development. All phases of energy are addressed—exploration, exploitation, and conservation. Energy is linked quite closely to other Chinese infrastructure and economic development issues. For example, energy availability influences locational decisions. Thermal power plants will be located in coal-rich regions, and energy-intensive industries (metallurgy is one specifically mentioned) will be located in areas where some of China's massive hydroelectric potential can be tapped. Additionally, since so much of the coal is moved by rail, energy considerations also dominate transport. Energy emphasis extends to such areas as the need to construct improved machinery to exploit available resources, including coal and offshore oil, as well as better instrumentation to promote such disparate objectives as offshore exploration, and energy conservation.

Coal production is projected to increase by 80 million tons over the five-year period. Half of the growth will occur in the Shanxi province. The plan also projects output targets for the subsequent five-year period. The importance of Shanxi province is again emphasized, as over the two planning periods the cumulative share of total new output to be located in Shanxi is approximately one-third. Shanxi's reserves, existing infrastructure, and position as the number one producer of coal allow it to be developed more rapidly than other provinces. Some strains on this region should be anticipated as a result of this emphasis. Although the trend will slow down considerably over the next five years, there will be a disproportionate development of this province. Specific targets are set for the success of coal exploration—it is anticipated that some 48

billion tons will be found during the planning period.

Railways improvement objectives illustrate the linkage of coal and railroads in China. For example, railway improvements are critical to the ability of three provinces, including Shanxi, to increase export capability by about 48 million tons. Railroads in southern China will increase the export capabilities by a much smaller amount. Railway improvements will include construction/addition of new capacity and the upgrading of the existing system. Anticipated capacity increases are shown in Table V-3-2.

INSTITUTIONAL ARRANGEMENTS FOR FOREIGN INVOLVEMENT

Institutional arrangements for foreign involvement in Chinese resource development must be sufficiently attractive to outsiders, while meeting the financial and technical acquisition needs for the Chinese. Additionally, they must be sufficiently flexible to allow accommodation of the quite dissimilar Chinese bureaucratic and outside corporate organizational structures. The primitive stage of Chinese commercial law development is a particularly important constraint when attempting to attract outside investment.

Pattison points to an "absence of a coherent, unified body of law . . ."[7] as the major hindrance to outside investment in China. The Chinese have set out to make deliberate improvements. The new constitution provides for protection of foreign

TABLE V-3-2 Sixth Five-Year Plan of the PRC: Capacity Increases, Energy/Minerals/Infrastructure

Coal
101 new pits
80 million tons new capacity
Power
37 new power stations
Additional 12.9 million kilowatts[a]
Harbors
54 deep-water berths
Additional 100 million tons capacity
Railways
2067 km new tracks
2511 km electrified
1689 km double-tracked
Metallurgical industry
First stage Baoshan steel construction
Add following production capacity by 1985
3 million tons iron
3 million tons steel
Petroleum industry: 35 million tons capacity added

[a]Includes new plants plus completion of work in progress.
Source: Ref. 5.

investment, and legislation has been passed addressing taxation, foreign exchange, joint ventures, offshore oil, and settlement of disputes.[8] However, there are still gaps in the overall legal framework, an insufficient supply of lawyers, and a lack of experience and precedent in implementing the laws.

The Chinese wish to use certain techniques that will allow them to utilize a full range of technical and material contributions, as well as financial assistance, in their dealings with foreign corporations. Various organizational structures and trade relationships have been developed to achieve these objectives. One is outright trade of Chinese goods for foreign goods, or barter trade. In an alternative form of compensation trade, the Chinese acquire technology from a foreign supplier in exchange for a portion of the production provided free or at a discount. Next to direct investment, some observers feel that compensation trade is the riskiest type of venture in China.[9]

Joint ventures are an approach that also appeals to the Chinese. In such a venture important Chinese contributions could include land, infrastructure, and labor. The foreign participant could provide technology, management, marketing assistance, and additional capital.[10] The key components of a joint venture are pooled assets, joint management, and shared profit and losses.[7] Taxation of foreign enterprises will be set at 20 to 40%, dependent on income, with a 10% local tax in addition.[11]

The implications for Chinese financing should be examined. Different types of financing generate different types of returns. For example, infrastructure generates minimal or no revenues. Export projects generate foreign exchange.[12] A problem with Chinese resource development is that generally, improvements to both the resource producing sector and infrastructure must be made for a feasible project. This increases the overall scope of these projects, with implications such as a longer time to complete the project and a higher project cost. Under compensation trade, the higher price tag will result in a larger amount of the output that must be provided as reimbursement to the foreign partner. The key point, then, is that the techniques that minimize pressure on China's finances can increase pressure on resource production and infrastructure. The size of natural resource projects involving foreign interests introduces complexities related to large capital expenditures, the need for detailed planning studies, and consequently difficult project negotiations.[13] While such arrangements can be expected to have major impacts on the Chinese economy, quick results cannot be realistically anticipated.

The response of some of the foreign companies to compensation trade should be noted. There is the problem of what the company that gets a share of goods under the compensation trade deal will do to turn this into cash. These difficulties have caused certain West German firms reluctance to enter into such agreements.[14] Additionally, such approaches may discourage U.S. involvement, as U.S. firms are seen as being most interested in entry into the China market, rather than producing for export.[15]

BAOSHAN: TECHNOLOGY IMPORT FOR METALS PROCESSING

The Baoshan steel mill project provides an example of the range of problems that the Chinese face in large-scale imports of technology in general, and in the metals processing sector specifically. Baoshan has proven to be the realization of a worst-case scenario—an ill-fated component of an ill-fated plan. The intention was to construct a modern, up-to-date 6-million ton per year facility, to be constructed in two phases, that would allow substitution of Chinese production for imported production. Steel imports, specifically finished products, have been the largest single import by the Chinese in recent years. The description of the problems of Baoshan is based on a detailed account of this project by Weil.[16]

The following is a partial list of some of the concerns that were expressed about the project:

Baoshan was a power-intensive project in an energy-constrained economy.

A poor site location increased costs, as the low-lying area required pilings for the foundation.

The site was unsuitable for a port, so one had to be constructed nearby.

A firm source of coal was not established until late in the project, despite expected large-scale consumption.

The project was politically vulnerable because it was dependent on imported technology and raw materials.

Possible environmental problems were associated with the project.

A few of the major milestones in the project are noted in Table V-3-3. The project was politically controversial from the beginning. Not long after its initiation, there was a recognition that the Chinese had overextended their capabilities, not only with this project but in the entire plan, and a period of readjustment began. Heavy industry, import of entire plants, and steel were no longer emphasized. Cancellation of phase II brought more controversy than China had anticipated. Contractors demanded

TABLE V-3-3   Baoshan Steel Plant: Major Events

Included in economic plan, 1978, largest complete plant
  import project
Nipon Steel–China National Technical Import Corporation
  sign contract, 1978
Contract suspended February 1979, financial discussions
New agreement concluded, June 1979
Phase II suspended, November 1979
Phase II terminated, contracts suspended January 1980

Source: Ref. 16.

payments to cover costs of work already begun.
Japan was particularly hard hit. China canceled
projects with Japanese manufacturers, then was
forced to request loans to cover some of the costs
associated with the cancellations. The impression
was that China was not a reliable trade partner—not
the image China wished to cultivate with Japan or
the world at large.

Current thought is that the plant was ill conceived from the beginning and should never have
been constructed. Weil[16] has noted the political as
well as economic complexities involved. Baoshan is
indicative not only of poor planning, but covers
issues such as the role of foreign interests in China
and the periodic policy reversals that occur. Weil
has also noted, however, that there has been a positive side of this controversy. A better understanding
of the role of technology in modernization and emphasis on better feasibility analysis are some of the
expected results. Additionally, in the future, emphasis will probably be on selective imports rather
than on whole-plant imports.

Imports of whole-plant technology in the nonferrous metals processing sector have been more
modest than the massive imports for improving the
ferrous industry. Additionally, more attention has
been devoted to those areas where China's processing capability is lacking, indicating at least an initial
effort at import substitution rather than export.
From 1974 to 1979, China contracted for $130
million worth of nonferrous whole-plant imports
compared with $3881 million worth of ferrous
whole-plant imports. Due to readjustment, some of
these contracts were suspended.[17]

Specific examples of nonferrous processing imports included an aluminum smelter construction
by the Japanese at Guiyang, Guizhou, and a copper
smelter to be built by the Japanese at Guixi,
Jiangxi.[11] The results of technology imports have
often been disappointing. Whole-plant imports
have encountered problems of delays in plant completion, operations at less than full capacity, and
poor return on investment.[17]

JAPANESE-CHINESE RESOURCE DEVELOPMENT

The trade relationship between Japan and China
deserves special mention. Japan is China's largest
trade partner, capturing almost one-fourth of the
total trade in 1981.[18] In 1980, Japan was one of
the most significant importers of 41 of China's
mineral products[11] (Table V-3-4). However, many
of these imports still represent small quantities.
Japan has exhibited a particularly strong interest
in Chinese energy resources. For example, in 1979,
slightly less than half of Chinese oil exports were
shipped to Japan.[19] As part of the policy of assuring a stable source of supply, Japan has contributed
to Chinese resource development ventures.

There are a number of reasons why Japan and
China would be active trade partners. These include
geographical proximity, a common cultural heritage, and the historical relationship between the
two countries. From a trade standpoint, the import needs of the two countries are complementary.
China has a strong resource base, but needs technology. Japan provides a market for the raw materials and is a source of Chinese technological needs.

Despite these advantages, Japan is in some
ways reluctant to invest heavily in China. Perceived
problems include a potentially unstable economic
policy, as well as the problems of the Chinese
bureaucracy, legal system, and infrastructure.[15]
This concern notwithstanding, Japan and China
negotiated a long-term trade agreement in 1978.
Under the agreement, China is to provide coal,
coking coal, and petroleum to Japan. Japan is to
export technology, plants, construction materials,
and machinery to China. The objective of the agree-

TABLE V-3-4   Selected Japanese Mineral Imports
from China

	Quantity to Japan[a]	Percent of total Chinese exports
Bauxite	38,834	9
Antimony, elemental	1,432	58
Chromite	2,041	99
Manganese ore/concentrates	21,290	84
Niobium/columbium	14	100
Titanium oxides	483	29
Feldspar/fluorspar/leucite/ nepheline	225,707	63
Anthracite/bituminous coal	2,071	74
Crude oil	57,530	79

[a]All quantities expressed in metric tons, except coal (thousand tons) and oil (thousand 42-gallon barrels).

Source: Ref. 11.

ment is that the partners keep trade in balance.[20] However, China has had problems in meeting its end of the agreement in terms of export targets.

In addition to being a source of technology, Japan provides an important source of capital for the Chinese. A $1.5 billion loan from the Overseas Economic Cooperation Fund was provided for six infrastructure projects, related primarily to coal exports. Subsequently, some of the funds were diverted to the Baoshan steel mill and petrochemical projects, and $2 billion credit was provided for coal and oil development through the Japanese Ex-Im Bank.[21]

The infrastructure projects financed by Japan included the following:[20,22]

Railway construction, Yanzhou-Shijiusuo
Railway expansion/electrification, Beijing-Qinhuangdao
Port construction, Shijiusuo
Port expansion, Qinhuangdao
Power plant construction, Wuqiangxi
Railway expansion, Hengyang to Guangzhou

One of the most significant of these projects is the Yanzhou-to-Shijiusuo railway, which will allow use of unit trains. These projects were to be completed by 1985, but appear to be behind schedule.[23]

Despite the difference in import composition, China and Japan exhibit convergence in their policy toward diversification of imports. For example, the Japanese have stated that they would like to import 25% of their coal from each of four sources, including China. The Chinese are reluctant to become overly dependent on a single source for technology and technical assistance. This stems from disruptions to the Chinese economy when the Soviets withdrew their assistance in the 1960s. China could be expected to benefit from the policy of diversification, in being able to shop for the best technology for her needs. However, there are some limitations. Use of diverse technology sources minimizes the ability to standardize parts replacement and training programs. Meanwhile, as long as specifications are met, Japan can look to a variety of sources for raw materials. Perhaps the biggest problem for China is that Japan's commitment to diversification effectively limits the amount of Japanese involvement in and assistance to the Chinese economy.

COAL DEVELOPMENT

China's current energy consumption is 70% dependent on coal, and this dependence is expected to continue as oil production increases are diverted to export markets and industry and power reliance on coal increases. China is attempting to modernize older mines and to construct new mines that use current technology. China anticipates that outside assistance will play an important role in this process as a source of investment funds, for mining technology, and for mine planning assistance.

Chinese Policies and Plans

Energy and coal investment targets and anticipated output increases have been described previously in the discussion of the 1981–1985 plan. Additionally, the Chinese have prepared a plan specifically addressing coal development needs to the end of the century. This was presented by Yu Hongen, Vice Minister of Coal Industry.[24] Areas of stength in the coal industry, based on the Chinese perspective, include a large reserve base, a high priority accorded to this sector, and a good foundation for subsequent improvements. In support of the objective to develop coal production, the following policies will be pursued:

Increased mechanization
Improved safety in coal operations
Development of a broader range of products from coal
Improved transport for coal
Additional use of economic incentives, in addition to administrative mechanisms promoting production
Institutional arrangements promoting coordination of the coal industry with related industries

In addition to large-scale production, improvements must be made in the area of processing and utilization of coal. Export of coal will often require additional processing, in order to meet specifications of foreign customers.[25] Given the large reserves of Chinese coal, additional uses of coal are also to be researched, including coal for gasification and liquefaction. China is in many ways in a similar situation to the United States, having ample reserves of coal while future oil production prospects are problematic.

One problem faced by the Chinese is that bureaucracies and provinces tend to operate in isolation. Decisions are made without adequate consideration of how these actions will affect related organizations or the national interest. China has devised institutional arrangements to promote coordination. One example is the National Coal Import and Export Corporation.[26] This organization has a range of functions relating to its role as a link between the Chinese coal industry and foreign entities, including coal export, equipment import, compensation trade, joint-venture development, and utilization of foreign funds for coal development.

A second example of a coordinating mechanism is the General Corporation for the Joint Development of Energy Resources in Southwest China.[27] This organization is intended to transcend bureaucratic and provincial boundaries for a specific purpose. Participants in the scheme are ministries (coal industry, communications, and railways), banks (Bank of China, China Construction Bank, Hong Kong Bank of China), provinces (Guizhou, Yunnan, Guangdong), and the Zhuang Autonomous Region. The corporation was provided an initial capital allocation and is expected to seek other sources of funding. Although coal and transport are the initial priorities, the corporation will eventually become involved in nonferrous metals development as well.

Other approaches are financial rather than organizational in nature. Provinces without sufficient energy reserves have been encouraged to invest in energy projects in adjacent provinces with good reserve bases, in exchange for a share of production and some control.[23] Additionally, the central government will, in certain cases, utilize surplus funds generated at the local governmental and collective level to finance certain high-priority projects, such as energy and transport.[28] It is not anticipated that these funds will be returned.

Achieving an increasingly mechanized coal-mining system is a Chinese objective. Based on Chinese accounts, there were 138 fully mechanized coal work faces in major mines that produce one-fifth of the production from the major mines. China maintains that its mechanization is as productive as similar operations in other countries.[29] Although it increases productivity, there are problems that accompany this approach—maintenance and trained workers being the foremost. China has recognized the need for additional workers trained to maintain equipment, as well as management that is more aware of the problems of mechanized mining.[30] An additional problem with the move toward more mechanization is that high-technology equipment requires additional electrical power, an important consideration in a power-constrained country such as China.[23]

Outside Assistance: Coal Development

In March 1982, Occidental Petroleum Corporation agreed to carry out a feasibility study for a joint venture involving coal mining in the Pingshuo mining area, Shanxi province.[31] The anticipated production was 15 million tons per year. Essen-

tially, Occidental would provide technology and China would provide infrastructure. Anticipated investment by Occidental would be some $230 million. The share of the profits would be 50:50 until the Occidental investment was recouped. Subsequently, the Chinese would get a 60% share of the profit. A preliminary agreement to develop the mine was signed in March 1983.[32] Anticipated coal exports from the mine would exceed the total current level of Chinese coal exports.

An intriguing venture, involving multiple entities on both sides, has been proposed between the European Economic Community and the China Southwest Energy Resources United Development Corporation.[33] Foreign interests would finance a railway, mines, a power station, and port improvements. Overall outside investment is anticipated to be up to $6 billion; the time frame for completion will be 10 to 15 years. This project summarizes some of the characteristics of resource development in China:

Long lead times
Necessity of infrastructure improvements
Importance of outside assistance
Reliance on multiple sources of outside assistance

The complexity of such a development program could result in coordination problems beyond the capabilities of current Chinese management practice.

Another U.S. firm involved in Chinese coal mines development is Fluor Mining and Metals, Inc. Their work with the Fushan mine is the initial example of an outside engineering service contract in the Chinese coal industry.[34] This could be considered part of the Chinese movement toward import of technical expertise, rather than being overly reliant on whole-plant import. It could also be considered part of the Chinese response to inadequate numbers of technically trained personnel. Such an approach can result in the training of Chinese personnel. As part of this project, Fluor personnel presented technical seminars to the Chinese.[35] An added advantage is that there is the possibility of transferability of the findings of this study to similar areas.

OUTSIDE ASSISTANCE: HARD MINERALS

The degree of outside assistance that China is acquiring for the hard minerals sector is modest in comparison to the energy sector. The Chinese have indicated interest in the use of innovative institutional arrangements in nonferrous, as well as the energy minerals, as a method of utilizing foreign assistance.[36] The Chinese have also utilized outside firms in helping to develop mineral resources.

One of the more interesting examples is the project involving the Fluor Mining and Metals Corporation[37] in developing copper resources. One of the key features of this assistance is that the outside help was brought in at an early stage. The steel project at Baoshan suffered, as outside assistance was brought in after the feasibility studies had been completed.[16] The disadvantage of such an approach is that the outside firm would have to operate within the constraints of previous decisions.

Nonetheless, the project still has been hampered by the Chinese tendency to try to address too much of the backlog of needs at one time. Recognizing the monumental needs that they face, the Chinese have been reluctant to address this through incremental increases in capacity. As an idea of the ambitious nature of this project as proposed, the intent was to increase an initial operation of 10,000 tons per day to the world's largest. Subsequently, the project was scaled down to 135,000 tons of ore per day,[37] although this is an uncertain figure.[11] Some of the services performed by Fluor included the following: project cost calculations, metallurgical testing, mine planning, equipment selection, reserve assessment, ancilliary equipment design, and logistics considerations.[36] In addition to the assistance from Fluor, the Chinese were to have a copper smelter constructed by the Japanese in conjunction with the project. However, this was subsequently delayed due to readjustment. An important point regarding this project is that it is at least initially directed toward import substitution.

Regarding other sources of outside assistance in Chinese mineral development, there have been negotiations to set up a compensation trade arrangement in tungsten by Bechtel. However, no agreement has been developed to date.[38] In exchange for a share of production, an Italian firm agreed in principle to construct a plant to produce calcined bauxite in Henan.[11]

THE CASE OF OFFSHORE OIL

Development of offshore oil in China has generated a great deal of interest from outside corporations, as this is arguably the largest mostly unexplored coastal area in the world. A recent study of locations where exploration activity was expected to be most intense in 1983 indicated that exploration of the South China Sea was one of the highest-ranked target areas.[39] The interest of the Chinese in exploiting the offshore resource, and the corresponding interest demonstrated by international oil firms, has resulted in Woodard noting that offshore oil and related development potentially could result in the largest-scale economic involvement by foreign corporations in Chinese enterprises.[40] From an international perspective, the offshore development in China shows potential for being the largest such development since the North Sea activity.[41]

It would be interesting initially to establish why the offshore development has generated more interest than the Chinese onshore oil production potential. One reason is that the relatively shallow depths of the areas to be worked will allow development at costs comparable to new areas onshore.[42] Offshore production will be near to both Chinese and export markets. New infrastructure will have to be constructed for offshore development. However, the coastal areas of China already have much of the necessary infrastructure in place. In comparison, new onshore production presents problems related to the distance of prospective producing areas, lack of infrastructure, and difficult terrain and climate. As an indication of the contrast, whereas 33 companies remain interested in offshore leasing, in addition to the existing U.S., Japanese, and French companies already in operation, a single Japanese firm is currently involved in onshore oil production possibilities.[41]

China wishes to develop offshore energy rapidly. Failure to solve its energy problems can be damaging economically and politically. Essentially, in the absence of offshore oil production increases, the options for China necessitate trade-offs between oil used for domestic development or for foreign exchange earnings.[43] Given this incentive and the fact that China recognizes that offshore oil is the biggest single lure to attract outside assistance, China has linked the fate of her overall economic development scheme quite closely to oil. Accordingly, China expects a great deal more to be realized from offshore oil than simply revenue. These expectations are best indicated in China's regulations affecting offshore development.[44] Oil development is not an industry that automatically gives rise to numerous economic spinoffs. That is, oil development does not necessarily stimulate other enterprises. However, the Chinese are trying to ensure that they capture as many of the benefits generated by offshore development as possible. Therefore, preference is to be given to Chinese goods for offshore operations, including design, construction, materials, and services, provided that these are comparable and competitive.

Chinese rather than foreign bases are to be

used for offshore operations. Chinese personnel are to be given preference over foreign workers and are to be trained. Also, provisions written into the regulations call for technology transfer and the use of appropriate and advanced technology. A possible problem relates to the operations of a Joint Management Commission composed of Chinese and foreign firm representatives who direct the venture. The Chinese will appoint the director of this committee. An issue relates to whether the committee's role will involve long-term strategic decision making as well.[43] The Joint Management Committee will provide not only Chinese control but help to familiarize the Chinese personnel with Western management techniques.

The proposed contract is another area where the Chinese have stipulated tough provisions for Western entry into resource development. Three areas will be bid upon. These include the work program specifying scheduling and number of wells, the factor bid determining the company's profit, and other contributions to offshore petroleum exploitation or economic development.[43] Other contractual terms may also be negotiable.

The initial contracts for offshore development were negotiated with French and Japanese companies. Exploration costs are to be covered by the foreign firms and recovered only if development occurs. The foreign companies are responsible for 5% of development costs and the Chinese for the remainder. Of the oil produced, 42.5% is royalty oil to the Chinese, 15% of the oil is earmarked for production and development cost recovery, and the remainder is the share of the foreign firm. A portion of the foreign share will be considered reimbursement for exploration costs, and a portion is to be purchased by the foreign firm.[45] China has a deserved reputation as a tough trading partner. Initially, the Chinese succeeded in having Western firms provide exploration assistance as an entry requirement into the bidding process. Some $200 million worth of geophysical exploration activity was provided by the firms to the Chinese. As noted above, the Chinese are trying to secure agreements advantageous to their interests. However, their deliberate pace has hampered them in this respect. Since initiation of the negotiations, the oil market has weakened substantially.[46] Therefore, the Chinese negotiating position is not as strong as it was. Some observers feel that the Chinese bureaucracy is typically slow to respond to dynamic market situations, and this may be such an example.

Arco, operating in the Hainan Island area in conjunction with Sante Fe International, is the first American firm to drill offshore China. Arco benefited from early entry into the China market.

The particular deal negotiated by Arco may not be characteristic of later deals, as Arco may have made certain concessions in return for the grant of a larger area with a greal deal of geologic potential.[47]

One indication of the importance that the Chinese attach to outside assistance is the recent decision by China to continue to allow U.S. firms to engage in oil exploration despite other trade disagreements between the Chinese and the United States. Specifically, disagreement has occurred because of the restricted entry of certain Chinese textiles, and China retaliated against the United States by refusing to allow imports of certain agricultural goods. The fact that China is allowing the continuation of oil contracts is seen as an indication of Chinese pragmatism, as well as an indication of their need for technology.[48] The status of China offshore operations as of December 1982 was as follows:[41] Gulf of Bohai, six discoveries, one of which is under development; Gulf of Beibu, three discoveries; East China Sea, one gas discovery.

OTHER SOURCES OF ASSISTANCE

World Bank

Since 1980, China has been granted representation in the World Bank. Significant benefits have been realized by China as a result. A study of the Chinese economy by the World Bank has not only provided a basis for World Bank programming, but also provided China with a detailed outside perspective on the problems faced. Additionally, the World Bank has provided economic development training to Chinese planners, a valuable contribution given some of the recent Chinese planning failures.[49] Assistance has been extended to physical and social infrastructure improvements, including scientific and engineering personnel training and port improvements.[50] The ports to be improved are Huangpu, Shanghai, and Tianjin. Huangpu is a coal-handling port.

An initial study of China by the World Bank noted the inefficiency of energy use and the need for massive investment in this sector. This confidential report estimated that investment needs for energy-related improvements could necessitate as much as $400 billion in expenditures over the next 20 years.[51] One area where the World Bank has assisted is in studies of transportation linkages to small coal mines in certain Chinese provinces.[51] Additionally, the World Bank is considering providing financing for China's coal mining.

The stagnation of Chinese oil production has also been addressed by the World Bank. A loan to China of $162.4 million has been provided by the International Bank for Reconstruction and Development for a program to make improvements at the Daqing oil field. The overall project anticipates $674.3 million in expenditures, including possible foreign funding. The project will include application of improved oil recovery methods, seismic studies, technical training, and support of research, all intended to boost production in Daqing.[52,53]

United Nations

Indicative of their more international outlook is the Chinese participation in the United Nations. Technical transfer and financial assistance are the primary benefits realized by the Chinese as a result of this association. China is involved in the United Nations Development Programme and will be funded for a number of projects under this program, including energy-related improvements.[54] Included in this program were a number of energy development and conservation measures. A seismic exploration team for oil and gas exploration in southwestern China is to be funded for $1.2 million. Some $400,000 is proposed to finance research into technology for thermal energy utilization. Efforts to control sand in oil production in the Shengli oilfield are to be funded by a $500,000 grant. Additionally, addressing the problem of inadequately trained personnel, a center was to be established to provide such training for large-scale thermal power plants, requiring $600,000 of United Nations funding. Other projects include geothermal, biogenic, solar, and wind-power projects.

In addition to the funding, the United Nations has provided other services to China in the area of natural resources.[55] For example, a technical session on oilfield development with representatives of 23 countries was sponsored by the United Nations in Daqing, China, in 1982.[55]

REFERENCES

1. "Metals and Minerals in China." *Canadian Mining Journal* 102 (August 1981), pp. 20-21.
2. "Foreign Trade Booming." *Beijing Review* 25 (8 November, 1982), p. 7.
3. "On China's Economic Relations with Foreign Countries." *Beijing Review* 25 (31 May 1982), pp. 13-16.
4. *Xinhua*, 20 October 1983. Foreign Broadcast Information Service (FBIS), *China Daily Report*, 22 October 1982, pp. K14-K15.
5. "The Sixth 5-Year Plan of the PRC for National Economic and Social Development (1981-1985)," excerpts. *Renmin Ribao*, 13 December 1982, pp. 1-4. Foreign Broadcast Information Service (FBIS), *China Daily Report*, 20 December 1982, pp. K5-K42.
6. "Foreign Exchange." *The Wall Street Journal*, 5 May 1983, p. 44.
7. Pattison, Joseph E. "China's Developing Legal Framework for Foreign Investment: Experience and Expectations." *Law and Policy in International Business* 13 (1981), pp. 80-175.
8. Cohen, Jerome A. "United States–China Trade Still Needs Nurturing." *The Asian Wall Street Journal*, 1 March 1983, p. 6.
9. Japanese External Trade Organization. *China: A Business Guide*. Tokyo: Japanese External Trade Organization and Press International Ltd., 1979.
10. U.S. Department of Commerce, International Trade Association. *Doing Business with China*, November 1980.
11. U.S. Department of the Interior, Bureau of Mines. *The Mineral Industry of China*. Preprint from the 1981 Bureau of Mines' *Minerals Yearbook*. By Edmond Chin.
12. Tomlinson, Alexander C. "Some Perspectives on Financing China's Projects." In *Business with China*. Edited by N. T. Wang, Elmsford, N.Y.: Pergamon Press, Inc., 1980, pp. 61-66.
13. Ching, Frank. "Foreign Investment Slowed in China in '82 As Peking Altered Goals; '83 Rise Forecast." *The Wall Street Journal*, 7 March 1983.
14. Heuser, Robert. "Sino-German Trade: A Summary of Economic and Legal Data." In *China Trade: Prospects and Perspectives*. Edited by David C. Buxbaum, Cassondra E. Joseph, and Paul D. Reynolds. New York: Praeger Publishers, 1982, pp. 217-248.
15. Kraus, Willy. *Economic Development and Social Change in the People's Republic of China*. Translated by E. M. Holz. New York: Springer-Verlag New York, Inc., 1982.
16. Weil, Martin. "The Baoshan Steel Mill: A Symbol of Change in China's Industrial Development Strategy." In *China under the Four Modernizations, Part 1*. Joint Economic Committee, U.S. Congress, 13 August 1982, pp. 365-393.
17. Simon, Denis Fred. "China's Capacity to Assimilate Foreign Technology: An Assessment." In *China under the Four Modernizations, Part 1*. Joint Economic Committee, U.S. Congress, 13 August 1982, pp. 514-552.
18. U.S. Department of Commerce, International Trade Association. *China's Economy and Foreign Trade, 1979-1981*, by Dr. Nai-Ruenn Chen, May 1982.
19. "China's Petroleum Surplus May Vanish in the 1980's." *Oil and Gas Journal* 78 (6 October 1980), pp. 27-32.
20. Payne, Douglas B. "Sino–Japanese Trade: Does It Hold Lessons for American Traders?" In *China Trade: Prospects and Perspectives*. Edited by David C. Buxbaum, Cassondra E. Joseph, and Paul D. Reynolds. New York: Praeger Publishers, 1982, pp. 249-278.
21. Larkin, Bruce D. "Sino–Japanese Relations: Economic Priorities." *Current History* 81 (September 1982), pp. 268-271, 281.
22. Feintech, Lynn. *China's Modernization Strategy and the United States*. Washington, D.C.: Overseas Development Council, August 1981.
23. Weil, Martin. "China's Troubled Coal Sector." *The China Business Review* 9 (March-April 1982), pp. 23-34.

24. *Xinhua*, 7 October 1982. FBIS, *China Daily Report*, 12 October 1982, pp. K14-K15.

25. *China Daily*, 3 November 1982, p. 3. FBIS, *China Daily Report*, 5 November 1982, p. K18.

26. *Xinhua*, 22 June 1982. FBIS, *China Daily Report*, 24 June 1982, p. K6.

27. "Exploring Energy Resources in Southwest China." *Beijing Review* 25 (1 March 1982), pp. 7-8.

28. Ching, Frank. "Chinese Energy, Transport to Get $10.3 Billion." *The Wall Street Journal*, 14 December 1982, p. 30.

29. *Xinhua*, 13 January 1983. FBIS, *China Daily Report*, 14 January 1983, p. K14.

30. "Close Attention Should Be Paid to Strengthening Mechanical and Electrical Management in Coal Mines." *Renmin Ribao*, 10 July 1982, p. 3. FBIS, *China Daily Report*, 15 July 1982, pp. K18-K19.

31. "Occidental Joint Venture Accents China's Commitments to Coal." *World Coal* 8 (May-June), p. 13.

32. "China Mine Pact Set by Occidental Petroleum Corp." *The Wall Street Journal*, 9 March 1983, p. 29.

33. China Daily, 24 June 1982, p. 2. FBIS, *China Daily Report*, 24 July 1982, p. G1.

34. "Developments in China Signal Serious Efforts for Coal." *World Coal* 8 (November-December 1982), pp. 92-93.

35. Brady, Steve C., and John Bassarear. "Fluor and the Chinese Government." *Mines Magazine* 74 (September 1982), pp. 16-20.

36. "Progress in Nonferrous Metals." *Beijing Review* 25 (15 February 1982), pp. 7-8.

37. Argall, George O., Jr. "Huge Chinese Porphyry: Dexing Operations and Plans." *World Mining* 34 (January 1981), pp. 24-28.

38. Brady, E. Sabrina. "China's Strategic Minerals and Metals." *The China Business Review* 8 (September-October 1981), pp. 55-69.

39. "Experts Agree Upturn in Sight." *Offshore* 42 (December 1982), pp. 45-46.

40. Woodard, Kim. "China and Offshore Energy." *Problems of Communism* 30 (November-December 1981), pp. 32-45.

41. "Chinese Shelf May Generate Biggest Offshore Flurry since North Sea." *Oil and Gas Journal* 80 (13 December 1981), pp. 55-60.

42. Yuan, Paul C. "China's Energy Image Slightly Distorted." *Offshore* 41 (September 1981), pp. 97-99.

43. Brown, Chris. "Tough Terms for Offshore Oil." *The China Business Review* 9 (July-August 1982), pp. 34-37.

44. "Regulations of the People's Republic of China on the Exploitation of Offshore Petroleum Resources in Co-operation with Foreign Enterprises." *Beijing Review* 25 (22 February 1982), pp. 14-18.

45. "China to Begin Awarding Exploration Pacts Soon." *Houston Chronicle*, 27 February 1983, Section 4, p. 10.

46. Walker, Tony. "Falling Oil Prices Delay Offshore China Drilling." *Houston Chronicle*, 13 March 1983, Section 4, p. 10.

47. "Arco's Deal with China Is a Tough Act to Follow." *Business Week*, 4 October 1982, pp. 43-44.

48. "China Won't Bar Oil Exploration by United States' Concerns." *Wall Street Journal*, 21 January 1983, p. 18.

49. Ludlow, Nicholas H. "China and the World Bank." The *China Business Review* 9 (November-December 1982), pp. 12-13.

50. World Bank. "China Borrows $199.4 Million for Port Development and Agricultural Research." Bank News Release 83/16, 3 November 1982.

51. Mofsan, Steve. "China Making Massive Energy Push." *The Wall Street Journal*, 12 December 1982, p. 19.

52. World Bank. "Petroleum Project in China to Use $162.4 Million World Bank Loan." Bank News Release 83/38, 27 January 1983.

53. "International Bank to Aid Daqing Flow." *Oil and Gas Journal* 81 (7 February 1983), p. 29.

54. United Nations Development Programme. *First Country Programme for the People's Republic of China.* DP/CP/CPR/2, 18 February 1982.

55. *Xinhua*, 7 September 1982. FBIS, *China Daily Report*, 8 September 1982, p. A1.

Limitation of Current Chinese Export Policies and Anticipated Results

LIMITATIONS

One of the most significant trends in China's foreign trade in the 1970s was the increasing share of natural resources exports as a proportion of overall exports.[1] Much of this can be attributed to China's emergence as a petroleum exporter. Chinese policy statements and investment commitments provide indications that the Chinese wish to increase mineral and energy resource exports as part of an overall policy to increase exports of selected products. Although this could provide additional income to the Chinese, as well as other benefits, there are limitations to this policy.

Increasing Indebtedness

China has departed substantially from the traditional policy of self-reliance. Cognizant of the numerous constraints to resource development that must be addressed, the Chinese have chosen to rely on outside assistance to help speed up this process. Due to conservative policies in the past, China is not burdened by a large external indebtedness; reliance on outside assistance will increase this indebtedness. The debt may not necessarily be expressed in monetary terms. Rather, it could be in the form of a claim on a portion of future production, based on the institutional arrangements the Chinese have chosen to attract foreign assistance. This does have an economic component, represented by opportunity cost, expressed as the value of the alternative uses of these resources for export revenue, or as inputs into the Chinese economy.

Benefits of Mining Ventures Not Realized Locally

It is not unusual for large-scale mining ventures in developing countries to be reliant on outside technical assistance and finance. China will require such assistance, particularly technology and technical expertise. Advantages realized from mineral development ventures include direct employment, stimulation of other ventures, and the government's share in the project. However, in recent years there has been a tendency for much of the benefits from mineral development projects to be realized outside the region, and even the country, where the project is located.[2] Rather than be concerned about export earnings, the Chinese may use mineral exports to reimburse outside financial and technical assistance in resource development.[3]

Failure to Increase Processing

Mineral processing is an approach to capturing additional value added associated with mineral development ventures. Some of the benefits expected from processing include anticipated increases in employment, income, and tax revenue.[4] Mineral-importing countries' discriminatory duties on processed materials represent limitations on attempts to improve the level of processing by the Chinese. At this point, China tends to export mostly unprocessed materials. For example, exports of tungsten—an area where China enjoys a clear comparative advantage in terms of reserves—were as follows in 1980: 6207 metric tons of ore and concentrate and 23 metric tons of metal. As another example, China exports relatively low value bauxite, but imports primarily higher-value unwrought aluminum metal.[5] Problems related to the failure of previous imports of processing technology to work, as well as anticipated and possible electrical shortages, would tend to discourage the Chinese from rapidly expanding processing for the purpose of export.

Problems of Oil-Exporting Nations

China has based its future development strategy heavily on the expectations of relatively rapid offshore oil development. Amuzegar[6] has observed that oil-exporting countries tend to experience similar disruptive influences, regardless of their economic and social characteristics. Generally, the expectations that oil wealth will guarantee rapid economic development have not been realized. The tendency has been for oil revenues to fail to keep pace with investment commitments. Governments have generally become increasingly dependent on oil revenue for current as well as future expenditures. This is dangerous, as oil revenue is uncertain, and dependent on forces outside the control of the government. Benefits are not dispersed through the economy, as oil is not characterized by extensive backward or forward linkages. The hope for quick improvements often result in inappropriate development strategies. Oil exporters have tended to rely on large-scale industrial projects to provide momentum to development. This has been accompanied by a tendency to pay inadequate attention to the rural and agricultural sectors.

It will be interesting to see if the results documented by Amuzegar will also occur in China. China is a centrally controlled society as well as a latecomer to the oil export business. The Chinese have the opportunity to learn from previous mistakes and to control information and resource allocation. Perhaps societal expectations about the benefits to be realized from oil development can be kept within reason. As has been noted, the Chinese are attempting to maximize benefits to be realized from oil development. The failure of previous emphasis on heavy industrialization to promote the desired results has encouraged a strategy accommodating light industry for quicker investment returns and consumer welfare. Despite these adjustments, it is difficult to avoid the issue of uncertain revenue from oil exports. Substantial changes have occurred in the oil market from the time that the negotiations with outside firms began to the time that the first American firm began operating in Chinese waters. It is not clear if Chinese expectations regarding this resource will change accordingly.

SCENARIOS

When examining the Chinese export potential, data that would support projections are limited. However, scenarios regarding possible actions that would support or limit exports are possible. For China to become a major energy/mineral exporter, a number of conditions would have to be met. There would have to be a continuation of policy in support of resource exports, accompanied by financial commitment and improvements to certain basic infrastructure. Four scenarios are briefly considered:

1. Political disruptions
2. Inadequate infrastructure development
3. Incremental increases internally consumed
4. China as a major exporter

Political Disruptions

Political and economic disruptions are closely linked. Political shifts could range from changes in policy to full-scale societal disruption, as has previously occurred in China. The move to attract outside assistance was based on pragmatic considerations. Failure to achieve expected results could result in a move back to predominance of ideological considerations, rather than economic development concerns. Reconsideration of the value of outside assistance could diminish the role of exports. The effects of previous political disruptions included drops in production levels of certain key materials as well as transportation problems. Despite recent policy stability, the potential for disruption remains a concern. Questions remain about the reliability of China as a supplier. Political disruptions, in short, could prevent China from becoming a *secure* supplier of resources.

Inadequate Infrastructure Development

Lack of adequate infrastructure (social and physical) is perhaps the single most significant limitation on China's ability to become a large mineral exporter. Under this scenario, despite commitments by the Chinese to make the necessary improvements, available finances and expertise would be insufficient for all necessary improvements. This could, in many cases, result in continued bottlenecks, such as congested railroads with insufficient capacity to move raw materials for domestic and export needs. Additionally, inadequate infrastructure could hamper China's competitiveness as an exporter. The lack of deep-draft ports is an example. These provide large economies of scale in transport costs. Such infrastructure limitations also limit China's attractiveness for would-be outside investors.

Even if physical infrastructure is improved, social infrastructure also presents certain limitations. An emerging legal system, with an uncertain accommodation to Western corporations, is a major problem area. An educational system that cannot provide adequate numbers of trained personnel for modernization efforts is another. China provides a challenge, due to the range of problems that must be addressed. In short, infrastructure improvements will have to be successful across the board to permit China's emergence as a large exporter of metal and energy products.

Incremental Production Increases Internally Consumed

The magnitude of material needs that must be met by the Chinese provide powerful limitations on exports. Success in China's efforts to improve output through indigenous and outside assistance may still be unable to stay even with rising domestic demand. China faces the largest absolute population pressures of any country. At the same time, transitions in the economy are also important. China has experienced modest success in raising per capita incomes. Larger per capita consumption of resources is an expected result of a growing economy, until the point that service industries become dominant. There is a good possibility that China could make substantial gains in resource production and still be unable to keep even with growing internal demand.

China as a Major Exporter

China could emerge as a major exporter of a variety of minerals provided that certain conditions are met, specifically including political stability, infrastructure improvements, and production increasing more rapidly than internal demand. The difficulties involved make this highly unlikely before the end of the century, if at all. Realistically, this is the least likely scenario.

At this time, a summation of the range of viewpoints regarding the role of China as a potential resource supplier is appropriate. Feintech,[7] for example, indicates that there could be sizable increases in Chinese nonferrous exports with foreign assistance in mining in return for a share of the production. On the other hand, Wu[8] does not see China emerging as a major resource exporter, based on recent production/export data. Another observer predicts that incremental increases in mineral production are likely to be internally consumed.[3]

Regarding oil export potential, observers such as Meyerhoff[9] and an oil executive in Asia[10] do not anticipate large oil exports from China. However, Barnett[11] believes that oil and coal exports might increase, due to the importance of these resources as generators of foreign exchange. Barnett[11] observes that exports might increase even if this creates domestic supply problems. Smil anticipates small exports having little impact on the world market or prices.[12]

CHINA'S ROLE AS A LIMITED EXPORTER

China might increase exports of selected commodities and still be characterized as a limited exporter. This appears to be the most likely scenario. Walsh[13] has documented the increasing emphasis on develop-for-import (DFI) projects financed by West Germany, France, and Japan. These countries are willing to provide assistance to developing countries in mineral resource development in return for a guaranteed share of the production. Such an approach is in line with the Chinese policy of payback in production, rather than monetarily, where possible. In this case, the interests of the producing countries coincide with those of consuming countries.

An advantage of such an approach is that the Chinese will gain technology and other assistance up front for a future share of production. A dis-

advantage is that this strategy, if broadly applied, will result in increased pressure on the resource and infrastructure sectors. The DFI approach was a response to failure of long-term supply agreements to hold up under the stress of severe price fluctuations.[13] In return for more assistance from the outside, more stringent requirements must be met.[13] Problems with this approach can be anticipated further down the line. Mineral resources might have to be exported to meet DFI commitments, despite the country not being able to satisfy domestic needs. Alternatively, use of overburdened infrastructure, such as railroads to move export products, could delay domestic shipments. These factors could have political implications in the future. Nevertheless, despite the problems with this approach, it seems the best hope for China to become an important exporter of selected commodities.

In short, China has potential as an exporter, although the magnitude of this potential is uncertain. Those factors hampering China as a large exporter are the possible political instability, massive social and physical infrastructure constraints, and increasing domestic demand. However, Chinese policies and those of importing countries are increasingly in line with one another. Future production can be exchanged for current assistance. This is in accordance with the preferred Chinese approach to attract development. Therefore, China can be expected to play a role as an important supplier of mineral resources. However, the exports will be limited to a small number of commodities, and exports will predominately go to those countries that grant China financial and technical assistance.

The extreme cases that have been considered include China as a large mineral exporter and China as a minor influence on world mineral trade. The last case is essentially equivalent to the existing situation. From the Chinese viewpoint, becoming a major exporter would provide an additional source of foreign exchange. Large increases in Chinese mineral exports, particularly in those commodities for which they have massive reserves, would exert downward pressure on world commodity prices and erode the market shares of certain other mineral-producing countries. For countries that are dependent on mineral imports, China as a large exporter would provide an opportunity to diversify sources of supply further.

China might increase its role in mineral exports primarily by attracting additional foreign involvement in Chinese mineral projects. The key implications relate to which countries become involved in the Chinese economy and the potential for future trade as a spin-off of such activity. Investing in a mineral project, given the scale and scope of such projects, could be considered a vote of confidence in the Chinese economy. More important, it could lead to additional trade between China and the investing country, including mineral trade not linked to a specific project. China's major trade partners are Japan, the United States, and West Germany. Despite considerable investment to date, Japan sometimes shows an ambivalence toward involvement in China. U.S. firms are particularly interested in opportunities in the Chinese oil industry. West Germany has invested in develop-for-import mineral projects in other countries, but no such arrangement has been negotiated with the Chinese to date.

REFERENCES

1. Japanese External Trade Organization. *China: A Business Guide*. Tokyo: Japanese External Trade Ogranization and Press International Ltd., 1979.
2. Radetzki, Marian. "Regional Development Benefits of Minerals Project." *Resources Policy* 8 (September 1982), pp. 193-200.
3. "Metals and Minerals in China." *Canadian Mining Journal*, August 1981, pp. 20-21.
4. Odingo, R. S. "New Perspectives on Natural Resource Development in Developing Countries." *GeoJournal* 5 (1981), pp. 521-530.
5. U.S. Department of the Interior, Bureau of Mines. *The Mineral Industry of China*, Preprint from the 1981 Bureau of Mines' *Minerals Yearbook*. By Edmond Chin.
6. Amuzegar, Jahangir. "Oil Wealth: A Very Mixed Blessing." *Foreign Affairs* 60 (Spring 1982), pp. 814-835.
7. Feintech, Lynn. *Chinese Modernization Strategy and the United States*. Washington, D.C.: Overseas Development Council, August 1981.
8. Wu, Yuan-li. "Resources, Competition, Economic Warfare and East-West Relations." In *China, the Soviet Union and the West: Strategic and Political Dimensions in the 1980's*. Edited by Douglas T. Stuart and William T. Tow. Boulder, Colo.: Westview Press, Inc., 1982, pp. 87-98.
9. Meyerhoff, A. A. "China's Petroleum Industry: Geology, Reserves, Technology and Politics." In *China Trade: Prospects and Perspectives*. Edited by David C. Buxbaum, Cassondra E. Joseph, and Paul D. Reynolds. New York: Praeger Publishers, 1982, pp. 94-112.
10. "China Not Seen Selling Much Offshore Oil." *Asian Wall Street Journal*, 15 January 1983, p. 5.
11. Barnett, A. Doak. *China's Economy in Global Perspective*. Washington, D.C.: The Brookings Institution, 1981.
12. Smil, Vaclav. Energy Development in China: The Need for a Coherent Policy. *Energy Policy* 9 (June 1981), pp. 113-126.
13. Walsh, James I. "The Growth of Develop-for-Import Projects." *Resources Policy* 8 (December 1982), pp. 277-284.

Chapter V-5

Conclusions

The potential for China to emerge as a significant diversified mineral and energy commodity supplier has been examined. Much is dependent on China's ability to meet domestic needs, which is linked to the success of modernization efforts. The Chinese must overcome certain constraints, such as inadequate power and transport systems, that prevent rapid natural resource development. Insufficient finances, an inadequate educational system, and a shortage of skilled workers are additional problems.

The Chinese resource position has been described. Essentially, China has a large reserve base in absolute terms that could be used for domestic inputs or for export purposes. Like many developing countries, China has yet to investigate its resource potential fully; however, impressive strides have been made in exploration efforts. For a number of mineral commodities, including tin, antimony, coal, rare earths, and tungsten, the Chinese have some of the largest reserves in the world. A comparison of the reserve base with the population base is perhaps a more useful evaluation criterion. When expressed in these terms, the Chinese resource position is less promising. China plans a minor role in mineral trade with the exception of those commodities where China has a significant advantage in terms of reserves.

Given a generally advantageous resource position and an inadequate social and physical infrastructure, there is a basis for the Chinese to develop programs that would permit them to meet domestic needs and export targets. Chinese economic policy utilizes five-year plans to address economic development issues. A plan prepared in 1978 emphasized heavy industrial and resource development projects. However, the scope of this plan proved to be overly ambitious. A period of readjustment to bring the plan into line with Chinese capability was therefore necessary. The current plan appears to be a more realistic document. Much of the emphasis is on energy and infrastructure development. Natural resource development continues to be a priority.

In a departure from previous isolationism, much future development will rely on significant foreign contributions. The Chinese have sought to develop institutional arrangements capable of accommodating outside participation. This has necessitated significant changes in the Chinese legal system. The Chinese have tried to emphasize techniques that minimize their own financial contributions in favor of in-kind contributions. Foreign participation is already significant in offshore oil and coal development.

It is difficult to isolate energy and mineral commodity trade from the larger issues of trade and foreign affairs. Japan and the United States are two countries that have exhibited great interest in trade with China. China provides an opportunity for Japan to export technology and to secure raw materials. It is expected that this trade will continue to be important; however, there are some limitations. Japan's raw material policy calls for diversifying sources of supply. Should Japan continue to adhere to this policy, this would effectively limit the Japanese contribution to Chinese resource development projects. The United States and China also have certain complementary needs. U.S. firms' interest in Chinese trade is often directed toward gaining access to the Chinese consumer market. The Chinese are particularly interested in those goods and services in which the United States has superior capabilities, notably offshore oil development. Certain policy issues complicate U.S.-China trade. The issue of Taiwan continues to be a difficult one. Additionally, U.S. protectionism directed against Chinese textiles have hurt trade relations. For these reasons, the perceived potential for trade opportunities between the two countries has never been realized.

China's relationship with the USSR is also problematic. Any departure from the status quo raises questions. It makes little difference whether the relationship improves or if it deteriorates to the point of conflict. In either case, questions are raised about China as a secure source of strategic materials and about the security of investment in China in such capital-intensive, long-term projects as energy or mineral development.

Based on current conditions and apparent short-term improvements, it is unlikely that China will emerge as a significant, diversified exporter of mineral commodities. A limited role, somewhat larger than currently is the case, can be anticipated. As mining, processing, and transport are improved through import of technology and Chinese investment, China can be expected to become a more reliable supplier. Security of supply is questionable over the long term, subject to domestic political stability and Chinese-Soviet relations.

For China to become a significant exporter, certain conditions have to be met. Chinese commitment to increase resource exports has to be accompanied by improvements to eliminate bottlenecks. Favorable market conditions would have to prevail. Foreign involvement in Chinese development is particularly important in increasing the role of China as an exporter. Infrastructure improvements will generally be linked to joint mineral resource development projects, either as part of the Chinese contribution or an essential portion of the foreign investment. With long-term share-of-production agreements, where future production is exchanged for up-front improvements, market fluctuations can be bypassed to some degree. Finally, once such

agreements are negotiated, it would appear that they would be less subject to disruptions due to Chinese policy shifts than would straight market transactions. Contractual remedies would probably be quite costly to the Chinese if they failed to meet export targets under such arrangements.[1] Additionally, despite Chinese difficulties in meeting export targets specified in the Japan-China Long Term Trade Agreement, the Chinese seem concerned with developing an image as a reliable trade partner.

Among Chinese leaders there is an awareness that the extensive resource base of the country must be more fully utilized to satisfy domestic and foreign trade demands. The implication is that these resources will be developed within the limitations of Chinese administrative, financial, and technical capabilities.

REFERENCE

1. Walsh, James I. "The Growth of Develop-for-Import Projects." *Resources Policy* 8 (December 1982), pp. 277-284.

PART VI

THE ONE-CROP-ECONOMY COUNTRIES

Chapter VI-1

Introduction

Complementing the roster of major world mineral producers are nations, typically less developed, that rank as important suppliers for only one, or a few, minerals. Particularly in the cases of tin, bauxite, copper, and most notably petroleum, the developing countries play an important role. These less developed countries (LDCs) can be characterized by two attributes. First, capital, technical expertise, and frequently labor, have until recently been provided mainly by large foreign corporations. In many cases this foreign dominance continues. As a result, governments are striving to gain more control over their own resources, particularly in terms of revenue sharing. The Organization of Petroleum Exporting Countries (OPEC) has proven inspirational in this regard. Second, while production is often mainly from LDCs, consumption is chiefly in the industrial nations. Tin is an excellent example of this characteristic. In addition, not only are minerals produced mostly for export, but heavy reliance is often placed on only one or two minerals for foreign exchange earnings.

On the one hand, large export levels for a key mineral would seem to provide favorable conditions for aiding development, by stimulating cash flows into the country. Often, however, a commodity represents so large a share of the nation's total export earnings or total GNP that the producer is quite vulnerable to external market conditions. Such vulnerability is exacerbated by rising production costs, vacillating commodity prices, massive debt, high energy import bills, and economic recession in the world economy. This problem becomes most serious where few possibilities exist to broaden the range of goods exported. These nations are in effect "single-commodity" producers.

The importance of mineral exports to a national economy varies from country to country, as seen in Table VI-1-1. The Brandt Report[5] concluded that in 1978, 57% of the Third World's export earnings came from primary commodities, and 80% if petroleum was included. In some instances, commodities contributed more than 50% to national GNP. The growing realization of the consequences of excessive reliance on a single export for LDCs during the past decade resulted in the North-

TABLE VI-1-1 Examples of Countries Heavily Dependent on Exports of Nonfuel Minerals

	Commodity	Percent share of total exports
Gilbert Islands (Nauru)	Phosphate	98
Guinea	Bauxite	95
Zambia	Copper	95
Thailand	Tin	80
Mauritania	Iron ore	80
Zaire	Copper	65
Liberia	Iron ore	62
Bolivia	Tin	40
Chile	Copper	40
Guyana	Bauxite	40
Morocco	Phosphate	30
Peru	Copper	15

Sources: Refs. 1–4.

South debate, the demands of the New International Economic Order (NIEO) for a restructuring of the world economy, the resource nationalism manifested by expropriation of mining facilities, and forced renegotiation of agreements between corporations and governments. There were attempts to emulate the remarkable success of OPEC, through formation of other producer associations, such as those found today in the copper industry (CIPEC) and the bauxite industry (International Bauxite Association). Due to the widespread geographical distribution of most mineral reserves, and the impediments to organizational unity which typically preclude the successful operation of a cartel, these associations have had very limited impact.

Several characteristics seem to mark single-commodity producers. First, owing largely to corporate uncertainty as to future host government attitudes, there has been a steady decline in Western investment in Third World mineral ventures. This not only inhibits attempts to widen the economic base to support the growth of nonmineral sectors, but also impedes infrastructure development, exploration, and other mineral activities dependent on major financing. Second, as a result of the oil price hikes of the 1970s, the strong dollar of the early 1980s, and depressed commodity prices, many LDCs have accumulated a huge debt obligation. This "debt bomb" frequently requires major portions of export earnings just to meet interest payments. (It is ironic that one of the main causes of burdensome Third World debt can be attributed to rising energy prices instigated by OPEC, yet the oil cartel remains a role model for other producer nations. High energy prices will prove a continuing difficulty since, in contrast to projected flat demand in industrial nations, a 23% rise in the demand for oil is seen for LDCs by 1990.)[6]

Third, despite attempts to maximize value added before export, processing capability remains at a relatively low level in most producer nations. Although comparatively low energy and labor costs provide a competitive processing advantage for Third World countries, a dearth of available capital to construct necessary infrastructure forces continued export of ores to the industrial nations, where most processing still occurs. Particularly in the case of bauxite, where shipping cost differentials between it and alumina are so apparent, this situation will slowly improve from the producers' perspective.

Fourth, the leverage available to producers to modify existing conditions is at best marginal, especially on an individual basis. Production, while vital from a national standpoint, is normally a small enough percentage of total world supply for a given mineral to preclude a negotiating advantage. This is one of the primary reasons that producer associations have proven so weak—substitutes typically are available, not only geographically, but even generically. The projected total Third World's share of world exports for several minerals is illustrated in Table VI-1-2. Only in the cases of copper, bauxite, and tin does the developing group of producers represent a majority. Even this does not

guarantee a producer influence, however, as buffer stocks held by consuming nations, weak world demand induced by economic recession, and economic and political disparity among producers themselves limit the potential strength of either an existing or a potential united producer organization.

The extreme economic and political diversity extant for the entire Third World has in fact proven a major hurdle to effective collective action bargaining with the industrial consumers. Thus the demands of the Group of 77 and the NIEO during the 1970s have resulted in little real improvement. Moreover, Brewster[8] has argued the inevitability of this fact. Once past the loud rhetoric, few resources were ultimately devoted to the North-South negotiations by the Third World, seemingly indicating a significantly lower level of actual priority on this agenda than was anticipated by the developed countries. Such diminished interest supports a contention by some that mineral exports are not as truly vital for LDC producers as once thought, and more particularly that export of manufactured goods and other items is rapidly increasing in importance. For example, phosphate as a percent of Moroccan export revenue has dropped from 50% in the mid-1970s to 30% by 1983.[2] It is an unfortunate fact, however, that many of the poorest LDCs are still very dependent on exports of one or a few commodities. These nations, the ones most in need of assistance and with the least amount of political influence in the world, are also the most susceptible to drops in export earnings caused by falling demand and/or prices. Although agricultural exports constitute the main cash commodity for these deprived countries, mineral exports are in some instances quite important. Of all mineral commodities, the one of most importance on a value basis to LDCs, and of primary strategic significance worldwide, is oil.

TABLE VI-1-2 Projected Third World Share
of World Mineral Exports (Percent)

	1985	1990
Tin	85.0	85.9
Bauxite	77.0	75.4
Aluminum	32.2	34.3
Copper	60.5	63.7
Manganese ore	47.1	48.5
Iron ore	43.9	43.5
Nickel[a]	43.3	44.3
Lead[a]	28.6	25.3
Zinc[a]	26.9	29.0

[a]Excludes exports from centrally planned economies.
Source: Ref. 7.

REFERENCES

1. Cobbe, James H. *Governments and Mining Companies in Developing Countries.* Boulder, Colo.: Westview Press, Inc., 1979, p. 42.
2. "Morocco's Debt, War in Desert." *U.S. News and World Report*, 12 December 1983, pp. 51–52.
3. U.S. Department of the Interior, Bureau of Mines. *Minerals Yearbook Volume III. Area Reports: International, Centennial Edition 1981.* Washington, D.C.: U.S. Government Printing Office, 1983.
4. U.S. Department of the Interior, Bureau of Mines. *Minerals Yearbook. Volume III. Area Reports: International, 1982.* Washington, D.C.: U.S. Government Printing Office, 1984.
5. "Commodity Trade and Development." Chapter 9 in *North-South. A Program for Survival. The Report of*

the *Independent Commission on International Development Issues under the Chairmanship of Willy Brandt*, pp. 141-157.

6. Meloe, Dr. Tor. "Foreign Oil Demand to Increase 23% in LDCs, Hold Level in Industrial Countries to 1990." *Oil and Gas Journal*, 14 November 1983, pp. 152-158.

7. Grilli, Enzo R. "Demand for Raw Materials and Requirements for the Financing of Supply." *Probleme der Rohstoffsicherung, (Primary Commodities: Security of Supply.)* Bonn, West Germany: Friedrich Ebert Stiftung, 1981, pp. 7-17.

8. Brewster, Havelock. "Interests and Strategies in International Conferences on Commodities." *Probleme der Rohstoffsicherung. (Primary Commodities: Security of Supply.)* Bonn, West Germany: Friedrich Ebert Stiftung, 1981, pp. 45-54.

Chapter VI-2

OPEC

The Organization of Petroleum Exporting Countries is by far the most prosperous of the producers' associations; it epitomizes the single-commodity-producing nations. By its very longevity, impact on the world economy, and success in redistributing wealth from other countries to its members, OPEC has defied classical economic cartel theory. Yet in its early years, OPEC wielded relatively little influence.

HISTORY

Prior to the existence of OPEC, two rather unorthodox producer arrangements evolved in the international oil market.[1] The first of these was the group of major multinational oil corporations, as delimited by the Red Line Agreement of 1914 and the As-Is Agreement of 1928. These essentially divided the world market among the giant oil-producing companies of that era, especially the "Seven Sisters" [Gulf, Texaco, Shell, British Petroleum, Standard Oil of New Jersey (now Exxon), Standard Oil of New York (now Mobil), and Standard Oil of California]. In reality, these corporations formed a cartel. Invariably, they were perceived as being callously indifferent to the needs and interests of the host governments.

The second arrangement in existence before OPEC which contributed not only to its formation, but also its structure, was prorationing. This system was established in the United States during the early 1930s, first in Oklahoma and later in Texas. Its purpose was to limit (prorate) production to market demand by allocating output among the various producers through quotas, to help establish a price floor. Eventually, oil companies themselves began voluntarily to restrict production outside the United States to mitigate oil price declines. In the Middle East, however, production constraints were mistakenly viewed by governments as a lack of diligence by the corporations in developing Eastern reserves. Consequently, these nations opened up new areas for bidding to independent oil companies. Two aspects of this action were significant. First, it indicated a basic weakening of the domination of the Seven Sisters. Second, it ultimately increased the supply of oil, thus frustrating the intent of prorationing by the majors, and driving down prices.

In addition to the above, Danielson[1] cites a final factor—profit sharing—as critical in promoting an environment conducive to the emergence of OPEC. The system of profit sharing was conceptually a scheme to increase the host nation's portion of the revenues from its oil reserves. Profit sharing originated in Venezuela in the mid-1940s, with a 50:50 split mandated between the state and oil corporations operating with the country. As a consequence, corporations immediately shifted production to the Middle East, where lift costs, unimpeded by profit-sharing constraints, were quite low. In response, Venezuela opened up vast new areas for bid to independent producers. Although this helped to increase Venezuelan production, it again had the added effect of depressing prices. In addition, Venezuela lobbied the Mideast nations to promote the establishment of their own profit-sharing system. This was quite attractive to Persian Gulf countries, who were concerned about their own falling revenues due to price decreases. By 1952, therefore, profit sharing was universally applied. The end results of profit sharing and of prorationing, ironically, were the opposite of that intended—an oil glut, a concomitant downward pressure on prices, and worsening relations between producing companies and host governments. The nationalization of the Iranian oil industry in 1951 only exacerbated the latter condition.

In particular, falling oil prices fomented OPEC's formation. Implementation of the Mandatory Oil Import Control Program in the United States in 1959 and the continuing presence of an oil oversupply prompted the majors to lower the posted price again, first in February 1959 and then again in August 1960. This was, typically, a unilateral action, advanced without prior consultation with producer nations. In response to these final cuts, the first OPEC meetings were held in September 1960, attended by Iran, Iraq, Kuwait, Saudi Arabia, and Venezuela. The initial intent was to resist continued price cuts and, if possible, to restore former price levels. Throughout the 1960s, however, prices continued to fall. To offset this, OPEC pursued production increases. This was in direct contrast to the major oil companies' concept of profit, which was a reliance on the constraint of production to maintain prices. Expansion of output, and the continual issuance of new concessions to smaller independents, underscored the loss of control by the majors. Domination of the oil market slowly passed to OPEC; by 1973, it was complete.

THE CARTEL

OPEC remains a price-fixing cartel, with the amount buyers will purchase determining the level of production. Since 1973, the organization has been more willing to drop production, rather than prices, although weak demand precipitated a price cut in March 1983. Although overall demand for petroleum is regarded in the short term as relatively inelastic, OPEC, as the residual or "swing" supplier, has a much more elastic demand curve than do other producers. It thus typically absorbs the variance in the market.

OPEC's success can be ascribed to four main features. First, OPEC nations possess a very high percentage of world reserves; current figures place it at more than two-thirds. This concentration makes interference with the organization's control of the world market by outside producers much more difficult. Second, OPEC is not truly a pure cartel but rather a "dominant producer" cartel. Saudi Arabia, with over 40% of total OPEC producing capacity, is the clear leader in efforts to maintain production and price levels. Any neces-

sary cuts in production, for example, to support the OPEC contract price are normally assumed mainly by the Saudis. In addition, Saudi Arabia sets the benchmark price, with other members following according to transport and oil quality differentials. Third, and probably most important, is that no single perfect substitute exists for petroleum in its major applications, most particularly in the short run. This inelasticity of world demand for oil has proven a boon to OPEC's price hikes. Finally, extremely low marginal production costs inside the cartel, believed as low as 15 cents per barrel for Saudi Arabia, have given the organization a clear advantage.

This is not to say that classical theory does not apply to OPEC. Certainly, the organization cannot be construed as a group of 13 homogeneous nations. Rather, members represent a wide spectrum of demographic and economic characteristics. This is illustrated in Tables VI-2-1 and VI-2-2. The most important feature apparent here is that OPEC can be divided into two groups: those with a large population and small oil reserves on the one hand, and those with few people and large reserves on the other. Clearly, the interests of the former will come into direct conflict with those of the latter, particularly regarding OPEC pricing and production strategy. For example, Saudi Arabia would be more inclined to favor moderate prices, thus assuring demand far into the future, due to its massive reserves. Conversely, Indonesia, with more than 150 million people and limited reserves, would

TABLE VI-2-1 OPEC Oil Reserves and 1983 Production

	Oil reserves (billions of barrels)	Percent of total OPEC reserves[a]	1983 production (thousand barrels/day)
Saudi Arabia	166.0	37	4,872
Kuwait	63.9	14	912
Iran	51.0	11	2,606
Iraq	43.0	10	905
United Arab Emirates (UAE)	38.0	8	1,517
Venezuela	24.9	6	1,791
Libya	21.3	5	1,020
Nigeria	16.6	4	1,232
Algeria	9.2	2	686
Indonesia	9.1	2	1,292
Qatar	3.3	1	270
Ecuador	1.7	0.4	236
Gabon	0.5	0.1	150
	448.5, 67% of world reserves		17,489

[a]May not add to 100 due to rounding.

Source: Ref. 2.

TABLE VI-2-2   Estimated 1983 Production for OPEC Nations and Production per Capita

	1983 population (millions)	Per capita production (barrels/person/year)
Indonesia	154.3	3
Nigeria	89.1	5
Iran	40.5	23
Algeria	20.0	13
Venezuela	14.7	44
Iraq	14.0	24
Ecuador	9.3	9
Saudi Arabia	8.9	200
Libya	3.4	110
Kuwait	1.6	208
Gabon	1.4	39
UAE	1.1	503
Qatar	0.2	493

Sources: Refs. 2 and 3.

favor a plan maximizing present revenues, and hence high prices, to speed current development efforts for its large population. This disparity among members, especially regarding per capita wealth, is a decidedly major impediment to cohesiveness.

During the early 1980s quotas on production have proven essential to maintain high prices. Consequently, output has dropped to approximately 17.5 million barrels per day, a decline of more than 40% from the peak of 31 million barrels per day set in 1977. World recession, improvements in fuel usage, and substitution by other fuels prompted by its own price hikes have resulted in a fall in OPEC's share of world oil output from 56% in 1973 to 35% by late 1983.[4] As an illustration of the decrease in the importance of OPEC oil, its decline as a percentage of U.S. consumption is shown in Figure VI-2-1. U.S. dependence on OPEC crude peaked at 34% of total consumption in 1977; by the end of 1983 this figure was less than 17%.

To counter declining demand, OPEC lowered its market price in early 1983 by $5, to $29 per barrel. Combined with production ceilings, the resultant drop in oil revenues has forced delay or outright cancellation of major development projects in virtually all OPEC countries. For those members with previous economic problems, such as Nigeria, Libya, Gabon, Ecuador, and Indonesia, the loss of petrodollars is particularly devastating. The estimated balance of payments deficit for OPEC in 1983 was put in excess of $30 billion.[6] Moreover, the organization's own pricing policy can be held largely accountable for the drop in demand for its oil. In addition, prices have made feasible the development of high-cost petroleum in areas such as Alaska and the North Sea. OPEC's pricing policy has thus prompted lower prices by increasing non-OPEC supplies and reducing overall consumption.

In the face of such pressures, it is uncertain how long all members can be compelled to maintain assigned production levels at the benchmark price. Indeed, as many as eight different producers were accused during 1983 of production above quota.[7] Charges of price skimming and hidden discounts have also surfaced.[8] Such discord during a time of reduced demand is only exacerbated by the bitter Iran-Iraq war. To make matters worse, the prospect is bleak for any significant demand spurt for the immediate future; some predict an oil glut lasting into the 1990s.[6]

Such projections are of course moot. The Iran-Iraq conflict, for example, could produce shortages, particularly if oil shipments through the Straits of Hormuz are halted. On the other hand, a settlement of this dispute could only increase oil supplies, as these nations resume prewar production. The possibility of future price cuts thus cannot be excluded.

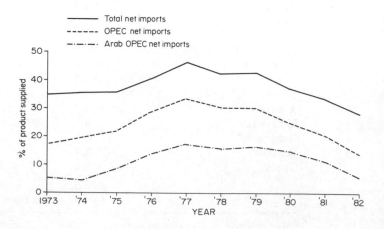

FIGURE VI-2-1   U.S. dependence on petroleum net imports. (From Ref. 5.)

Such action would only magnify existing stress on organizational ties. However, OPEC cannot be construed as no longer being influential or powerful. In retrospect, OPEC has engendered scrutiny of individual energy policies by all importers, and fostered more pervasive conservation and substitution programs. It has been able to support a very high price in the face of a severe recession and low demand. In addition, OPEC still supplies more than 45% of noncommunist oil for the world, and possesses in excess of 70% of noncommunist reserves. A strong worldwide economic recovery, and/or a harsh winter or two, could revive sluggish demand. In fact, some experts have concluded that oil markets are stabilizing, and that non-OPEC producers are now helping support the market price.[9] Thus, although certainly faced with difficult problems, especially regarding internal unity, OPEC is not yet "yesterday's cartel."

REFERENCES

1. Danielson, Albert L. *The Evolution of OPEC*. New York: Harcourt Brace Jovanovich, Inc., 1982.
2. "Worldwide Report." *Oil and Gas Journal*, 26 December 1983, pp. 80-81.
3. *Encyclopaedia Britannica. 1983 Book of the Year*. Chicago: Encyclopaedia Britannica, Inc., 1983.
4. "Is OPEC Going the Way of All Cartels?" *U.S. News and World Report*, 24 October 1983, pp. 30-31.
5. U.S. Department of Energy, Energy Information Administration. *Monthly Energy Review*, December 1983.
6. "Yesterday's Oil Cartel?" *The Economist*, 3 December 1983, pp. 13-14.
7. "OPEC Crude Exports, Revenue Down in '82." *Oil and Gas Journal*, 21 November 1983, p. 51.
8. Ibrahim, Youssef M. "OPEC Chiefs Try to Stop Price Erosion, If They Can; Iran-Iraq War Is Wild Card." *The Wall Street Journal*, 6 December 1983, p. 33.
9. Ibrahim, Youssef M. "Firming Oil Prices Strengthen OPEC." *The Wall Street Journal*, 8 March 1984, p. 30.

Chapter VI-3

LDC Nonfuel Mineral Production

Petroleum remains by far the most important commodity exported by the developing countries. However, LDC producers also represent more than 50% of the world's supply of tin, bauxite, and copper. Other high-tonnage LDC production includes phosphate, iron ore, nickel, chromite, and manganese.

BAUXITE

The Third World is the principal supplier of bauxite, the raw material from which aluminum is typically derived. This is in keeping with the tropical geological occurrence of the mineral. While Australia continues as the output leader, the importance of the LDC producers is obvious, all the more so if Eastern Bloc production, which is normally earmarked for Soviet consumption, is excluded. As seen from Table VI-3-1, the major LDC producers of bauxite are Guinea, Jamaica, Brazil, Suriname, India, and Guyana.

Historically, bauxite production has been dominated by an oligopoly made up of six major international companies (Alcoa, Pechiney Ugine Kuhlmann, Alcan, Alusuisse, Reynolds, and Kaiser). Presently, these firms control 35% of world bauxite capacity, 50% of alumina capacity, and 40% of aluminum capacity.[2] The availability of capital for LDC nations, coupled with the aluminum companies' control of most downstream operations in the industry and with depressed world demand for aluminum, has worked to hinder free access to the market for many new LDC deposits. However, prospects shall improve for increasing processing capacity in LDCs (especially alumina refineries), in response to comparative transport and energy cost advantages in these nations. Such advantages, however, are often offset by a dearth of infrastructure in many African and South American producing countries.

The aluminum companies' dominance is further challenged by other LDC actions. For instance, in 1974 the International Bauxite Association (IBA) was formed by Guinea, Australia, Guyana, Jamaica, Sierra Leone, Suriname, and Yugoslavia. These nations were subsequently joined by the Dominican Republic, Haiti, Ghana, and Indonesia. IBA producers now account for approximately 75% of all world bauxite output. While attempts by this organization to set a price floor have had limited success, individual members have promulgated nationalization of mines and much higher tax rates (for example, a 700% increase in Jamaica).[3] High levies on bauxite, and increasing government equity participation in mining revenues, were largely caused by the oil price hikes of 1973, which greatly boosted energy import costs. The IBA's potential for future large price increases appears thwarted by Australia's presence in the association. Australia's huge bauxite reserves, combined with some remaining production capacity outside the organization (e.g., in Brazil), constrains price elevations. The impact of the IBA in the future thus appears limited.

Guinea

The centrally planned West African nation of Guinea ranks second in world bauxite production, behind Australia and just ahead of Jamaica. Mining operations began there in 1952. By 1982, output stood at 10.2 million tons per year.[1] More important, Guinea holds the largest bauxite reserves in the world. This figure is estimated to 5.9 billion tons, representing more than 25% of total global reserves (see Table VI-3-2). Moreover, much of the bauxite in Guinea is of very high quality, particularly the massive gibbsite-boehmite deposits at Boke.

TABLE VI-3-1 1982 Bauxite Production
(Thousand Metric Tons)

Australia	23,000
Guinea	10,200
Jamaica	8,000
Brazil	5,400
USSR	4,600
Greece	3,400
Yugoslavia	3,300
Hungary	3,000
Suriname	2,900
India	2,300
Guyana	1,500

Source: Ref. 1.

TABLE VI-3-2 World Bauxite Reserves

	Bauxite reserves (million metric tons)
Guinea	5900
Australia	4600
Brazil	2300
Jamaica	2000
India	1200
Guyana	700
Greece	650
Suriname	600

Source: Ref. 1.

Not surprisingly, minerals account for approximately 95% of all export earnings, with bauxite providing most of this. More than three-fourths of all bauxite mined is exported. What is not immediately traded is refined into alumina at Fria and then exported. At present, however, alumina capacity remains under 1 million tons. Ninety percent of all bauxite exported is destined for the USSR, with the remainder bound for other Eastern Bloc nations. Importantly, the Soviets both finance and operate the Guinean bauxite company, Compagnie des Bauxites de Guinée (CBG). This is in keeping with the keen Soviet interest in foreign sources of bauxite, this mineral being one of the few the USSR cannot produce in sufficient amounts internally. (Other LDC producers supplying Soviet bauxite needs include Brazil and Jamaica.) Given the magnitude of its reserves, Guinea's ties with the USSR can only be expected to increase in importance, as expansion by the Soviet aluminum sector boosts its requirements for raw materials.

Also anticipated to rise is the Guinean government's reliance on revenues from its bauxite operations. Not only do taxes on exports and mining comprise a major income to the state, but the government is also itself a partner in CBG. These revenues are vital to meet interest requirements on a foreign debt in excess of $1.5 billion.

Suriname

In Suriname, the health of the entire economy rests on its bauxite exports. Production slipped to 2.9 million tons in 1982, a decrease of more than 20% from the preceding year.[1] Production of alumina, however, does exceed 1 million tons. Reserves are at the low end of the major producing nations and are listed at 600 million tons.

Suriname's coastal plain has been a primary source of high-grade gibbsite since the 1920s. Rich deposits in the interior remain undeveloped. This is particularly true of the Bakhuys region in western Suriname, where development for the entire area hinges largely on projected bauxite mining operations and a planned alumina refinery. At present, three companies are active in the aluminum industry in Suriname: Suriname Aluminum Co. (a subsidiary of Alcoa); Suriname NV, Billiton Maatschappij (a subsidiary of Royal Dutch/Shell); and NV Grassalco, a state-owned company. Grassalco owns the sole currently operating refinery and smelter.

The United States is the major recipient of most of Suriname's bauxite and alumina exports. Aluminum metal production remains limited, with 90% of alumina refined being exported. Several problems have inhibited any recent increments to bauxite production. In addition to the depressed international economic climate of the early 1980s (which has greatly affected all producers), competition from more low-cost operations (particularly in Australia), coupled with demands for much higher wages from the bauxite labor union, have worked to stifle growth in Suriname's industry.[4]

India

Despite many substantial operational problems, production of Indian bauxite continues to grow; it reached 2.3 million tons in 1982. Reserves of bauxite are estimated at 1.2 billion tons. Alumina and aluminum metal output, however, is facing slow or no growth.

The failure of the electrification plan for the country to meet targeted capacity has been responsible for two persistent difficulties in the Indian aluminum sector.[4] First, the ability of refineries and smelters to maintain operation is hindered. Second, due to the mandate requiring that electrical-conductor-grade aluminum make up a high percentage of total aluminum output, delays in meeting electrification plans prevented purchases of this grade, causing a massive surplus. Compounding the problem were high aluminum prices, exacerbated by an exorbitant ad valorem tax rate in the early 1980s, as high as 44%. In spite of these difficulties, however, there is considerable room for continuing expansion of bauxite, alumina, and aluminum metal production, particularly in light of the size of India's reserve base. Financial assistance from the USSR, India's chief trading partner, can only aid this growth.

Guyana

Guyana, primarily a producer of agricultural commodities, is nonetheless dependent on bauxite for 20% of its GNP. In addition, bauxite accounts for about 40% of all export earnings. Reserves are believed to be 700 million tons, with much bauxite containing very little iron. Alumina content is estimated to average 59%. Guyana has been a major supplier of high-grade gibbsite since 1971. In 1982, production was 1.5 million tons, a decline from the previous year of more than 10%. This decrease can be attributed to many factors beyond depressed world demand, including labor unrest, technological shortcomings, and shortages of equipment and spare parts.[4] In keeping with the traditional LDC problem of limited refining capacity, alumina production is under 250,000 tons annually.

TIN

LDC producers provide more than half the world's supply of tin. In particular, Malaysia, Indonesia, Thailand, and Bolivia are not only major producers, but also represent a significant portion of world reserves (see Table VI-3-3). These four nations constitute the chief suppliers in the Sixth International Tin Agreement, which became effective in July 1982. Tin is, in fact, the only mineral for which an international commodity agreement has been concluded. Existing trade arrangements, import quotas, and buffer stocks have interacted to hinder agreements for other minerals.[5] Originally established in 1956 in response to a long history of market instability, the Tin Agreement is now composed of many fewer producing and consuming nations than prior pacts. Neither the USSR nor the United States, for example, are participants in the present accord. This may erode the ability of producers to maintain price stability. Particularly in view of substitution by other commodities, and to supply surpluses as production exceeds world demand, current export quotas imposed on member producers could prove inadequate.

Malaysia

Although Malaysia is the number one producer of tin, exports of the mineral account for less than 10% of total export earnings. This reflects the importance to the nation of petroleum exports. Despite recent drops in output, Malaysia still provides approximately 30% of world tin concentrates and 35% of all tin metal. The world's largest tin company operates from the country, the Malaysia Mining Corporation Bhd.; it is 56% state owned.

As seen from Table VI-3-3 Malaysia ranks third in the size of tin reserves. The states of Perak and Selangor provide nearly 90% of all tin concentrates. More than half of the concentrate is derived from gravel pumping, with dredging operations accounting for another 30%.[4]

In view of the capital available to the Malaysia Mining Corporation (which exceeds $1 billion), its ability to establish an integrated operation, and the size of the country's reserves, Malaysia should remain a significant producer and exporter of both tin concentrate and metal. Currently, 40% of American tin metal imports are supplied by the country. Other important consumers of Malaysian tin include Japan and the Netherlands.

Indonesia

Despite possessing the largest tin reserves in the world, estimated at 1,550,000 tons, the Indonesian economy is dominated by the oil and gas sector. This industry provides more than 20% of the nation's GNP and more than four-fifths of total exports by value. In contrast to some of Indonesia's competitors, price downturns throughout the early 1980s did not prevent continued expansion of domestic tin output. This enabled Indonesia to pass Thailand in 1981 and establish itself as the third-leading producer. Output in 1982 was 32,000 tons of ore (metal content). Tin exports now rank in importance behind petroleum, timber, and rubber.

In contrast to Malaysia, approximately 50% of Indonesian concentrate is derived from dredging operations, reflecting the location of about two-thirds of the reserve base offshore. Gravel pumping methods provide another 40% of production. Approximately 75% of all tin is supplied by the state-owned mining company, P.T. Tambang Timah. In light of the extent of existing reserves, and the continuing increments to this base from large offshore discoveries, Indonesian output could well move into second place sometime in this decade. Singapore, the Netherlands, Japan, and the United States are the primary recipients of Indonesian tin exports.

Thailand

Thailand is an excellent example of a developing country highly dependent on production and export of a single mineral commodity, tin supplying more than 80% of all Thai export earnings. During the early 1980s, increases in the price of its energy imports, high domestic inflation, and falling tin prices all worked to worsen Thailand's balance

TABLE VI-3-3   Major Tin-Producing Countries
(Metric Tons Tin Content)

	1982 production	Reserves
Malaysia	55,000	1,200,000
USSR	37,000	1,000,000
Indonesia	32,000	1,550,000
Thailand	30,000	1,200,000
Bolivia	26,000	980,000
China	15,000	1,500,000

Source: Ref. 1.

of trade, ultimately causing a currency devaluation. Although possessing substantial tin reserves, several factors have forced declines in Thai tin production since 1979. Downward price trends with a con-comitant world supply glut have discouraged output increments. This malady, however, has been common to all market economy producers. Thai tin mining also faces the problems of serious declining ore grades and of rising production costs.[4] In the case of ore depletion, some of the difficulty can be attributed to "high-grading" of large areas by poachers and small operators, rendering significant zones uneconomic for large dredges. New discoveries in deeper water, however, are offering the potential not only to reverse the ore-grade decline, but to technologically preclude poaching, and government tax reductions are alleviating the cost increase problem. Construction of the third Thai tin smelter indicates that state reliance on this metal as a source of revenues could intensify. The Netherlands, the United States, and Japan are the main importers of Thai tin.

COPPER

In terms of value, copper remains the major non-ferrous mineral traded internationally. LDC producers are again important suppliers of this commodity. The copper producer association, CIPEC, comprised of Chile, Indonesia, Papua New Guinea, Peru, Yugoslavia, Zaire, Zambia, and Australia, provides more than 40% of total world output. Table VI-3-4 lists major copper producers and their estimated reserves.

Historically, CIPEC was formed in 1967 by Chile, Peru, Zambia, and Zaire, in the first explicit attempt by LDC producers to emulate OPEC. These four nations are today still quite dependent on copper exports. Copper trade provides more than 40% of Chile's foreign currency earnings, 15% of

TABLE VI-3-4   Major Copper-Producing Countries
(Hundred Metric Tons of Copper Content)

	1982 production	Reserves
Chile	1,241	97,000
U.S.	1,140	90,000
USSR	1,000	36,000
Canada	606	32,000
Zambia	530	34,000
Zaire	495	30,000
Peru	369	32,000
Poland	338	13,000
Philippines	280	18,000
Australia	245	16,000

Sources: Refs. 1 and 6.

Peru's, about 67% of Zaire's, and 95% of Zambia's. In addition, copper is the major export commodity for both the Philippines and Papua New Guinea, and is a significant export for many other LDCs.

Competition among copper producers super-sedes national boundaries, due to relatively few import barriers in most consuming nations. Thus, in a time of depressed demand, low-priced imports can "squeeze out" domestic high-cost production. Under most conditions, low demand includes decreases in output from all producers in an attempt to prevent surpluses and precipitous price declines. In actuality, however, due to the heavy reliance on copper for export earnings and domestic employment, nations such as Chile continue to maintain high production levels, often through government subsidy. The burden of absorbing market imbalance then falls upon free market producers, such as Canada, the Philippines, or the United States.

American producers in particular face a double-edged sword. If U.S. mines are closed in response to low demand, foreign producers make up the difference through increased output. On the other hand, if U.S. producers attempt to remain open, they are undercut by foreign copper prices. Due to high U.S. production costs (caused largely by the world's highest environmental control costs) and to very low grades of porphyry deposits exploited, the U.S. copper industry has a much higher break-even point than do foreign producers. This explains the often-unfavorable capacity utilization rate of the United States compared to other nations. In 1982, the U.S. copper industry operated at 65% of capacity, while foreign production ran at 85%.[6] Moreover, the latter figure does not indicate the significant production increments from Chile and Peru. These two nations, in particular, can be expected to continue their expansion. CIPEC believes as much as one-fourth of the world's total reserves lie within their borders, and that they could supply almost 33% of world output by 2000.[6] As a consequence, U.S. smelters are increasingly reliant on foreign concentrates, engendering growing protectionism sentiments among the U.S. industry. Such sentiments could see the end of duty-free copper trade for the LDC producers, at least in the short term.

IRON ORE

As a result of the continuing industrialization extant in all parts of the globe, iron ore has undergone the fastest growth in international trade of any mineral commodity since the 1950s. An increasing proportion of total world production is exported. In 1950, 16% of world output was traded. By 1983, this figure was 40%.[5,6]

India

Iron ore production in India rose slightly in 1982 to 41 million tons.[1] Although approximately 50% of all output is exported, iron ore accounts for only 5% of the nation's foreign currency earnings. Indian reserves are estimated at 9 billion tons of ore.[1] The depression in the world steel market, and postponement of a large Iranian contract for Indian iron due to the war with Iraq, have hindered export growth potential. At present, Eastern Bloc nations constitute the main consumers of Indian iron ore. However, despite the world market slump, the Indian steel sector has continued to expand. In order to meet domestic demand, therefore, while still meeting export requirements, sound development plans will be essential. Sensible development should witness notable increases to iron ore output, although, again, domestic usage could be deemed as having a higher economic value than that of exported ore.

Liberia

Iron ore is very important to the Liberian economy, accounting for more than 60% of total export earnings in 1981. Only in Mauritania does iron ore play a more dominant economic role. The United States and the European Economic Community (EEC) are the major purchasers of Liberian iron; they are also a vital source of funds and technical assistance. As might be expected, depressed steel markets in the United States and Europe have had a heavily stagnating effect on Liberian production, virtually all of which is exported. Liberian output in 1982 was 18 million tons,[1] a fall of 7% from the previous year. Despite a recent $20 million loan from the World Bank to aid in future iron ore development, Liberia's reserves, at 1.4 billion tons,[1] may prove too small a foundation on which to base the entire health of the nation's economy.

PHOSPHATE

Morocco

Morocco's phosphate reserves are massive. At 20 billion tons, they represent 50% of total world reserves.[1] Phosphate trade provides approximately 33% of the nation's export revenue, mainly from Western Europe. Proximity to its markets gives Morocco a decided transportation cost advantage. These two factors, magnitude of reserves and importance as an export commodity, indicate a tendency to rely increasingly on phosphate as the cornerstone of the Moroccan economy.

In 1982, production stood at 17.2 million tons,[1] second only to the United States. This was a decrease of about 2.5 million tons from 1981. Declining revenues from phosphate, the price of which has fallen more than 50% in seven years, is but one of many serious problems Morocco faces.[7] Oil import prices remain high. Most of any expected increase in phosphate demand in the immediate future is expected to occur mainly in the United States, which has huge reserves of its own. Additionally, severe drought conditions in northern Africa continue to restrict revenue potential from agricultural exports. An enormous debt of over $11 billion stifles growth possibilities; annual interest payments alone exceed $1.5 billion. Moreover, the continuing war with Polisario rebels in the western Sahara puts added burdens on an already strained economy. Although IMF loan assistance is providing some cash, austerity measures stipulated by the loaner have raised food prices, prompting riots in some areas in 1983.

Despite the present economic climate, which has forced postponement or cancellation of new projects in the U.S. phosphate industry, expansion plans continue unabated in Morocco. The Moroccan phosphate monopoly, Office Chérifien des Phosphates (OCP), has initiated a $6 billion five-year plan whose goal is to double production by 1990, to process 30% of all output by that year, and to develop new port facilities. Aiding this development is a contract with the USSR, under which Soviet financing of $20 million will help exploit the huge Meskala deposits of southern Morocco.

While present conditions thus appear unfavorable, Morocco is relying on the essential role of phosphate in agriculture to boost demand again in the future. When this occurs, Morocco's intent is clearly to be in a position to take advantage of such growth. Huge reserves and ambitious development plans underscore Morocco's potential for surpassing the United States as the number one producer of phosphate.

MANGANESE

Production of manganese, essential in the manufacture of steel, remains largely under the control of the Soviets and the South Africans. The reserves of these two nations dwarf all others, as does their output (see Table VI-3-5). Gabon, India, and Brazil rank as the LDC countries producing more than 1 million tons of manganese ore annually.

Gabon

Gabon is the fourth largest producer of manganese ore. However, manganese represents only about 5% of export earnings, reflecting the impor-

TABLE VI-3-5 Principal Manganese Producers
(Thousand Metric Tons)

	1982 production	Reserves
USSR	10,140	2,400,000
South Africa	5,750	2,200,000
China	1,760	50,000
Gabon	1,667	159,000
India	1,596	50,000
Brazil	1,433	95,000
Australia	1,248	330,000

Sources: Refs. 1 and 6.

tance of oil exports to the nation. The world depression in steel during the early 1980s has cut output in Gabon by nearly 30%. Production averages 51% manganese content,[6] with most metallurgical-grade ore exports going to Western Europe. Future plans include increasing state ownership in the Compagnie Minière de l'Ogooue S.A. (Comilog), the manganese mining company operating in Gabon. Development of the Trans-Gabon Railroad continues, with French assistance. Completion of this rail network will have a profound impact on subsequent development of manganese deposits in the interior of the nation, as well as on development of other raw materials, notably timber and iron ore. Gabon is the main supplier of U.S. manganese ore requirements, accounting for nearly 33% of American imports. France is also a major consumer.

India

Approximately 50% of Indian manganese ore produced is exported. Japan receives 67% of this trade, although Eastern Europe is increasing in importance as a customer. India's ambitious industrialization plans, with expansion of its own steel sector, will cause a projected doubling of domestic ore requirements by 1990.[4] This will strain existing export capabilities, barring major new discoveries. To this end, the state-owned company Manganese Ore (India) Ltd., which produces most of India's high-grade ore, has initiated

a large-scale exploration program to meet this internal growth in demand.

NICKEL

New Caledonia

Cutbacks in nickel output of more than 50% for 1982 in New Caledonia, induced by three consecutive years of depressed market conditions, dropped the nation from third to fourth place as a principal producer. However, nickel reserves are believed to be the largest in the world, estimated at 15,000,000 tons of nickel (see Table VI-3-6). Japan could increase in importance as a user of New Caledonian nickel, due to the failure of the French steel industry, presently the major export recipient, to recover from the world recession. Beyond depressed world demand, energy needs remain a serious problem in New Caledonia, as all energy must be imported.[4]

The Philippines

Nickel exports rank sixth in importance as an earner to the Philippines of foreign currency. Low prices and depressed demand have worked to hinder any substantial production increases from either of the nickel producers in the country, Marinduque Mining and Industrial Corporation and Rio Tuba Nickel Mining Corporation. Most production occurs in either Nonoc or Palawan Islands. Virtually all ore exported is shipped to Japan for processing, with most nickel metal exports going to the United States or the Netherlands. Philippine nickel export earnings have declined more than 30% in recent years, indicative of the production cutbacks occurring as market economy producers respond to soft demand. Under such conditions, only those producers subsidized by state funding, or unfettered by profit constraints (e.g., the centrally planned economies) can afford to maintain high levels of output. The present Philippine foreign debt, which exceeds $15 billion, coupled with large reserves of nickel ore, are strong inducements to remain a major exporter for this mineral. The Philippines

TABLE VI-3-6 World Nickel Producers (Tons of Nickel Content)

	1982 production	Reserves	Ore grade (%)
USSR	170,000	8,100,000	1.5
Canada	90,000	8,600,000	1.5–3
Australia	70,000	5,600,000	1.5–3
New Caledonia	35,000	15,000,000	1–3
Cuba	35,000	3,400,000	1.4
Philippines	35,000	5,700,000	0.4–2.2

Source: Ref. 1.

should thus remain an important supplier of nickel for the noncommunist world.

Cuba

As a centrally planned economy, Cuba represents a producer minimally affected by international market forces. This is because most output is destined for communist countries. In return for Soviet financial assistance to its nickel industry, Cuba supplies the USSR with approximately 20,000 tons of nickel-cobalt concentrate each year. Such exports amount to less than 10% of Cuba's total exports to the USSR. The Cuban nickel industry is pursuing a major expansion program, including projects at Las Camoriocas and Punta Gorda. Continued funding from the Council for Mutual Economic Assistance is anticipated to help maintain Cuba's role as a vital supplier of nickel to Eastern Bloc nations.

CHROMIUM

As 97% of world chromite reserves lie in South Africa and Zimbabwe, these two nations dominate world production. Despite their insignificant reserve base, however, the Philippines and Turkey have been important chromite producers. Since implementation of the AOD process, permitting usage of the vast stratiform chemical-grade deposits of southern Africa in metallurgical applications, the importance of these LDC producers has greatly diminished.

The Philippines

Zambales Province northwest of Manila remains the main chromium area in the Philippines, with the Coto podiform deposit being a particularly large source of refractory-grade ore. Output of both refractory and metallurgical grades has declined since 1979; 1982 production stood at approximately 400,000 tons.[1] The United States is the primary importer of refractory-grade chromite, and Japan is the chief user of Philippine metallurgical ore. Reserves are estimated at 3.3 million tons.[1]

Turkey

Chromite occurs in numerous podiform bodies throughout Turkey, which has been one of the world's major sources of the mineral for more than 60 years. Large deposits are found in Elazig, Bursa, Mugla, Eskisehir, Kutahya, and Kayseri Provinces. Reserves are listed at 5.5 million tons.[1]

Production in 1982 was 400,000 tons,[1] over half of which was mined by the government-owned mining company, Etibank. This enterprise also sup-

plies all Turkish ferrochrome output. Approximately 33% of all chromite production is processed into ferrochrome before export, with European markets being the primary destination for both ore and refined chrome. A continuing problem for the entire Turkish minerals industry revolves around the availability of adequate investment capital. This factor can be expected to continue to constrain future chromium production.

OTHER MINERALS

In addition to the LDC producers noted above, other developing nations are significant suppliers for some minerals. Zambia and Zaire are the principal producers of cobalt. Chile is a major producer of molybdenum, and Algeria of mercury. Bolivia and Korea have important tungsten production. Brazil, the Philippines, and Papua New Guinea produce more than 500,000 troy ounces of gold yearly. Papua New Guinea is also a supplier of copper, the Philippines of cobalt, and Mauritania of iron ore. Peru has notable silver, lead, and zinc output, as does Mexico. Mexico is also the leading producer of fluorspar. Generally, however, production represents a minor share of total world supplies, usually considerably under 10%. Moreover, individual reserves are typically a quite small proportion of global reserves as well.

REFERENCES

1. U.S. Department of the Interior, Bureau of Mines. *Mineral Commodity Summaries 1983.* Washington, D.C.: U.S. Government Printing Office, 1983.
2. U.S. Department of the Interior, Bureau of Mines. *Aluminum Availability—Market Economy Countries.* Washington, D.C.: U.S. Government Printing Office, 1983.
3. van Rensburg, W. C. J., and Susan Bambrick. *The Economics of the World's Mineral Industries.* Johannesburg, South Africa: McGraw-Hill Book Company, 1978.
4. U.S. Department of the Interior, Bureau of Mines. *Minerals Yearbook. Volume III. Area Reports: International. Centennial Edition 1981.* Washington, D.C.: U.S. Government Printing Office, 1983.
5. McCaskill, Joseph C. "Minerals and International Trade." In *Economics of the Mineral Industries.* Edited by William A. Bogely et al. New York: American Institute of Mining, Metallurgical, and Petroleum Engineers, Inc., 1976, pp. 73-107.
6. U.S. Department of the Interior, Bureau of Mines. *Minerals Yearbook. Volume III. Area Reports: International. 1982.* Washington, D.C.: U.S. Government Printing Office, 1984.
7. "Morocco's Debt, War in Desert." *U.S. News and World Report,* 12 December 1983, pp. 51-52.

Chapter VI-4

Future Prospects for LDC Producers

LDC producers will continue to be essential sources of certain key raw materials, such as bauxite and tin (see Table VI-1-1.) However, as emphasized by Govett and Govett,[1] most of the world's mineral production remains concentrated in the hands of a few developed and centrally planned nations. This group is dominated by Canada, the USSR, South Africa, Australia, and the United States. Moreover, mineral trade has been increasingly important between the developed countries, as opposed to between developed and developing nations. This swing away from LDC supplies by the industrial world has also been stressed by Page[2] in an analysis of Western European trade patterns.

These facts, combined with projected small increases in mineral demand for the near future, provide little leverage for LDC producers to improve their bargaining position, a position weakened first by the diatribe of the 1970s as voiced by the Group of 77, and later by the world recession of the early 1980s. Political invectives have now largely diminished, as it became increasingly apparent that LDC demands were detrimental to individual goals, being in effect a "beggar thyself" stand. Clear trends are now emerging in many LDCs of an intent to reestablish an environment conducive to attracting investment from industrial nations and multinational corporations. These changing attitudes reflect the realization that inflexible stipulations regarding North-South wealth transfer ultimately provide much less economic benefit than can a cooperative posture. While the importance of some LDCs as major suppliers of certain minerals often promotes negotiating advantage regarding pricing policy, it also helps guarantee a continuity of available capital. Such continuity is not assured for other LDC producers. Adequate investment remains a major hindrance to mineral, and hence industrial, development for most of the Third World.

This offers small hope for those LDCs whose economies are highly dependent on mineral sales but whose production is minor in relation to total world supplies. As the Brandt Report[3] accented, these nations are forced to accept market fluctuations, often with serious negative consequences on their export earnings. However, as Govett and Govett[1] have observed, relatively few LDCs are in fact heavily dependent solely on mineral trade. Exports from most of the developing world are more usually agricultural (e.g., sugar, rice, rubber, cotton, and cocoa), indicating the agrarian nature of these countries. Regrettably, agricultural commodities are themselves volatile.

Those LDCs that are intensely export reliant on minerals are, unfortunately, typically "Fourth World" nations, the poorest of the poor. Prospects for these countries remain grim, captives of an often capricious marketplace. This is not to diminish the vulnerability of all LDC mineral producers to market forces, to world recession, to high energy import bills, to huge debt burdens, and to loss of customers due to substitution. However, growing contributions to GNP from other sectors, such as manufacturing, construction, commerce, or transportation, will increasingly expand the economic base of industrialization in these nations, decreasing their vulnerability from fluctuating mineral export earnings. Meanwhile, the essential nature to world supplies of the truly large LDC producers, such as Malaysian and Indonesian tin, Chilean copper, or Jamaican and Guinean bauxite, will endure.

REFERENCES

1. Govett, M. H., and G. J. S. Govett. "The New Economic Order and World Mineral Production and Trade." *Resources Policy* 4, no. 4 (December 1978), pp. 230-241.
2. Page, William. "UK Balance of Payments in Raw Materials." *Resources Policy* 5, No. 3 (September 1979), pp. 185-196.
3. "Commodity Trade and Development." Chapter 9 in *North-South. A Program for Survival. The Report of the Independent Commission on International Development Issues under the Chairmanship of Willy Brandt*, pp. 141-157.

Bibliography

PART I: CANADA

American Embassy, Ottawa. "Canada." In *Foreign Economic Trends and Their Implications for the United States*. Washington, D.C.: U.S. Department of Commerce, March 1982.

Anderson, Ian. "Metals That No Longer Glitter." *Maclean's*, 15 November 1982.

Anderson, Ian. "The World According to Yamani," *Maclean's*, 7 March 1983.

"Annual Mineral Review and Forecast." *Canadian Mining Journal* 103 (February 1982).

Armstrong, Willis C., et al. "U.S. Policy toward Canada: The Neighbor We Cannot Take for Granted." *The Atlantic Community Quarterly* 19 (Fall 1981).

Baker Fox, Annette. "The Range of Choice for Middle Powers: Australia and Canada Compared." *The Australian Journal of Politics and History* 26 (1980).

"Balkanizing Canada: The Cost of Provincial Barriers." *Business Week*, 15 September 1980.

Barry, Donald. "Retrospective on Canada-U.S. Relations." *Foreign Investment Review* 6 (Autumn 1982).

Bayless, Alan. "Cheap Oil Is Causing Problems in Ottawa As Government Faces Decline in Revenue." *The Wall Street Journal*, 22 April 1982.

Bayless, Alan, "Price of Coal under Pressure." *The Globe and Mail*, 27 January 1983.

Bayless, Alan, "Canada Likely to Cut Natural Gas Price to U.S., but Rise in Canadian Exports Isn't Expected." *The Wall Street Journal*, 18 March 1983.

Beard, R. C. "Ontario Ministry of Natural Resources Incentive Programs to Encourage Exploration and to Facilitate the Development of Small Mines in Northwestern Ontario." *CIM Bulletin* 75 (September 1982).

Beckman, Christopher C. *Canada's International Trade Performance: A Survey of Recent Trends*. Executive Bulletin 23. Ottawa: The Conference Board of Canada, 1982.

Best, Dunnery, and Giles Gherson. "One Hurdle Keeps Gas Firms from Joy." *The Financial Post*, 5 February 1983.

Birnbaum, Jeffrey H., and Claudia Waterloo. "Price-Support Program for Grain, Cotton Draws Extensive Response from Farmers." *The Wall Street Journal*, 23 March 1983.

Bonus, John L., Managing Director, The Mining Association of Canada. Personal communication, 5 November 1982.

Booth, Amy. "Hydro Sweetener Helps Giant Smelter Deal." *The Financial Post*, 22 January 1983.

Bott, Robert, "The Megaproject Scorecard," *Canadian Business*, April 1981.

BP Statistical Review of World Energy 1981. London: The British Petroleum Company, 1981.

Bradbury, John H. "State Corporations and Resource Based Development in Quebec, Canada: 1960–1980." *Economic Geography* 58 (January 1982).

Branch, Stephen N. "Eastern Potash Industry Born As PCA Readies First Producer in NB." *Canadian Mining Journal* 103 (June 1982).

Brault, M. A. "Canada's Advantage in Francophone Africa." *The Canadian Business Review*, Summer 1982.

Brecher, Irving, and S. S. Reisman. *Canada-United States Economic Relations*. Ottawa: Royal Commission on Canada's Economic Prospects, July 1957.

Brown, Robert Douglas. "The Impact of Canadian Federal and Provincial Mining Taxation." Proceedings of the Council of Economics, AIME Annual Meeting, Denver, Colo., 26 February–2 March 1978.

Byleveld, Herbert C. "Foreign Investment in Canada: What's the Score?" *The Canadian Business Review* 9 (Summer 1982).

Cairns, Robert D. "A Reconsideration of Ontario Nickel Policy." *Resources Policy* 7 (Autumn 1981).

Cameron, John I. "Nickel." In *Natural Resources in U.S.-Canadian Relations, Volume II: Patterns and Trends in Resource Supplies and Policies*. Edited by Carl E. Beigie and Alfred O. Hero, Jr. Boulder, Colo.: Westview Press, Inc., 1980.

"Canada—Crisis of Confidence." *Mining Journal*, 12 December 1982.

"Canada's Jobless Rate Rose to 12.5% in Month." *The Wall Street Journal*, 14 March 1983.

"Canada's Potash Corp. to Resume Operations." *The Wall Street Journal*, 9 February 1983.

Canadian-American Committee. *Improving Bilateral Consultation on Economic Issues*. Montreal: C. D. Howe Research Institute/Washington, D.C.: National Planning Association, October 1981.

"The Canadian Economy Is in Crisis." *Business Week*, 28 June 1982.

Clancy, John, and Andrew J. Freyman. "The Pacific Perspective: Coal Development in British Columbia, 1981–2000." *CIM Bulletin* 75 (December 1982).

Cook, Peter, "OPEC Standoff Clouds Canadian Price Outlook." *The Globe and Mail*, 13 July 1982.

Cotter, Nicholas. "Anticipate Federal Clearance for Roche Bay Magnetite Mine." *The Northern Miner*, 17 March 1983.

Crabb, Peter. "There Is More to Canada's Constitutional Problems Than Albertan Oil." *Energy Policy* 9 (December 1981).

Creighton, Donald. *Towards the Discovery of Canada*. Toronto: Macmillan of Canada, 1972.

"Crumpled Maple Leaf." *The Economist*, 7 August 1982.

"Curing Our Ills." *The Globe and Mail*, 19 August 1982.

Curtis, John. "The China Trade." *Policy Options* 3 (January-February 1982).

Daly, D. J. "Mineral Resources in the Canadian Economy: Macro-economic Implications." In *Natural Resources in U.S.–Canadian Relations, Volume I: The Evolution of Policies and Issues*. Edited by Carl E. Beigie and Alfred O. Hero, Jr. Boulder, Colo.: Westview Press, Inc. 1980.

Darisse, Alan. Chief, Communications Division, Foreign Investment Review Agency. Personal communication, 14 October 1982.

Dauphin, Roma. "Asbestos." In *Natural Resources in U.S.–Canadian Relations, Volume II: Patterns and Trends in Resource Supplies and Policies*. Edited by Carl E. Beigie and Alfred O. Hero, Jr. Boulder, Colo.: Westview Press, Inc., 1980.

Derry, Duncan R. "Canada's Mineral Resources." *Engineering and Mining Journal* 181 (November 1981).

DeYoung, John H., Jr. "Effect of Tax Laws on Mineral Exploration in Canada." *Resources Policy* 3 (June 1977).

"Dome Petroleum Proceeds with Plan to Export LNG." *The Wall Street Journal*, 25 March 1983.

Dorr, Andre L., and John E. Tilton. "Bauxite and Aluminum." In *Natural Resources in U.S.–Canadian Relations, Volume II: Patterns and Trends in Resource Supplies and Policies*. Edited by Carl E. Beigie and Alfred O. Hero, Jr. Boulder, Colo.: Westview Press, Inc., 1980.

Douglas, R. J. W., ed. *Geology and Economic Minerals of Canada*. Ottawa: Minister of Supply and Services Canada, 1976.

Dowsing, R. J. "Spotlight on Tungsten," *Metals and Materials*, November 1981.

Drolet, Jean-Paul. "Strategic Importance of Canada as a Mineral Supplier to the World." *Mining Congress Journal*, February 1980.

Drouin, Marie-Josée, and Harald B. Malmgren. "Canada, the United States and the World Economy." *Foreign Affairs* 60 (Winter 1981/82).

Eisler, Dale. "Bidding Farewell to the Old Crow." *Maclean's*, 14 February 1983.

Energy, Mines and Resources Canada. *Canadian Minerals Yearbook*. Ottawa: Minister of Supply and Services, various years.

Energy, Mines and Resources Canada. *Discussion Paper on Coal 1980*. Resource EP 80-1E. Ottawa: Minister of Supply and Services Canada, 1980.

Energy, Mines, and Resources Canada. *Mineral Policy: A Discussion Paper*. Ottawa: Minister of Supply and Services Canada, December 1981.

Energy, Mines and Resources Canada. *Mineral Policy Backgrounder*, No. 82/25(b) (8 March 1982).

Energy, Mines and Resources Canada. *Minerals*. SS 82-18. December 1982.

Energy, Mines and Resources Canada. *Principal Mineral Areas of Canada*. Map 900A. 32nd ed. Ottawa: Surveys and Mapping Branch, 1982.

Energy, Mines and Resources Canada, Mineral Policy Sector. *The Non-fuel Mineral Industry to 1991: A Quantitative Outlook*. Internal Report MRI 81/6. Ottawa, July 1982.

Erola, Judy. "Canada's Resource Policy." *American Mining Congress Journal* 69 (16 February 1983).

"Exploration-Changing Perceptions." *Mining Journal* 299 (26 November 1982).

External Affairs Canada. *Canada's Export Development Plan for Japan*. Ottawa: Government of Canada, August 1982.

External Affairs Canada, Bureau of United States Affairs. *Notes on the Management of Canada/USA Relations*, June 1982.

External Affairs Canada, Domestic Information Programs Division. *Canada and the North–South Dialogue*. Canadian Foreign Policy Text 82/8 (September 1982).

External Affairs Canada, Domestic Information Programs Division. *Relations between Canada and Mexico*. Canadian Foreign Policy Text 82/3 (September 1982).

Fayerweather, John. *Foreign Investment in Canada: Prospects for National Policy*. White Plains, N.Y.: International Arts and Sciences Press, Inc., 1973.

The Financial Post, 22 January 1983.

The Financial Post, 12 February 1983.

The Financial Post, 26 February 1983.

The Financial Post, 12 March 1983.

The Financial Post, 19 March 1983.

The Financial Post, 31 July 1983.

"FIRA Approvals at 86% in Year." *The Globe and Mail*, 20 November 1982.

Foreign Direct Investment in Canada. Ottawa: The Government of Canada, 1972.

Foreign Investment Review Agency. *Foreign Investment Review Act, Businessman's Guide*. Ottawa: Minister of Supply and Services Canada, 1977.

Freeman, Alan. "Canada Asbestos Firms Blame Problems Mainly on Recession, Rather Than Suits." *The Wall Street Journal*, 3 September 1982.

Freeman, Alan, "Quebec Pension Agency Fights Bill to Stop Any Increase in Transportation Holdings." *The Wall Street Journal*, 18 November 1982.

Freyman, A. J. *The Role of Smaller Enterprises in the Canadian Mineral Industry with a Focus on Ontario*. Toronto: Ministry of Natural Resources, 1978.

Garreau, Joel. "Quebec." In *The Nine Nations of North America*. Boston: Houghton Mifflin Company, 1981.

"Gas Exports by Canada Seen Doubling." *The Wall Street Journal*, 28 January 1983.

Gaudet, Gerard. "Forces Underlying the Evolution of Natural Resource Policies in Quebec." In *Natural Resources in U.S.–Canadian Relations, Volume I: The Evolution of Policies and Issues*. Edited by Carl E. Beigie and Alfred O. Hero, Jr. Boulder, Colo.: Westview Press, Inc., 1980.

Gherson, Joan. "Japanese Investment in Canada." *Foreign Investment Review* 3 (Autumn 1979).

Gilpin, Robert. "American Direct Investment and Canada's

Two Nationalisms." In *The Influence of the United States on Canadian Development: Eleven Case Studies.* Edited by Richard A. Preston. Durham, N.C.: Duke University Press, 1972.

The Globe and Mail, 17 January 1983.

Globerman, Steven. *U.S. Ownership of Firms in Canada: Issues and Policy Approaches.* Montreal: C. D. Howe Research Institute, 1979.

Goldstein, Walter. "Canada's Constitutional Crisis: The Uncertain Development of Alberta's Energy Resources." *Energy Policy* 9 (March 1981).

Golt, Sidney. *Trade Issues in the Mid-1980's.* London: British–North American Committee, October 1982.

Govett, M. H., and G. J. S. Govett. "The Canadian Minerals Industry." *Resources Policy* 2 (March 1976).

Gray, John. "Power and Price Peckford Foes." *The Globe and Mail*, 30 September 1982.

Greenspan, Edward. "Potash Producers Cast Worried Eyes South." *The Financial Post*, 15 January 1983.

Greenwood, Ted, and Alvin Streeter, Jr. "Uranium." In *Natural Resources in U.S.-Canada Relations, Volume II: Patterns and Trends in Resource Supplies and Policies.* Edited by Carl E. Beigie and Alfred O. Hero, Jr. Boulder, Colo.: Westview Press, Inc., 1980.

Harrison, Fred. "U.S. Clashes Threaten Natural Gas Markets." *The Financial Post*, 29 January 1983.

Hatfield, Scott. "The Australian Parallel." *The Canadian Business Review*, Winter 1982.

Hay, Keith A. J. "Canadian Trade Policy in the 1980's." *International Perspectives*, July–August 1982.

Hero, Alfred O., Jr., and Roderick M. Logan. "Other Minerals and Deep-Sea Nodules." In *Natural Resources in U.S.-Canada Relations, Volume II: Patterns and Trends in Resource Supplies and Policies.* Edited by Carl E. Beigie and Alfred O. Hero, Jr. Boulder, Colo.: Westview Press, Inc., 1980.

Hjorleifson, G. R. "A National Minerals Policy." *Resources Policy* 2 (March 1976).

Hogan, William T. "Iron Ore." In *Natural Resources in U.S.-Canada Relations, Volume II: Patterns and Trends in Resource Supplies and Policies.* Edited by Carl E. Beigie and Alfred O. Hero, Jr. Boulder, Colo.: Westview Press, Inc., 1980.

"How FIRA Works." *Foreign Investment Review* 6 (Autumn 1982).

"IMF Sees 0.9% Growth in GNP during 1983." *The Globe and Mail*, 1 February 1983.

Janigan, Mary. "Taking the Bite-Out of FIRA." *Maclean's*, 27 September 1982.

Jeremic, M. L. "Coal Resources Are Plentiful but Unevenly Scattered." *World Coal* 7 (July–August 1981).

Joyce, Randolph. "Brian Peckford on the Rocks." *Maclean's*, 28 February 1983.

Kanabayashi, Masayoshi. "Japan's Economy Expected to Stay Weak in 1983, with Some Recovery Late in Year." *The Wall Street Journal*, 1 February 1983.

Kaplan, Jacob. "U.S. Resource Policy: Canadian Connections." In *Natural Resources in U.S.-Canada Relations, Volume I: The Evolution of Policies and Issues.* Edited

by Carl E. Beigie and Alfred O. Hero, Jr. Boulder, Colo.: Westview Press, Inc., 1980.

King, John. "U.S. Investments' Book Value Up, but Equity Outlay Off." *The Globe and Mail*, 26 August 1982.

King, Robert E. "Canada: Offshore Search Continues." *World Oil*, May 1982.

Kostuik, John, "Mining in the Next Decade." *CIM Bulletin* 73 (November 1980).

Kundig, Konrad. "The Tungsten Market—From Chaos to Stability." *The Journal of Metals*, May 1981.

LaForest, Gerard V. *Natural Resources and Public Property under the Canadian Constitution.* Toronto: University of Toronto Press, 1969.

Laux, Jeanne Kirk, and Maureen Appel Molot. "Asbestos." In *Natural Resources in U.S.-Canada Relations, Volume II: Patterns and Trends in Resource Supplies and Policies.* Edited by Carl E. Beigie and Alfred O. Hero, Jr. Boulder, Colo.: Westview Press, Inc., 1980.

Lewington, Jennifer. "NEB Approves Record in Natural Gas Exports." *The Globe and Mail*, 28 January 1983.

"Loss of By-product Support Is Critical to Copper and Base Metals, Study Shows." *Engineering and Mining Journal* 182 (April 1982).

"Low Price, Write-Downs Hit Canada Tungsten." *The Northern Miner*, 7 April 1983.

Lucas, A. R. "Natural Resources and the New Constitution." *Resources*, No. 2 (September 1982).

McCallum, Anthony. "Oil Exports Urged to Offset Slump." *The Globe and Mail*, 15 February 1983.

Mackenzie, Brian W. "The Competitive Position of Canadian Copper Supply." *Resources Policy* 7 (December 1981).

Mackenzie, Brian W., and Geoffrey G. Snow. "The Environment of Exploration: Economic, Organizational, and Social Constraints." *Economic Geology*, 75th Anniversary Volume (1981).

Miller, James A. "Ominous Trends in Canada's Resources Sector: Canada Development Corporation, Collapsing Megaprojects, The New Mineral Policy." *Alert Letter on the Availability of Raw Materials* No. 13 (June 1982).

Mitchell, Bruce. "The Natural Resources Development Debate in Canada." *Geoforum* 12 (1981).

Mulgrew, Ian. "B. C.'s Coal Superproject: Dream or Nightmare?" *The Globe and Mail*, 4 September 1982.

Mulholland, W. D., "Toward a Mineral Industries Policy for Canada." *CIM Bulletin* 73 (October 1980).

Mulvihill, R. P. *Coal: The Canadian Industry—1982.* Ottawa: Department of Industry, Trade and Commerce, Industrial Minerals Division, 1982.

National Research Council, Commission on Engineering and Technical Systems, National Material Advisory Board. *Tantalum and Columbium Supply and Demand Outlook.* Washington, D.C.: National Academy Press, 1982.

"Nickel Operations Are in Trouble." *World Mining* 33 (December 1982).

"The 1982 Country Credit Ratings." *Institutional Investor* 16 (March 1982).

"Noranda Closes Molybdenum Mine." *The Globe and Mail*, 16 December 1982.

Norrie, Kenneth H. "Natural Resources Development and

U.S.-Canadian Relations." In *Natural Resources in U.S.-Canadian Relations, Volume I: The Evolution of Policies and Issues*. Edited by Carl E. Beigie and Alfred O. Hero, Jr. Boulder, Colo.: Westview Press, Inc., 1980.

The Northern Miner, 7 April 1983.

Noumoff, S. J. "China: The Unexplored Prospects." *The Canadian Business Review*, Winter 1982.

"$1 Billion Slated for 15 Big Ports." *The Globe and Mail*, 12 July 1982.

Operating in a Changing Canada: A Rich Market, an Uncertain Future. New York: Business International Corporation, 1978.

Owens, O. E. "Canada and Mineral Exploration in the 1980's." *CIM Bulletin* 73 (August 1980).

Parsons, Robert B. "Time to Get Serious About Tax Simplification." *CIM Bulletin* 75 (December 1982).

Paterson, Donald G. *British Direct Investment in Canada 1890-1914*. Toronto: University of Toronto Press, 1976.

Patton, Donald J. "The Evolution of Canadian Federal Mineral Policies." In *Natural Resources in U.S.-Canada Relations, Volume I: The Evolution of Policies and Issues*. Edited by Carl E. Beigie and Alfred O. Hero, Jr. Boulder, Colo.: Westview Press, Inc., 1980.

Peters, W. C. "The Economics of Mineral Exploration." *Geophysics* 34 (August 1969).

Powers, Ned. "A Good Idea Gone Awry." *The Globe and Mail*, 30 October 1982.

Province of British Columbia, Ministry of Industry and Small Business Development. *Doing Business in British Columbia, No. 14: Tax Incentives* (November 1981).

"Quebec-Labrador Iron Ore/Steel Output: Hurt by U.S. Steel Slump." *Engineering and Mining Journal* 182 (September 1982).

"Quebec Premier Plans Renewed Bid to Turn Province into Nation." *The Wall Street Journal*, 17 March 1983.

"Report on Japan Business." *The Globe and Mail*, 10 December 1982.

Rose, Frederick. "Canada Gets Tougher on Wage and Price Increases." *The Wall Street Journal*, 19 August 1982.

Rosencranz, Armin. "The International Law and Politics of Acid Rain." *Denver Journal of International Law and Policy* 10 (Spring 1981).

Rotstein, Abraham. "Independence When Times Are Hard." *Policy Options* 3 (September-October 1982).

Russell, Allen S. "Pitfalls and Pleasures in New Aluminum Process Development." *Journal of Metals*, June 1981.

Salaff, Stephen. "Saskatchewan Uranium in the 1980's." *CRS Perspectives*, No. 12 (April 1982).

"Saskatchewan Requirement for Compulsory Crown Participation Repealed." *CIM Bulletin* 75 (December 1982).

Sassos, Michael P. "Modernization Extends to Most Alcan Plants." *Engineering and Mining Journal* 182 (January 1982).

Scales, Marilyn. "Canada Backs Asbestos Use, Third World Provides Markets." *Canadian Mining Journal* 103 (July 1982).

Scales, Marilyn. "CMJ Capital Spending Report." *Canadian Mining Journal* 103 (October 1982).

Schabas, Bill. "Canada Backs Ocean Mining, U.S. Refuses to Sign Treaty." *Canadian Mining Journal* 103 (October 1982).

Scheinberg, Stephen. "Invitation to Empire: Tariffs and American Economic Expansion in Canada." In *Enterprise and National Development*. Edited by Glenn Porter and Robert D. Cuff. Toronto: A. M. Haekkert Ltd., 1973.

Schiller, E. A. "Coal in Canada." *Mining Magazine*, July 1982.

Schmidt, Helmut, and Manfred Kruszona. *Regional Distribution of Mining Production and Reserves of Mineral Commodities in the World*. Hannover, West Germany: Federal Institute for Geosciences and Natural Resources, January 1982.

Schreiner, John. *Transportation: The Evolution of Canada's Networks*. Toronto: McGraw-Hill Ryerson Limited and Maclean Hunter Limited, 1972.

Shao, Maria. "U.S. Copper Industry is Being Devastated by Its High Costs As Well As by Recession." *The Wall Street Journal*, 16 September 1982.

Shuyama, Thomas K. "Canada-Japan/South Korea Trade Promises to Grow." *The Canadian Business Review*, Winter 1982.

Sigurdson, Albert. "Coal Production, Projects Suffer As Exports Decline." *The Globe and Mail*, 17 September 1982.

Sigurdson, Albert. "B. C. Coal Chief Lauds Proposed Growth Solution." *The Globe and Mail*, 18 February 1983.

Skea, J. "Electricity Supplies for the Primary Aluminum Industry." *Resources Policy* 6 (March 1980).

Smiley, Donald V. "The Political Context of Resource Development in Canada." In *Natural Resources Revenues: A Test of Federalism*. Edited by Anthony Scott. Vancouver: University of British Columbia Press, 1976.

Soganich, John. "New Golden Era Could Be Coming." *The Financial Post*, 22 January 1983.

Spector, Stewart R. "Price and Availability of Energy in the Aluminum Industry." *Journal of Metals*, June 1981.

"Spotlight on Canada's Resourceful Mining Industry." *Engineering and Mining Journal* 181 (November 1981).

Stanfield, Robert. "What to Do about Quebec's Isolation." *Policy Options* 3 (July-August 1982).

Statistics Canada. *Canada Year Book 1980-81*. Ottawa: Minister of Supply and Services Canada, 1981.

Stevenson, Garth. "The Process of Making Mineral Resource Policy." In *Natural Resources in U.S.-Canadian Relations, Volume I: The Evolution of Policies and Issues*. Edited by Carl E. Beigie and Alfred O. Hero, Jr. Boulder, Colo.: Westview Press, Inc., 1980.

Sultan, Ralph G. M. "Canada's Recent Experiment in the Repatriation of American Capital." *Canadian Public Policy* 8 (October 1982).

"Suneel, Koren Utility Set Contract for Coal." *The Wall Street Journal*, 21 January 1983.

U.S. Central Intelligence Agency, National Foreign Assessment Center. *The World Factbook—1981*. Washington, D.C.: U.S. Government Printing Office, April 1981.

van Rensburg, W. C. J. "The Classification of Coal Resources and Reserves." Mineral Resource Circular 65-1980. Austin, Tex.: The University of Texas—Austin, Bureau of Economic Geology, 1980.

Volpe, John. *Industrial Incentive Policies and Programs in the Canadian-American Context.* Montreal: C. D. Howe Research Institute/Washington, D.C.: National Planning Association, January 1976.

Von Riekhoff, Harald, John H. Sigler, and Brian W. Tomlin. *Canadian-U.S. Relations: Policy Environments, Issues, and Prospects.* Montreal: C. D. Howe Research Institute, 1979.

The Wall Street Journal, 3 March 1983.

The Wall Street Journal, 14 March 1983.

Watson, William. "We're Moving toward Two Nations in Quebec." *The Financial Post,* 2 October 1982.

Welsh, Lawrence. "Inco, Falconbridge Facing Soviet Market Threat." *The Globe and Mail,* 2 October 1982.

Welsh, Lawrence. "Value of Base Metals Mined Slides by $1.5 Billion." *The Globe and Mail,* 26 January 1983.

What Mining Means to Canada. Ottawa: The Mining Association of Canada, September 1981.

Whichard, Obie G. "U.S. Direct Investment Abroad in 1981." *Survey of Current Business.* U.S. Department of Commerce 62 (August 1982).

Whitney, John W. "Copper." In *Natural Resources in U.S.-Canadian Relations, Volume II: Patterns and Trends in Resource Supplies and Policies.* Edited by Carl E. Beigie and Alfred O. Hero, Jr. Boulder, Colo.: Westview Press, Inc., 1980.

Wojciechowski, Margot J. *Federal Mineral Policies, 1945 to 1975: A Survey of Federal Activities That Affected the Canadian Mineral Industry.* Working Paper 8. Kingston, Ont.: Centre for Resource Studies, Queen's University, May 1979.

Wojciechowski, Margot J., Assistant Director, Centre for Resource Studies, Queen's University, Kingston, Ontario. Personal communication, 14 February 1983.

Wojciechowski, M. J., and C. E. McMurray. *Mineral Policy Update 1981.* Kingston, Ont.: Centre for Resource Studies, Queen's University, November 1982.

Wonnacott, Ronald J. "Controlling Trade and Foreign Investment in the Canadian Economy: Some Proposals." *Canadian Journal of Economics* 15 (November 1982).

World Coal, July-August, 1981.

World Mine Production of Gold, 1981-1985. Washington, D.C.: The Gold Institute, January 1983.

Zehr, Leonard, et al. "Alsands Failure Is Major Loss for Canada, May Force It to Revise Economic Strategy." *The Wall Street Journal,* 3 May 1982.

"Zinc-Lead Enigmas." *Mining Journal* 297 (9 October 1981).

PART II: SOUTHERN AFRICA

Buchanan, D. L. Bureau for Mineral Studies. *Chromite Production from the Bushveld Complex and Its Relationship to the Ferrochromium Industry.* Report 3. Johannesburg, South Africa: University of the Witwatersrand, October 1978.

Buchanan, D. L. "Platinum—Great Importance of Bushveld Complex." *World Mining,* August 1980, pp. 56–59.

Cobbe, James H. *Governments and Mining Companies in Developing Countries.* Boulder, Colo.: Westview Press, Inc. 1979.

Coetzee, C. B., ed. *Mineral Resources of the Republic of South Africa,* 5th ed. Pretoria, South Africa: Government Printer, 1976.

Colcough, Christopher, and Stephen McCarthy. *The Political Economy of Botswana: A Study of Growth and Distribution.* New York: Oxford University Press, Inc., 1980.

Dayton, Stanley H., and John R. Burger. "Mining in South Africa." *Engineering and Mining Journal,* November 1982, pp. 52–139.

Embassy of Botswana, Washington, D.C. Information provided.

Evans, Anthony M. *An Introduction to Ore Geology.* New York: Elsevier North-Holland Inc., 1980.

Geertsema, A. "SASOL's Expansion Programme—Economic Implications for South Africa." Paper presented at the 13th AISEC Annual Economic Congress, Cape Town, 7 July 1981.

Government of Botswana, Department of Mines. *Annual Report,* 1981, p. 309.

Government of Botswana, Geological Survey Department. Areas Held under Prospecting Licenses and Mining Leases As at 1st July 1982. (Map.) Lobatse, Botswana.

Government of Botswana, Government Information Services. *Mining in Botswana.* Gaborone, Botswana: Government Printer.

Government of Botswana, Mines and Mineral Act, 1976. *Botswana Government Gazette,* Supplement A, 31 December 1976.

Government of Botswana, Ministry of Finance and Development Planning. *Statistical Bulletin* 5, No. 4. Gaborone, Botswana: Government Printer, December 1980.

Government of Botswana, Ministry of Mineral Resources and Water Affairs. *Mirwana News.* Gaborone, Botswana: Government Printer.

Government of South Africa, Department of Foreign Affairs and Information. *Official Yearbook of the Republic of South Africa,* 8th ed. Johannesburg, South Africa: Chris van Rensburg Publications, 1982.

Government of South Africa, Department of Mines, Minerals Bureau. *Alternative Routes to Steelmaking in South Africa.* Preliminary Report 2175. Braamfontein, South Africa, 1975.

Government of South Africa, Department of Mines, Minerals Bureau. *Reserve and Resource Analysis Non-metallic Minerals Andalusite, Sillimanite, Kyanite.* Internal Report 6. Braamfontein, South Africa, 7 February 1977.

Hodges, Tony. "Mozambique: The Politics of Liberation." In *Southern Africa: The Continuing Crisis,* 2nd ed. Edited by Gwendolyn M. Carter and Patrick O'Meara. Bloomington, Ind.: Indiana University Press, 1982.

Hull, Galen Spencer. *Pawns on a Chessboard: The Resource War in Southern Africa.* Washington, D.C.: University Press of America, Inc. 1981.

Hunter, D. R. *Some Enigmas of the Bushveld Complex.* Economic Geology Research Unit. Information Circular 92.

Johannesburg, South Africa: University of the Witwatersrand, 1974.

Interagency Task Force. *Report on the Issues Identified in the Nonfuel Minerals Policy Review.* Draft Report. Washington, D.C.: U.S. Government Printing Office, 1979.

Johnson, Charles J. "Mineral Objectives, Policies, and Strategies in Botswana—Analysis and Lessons." *Natural Resources Forum* 5, (1981), pp. 347–367.

"Jwaneng Diamond Mine, Botswana Officially Opened." *Mining Magazine*, October 1980, pp. 263–265.

Kowet, Donald Kalinde. *Land, Labour Migration and Politics in Southern Africa.* New York: Africana Publishing Company, 1978.

Landis, Elizabeth S., and Michael I. Davis. "Namibia: Impending Independence." In *Southern Africa: The Continuing Crisis*, 2nd ed. Edited by Gwendolyn M. Carter and Patrick O'Meara. Bloomington, Ind.: Indiana University Press, 1982.

Marcum, John. "Angola: Perilous Transition to Independence." In *Southern Africa: The Continuing Crisis*, 2nd ed. Edited by Gwendolyn M. Carter and Patrick O'Meara. Bloomington, Ind.: Indiana University Press, 1982.

Mufson, Steve. "Investors Unconvinced on Zimbabwe." *The Wall Street Journal*, 5 March 1982, p. 25.

O'Meara, Patrick. "Zimbabwe: The Politics of Independence." In *Southern Africa: The Continuing Crisis*, 2nd ed. Edited by Gwendolyn M. Carter and Patrick O'Meara. Bloomington, Ind.: Indiana University Press, 1982.

Pretorius, D. A. *The Depositional Environment of the Witwatersrand Goldfields: A Chronological Review of Speculations and Observations.* Economic Geology Research Unit. Information Circular 95. Johannesburg, South Africa: University of the Witwatersrand, March 1975.

Pretorius, D. A. *Gold and Uranium in Quartz–Pebble Conglomerates.* Economic Geology Research Unit. Information Circular 151. Johannesburg, South Africa: University of the Witwatersrand, January 1981.

Resource Investory of Botswana: *Metallic Minerals, Mineral Fuels and Diamonds*, 1977.

Ridge, John Drew. "Zaire". In *Annotated Bibliographies of Mineral Deposits in Africa, Asia (exclusive of the USSR) and Australia*, Elmsford, N.Y.: Pergamon Press, Inc., 1976, pp. 177–195.

Ridge, John Drew. "Zimbabwe". In *Annotated Bibliographies of Mineral Deposits in Africa, Asia (exclusive of the USSR) and Australia.* Elmsford, N.Y.: Pergamon Press, Inc., 1976, pp. 21–46.

Savastuk, David J. "Angola: U.S. Business Involvement Is Significant, Expanding." *Business America*, 9 August 1982, p. 25.

Stocks, Kevin. "Botha Regime Facing a Test in South Africa." *The Wall Street Journal*, 2 May 1982, p. 32.

Stoneman, Colin, ed. *Zimbabwe's Inheritance.* New York: St. Martin's Press, Inc., 1981.

Storrar, C. D., ed. *South African Mine Valuation.* Johannesburg, South Africa: Chamber of Mines in South Africa, 1977.

Tankard, A. J., et al. *The Crustal Evolution of Southern Africa.* New York: Springer-Verlag New York, Inc., 1982.

U.S. Central Intelligence Agency, National Foreign Assessment Center. *The World Factbook—1981.* Washington, D.C.: U.S. Goverment Printing Office, 1982.

U.S. Congress, Senate, Committee on Foreign Relations. *Imports of Minerals from South Africa by the United States and the OECD Countries.* 96th Congress, 2nd Session, 1980. Washington, D.C.: U.S. Government Printing Office, September 1980.

U.S. Congress, Senate Committee on Foreign Relations. *U.S. Minerals Dependence on South Africa.* 97th Congress, 2nd Session, 1980. Washington, D.C.: U.S. Government Printing Office, October 1982.

U.S. Congress, Senate, Committee on the Judiciary, Subcommittee on Security and Terrorism. *Soviet, East German and Cuban Involvement in Fomenting Terrorism in Southern Africa.* 97th Congress, 2nd Session. Washington, D.C.: U.S. Government Printing Office, 1982.

U.S. Department of Commerce. "Marketing in Swaziland." *Overseas Business Report.* OBR 77-36, July 1977.

U.S. Department of Commerce. "Kingdom of Lesotho." *Foreign Economic Trends and Their Implications for the United States.* FET 81-086, Washington, D.C.: U.S. Government Printing Office, August 1981.

U.S. Department of Commerce. "Kingdom of Swaziland." *Foreign Economic Trends and Their Implications for the United States.* FET 81-113. Washington, D.C.: U.S. Government Printing Office, September 1981.

U.S. Department of Commerce. *Botswana.* Prepared by American Embassy, Gaborone, December 1982.

U.S. Department of Commerce, Industry and Trade Administration. "Marketing in Zambia." *Overseas Business Reports.* OBR 78-24, June 1978.

U.S. Department of Commerce, Industry and Trade Administration. "Marketing in Zaire." *Overseas Business Reports.* OBR 79-44, December 1979.

U.S. Department of Commerce, International Trade Administration. "Marketing in South Africa." *Overseas Business Reports.* OBR 81-03. Washington, D.C.: U.S. Government Printing Office, March 1981.

U.S. Department of Commerce, International Trade Administration. "Marketing in Zimbabwe." *Overseas Business Reports.* OBR 81-21, August 1981.

U.S. Department of Commerce, International Trade Administration. "Zaire." *Foreign Economic Trends and Their Implications for the United States.* FET 81-143. Washington, D.C.: U.S. Government Printing Office. December 1981.

U.S. Department of Commerce, International Trade Administration. "Zambia." *Foreign Economic Trends and Their Implications for the United States.* FET 82-023, March 1982.

U.S. Department of Commerce, International Trade Administration. "Zimbabwe." *Foreign Economic Trends and Their Implications for the United States.* FET 82-041, June 1982.

U.S. Department of State. "Kingdom of Lesotho." *Back-*

ground Notes. Washington, D.C.: U.S. Government Printing Office, December 1981.

U.S. Department of State. *Industrial Outlook Report: Minerals South Africa—1981*. Prepared by the American Embassy in Pretoria, 11 June 1982.

U.S. Department of State, Bureau of Public Affairs. Namibia/South-West Africa. *Background Notes*, September 1980.

U.S. Department of the Interior, Bureau of Mines. *Mineral Industries of Africa*. Washington, D.C.: U.S. Government Printing Office, 1976.

U.S. Department of the Interior, Bureau of Mines. *Cobalt Availability—Domestic*. IC 8848. Washington, D.C.: U.S. Government Printing Office, 1981.

U.S. Department of the Interior, Bureau of Mines. *Critical and Strategic Minerals in Alaska*. IC 8869. Washington, D.C.: U.S. Government Printing Office, 1981.

U.S. Department of the Interior, Bureau of Mines. "Malawi." *Minerals Yearbook 1978/79*. Washington, D.C.: U.S. Government Printing Office, 1981.

U.S. Department of the Interior, Bureau of Mines. "The Mineral Industry of Angola." *Minerals Yearbook 1978/79*. Washington, D.C.: U.S. Government Printing Office, 1981.

U.S. Department of the Interior, Bureau of Mines. "The Minerals Industry of Botswana." *Minerals Yearbook 1978/79*. Washington, D.C.: U.S. Government Printing Office, 1981.

U.S. Department of the Interior, Bureau of Mines. "The Mineral Industry of Mozambique." *Minerals Yearbook 1978/79*. Washington, D.C.: U.S. Government Printing Office, 1981.

U.S. Department of the Interior, Bureau of Mines. "The Mineral Industry of the Republic of South Africa." *Minerals Yearbook 1978/79*. Washington, D.C.: U.S. Government Printing Office, 1981.

U.S. Department of the Interior, Bureau of Mines. "The Mineral Industry of the Territory of South-West Africa." *Minerals Yearbook 1978/79*. Washington, D.C.: U.S. Government Printing Office, 1981.

U.S. Department of the Interior, Bureau of Mines. "The Mineral Industry of Zaire." *Minerals Yearbook 1978/79*. Washington, D.C.: U.S. Government Printing Office, 1981.

U.S. Department of the Interior, Bureau of Mines. "The Mineral Industry of Zambia." *Minerals Yearbook 1978/79*. Washington, D.C.: U.S. Government Printing Office, 1981.

U.S. Department of the Interior, Bureau of Mines. "Swaziland." *Minerals Yearbook 1978/79*. Washington, D.C.: U.S. Government Printing Office, 1981.

U.S. Department of the Interior, Bureau of Mines. "Zimbabwe." *Mineral Perspectives*, August 1981.

U.S. Department of the Interior, Bureau of Mines. *Chromium Availability—Domestic*. IC 8895. Washington, D.C.: U.S. Government Printing Office, 1982.

U.S. Department of the Interior, Bureau of Mines. "Lesotho." *Minerals Yearbook 1980*. Washington, D.C.: U.S. Government Printing Office, 1982.

U.S. Department of the Interior, Bureau of Mines. "Malawi."

Minerals Yearbook 1980. Washington, D.C.: U.S. Government Printing Office, 1982.

U.S. Department of the Interior, Bureau of Mines. *Manganese Availability—Domestic*. IC 8889. Washington, D.C. U.S. Government Printing Office, 1982.

U.S. Department of the Interior, Bureau of Mines. "The Mineral Industry of Angola." *Minerals Yearbook 1980*. Washington, D.C.: U.S. Government Printing Office, 1982.

U.S. Department of the Interior, Bureau of Mines. "The Minerals Industry of Botswana." *Minerals Yearbook 1980*. Washington, D.C.: U.S. Government Printing Office, 1982.

U.S. Department of the Interior, Bureau of Mines. "The Mineral Industry of Mozambique." *Minerals Yearbook 1980*. Washington, D.C.: U.S. Government Printing Office, 1982.

U.S. Department of the Interior, Bureau of Mines. "The Mineral Industry of Namibia (Terrotory of South-West Africa)." *Minerals Yearbook 1980*. Washington, D.C.: U.S. Government Printing Office, 1982.

U.S. Department of the Interior, Bureau of Mines. "The Mineral Industry of the Republic of South Africa." *Minerals Yearbook 1980*. Washington, D.C.: U.S. Government Printing Office, 1982.

U.S. Department of the Interior, Bureau of Mines. "The Mineral Industry of Zaire." *Minerals Yearbook 1980*. Washington, D.C.: U.S. Government Printing Office, 1982.

U.S. Department of the Interior, Bureau of Mines. "The Mineral Industry of Zambia." *Minerals Yearbook 1980*. Washington, D.C.: U.S. Government Printing Office, 1982.

U.S. Department of the Interior, Bureau of Mines. "The Mineral Industry of Zimbabwe." *Minerals Yearbook 1980*. Washington, D.C.: U.S. Government Printing Office, 1982.

U.S. Department of the Interior, Bureau of Mines. *Platinum Availability—Market Economy Countries*. IC 8897. Washington, D.C.: U.S. Government Printing Office, 1982.

U.S. Department of the Interior, Bureau of Mines. "Platinum-Group Metals." *Mineral Facts and Problems 1980*. Washington, D.C.: U.S. Government Printing Office, 1982.

U.S. Department of the Interior, Bureau of Mines. "Recent Government Actions Affecting the Minerals Industry of Zimbabwe." *Minerals and Materials*, August/September 1982, pp. 42–53.

U.S. Department of the Interior, Bureau of Mines. "Swaziland." *Minerals Yearbook 1980*. Washington, D.C.: U.S. Government Printing Office, 1982.

U.S. Department of the Interior, Bureau of Mines. "Chromium." *Mineral Commodity Profiles 1983*. Washington, D.C.: U.S. Government Printing Office, 1983.

U.S. Department of the Interior, Bureau of Mines. "Lesotho." *Minerals Yearbook 1981*. Washington, D.C.: U.S. Government Printing Office, 1983.

U.S. Department of the Interior, Bureau of Mines. "Malawi."

Minerals Yearbook 1981. Washington, D.C.: U.S. Government Printing Office, 1983.

U.S. Department of the Interior, Bureau of Mines. "Manganese." *Mineral Commodity Profiles 1983*. Washington, D.C.: U.S. Government Printing Office, 1983.

U.S. Department of the Interior, Bureau of Mines. *Mineral Commodity Summaries 1983*. Washington, D.C.: U.S. Government Printing Office, 1983.

U.S. Department of the Interior, Bureau of Mines. "The Mineral Industry of Angola." *Minerals Yearbook 1981*. Washington, D.C.: U.S. Government Printing Office, 1983.

U.S. Department of the Interior, Bureau of Mines. "The Mineral Industry of Botswana." *Minerals Yearbook 1981*. Washington, D.C.: U.S. Government Printing Office, 1983.

U.S. Department of the Interior, Bureau of Mines. "The Mineral Industry of Mozambique." *Mineral Yearbook 1981*. Washington, D.C.: U.S. Government Printing Office, 1983.

U.S. Department of the Interior, Bureau of Mines. "The Mineral Industry of Namibia." *Minerals Yearbook 1981*. Washington, D.C.: U.S. Government Printing Office, 1983.

U.S. Department of the Interior, Bureau of Mines. "The Minerals Industry of the Republic of South Africa." *Minerals Yearbook 1981*. Washington, D.C.: U.S. Government Printing Office, 1983.

U.S. Department of the Interior, Bureau of Mines. "The Mineral Industry of Zaire." *Minerals Yearbook 1981*. Washington, D.C.: U.S. Government Printing Office, 1983.

U.S. Department of the Interior, Bureau of Mines. "The Mineral Industry of Zambia." *Minerals Yearbook 1981*. Washington, D.C.: U.S. Government Printing Office, 1983.

U.S. Department of the Interior, Bureau of Mines. "The Mineral Industry of Zimbabwe." *Minerals Yearbook 1981*. Washington, D.C.: U.S. Government Printing Office, 1983.

U.S. Department of the Interior, Bureau of Mines. "Platinum-Group Metals." *Mineral Commodity Profiles 1983*. Washington, D.C.: U.S. Government Printing Office, 1983.

U.S. Department of the Interior, Bureau of Mines. "Swaziland." *Minerals Yearbook 1981*. Washington, D.C: U.S. Government Printing Office, 1983.

U.S. Department of the Interior, Bureau of Mines. *Minerals Yearbook 1982*. Washington, D.C.: U.S. Government Printing Office, 1984.

U.S. Department of the Interior, Office of Minerals Policy and Research Analysis. *Implications of Current Events in Zaire for World and U.S. Cobalt Markets*, 27 April 1977.

U.S. Department of the Interior, Office of Minerals Policy and Research Analysis. *The Future Role of Central and Southern Africa in the Supply of Nonfuel Minerals to the U.S.—Qualitative Report on Zimbabwe*, 20 June 1980.

U.S. Department of the Interior, Office of Minerals Policy and Research Analysis. *The Future Role of Central and Southern Africa in the Supply of Nonfuel Minerals to the U.S.—Qualitative Report on South Africa*, 20 June 1980.

U.S. Department of the Interior, Office of Minerals Policy and Research Analysis. *The Future Role of Central and Southern Africa in the Supply of Nonfuel Minerals to the U.S.—Qualitative Report on Zambia*, 11 July 1980.

van Rensburg, W. C. J., and D. A. Pretorius. *South Africa's Strategic Minerals—Pieces on a Continental Chess-Board*. Johannesburg, South Africa: Valiant Publishers, 1977.

Vermaak, C. F. *The Global Status of the South African Minerals Economy and Data Summaries of Its Key Commodities*. Review Paper 1. Johannesburg, South Africa: The Geological Society of South Africa, 1979.

Vilakazi, Absolom. "Swaziland: From Tradition to Modernity." In *Southern Africa: The Continuing Crisis*, 2nd ed. Edited by Gwendolyn M. Carter and Patrick O'Meara. Bloomington, Ind.: Indiana University Press, 1982.

"Wanted: Capitalist Comrades." *The Economist*, 24 July 1982, pp. 65–67.

Weisfelder, Richard. "Lesotho: Changing Patterns of Dependence." In *Southern Africa: The Continuing Crisis*, 2nd ed. Edited by Gwendolyn M. Carter and Patrick O'Meara. Bloomington, Ind.: Indiana University Press, 1982.

White, Lane. "Zaire." *Engineering and Mining Journal*, November 1979, pp. 188–206.

White, Lane. "Zambia." *Engineering and Mining Journal*, November 1981, pp. 146–183.

World Bank, International Bank for Reconstruction and Development. *World Development Report 1982*. New York: Oxford University Press, Inc., 1982.

"Zaire: Problems Mount." *Mining Journal*, 22 October 1982.

"Zambia." *Mining Annual Review*, June 1982, pp. 437–441.

"Zambia: World's Second Largest Copper Producer Is Born." *World Mining*, May 1982, pp. 71–76.

"Zimbabwe: 18 Months After." *Mining Journal*, 30 October 1981, pp. 329–330.

"Zimbabwe Leader Tries to Ignore Them, but Massacre Charges Keep Multiplying." *The Wall Street Journal*, 28 February 1982, p. 25.

PART III: AUSTRALIA

"Aboriginal Land Rights: The Search for a National Consensus." *Mining Review*, March 1982.

"AJV Agreements—Northern Pursues Its Own Markets." *World Mining*, December 1982.

Alexander, J., and R. Hattersly. *Australian Mining, Minerals, and Oil*. Sydney: The David Ell Press Pty. Ltd., 1980.

Aluminum. Perth: Geological Survey of Western Australia, 1982.

"Argyle Mine, Australia Will Be One of the World's Major Diamond Producers." *Mining Magazine*, October 1982.

"Aussie Aluminum: Chilled by the Recession's Pall." *Chemical Week*, 15 September 1982.

"Aussie Diamonds May Alter the Market." *The Wall Street Journal*, 22 February 1982.

"Australia Looks to Labor." *Mining Journal*, 11 March 1983.

Australian Coal Association. "The Present State and Prospect of the Australian Export Coal Industry with Particular Reference to Its Contribution to the Economy: Taxation Policy." Submission to the Senate Standing Committee on Trade and Commerce, Canberra, November 1982.

Australian Department of National Development and Energy. *Australian Energy Stastistics: 1982*. Canberra: Australian Government Publishing Service, 1982.

Australian Department of Trade. "ABS Overseas Trade Statistics 1970-71 through 1981-82." Canberra, 1983.

Australian Department of Trade and Resources. *Australia's Mineral Resources: Bauxite/Alumina/Aluminum*. Canberra: Australian Government Publishing Service, 1980.

Australian Department of Trade and Resources. *Australia's Mineral Resources: Mineral Sands*. Canberra: Australian Government Publishing Service, 1980.

Australian Department of Trade and Resources. *Australian Coal*. Canberra: Australian Government Publishing Service, 1981.

Australian Department of Trade and Resources. *Australia's Mineral Resources: Development and Policies*. Canberra: Australian Government Publishing Service, 1981.

Australian Department of Trade and Resources. *Australia's Mineral Resources: Iron Ore*. Canberra: Australian Government Publishing Service, 1981.

Australian Department of Trade and Resources. *Australia's Mineral Resources: Manganese*. Canberra: Australian Government Publishing Service, 1981.

Australian Department of Trade and Resources. *Australia's Mineral Resources: Metallurgical Coal*. Canberra: Australian Government Publishing Service, 1981.

Australian Department of Trade and Resources. *Australia's Mineral Resources: Nickel*. Canberra: Australian Government Publishing Service, 1981.

Australian Department of Trade and Resources. *Australia's Mineral Resources: Steaming Coal*. Canberra: Australian Government Publishing Service, 1981.

Australian Department of Trade and Resources. *Australia's Mineral Resources: Tin*. Canberra: Australian Government Publishing Service, 1981.

Australian Department of Trade and Resources. *Australia's Mineral Resources: Tungsten*. Canberra: Australian Government Publishing Service, 1981.

Australian Department of Trade and Resources. *Australia's Mineral Resources: Uranium*. Canberra: Australian Government Publishing Service, 1981.

Australian Department of Trade and Resources. *Energy Australia*. Canberra: Australian Government Publishing Service, 1981.

Australian Department of Trade and Resources. *Australian Resources in a World Context*. Canberra: Australian Government Publishing Service, 1982.

Australian Department of Trade and Resources. *Australia's Mineral Resources: Gemstones*. Canberra: Australian Government Publishing Service, 1982.

Australian Department of Trade and Resources. *Exports Australia*. Canberra: Australian Bureau of Statistics, 1982.

Australian Department of Trade and Resources. *Survey of Major Western Pacific Economies*. Canberra: Australian Government Publishing Service, 1982.

Australian Department of Trade and Resources. *Australia's Mineral Resources: Copper*. Canberra: Australian Government Publishing Service, 1983.

Australian Department of Trade and Resources. *Australia's Mineral Resources: Lead and Zinc*. Canberra: Australian Government Publishing Service, 1983.

Australian Information Service. *Fact Sheet on Australia: Aluminum Industry*. Canberra: Department of Administrative Services, February 1980.

Australian/Japan Joint Study Group on Raw Materials Processing. *Australian and Japanese Aluminum Smelting Industries: Future Development and Relationship*. Canberra: Australian Government Publishing Service, 1980.

"Australian Mining Analysed." *Mining Journal*, 14 January 1983.

Australian Mining Industry Council. *Mineral Industry Survey 1982*. Canberra, 1982.

Australian Trade Development Council. *Minerals Processing, a Comparative Study*. Canberra: Australian Government Publishing Service, 1980.

"Australia's Key Issue." *London Financial Times*, 4 February 1983.

Bain, G. W. "Resources of Tungsten." *Institution of Mining and Metallurgy*, April 1980.

Bambrick, Susan. *Australian Minerals and Energy Policy*. Canberra: Australian National University Press, 1979.

Bambrick, Susan. "Australian Energy Policy." Paper presented at the Conference on New Developments in Foreign Trade and Investment, San Francisco, 12 August 1982.

Bambrick, Susan. "Australian Mineral Resources Trade and Policies." Paper presented to the Energy and Mineral Resources seminar, The University of Texas—Austin, Austin, Tex., 29 November 1982.

Bambrick, Susan. "Energy Resource Development: Economic Growth and Socio-political Priorities." Paper presented at the Conference on Australian Energy Resource Development, State College, Pa., 5-7 December 1982.

Barnett, D. W. *Minerals and Energy in Australia*. Melbourne: Cassell Australia Ltd., 1979.

Bayless, A. "Japan Wants Coal Producers to Cut Prices and Deliveries As Steel Market Declines." *The Wall Street Journal*, 21 January 1983.

Bell, R. "Problems in Australian Foreign Policy July–December 1979." *The Australian Journal of Politics and History*. 26, No. 1 (1980).

Blainey, G. *The Rush That Never Ended*. Melbourne: Melbourne University Press, 1963.

Blayden, I. D. "Geology of Australian Coalfields." In *Australian Black Coal*. Edited by A. C. Cook. Illawarra, Australia: Australasian Institute of Mining and Metallurgy, 1975, pp. 19-30.

Bowen, K. G. "Brown Coal: Victoria's Prime Energy Resource." *World Coal*, July 1978.

British Petroleum Company. *BP Statistical Review of World Energy: 1982*. London: Dix Motive Press Ltd., 1982.

Broinowski, R. P. *Australian Energy Resources; Their Marketability in Japan and the Pacific Region.* Paper presented at the Conference of Australian Energy Resource Development, Canberra, 5-7 December 1982.

Brown, G. R., N. E. Guernsey, and T. M. Li, eds. *Mineral Resources of the Pacific Rim*. New York: American Institute of Mining, Metallurgical, and Petroleum Engineers, Inc., 1982.

Carr, Bob. "When Will the Waterfront Mess Be Cleaned Up?" *The Bulletin*, 26 January 1982.

Carrick, John. "Outlook for Energy Source Development Promising." *Mining Review*, December 1982.

Chadwick, J. R. "Tin, What of the Future?" *World Mining*, November 1981.

"Coal: Focus on Queensland." *Mining Journal*, 6 August 1982.

"Commodity Review 1982." In *Australian Mineral Industry Review*. Canberra: Australian Bureau of Mineral Resources, 1983.

Commonwealth/State Joint Study Group on Raw Materials Processing. *Discussion Paper I: The Impact of Fiscal Measures on the Cost and Availability of Capital on Raw Materials Processing in Australia*. Canberra: Australian Government Publishing Service, 1981.

Commonwealth/State Joint Study Group on Raw Materials. *Processing Discussion Paper II: The National Benefits Arising from Raw Materials Processing*. Canberra: Australian Government Publishing Service, 1981.

"DeBeers May Be Cutting Diamond Output in Renewed Bid to Support Ailing Market." *The Wall Street Journal*, April 19, 1982.

"Dispute over New South Wales' Third Coal Loader." *Mining Journal*, 22 January 1982.

Durie, R. A., and I. W. Smith. "Production of Gaseous and Liquid Fuels from Coal." In *Australian Black Coal*. Edited by A. C. Cook. Illawarra, Australia: Australasian Institute of Mining and Metallurgy, 1975, pp. 161-172.

"Estimates of Olympic Dam Reserves Finally Released." *World Mining*, July 1982.

"Exploration Roundup." *Engineering and Mining Journal*, March 1982.

Fitzgerald, M. D., and G. Pollio. "Aluminum: The Next Twenty Years." *Journal of Metals*, December 1982.

Ford, J. "New Energy Sources to Restore the Balance of Power." *Australia Now*, 1982.

"Foreign Investment Statistics: Need for Objectivity in New Definitions." *Mining Review*, August 1982.

Galante, S. "Australia, New Zealand Near Accord to Cut Trade Barriers, Merge Markets." *The Wall Street Journal*, 20 October 1982.

Govett, G. J. S., and M. H. Govett. "The Role of the Australian Mineral Industry in World Mineral Supplies." *Resources Policy*, June 1980.

Govett, G. J. S., and M. H. Govett. "Australian Minerals." *Resources Policy*, December 1982.

Gregory, R. G. "Some Implications of the Growth of the Mineral Sector." *The Australian Journal of Agricultural Economics*, August 1982.

Hampson, D. C. "Australia's Uranium." *Resources Policy*, June 1980.

Harris, Stuart. "Resource Policy in Australia." *Resources Policy*, June 1980.

Ho. C. E. "Tungsten." *Mining Annual Review*, June 1982.

Hough, R. A., and L. E. Lynd. "Titanium." In *Minerals Yearbook 1981*. Washington, D.C.: U.S. Department of the Interior, 1983.

"How Labor Plans to Tax the Miners." *The Bulletin*, 3 May 1983.

The International Tin Council. "Tin." *Mining Annual Review*, June 1982.

"In the Pacific: Australia." *Engineering and Mining Journal*, July 1983.

Iron Ore. Perth: Geological Survey of Western Australia, 1983.

Joint Coal Board. *Black Coal in Australia 1981-82*. Sydney: Joint Coal Board, 1982.

Knight, C. L., ed. *Economic Geology of Australia and Papua New Guinea, Volume 1: Metals*. Victoria: Australian Institute of Mining and Metallurgy, 1975.

Krause, L. B., and H. Patrick, eds. *Mineral Resources in the Pacific Area*. San Francisco: Federal Reserve Bank of San Francisco, 1978.

Lancer, John. "New Concessions on Taxation System." *Mining Review*, May 1983.

Lawriwsky, M. "Some Issues of Foreign Relations and Control of Australia's Mineral Resources." *The Australian Quarterly*, Winter 1982.

Lewis, John. "Water Resources and Coal Mining." *Mining Review*, July 1983.

"Light Metals." *Mining Annual Review*, June 1982.

"Loy Yang: A Major Brown Coal Project." *Mining Magazine*, October 1982.

Lyons, L. A. "Worldwide Survey: Australia." *World Mining*, August 1982.

McCawley, F. X., and P. A. Stephenson. "Aluminum." In *Minerals Yearbook 1981*. Washington, D.C.: U.S. Department of the Interior, 1983.

"Markets Available to Black Coal." In *Australian Black Coal*. Edited by A. C. Cook. Illawarra, Australia: Australian Institute of Mining and Metallurgy, 1975, pp. 102-106.

Matusuoka, H. "Requirements for Coals in Japanese Coking Blends." In *Australian Black Coals*. Edited by A. C. Cook. Illawarra, Australia: Australasian Institute of Mining and Metallurgy, 1975, pp. 252-263.

Miline, J. "Uranium to Fuel the World's Reactors." *Australia Now*, 1982.

Milligan, E. N. "The Geology of the Bowen and Galilee Basin Coalfields." In *Australian Black Coal*. Edited by A. C. Cook. Illawarra, Australia: Australasian Institute of Mining and Metallurgy, 1975, pp. 10-18.

"Miners Focus on Stability." *Mining Review*, June 1982.

Nahai, L., and C. Wyche. *Future Trends and Prospects for the Australian Minerals Processing Sector*. Washington D.C.: U.S. Department of the Interior, 1982.

New South Wales Department of Mineral Resources. *New South Wales Mineral Industry 1981 Review*. Sydney: Government of New South Wales, 1982.

"Olympic Dam/Roxby Downs Mineralization Valued at $140 Billion." *Engineering and Mining Journal*, 21 January 1983.

Petroleum in Western Australia. Perth: Government of Western Australia, Department of Mines, Petroleum Branch, January 1982.

Pritchard, Chris. "Australian Labor Scene, Now Tranquil, Shows Signs of Growing More Turbulent." *The Wall Street Journal*, 5 January 1983.

Prospecting for Uranium. Perth: Geological Survey of Western Australia, 1978.

"Queensland Has Integrated Aluminum Industry." *World Mining*, November 1982.

"Ranger Mine Produces above Design Capacity." *World Mining*, November 1982.

Regan, R. J. "A Leaner Aluminum Industry Searches for the Comeback Trail." *Iron Age*, 4 October 1982.

"Restraint Needed." *Mining Review*, July 1983.

Rich, P. J. H. "The Future of Tin as a Tonnage Commodity." *Institute of Mining and Metallurgy*, January 1980.

Ridge, J. *Annotated Bibliographies of Mineral Deposits in Africa, Asia (Exclusive of the USSR), and Australia*. Oxford: Pergamon Press Ltd., 1976.

Rodrick, D. "Managing Resource Dependency: The United States and Japan in the Markets for Copper, Iron Ore, and Bauxite." *World Development*, July 1982.

"Roxby Will Cost $1400 Million to Develop." *Engineers Australia*, 21 January 1983.

Scambary, R. "Brown Is Bountiful for Local Use." *Australia Now*, 1982.

Scambary, R. "Oil and Gas: High Hopes in Ever-Deeper Waters." *Australia Now*, 1982.

Schmidt, Helmut, and Manfred Kruszona. *Regional Distribution of Mining Production and Reserves of Mineral Commodities in the World*. Hannover, West Germany: Federal Institute for Geosciences and Natural Resources, January 1982.

South African Department of Foreign Affairs and Information. *Official Yearbook of the Republic of South Africa: 1982*, 8th ed. Johannesburg, South Africa: Chris van Rensburg Publications, 1982.

Stewart, G. N. "Australia." *Mining Annual Review*, June 1982.

Thompson, Herb. "Normalisation: Industrial Relations and Community Control in the Pilbara." *The Australian Quarterly*, Winter 1982.

"Uranium Enrichment: Report Outlines Details." *Mining Review*, October 1982.

U.S. Department of the Interior, Bureau of Mines. *Mineral Commodity Summaries 1982*. Washington, D.C.: U.S. Government Printing Office, 1983.

"Use of Aluminum in Auto Wheels Expected to Triple by 1983." *Modern Metals*, 16 August 1982.

Vardabasso, Monica. "Mining Towns: An Extra Dimension to Regional Development." *Mining Review*, June 1983.

Walsh, M. "BHP's Threat: We'll Close the Steel Industry." *The Bulletin*, 28 December 1982.

Walsh, M. "How Hawke and the ACTU Conned Big Business." *The Bulletin*, 26 April 1983.

Western Australian Department of Resources Development. *Minerals and Mineral Development 1981*. Perth: Government of Western Australia, 1982.

Woodward, O. H. *A Review of the Broken Hill Lead-Silver-Zinc Industry*. Sydney: West Publishing Corporation, 1965.

Wyche, C. "The Mineral Industry of Australia." In *Minerals Yearbook 1980*. Washington, D.C.: U.S. Department of the Interior, 1980.

PART IV   LATIN AMERICA

"Araxa Niobium Mine." *Mining Magazine*, February 1982.

Banco Minero del Perú. *Metallic Gold Mining Law Regulation*. Lima: Banco Minero del Perú, 1979.

"Brazil Battles the Jungle to Mine Carajas Minerals." *World Mining*, January 1983.

"Brazilian Leaders Discuss the Economy." *The Wall Street Journal*, 31 January 1983.

"Brazil is Making Massive Minerals Investments." *The Wall Street Journal*, 28 January 1983.

"Brazil: Where Growth Is a Necessity." *The Wall Street Journal*, 27 January 1983.

Business Week 53 (6 July 1965).

Business Week 69 (6 July 1981).

"Chilean Copper: Major Expansion." *Mining Journal* 292 (29 August 1980).

"Chilean Copper: Plans for the Future." *Mining Magazine*, October 1980.

"Chile Will Maintain Policy of Stepped Up Copper Production during 1982." *Engineering and Mining Journal* 182 (June 1982).

"Codelco-Chile Meeting the Challenge." *Mining Journal* 248 (18 June 1982).

"Codelco Costs Now Seen As Topping Current Cu-Mo Prices." *Engineering and Mining Journal* 182 (July 1982).

"Colombia Unveils Plans to Step Up Mineral Development." *Engineering and Mining Journal* 180 (November 1980).

"Cuajone." *Mining Magazine*, August 1976.

"CVRD Obtains Mining Concession for Carajas Manganese Reserves." *Engineering and Mining Journal* 182 (November 1982).

Dayton, Stan. "Chile." *Engineering and Mining Journal* 179 (November 1982).

"Decisive Year for Brazil." *Mining Journal* 298 (1 January 1982).

De Shields, Anne P. Council of the Americas. "Colombia." *Engineering and Mining Journal* 182 (August 1982).

Evans, Anthony M. *An Introduction to Ore Geology*. New York: Elsevier, 1980.

"Eyes On Mexico." *Mining Journal* 296 (15 May 1981).

First International Bancshares, Economics Division. "Chile." April 1980.

Fishlow, Albert. "The United States and Brazil: The Case of the Missing Relationship." *Foreign Affairs* 60 (Spring 1982).

Harris, Robert. "ROFOMEX Moves Mexico toward Phosphate Self-sufficiency." *Engineering and Mining Journal* 181 (July 1981).

"Importing Oil to Export Bauxite." *The Courier* 72 (March–April 1982).

Knakal, Jan. "Transnationals and Mining Development in Bolivia, Chile, and Peru." *CEPAL Review*, August 1981.

Loprete, Carlos A., and Dorothy McMahon. *Iberoamerica*, 2nd ed. New York: Charles Scribner's Sons, 1974.

Martin, Everett G. "A Little Town in Bolivia Is Thriving as a Financial Center." *The Wall Street Journal*, 17 February 1983.

"Mexico: Reasons for Optimism." *Mining Journal* 290 (2 June 1978).

Ministerio de Recursos Naturales y Energéticos, Dirección General de Geología y Minas. *Informativo Geológico-Minero*, December 1981.

National Association of Medium Miners. *Annual Report 1981*. La Paz: Inderpa Ltd. 1981.

Newsweek 89 (6 July 1981).

"Nickel Mine Dedicated in Northern Colombia." *Mining Engineering* 34 (August 1982).

"Peru Adopts Measures for Small/Medium Mines." *Engineering and Mining Journal* 182 (September 1982).

"Peru: Taking the Long View." *Mining Journal* 299 (30 July 1982).

"Peruvian Promise." *Mining Journal* 296 (13 February 1981).

"Production Starts at Minero Peru's Cerro Verde Copper Mine." *Engineering and Mining Journal* 177 (July 1977).

Purcell, Susan Kaufman. "Mexico–U.S. Relations: Big Initiatives Can Cause Big Problems." *Foreign Affairs* 60 (Winter 1981-1982).

República de Venezuela, Ministerio de Energía y Minas. *Hierro y Otros Datos Estadísticos Mineros*. Caracas, December 1980.

República Federativa do Brasil, Ministério das Minas e Energia. *Anuário Mineral Brasileiro*, 1981.

República Federativa do Brasil, Ministério das Minas e Energia, *Contribução da CPRM ao Sector Mineral*. May 1982.

Ridley, Regina Starr. "Peru's Cobriza Brings New Capacity on Stream." *World Mining*, September 1982.

Rodado Noriega, Carlos. *Transición Energética y Minera*. Bogotá: Ministerio de Minas y Energía, June 1982.

Scheiderman, Stephen J. "Colombia Joins Latin American Mining's Big League." *World Mining*, September 1981.

Schiller, E. A. "Mineral Exploration and Mining in Brasil—1977." *Mining Magazine*, September 1977.

Schiller, E. A. "Mineral Exploration and Mining in Colombia." *Mining Magazine*, January 1980.

Sillitoe, R. H. "Regional Aspects of the Andean Porphyry Copper Belt in Chile and Argentina." *Applied Earth Science* 90 (February 1981).

Sisselman, Robert. "Peru." *Engineering and Mining Journal* 179 (October 1979).

Spooner, John, ed. *Mining Journal Annual Review 1980*. London: Mining Journal, June 1980.

Spooner, John, ed. *Mining Journal Annual Review 1982*. London: Mining Journal, June 1982.

Sutulov, Alexander. *Chilean Mining*. Santiago: University of Concepción Press, 1978.

Todd, Joan C. "Peru's Tintaya Copper Project to Go on Stream in September 1984." *Engineering and Mining Journal* 182 (September 1982).

U.S. Central Intelligence Agency, National Foreign Assessment Center. *The World Factbook—1981*. Washington, D.C.: U.S. Government Printing Office, 1982.

U.S. Department of the Interior, Bureau of Mines. *Mineral Commodity Summaries 1983*. Washington, D.C.: U.S. Government Printing Office, 1983.

U.S. Department of the Interior, Bureau of Mines. *Mineral Industries of Latin America*. Mineral Perspectives Series. December 1981.

U.S. Department of the Interior, Bureau of Mines. *Minerals and Materials*, August-September, 1982.

U.S. Department of the Interior, Bureau of Mines. *Minerals and Materials*, October-November, 1982.

U.S. Department of the Interior, Bureau of Mines. *Minerals Yearbook 1980*. Washington, D.C.: U.S. Government Printing Office, 1982.

U.S. Department of the Interior, Bureau of Mines. *Minerals Yearbook 1981*. Washington, D.C.: U.S. Government Printing Office, 1983.

U.S. Department of State. *Mineral Trade Notes* 76, No. 3 (March 1979).

U.S. Department of State. *Mineral Trade Notes* 76, No. 4 (April 1979).

U.S. Department of State, *Mineral Trade Notes* 76, No. 5 (May 1979).

"Venezuela's $200 Million Carbozulia Coal Project." *Mining Magazine*, May 1982.

Welna, David. "Argentina Meeting with Success in Securing Foreign Funds for Mineral Development." *Minerals Week*, 9 January, 1981.

White, Lane. "Mining in Mexico." *Engineering and Mining Journal* 180 (November 1980).

Wippman, Larry. "Centromin Invests $261 Million to Expand Cobriza Production." *Engineering and Mining Journal* 182 (August 1982).

"Worldwide Survey: Latin America." *World Mining*, August 1981.

"Worldwide Survey: Latin America." *World Mining*, August 1982.

Zozaya, F. Dionisio. *La Industria Minera en Venezuela: Situación y Perspectivas para el Desarrollo Nacional*. División de Información Técnica Geológica-Minera, Ministerio de Energía y Minas, 1978.

PART V: CHINA

American Petroleum Institute. *Basic Petroleum Data Book, Petroleum Industry Statistics, Volume III, No. 1*. Washington, D.C.: API, January 1983.

Amuzegar, Jahangir. "Oil Wealth: A Very Mixed Blessing." *Foreign Affairs* 60 (Spring 1982), pp. 814-835.

"Arco's Deal with China Is a Tough Act to Follow." *Business Week*, 4 October 1982, pp. 43-44.

Argall, George O., Jr. "Huge Chinese Porphyry: Dexing Operations and Plans." *World Mining* 34 (January 1981), pp. 24-28.

Barbier, Claude. *The Economics of Tungsten.* London: Metal Bulletin Books Ltd., 1971.

Barnett, A. Doak. *China's Economy in Global Perspective.* Washington, D.C.: The Brookings Institution, 1981.

"The Battle for Resources." *Beijing Review* 24 (4 May 1981), pp. 14-15.

Brady, E. Sabrina. "China's Strategic Minerals and Metals." *The China Business Review* 8 (September–October 1981), pp. 55-69.

Brady, Steve C., and John Bassarear. "Fluor and the Chinese Government." *Mines Magazine* 74 (September 1982), pp. 16-20.

Brown, Chris. "Tough Terms for Offshore Oil." *The China Business Review* 9 (July–August 1982), pp. 34-37.

Bunge, Fredericka M., and Rinn-Sup Shinn, eds. *China: A Country Study.* Washington, D.C.: U.S. Government Printing Office, 1981.

Business International Asia/Pacific Ltd. *Business Strategies for the People's Republic of China.* Hong Kong: Business International Asia/Pacific Ltd., 1980.

Cannon, Joseph G. "Lanthanides (Rare Earths): Shipment Lower in 1982, but Market Growth Potential Is High." *Engineering and Mining Journal* 189 (March 1983), pp. 136-137.

Chang, Raymond. *Chinese Petroleum: An Annotated Bibliography.* Boston: G. K. Hall & Co., 1982.

Chieh, Hsu. "China's Mineral Resources." *China Reconstructs* 30 (April 1981), pp. 62-64.

Chin, Edmond. "China." *Mining Annual Review*, 1982, pp. 385-391.

"China: A Major U.S. Supplier of Calcined Refractory Bauxite." *Engineering and Mining Journal* 183 (October 1982), p. 158.

China Daily, 24 July 1982, p. 2. FBIS, *China Daily Report*, 24 July 1982, p. G1.

China Daily, 3 November 1982, p. 3. FBIS, *China Daily Report*, 5 November 1982, p. K18.

"China Debates Foreign Involvement in Energy Development: Energy Planners Want Self-Reliance but Need Foreign Technical Assistance." *World Business Weekly*, 13 April 1981, pp. 15-16.

"China Mine Pact Set by Occidental Petroleum Corp." *The Wall Street Journal*, 9 March 1983, p. 29.

"China Not Seen Selling Much Offshore Oil." *Asian Wall Street Journal*, 15 January 1983, p. 5.

"China Poised to Become the World's New Supplier of Platinum Group Metal." *Metals and Materials* December 1980, pp. 27-30.

"China Says It Won't Cancel Any More Projects in the Next Few Years As Retrenchment Is Over." *The Wall Street Journal*, 27 July 1982, p. 32.

"China's Petroleum Surplus May Vanish in the 1980's." *Oil and Gas Journal* 78 (6 October 1980), pp. 27-32.

"China to Begin Awarding Exploration Pacts Soon." *Houston Chronicle*, 27 February 1983, Section 4, p. 10.

"China Trying to Maintain Slow Growth." *Asian Wall Street Journal*, 2 September 1982, p. 3.

"China Won't Bar Oil Exploration by U.S. Concerns." *The Wall Street Journal*, 21 January 1983, p. 18.

"Chinese Envisage Big Project for Production of Aluminum." *Japanese Economic Journal*, 25 January 1983, p. 7.

"Chinese Shelf May Generate Biggest Offshore Flurry since North Sea." *Oil and Gas Journal* 80 (13 December 1982), pp. 55-60.

Ching, Frank. "Chinese Energy, Transport to Get $10.3 Billion." *The Wall Street Journal*, 14 December 1982, p. 30.

Ching, Frank. "Foreign Investment Slowed in China in '82 As Peking Altered Goals; '83 Rise Forecast. *The Wall Street Journal*, 7 March 1983, pp. 22.

Clarke, William. "China's Electric Power Industry." In *Energy in the Developing World: The Real Energy Crisis.* Edited by Vaclav Smil and William E. Knowland. Oxford: Oxford University Press, 1980, pp. 145-166.

"Close Attention Should Be Paid to Strengthening Mechanical and Electrical Management in Coal Mines." *Renmin Ribao*, 10 July 1982, p. 3. FBIS, *China Daily Report*, 15 July 1982, pp. K18-K19.

Cohen, Jerome A. "U.S. China Trade Still Needs Nurturing." *Asian Wall Street Journal*, 1 March 1983, p. 6.

Copper, John F. *China's Global Role.* Stanford, Calif.: Hoover Institution Press, 1980.

Cox, Madeline. *China Economic Overview: A British Columbia Perspective.* Province of British Columbia, Ministry of Industry and Small Business Development, no date.

Curtis, John. "The China Trade." *Policy Options* 3 (January–February 1982), pp. 29-32.

"Developments in China Signal Serious Efforts for Coal." *World Coal* 8 (November–December 1982), pp. 92-93.

Earney, Fillmore C. F. "China's Offshore Petroleum Frontiers: Confrontation? Conflict? Cooperation? *Resources Policy* 7 (June 1981), pp. 118-128.

"Electricity for the P.R.C." *Electric Light and Power* 60 (February 1982), pp. 11-12.

"Experts Agree Upturn in Sight." *Offshore* 42 (December 1982), pp. 45-46.

"Exploring Energy Resources in Southwest China." *Beijing Review* 25 (1 March 1982), pp. 7-8.

Feintech, Lynn. *China's Modernization Strategy and the United States.* Washington, D.C.: Overseas Development Council, August 1981.

Field, Robert Michael. "Growth and Structural Changes in Chinese Industry: 1952-79." In *China under the Four Modernizations, Part 1.* Joint Economic Committee, U.S. Congress, 13 August 1982, pp. 303-333.

Field, Robert Michael, and Judith A. Flynn. "China: An Energy-Constrained Model of Industrial Performance through 1985." In *China under the Four Modernizations, Part 1.* 364. Joint Economic Committee, U.S. Congress, 13 August 1982, pp. 334-364.

"Foreign Exchange." *The Wall Street Journal*, 5 May 1983, p. 44.

"Foreign Trade Booming." *Beiging Review* 25 (8 November 1982), p. 7.

Heuser, Robert. "Sino-German Trade: A Summary of Economic and Legal Data." In *China Trade: Prospects and Perspectives.* Edited by David C. Buxbaum, Cassondra E.

Joseph, and Paul D. Reynolds. New York: Praeger Publishers, 1982, pp. 217-248.

Hongren, Zhang. "Mineral Exploration and Development in China." Paper presented at a workshop on Mineral Policies to Achieve Development Objectives, East-West Center, Honolulu, 9-13 June 1980.

"International Bank to Aid Daqing Flow." *Oil and Gas Journal* 81 (7 February 1983), p. 29.

Japanese External Trade Organization. *China: A Business Guide*. Tokoyo: Japanese External Trade Organization and Press International Ltd., 1979.

Jensen, Mead L., and Alan M. Bateman. *Economic Mineral Deposits*, 3rd ed. New York: John Wiley & Sons, Inc., 1981.

Jones, Dori. "Breaking the Bottlenecks." *The China Business Review* 7 (May-June 1980), pp. 47-48.

Jones, Sam L. "Samarium-Cobalt Magnets Spur Electronic Advances." *American Metal Market Metalworking News*, 28 June 1982, p. 27.

Jones, Sam L. "Rare Earths' Development Said to Be in Its Infancy." *American Metal Market Metalworking News*, 8 August 1982, p. 29.

Kraus, Willy. *Economic Development and Social Change in the People's Republic of China*, translated by E. M. Holz. New York: Springer-Verlag New York, Inc., 1982.

Larkin, Bruce D. "Sino-Japanese Relations: Economic Priorities." *Current History* 81 (September 1982), pp. 268-271, 281.

Ludlow, Nicholas H. "China and the World Bank." *The China Business Review* 9 (November-December 1982), pp. 12-13.

"Metals and Minerals in China." *Canadian Mining Journal* 102 (August 1981), pp. 20-21.

Meyerhoff, A. A. "China's Petroleum Industry: Geology, Reserves, Technology and Politics." In *China Trade: Prospects and Perspectives*. Edited by David C. Buxbaum, Cassondra E. Joseph, and Paul D. Reynolds. New York: Praeger Publishers, 1982, pp. 94-112.

Mufson, Steve. "China Making Massive Energy Push." *The Wall Street Journal*, 12 December 1982, p. 19.

"Occidental Joint Venture Accents China's Commitment to Coal." *World Coal* 8 (May-June 1982), p. 13.

Odingo, R. S. "New Perspective on Natural Resource Development in Developing Countries." *GeoJournal* 5 (1981), pp. 521-530.

"On China's Economic Relations with Foreign Countries." *Beijing Review* 25 (31 May 1982), pp. 13-16.

Pattison, Joseph E. "China's Developing Legal Framework for Foreign Investment: Experience and Expectations." *Law and Policy in International Business* 13 (1981), pp. 89-175.

Payne, Douglas B. "Sino-Japanese Trade: Does It Hold Lessons for American Traders?" In *China Trade: Prospects and Perspectives*. Edited by David C. Buxbaum, Cassondra E. Joseph, and Paul D. Reynolds. New York: Praeger Publishers, 1982, pp. 249-278.

Perkins, Dwight H. "The Chinese Economy in the 1980's." In *China among the Nations of the Pacific*. Edited by Harrison Brown. Boulder, Colo.: Westview Press, Inc., 1982, pp. 1-14.

Peterson, Albert S. "China: Transportation Developments, 1971-80." In *China under the Four Modernizations, Part 1*. Joint Economic Committee, U.S. Congress, 13 August 1982, pp. 138-170.

"Progress in Nonferrous Metals." *Beijing Review* 25 (15 February 1982), pp. 7-8.

Radetzki, Marian. "Regional Development Benefits of Minerals Projects." *Resources Policy* 8 (September 1982), pp. 193-200.

"Regulations of the People's Republic of China on the Exploitation of Offshore Petroleum Resources in Cooperation with Foreign Enterprises." *Beijing Review* 25 (22 February 1982), pp. 14-18.

Schmidt, Helmut, and Manfred Kruszona. *Regional Distribution of Mining Production and Reserves of Mineral Commodities in the World*. Hannover, West Germany: Federal Institute for Geosciences and Natural Resources, January 1982.

Simon, David R. "Taxation of Joint Ventures in the People's Republic of China: A Legal Analysis in the Context of Current Chinese Economic and Political Conditions." *Vanderbilt Journal of Transnational Law* 15 (Summer 1982), pp. 513-582.

Simon, Denis Fred. "China's Capacity to Assimilate Foreign Technology: An Assessment." In *China under the Four Modernizations, Part 1*. Joint Economic Committee, U.S. Congress, 13 August 1982, pp. 514-552.

"The Sixth 5-Year Plan of the PRC for National Economic and Social Development (1981-1985)," *Renmin Ribao*, 20 December 1982, pp. K5-K42.

"Sleeping Giant, but Not in Mining." *The Northern Miner*, 16 September 1982, p. 2.

Smil, Vaclav. "Energy Development in China: The Need for a Coherent Policy." *Energy Policy* 9 (June 1981), pp. 113-126.

Stuart, Douglas T., and William T. Tow. "China's Military Modernization: The Western Arms Connection." *China Quarterly*, June 1982, pp. 253-270.

Szuprowicz, Bohdan O. *How to Avoid Strategic Materials Shortages: Dealing with Cartels, Embargoes and Supply Disruptions*. New York: John Wiley & Sons, Inc., 1981.

Szuprowicz, Bohdan O., and Maria R. Szuprowicz. *Doing Business with the People's Republic of China*. New York: John Wiley & Sons, Inc., 1978.

Tan, C. Suan. "China's Tungsten Pricing Strategy: Why the PRC Has Reduced Its Tungsten Supplies." *The China Business Review* 6 (March-April 1979), pp. 28-33.

Taylor, Robert P. *Rural Energy Development in China*. Washington, D.C.: Resources for the Future, 1981.

Tomlinson, Alexander C. "Some Perspectives on Financing China's Project." In *Business with China*. Edited by N. T. Wang. New York: Pergamon Press, 1980, pp. 61-66.

United Nations Development Programme. *First Country Programme for the People's Republic of China*. DP/CP/CPR/1, 18 February 1982.

U.S. Central Intelligence Agency. Research Aid: *China's Minerals and Metals Position in the World Market*. ER76-10150, March 1976.

U.S. Central Intelligence Agency, National Foreign Assessment Center. *Chinese Coal Industry: Prospects over the Next Decade*, February 1979.

U.S. Central Intelligence Agency, National Foreign Assessment Center. *China: The Continuing Search for a Modernization Strategy*. ER80-10248, April 1980.

U.S. Central Intelligence Agency, National Foreign Assessment Center. *Handbook of Economic Statistics 1981*.

U.S. Department of Commerce, International Trade Association. *Doing Business with China*, November 1980.

U.S. Department of Commerce, International Trade Association. *China's Economy and Foreign Trade 1979-81*, by Dr. Nai-Ruenn Chen, May 1982.

U.S. Department of the Interior, Bureau of Mines. *Minerals Yearbook*. Various editions, 1962-1980, inclusive.

U.S. Department of the Interior, Bureau of Mines. *Minerals Facts and Problems, 1980 Edition*. Washington, D.C.: U.S. Government Printing Office, 1980.

U.S. Department of the Interior, Bureau of Mines. *The Mineral Industry of China*. Preprint from the 1980 Bureau of Mines' *Minerals Yearbook*. By Edmond Chin.

U.S. Department of the Interior, Bureau of Mines. *Mineral Commodity Summaries 1983*. Washington, D.C.: U.S. Government Printing Office, 1983.

U.S. Department of the Interior, Bureau of Mines. *The People's Republic of China: A New Industrial Power with a Strong Mineral Base*, by K. P. Wang. 1975.

Walker, Tony. "Falling Oil Prices Delay Offshore China Drilling." *Houston Chronicle*, 13 March 1983, Section 4, p. 10.

Walsh, James I. "The Growth of Develop-for-Import Projects." *Resources Policy* 8 (December 1982), pp. 277-284.

Wang, K. P. *Mineral Resources and Basic Industries in the People's Republic of China*. Boulder, Colo.: Westview Press, Inc., 1977.

Wang, K. P. "China's Mineral Economy." In *Chinese Economy Post-Mao*. Joint Economic Committee, U.S. Congress, 9 November 1978, pp. 370-402.

Wang, K. P. "Mineral Industries of China." U.S. Department of the Interior, Bureau of Mines. Reprinted from *Mining Annual Review 1979*.

Wang, K. P. "China." *Mining Annual Review*, 1980, pp. 435-456.

Weil, Martin. "China's Troubled Coal Sector." *The China Business Review* 9 (March-April 1982), pp. 23-34.

Weil, Martin. "The Baoshan Steel Mill: A Symbol of Change in China's Industrial Development Strategy." In *China under the Four Modernizations, Part 1*. Joint Economic Committee, U.S. Congress, 13 August 1982, pp. 365-393.

Willrich, Mason. *Energy and World Politics*. New York: The Free Press, 1975.

Wilson, Thomas W., Jr. *Science, Technology and Development: The Politics of Modernization*. New York: Foreign Policy Association, 1979.

Woodard, Kim. "China and Offshore Energy." *Problems of Communism* 30 (November-December 1981), pp. 32-45.

World Bank. "China Borrows $199.4 Million for Port De-

velopment and Agricultural Research." Bank News Release 83/16, 3 November 1982.

World Bank. "Petroleum Project in China to Use $162.4 Million World Bank Loan." Bank News Release 83/38, 27 January 1983.

Wu, Yuan-li. "Resources, Competition, Economic Warfare and East West Relations." In *China, the Soviet Union, and the West: Strategic and Political Dimensions in the 1980's*. Edited by Douglas T. Stuart and William T. Tow. Boulder, Colo.: Westview Press, Inc., 1982, pp. 87-98.

Xinhua, 22 June 1982. FBIS, *China Daily Report*, 24 June p. K6.

Xinhua, 7 September 1982. FBIS, *China Daily Report*, 8 September 1982, p. A1.

Xinhua, 7 October 1982. FBIS, *China Daily Report*, 12 October 1982, pp. K14-K15.

Xinhua, 17 October 1982. FBIS, *China Daily Report*, 21 October 1982, p. K12.

Xinhua, 20 October 1982. FBIS, *China Daily Report*, 22 October 1982, pp. K14-K15.

Xinhua, 9 December 1982. FBIS, *China Daily Report*, 9 December 1982, p. K26.

Xinhua, 13 January 1983. FBIS, *China Daily Report*, 14 January 1983, p. K14.

Yuan, Paul C. "China's Energy Image Slightly Distorted." *Offshore* 41 (September 1981), pp. 97-99.

PART VI: THE ONE-CROP-ECONOMY COUNTRIES

Brewster, Havelock. "Interests and Strategies in International Conferences on Commodities." *Probleme der Rohstoffsicherung. (Primary Commodities: Security of Supply.)* Bonn, West Germany: Friedrich Ebert Stiftung, 1981, pp. 45-54.

Cobbe, James H. *Governments and Mining Companies in Developing Countries*. Boulder, Colo.: Westview Press, Inc., 1979, p. 42.

"Commodity Trade and Development." Chapter 9 in *North-South. A Program for Survival. The Report of the Independent Commission on International Development Issues under the Chairmanship of Willy Brandt*. pp. 141-157.

Danielson, Albert L. *The Evolution of OPEC*. New York: Harcourt Brace Jovanovich, Inc., 1982.

Encyclopaedia Britannica. 1983 Book of the Year. Chicago: Encyclopaedia Britannica, Inc., 1983.

Govett, M. H., and G. J. S. Govett. "The New Economic Order and World Mineral Production and Trade." *Resources Policy* 4, No. 4 (December 1978), pp. 230-241.

Grilli, Enzo R. "Demand for Raw Materials and Requirements for the Financing of Supply." *Probleme der Rohstoffsicherung. (Primary Commodities: Security of Supply.)* Bonn, West Germany: Friedrich Ebert Stiftung, 1981, pp. 7-17.

Ibrahim, Youssef M. "OPEC Chiefs Try to Stop Price Erosion, If They Can; Iran-Iraq War Is Wild Card." *The Wall Street Journal*, 6 December 1983, pp. 33.

Ibrahim, Youssef M. "Firming Oil Prices Strengthen OPEC." *The Wall Street Journal*, 8 March 1984, p. 30.

"Is OPEC Going the Way of All Cartels?" *U.S. News and World Report*, 24 October 1983, pp. 30–31.

McCaskill, Joseph C. "Minerals and International Trade." In *Economics of the Mineral Industries*. Edited by William A. Bogely, et al. New York: American Institute of Mining, Metallurgical, and Petroleum Engineers, Inc., 1976, pp. 73–107.

Meloe, Dr. Tor. "Foreign Oil Demand to Increase 23% in LDCs, Hold Level in Industrial Countries to 1990." *Oil and Gas Journal*, 14 November 1983, pp. 152–158.

"Morocco's Debt, War in Desert." *U.S. News and World Report*, 12 December 1983, pp. 51–52.

"OPEC Crude Exports, Revenue Down in '82." *Oil and Gas Journal*, 21 November 1983, p. 51.

Page, William. "UK Balance of Payments in Raw Materials." *Resources Policy* 5, No. 3 (September 1979), pp. 185–196.

U.S. Department of Energy, Energy Information Administration. *Monthly Energy Review*, December 1983.

U.S. Department of the Interior, Bureau of Mines. *Aluminum Availability—Market Economy Countries*. Washington, D.C.: U.S. Government Printing Office, 1983.

U.S. Department of the Interior, Bureau of Mines. *Mineral Commodity Summaries 1983*. Washington, D.C.: U.S. Government Printing Office, 1983.

U.S. Department of the Interior, Bureau of Mines. *Minerals Yearbook. Volume III. Area Reports: International. Centennial Edition 1981*. Washington, D.C.: U.S. Government Printing Office, 1983.

U.S. Department of the Interior, Bureau of Mines. *Minerals Yearbook. Volume III. Area Reports: International. 1982*. Washington, D.C.: U.S. Government Printing Office, 1984.

van Rensburg, W. C. J., and Susan Bambrick. *The Economics of the World's Mineral Industries*. Johannesburg, South Africa: McGraw-Hill Book Company, 1978.

"Worldwide Report." *Oil and Gas Journal*, 26 December 1983, pp. 80–81.

"Yesterday's Oil Cartel?" *The Economist*, 3 December 1983. pp. 13–14.

Index

DATE